Applied Instrumentation in the Process Industries

Volume I

Second Edition

A Survey

Applied Instrumentation in the Process Industries

Volume I

Second Edition

A Survey

W.G. Andrew
H.B. Williams

Gulf Publishing Company
Book Division
Houston, London, Paris, Tokyo

*To my wife, Edna Williams,
and children, Raymond, Rhonda, and Sandra
for their encouragement and loving kindness.*

**Applied Instrumentation
in the Process Industries
Volume I/Second Edition**

A Survey

Library of Congress Cataloging in Publication Data

Andrew, William G
Applied instrumentation in the process industries.

Includes index.
CONTENTS: v. 1. A survey.
1. Process control. 2. Automatic control.
3. Engineering instruments. I. Williams, H.B.,
1939- joint author. II. Title.
TS156.8.A534 1979 660.2'81 79-9418
ISBN 0-87201-382-0

Contents

Preface

The word *automation* has become a familiar sound to the average American. In the process industries, the somewhat synonomous term *instrumentation* is no less familiar. It is ironic that in a day of automation (or instrumentation) there still exists a partial void in efforts to train people engaged in a field that involves both art and science. The scientific aspect has often lagged the state of the art. Formal education for the field has been nil; practitioners have come from other disciplines, primarily electrical, chemical, and mechanical. The training afforded those entering the profession has come through formal and informal communication within the small but closely knit group. Trainees are often bewildered by the slow rate at which they grasp the entire field and achieve the level of competence and confidence of a professional.

These four volumes present the combined knowledge and experience of many people by providing a survey of instruments used in the chemical, refining, and petrochemical industries; a compilation of practical guidelines widely accepted throughout the industries; and engineering information and resource material that provides shortcuts to the practicing profession, for engineers, designers, and operating people.

The scope of the books and their presentation make them profitable library additions for experienced as well as inexperienced people. Instrument engineer trainees, process engineers, project engineers, and managers who desire easily referenced sources to confirm control concepts and methods will use this source confidently.

Technicians and maintenance men who aim at proficiency, operators who want to know how their processes are controlled, ambitious designers who want to progress in the instrument field will find the coverage to fit their needs.

Instrument manufacturers and sales personnel will find it interesting and profitable to understand more fully the user's view of the industry they serve.

Experienced instrument people will also find it useful as a thought-tickling reference for many applications. The experienced professional, more aware of the need for thorough analysis than are the novices and inexperienced, still seeks authoritative references for the control problems of today's industries. These four cross-referenced volumes serve a very useful tool for the accomplished professional in the field of instrumentation.

Volume 1, "A Survey," provides a comprehensive coverage of measurement and control devices in the processing industries. A high percentage of devices used in industry are described briefly and their operating principles are explained. Many of these are pictured or illustrated so that their use and function are easily understood.

The book provides an excellent introduction to the instrument profession, including a brief historical treatment that sets the stage for an appreciation of the measurement and control concepts and methods prevalent today. An interesting trail of progress is reviewed.

Volume 1 presents the basic fundamental concepts and theories of automatic control. The presentation is a nonmathematical, layman type treatment easily understood by the novice. The clarity and conciseness with which these theories and concepts are given provides an excellent reference for experienced and inexperienced readers.

Volume 2, "Practical Guidelines," is unique in the field of instrumentation literature. For the first time a work is available that formally sets forth information, suggests methods, makes comparisons, issues principles, and provides guidelines for those who want to master instrumentation quickly and thoroughly. It provides the accumulated knowledge and experience of instrument practitioners who are applications-minded.

Volume 3, "Engineering Data and Resource Material," is yet another unusual addition to the instrument field. It is filled with resource material necessary for people involved in instrument application, engineering, design, and operation.

This volume includes a thorough treatment of fluid flow problems. Physical properties of fluids are discussed, the nature of liquid and compressible fluid flows are treated clearly and concisely. There are charts, tables, nomographs, formulas, and symbols that instrument people need in their work. The information is arranged topically and indexed for easy reference. There is an abundance of information on the physical properties of fluids, flow data, conversion data, mathematical functions, piping information, and electrical data that are essential to instrument engineering.

Other features in Volume 3 are a listing of formulas needed for calculations in engineering work, typical installation details for instrument devices and sample calculation problems that are helpful to novices in the field. Sample problems include fluid flow problems, orifice calculations, control valve and relief valve calculations, and other problems involving the use of charts, tables, etc., given in Volume 3.

Volume 4, "Control Systems: Theory, Trouble-shooting, and Design," ties the practical to the theoretical in an easily understandable self-study text. Process dynamic responses are analyzed mathematically with respect to operating parameters. Study questions and problems are incorporated to give the student a working knowledge of transfer function development.

These four volumes are directed at hastening the development of new instrument trainees, bridging the gap between instrument people and engineers and designers in other engineering disciplines, and aiding experienced instrument engineers in their jobs.

Acknowledgments

Any technical book draws material from a large number of sources. Although many of these are referenced in the text, it is not feasible to include all the contributors to whom the authors are indebted. Data and information were furnished by many industrial companies.

The authors were encouraged to undertake the work by A.C. Lederer, former president of S.I.P., Inc. The cooperation of W.L. Hampton, Manager of Engineering, is gratefully acknowledged in producing the work on schedule.

Appreciation is extended particularly to B.J. Normand and K.G. Rhea for time spent in reviewing and criticizing many chapters and sections of the manuscript. Others who contributed in this area include L. Ashley, W.E. DeLong, D.M. Dudney, L.C. Hoffman, T.E. Lasseter, and J.G. Royle.

In addition, the authors are deeply appreciative to S.I.P., Inc. and its staff in providing the environment and materials for producing this work.

On the second edition: Mr. H.B. Williams gives his profound appreciation to all who have assisted in preparing this revised text—those who have contributed technical information; those who have typed and proofread; those who willingly provided illustrations, suggestions, and encouragement; and to all who have, by their confidence, inspired this effort.

1 Introduction

William G. Andrew

Necessity has dictated many of the advances of technology. In no other field is this more apparent than the discipline known as *instrumentation*—a word unknown 50 years ago.

In this 50-year development period, instrumentation has evolved from a series of devices, developed to fill specific needs of measurement and control, to a science in itself, where the premises and economics of entire plants and processes are based on suitable control strategies and instrumentation systems.

Initially, instrumentation had no separate status. Primarily, it consisted of mechanisms devised to fill the needs of specific localized control problems. Its development was relatively slow until the 1940s.

The critical demands of wartime production schedules during World War II pulled the instrument industry into the twentieth century. New control techniques were developed and used to keep pace with the accelerated output of America's wartime industrial complex. New control system technologies were developed for weapons control systems and for chemical and petroleum processes developed and/or expanded by wartime demands.

In the postwar period, the wartime technology continued to infuse the industrial community so that when the economic boom period of the 1950s and 1960s occurred, rapid expansions in petroleum, chemical, pulp and paper, automotive, machine tool and space industries provided a terrific impetus to instrumentation, producing great advancements that have brought the discipline to the respected status it commands today. These industries and most other process and assembly-type industries would fail to survive today if they were to be deprived of modern control techniques.

A technology as dynamic as instrumentation requires a constant inflow of new people who need to know not only the latest analytical control techniques from educational institutions but also the more practical aspects of hardware knowledge and application philosophies and guidelines from industry. The latter function is usually obtained through experience—frequently slowly and expensively.

This four-volume series presents a practical source of information and background on instrumentation based on experience in selection, application and maintenance. It is toward this goal that this series is aimed: Volume 1 is a survey of the commonly used instruments; Volume 2 presents practical guidelines for application; Volume 3 presents engineering data and resource material; and Volume 4 presents theory, troubleshooting and system design.

The material in these volumes has been selected for applications-oriented people. People who need a broad understanding of the many factors which enter into instrument selection—process operating demands, maintenance requirements, economics, operator interface, cost of downtime and innumerable other factors which must be weighed and considered—will find them a valuable asset to their library of references.

The remaining portion of this chapter reviews early instrument development, examines the status of instrumentation as it stands today and projects future trends. The industry is dynamic and ever-changing, and its scope is difficult to define because of its changeable nature.

Early Instrument Development

In the 1920s and 1930s industrial instruments were rather crude compared to today's standards. Local control was prominent. Efforts to centralize controls were confined primarily to units or subunits of processes, even for the more sophisticated plants. Indicators and recorders revealing the basic measurements of temperature, pressure, level and flow were used in many cases, with the operator observing measurements and adjusting valves as necessary to achieve desired results.

Prior to 1940, few people were working on control theory as it is now defined. Instead, empirical methods were used in attempts to apply control methods to process problems. It was a period of cut-and-try. Applications were founded primarily on experience and not on a strong theoretical base. This point is emphasized when one considers that it has been estimated that probably more than 90% of all control engineers who ever practiced are still active today.

From the beginning of World War II through the 1950s, control theorists produced a rather complete and general body of knowledge for the analysis, synthesis and design aspects of linear control systems. This has been referred to as the "classical" period in control theory development. Subsequent work is referred to as being in the "modern" period. This does not imply that the work of the 1940s and 1950s is not correct or appropriate; in fact, most control systems used today are based on the concepts of the classical period theorists.

The 1950 period also witnessed the introduction of electronic instruments. Among the earliest were Swarthout Instruments using AC voltage signals and Manning, Maxwell and Moore's Microsen line using DC milliamp signals. The increasing use of electronic transmitters and controllers, no doubt, has hastened the use of computer techniques, and the adoption of the more advanced control methods such as adaptive and optimal control.

This same period saw the beginning of miniaturization in packaging of control components, both pneumatic and electronic. Large case indicators, recorders and controllers that had occupied front-of-panel areas of approximately 180 square inches (12 by 15 inches) soon took no more than 36 square inches (6 by 6 inches). These were subsequently reduced to 18 square inches (3 by 6 inches) and even smaller. Not only have instrument sizes been reduced, but the density of instruments has also increased to provide much more process information in a given area.

While new control theories evolved and new and improved equipment was introduced, improved signal transmission methods also came into use. Multiconductor tubing cables of copper, aluminum and plastic began to replace the single tubing runs installed in the 1940s and 1950s in 50-foot lengths. Single tubing runs required stretching to attain rigidity and stiffness and rather expensive supports to provide for proper installation. Joints were made with connectors (fittings) or by soldering. Installation and testing methods were long, laborious and costly.

These brief highlights of control theory development and hardware use provide an interesting perspective for viewing instrumentation and control system concepts used in today's industrial plants.

Instrumentation Today

Instrumentation is now in a state of flux. There exist today plants with old but operable control systems, just as there exist processes which are old but still profitable. If they were being built today, changes would be made, but a time for change has not been deemed appropriate in these plants.

Today, plants are being built using the latest available electronic hardware, computer controls and advanced control concepts; others are built with appropriate hardware for future conversion to computer control; and still others are built with conventional hardware (pneumatic or electronic) that would require major, expensive modifications to convert to computer control.

In the last two decades, there has been much debating among those involved in instrumentation and control system technology about the relative merits of local versus central control, pneumatic versus electronic instrumentation, computer versus conventional control systems and direct digital control versus supervisory control. A rationale has emerged from these seemingly conflicting approaches which dictates that the application engineer must judiciously select from among these alternatives the system(s) which best fit the criteria under which he is working. Each system has its own merits, yet none has been demonstrated universally superior in *all* applications.

A dominant factor in current instrumentation is the impact of computer applications. Computers are being used to control directly, to perform economic optimization calculations, to make heat and material balance calculations or simply to perform the conventional monitoring, logging and alarm functions so essential in today's industrial processes. Computer applications have penetrated all the major industries as well as specialized applications in many smaller industries.

Large computer systems are being designed and built, having fully redundant computers, with complete, fully automatic transfer of data and controls in the event that the on-line computer fails. These systems include sophisticated display features and require complex interface hardware and software techniques.

Mini-computers are finding increasing usage in control systems as their costs decrease and as reliability and improved packaging of integrated circuits improve. Prices for a small system with standardized programming capability are competitive with systems utilizing conventional control systems. Dedicated computer systems with a mini-computer used exclusively for controlling a single unit or portion of a process are becoming popular for several reasons.

1. Attractive economic benefits can be realized by controlling and optimizing only a few critical loops in a process.
2. Hardware and software costs are reduced.
3. Operating losses from downtime are reduced.
4. Manpower for design, operation and maintenance is small compared to that required for a large computer system.

Microprocessor-based computers, a recent innovation brought about through advances in microminiaturization, are finding a niche in the marketplace. Able to concentrate many of the features of computers and mini-computers into a fraction of their required space and using much less energy, they are bringing a new degree of flexibility to the control room operator. Their low initial cost has appeal in both operational and managerial budgetary circles.

Dedicated microprocessors, have been immediately recognized for their ability to solve a myriad of OEM controls problems, have caused a wave of industrial interest, ranging from children's toys to washing machines, automatic machine tool programmers, automobile fuel optimization, industrial lab equipment, process data scanners, local mass flow calculation hardware, and basic control room instrumentation.

Programmable logic controllers now provide versatile alternatives to relay-type motor control systems. Allowing easy reprogramming in the field, entire control schemes can be reordered without costly rewiring.

The increased frequency of computer applications has not stopped the expansion and improvement of conventional control hardware. One major instrument manufacturer has completely redesigned a pneumatic panel instrument control line in the last few years. Another instrument manufacturer has redesigned its transmitter line.

Bumpless transfer when switching from automatic to manual control and vice versa, anti-reset windup and plug-in capability for adding alarm functions, output limiters and feedforward units are features that are offered by various controller manufacturers. Following are other facets and features of today's changing control technology.

Implementation of New Hardware and Control Techniques

Even in conventional control systems, new and improved hardware and useful control features have been added. Electrohydraulic valve operators with extremely fast (compared to pneumatic or electric) operating speeds are available for processes demanding fast response speeds. Electric operators that respond to the low-level electronic signals of modern controllers are offered, but they do not yet meet the speed, power and cost requirements to be competitive with pneumatic or hydraulic operators.

Equipment Packaging

There have been many recent improvements in this area. The size of both pneumatic and electronic panel instruments has been reduced; instrument density in control panel arrangement has increased; and field connections have been simplified (for panel mounted instruments). Electronic instruments increasingly use solid-state components for increased reliability and improved maintainability. One of the most significant improvements introduced is the modular concept of construction. Plug-in assemblies and subassemblies have made replacement and repair much easier. Elimination of vacuum tubes and the use of solid-state devices have given components longer life expectancy and more freedom from vibration, heat and other environmental problems.

Several instrument manufacturers now offer control room instruments using "split architecture," which locates the functions needing continuous operator attention in a console assembly but locates terminations, calibration and maintenance functions in remote cabinets or racks, perhaps in an adjacent room. The separate racks and consoles can be prewired and tested at the factory. While allowing smaller control rooms, split architecture adds space requirements for the auxiliary racks.

Analytical Equipment

Analytical techniques for on-line measurement and control are changing rapidly, particularly in gas chromatography. Reliability and reproducibility have been recognized problems. These are improving with better sampling systems, more reliable programming techniques and the ever-increasing storehouse of applications data. Memory units have been developed allowing the periodically sampled signal levels to be used on a continuous basis. Other significant analytical techniques and devices include:

1. pH measurement and control for product quality improvement, corrosion inhibition and neutralization of waste effluents
2. Oxygen analysis for its essential role in oxidation, combustion and processing applications (the analysis for dissolved oxygen is highly pertinent to the ecological field)
3. The measurement of total hydrocarbon content in waste streams
4. Numerous methods of moisture analysis important to many processes in the refining, chemical and petrochemical industries as well as other processing applications
5. A variety of other physical and composition measurements including conductivity, density, turbidity, refractometry, colorimetry and viscosity

These type measurements are being refined and improved to provide essential characteristics of processes which determine quality and reproducibility. These, of course, are prerequisites to product profitability.

Basic Measurements of Flow, Temperature, Pressure and Level

The accuracy of these basic process measurements have not increased significantly in recent years. Accuracies to within ±½ to ±1% have been attainable for a long time. Many devices accurate to within ±¼% (or better with proper calibration) are available now, but this cannot be claimed as an outstanding achievement. Several new principles and techniques for flow measurement have been introduced and find increasing use. These include electromagnetic meters, turbine meters, magnetic reluctance and capacitance change differential pressure transmitters, swirlmeters, vortex shedding meters, several variations of mass flow gas meters which utilize gas flow to remove heat from constant power/heat sources (in this type mass flow meter, the amount of heat removed depends on the thermal properties of the gas which either must be known or else calibration must be made to determine the thermal properties), and sonic and ultrasonic sensors which are mounted through the pipe wall or simply clamped on the outside.

The swirlmeter and the vortex shedding meter use flow disturbance in the line to provide pulse signals detected by resistance elements. These meters are new enough that their use is not yet widespread.

Methods of temperature measurement have not changed appreciably in the past few years. Filled systems and thermocouples are still predominant in use. Radiation and optical pyrometers are suitable for certain applications. Resistance elements are used when high accuracies are needed. There probably have been fewer new techniques introduced in temperature measurement than in any other of the basic process measurements.

Few changes have been made recently in pressure measuring techniques. Bourdon tubes, spiral elements, diaphragms and bellows remain most often used. Probably the greatest change in pressure measurement has been the increased use of strain gauges. They are particularly adaptable for high-pressure measurement where high accuracy and fast response are necessary.

There have been many new techniques introduced for level measurement. As a complement to standard float, differential pressure, gas bubbler and displacement types, there is increasing use of newer capacitance, radiation and ultrasonic type devices. The newer devices are more likely to be used for special applications where the more standard types are not suitable.

Thus, when the question is asked about the status of instrumentation today, the answer depends largely on the experience of the person queried. People who are researching and developing new equipment, studying advanced control theories and techniques and applying them to new process installations have vastly different concepts of "where we are" than the people whose duties relate to 15- or 20-year-old plants that are still functioning economically with control hardware of the same vintage.

"Where we are" is somewhere between the two extremes. There have been significant advances in the application of advanced control methods, in improved equipment designs and in the acceptance of sophisticated data acquisition, monitoring and control systems. However, recognition of these advances and improvements does not warrant scrapping existing operable controls. They will be replaced by better and more sophisticated controls only when replacement can be economically justified.

Looking to the Future

A historical review of instrumentation and an analysis of "where it stands today" lead to the question, "What will instrumentation be like tomorrow?" Some needs, trends and predictions for the future can be enumerated.

One of the greatest needs is for new and improved measuring sensors. While the technology for processing, displaying and using sensor-produced signals has increased tremendously, there has been a notable lack of progress in introducing new sensors or improving accuracies and realibilities of those existing.

In the basic process measurements of flow, temperature, level and pressure, the majority of measurements have accuracies of ±½ to ±2%. Most flow measurements are still made by the restriction orifice method where many errors may exist because of the condition of the element, the installation method, the calibration of the instrument, etc. The other basic measurements have similar problems in application. There needs to be an order-of-magnitude increase in accuracies. Instead of ±½ and ±2%, there should be ±0.05 and ±0.2%, or even ±0.01 and ±0.02%, which would be much closer to matching accuracies available in some of the signal processing devices.

In addition to accuracy improvements, sensors should be unaffected by ambient temperature variations, vibration, power supply fluctuations and corrosive environments. A better understanding is needed to identify applications which are critical and utilize the highest levels of accuracy and dependability for those cases.

Because of the cost of handling and installing field devices, they should be small, compact, simple and easy to install and calibrate.

There is a need now for new insights into processes. Instead of the standard flow-temperature-pressure-level measurements, there are more meaningful physical functions and process conditions that might be used. For example:

1. Heat transfer coefficient is preferable to steam pressure alone.
2. Mass flow is better than orifice ΔP.
3. Knowledge of complete chemical composition is more valuable than proof of the presence or absence of a particular component.
4. Sensors to measure thermal conductivity, sound, speed, shear modulus, color, odor and other physical properties of products are needed.

The increasing use of computers dictates the need for digital transmitters and transducers outputting directly into the computer. Final control elements must be capable of receiving directly the digital outputs from the computer. A digital pressure transducer has recently appeared on the market, as well as a digital valve positioner, but widespread usage of these elements has not been evident.

Computer usage will also permit redundancy not presently possible. With the present emphasis on large, single train processes, instrument failures that cause shutdowns are very expensive. Multiple measurements of critical process variables may be feasible where action may be taken on the best two-out-of-three measurements, with alarms functioning at any time that one of the three measurements deviates from the other two, outside of prescribed limits.

Predicting sizes and configurations of computer systems is difficult. As previously mentioned, large sophisticated systems are being installed, on line, with full backup capability, and it appears certain that these systems will continue.

A new trend seems definitely established in the control industry—the application of mini-computers or microprocessor-based systems to perform many of the logic and control functions normally associated with conventional instrumentation. These computers are frequently limited or dedicated to a single unit or piece of equipment critical to the process to monitor, to perform logical decisions, to control and sometimes to optimize operation of the unit.

This trend will likely increase because:

1. Programming becomes standardized for various process units (i.e., fractionation).
2. Hardware costs decrease as the integrated-circuit impact in mini-computer manufacturing increases.
3. These systems will compete cost-wise with conventional instrument controls as overall costs decrease.
4. Many of the advantages of small computer systems can be obtained with far less capital and operating risk than for a full-blown system.
5. These systems may be linked to an overall hierarchial computer management system when the need arises.

Microprocessors are now being used in individual instruments. Ideally suited for dedicated measurement and control tasks, "computers on a chip" are now economical and versatile enough to perform functions that range from simple logic decisions to complex analog control.

It has been estimated as early as mid-1978 that 10-15% of current chromatographs had microprocessors within their programmer-controllers. Microprocessors provide versatility that was never before possible in analyzer control, and will certainly be found in nearly all of the chromatograph product lines within a very short time. Analyzers currently microprocessor-based include chromatographs, hydrogen analyzers, carbon dioxide-, sulfur-, air-, and other stack-analyzers, mass spectrometers, corrosometers, and particle size analyzers.

Data logging equipment of the future will continue the trend towards sophistication. Recent equipment features programmable functions as well as internal signal conversion, linearization, and engineering units readouts.

Conversion of existing plants to computer control will continue as they are economically justified, depending on several factors.

1. Careful studies of the process to assure that the conversion will increase profits.
2. The availability of personnel for programming, operating and maintaining a more complex system—long-range planning for acquisition and training of these people can be difficult and expensive.
3. An analysis of the impact of computer downtime on the process and plant. Reliability of computer systems has increased greatly, but the cost of downtime may dictate redundant or backup systems which would push total systems cost out of economic range.

Configurations of computer systems in older plant conversions depend on the type of existing instrumentation, the size of the plant to be converted and whether or not concurrent control room consolidations are occurring. Depending on these circumstances, all-new electronic transmitters may be selected, or pneumatic to electronic transducers may be used.

Microprocessors are now forming the base for a number of control systems that provide the operator in the control room fingertip access to the process data. Video readouts show overall process parameters, a group of control parameters and their set points, and/or the detailed data surrounding an individual loop. Using color CRTs, data can be displayed with priorities arranged by color. Excellent high-resolution B/W CRTs also display data using black, white, gray, inverse, flashing, various character and line widths and sizes, and even characters from other languages. Programming can be accomplished at more than one level of expertise so that operators with varying degrees of experience can control the process without being intimidated by its electronic control system.

Graphic display, particularly the interactive cathode-ray tube devices, is finding increasing use for operator interfacing—for display of trends, profiles, tabulated data, as well as diagrammatic display with real time process data information. An associated keyboard allows additional data entry or changes.

The use of analytical instrumentation will continue to increase. On-line control of processes has been achieved in many other industries. The food industry, for example, is using gas chromatography to describe or identify compounds which are partly or largely responsible for good and bad odors and flavors in food products. In the beverage line, pH values are used to determine the condition of the finished product.

Pollution control regulations are introducing a complete new era in analytical instrumentation. Because of legal implications, the design of analytic systems must yield not only qualitative and quantitive information but also the time analysis occurred, the duration of the analysis and the source of the various contaminants. Many of the measurements required cannot be continuously measured by existing instruments. Sample systems themselves become a difficult design problem.

Signal transmission techniques are changing. The use of telemetry has made remote control of pipeline compressor stations possible. Greater exploitation of techniques are likely for other industries. Cost reductions are possible through use of time-division multiplexing, analog-frequency channels and audio-tone matrixing. The use of remote multiplexing and analog-to-digital conversion units with serial digital transmission to a central area will be common. Eventually, the use of microwave or radio transmission may eliminate hard-wire transmission altogether. A system has already been demonstrated that can transmit and receive 248 points of control over a single pair of wires. The use of dedicated microprocessors in the field that report only data which has changed will allow quicker control action or coverage of more field points.

The increase of electronic instruments has caused some people to predict the phase-out of pnuematic instruments. Others predict the use of pneumatics into the 1980s at least. A significant contributor of the demise of pneumatics could be the use of wireless transmission techniques which would drastically reduce signal transmission costs.

Miniaturization of equipment and the high density arrangement of readout and control devices have been evident for several years. There is little indication that the ultimate has been reached in this respect. As this trend continues, the need for more alert and better trained operators will be evident. The present limitation on this type display is the physical size of the meters, knobs and switches required for adequate observation and operation by the operator.

The next step was limited display/operating function with blind controllers. In such a system, the controllers function normally within their prescribed limits, but with no display or available controls. On operator demand or on out-of-limits operation, the control loop and its necessary operating functions are selected through logic circuits to be displayed on the control panel, permitting the operator to view and control the loop as required, thus minimizing the requirements for panel face area. The next step in development would appear to be advanced control strategies, in which empirically-determined process equations are solved by dedicated minicomputers or microcomputers which then control operating unit parameters to provide higher production rates, high quality product, or reduced energy requirements.

Preparing for Now

As one contemplates the changes of the recent past and the possibilities of the future with its advancing technology, it is easy to overlook the problems of today. This must not happen. Equipment and plant facilities worth billions of dollars are being built today and planned for tomorrow. It is essential that the best equipment be selected for the job consistent with the control philosophies for each of these facilities.

It seems ironic that a recognized study discipline has not been assigned to the technology of instrumentation. Few schools offer degree courses in the field, yet no discipline covers such a wide area of application and none is more important to today's industrial world.

There is a need for expertise at every level involved in instrumentation—in the design and development of the instruments, in the application and design of control systems and in their maintenance. The people needed are difficult to find and hire; they are difficult to keep. Some can be bought, but in general they must be reared and trained.

The ultimate success of any plant control system rests on the ability of instrument experts to make proper application of components and systems and on the ability of maintenance people to keep them calibrated and working properly. The purpose of this volume and the three volumes to follow is to prepare people for these tasks.

2 Fundamentals of Automatic Control

William G. Andrew

The study of automatic control is a complex subject. Exact solutions to particular control problems require detailed process knowledge, not only of the physical and chemical characteristics of the fluids, but also of the mechanical aspects of the process—equipment (pumps, mixers, reactors, heat exchangers, etc.), piping systems and the control loop itself. Fortunately, good control usually is achieved with limited knowledge of these physical, chemical and mechanical conditions, and control applications are made quickly and economically with a high degree of confidence in their successful operation. The next few paragraphs discuss some fundamental principles underlying all control problems. A review of these principles makes it easy for the engineer and designer to cope with control problems when the occasion arises.

Control Defined

Automatic control may be defined as the technique of measuring the value of a variable and producing a counter response to limit its deviation from a selected reference. Many other equally suitable definitions could be given, but automatic control can be understood better in terms of why it is needed, the advantages it offers and the forms of energy it controls. A really simple approach is taken in the discussion of these facets, for a complete understanding of basic principles is necessary in comprehending and evaluating control problems.

A closed control loop may use several devices to accomplish the control of a process variable. Regardless of the number of elements used, the loop will contain at least four basic elements: (a) detecting, (b) measuring, (c) controlling and (d) final control element. Often, a transmitting element is added to these. More than one element is often designed into the same housing so that a loop does not always contain four separate units.

Purpose

The one basic purpose for using automatic controls is that production is achieved more economically. Some processes would not be possible except through the use of automatic controls. Economy is achieved in several different ways:

1. Lowering labor costs
2. Eliminating or reducing human errors
3. Improving process quality
4. Reducing the size of process equipment and the amount of space it requires
5. Providing greater safety in operation
6. Minimizing energy consumption

Material Properties and Energy Forms Controlled

The control of any chemical process reduces to a balance of materials and energy related to that process. It involves operating conditions peculiar to its requirements, such as the ratio of material flows, and the correct temperatures, pressures, fluid levels and material composition. The conditions mentioned are interrelated, and normally the more successful processes are those in which these interrelationships are known and controlled. The next few paragraphs discuss various control parameters peculiar to almost all processes.

Flow

Any continuous chemical process requires the control of material flows. Flow control loops are more common than any other single process variable. The control of other variables usually depends on flow regulation to achieve stability. For example, most temperature control systems are achieved by regulating the flow of the heating or cooling mediums.

Many chemical processes are sensitive to the ratio of ingredients for proper reaction, and nearly all of them require correct proportioning of materials to achieve quality products. Product properties are affected by varying the ratio of raw materials and production rates vary with catalyst addition; so whether the fluid is liquid, gas or solid, accurate flow control is nearly always essential. The various methods of achieving good flow control are discussed in Chapter 4 of this volume.

Temperature

Energy in the form of heat is another important variable which is controlled in most chemical reactions. In reactions where heat in uncontrolled, it must still be removed from the process in some way. Maximum yields, completeness of reaction and desirable product characteristics—all are functions of temperature in most processes. Some processes operate close to their critical temperatures. When critical temperature is exceeded, the process "runs away," it reaches an uncontrollable state. Good automatic control techniques normally prevent that from happening and are considered economic necessities. Although most processes have more flow than temperature loops, the need for good temperature control cannot be overemphasized.

Pressure

Pressure control is necessary in all chemical processes. Many reactions are functions of pressure (positive or vacuum) as well as temperature. In addition to its control to obtain the desired condition for a reaction, pressure is also necessary to the delivery of materials through equipment and piping systems at desired flow rates.

Level

Still another important function in controlling continuous processes is level control. It is used (a) to obtain process material balances when variations occur in raw material flow rates, (b) for proper functioning of fractionating towers, settling tanks, reactors and other equipment and (c) to regulate the flow of intermediate and finished products to and from temporary storage facilities. Like temperature control, level control is often closely associated with material flow control.

Other Controllable Properties

The four variables previously mentioned comprise a high percentage of the control loops of most processes. Some may be clearly defined as a form of energy control while others do not readily fall into this definition. Other variables measured and controlled include the following list.

1. Component analysis. There are several types of analytical devices used to measure process stream components or otherwise determine the stream mixture. Sometimes the need is to determine and maintain the presence or absence, within prescribed limits, of one or more components. These analyses are made with chromatographs or other special analytical devices, many of which are described in Chapter 9.
2. Physical properties. Physical properties which often must be controlled include viscosity, specific gravity, melt index, melt flow, haze, turbidity, initial boiling point, end point and color. Some of these properties are peculiar to only a few processes, and some are functions of one or more other variables.
3. Chemical properties. For many processes chemical properties which need to be controlled include pH, BOD, TOC, conductivity and redox.
4. Miscellaneous energy forms. Still other variables which are sometimes controlled include speed, frequency, power, voltage and current.

Responses of Detecting Elements

A time lag always occurs between a change in a process condition and the notice of change by a process operator or an automatic controller. This time lag is due to three distinct causes: the lag inherent in the process, the response lag of the detecting element and the signal transmission lag. In some control loops all three lags are negligible; in others, none are negligible. In most cases one of the three sources has a much greater effect than the other two. In attempting to minimize lag effects, detection response should be considered first and its lag effect reduced if possible. The purpose here is to discuss the lags of detecting elements; the other two lags are discussed in Chapter 3.

Measurement lags imply errors whenever a process is changing. The measurement is not only late but also inaccurate because it is continuing to change even as the reading is made available. The slower the response, the more inaccurate may be the measurement when received. A large disturbance of short duration, however, may go completely unnoticed if its time duration is short compared to the measuring lag. In such instances the disturbance probably has little effect on the process.

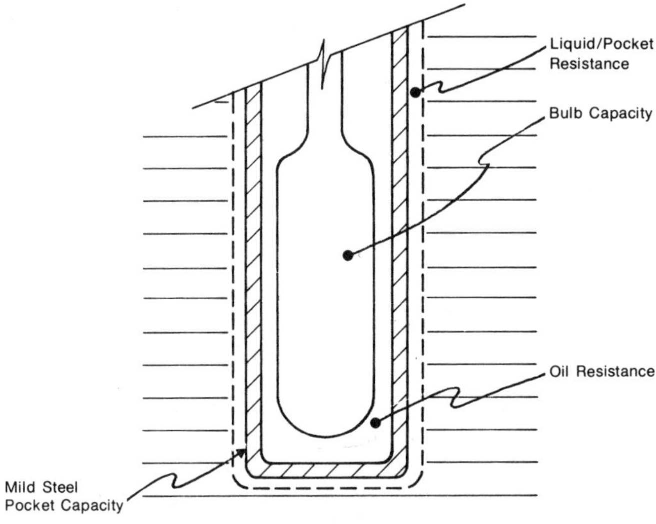

Figure 2.1. Temperature response time to a step change in temperature depends on the capacity of the bulb, the difference in temperature and the resistance to change which includes the boundary effect, the capacity of the well material and the medium between the well and the bulb.

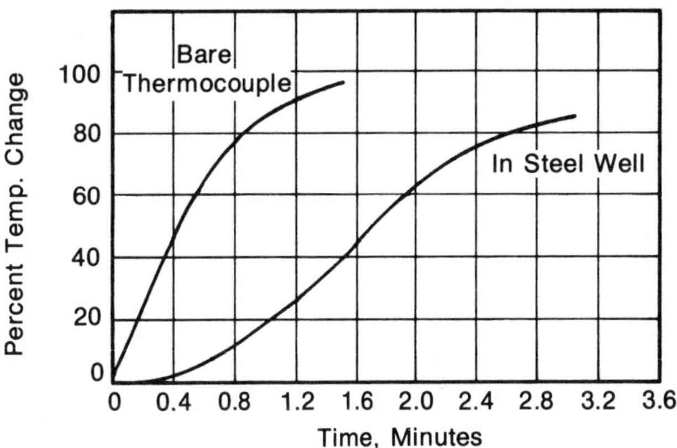

Figure 2.2 Response curves of a bare thermocouple and a well-immersed couple to a sudden sustained temperature change.

Temperature

A classic example of detecting element lag is a temperature measurement using a filled system for temperature detection (Figure 2.1). The response to a step change in temperature at the element depends on the thermal capacity of the element and on the rate at which heat is exchanged between the element and its surrounding fluid. The rate of heat exchange, in turn, depends on the resistance of the element to heat flow across its boundary and the difference in temperature between the element and the fluid.

The thermal capacity of the detecting element is a function of its size, shape and material. The resistance to heat flow, however, depends on the nature of the flowing medium and its velocity. The resistance of the boundary layer increases with thermal conductivity, specific heat and density of the fluid; it decreases with fluid velocity. The boundary layer resistance for gases is much greater than for liquids; therefore, rapid responses for gases are more difficult to obtain.

Figure 2.2 shows the response curve of a bare thermocouple placed in a fluid of higher temperature. It is an exponential curve which reaches 83% of its final amplitude in about one minute. Another curve shows the response of a thermocouple element in a well immersed in the same fluid. The delayed temperature rise caused by the capacity (or amount of mass) whose temperature must change should be noted.

Of the common methods for detecting temperature changes, the filled system has the slowest response speed. The response of resistance elements and thermocouples are faster and depend more on the thermal capacity of the protecting well used for insertion into the fluid than on the detecting elements themselves. To ensure quicker responses, the following precautions should be observed.

1. A close fitting should be provided between the element and its protecting well.
2. The protecting well surface should always be clean.
3. A high velocity of the fluid to be measured must be maintained. (One ft/sec or more for liquids and 10 ft/sec minimum for gases.)
4. Protecting wells with thicknesses as small as possible consistent with plant specifications should be selected. (Pressure requirements must necessarily be met.)
5. Protecting wells should be immersed deeply into lines and vessels.

Flow

The most widely used flow elements respond almost instantaneously to flow changes, including differential type devices, rotameters, turbine meters and magnetic meters. One type flow device that theorectically is subject to measurement lag is the thermal flowmeter. There are several designs of this type, most of which use heated elements. Measurement lag is introduced by temperature sensing devices which maintain a constant temperature differential across a section of the flowmeter. The lag in these devices is usually only a few seconds—even in extreme cases.

Pressure

Pressure elements, like flow elements, respond almost instantaneously to step changes in pressure. Detection elements such as bourdon tubes, other spiral elements, diaphragms and bellows respond quickly to pressure changes, as do strain cells and most other pressure measuring elements. Element response time for pressure devices normally presents no problem unless the volume of the system to be controlled is small.

Level

Most level devices, such as float and displacement types, capacitance and ultrasonic types, have very fast response times. One device that does respond a little more slowly is the radioactive device operating on the principle of gamma ray absorption. Its lag time may range from a fraction of a second to a second or two. The reason for the delay is the characteristically low level signal base where background noise must be filtered from the true signal base. The time constant is variable.

Analytical

The response time for analytical devices ranges from instantaneous, for pH or conductivity elements, to several minutes for chromatographic analyses. The analysis time for multiple streams or for multiple components in single streams depends on the components to be measured, those which are to be discarded and the flushing and backwashing procedures which are sometimes required for a careful analysis.

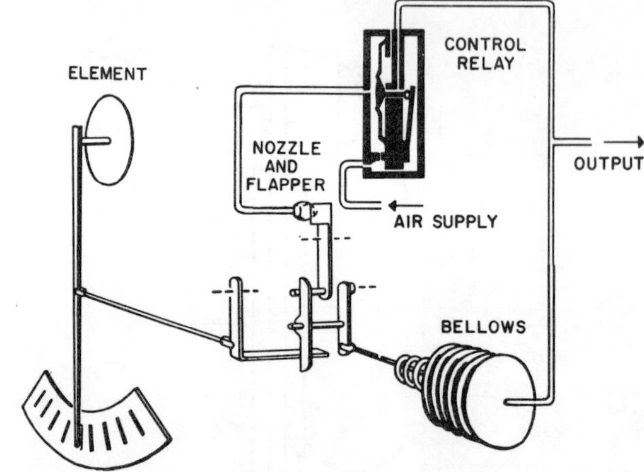

Figure 2.3. The expansion or contraction of the pressure bellows exerts, through linkages, pen movement along the scale for local indication or flapper movement at the flapper-nozzle assembly to produce a pneumatic output signal proportional to bellows movement. (Courtesy of Foxboro Co.)

Measuring Methods

After automatic control is defined, its purpose is discussed, controlled energy forms are analyzed and response times of detecting elements are briefly examined, the next logical step in understanding automatic control systems is to analyze measuring methods and the conversion of detected responses into readable signals.

Pressure and differential pressure devices are constructed so that increasing or decreasing pressures produce linear movements on pointers or dials which move over calibrated scales (Figures 2.3 and 2.4). At the same time that they provide a local indication, they may also provide a movement to a flapper in a flapper-nozzle relationship to give a proportional output signal for remote use. The same principle applies to filled temperature systems where the pressure of constant volume systems changes with the contraction and expansion of the fluid in the system (Figure 2.5). Indication or transmission of these measured values is accomplished primarily by balancing systems commonly referred to as motion balance or force balance measuring systems.

Motion Balance

In the early days of process control, among the first control tools to be used were such simple indicators as glass bulb and dial thermometers for temperature, dial pressure gauges for process pressure measurement and U-tube manometers for differential pressure. As processes became more sophisticated and the desirability for automatic control increased, the need not only to measure but also to transmit process variables to remote locations became more

Figure 2.4. Movement of the differential bellows moves the pointer over a 270° arc through the torque tube whose motion is multiplied by a gear and pinion. (Courtesy of Barton Instrument Co.)

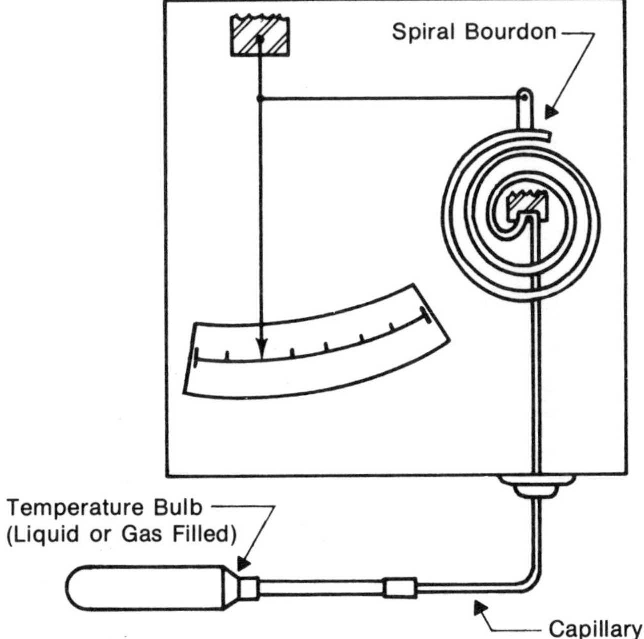

Figure 2.5. Fluid expansion or contraction with temperature changes produces bourdon movement that is read on a calibrated dial.

intense. The early indicators mentioned above were (and some still are) motion balance devices. So were the first transmitters.

A typical motion balance pressure transmitter is shown in Figure 2.6. The process pressure acting on the bellows moves a beam or spindle. The amount of beam motion is determined by the force of the restraining spring (or weight) as depicted in the illustration. The action of the process pressure on the bellows at the left forces the beam in a counterclockwise direction. The motion is restrained by two forces: (a) the spring rate of the measuring bellows itself and (b) the spring at the right hand side of the pivot beam. Instruments are usually designed so that process pressure changes result in motions which typically range from one-quarter to three-quarters of an inch. This usually provides a sufficient motion to indicate accurately the process variable.

For transmission of pneumatic signals, the beam position is sensed by a flapper and nozzle arrangement (Figure 2.6). The beam itself serves as the flapper, and as it moves toward the nozzle, the back pressure from the air supply increases linearly with decreasing distance between nozzle and flapper, providing the increasing signal from the transmitter. The output is also sensed by a feedback bellows which repositions the nozzle with respect to the flapper, bringing the transmitter into servo balance as dictated by the new beam position. The only restraining force exerted by the transmitter itself on the beam is a small force generated by the nozzle pressure. The primary restraining force is exerted by the spring.

The following are the main advantages of motion balance transmitters.

1. Normally they are slightly more economical than force balance transmitters in terms of initial cost.
2. They provide direct indication of the process variable.
3. Direct, correct local indication may be available even though the transmitter portion is not operating properly.

Disadvantages of motion balance transmitters include:

1. A large number of moving parts, linkages and pivots which may result in friction and wear
2. Limited span and zero adjustments compared to force balance transmitters

Although many simple motion balance indicators are employed, their use as transmitting devices has declined appreciably in the last few years.

Force Balance

The force balance transmitter was introduced in the late 1940s. It is similar in construction to the motion balance transmitter. In pneumatic types, the process pressure is sensed by the measuring element which pushes against a beam (Figure 2.7) like the motion balance device. Instead of being opposed by a restraining spring, however, the output signal itself pushes back and generates the force necessary to restore the beam to its servo balanced positon. As the variable changes in any direction, the beam assumes a new position, changing the flapper-nozzle relationship and, thus, the output signal. The output signal is also sensed by the feedback bellows which acts as a restoring force on the beam to counteract the force generated by the process variable.

The force balance transmitter has the following advantages.

1. It is simpler than the motion balance type.
2. It requires less maintenance.
3. Servicing is usually easier.
4. Adjustments for span and zero are usually easier to make.
5. Moving parts are fewer.
6. It normally has no pivots or bearings.

Force balance transmitters are usually considered to be motionless, but they do undergo a slight motion of a few thousandths of an inch.

The electronic force balance transmitter (Figure 2.8) is similar to its pneumatic counterpart. The flapper and nozzle arrangement is replaced with some type of electrical detector, often a coil whose inductive reactance is modified by the proximity of the beam. Electronic amplifiers (normally transistorized or with integrated circuits) convert the

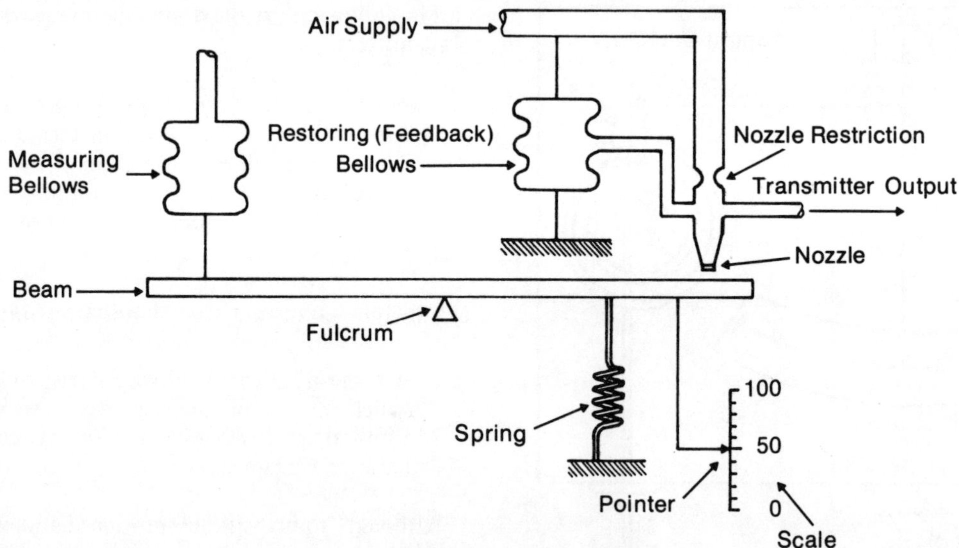

Figure 2.6. Pressure from measuring bellows imparts motion to beam for indication and provides a flapper motion at nozzle to provide a linear output pressure in this motion balance indicating transmitter.

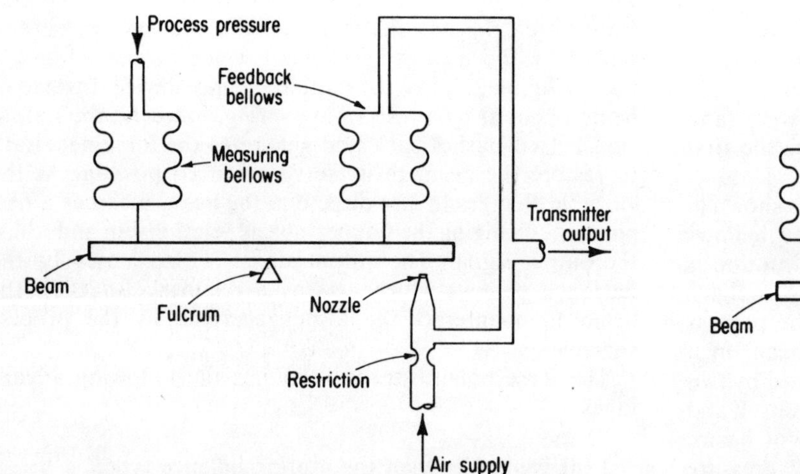

Figure 2.7. Force balance transmitters utilize the signal output pressure to provide a restoring force to oppose the measuring bellows and provide a servo balanced system.

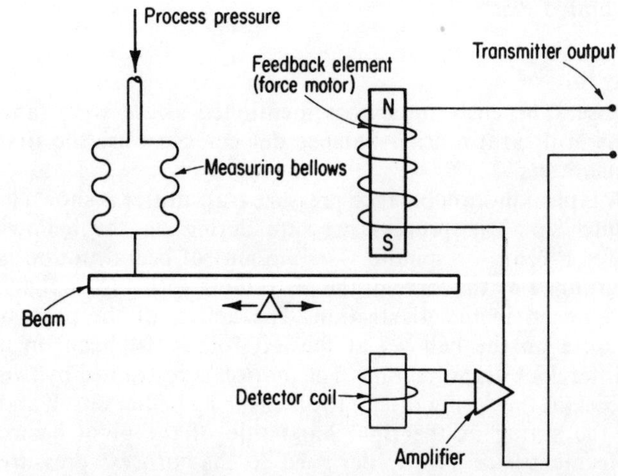

Figure 2.8. This electronic force balance transmitter uses a force motor (feedback element) powered by the output signal to oppose the measuring bellows and provide the servo balanced system inherent in the force balance principle.

detected signal to a DC current which serves as the transmitted signal. The feedback force (as in the pneumatic system) is provided by the output signal and is used to counterbalance the force generated by the process variable. The feedback element (called a force motor) consists of a coil within the permanent magnet field .

The span of the transmitter is changed by simply adjusting the position of the beam's fulcrum. Range adjustments of 10:1 are typical for force balance transmitters.

Temperature and differential pressure transmitters are similar to the pressure type just described. Small mechanical movements of primary sensors are used to set up self-balancing systems which are accurately calibrated against values of the measured variables. Differential transmitters are capable of measuring a very small differential pressure (in inches of water range) above a static pressure that may be as high as several thousand pounds. A typical range is 20 inches H_2O in a service that may go as high as 6,000 pounds per square inch static pressure.

Chemical Reaction

From the previous paragraphs discussing motion and force balance methods of indicating and transmitting measured variables, it is easily recognized that the detecting element relies on dynamic or static sources of energy for actuation—flowing fluids and pressure sources or static fluid heads (in the case of level measurement).

Another energy source used for detecting desired variations in processes is that of chemical reactions. Examples are found primarily in analytical devices such as oxygen analyzers (catalytic-combustion type, microfuel cell, polarographic cell, galvanic cell and the thallium electrode), total organic carbon analyzers, certain types of moisture analyzers and pH analyzers, to mention a few. Conductivity measurements and pH analyzers may be thought of primarily as electrical energy devices, but the electrical current that is produced is a result of chemical reactions.

Electrical Energy

In addition to voltages and currents produced by chemical means, there are electrical energy sources by other means. These include current, voltage, power, phase angle, and frequency measurements made using current and potential transformers; voltages developed at thermocouple junctions; currents generated by magnetic flow meters and turbine flow meters; capacitance level measurements; sonic level and flow transducers; light emitting diode (LED) and photodiode optoisolators; and many other devices powered by external AC or DC sources but using electrical components for detection (resistances, inductances, capacitances, balanced bridge circuits, etc.).

Physical Change or Separation

Another method used to detect and measure process variable characteristics is to note physical changes or cause physical changes to occur and at the same time measure the desired characteristic. A notable example is the chromatograph which physically separates individual components of a stream and uses various methods to measure them quantitatively. Other examples include electrolytic hygrometers, infrared analyzers, refractometers, turbidity analyzers, etc. The turbidimeter, for example, measures the amount of light reflected by suspended particles in the fluid sample. Any change in concentration of suspended particles allows more or less light to be reflected to a photocell, and the change is converted to an electrical signal which is proportional to the suspended particle concentration.

Controllers

For review, the following are fundamental elements of any control loop:

1. Detecting element
2. Measuring element
3. Controller element
4. Final control element

Two or more of these functions may be combined or placed in a single enclosure, but this is not usually the case. Detecting elements and measuring methods have been discussed in some detail, and now the control element is considered. Since the next chapter treats the subject of automatic control methods in great detail, this section gives only some basic concepts to provide continuity to the fundamentals covered in this chapter.

A controller may be defined as a device which compares the value of a measured variable (signal input) to its desired value (set point) to produce an output signal that maintains the desired value. Another way to state its function is that a controller senses the difference between the actual and desired value of a variable and uses this difference to manipulate the controlled variable. Controllers are classified as pneumatic, electronic or hydraulic. The hydraulic type constitutes only a very small percentage of all controllers and for practical purposes is given little consideration.

Figure 2.9 shows a typical pneumatic control loop with the four fundamental elements previously mentioned. They are designated as (a) detecting, (b) measuring, (c) controlling and (d) final control elements. The controller section of the loop consists basically of a flapper-nozzle arrangement (Figure 2.10) in which a signal representing the measured variable moves the flapper toward or away from the nozzle to change its back pressure. The controller shown is a force balance type controller, and the set point signal bellows is shown opposing the measuring signal bellows. (The controller is designed to manipulate in response to a difference between the measured and the desired or set point signal.) Figure 2.10 also shows a feedback bellows whose purpose is to "feed back" into the controller mechanism to restore balance and provide stability.

Figure 2.11 shows a motion balance type controller schematic in which a mechancial linkage receives a motion from the measuring device to actuate the flapper-nozzle arrangement. Regulation of the controlled variable utilizing feedback control, as shown in the illustrations, is the predominant method used in control devices at this time. Feedback and feedforward control concepts are discussed in the next chapter, and their relative merits are considered.

Final Control Elements

In discussing the fundamental elements of control loops, the final control element (in a majority of cases, a control valve) is logically the last of the major items to be discussed. *Final control element* is the term that has been applied to the item that finally responds to make a change in the measured variable.

Control Valves

Most process variables are controlled by the operation of valves (Figure 2.12) in the piping system, whether the controlled variable is flow, pressure, level, temperature or com-

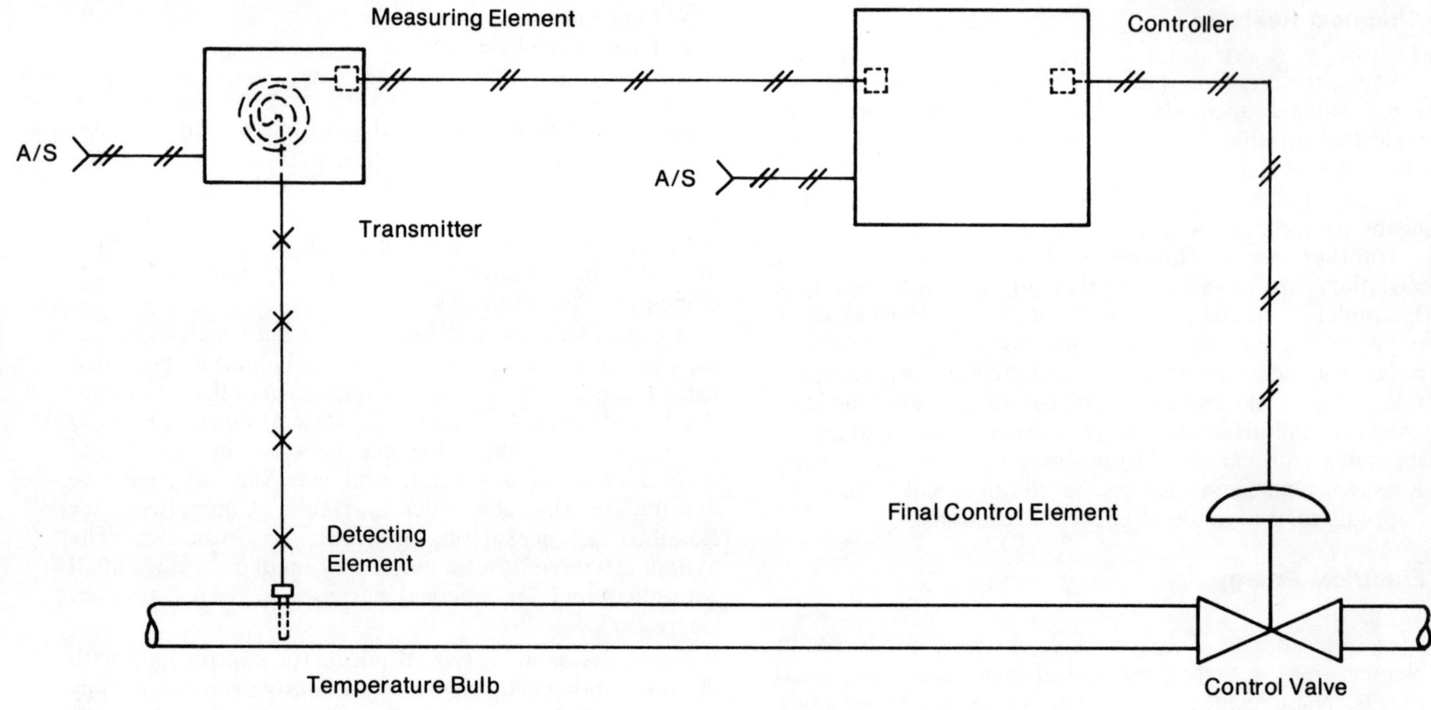

Figure 2.9. This pneumatic loop contains the four elements common to any closed control loop: detecting element, measuring element (and transmitting element in this case), controller and final control element.

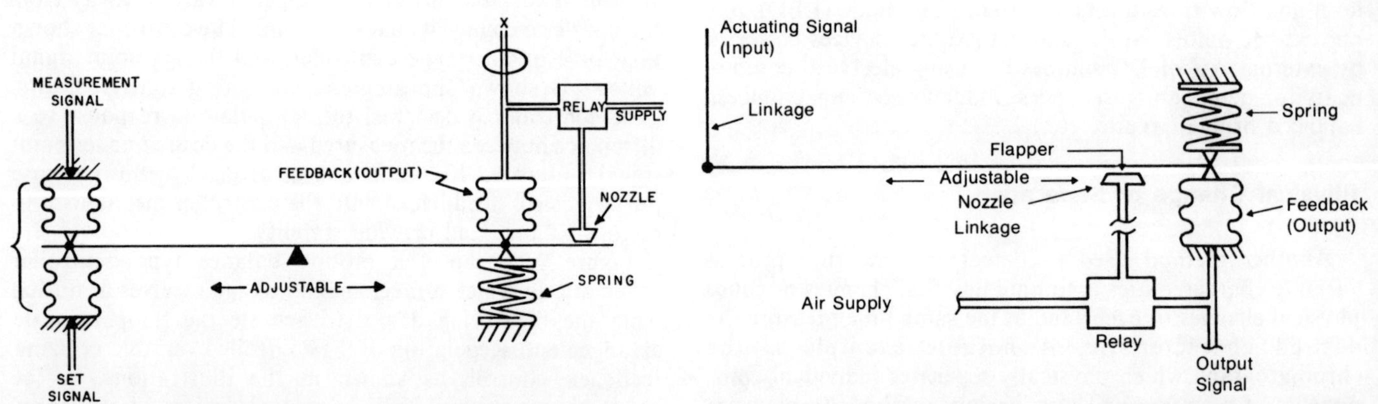

Figure 2.10. Schematic of a force balance control unit in which the force balance is achieved by balancing the measured and set point signals. The feedback bellows provides balance and stability to the system.

Figure 2.11. Motion balance control schematic showing the flapper-nozzle relationship resulting from the actuating signal and producing an output signal in proportion to that input. Proprotional only control is shown.

ponent mixture. Most fluid flows are valve controlled; the primary exceptions being metering pumps for some liquid services, louvers and dampers for some gas services and feeders and speed controlled conveyors for solids.

Pressures and levels generally are controlled indirectly by controlling flows from one part of a process system to another. Temperatures are often controlled directly or indirectly by mixing fluids of different temperatures, by controlling the flow of cooling or heating mediums or by controlling catalyst flows or other factors which affect the reaction of exothermic processes.

Other facets concerning control valves—types available, characteristics, selection, etc.—are discussed in Chapter 11 of this volume and in Chapter 4 of Volume 2.

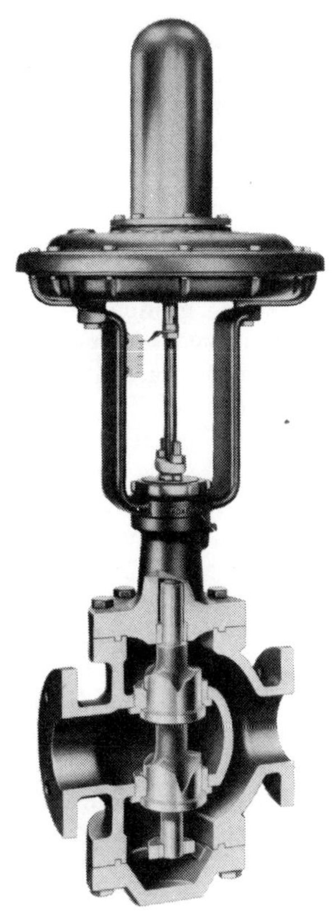

Figure 2.12. The final control element in most control loops is a control valve. This one is a double port valve using equal percentage trim characteristic. (Courtesy of Foxboro Co.)

Other Final Control Elements

Though the ratio of airplanes to automobiles is extremely small, modern transportation systems would be seriously hampered if there were no airplanes or if automobiles were the only method of transporting people. Similarly, automatic control would be seriously hampered if control valves were the only available final control element. Other final control elements used include dampers, louvers, governors, pumps, feeders and variable resistors.

Dampers and Louvers

Dampers and louvers are used for throttling gas flows, primarily where the control quality is not critical or exacting and normally where pressures are low. Typical applications include air conditioning systems and flue gas ducts where fan and blower discharges must be regulated. Leakage rates are usually high but can be reduced by using soft edging on the damper blades.

Pumps

Flow control of liquids and slurries is often accomplished by metering pumps as final control elements. The method of accomplishment varies with the type of pump.

Reciprocating pumps consist of a plunger and cylinder complete with stuffing box, packing, suction and discharge valves and other appropriate accessories. A rotary driven shaft imparts a linear motion to the plunger inside the cylinder to provide a reciprocating motion. Each stroke displaces a known fixed volume of fluid.

Control is established by adjusting the stroke length or by varying the speed of the motor driven shaft. A combination of these two methods is often used to provide an overall rangeability of 100 to 1 or greater.

An undesirable feature of reciprocating pumps is the pulsating flow pattern inherent in the pump action. This feature can be reduced somewhat by using a unique drive arrangement that allows staggered discharges which tend to smooth out flow pulsations. Single ended pumps are called simplex pumps. Duplex and triplex arrangements (Figure 2.13) are available and arranged so that the discharge strokes are phased 180° (for duplex) or 120° (for triplex) apart. Figure 2.14 shows characteristic curves for simplex, duplex and triplex pumps that reveal the comparative pulsation characteristics of the three arrangements. These pumps can be used in clean services for pressures up to 4,000 psig and higher. They are not recommended for slurries.

Diaphragm pumps use a flexible diaphragm (Figure 2.15) to achieve pumping action. The drive in this instance is a three-dimensional space crank mechanism. The worm and worm gear revolve the space crank in a conical path to provide the rod movement which flexes the diaphragm for fluid pumping.

Flow control is achieved in the same manner as for plunger pumps, by speed variation and stroke adjustment. Diaphragm pumps cost less than plunger pumps, but because of strength limitations of diaphragm materials, they are limited to much lower operating pressures. Their rangeability is considerably less than plunger pumps, but

Figure 2.13. A unique drive arrangement allows from one to three liquid ends to be pumped by a single motor in this reciprocating pump. (Courtesy of Wallace and Tiernan, Division of Pennwalt Corp.)

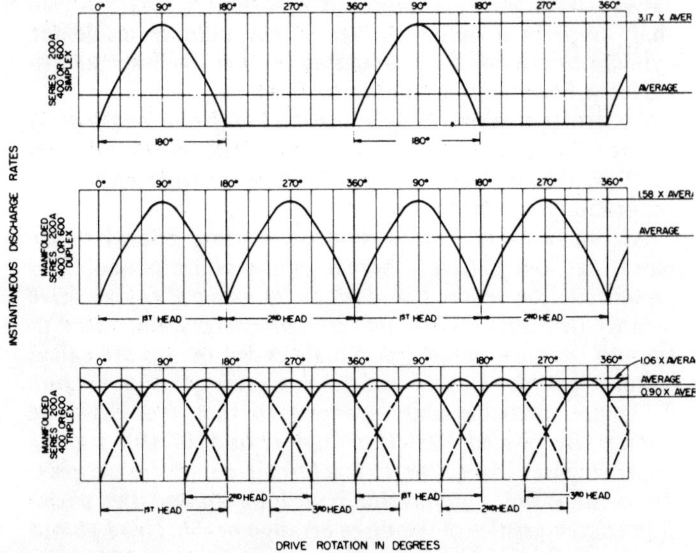

Figure 2.14. The discharge curves of simplex, duplex and triplex pumps reveal the smoothing effect on pulsations when multiple, out-of-phase discharge ends are used. (Courtesy of Wallace and Tiernan, Division of Pennwalt Corp.)

they are more adaptable to slurry services because their flow path is straight with few obstructions and no cavities.

Variable Speed Drives

Many devices use some kind of variable speed control for adjusting feed rates, for varying electrical power or for varying motor speed for various applications. Typical of these uses are belt type gravimetric feeders, rotary feeders and piston and diaphragm pumps (discussed in the previous section).

Variable-speed DC motors are used for many control applications. Variable speed is obtained by adjusting field flux, armature voltage or armature resistance. Field flux can be adjusted by a field rheostat or a silicon controlled rectifier (SCR). Armature voltage adjustment is made by motor-generator sets, SCRs, ignitrons, saturable core reactors and magnetic amplifiers. Armature resistance adjustment is obtained by inserting a variable external resistor in the armature curcuit. This method is seldom used, however, because power consumption in the resistor makes it inefficient. All these requirements for variable speed drives can be met by using open or closed loop controllers furnishing standard signals to actuate the final control element.

Other Elements

Previous sections of this chapter have listed and discussed briefly the fundamental functions of any control loop—detection, measurement, the controller and the final control element. In almost all control loops, one or several other functions are used, and some are listed and described briefly. These include recorders, indicators, transducers or converters, integrators and alarm and shutdown functions.

Indicators

Indicating elements include pointers, floats, liquid columns, light beams, etc. The magnitude of the indicated measurement is determined by providing graduated scales from a reference point to the maximum measurement value the element is expected to make. Indicating scales are usually laid out in straight lines or on a circular arc. They may be uniform, square root or nonlinear (see Figure 2.16).

Normally ranging from 2 to 6 inches, scale lengths depend largely on indicator use. Most instruments utilize fixed scales with moving pens or pointers to show the measured variable, but some have fixed reference points or reference lines with movable scales to show the measured values.

Digital indicators (Figure 2.17) that present measured variables in numerical form are becoming more commonplace. They are easy to read with little likelihood of errors that occur when deciphering scale indications. Presently temperature is the variable most commonly displayed digitally in the process industries, but almost any variable may be thus displayed if desired. Increased use of digital displays is almost a certainty.

Digital indicators use several methods to indicate data. Glowing filament "Nixie" tubes were widely used and gave good service, but they are best read from directly in front of the device and use more power than newer types of indicators. LED, gas discharge, and liquid crystal displays use much less energy and therefore give off less heat.

LED displays now come in several colors, although red should continue to be more in evidence due to visibility and lower cost. Individual LEDs are used to indicate logic, power "ON", high and low limit visual alarm, and, when arranged in a row and coupled with proper circuitry, give a bar graph type indication.

Gas discharge displays give bright display characters. Light blue-green is a usual color and can be produced in fairly large sizes.

Liquid crystal displays (Figure 2.18), by virtue of their minimal power requirements, are seeing much more use today. From wrist watches to 6-inch-high digital indicators, liquid crystal displays have entered the control room. Technicians now carry test instruments with liquid crystal readouts. Operating modes are reflective, in which incident rays of light are reflected away, as well as transmissive, in which the liquid crystals are transparent to light rays. Since a reflective mode display is dependent on ambient light reflections, readout is impaired in poorly lit areas.

Recorders

Recorders provide continuous records of measured variables with respect to time. Recorder charts use essentially the same scales that are used for indication but with an

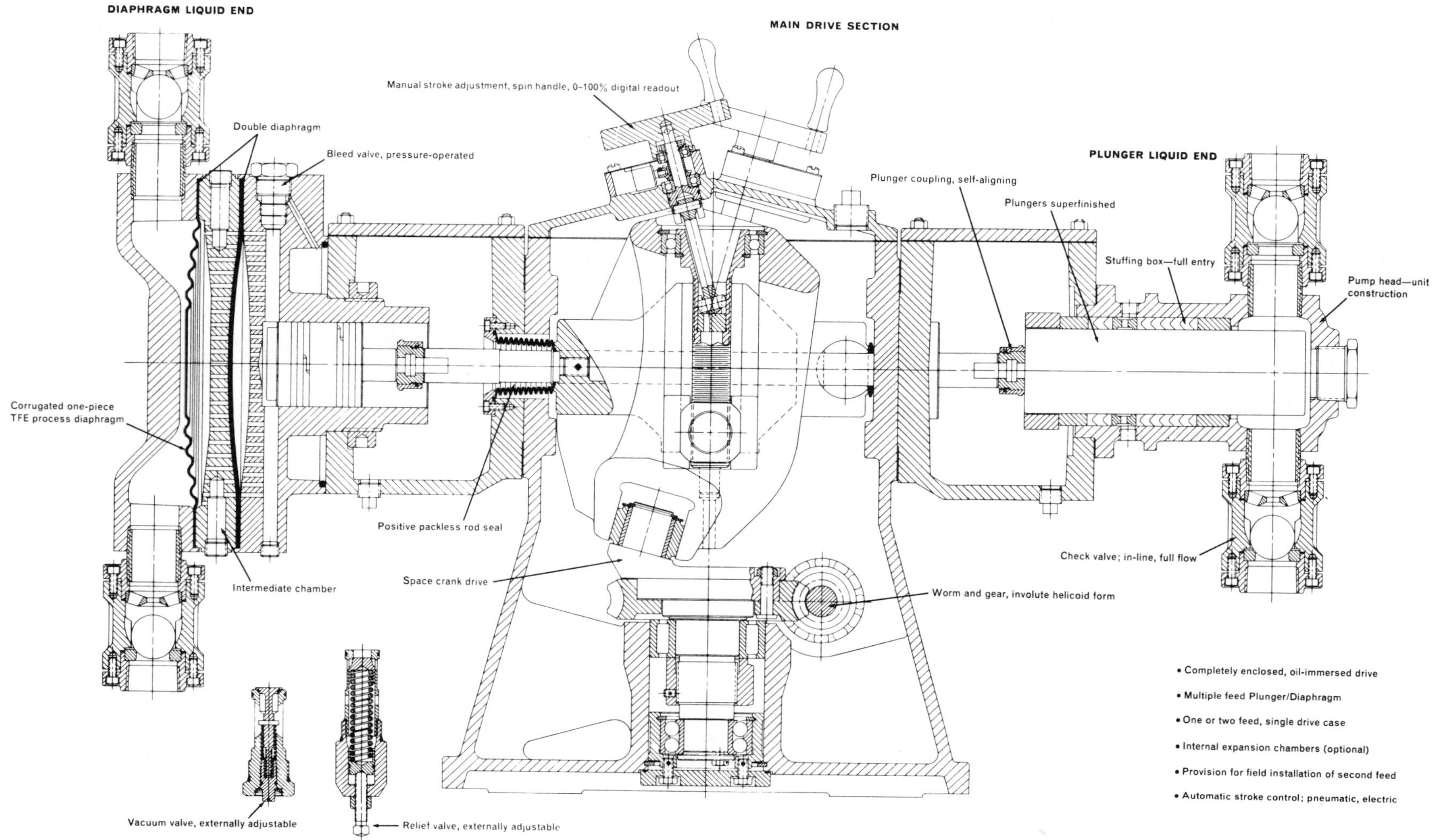

Figure 2.15. Diaphragm pumps isolate many pump parts from the process fluid and provide effective seal by flexing a diaphragm to achieve pumping action. (Courtesy of Hills-McCanna)

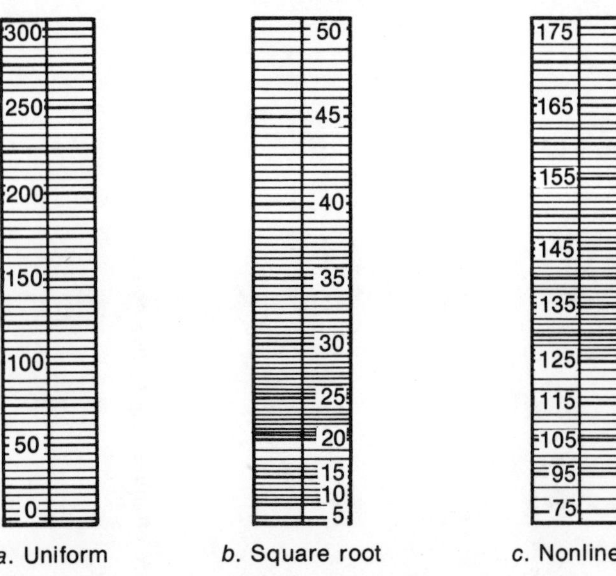

Figure 2.16. Scales vary in shape and length and calibration is uniform (a), square root (b) or non-linear (c).

Figure 2.17. Digital indicators are increasing in popularity because of accuracy and ease of reading. (Courtesy of Ircon, Inc.)

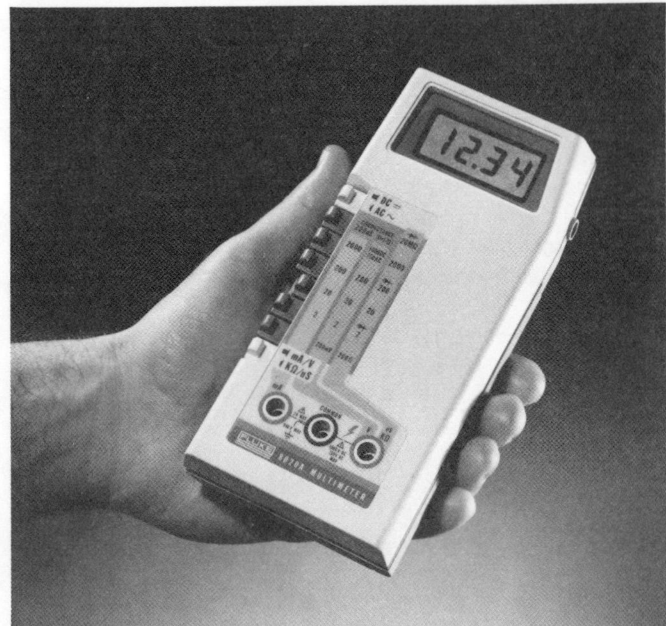

Figure 2.18. Liquid crystal displays use very little energy. (Courtesy of John Fluke Manufacturing Company, Inc.)

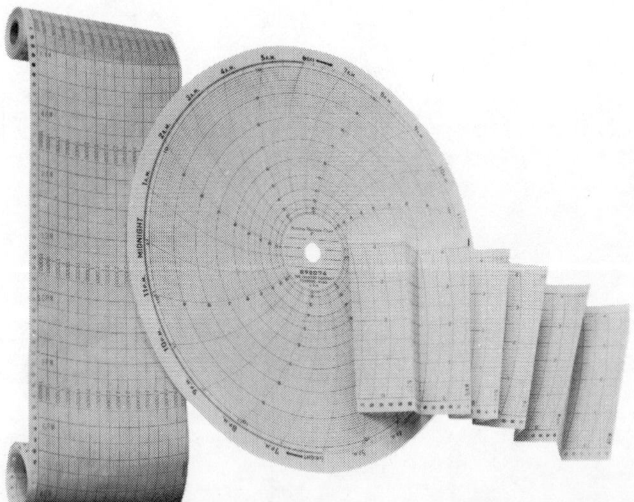

Figure 2.19. Circular charts were used almost exclusively in the early days of instrumentation, but strip charts (roll and fold types) are more practical in most cases because they are adjustable to longer records and require less attention. (Courtesy of Foxboro Co.)

added coordinate to designate time. Recorder types most frequently used are circular charts and strip charts (Figure 2.19) so designated because of their shape. Circular chart recorders were popular in the early days of process instrumentation, but strip charts are now much more common, especially in miniature sizes. Circular charts are usually 24-hour charts and must be replaced daily. Strip charts usually last for 30 days—at standard speeds of ¾ or 1 inch per hour.

Special purpose recorders such as X-Y recorders, event recorders, magnetic tape recorders and digital printers are available but will not be discussed in detail here.

Both circular and strip charts are normally calibrated or selected to read directly in process variable units that are familiar to operators or other people handling the data. Standard circular chart diameters include 6, 8, 10 and 12 inches, and other sizes are available. Circular chart speeds include ¼, 1, 2, 3, 4, 6, 8, 12 and 24 hours per revolution and 2, 7, 8, 14 and 28 days per revolution. Most charts are driven electrically, but mechanically and pneumatically driven charts are also available.

Strip charts may be furnished in sheets or in strips; the latter may be rolled or folded. Standard lengths range from 100 to 250 feet. Standard chart speeds range normally from ¾ inch per hour to 1 inch per minute. The most common speeds are ¾ and 1 inch per hour. Speeds of several feet per second are also available on many recording devices.

Multi-point recorders are often used when more than one variable is needed on a common time and range base. The greatest need for such recorders is for temperature measurements relating to the same equipment or system or at various locations. These records are needed for comparison, for future reference and as checkpoints to guide operators within specified tolerance limitations.

Transducers or Converters

The development of miniature electronic controllers in the early 1950s created a need for a group of instrument devices—transducers or converters—whose use continues to grow. Their main purpose is to convert signals from one energy form to another or from one signal level to another. Electronic transmitters and controllers were introduced about the same time. Because there were no electronically actuated valves (final control elements) available, a transducer was necessary to convert the electronic current signal to the pneumatic signal needed at the valve. Figure 2.20 shows a device used for that purpose. These current-to-air converters were developed along with the transmitters and controllers and are used in most electronic closed loop control circuits.

About the same time or earlier, another transducer was introduced that was widely used for several years—a voltage-to-air converter. The signal range commonly used was a few millivolts DC to a 3 to 15 psi signal. These devices are used on primarily pneumatic systems, and the converters are used to change thermocouple generated DC signals to appropriate pneumatic signals. The use of E.M.F./air converters has declined with increasing use of electronic systems which use E.M.F./current converters which are compatible with electronic signal systems.

Often there are requirements for signal conversion from one signal system to another because of the diversity of signal levels still in use. For example, electronic systems may use 1 to 5 volts DC, 0 to 5 volts AC, 1 to 5, 4 to 20 or 10 to 50 milliamps DC. Special instruments such as analyzers may operate at 0 to 1 or 0 to 10 millivolt DC signal levels. Some devices may have even smaller (and sometimes larger) DC or AC signal ranges which must be made compatible with the standard signal ranges for the control systems used. All these devices—air to-current, voltage-to-current, current-to-current, resistance-to-current and current-to-air tranducers—are used as components in control systems to provide needed compatibility and economy.

It should also be noted that analog-to-digital and digital-to-analog converters are necessary to interface with the increasing use of computers in industry today. Since many models of computers are now available and in service, the computer industry has developed a standard specification for the devices which interface computers to each other. Electronic Industries Association (EIA) specification No. 232C, better known as RS232, specifies grounds, control signals, timing signals, data, and other information pertinent to digital data transmission. Not all manufacturers furnish RS232C-compatible equipment, so specifications must certainly be reviewed to assure compatibility with appropriate external and remote computer equipment.

Integrators (Figure 2.21) are instruments that receive continuous rate signals (either pneumatic or electronic) and automatically provide running totals or integrated values of these rates. They are used primarily for flow measurement and are necessary for accounting purposes, purchasing, material balancing and quality control.

Figure 2.20. This current-to-air transducer with cover removed is mounted on valve yoke. Transducers became necessary when electronic control systems came into use, but pneumatics continued to be the power source for valve operators. (Courtesy of Fisher Controls)

Figure 2.21. Integrators receive continuous rate signals and automatically provide running totals or integrated values of the instantaneous rates. (Courtesy of Foxboro Co.)

Units are available with six-digit or eight-digit counters normally and may or may not have reset functions. An integrator with a reset function may be reset to a zero position at any time.

Integrators may also be provided with other functions such as switching so that when a predetermined value is reached, initiation of appropriate actions occur (such as valve closures or pump shutdowns).

Alarm and/or Shutdown Functions

Alarm and shutdown functions are described at the same time because both conditions represent abnormal situations and both are actuated by the same type devices. Ordinarily an alarm condition is one that warns that a shutdown may be imminent. Conversely, a shutdown is a situation in which an alarm condition existed that was so serious that no time was available for remedial action. It was predetermined that a shutdown was preferable to a course of correction. Of course, there are instances when time exists for remedial action, but for some reason the action is not taken. In either case a switching action occurs to initiate the alarm or the shutdown. The switch, in this case, is a device which "measures" the variable at a particular value and operates (opens or closes) when the preset value is reached.

Switches which actuate alarms or shutdowns may be piped directly to the process (process actuated), or they may be actuated by transmitted signals. The latter method is usually cheaper and in many situations is used for alarms. Many companies insist that shutdown functions be initiated directly by process mounted devices rather than secondary (transmitted signal) devices. Figure 2.22 shows a pressure switch, a temperature switch, a flow switch and a level switch all of which mount directly in the process and may be used for alarm or shutdown.

Figure 2.23 shows a typical application of a pressure switch used on pneumatic signal lines and is applicable to whatever variable the system is measuring—flow, pressure, level or other. Figure 2.24 shows an electronic counterpart of the pneumatic pressure switch which is used in a similar manner. Most manufacturers package dual-alarm trip units so that "high-low," "high-high high," etc., functions can be purchased together (Figure 2.25). Set points are individually adjustable by screwdriver or knob, and output contacts are available usually in SPDT form. Other switch actions may come from relays or auxiliary contacts in motor starters or other electrical or electronic apparatus.

Alarm units which provide the visual alarm indication include units such as a light (bullseye) or it may be part of a multiunit system (Figure 2.26). Alarm units are discussed in Chapter 10. The devices and equipment that may be shut down include pumps, compressors and other types of rotating machinery that could be damaged under adverse operating conditions. Solenoid valves may be operated to open or close piping and equipment systems for safety precautions, for material conservation, to maintain product quality or for some other economic consideration.

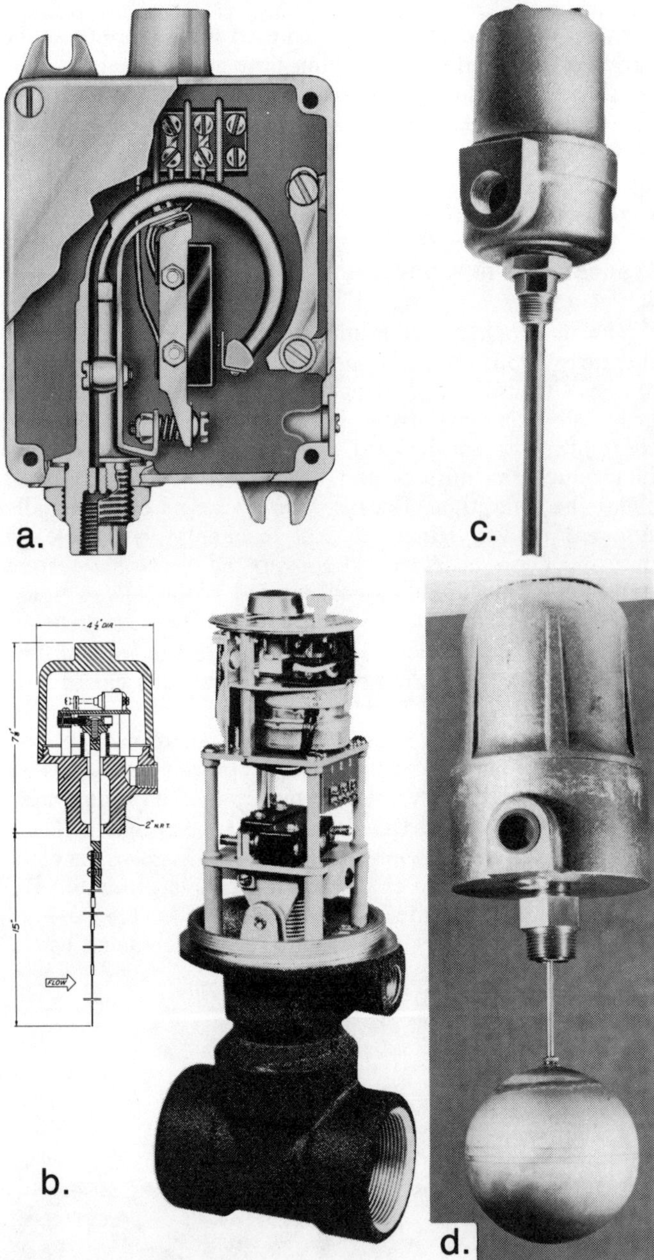

Figure 2.22. Direct mounted switches for pressure (a), Flow (b), Temperature (c) and Level (d) are used for alarm and/or interlock functions. (Courtesy of Delaval, Barksdale Control Div. and Power and Equipment Company, Inc. and Burling Instrument Co. and Jo-Bell Products, Inc.)

Miscellaneous

Other items which frequently are elements of control systems include relays, solenoid valves, timers and other devices which perform special functions. These are not discussed in detail but are mentioned or described briefly to provide a wider concept of the variety of units that may comprise a control system.

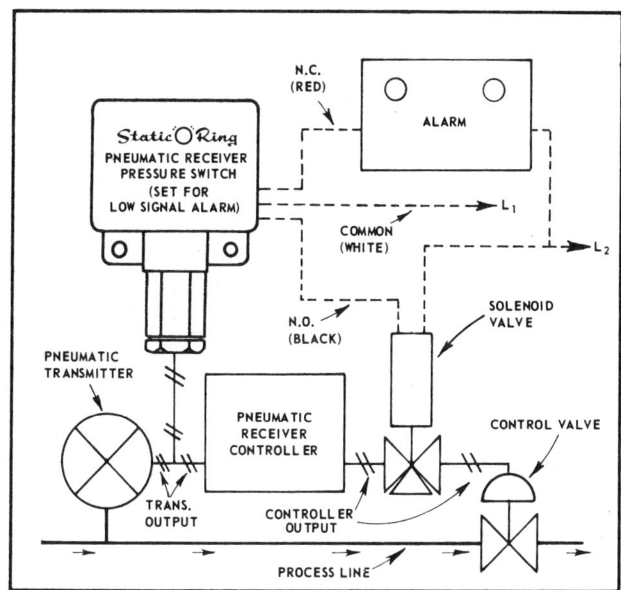

Figure 2.23. Pressure switches with 3 to 15 psig receiver elements are used as control devices or as alarm devices, taking their signals directly from the transmitter in the loop. (Static "O" Ring Pressure SW. Co.)

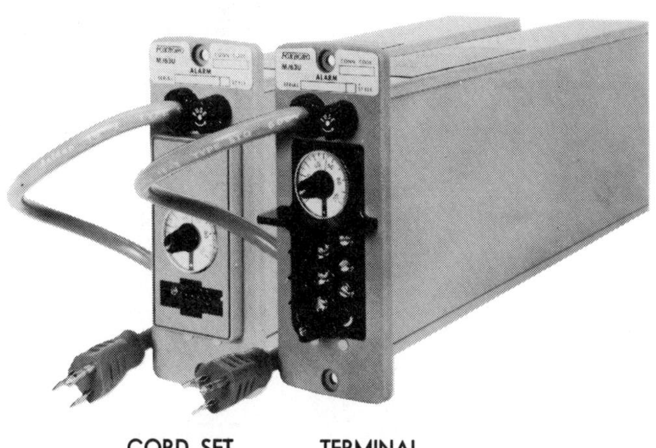

Figure 2.24. This electronic alarm unit is used with any compatible electronic system to set alarm or interlock functions for any variable desired. (Courtesy of Foxboro Co.)

Relays

Relays fall into two categories—pneumatic and electronic (Figure 2.27). Pneumatic relays may be classified under another heading, *special functions*, and are discussed briefly in a later paragraph.

Electronic relays are used in all types of instrument loops such as pressure, flow, level and temperature and, like pneumatic units, are used to perform a number of functions—multiplying, dividing, averaging, switching, etc., to

Figure 2.25. Dual electronic trip units can provide "high-low" alarm functions in a single package. (Courtesy of Moore Industries.)

Figure 2.26. Visual alarm units may be single bullseye lamps (a) or multipoint units using lighted window displays as in b. (Courtesy of Panalarm Division, Riley Co.)

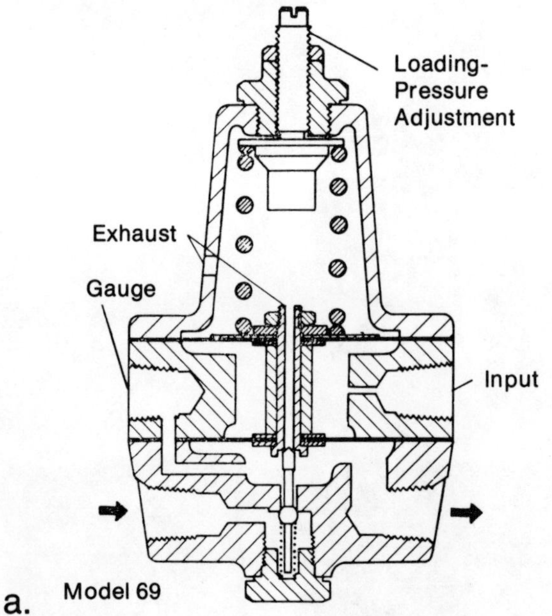

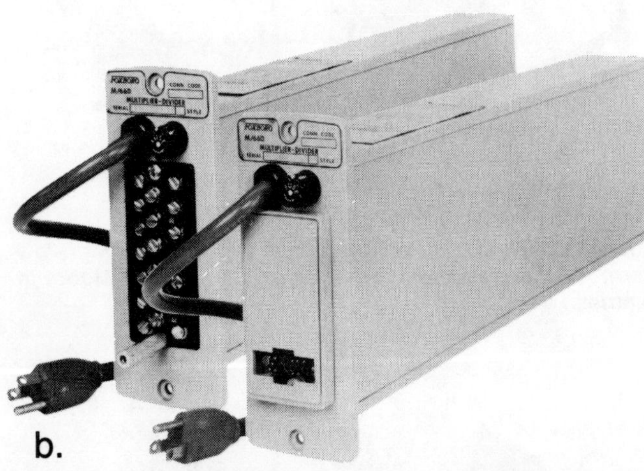

Figure 2.27. Pneumatic and electronic relays provide a wide variety of functions—multiplying, dividing, averaging, scaling and reversing. The pneumatic relay (a) performs two functions—ratioing and reversing. The electronic relay (b) can multiply, divide, square, take the square root or perform multiple functions as specified. (Courtesy of Moore Products Co. and The Foxboro Co.)

accomplish the desired control actions. These special type relays form an integral part of the control system, and its signal must be compatible with the system in which it is used.

Solenoid Valves

A solenoid valve is a combination of two basic functional units—a valve and a solenoid (electromagnetic device). There are several types—direct action, internal pilot operated, external pilot operated, two-way, three-way and four-way. Figure 2.28 shows several of these types. They come in different enclosures to suit various electrical

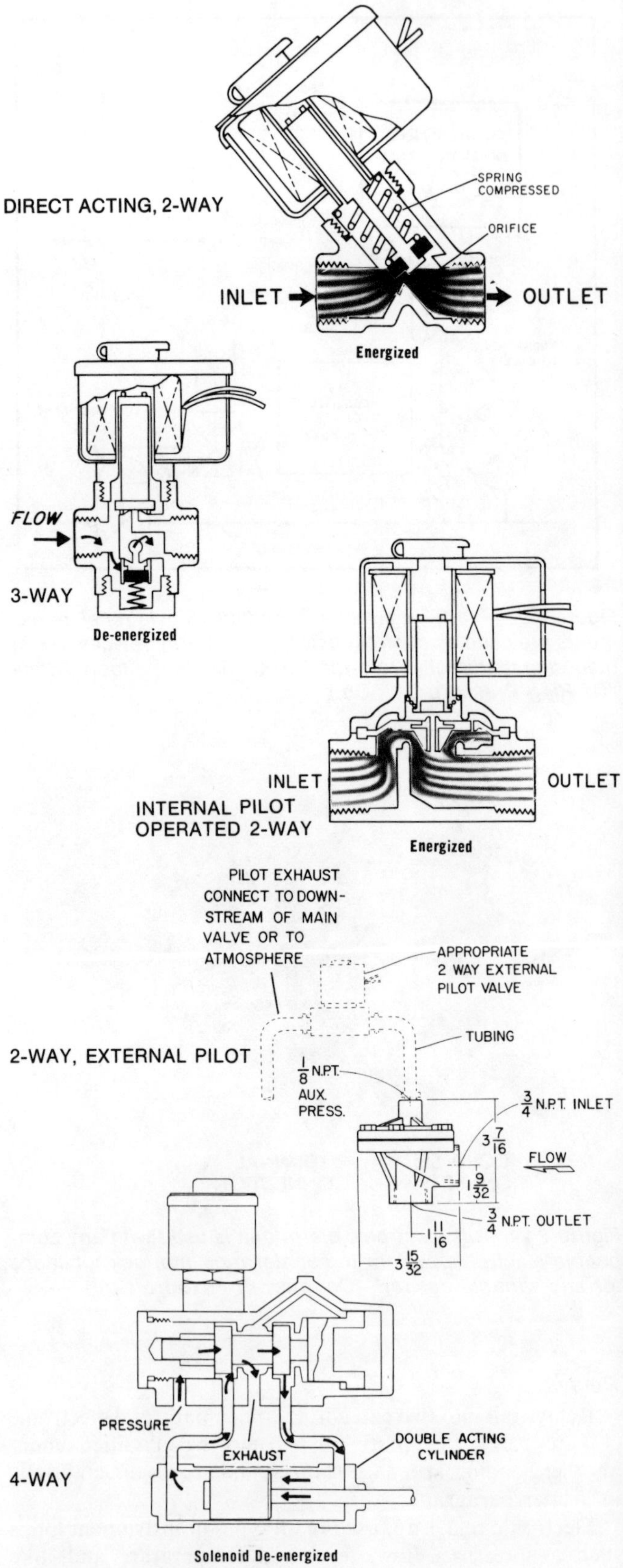

Figure 2.28. Solenoid valves may have two, three or four ports and may be classified as direct acting, internal or external pilot-operated. (Courtesy of ASCO)

classifications, and there are many variations relative to electrical characteristics as well as mechanical and operating characteristics. The function of solenoid valves is to provide an on-off switching option in the system. They accomplish in a pneumatic or liquid system what an electrical relay accomplishes in an electrical system. They are frequently used in conjunction with control valves to open or close the valve at predetermined conditions or limits. The limits are determined by measured variables pertinent to the system.

Solenoid valves are discussed in greater detail in Chapter 10.

Special Functions

The quantity of special functions available to control systems has increased manifold in the last few years. The purpose here is to mention them briefly. Their application or operation is not discussed in detail.

Square root extractors are used to linearize flow signals available from differential flow transmitters (Figure 2.29). Computing relays for totalizing, multiplying, reducing and averaging are available in both pneumatic and electronic types. Other relay functions include limiting, damping, accelerating, throttling and characterizing. Other special function devices include timers, printers, scanners, etc.

Figure 2.29. Square root extractors are often used on the output of flow d/p cells to linearize the square root signal and present a linear output for better definition of readout or control. (Courtesy of Moore Products Co.)

Most of these instruments belong to specialty areas, and detailed discussions on their application and selection are not given.

3 Automatic Control Principles

Tom W. Clark, Laurence C. Hoffmann,
William G. Andrew

The definition and purpose of automatic control were given in Chapter 2, along with a discussion of many fundamental concepts and elements involved in control loops. Various material properties of processes were listed and described; energy forms related to these processes were considered; the responses of various types of detecting elements used for measurement were analyzed; measuring methods were outlined; and many of the instruments used for open and closed loop control were described and their functions outlined briefly.

This chapter presents in simple, layman language the control responses used to accomplish automatic control and some of the underlying principles involved in these responses. For the reader who desires a precise mathematical analysis of automatic control theory and a derivation of the laws and principles used, there are several good texts which he may study and to which he may refer. The simplified approach presented here should provide a basic knowledge of control theory which will help in understanding the measurements and control concepts discussed throughout this volume.

Prior to discussing the control responses and control methods and schemes used, definitions of commonly used terms are given, and process and measurement control problems which need solutions are defined and discussed briefly.

Automatic Control Terminology

The following terms are frequently encountered in the field of instrumentation. Some will not appear in this chapter, but many will. The reader will likely see and use most of them repeatedly and needs to be fully aware of their definition.

Accuracy: the capacity of a device or system to render a true value of a measured variable under reference conditions. (See Chapter 11, Volume II.)

Amplification: the dimensionless ratio output to input in a device intended by design to increase that ratio.

Attenuation: a decrease in the magnitude of a signal value between two points.

Automatic controller: a device which measures the value of a variable and operates to correct or limit the deviation of this measured value from a selected (set point) reference.

Automation: the act or method of making a process perform without the necessity of operator intervention.

Calibrate: to ascertain the error (if any) in a device by checking it against a standard; to adjust the output of a device to bring it to a desired value (within a specified tolerance) for a particular value of input.

Cascade control system: a control system in which the output of one controller is the input for another.

Closed loop (feedback loop): several automatic control units connected so as to provide a signal path which includes a forward path, a feedback path and a summing point. The controlled variable is constantly measured, and if it deviates from a prescribed set point, corrective action is applied to the final element so as to return the controlled variable to the desired value.

Control point: the point at which the controller actually regulates the process; it may not agree with the set point applied to the controller.

Cycling: a periodic change in the variable under control resulting in excursions above and below the control point.

Dead band: the change through which the input to an instrument can be varied without initiating instrument response.

Derivative action: a controller response which is proportional to the rate of deviation change.

Detecting element: that part of the measuring system which responds directly to the controlled variable.

Deviation: the departure from the desired value—the difference between the set point and the control point.

Device: an apparatus for performing a prescribed function.

Drift: an undesired change in output over a period of time, which is unrelated to input, operating conditions or load.

Error: the difference between the indication and the true value of the measured signal, often expressed as a percentage of span or full-scale value.

Feedback: in a control loop, it is information about the status of the controlled variable which is compared with that which is desired, in the interest of making them coincide.

Final control element: component of a control system (such as a valve) which directly regulates the flow of energy or material to the process.

Flapper-nozzle: pneumatic components which generate a signal proportional to the detected signal.

Force balance: a pneumatic or electronic control mechanism which generates a signal by opposing torques, with little palpable motion.

Gain: the ratio of output to input—a dimensionless term since both output and input are expressed in the same terms.

Hunting: oscillation or cycling of a measured variable caused by the system's overzealous effort to achieve a prescribed level of control.

Hysteresis: difference between upscale and downscale results in instrument response when it is subjected to the same input approached from opposite directions.

Input: incoming signal to a measuring instrument, control unit or system.

Instrument: in process measurement and control, this term is used boradly to describe any device that performs a measuring or controlling function.

Instrumentation: the application of instruments to an industrial process for measuring or controlling its acitivity.

Integral control action: action in which the controller's output is proportional to the time integral of the error input; when used in combination with proportional action, it is often called reset action.

Interference, common mode: a form of interference which appears between measuring circuit terminals and ground.

Linearity: the nearness with which the plot of a signal or other variable plotted against a prescribed linear scale approximates a straight line.

Manipulated variable: that which is altered by the automatic control equipment so as to change the variable under control and make it conform with the desired value.

Measuring element: the element which converts a measurable variable into a form or language that the controller can understand.

Measurement: any instantaneous value of the process variable.

Motion balance: a control mechanism which generates a signal by palpable motion of its part.

Noise: unwanted signal components that tend to obscure the genuine signal information that is being sought.

Offset: the difference between the value of the measured signal and the set point that is a characteristic function of proportional action.

On-off control: a control mechanism having only two discrete values of output, fully on or fully off.

Open loop: control without feedback.

Overdamped: damped so that overshoot cannot occur.

Overshoot: the overzealous effort of a control system to reach the desired level which frequently results in going beyond (overshooting) the mark.

Primary element (detector): the first system element that responds quantitatively to the measured variable and performs the initial measurement operation.

Proportional action: a controller response which is proportional to deviation.

Proportional band: the reciprocal of gain expressed as a percentage; refers to the percentage of the controller's measurement span over which the full travel of the control element is divided.

Range: the region between the limits within which a quantity is measured, received or transmitted, expressed by stating the lower and upper range values.

Rate action: that portion of a controller output which is proportional to the rate of change of input (output on some controllers); it is also called derivative action.

Repeatability: the closeness of agreement among a number of consecutive measurements of an output for the same value of the measured signal.

Reproducibility: the exactness with which a measurement or other condition can be duplicated over a period of time.

Reset action: a controller response which is proportional to the extent and duration of deviation.

Reset windup: the overcharging, in the presence of a continuous error, of the reset capacitor (bellows in a pneumatic controller) which must discharge through a long time constant discharge path and prevents a quick return to the desired control point.

Response: reaction to a forcing function applied to the input, e.g., the variation in measured variables which occurs as the result of input changes.

Sensitivity: the minimum change of input to which the system is capable of responding.

Set point: the point at which the controller is adjusted to regulate the process.

Signal: information in the form of a pneumatic pressure, an electric current or mechanical position that carries information from one control loop component to another.

Span: the algebraic difference between the upper and lower range values.

Steady state: that state in which the static conditions prevail and all dynamic changes may be assumed completed.

Step change: a change from one level to another in supposedly zero time.

Time constant: for a first-order system, the time required for an output to complete 63.2% *of the total rise or decay as a result of a step change of the input.*

Time, dead: the interval of time between the initiation of an input change or stimulus and the start of the resulting response.

Transducer: a device which receives information in the form of one physical quantity and converts it to information in the form of the same or other physical quantity.

Transmitter: a device which responds to a measured variable by means of a sensing element and converts it to a standardized transmission signal which is a function only of the measurement.

Variable: a level, quantity or value which is subject to change; this may be regulated (e.g., the controlled variable) or simply measured (e.g., a barometer measuring atmospheric pressure).

Zero: the lower end of the measuring instrument's scale; zero scale and zero may not coincide.

Zero shift: any parallel shift of an input-output curve.

Control Problems

All physical devices and processes have energy storage capabilities and are resistant to change. In this respect they are similar and may be compared to the electrial properties of capacitance and resistance. Like an electrical RC circuit in which the output is distorted with respect to the input, the response of a process to a parameter change is likely to be distorted to some degree. This distortion or attenuation is referred to as lag.

There are three broad classifications of lag inherent in process control systems:

1. Process lags
2. Measurement lags
3. Transmission lags

Process Lags

Process lags result from the inability of a process to accept or give up energy instantaneously. These are sometimes referred to as velocity-distance lags or dead time. Some examples of these delays follow:

1. The time required for gas to flow from one point to another to produce a pressure change—a function of velocity, distance and capacity
2. The time required for liquid to flow from one vessel to another in a process to produce a level change—same three factors
3. The time required for heat to be transferred from one process stream to another to produce a temperature change—same three factors plus a consideration of the heat transfer surface and the heat transfer coefficient
4. The time required for gases or liquids to combine or react to produce a composition change—includes one or more of the factors previously metioned

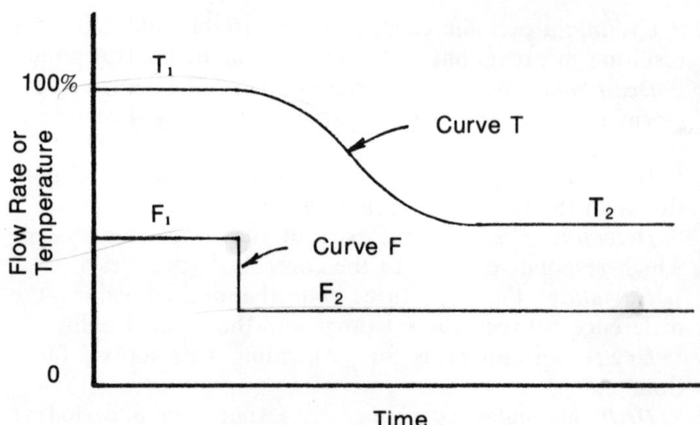

Figure 3.1. Curve T shows how a process temperature responds to a change in cooling water flow rate (Curve F) *where subscript 1 represents initial conditions and subscript 2 represents altered conditions.*

These examples are further emphasized by graphical illustrations. A typical response curve for a heat exchanger is shown in Figure 3.1. When the flow rate of cooling water to an exchanger is suddenly changed, temperature at the fluid outlet does not respond as a step function nor does its change follow a straight line relationship. Because the mass of metal in the exchanger serves as a heat sink, its temperature changes slowly and causes the process temperature to respond as indicated by the curve.

In the example shown, it was assumed that the cooling water was changed as a step function although this is not strictly true. A step change cannot be expected since the final control element has inertia and its reaction time, though short, involves some lag time.

If the cooling water flow rate had remained constant and the water temperature were suddenly changed, the fluid outlet would possibly respond along a different curve from that of the example shown even though the change in btu/hour input was the same in both cases.

Another example of process lag is illustrated by the level application shown in Figure 3.2. In the initial case the flow into the vessel equals the flow out while the head, h_1 is constant. If the inlet flow rate is increased and the outlet valve remains unchanged, h will increase until its head produces a pressure high enough to cause the inlet and outlet flows to stabilize at a new head value h_2.

Since the outlet flow is a function of the square root of h, the flow out increases, but not at the same rate as the inlet flow. If the tank is tall enough, the increase in h will eventually allow the flow out to equal the flow in at h_2, and the level will stabilize.

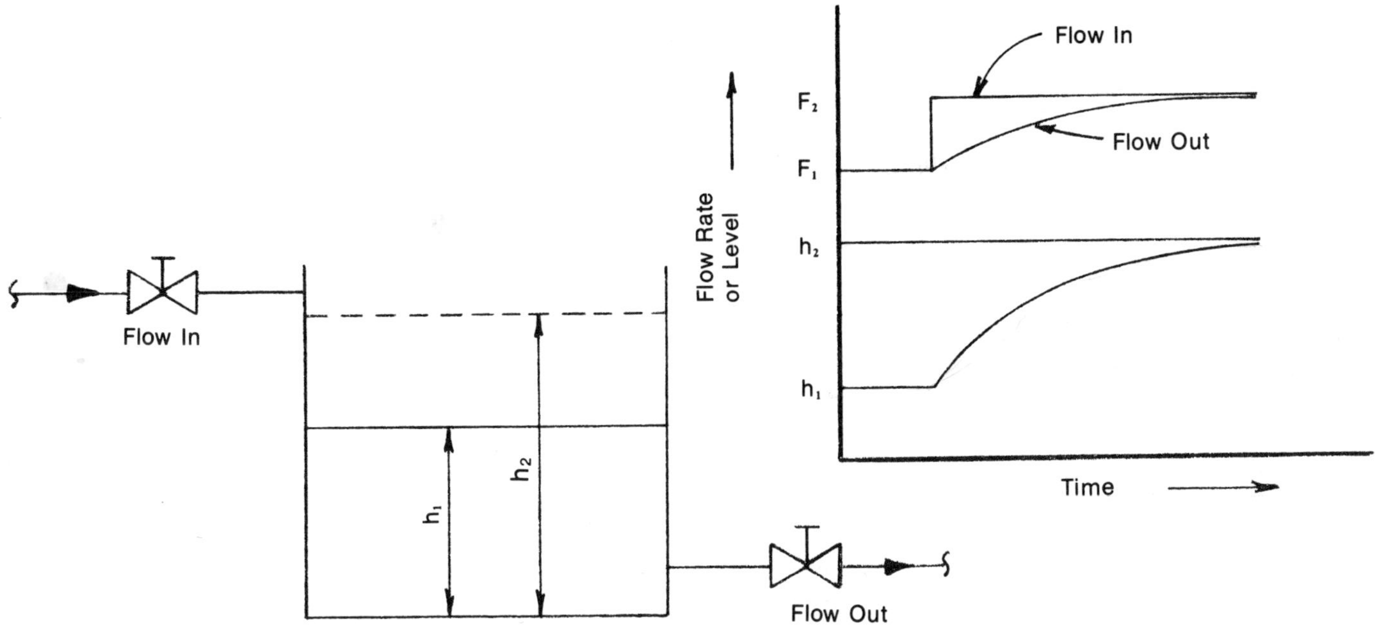

Figure 3.2. Process lag is illustrated by increasing the flow rate from F_1 to F_2. Since the flow rate out was stabilized at liquid head h_1, time elapses before the new flow out is stabilized with the flow in at h_2.

Measurement Lags

Measurement lags were discussed in considerable detail in Chapter 2 and are mentioned again briefly. In physical measurements such as temperature, for example, the lag measurement or delay is due to the thermal capacities of the equipment, of the temperature element well and of the measuring element itself. It is necessary to heat or cool the metals of a heat exchanger, of the well and of the thermocouple before the thermocouple can respond fully to a step change in temperature. In a filled system the liquid or vapor must be heated or cooled before it can sense a change.

Bellows, spirals, bourdons or diaphragms must be moved before an instrument can detect a change in pressure or differential pressure. In these measurements, physical and mechanical inertia must be overcome.

In many analytical measurements, samples must be taken and conditioned before an analyzer can detect changes of composition. Because of the batch nature of these measurements, lags of several minutes are sometimes encountered. For a noncontinuous sampling technique, such as is inherent in a chromatograph, it is necessary to store the output of the analyzer in a buffer in order to supply a continuous signal to a controller. As new samples are analyzed, the results are used to update the memory buffer. A problem that arises when using this method is that a gradually changing analysis of the process is interpreted by the controller as a series of step changes. This sometimes causes the controller to overcorrect and drives the process into oscillation. When this occurs, this apparent step measurement function may be converted to a smooth curve by introducing suitable amounts of resistance and capacitance between the analyzer output and the controller input. This type of control difficulty may also be minimized by properly tuning the control system.

Transmission Lags

The other lag source that must be considered in addition to process lags and measurement lags is transmission lag. These lags are more pertinent to pneumatic systems than to electronic systems for electronic signal transmission lags are negligible in most cases. In pneumatic systems, however, they can present problems, especially in fast acting processes.

A typical pneumatic control system may have a detector located some 300 feet from the central control room. A measured variable value is transmitted the 300 feet to its controller for comparison to a set point. After the controller responds to the information, its corrective signal travels approximately 300 feet back to a regulator (probably a control valve) which manipulates the variable. If the valve actuator is large, a considerable volume must be filled (or bled off) to reflect the difference in the old and corrected signal value. The time usually is only a few seconds, but this is sometimes too slow for fast acting processes. Corrective actions for this type problem include:

1. Larger diameter transmission tubing to reduce resistance to signal medium flow.
2. Booster relays in long transmission lines to decrease signal response time.

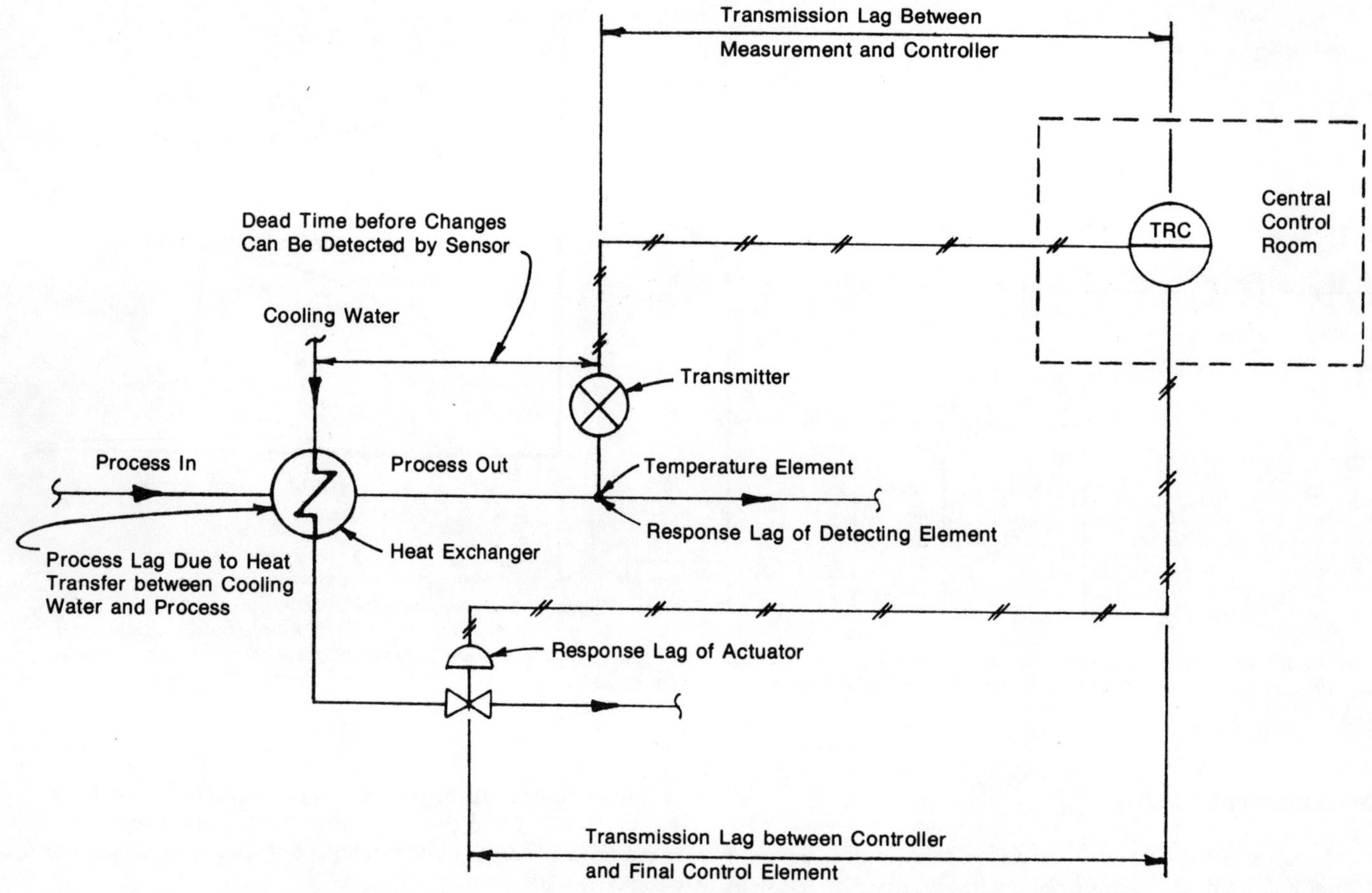

Figure 3.3. Process measurement and transmission lags are illustrated in this control system.

3. Location of controller near the regulator to reduce the distance that transmission and control signals must travel. Several years ago this was a frequent practice for fast acting systems. Now when this kind of response seems necessary, electronic systems are likely to be used.

Signal transmission lags are similar in some respects to certain types of process lags. There is an unavoidable delay between measurement time and action time due to physical restraints. This compares to what has been termed the velocity-distance lag of a process.

Figure 3.3 illustrates a control system in which the various lags discussed are labeled. It is beyond the scope of this work to place finite times on the lags of such a system, but the principle involved can be readily understood.

Control Responses

During the 1940s and 1950s, much of the control theory was developed that governs the makeup of automatic controllers today. In this period, the nature of processes was studied and analyzed rather extensively. Their dynamic characteristics were defined by predicting and observing their behavior with respect to time. Their behavior under steady state conditions was noted. These conditions were also related to the mechanics of the system—resistances and capacities—and control principles were advanced and used which are so common today.

Application of these principles were backed not only by mathematical analyses but also by actual experimentation, and both confirmed the soundness of the principles. During this period the control responses (or modes) that are still recognized as meeting most of the requirements of today's sophisticated processes were put into use. In addition to the two-position or on-off control response, there are proportional, reset (integral), rate (derivative) and combinations of proportional, reset and rate responses. Since these responses are prevalent today and basic to the understanding of automatic control principles, it is imperative that they be understood.

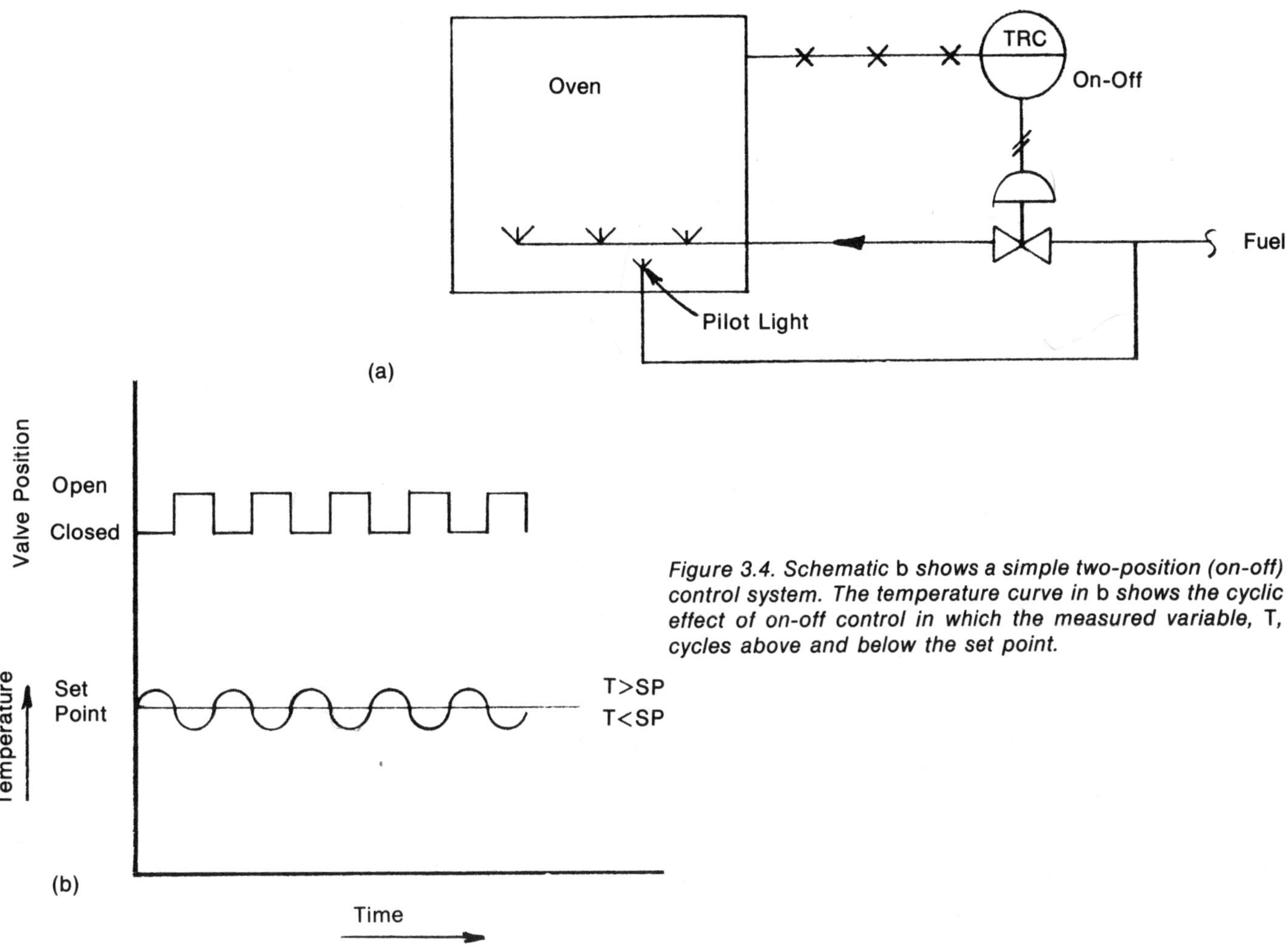

Figure 3.4. Schematic b *shows a simple two-position (on-off) control system. The temperature curve in* b *shows the cyclic effect of on-off control in which the measured variable,* T, *cycles above and below the set point.*

Two-Position Control

The simplest form of automatic control used is the two-position or on-off type control. The term *on-off* may be a misnomer in that it is possible to have two-position control without having the positions either *on* or *off.* However, since most two-position control systems are *on* or *off,* the term is normally applicable.

Two-position control is normally used when the controlled process variables need not be maintained at precise values. Typical examples are oven controls or alarm and shutdown functions. Another familiar example is the common house furnace which is set at a desired temperature. When the house gets cold, the thermostat turns the heat on. When enough heat is supplied to warm the house, the thermostat turns the heater off. Temperature variations from the cyclic effect of such a system usually go unnoticed because their magnitude is small (compared to the thermal capacity of the house).

The same is true in industrial systems; two-position control is often used when large capacities are involved and energy inflows or outflows are small compared to the system capacity. Controller outputs that have valves, heater elements, etc., in the fully on or fully off positions provide control close enough to be acceptable in many cases.

Figure 3.4*a* shows a gas fired oven with an on-off temperature controller. Theoretically, the fuel valve will be open when the measured variable is below the set point and the error signal is positive. Figure 3.4*b* shows a plot of temperature versus time and valve position. The fuel valve opens and remains open until the error signal (e) becomes zero at which time the valve closes, but because capacitance exists in the system, the temperature continues to rise and the measured variable rises above the set point, creating a negative error signal. After peaking, the temperature falls back below the set point and the valve opens again. The temperature does not respond immediately but continues to drop, reaching a low point, then rises to the set point starting another cycle.

Figure 3.5 shows a variation in the on-off controller scheme. It represents a system in which both heating and cooling may be required. A dead band exists around the set point in which neither heating nor cooling is used. When the temperature rises above the set point exceeding ½ the dead

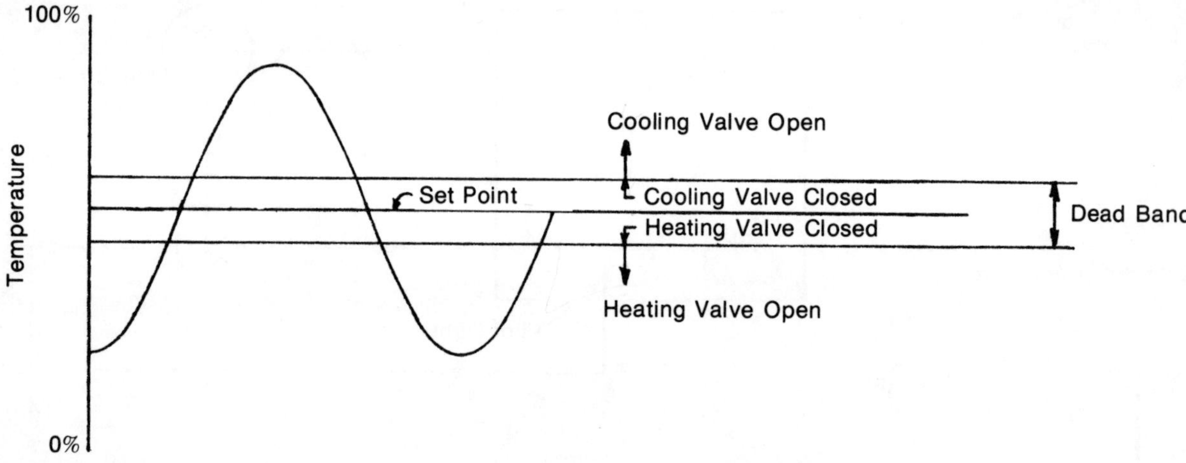

Figure 3.5. Multiposition on-off control is illustrated by a process temperature curve using both heating and cooling actions—but both actuated by switch actions.

band, the cooling valve opens. When it falls below the set point exceeding the dead band, the heating valve opens. This type controller is referred to as a multiposition on-off controller.

In a system with large capacity, the cycle frequency is rather low, providing longer equipment life. This is especially true if electrical heaters are used in the heating cycle. Slower cycle frequencies produce greater deviations from set point.

Many systems can tolerate the continuous cycle and lack of precise control, and the dead band is often adjustable to allow the frequency to be optimized. If closer control is required, other forms of control should be used.

Proportional Control

Two-position control produces a variation and a continuous cycle which many processes cannot tolerate. Proportional control modulates the final control element continuously between the "on" and "off" limits and is the basis for most indudstrial controllers today. Its name is derived from the fact that controller output is proportional to the difference between the measured variable and the set point, i.e., the error signal (*e*). Proportional band (*PB*) is defined as the percentage of full-scale change in input required to change the output from 0 to 100%. The relationship between input and output can be described by the following formula.

$$O_i = (100/PB)\, e + O_o \qquad (3.1)$$

Where:

O_i is the output at any given time
O_0 is the output at zero error
e is the error signal
PB is the proportional band in %

Proportional band is adjustable on most controllers and is field tuned to give optimum response to the process changes expected. Figure 3.6*a* shows the relationship between the measured span of a variable and a controller output when the proportional band is 100%. Figure 3.6*b* shows the relationships between measured spans and outputs for band settings of 20, 50, 200 and 500%. It can be noted that at a 20% *PB* setting, full output is achieved between 40 and 60% of the measured span, while at 500% PB, the output varies only from 40 to 60% over the entire measurement span.

Many manufacturers calibrate their adjustment in gain rather than proportional band. Gain is defined as proportional band divided into 100, i.e., 50% *PB* is equal to a gain of 2. In tuning controllers it must be remembered that increasing gain is decreasing the proportional band.

The float and lever operated valve in Figure 3.7 illustrates a simple proportional controller. In this case the instrument range is limited by the mechanical movement of the float and is 4 inches. The valve stroke is 4 inches from fully closed to fully open. If dimensions *A* and *B* are equal, the controller has a 100% proportional band. When the level is 2 inches above the set point, the control valve is wide open. Further increase in level will not affect valve position and will only increase the outlet flow rate by a change in static head. When the level drops 2 inches below the set point, the control valve will close.

If the fulcrum is moved toward the float so that $B = 2A$, the controller has a 50% *PB* setting or a gain of 2.0. This means that the control valve will be wide open when the level is 1 inch above the set point and closed when the level is 1 inch below the set point. The valve operates throughout its range when the float moves through only 50% of its range, i.e., 50% *PB*. Now if the *PB* were to be set at 200% ($A = 2B$), a 4-inch float movement would move the valve only 2 inches—it would stroke through only half of its range.

It may be noted that for a given *PB* setting there is only one valve position for each level position. In this case the turnbuckle is set so that the output error is zero when the

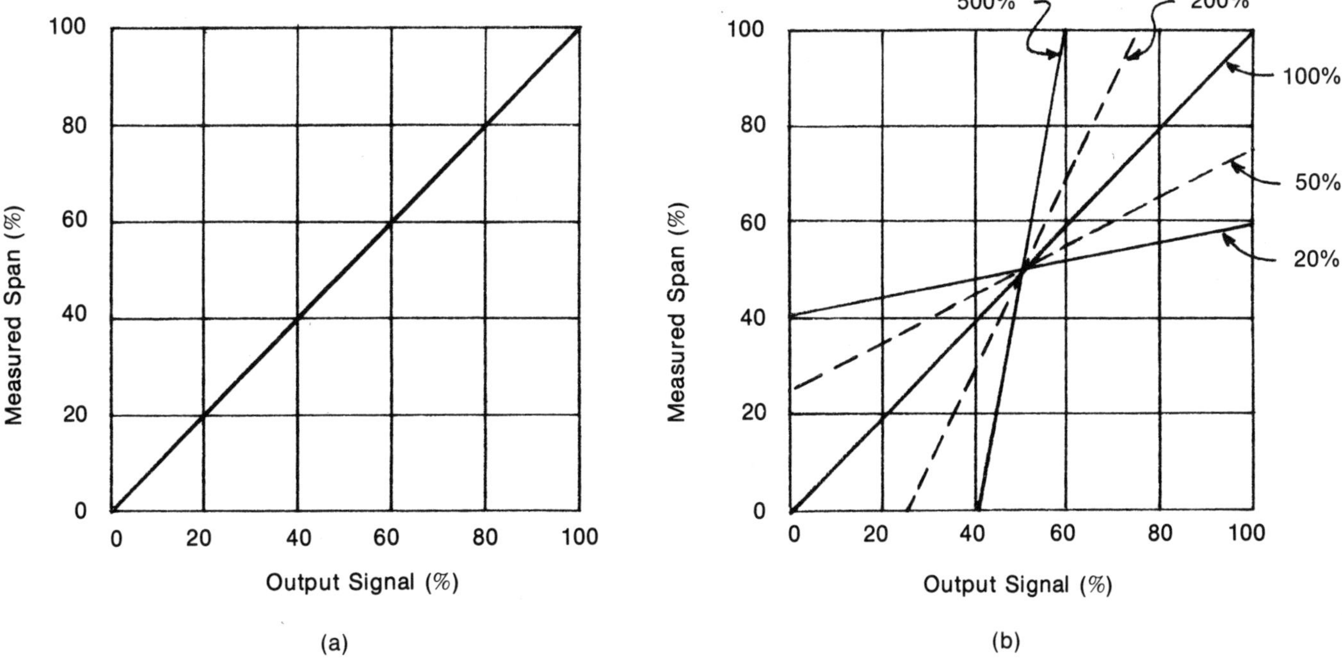

Figure 3.6. Proportional band is the ratio of input to output. Figure 3.6a shows a proportional band of 100% where 100% measurement change produces a 100% output change. Figure 3.6b shows various proportional band settings between 20% and 500% with relationships between input and output given.

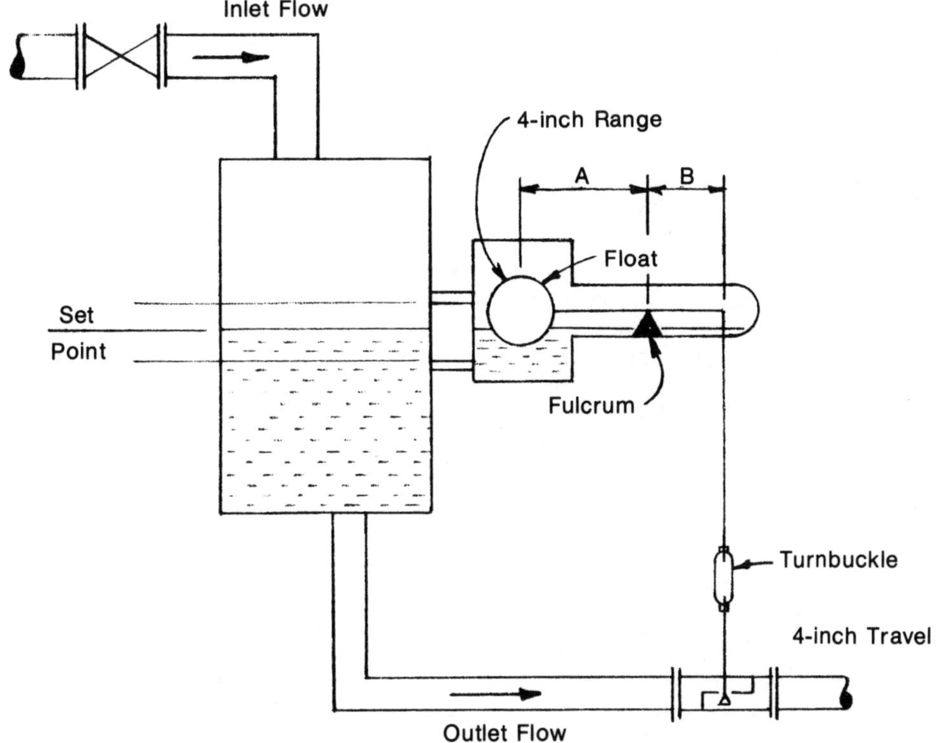

Figure 3.7. The float and lever system illustrates a simple proportional controller. The proportional band is adjusted by varying the fulcrum point.

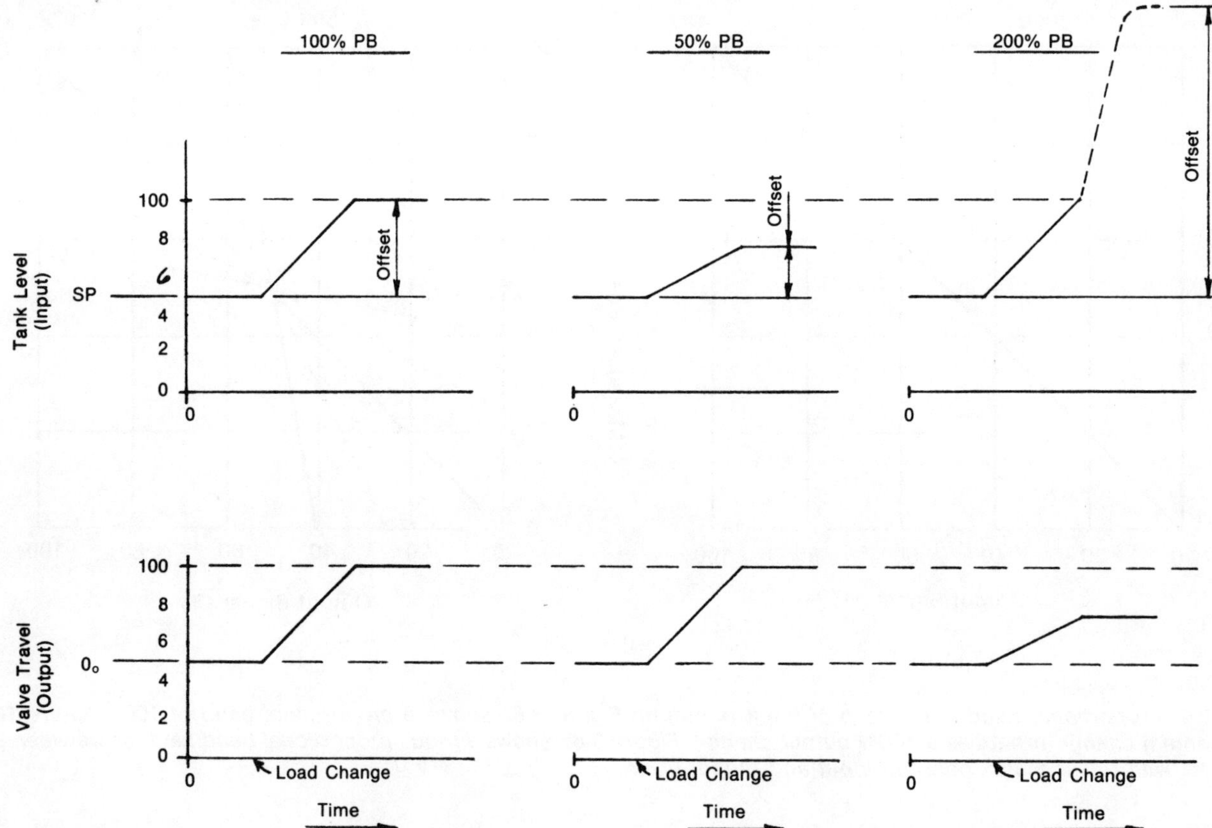

Figure 3.8. These input-output curves for load changes at various proportional band settings reveal the effect of proportional band on offset. The dynamics of the process limit the reduction of the proportional band settings.

valve is stroked 2 inches from the closed position. For any other valve opening there is an error due to the nature of the proportional controller. This error is called "offset" and is an important characteristic of proportional control.

Figure 3.8 shows input and output curves of this system for load changes at different *PB* settings. The load change in this case is an increase in the inlet flow rate. For a 100% *PB* (or gain of 1.0) the output is identical to the input. The 50% *PB* curve shows that decreasing *PB* (increasing gain) reduces the amount of offset. Reductions in *PB* settings are not always feasible, however, because in real systems the amount the *PB* can be reduced is limited by the process dynamics, and a point is reached where oscillation occurs, producing little or no improvement over on-off control. In the case shown for 200% *PB,* the point of maximum range on increasing level is reached with the valve only 75% open. If the tank were infinitely tall so that the hydrostatic head would continue to increase, a head would eventually be reached that would enable the valve to handle the increased flow rate.

In the hypothetical case cited, the assumption was made that a 50% valve opening would handle the normal flow rate before the load changes and would hold the 50% level set point. These values were selected only to illustrate a point. It is desirable, however, to design the system so that the set point is around 50% of the input signal range. Often the normal load occurs at valve openings as low as 30% or as high as 80% of full stroke. This is where the term, *output at zero error* (O_0), comes into effect. In Figure 3.7 the turnbuckle in the linkage between the level controller and the control valve should be noted. If the normal load condition takes a valve opening of 75% (or a 3-inch stroke) to handle the normal flow, the controller can be "manually reset" by extending the turnbuckle 1 inch. There is now a zero error output at 75% of valve stroke. The linkage can be adjusted to put the zero error output at any load condition the system is capable of handling, and the set point can be anywhere within the range. Regardless, however, of where the turnbuckle is moved, there can be only one point of zero error, i.e., one point without offset. This offset can be reduced by overall system design or reduced *PB,* but the amount it can be reduced is limited by the system dynamics.

Reset (INTEGRAL)

Reset action is defined as a controller response which is proportional to the extent and duration of a signal deviation. Proportional controllers always deviate from the set point when subjected to load changes. This is objectionable in most industrial control systems, and the reset mode is often combined with proportional control to eliminate this inherent offset. This is its primary purpose. The reset control mode may also be combined with both proportional and rate to form 3-mode controllers.

Reset is often called integral control because of its approximate mathematical relationship to the error signal. The amount of reset is proportional to the area under the error curve and is expressed mathematically by the equation:

$$O_i = f \int_{t=0}^{t=i} e(dt) + O_o \tag{3.2}$$

where

O_i is the output at any given time
O_o is the output at time zero or zero error
e is the error signal
t is time
f is the reset rate in repeats per minute

The output from a reset mode controller is constantly changing as long as there is an error signal. The rate of change of the output is dependent not only on the magnitude of the error signal but also on the length of time the error has existed.

The reset rate in repeats per minute means the number of times per minute the reset action produces a correction equal in magnitude to the correction from proportional action. Some manufacturers choose to express reset in the inverse unit, time, and it thus becomes the time required, in minutes, to repeat the proportional correction. Figure 3.9 shows a reset-only controller. If a step change is made in the input (I) at time, t_0, there will be a step change in the error signal (1). The output, O_i, will begin to change at a constant rate determined by the reset rate setting. This will produce a constant change in the output signal until time, t_i, when the measured variable returns to the set point.

Figure 3.10 illustrates the primary purpose of the reset function by showing a typical measured variable controlled with and without reset. The reset function is necessary if close control is needed and wide deviations from the set point are undesirable.

The reset function has a disadvantageous side effect. When a sustained deviation occurs, a controller containing reset will eventually be driven off scale. This will happen when the loop is opened—for shutdowns, transfers to manual control, etc. When a process is shut down, for example, by closing shutoff valves, reset action begins to force the proportional band of the controller upward to its limit, trying to change the measurement. This characteristic is known as "windup." It can be eliminated by automatically disabling the reset circuit.

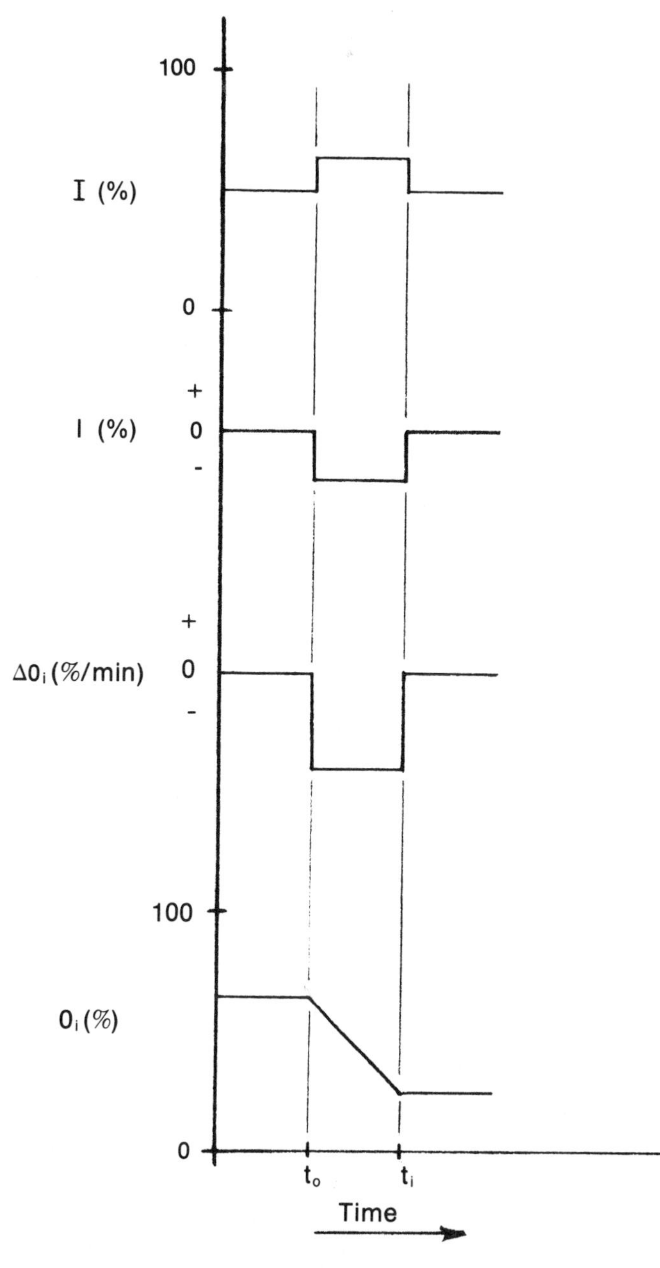

Figure 3.9. Reset action is a function of the magnitude and length of deviation from the set point.

Rate (DERIVATIVE)

The third control mode to be discussed is rate or derivative action. It is defined as that part of a controller response which is proportional to the rate of change of input (in many controllers, the output from the proportional amplifier). Rate may be used along with proportional control to form a two-mode controller, or it may be used with proportional control and reset to form a three-mode con-

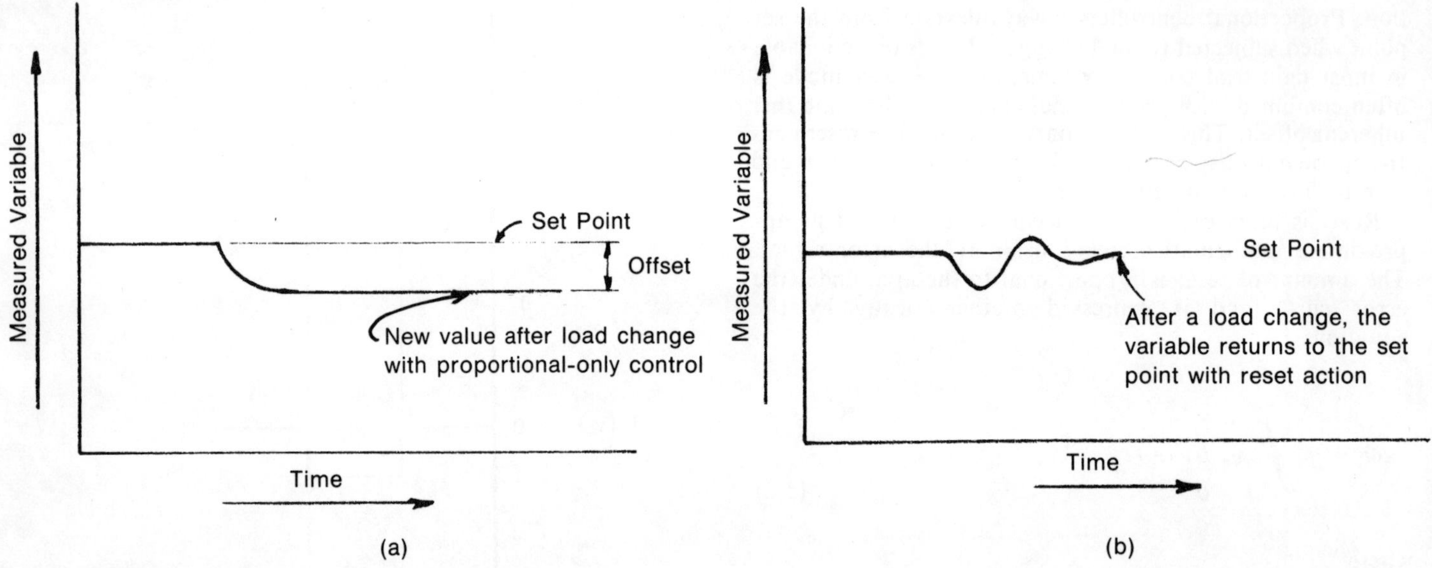

Figure 3.10. Curves A *and* B *demonstrate the use of the reset control mode to return a control variable to its set point after a disturbance.*

troller. Rate control is often useful on systems with large amounts of inertia or lag such as temperature. Since the change in output from a rate controller depends on the rate of change of the error signal, it gives a large amount of correction to a rapidly changing error signal while the error is still small. Because of this characteristic, it sometimes appears that rate action actually anticipates changes; hence, it is sometimes referred to as anticipatory action. Mathematically the rate mode can be expressed as follows:

$$O_i = r(de/dt) + O_o \tag{3.3}$$

where r is the rate time, and other symbols are as defined for Equation 3.2.

Figure 3.11 shows a rate-mode-only controller. A step change in the measured variable (I) would theoretically produce an infinite output; therefore, the measured variable is changed at a finite rate to show a meaningful change in the output. As the input increases, the error signal changes. Rate action is produced only by a changing error signal. A large constant error will produce no output change. The rate time (often called derivative time) is the time interval by which rate action advances the position of proportional action of the output. This is the time difference in getting a particular output change with proportional only action as compared to proportional plus derivative action.

Combination Control Responses

Control responses have been discussed individually; combinations of these modes or responses can now be considered. It is not quite true that these functions have been discussed entirely separately, for reset and rate responses relate quite closely to the proportional response.

In discussing combination modes, the commonly used terminology is single mode (proportional only), two-mode (proportional plus reset or proportional plus rate) and three-mode control (proportional plus reset plus rate). The purpose here is to define the three terms and to discuss briefly their application to process system.

Proportional Only

The listing of single mode control may seem out of place under the heading of "Combination Control Responses." However, the proportional response is the mode with which the other two prominent modes so often are combined. It is necessary, then, to understand that the proportional mode is the basic control response of automatic controllers. When single mode control is mentioned, proportional only control is understood.

Proportional control is adequate for systems which have small capacitances and therefore need fast responses to load changes. This type system normally requires narrow proportional control bands.

Proportional-only control is also adequate where load changes are relatively small and where close control is not critical. If close control is needed, the offset phenomenom inherent in proportional control dictates the use of additional control functions to provide better control.

Proportional Plus Reset—Two-Mode Control

Processes that use wide proportional band control to prevent cycling are subject to offset (see Figure 3.12). The greater the proportional band setting and/or the greater the

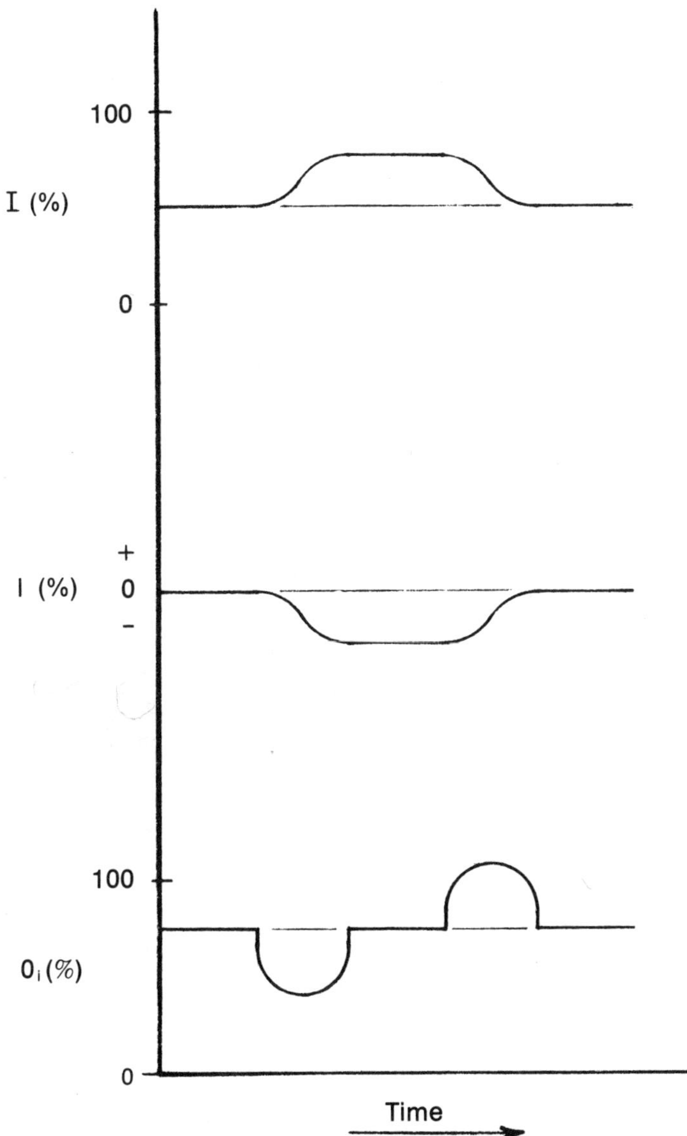

Figure 3.11. Rate or derivative action is proportional to the rate of change of the measured variable. Its effect is quick and is useful on systems with long lags or high capacities.

load change of the process, the greater is the offset. Flow control loops are typical examples requiring wide proportional settings.

For such loops the reset mode needs to be added to the controller. Its addition has the effect of reducing the proportional response to some extent, but this objection is far outweighed by its ability to return the measured variable to its set point, eliminating the offset so objectionable to proportional-only control. (For the addition of reset to the proportional response, Figure 3.10 can be noted.)

When the reset function is added to the controller, it continues to act on the output as long as the deviation from the control point exists; hence, it drives the deviation to zero (returns the variable to the set point).

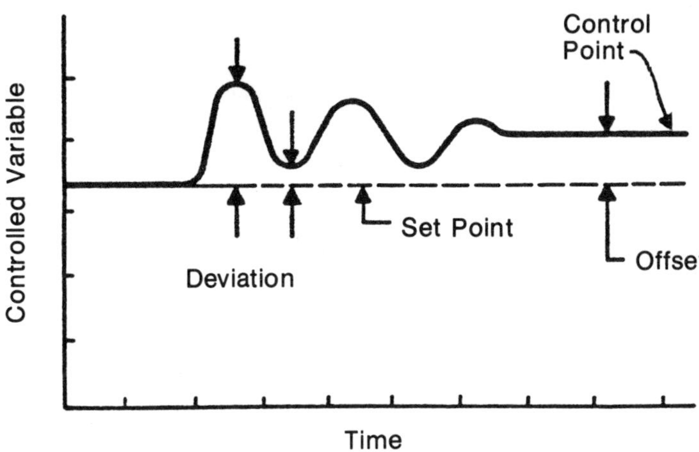

Figure 3.12. When proportional only control is used, a load change produces an offset from the set point.

Proportional Plus Rate—Two-Mode Control

The purpose of derivative or rate action is to speed the response of the closed loop. On difficult-to-control processes (multicapacity systems), the addition of rate or derivative action to the proportional mode is often preferable to adding reset. It enhances both the speed and stability of control responses, particularly on slow-responding systems. Its action is the reverse of reset in that it leads rather than lags the proportional action.

Rate action added to the proportional mode is not desirable for systems such as flow loops that are subject to noise problems (caused by flow turbulence or pumping action), for its derivative action amplifies the noisy error signals and produces instability.

In most controllers, rate action is physically introduced in the feedback path around the controller amplifier. Therefore, if the controller output is constant, no derivative action takes place, regardless of what the controlled variable may be doing. When rate action is applied to the controller output, its action amplifies the fluctuations caused by set point changes—an objectional feature. It also reduces the effectiveness of rate control in eliminating overshoot. Controllers are available, however, in which derivative action is added to the measurement signal before it reaches the controller. This eliminates the objectionable feature of rate action which occurs when instantaneous changes are made to process set points.

Proportional plus rate action is often used on discontinuous or batch control systems where periodic shutdowns present reset windup problems when the reset mode is used.

Proportional Plus Reset Plus Rate—Three-Mode Control

On many control loops, particularly those which are most difficult to control, it is desirable to use proportional plus reset plus rate—three-mode control.

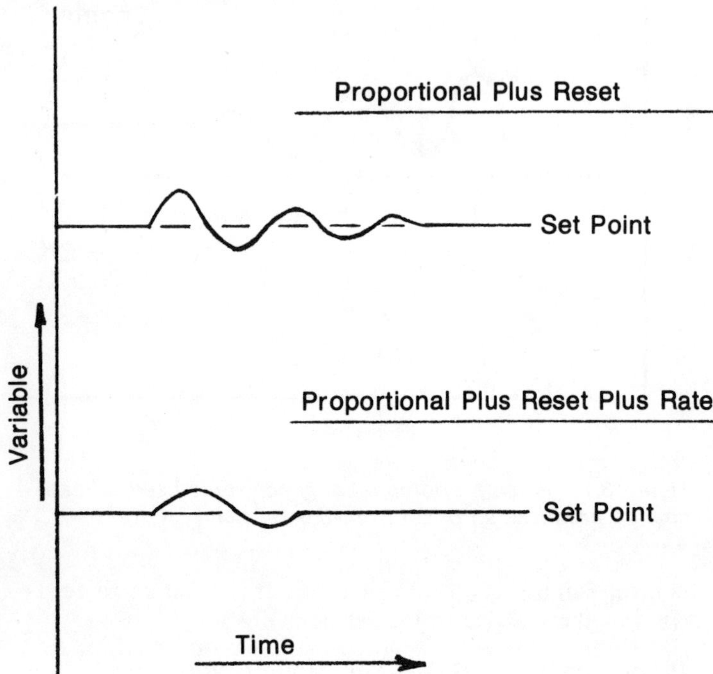

Figure 3.13. Reset added to proportional action sometimes has a tendency to overshoot the set point as it corrects for offset. The addition of rate action helps prevent or reduce the overshoot.

Rate action has the effect of reducing or stopping the overshoot that often occurs when reset is added to proportional action (Figure 3.13). Rate action also counteracts the lag characteristic introduced by reset action.

There is no doubt that a great deal of interaction exists among the control modes when all three are used. The adjustment of any one affects the other two. Tuning loops with all three modes can be difficult because of the interaction. This is much less of a problem if the rate action occurs on the measured signal rather than the controlled signal for interaction is practically eliminated in such cases.

Cascade Control

Cascade control is a technique that uses two *measuring and control* systems to manipulate a single final control element. Its purpose is to provide increased stability to particularly complex process control problems. The technique has been used for many years and is very effective in many applications.

The relationship that exists between controllers is referred to as a master-slave relationship or a primary-secondary relationship. Figure 3.14 illustrates schematically their hookup. The master or primary unit is the controller of the variable whose value is of primary inportance—in this instance, temperature. The slave or secondary unit is the controller of the variable whose value is important only as it affects the primary variable. Closer and more stable temperature control can be achieved by the cascade system shown in the lower schematic than by temperature control alone as shown in the upper schematic.

Cascade control accomplishes two important functions:(a) it reduces the effect of load changes near their source, and (b) it improves control by reducing the effect of time lags.

The second mentioned accomplishment is the more obvious one. Typically it occurs on temperature and analytical applications where measurement lag times are generally rather long.

The other effect is less obvious and perhaps is best explained by analyzing the illustration given in Figure 3.14. The upper schematic shows how control is accomplished directly with the temperature controller regulating steam flow through the heating coil. This system works very well except when disturbances occur in the feed rate or when steam pressure variations change the amount of flow through the heating coil. Because of the fluid capacity in the vessel and because of the measurement lag time, the temperature controller does not immediately detect the disturbances. By the time detection is made, the disturbance may have receded to its normal operation. Cyclic action probably occurs.

The lower schematic illustrates how the cascade system operates. The steam feed is placed on flow control so that the desired steam flow is maintained despite pressure fluctuations in the supply. The temperature controller is cascaded with the flow controller, however, so that long-range fluctuations such as feed rate, ambient temperature effects, etc., will be overcome, and the desired variable (temperature) is maintained as needed. It resets the flow control set point as necessary to maintain the correct temperature.

Another cascade system is illustrated in Figure 3.15 where measurement lag presents a problem. The feed stream to a process reactor needs to maintain a certain catalyst ratio. A chromatograph measurement in the upper schematic does a creditable job, but cycling often occurs because of the system lag. The analysis time takes about 5 minutes. A flow control loop is added to the system as shown in the lower schematic to provide a steady catalyst flow to the process, and its set point is changed by the analyzer as required to provide the correct catalyst ratio.

Cascade systems can be overemphasized; they are not a panacea for every unstable process condition encountered or for all measurement lag problems. However, they do provide satisfactory solutions to many application problems.

Ratio Control

As the name implies, ratio control is maintaining a fixed ratio between two variables. The most common application for ratio control is maintaining a fixed relationship between two flows, such as air-fuel ratios in furnaces, feed and catalysts ratios in reactors and mixtures of two or more raw materials in blending operations.

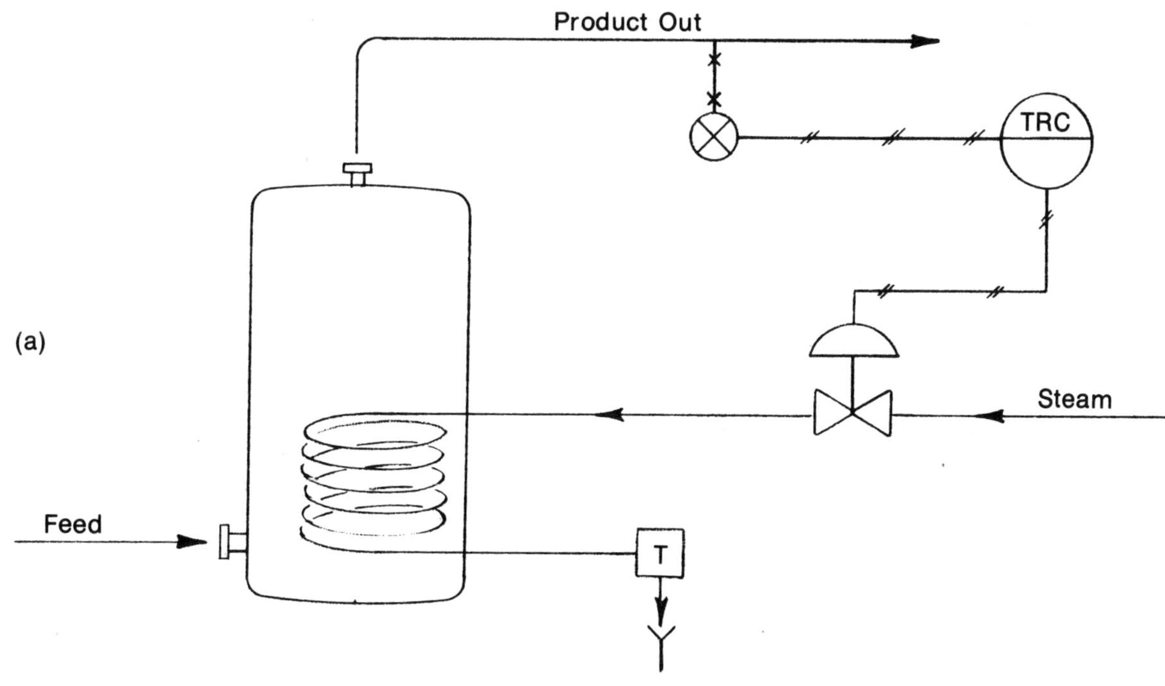

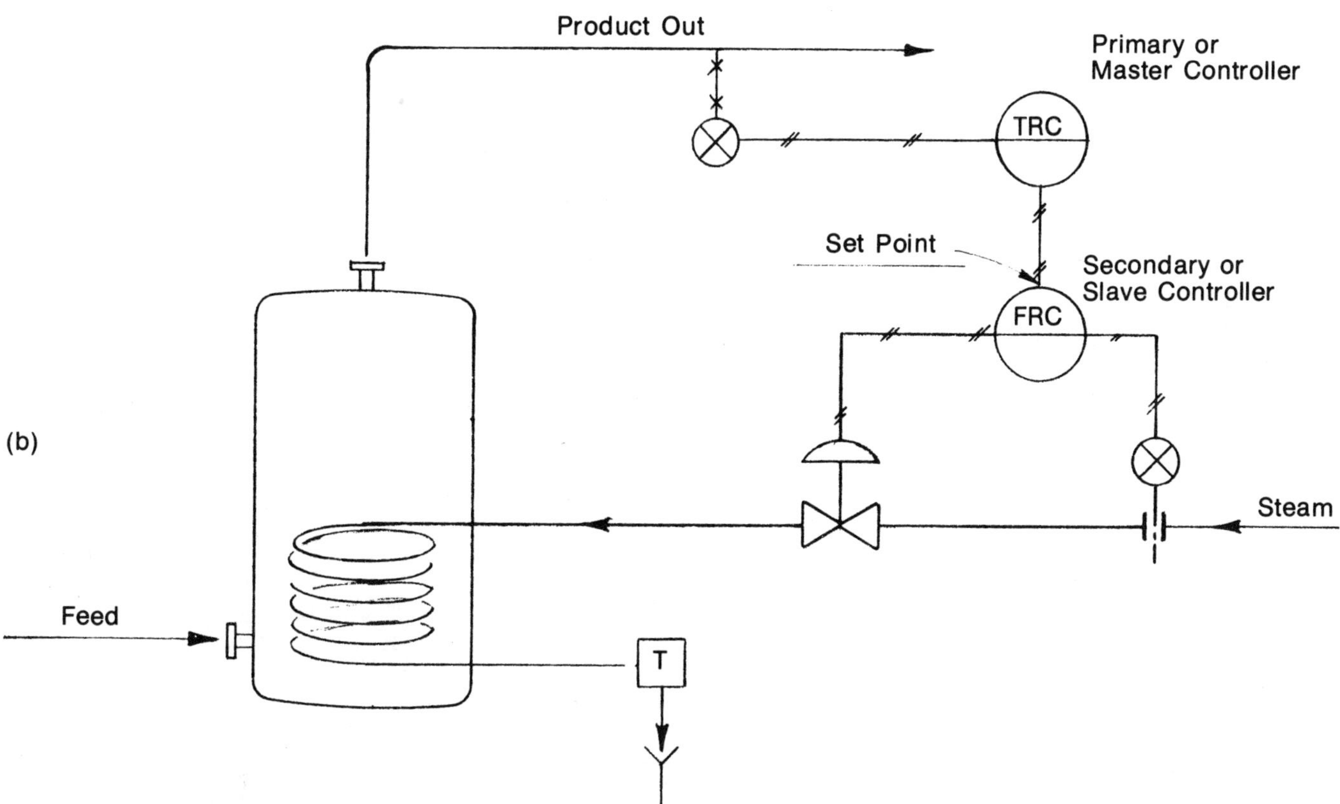

Figure 3.14. Schematic b, depicting a cascade system, achieves closer and more stable temperature control than is achieved by a single temperature controller as shown in the Schematic a.

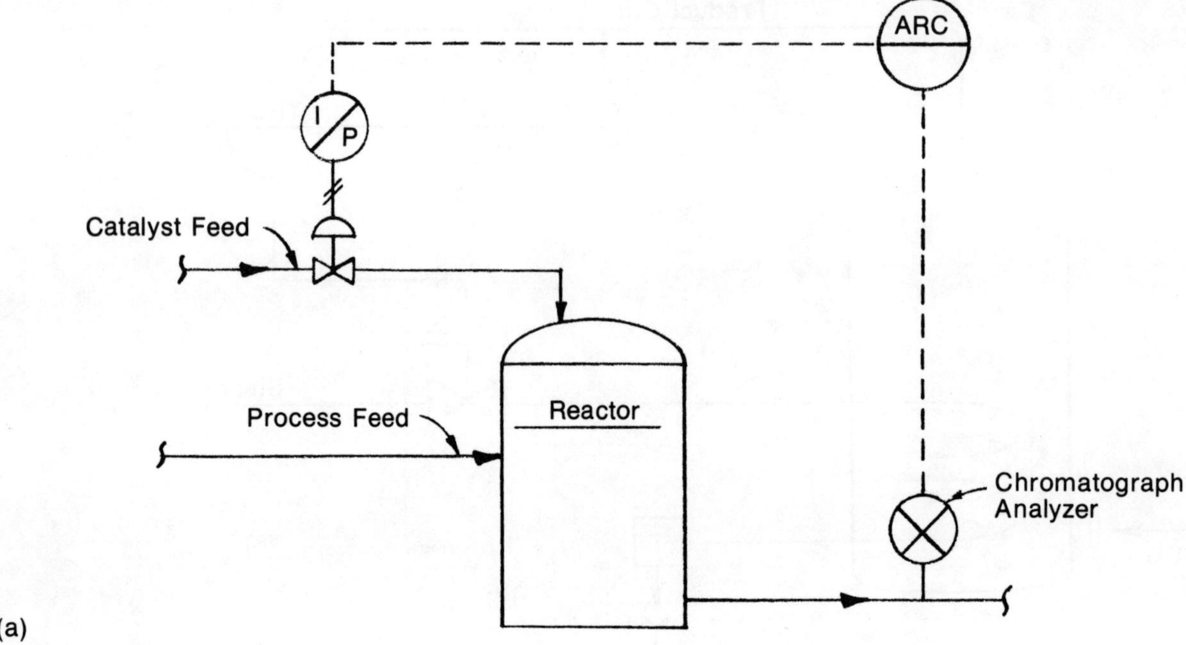

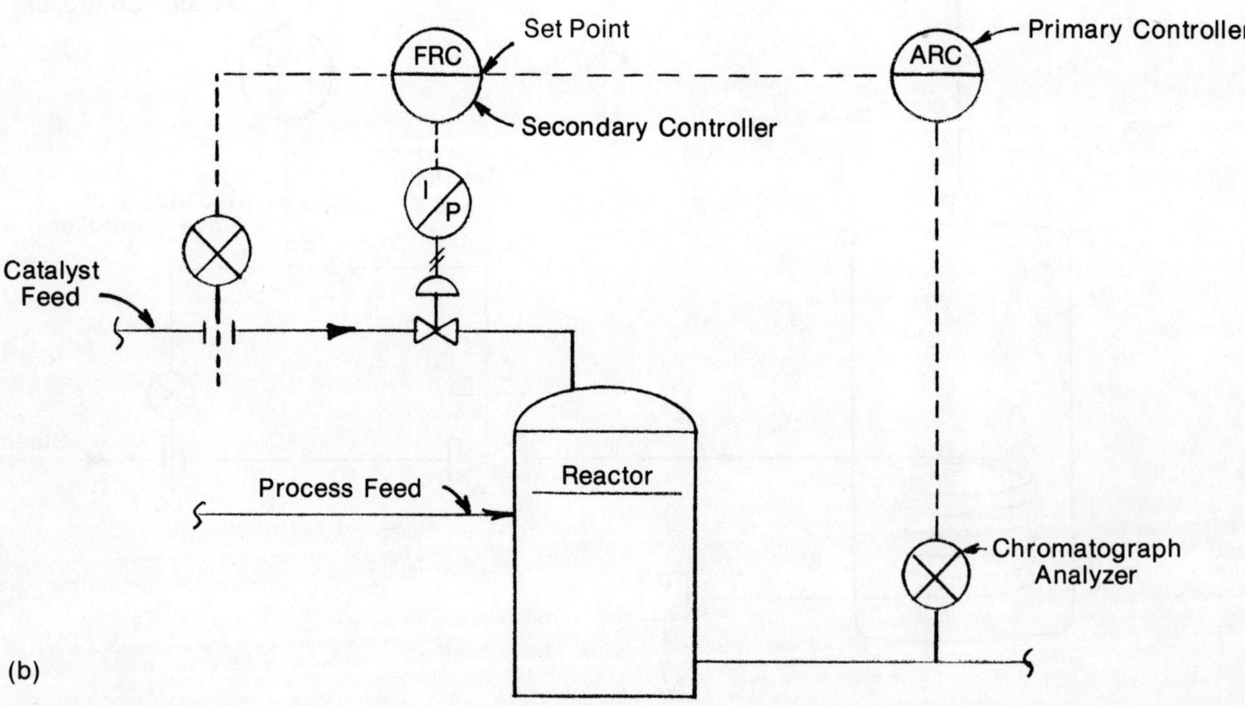

*Figure 3.15. The control scheme in Schematic a presents cycling problems due to measurement lag. The cascade control system depicted in Figure 3.15*b *provides close, stable control of catalyst requirement.*

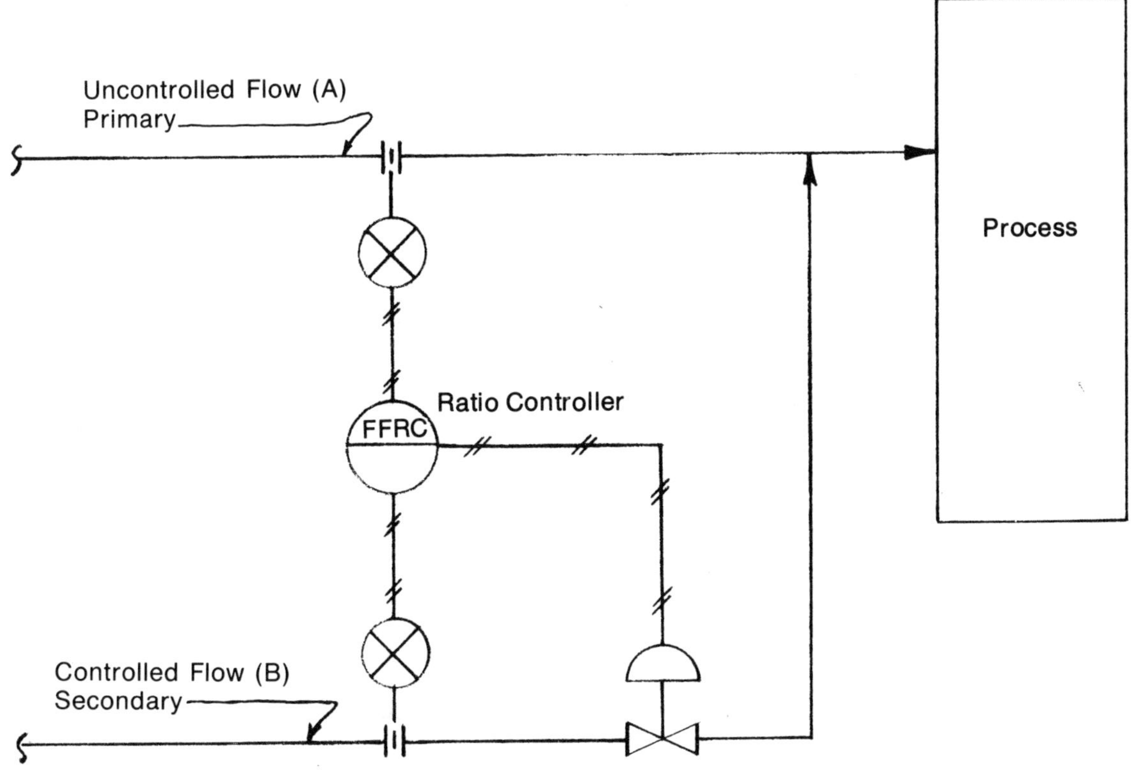

Figure 3.16. The uncontrolled flow (A) is measured and an adjustable ratio linkage on the controller is used to control flow B to the desired ratio between A and B.

There are several variations of control schemes used to obtain the ratio of two variables. Two of these are presented and discussed to illustrate the basic principles involved in ratio control.

The predominant scheme in the 1940s and 1950s (and still used to some extent today) used an adjustable ratio linkage between the primary or uncontrolled variable (sometimes called the "wild" variable) and the control index of the secondary or controlled variable. This type of linkage was used in large case pneumatic controllers. Figure 3.16 shows a typical loop of this type where the uncontrolled (primary) flow is measured and used to control the other (secondary) flow to maintain the desired ratio.

A more common method of ratio control is using separate units to provide the ratio system. Figure 3.17 shows the measurement of an uncontrolled flow transmitted to a ratio unit where it is multiplied by a ratio factor, and the output of the ratio unit becomes the set point of the seconday controller. The ratio unit normally has a manually adjusted scale to adjust the ratio between the two variables. The range of adjustment for standard ratio stations is normally 10:1 for linear scales (from 0.3:1.0 to 3.0:1.0) and 3:1 for square root scales (from 0.5:1.0 to 1.5:1.0).

The primary variable is not always uncontrolled. Very similar to the scheme shown in Figure 3.17 is one in which the primary flow is not only measured but also controlled at various rates. As in the previous case, the primary transmitter signal is fed to a ratio device whose output sets the control point of the secondary controller.

In using ratio control systems, one should ensure that both measurements are in the same units (i.e., gpm, pounds per hour, etc.) and that scales used are the same, either square root or linear.

Ratio control systems are available in pneumatic and electronic systems. In electronic systems, the analog or the digital system may be used. Digital systems are quite common for high accuracy blending where accuracy is ± one digital count.

Feedforward Control

Feedforward control is the application of a control action to a process before a deviation occurs in the controlled variable. To this point all discussions about control principles have centered around control modes that utilize the concept of feedback control in regulating process variables. Feedback control acts only after a deviation from target (set point) is sensed. Feedforward control theoretically prevents the deviation from occuring. It accomplishes this feat by measuring variables that cause load changes in the process

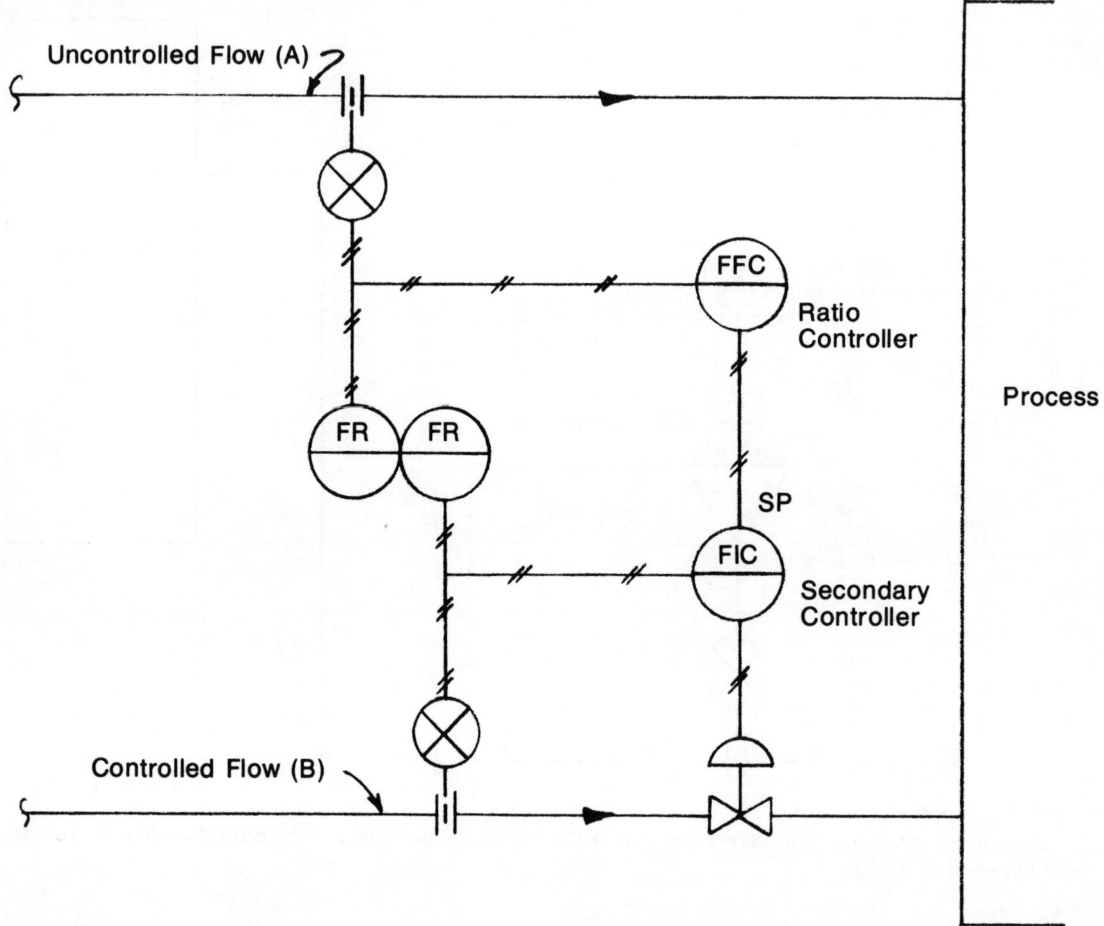

Figure 3.17. In this ratio control scheme, the flow feeds to a ratio controller whose output adjusts the set point of the secondary flow controller.

and manipulating other variables that cancel out load change effects before disturbances are created.

The concept of feedforward control has been realized for a long time. In the early 1960s, conventional electronic controls were made available with the option of feedforward action. The application of the technique in conventional control systems has never been widespread, however, probably because in most cases too little knowledge existed concerning the energy balances of the process. The technique certainly is applicable to computer controlled processes or to conventional control loops with added computing functions to augment its control capabilities.

Feedforward control is illustrated in Figure 3.18. Load changes to a heat exchanger are sensed in the liquid feed before the change reaches the exchanger. The computer calculates the amount of steam necessary to meet the load change requirement, and the steam flow controller is reset as necessary to meet the new demand. The heat change requirement is anticipated due to the load changes and the output temperature theoretically remains stable.

Feedforward action is more applicable to difficult-to-regulate processes. It also implies considerable knowledge of the process. Inherent in its concept is the knowledge that a load change will produce a known reaction which can be prevented by appropriate action.

The need for feedforward control is more likely to occur in loops where there is an appreciable lag time. In the example given in Figure 3.18, a conventional control system using temperature measurement to maintain stable control would deviate from the control point when load changes occur because that is the nature of a feedback control system and because of the inherent lag of a temperature measuring system.

A serious failing of the feedforward technique is its dependence on accuracy. To provide perfect control, the system must model the process exactly. This will not be accomplished precisely because measured loads and variables are inaccurate and information about the process is insufficient. Control errors are certain to occur and this means that some modes of feedback control are also necessary

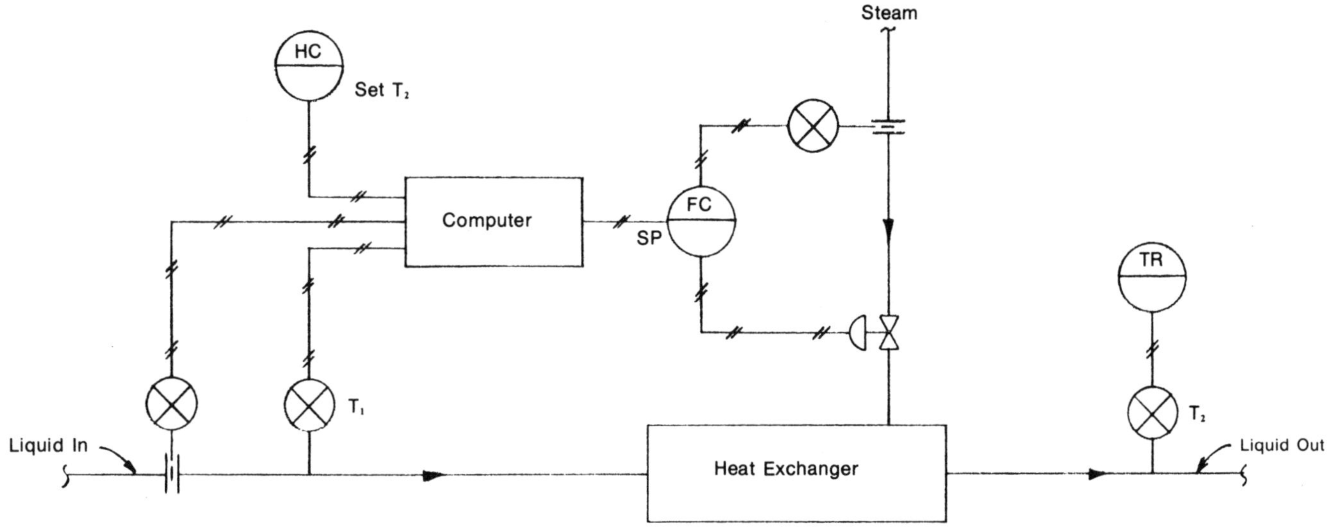

Figure 3.18. Feedforward control is illustrated by a control system for a heat exchanger. Measurement of load changes is made ahead of heat exchanger, and the calculated heat input requirements cause the steam control system to respond accordingly.

when feedforward is used. The role of feedback control may be somewhat reduced, however.

In summary, there is little question that feedforward control is an effective technique. Because additional control components are required, it is considerably more costly than conventional control systems. It requires a knowledge of the process to design the system and by its nature precludes mass production of fully adjustable feedforward systems.

Other Control Schemes

Several other control schemes or methods are described briefly which find application occasionally for specific problems. These include split-range control, override control, time-cycle control, end-point control and program control.

Split-Range Control

Split-range control normally involves two control valves operated by the same controller. In Figure 3.19, nitrogen makeup is used normally to maintain the correct vessel pressure. However, under certain conditions, the vessel pressure may become too high and the excess pressure must be vented off.

Under conditions such as these, it is often desirable to provide a small dead band between valve operations; that is, have the nitrogen makeup valve closing from 0 to 50% of signal range, leave a 5% dead band and have the vent valve opening from 55% to 100% of signal range. The percentage of dead band varies with application. In some cases there is no need for dead band at all, while some applications may require an overlap in action of the valves.

Override Control

In process control systems it often becomes desirable to limit a process variable to some low or high value to avoid damage to process equipment or to the product. This is accomplished by override devices. As long as the variable is within the limits set by the override devices, normal functioning of the control system continues; when the set limits are exceeded, the override devices take predetermined actions.

There are several options that may be taken when override occurs. In some cases, control may be switched to a secondary variable until operation of the primary variable returns within the prescribed limits. In other cases, the controller output may remain at the predetermined limit, and in some cases positive on-off actions may be taken at appropriate places in the process—actions such as closing a valve to stop steam flow to a reaction process, opening a valve for cooling purposes, opening a valve for quenching purposes, opening a vent valve, closing a catalyst feed valve, etc.

Time-Cycle Control

Time-cycle control involves one or more circuits, usually electrical, which activate on-off valves and other control devices to perform repetitive operations in process operations.

There are many process functions that require this type control for entire operational sequences. Some functions are simple, such as switching drying chambers in dessicant air dryers that have two dessicant beds used alternately in the drying and reactivation cycles.

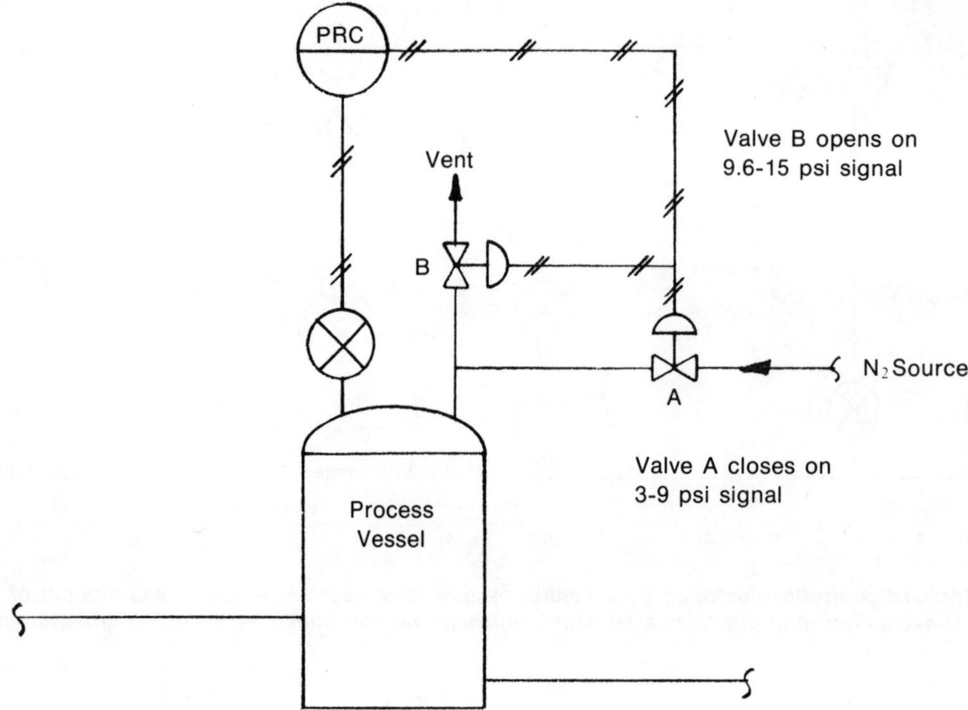

Figure 3.19. Split range control is illustrated by a pressure control system that normally requires N_2 makeup. A small dead band is provided between the closing of the makeup valve and opening of the vent when pressure becomes excessive.

Other more complex systems include absorbent type drying systems such as molecular sieves for moisture removal or other liquid or component separation. Such systems as these involve switching, furnace operation and other on-off functions that are accomplished on a pure time cycle or a combination of time cycle and end-point control.

End-Point Control

End-point control is a combination of control systems in which a primary variable automatically adjusts set points or ratios of controllers to achieve control of the primary variable. A typical application of end-point control is illustrated in Figure 3.20.

In the example a process must be neutral at the mixing tank to prevent unnecessary corrosion to process equipment downstream. A combination of cascade and ratio unit control systems is used for this purpose.

End-point analysis is made by a pH detector, and its controller adjusts the ratio of the neutralizing agent (secondary flow) to the acid stream (primary flow) to achieve a neutralized (basic) mixture. As acidity changes in the primary stream, the pH controller detects the deviation from set point and adjusts the ratio setting automatically to keep the mixture under control.

This technique can also be applied to controlling air-fuel ratio in furnaces by measuring the oxygen content of the exhaust gases.

Program Control

Program control is used in batch or continuous processes to control mixing operations, for sequencing events in process drying operations, for controlling batch processes or for other control schemes involving repeated sequencing of events. Program controllers vary in complexity and in adjustment flexibility. They may approach the simplicity of a time-schedule controller, but they usually employ end-point analysis in sequencing events. In some cases they consist of a combination of time-schedule and end-point analyses for sequencing direction.

Additional information and discussions of this topic are given in Chapter 10.

Analyzing Control Problems

The proper solutions to control problems are not always obvious. Solutions that seem obvious are not always the best. In order to evaluate the various alternatives, some pertinent questions may furnish clues to proper applications. Among the questions that need answering are these:

1. Which variable should be controlled?
2. Which detection method should be used for that variable?
3. Where should the detector be located?

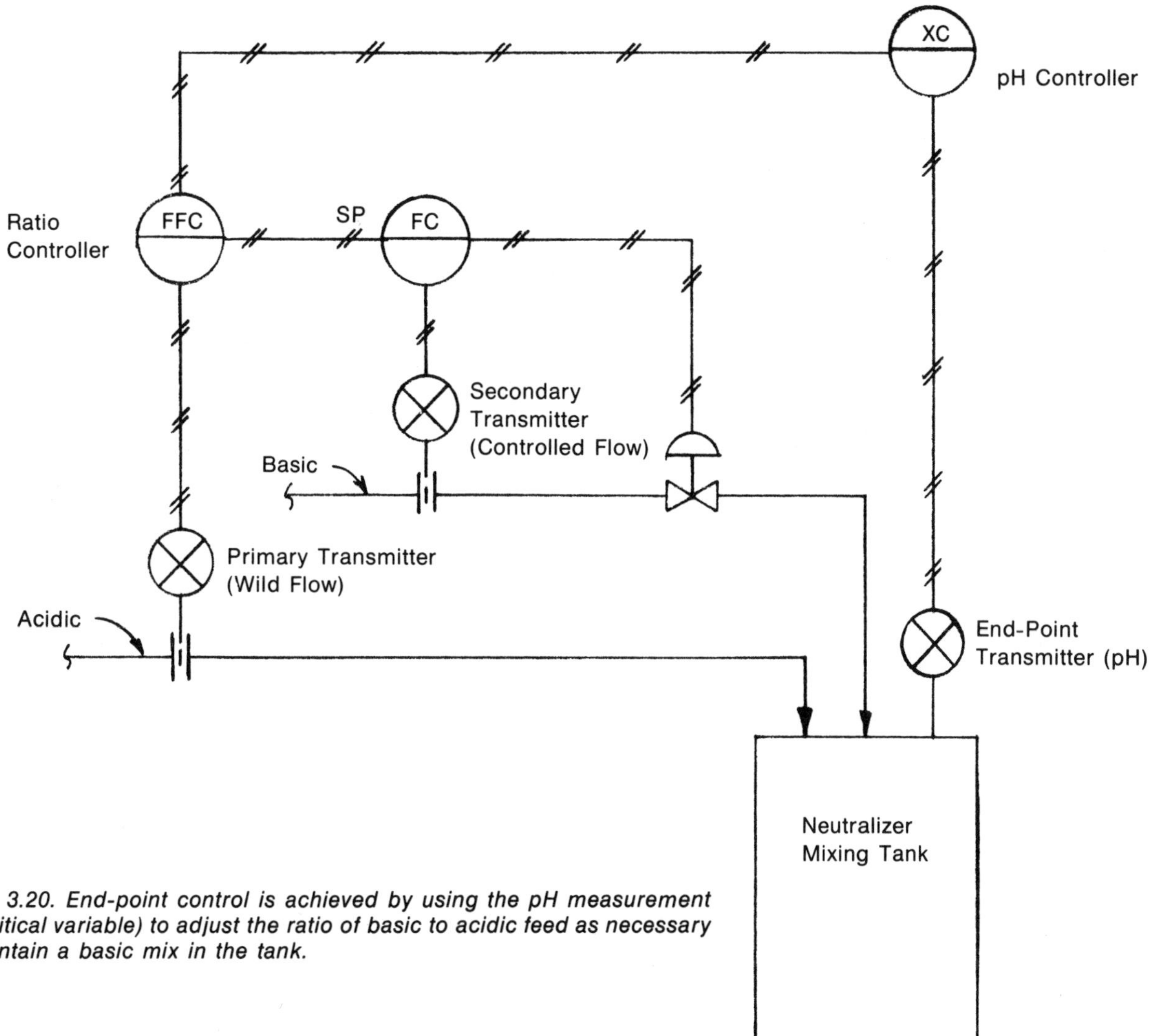

Figure 3.20. End-point control is achieved by using the pH measurement (the critical variable) to adjust the ratio of basic to acidic feed as necessary to maintain a basic mix in the tank.

4. How should system cost versus system efficiency be evaluated?

The following examples help illustrate the need for posing these questions.

Proper Variable

For many years in the processing industries, temperature, flow, pressure and level were essentially the only variables used to control processes. These four variables still comprise the major portion of control loops specified. However, in recent years truly significant gains in process control have been achieved by measurement and control of more meaningful physical and chemical properties of streams throughout the process—from the raw material stage to the finished product. Analytical devices for composition measurement have been particularly helpful. Many physical measurements now made routinely were seldom used only a decade ago.

Added to the four basic physical measurements so often used are the following: composition measurements by chromatograph or other analytical means, density or specific gravity, turbidity, viscosity, colorimetry, conductivity, moisture, consistency, odor, force, vibration and speed measurements, to mention a few. With these added methods for determining the physical and chemical characteristics of streams, process efficiency, quality and economy are often greatly improved.

Two hypothetical examples illustrate the need for selecting the proper variable in a control situation.

Example 1

A reaction process requires the addition of a catalyst whose weight percent normally must be 0.1 % of the reaction mixture to obtain maximum reaction. The catalyst flow can be ratioed to that of the main feed stream with creditable results; but a composition measurement of the catalyst component in the reactor allows closer control, resulting in a 3% increase in production rate.

Example 2

Flow control of a hydrocarbon feed stream into a reactor leads to a cyclic production rate and variations in quality level. Adding a specific gravity measurement and a computing device to provide mass flow measurement and control stabilizes production and improves quality. The problem results from variations in mass flow due to ambient temperature effects on the feed stream temperature.

Many other examples might be cited, but these emphasize that each control system should be analyzed to determine which variable measured presents the best information toward better control of the process.

Proper Detection

After the decision is made to use a particular variable for control, the best method for obtaining its measurement must be selected. In many cases this is not easy, particularly for new processes where the exact nature of the streams may not be well known. Again, the problem is illustrated by examples.

Example l

Liquid level needs to be measured in a vessel containing a polymer slurry. The vessel is 8 feet in diameter, about 20 feet high, and its contents are stirred by a top-entry agitator. Measurement range is 6 feet; design pressure is 500 psig; maximum temperature is 300°F. Density variations caused by polymer content and operating temperature are expected. Alternatives are listed with some appropriate comments.

1. External cages with stagnant legs would likely plug with polymer material so that a buoyancy type device would not work; polymer would also coat an internal displacer.
2. Differential or head type devices would require sealed systems or purges; in addition, it would be subject to errors caused by density changes.
3. A continuous capacitance type device might work satisfactorily, but the measurement range is high, the dielectric constant of the material has not been verified and it, too, is subject to a coating problem.
4. The nuclear radiation method might be used, but there still is the density variation problem, and calibration might be difficult to make.
5. Weighing the vessel contents is another approach, but this method, too, must take into account specific gravity variations caused by polymer content and operating temperature changes.

The solution chosen among the alternatives is a radioactive level measuring system, compensated by a density measurement. Other methods may also work satisfactorily.

Example 2

Flow measurement is desired in a 100 gpm polymer slurry stream where the solids content is about 5%. Pressure and temperature requirements present no problems.

Because of slurry, differential measurement would require liquid purging of lead lines which is undesirable from several standpoints. Elbow taps with sealed diaphragms might be used, but they are not accurate enough.

The solids content is too high for a rotameter, not to mention the high maintenance factor on rotameters.

A magnetic meter offers a good possibility except that the conductivity level is questionably low, and coating of the electrodes is a definite possibility.

A turbine meter is not seriously considered, but a target meter might work even though the stream contains an appreciable amount of solids.

A vortex meter might work, but extensive application of this device has not yet been made.

The choice boils down to three methods—the mag meter, the target meter or the Vortex meter. The low-conductivity model mag meter is chosen with a preamplifier to handle the low conductivity ranges.

Detector Location

Detector location normally is not a problem. The proper location of flow elements depends primarily on their isolation from disturbances from equipment, valves and fittings. These locations are fairly well defined in Chapter 4. Space requirements for straight run lengths of pipe sometimes present problems.

Pressure tap locations are not difficult to make. They might be erroneously placed on the wrong side of a control valve that develops a high pressure drop, but this does not occur very often.

Temperature elements can easily be mislocated. They may be located too close to the mixing point of convervging streams of different temperatures resulting in incomplete mixing and unstable readings. An element might not be immersed sufficiently into a stream, so that the indicated temperature is somewhere along the gradient between stream temperature and pipe skin temperature.

The location of temperature elements probably requires closer scrutiny than any other type measurement other than analytical measurements.

Sample points for analyzers must be located at the proper places in the streams to be analyzed and should also be located so that sample transporation time is minimized. In this situation it might be stated that the analyzer should be located close to the sampling point. This is not always easy if the area is crowded with equipment and/or piping. Some situations require compromises between the problems of sample lag time and convenience of analyzer location.

Cost Versus Efficiency

The cost of instrument control systems must be evaluated in terms of what they accomplish. An extensive treatment of the topic is beyond the intended scope, but some observations are in order.

A mistake often made in instrument application is the sacrificing of quality to minimize original cost. This may result in increased maintenance costs, and, as a by-product, loss of production time. The two factors, production loss and maintenance cost, are usually more important than original cost. This has been increasingly true in the past two decades, and there is no indication that the trend will reverse. As process plant sizes increase, as production rates increase, as labor costs continue to rise and as profit margins decrease the reduction of downtime and maintenance time are increasingly important.

Most instrument control systems are installed with a life expectancy of 10 years or more, yet many of these systems have been in use for more than 25 years. In many cases this minimizes the significance of original cost. Past experience shows that instrument costs may run from 7 to 10 percent (and sometimes more) of the total cost of a new plant. Their cost usually must come within some established guidelines just as other facets of the total cost picture.

As plant situations are analyzed, it sometimes becomes evident that control sophistication and complexity can work against plant efficiency as well as for it. A complex system that is highly advantageous in one plant where skilled operators and maintenance people are available might create severe production losses when operated and maintained with less skilled workmen. The experience level of operators and maintenance people must be considered. Consultants from plant operating groups can often furnish this type information.

4 Flow Measurement

William G. Andrew, B. J. Normand, Fred E. Edmondson

Flow measurements account for a high percentage of the process variables measured in the chemical processing industries. No other variable is more important to plant operation. Without flow measurements, plant material balances, quality control and even the operation of continuous processes would be almost impossible. It is important, then, to make a comprehensive survey to ascertain the types of flow measurements made and learn the good and bad features of each of these types. Only then will one be sure of the best method for a particular application.

Many accurate and reliable methods are available for measuring flow. Some are applicable only to liquids, some only to vapors and gases and some to both. Fluids measured may be clear or opaque, clean or "dirty," wet or dry, erosive or corrosive. Fluid streams may be multiphase, vapor, liquid or slurries. The flow may be turbulent or laminar. Viscosities may vary. Pressures may vary from near vacuum to many atmospheres; temperatures may range from cryogenic to hundreds of degrees. Flow rate requirements may range from a few drops per hour to thousands of gallons per minute. Rangeabilities (ratio of maximum to minimum flow requirements) may vary from essentially 1:1 to 100:1 or greater. With this possible assortment of conditions and requirements, it is essential to have a good grasp of the various flow measuring methods available and to know their applicability to a given set of conditions.

The most common method of measuring flow is the differential pressure or "head" device which utilizes restriction elements (orifices, venturis, etc.) in a line. For this method, flow rate is proportional to the square root of the differential pressure generated by flow through the restriction. Other flow rate measurements include magnetic meters, variable area meters (rotameters), turbine meters, gear meters, piston meters, nutating disc meters, vane meters, fluidic meters, anemometers, target meters, thermal flowmeters, ultrasonic flowmeters, swirlmeters and metering pumps.

This survey provides an insight into the operation of these meters, with salient features about their application.

Nature of Fluids

A knowledge of some basic characteristics of fluids and fluid flow is necessary for choosing the best method of flow measurement. The characteristics and physical properties which need to be considered include viscosity, density, specific gravity, compressibility, temperature and pressure. Many of these are direct functions of others. To know whether the flow is turbulent or laminar is also necessary, in most cases.

Viscosity

Viscosity is defined as a quantitative measure of the tendency of a fluid to resist the forces of shear or deformation. If the technical definition fails to register properly, then one should consider it as the tendency of the fluid to flow. Fluids that flow freely have low viscosities; fluids that seem to resist free flow are highly viscous. A good example of the highly viscous material is cold sorghum molasses, which does not flow easily or readily in comparison to water. Water at 68.0°F is a reference point for viscosity measurements; its value is 1.0 centipoise (1/100 of a poise) in the metric system of units.

The dimension of a centipoise is in grams per centimeter second. The poise or centipoise is referred to as absolute viscosity. Several other systems of units exist, making conversions often necessary. Other viscosity units used in the petroleum industry include kinematic viscosity, expressed in stokes or centistokes, and Saybolt Universal or Saybolt Furol viscosity expressed in seconds. These units are readily converted from one to the other when calculations require their use and values are given in other units.

The viscosity of a fluid depends primarily on temperature and to a lesser degree on pressure. Viscosities of liquids generally decrease with increasing temperature, and gas viscosities normally increase as temperature increases. The effect of pressure on viscosities of liquids is very small. Its effect on gases is normally significant only at high pressures, particularly in the vicinity of the critical pressure.

Density

The density of a substance is defined as its mass per unit volume. The density of water at 32°F and at atmospheric pressure is 62.42 pounds per cubic foot. Liquid densities change considerably with temperature, but only negligible amounts with pressure, except at extremely high pressures. Liquids generally are considered incompressible, but when pressures exceed a few thousand psi, liquid compressibilities should be considered in flow measurement applications.

The densities of gases and vapors are affected greatly by changes in pressure and temperature. Most flow measurements are made on the basis of volume measurements, so fluid densities must be known or measured to determine true mass flow, and, in most instances, mass flow is the desired measurement.

Specific Gravity

The specific gravity of a fluid is its weight (or mass) ratio to a standard. For liquids (or solids) the standard is water (s.g. = 1.0) at 4°C or 60°F (used by most engineers). It is a dimensionless term. The standard for gases and vapors is air (s.g. = 1.0) at 60°F and one atmosphere pressure (common reference—other references are sometimes used). The use of the term specific gravity is a matter of convenience since the characteristic that is important to fluid flow considerations is the fluid *density*. The use of specific gravity values referenced to a standard makes calculations simpler and important densities easier to remember.

Compressibility

It was noted earlier that liquids are considered incompressible except at high-pressure ranges. Its effect need not be considered for most liquid flow measurement methods. In the measurement of gas flows, however, compressibility is definitely a significant factor and must be considered. In differential measurements, the compressibility factor (Z) should be calculated in all cases, since the percentage error can be quite large. An example of this, for instance, is propane flowing at 100 psia and 300°F; thus $Z_{base} = 0.87$ and $Z_{flow} = 0.98$. The flow correction factor, F_{pv}, is equal to $Z_b/(\sqrt{Z_f}) = 0.888$. This is an 11.2% error.

Compressibility factors are a function of molecular weight, temperature and pressure.

Temperature

The effect of temperature changes has already been noted on viscosity, density, and compressibility. Few flow measurements are made that are entirely unaffected by temperature variations. The most common method of flow meaurement, the differential or head type, assumes that both temperature and pressure remain constant. This is seldom true. In many applications, the assumption is valid enough, however; for the inaccuracies involved are either insignificant to the process, or compensation is made through other manual or automatic adjustments.

For accounting applications or in other services where compensating means cannot be used, temperatures are measured separately and flow rates are adjusted to reflect mass rather than volume flow. As more processes are placed under computer control, the temperature problem decreases, for the computer often performs mass flow calculations at a much smaller differential cost than has been accomplished previously with measuring elements and complete signal transmission systems that comprise special purpose computers for that particular function.

Pressure

The effect of pressure variations has been well defined in its relation to density, specific gravity and compressibility. Its effect is not appreciable on liquids except for high-pressure ranges, but it definitely must be accounted for in gas and vapor measurements. When head flow measurements are made and conventional instrument hardware is used, pressure or density measurements are also made for mass flow calculations with special computing hardware or through manual computation. As in the case of temperature variations, this can be accomplished easily when process control is computerized.

Fluid Velocity

Fluids in motion through pipes move in two contrasting manners of flow depending to a high degree on their velocity. The two types of flow are called laminar and turbulent.

Laminar flow is sometimes referred to as viscous or streamline flow. It is distinguished by the fact that particular molecules of the fluid follow closely parallel paths as they proceed through the pipe or container.

Turbulent flow, on the other hand, is characterized by erratic patterns, as crosscurrents and eddies move the fluid molecules along irregular paths. The two types can be demonstrated easily by injecting a small jet of colored water slowly into a clear liquid stream in a transparent pipe. At low flow rates (laminar flow) the jet of colored water flows evenly and with little diffusion in the surrounding stream.

When a similar jet stream is released in a high velocity (turbulent) stream, diffusion is almost immediate and uniform across the entire pipe section.

The term velocity, when applied to fluid flow in pipes, refers to the average or mean flow. Average velocity must be used since the actual velocity of the fluid varies across the cross section of the pipe. For laminar flow, the velocity of the fluid adjacent to the pipe wall is theoretically zero. It increases rapidly a short distance from the wall and reaches a maximum value at the pipe centerline. The average velocity is about one half the maximum.

In turbulent flow, the velocity adjacent to the pipe wall is zero as in laminar flow. It increases much faster as it moves from the wall, however, and the average velocity is somewhat larger than one-half the maximum flow.

Turbulent Flow

Most flow measurements are made under turbulent flow conditions. This is not a necessary requirement for principle or accuracy but is imposed because of economic considerations. Larger pipes and slower velocities would require much more expensive piping systems. Turbulent flow is a necessary requirement to meet the empirical relationships that have been developed over the years for orifice flow calculations. The requirement is inherent in its development. When using these empirical relationships, it is essential to determine that flows are turbulent.

It has been determined that turbulent flow exists when the Reynolds number exceeds 4,000. The Reynolds number is an important factor in fluid flow calculations or in any analysis of flowing phenomena. A dimensionless term, its value is determined by several factors: pipe diameter, viscosity and density of the fluid and average fluid velocity. These four factors combined change the flow characteristics when the numerical value of the group reaches a certain magnitude. This dimensionless group is named in honor of Osborne Reynolds, an early researcher and observer of fluid flow phenomena.

Laminar Flow

Under ordinary conditions, fluid flow is considered to be laminar for Reynolds numbers up to 2,100. Turbulent flow exists for Reynolds numbers above 4,000. In the transition zone between 2,100 and 4,000, the flow can be either laminar or turbulent. The fluid velocity at which laminar flow ends and the transition zone begins is called the critical velocity.

Since laminar flow conditions do not occur very frequently, few flow measurements are made under these conditions. Laminar flow elements are made, however, for low flow differential measurement applications, and they offer a definite advantage over turbulent flow devices. Their rangeability exceeds that of the devices made for turbulent flow by a factor of about 25. Because of the square relationship between flow rate and the pressure differential across the flow element, the rangeability of an orifice type measurement is about 4:1. The rangeability when using a laminar flow element is about 100:1, mainly because the flow rate-pressure differential relationship is linear. Accuracy is just as good on the low end of the range as it is on the high end, whereas in turbulent flow measurements, accuracy (detection) is poor on the low end of a selected differential range.

Measurement Methods

Some specific methods of flow measurement were mentioned briefly in the introductory paragraphs of this chapter. These methods are divided into three broad categories as inferential types, discrete quantity types and mass measurement types. A brief discussion of these broad categories follows.

Most flow rates are determined by inferential measurements; that is, the rate is inferred from a characteristic effect of a related phenomenon. A classic example of an inference measurement is a body fever resulting from a serious sickness or body disability. A rise in body temperature—a fever—is not the cause of sickness, it is a characteristic phenomenon resulting from the body ailment.

Inferential flow measurements include:

1. The head type (inferred from differential pressure measurement across an engineered restriction)
2. The variable area or rotameter type (inferred from a position measurement resulting from a balance of weight force against a velocity force)
3. The magnetic meter (inferred from a velocity measurement)
4. Turbine meters (inferred from a velocity proportionality factor)
5. Target meters (inferred from a force measurement)
6. Thermal flowmeters (inferred from a thermal conductivity effect)
7. Swirl meters (inferred from temperature or pressure oscillations).
8. Sonic flowmeters (inferred from noise level or Doppler shift).

Discrete quantity flow measurement is used for a small percentage of industrial flow rate applications. Methods that fall into this category include positive displacement meters and positive displacement metering pumps. The principle of measurement is that with each cycle of operation (whether the method is a reciprocating piston, a rotating vane or lobe or a flexible diaphragm), a discrete volume of fluid is transferred through the meter or pump. It should be remembered that these are volume measurements only and provide mass rates only when fluid densities remain constant.

True mass flow devices are rarely found in industrial applications at present. Mass flow computations are being used with increasing frequency, however, through volume flow measurements and other necessary measurements such as temperature and pressure or density. Special purpose computational devices use these signals to provide mass flow information, or when processes are computerized, calculations are made by the process computer.

Mass flow devices that are available commercially include the impeller-turbine and twin-turbine types and one operating on the principle of a gyroscope. All of these utilize the principle of angular momentum. Industrial use is limited for these types because of their high cost and the relative ease and convenience of calculation by other methods. Broader descriptions and illustrations of all the methods referred to are given in following sections.

Operating Condition Factors

Many of the points discussed here are covered in greater detail in Volume 2. As this survey of available flow devices is made, it is important to keep in mind the operating conditions and fluid characteristics and how the various flow measuring methods are affected by these factors.

One of the first factors to consider is the range of flow rates required. It is essential to know what the normal flow rate is expected to be and also to have a fair idea of the maximum and minimum rates that can be expected. The most common flow measurement method, the head meter, is applicable only from a 3:1 to 5:1 ratio of flows. For record or indication only, a 5:1 ratio is acceptable for most applications. For closed loop control, however, the ratio should not exceed 4:1; 3:1 is preferable.

When a higher rangeability is needed, it is necessary to use two or more head devices in parallel or to use another method having a higher rangeability. Usually the latter choice is preferable. Most other devices have rangeabilities from 10:1 to 20:1. A few types, such as the swirlmeter and laminar flow devices, have rangeabilities of 100:1.

Temperature and pressure variations must be known so that their effects can be compensated for when measurements require it. Only the true mass flowmeter automatically takes these into consideration.

Corrosive effects must be known. It would be economically unjustifiable to install a Venturi tube and d/p cell into a line that required Hastelloy parts in contact with the fluid under many circumstances. A target meter would be far less expensive and might be used if its accuracy limitations were acceptable, and they can be calibrated almost as accurately as a d/p cell. When corrosive fluids require high cost alloys, special attention should be given to methods which minimize the cost.

Slurry fluids—liquids containing solids, polymers, etc.—require special consideration. There are several methods that can be used effectively and accurately. Various sealing methods can be used for head devices. Target meters can be used in some services. Magnetic meters can be used if the fluid is sufficiently conductive.

Viscosity effects must be borne in mind. Some fluids have, or can attain, viscosity values high enough that their flow is always laminar—in which case, a head measurement is not applicable except when using a laminar flow element. Where laminar flow exists, a differential measurement can be made accross a straight section of pipe, if a sufficiently long, straight section can be found to make accurate measurement. This is seldom practical, however.

Many methods of flow measurement are used today, and most of these are discussed in considerable detail. Although many different types of measurements are made, because most of them are relatively expensive, there are few areas in instrumentation where a new, accurate, economical method would be more welcomed than in the flow measurement area.

Variable Head or Differential Meters

Head metering is one of the oldest methods of measuring flow rates. Still used extensively in industry, it outstrips all other methods in use because it is simple, accurate, reliable and relatively inexpensive. It measures volume rather than mass flow rates, but mass rates can be calculated or computed easily by knowing or sensing temperature and pressure.

The principle of operation is that a restriction in the line of a flowing fluid produces a differential pressure across the element (restriction) which is proportional to the flow rate (Figure 4.1). The proportionality is not a linear one but has a square root relationship (Figure 4.2) in which the flow rate is proportional to the square root of the differential pressure. The restrictions or flow elements are made in a predetermined manner and there are several types. They are discussed in considerable detail later in this chapter.

An extensive amount of data has been compiled and published relative to head flow measurements. Most of it centers around orifice and Venturi elements and the con-

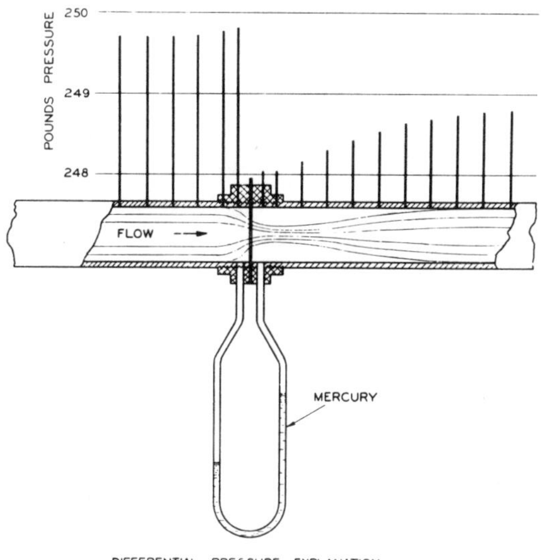

Figure 4.1. Flow through a restriction produces a differential pressure which is proportional to flow rate. (Courtesy of Foxboro Co.)

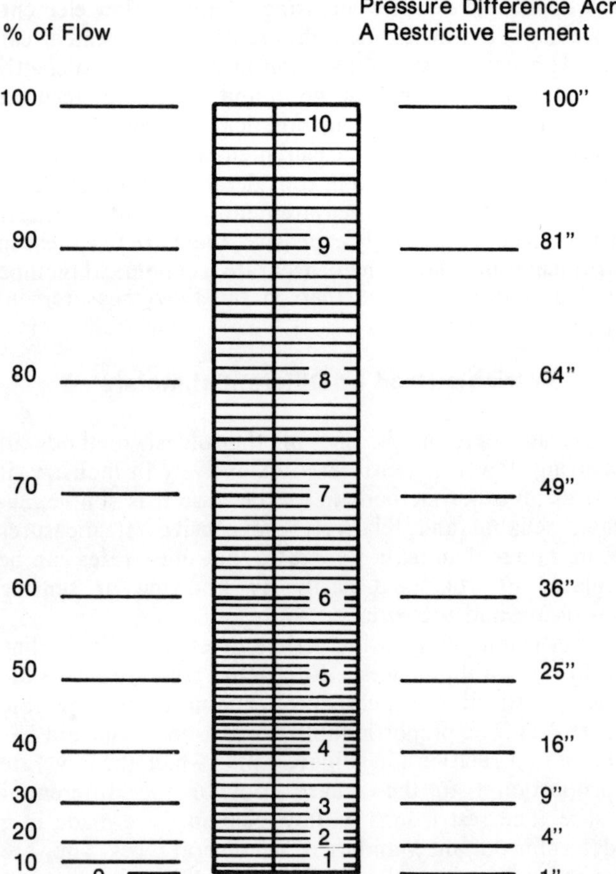

Figure 4.2. A square root relationship—flow is proportional to the square root of the differential pressure across the restrictive element. Pressure differential is shown for 100-inch range application.

figuration of the adjacent piping where the measurements are made. Equations have been developed, empirical in nature, which take into account almost all known factors which affect the measurements. The basic equations from which these were developed were derived from the Bernoulli Theorem established in the 18th century.

$$V = k\sqrt{(h/\rho)} \tag{4.1}$$

$$Q = kA\sqrt{(h/\rho)} \tag{4.2}$$

$$W = kA\sqrt{(h/\rho)} \tag{4.3}$$

where

V = velocity

Q = volume flow rate

W = mass flow rate

A = cross-sectional area of pipe

h = differential pressure across the element

ρ = density of the flowing fluid

k = constant which includes the ratio of the pipe cross-sectional area to the cross-sectional area of the element, units of measurement and correction factors.

The constant, k, contains many separately determined factors which affect the differential pressure measurement. These include the type differential device used, the flowing temperature and pressure, the expansion or contraction of the element at flowing temperature, the compressibility factor, velocity of approach, etc. These are discussed at length and sample calculations are given for a variety of conditions in Volume 3.

Refinement of empirical data has made it possible to calculate the bores of primary devices (orifices, Venturis, etc.) with great accuracy. The bores are then drilled to accuracies that are customarily within ±0.001 inch. The accuracy of head measurements is expected to be within ±½ to ±1% of span. Through good calibration techniques, even greater accuracies can be obtained.

The square root relationship (Figure 4.2) that exists between flow rate and differential pressure in head measurements has both good and bad consequences—for readout or control. A square root readout of the differential pressure expands the high end of the scale (good) but compresses the low end of the scale (bad). Fifty percent of the flow rate produces only 25% of the full differential pressure (hence, 25% of scale). At 10% of the flow, only 1% of the scale is reached. Accuracy and readability are poor on the low end. On closed loop control one must be cautious about relying on the low end for good control. For readout only (indication or recording), a rangeability of 5:1 (from 20 to 100%) is acceptable (Figure 4.3). For control applications a rangeability of only 3.5:1 (from about 27 to 95%) should be used. This rangeability leaves room for some control action at either end of the scale.

The difficulty that occurs at the low end of the scale is not because the differential pressure does not correspond to the flow rate but that low differential pressures are difficult to measure accurately when wide differential ranges are used. The solution in some instances is to use more than one differential range for measurements where high accuracies are needed.

Unlike most other types of flow measuring devices, the description of head meters breaks down quite naturally into two categories, primary and secondary elements. The primary devices (orifices, Venturis, flow nozzles, etc.) are inserted in pipelines to produce the differential pressures that are measured by the secondary devices (manometers, bellows, d/p cells, etc.). It seems appropriate to discuss primary elements first.

Primary Elements

Several different types of primary elements are used as restrictive devices for head flow measurements. The orifice

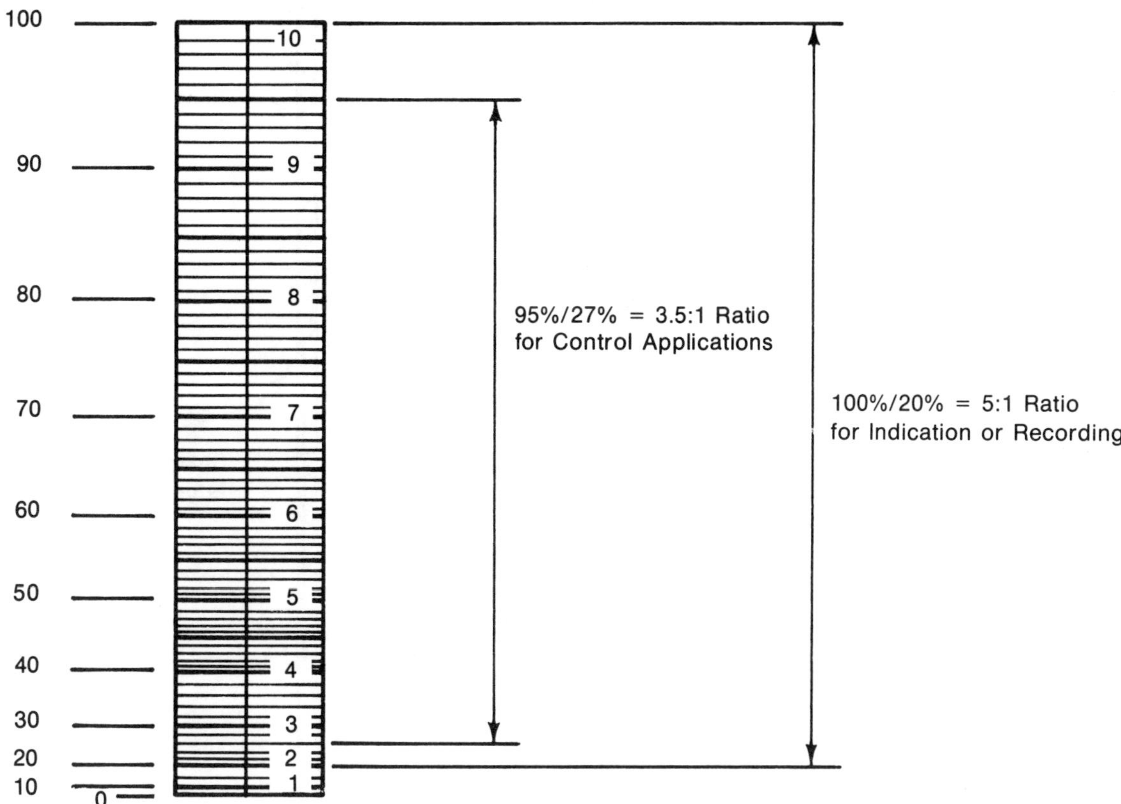

Figure 4.3. Practical application of rangeabilities is shown by observing readout or control points on a square root scale.

plate is used more frequently than all other types combined. There are four varieties of orifice plate: concentric, eccentric, segmental and quadrant edge. Other primary devices include Venturi tubes, flow nozzles, weirs, Pitot tubes and Annubars.

Orifice Plates

Concentric Plates. The orifice plate dates back to the days of Rome under the Caesars when they were used to measure residential water use by the Romans. In the early 1900s Thomas R. Weymouth began experimenting with thin plate, sharp edge, concentric orifices (Figure 4.4) for measuring large volumes of natural gas. He used flange pressure taps, 1 in. upstream and 1 in. downstream from the faces of the orifices, which were later to become the predominant standard for industries in the United States. Figure 4.1 illustrates the installed orifice, the flow pattern it causes and a typical pressure profile created by the restriction.

The standard concentric orifice plate is usually made of stainless steel from ⅛ to ½ inch thick, depending primarily on the line size. Other materials (nickel, Monel, Hastelloy, etc.) are used when needed to prevent corrosion or contamination. A tab is usually built onto the plate on which is stamped pertinent orifice plate data (identification, bore size, etc.).

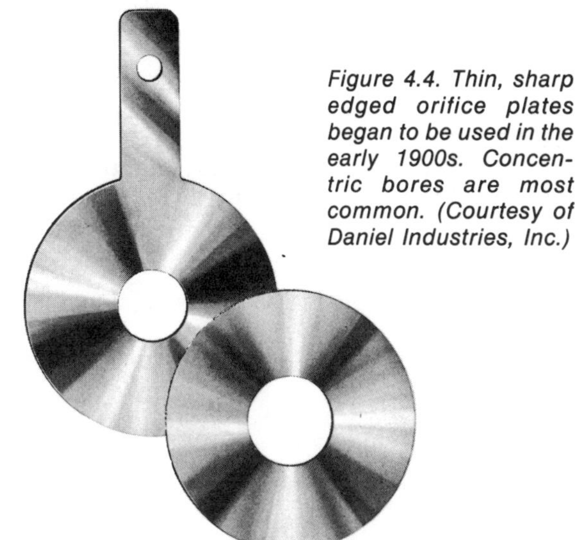

Figure 4.4. Thin, sharp edged orifice plates began to be used in the early 1900s. Concentric bores are most common. (Courtesy of Daniel Industries, Inc.)

Some flow orifices are made with one side of the hole bevelled at a 45° angle (Figure 4.5). The sharp or unbevelled edge should be mounted in the line facing the flow to

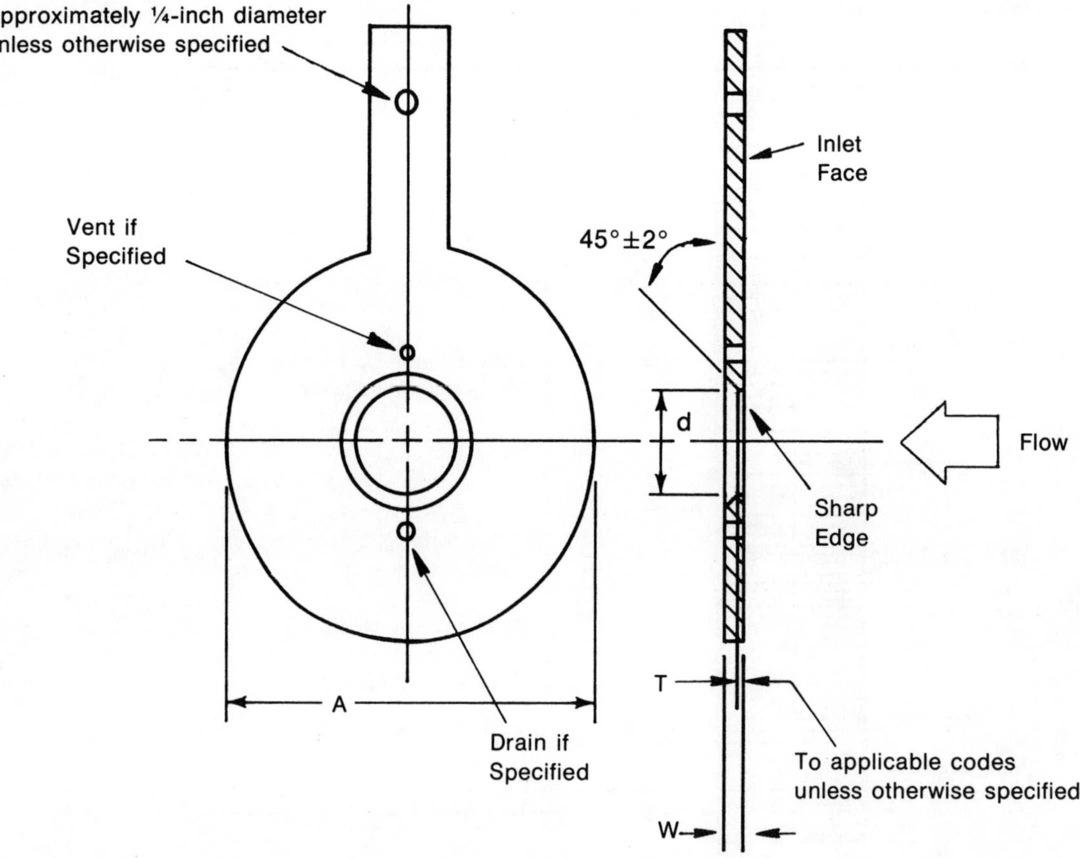

Figure 4.5. 45° beveled edges are often used to minimize friction resistance to flowing fluid.

minimize friction resistance as the flowing fluid passes through the restriction.

The plate thickness is usually a function of the line size. Its thickness at the orifice edge should not exceed any of the following:

D/50
d/8
(D-d)/8

where

D = the pipe ID
d = the orifice diameter or bore

The plate may be made thicker than indicated by the above criteria by bevelling the downstream edge at 45° or less to the required thickness. Square edge orifice plates can be used to measure reverse flow, but bevelled or counter-bored plates would produce measurement errors for flows in the reverse direction.

The ratio of the orifice bore to the internal pipe diameter, or the Beta (β) ratio, should fall within prescribed upper and lower limits to maintain measurement accuracy. Although data exists for d/D ratios beyond these limits, most applications should fall between values of 0.15 to 0.75 for liquids and 0.20 to 0.70 for gases, vapors, and steam. Best results occur between values of 0.4 and 0.6. Orifice runs are seldom installed in pipe sizes less than 1½ inches. When they are used for smaller sizes, prefabricated, specially bored pipe should be used to ensure accuracy.

Eccentric Orifices. Eccentric orifices (Figure 4.6) are bored tangent to a circle concentric with the pipe and of a diameter equal to 98% of that of the pipe. Other specifications relative to thickness, sharpness, smoothness and flatness are the same as for concentric orifices. Location of the bore prevents damming of solid materials or foreign particles and makes it useful for measuring fluids containing solids, oils containing water and wet steam.

Eccentric plates can use either flange or vena contracta taps, but the taps must be at 180° or 90° to the eccentric opening. The vena contracta locations are different from those for the concentric plates, and, in general, the possible error can be up to five times greater than on concentric plates.

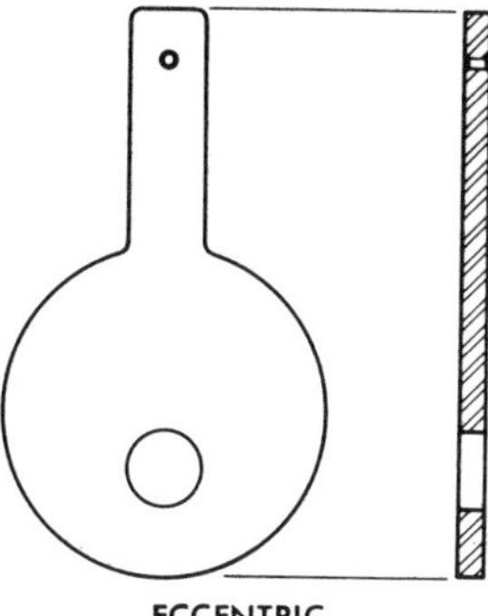

Figure 4.6. Eccentric orifices have the bore offset from center to minimize problems in services of solids-containing materials. (Courtesy of Daniel Industries, Inc.)

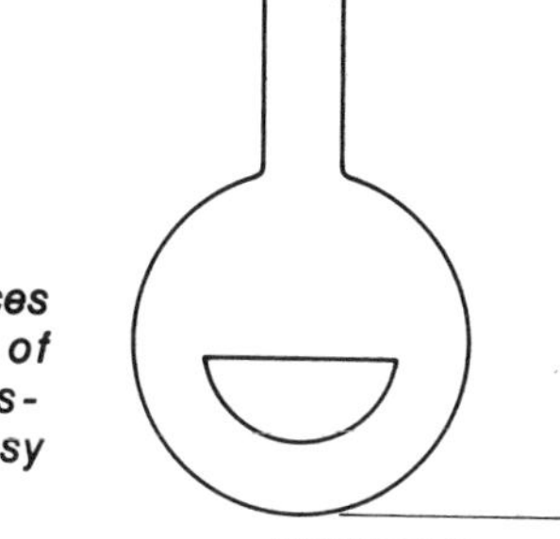

Figure 4.7. Segmental orifices provide another version of plates useful for solids-containing materials. (Courtesy Daniel Industries, Inc.)

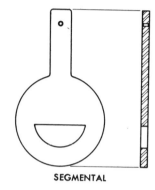

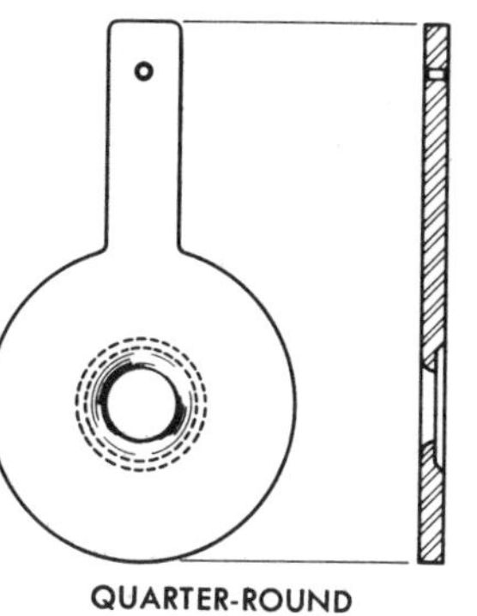

Figure 4.8. Quadrant edge orifices produce a relatively constant coefficient of discharge for services with low Reynolds numbers in the range from 100,000 down to 5,000. (Courtesy of Daniel Industries, Inc.)

Segmental Orifices. Segmentally bored plates (Figure 4.7) are used for the same type services as the eccentric orifice plate. The opening is a segment of a circle, the diameter of which is customarily 98% of the pipe diameter. The segmental opening may be placed either at the top or bottom of the pipe; however, it is generally used in services which require that it be placed at the bottom of the pipe. For the best accuracy, the tap location should be 180° from the center of tangency.

Quadrant Edge Orifice. The quadrant edge (Figure 4.8) or quarter-round orifice plate is constructed such that the edge is rounded to form a quarter-circle. The plate has a concentric opening with a rounded upstream edge rather than the sharp, square edge normally used. Tests indicate that this type plate produces a coefficient of discharge which is practically constant in the low Reynolds number or viscous flow region. The quarter-round orifice may be used when the line Reynolds numbers range from 100,000 or above down to 3,000 to 5,000 (depending upon beta ratio) with a coefficient accuracy of approximately 0.5%.

This plate is used for flows such as heavy crudes, syrups and slurries and viscous flows having Reynolds numbers below 100,000.

Types of Orifice Tap Connections. There are five common tap locations for taking differential pressure measurements across primary elements: flange, corner, vena contracta, radius and pipe.

Flange taps are the predominant type in use and are normally located 1 inch from the upstream face of the orifice plate and 1 inch from the downstream face. Flange taps are not recommended for pipe sizes less than 2 inches since the vena contracta may be less than 1 inch from the orifice plate. Figure 4.9 shows a cutaway of a set of orifice flanges using flange taps with the orifice plate installed.

Orifice taps are normally supplied with ½-inch taps in flanges with a minimum of 300-psig rating. These flanges have a minimum thickness of 1½ inches. Smaller size flanges are normally thicker than 1½ inches. In some cases

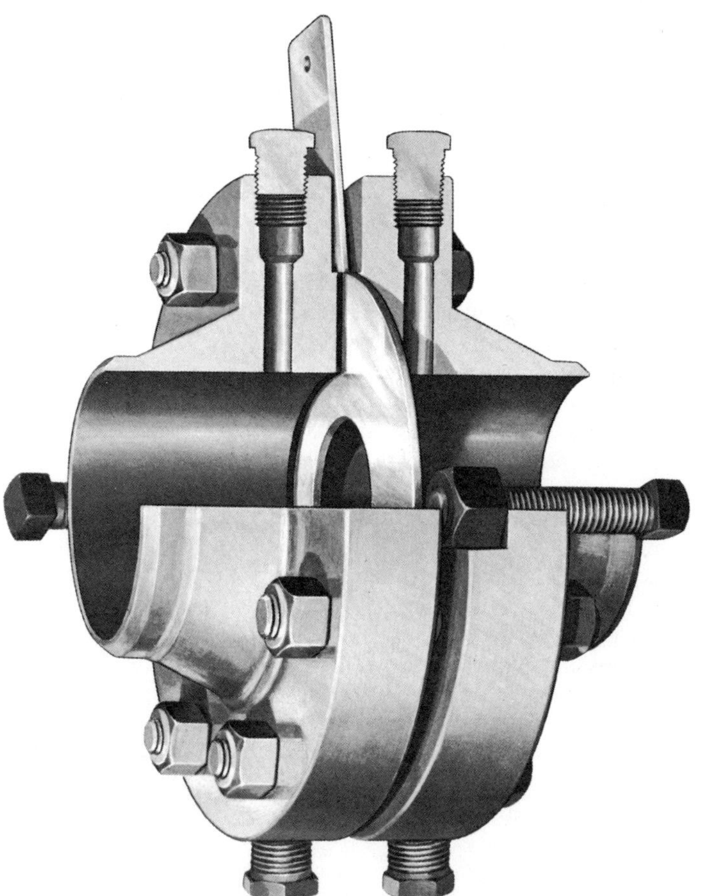

Figure 4.9. Orifice plate installed in a pair of flanges using flange taps. (Courtesy of Daniel Industries, Inc.)

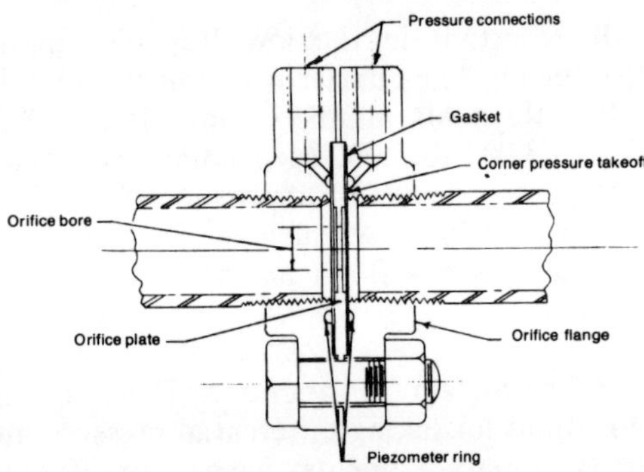

Figure 4.10. Corner taps are located adjacent to the plate and holes are drilled at an angle to the usual pressure connection location in the center of the flange. (Courtesy of Barton Instrument Corp.)

where ¾-inch taps are used, the flange thickness should be 1⅝ inches.

Corner taps are located directly at the faces of the orifice plate (Figure 4.10). They are in common use in Europe and are particularly useful for pipe sizes less than 2 inches where the vena contracta may occur inside the dimension of the standard flange tap. Possible difficulties in using corner taps include:

1. Small passages are vulnerable to plugging.
2. Tests have indicated pressure instability in the region of the orifice face.

Vena contracta taps are made at locations which theoretically take advantage of the highest ΔP available at the orifice. The upstream tap is located one pipe diameter from the face of the orifice while the downstream tap is located at the vena contracta, the point of minimum pressure caused by the restriction (Figure 4.11). However, the point of minimum pressure varies with the d/D ratio, thereby introducing an error if the plate bore is changed.

Radius taps are a close approximation of vena contracta taps, at locations one pipe diameter upstream and ½ pipe diameter downstream from the orifice face.

Pipe taps or full-flow taps (Figure 4.12) measure the permanent pressure loss across an orifice. The taps are located 2½ pipe diameters upstream (ahead of the pressure buildup near the orifice face) and eight pipe diameters downstream—where static pressure has reached its maximum following the vena contracta. There is more likelihood of measurement error when pipe taps are used, and they require longer runs of straight pipe. However, they are sometimes more economical in that they allow use of standard flanges (sometimes already existing), and they also have higher flow rate capabilities than flange or vena contracta taps.

Tap Orientation and Mounting of Secondary Device. Taps are usually drilled at the top, side or bottom of the orifice flanges and sometimes at 45° angles from these points. Wet gas taps are usually made from the top of the flange (Figure 4.13) to allow condensate drainage back into the line. Liquid taps are made at the side normally to prevent trapped gas bubbles from interfering with the measurement and to prevent sludge or foreign particles from entering the lead lines. Figure 4.14 shows a typical mounting of a d/p cell close-coupled to flange taps. Connecting pipe or tubing between orifice taps and secondary devices is usually sloped for drainage and vent purposes.

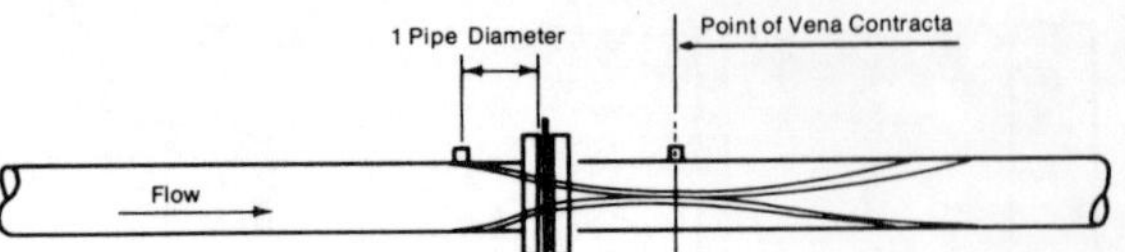

Figure 4.11. Vena-contracta taps are located to take advantage of the maximum pressure differential developed by the orifice.

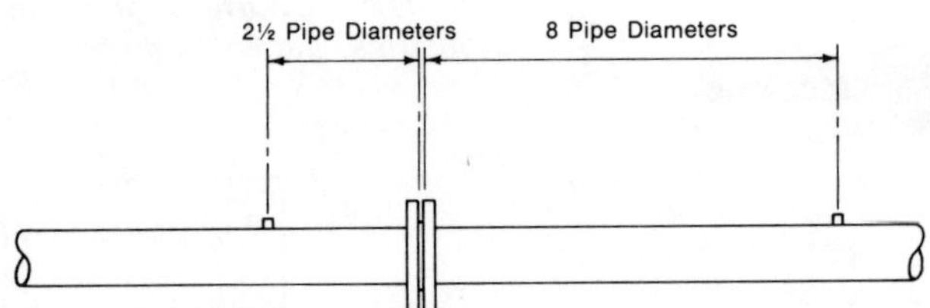

Figure 4.12. Pipe taps are installed to measure the permanent pressure loss across the orifice.

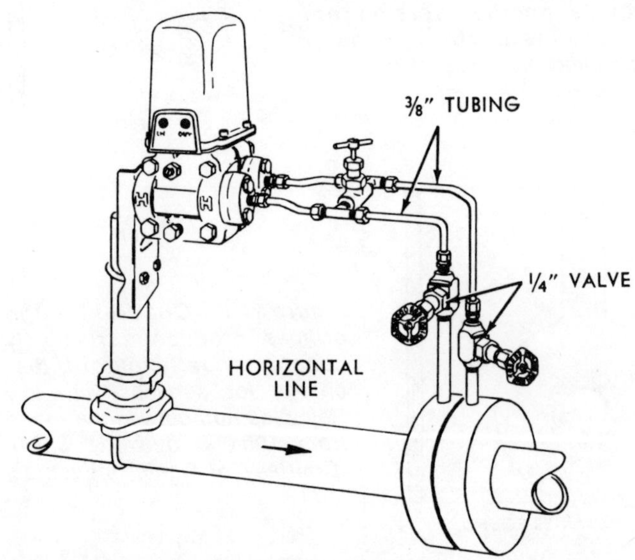

Figure 4.13. Wet gas taps are made from top of flange and d/p cell is mounted above takeoff to allow condensate drainage back into line. (Courtesy of Foxboro Co.)

Venturi Tubes

When permanent pressure loss is of primary importance in head flow measurement, the Venturi tube deserves strong

Figure 4.14. Close coupled d/p cell to flange taps at side of flange. Side mounting is normally for liquid service but close coupling is also used for steam and gas services. (Courtesy of Foxboro Co.)

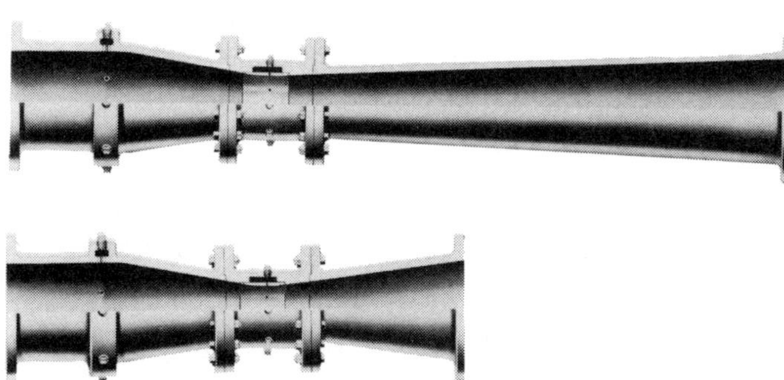

Figure 4.15. Venturi tubes possess the characteristic of a low permanent pressure loss. (Courtesy of BIF, a unit of General Signal Corp.)

consideration. The classical Venturi tube consists of a converging conical inlet section, a cylindrical throat and a diverging recovery cone (see Figure 4.15). Fluid velocity increases in the converging inlet section, increasing the velocity head and decreasing the pressure head. The flow rate remains static in the throat section where there is no cross-sectional dimensional change, but it decreases in the recovery section, and the decreased velocity head is recovered as pressure. The relatively large recovery at this point results in a permanent pressure loss of only 10 to 25% of the differential pressure across the tube. When large lines with high flow rates are involved, substantial savings in power requirements can be obtained.

The Venturi tube can be used to handle any fluid an orifice plate can handle, plus fluids that contain some solids because they contain no sharp corners nor project into the fluid stream. Slurries and dirty liquids that build up around other primary devices can be handled easily if the pressure taps are protected from plugging.

The overall accuracy of Venturi tubes is no better than orifice plates; it may range from ±¼ to ±3%. However, a Venturi coefficient is less affected by a decreasing Reynolds number than either a nozzle or an orifice and is, therefore, more accurate over wide flow ranges.

Flow calcuations based on dimensions of classical Venturi tubes are more uncertain than for orifice plates. The uncertainty becomes greater as dimensions and contours vary from the classical. Ultimately the accuracy and reliability depend primarily on the calibration facilities of the manufacturer.

To provide greater certainty of flow measurements involving Venturi tubes, the American Society of Mechanical Engineers adopted a Venturi design that used a flow coefficient value of 0.984 for all diameter ratios and sizes of Venturi tubes from 2- to 30-inch pipe.

Critical dimensions and contours of this design are shown in Figure 4.16. The pressure taps are located one-half pipe diameter upstream of the inlet cone and at the middle of the throat section. Standard designs are equipped with piezometer rings—multiple tap holes around the periphery of inlet and throat, surrounded by an annular ring. In slurry services piezometer rings are usually eliminated to permit purging of the pressure taps.

Eccentric Venturi tubes (Figure 4.17) with the throat flush with the bottom or top of the pipe, and rectangular Venturi tubes are also used occasionally. Eccentric tubes provide further assurance that buildup of solids will not occur and permit complete drainage of horizontal lines. Rectangular tubes are useful in air duct work. Both require individual calibration if the usual measurement tolerance is demanded.

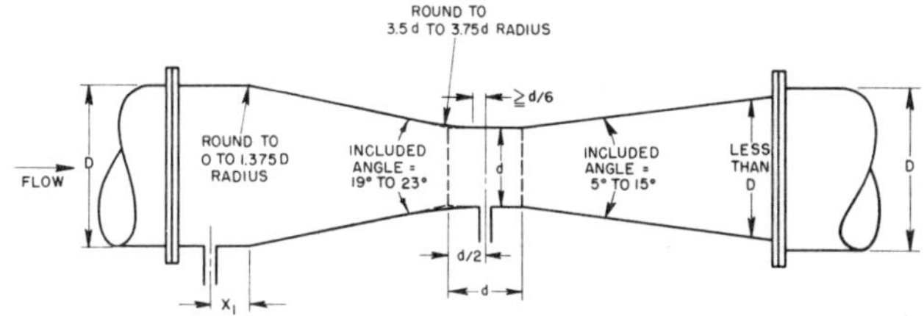

Figure 4.16. Critical dimensions and contours of the classical venturi design adopted by ASME. (Courtesy of Foxboro Co.)

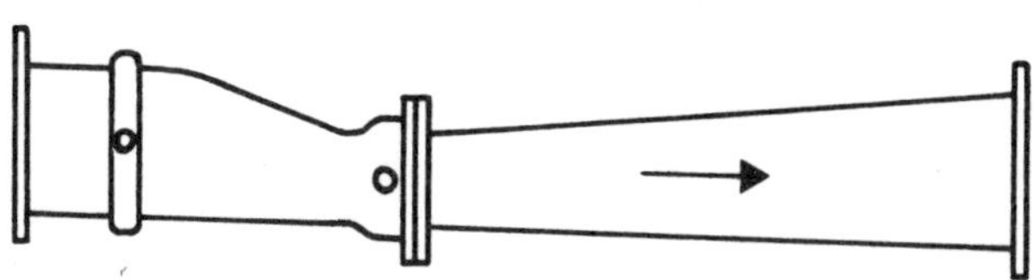

Figure 4.17. Eccentric Venturi tubes provide additional insurance against buildup of solids materials. (Courtesy of BIF, a unit of General Signal Corp.)

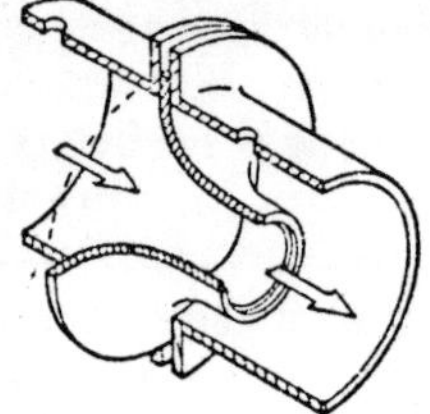

Figure 4.18. A standard flow nozzle. (Courtesy of Foxboro Co.)

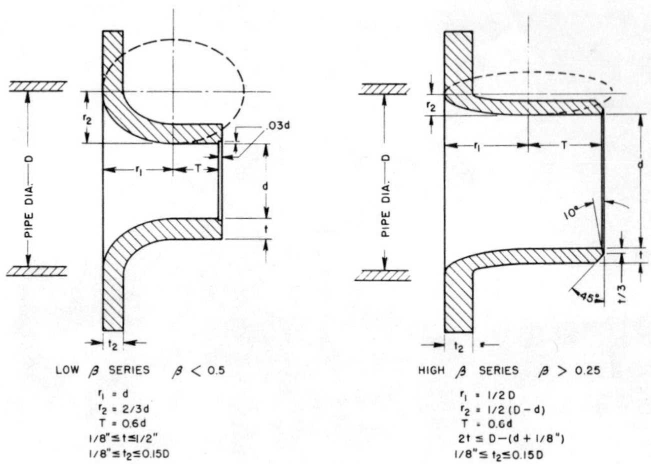

Figure 4.19. The long radius flow nozzle. (Courtesy of Foxboro Co.)

The high cost of Venturi tubes and the greater length required for their installation limit their use. Each application must be reviewed with all factors involved to determine the most suitable element for a particular case.

Low-Pressure Loss Tubes

There are several low pressure-loss tubes and flow nozzles available, including the Gentile patent flow tube, the Dall tube, the lo-loss tube and the twin-throat Venturi tube. They are designed with a combination of contour and tap locations to provide a high differential pressure with a low pressure loss. They are subject to coefficient change with variation in viscosity. Manufacturer's data is provided on the effect of Reynolds number on the coefficient.

The installed cost of these devices is usually less than for classical Venturi tubes, and they are commonly used for high liquid flow applications. Accuracy is dependent on the manufacturer's calibration data which is much less extensive than for classical Venturi tubes.

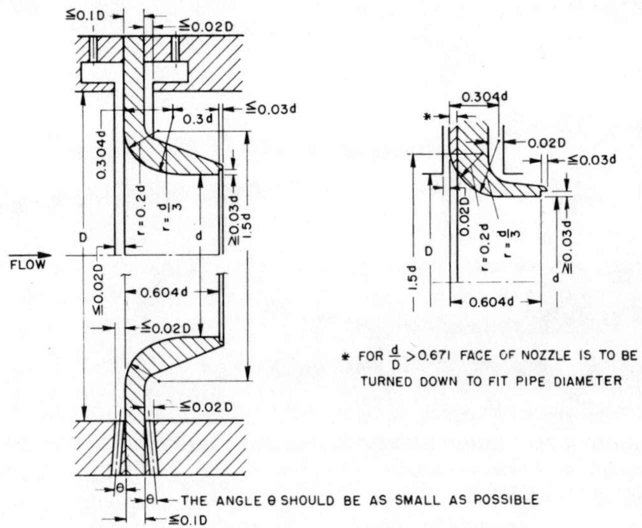

Figure 4.20 The I.S.A. flow nozzle. (Courtesy of Foxboro Co.)

Flow Nozzles

The first use of the flow nozzle (Figure 4.18) is difficult to trace. It was used in Germany in the nineteenth century. Several designs of flow nozzle contours are available: the long-radius flow nozzle (Figure 4.19), the I.S.A. flow nozzle (Figure 4.20) and the flow nozzle with the incomplete elliptical arc (Figure 4.21). Differential pressure taps are normally located one diameter upstream and one-half diameter downstream from the inlet faces of the nozzle.

Used for flow measurements at high fluid velocities, the flow nozzle is more rugged and more resistant to erosion than the sharp-edge orifice. For a given diameter and a given differential pressure, it will pass almost 65% more flow than the orifice. At high flow rates, it can be used with a higher d/D ratio without developing an excessively high differential pressure.

Because of its streamlined contour, it tends to sweep solids through the throat, but it should not be used if a high percentage of the total flow is solids. If possible, it should be installed in a vertical line with flow in the downward direction. It is more compatible for gas service than for liquids.

The cost of a flow nozzle is considerably higher than an orifice plate. Maintenance is higher since it is necessary to remove a section of pipe to inspect or install it. The initial cost may be many times higher than the orifice plate depending on the size and material.

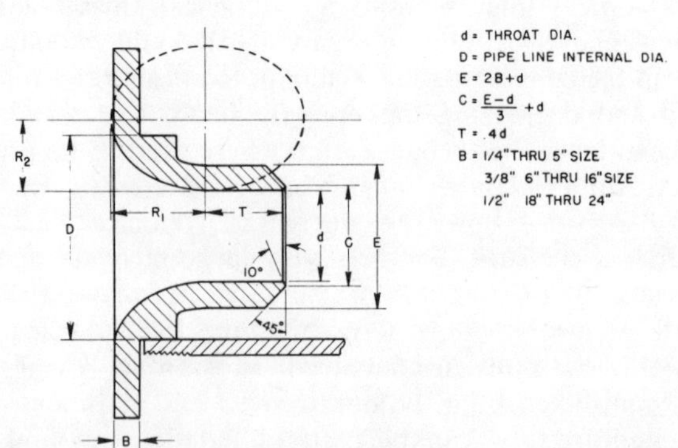

Figure 4.21. A flow nozzle with incomplete elliptical arc. (Courtesy of Foxboro Co.)

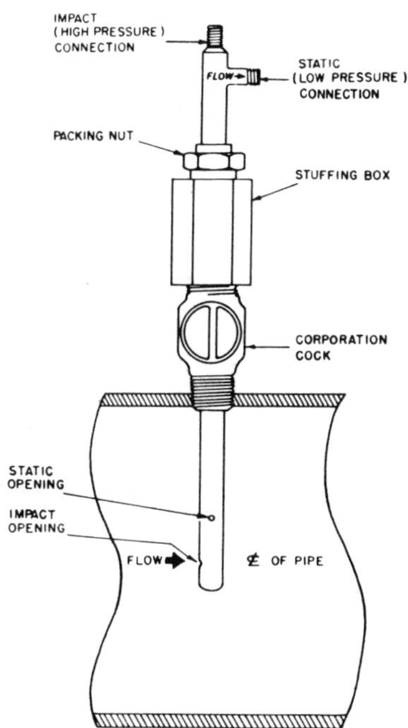

Figure 4.22. The Pitot tube measures fluid velocity at one point in a line. Line flow is determined by the ratio of that point to average velocity. (Courtesy of Foxboro Co.)

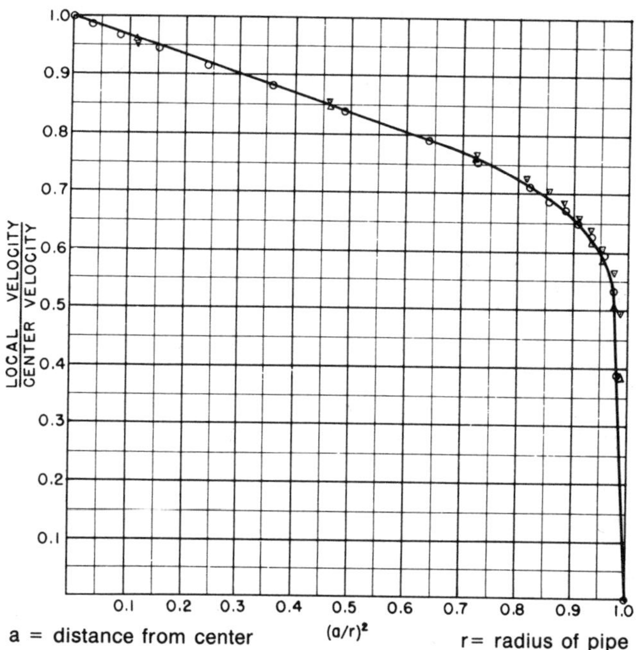

Figure 4.23. A typical velocity distribution curve for a Pitot tube with ratio of local to center velocity plotted against the square of the ratio of the distance from center to the pipe radius. (Courtesy of Foxboro Co.)

Pitot Tubes

The Pitot tube measures fluid velocity at one point in a pipe. Measurement is determined by placing the impact opening directly in the line of flow and the static opening at 90° from the impact opening (Figure 4.22). The differential pressure across these taps is proportional to velocity. Quantity rate measurement is calculated from the ratio of average velocity to the velocity at the point of measurement. If a more accurate measure of flow is desired, it is obtained by making a complete traverse—moving the Pitot tube across the entire diameter of the pipe to measure the velocity at several points—and computing from the true average velocity.

Figure 4.23 shows a typical velocity distribution curve for a Pitot tube when a sufficiently long straight run of pipe is installed upstream of the tube (50 or more pipe diameters). The average velocity is obtained at a location approximately 30% of the pipe radius from the pipe wall. The velocity distribution, however, changes with Reynolds number, which is a function of velocity.

Special applications of the Pitot tube include the Pitot-Venturi (Figure 4.24) and the double Venturi (Figure 4.25). The purpose of these designs is to provide a higher differential pressure than that produced by impact pressure alone. The need for these special devices occurs at low velocity flows where differential pressures would be small. They are still sensitive to Reynolds number changes and to dimension errors.

Pitot and Pitot-Venturis offer a distinct advantage in that they cause practically no pressure drop in the flowing

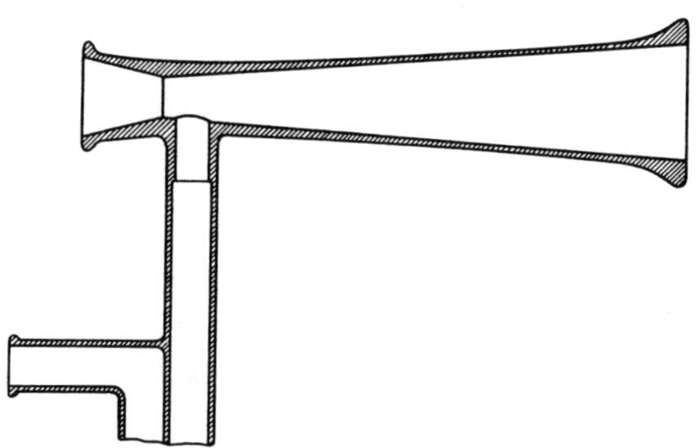

Figure 4.24. A Pitot Venturi also produces a higher differential pressure than the standard Pitot tube. (Courtesy of Foxboro Co.)

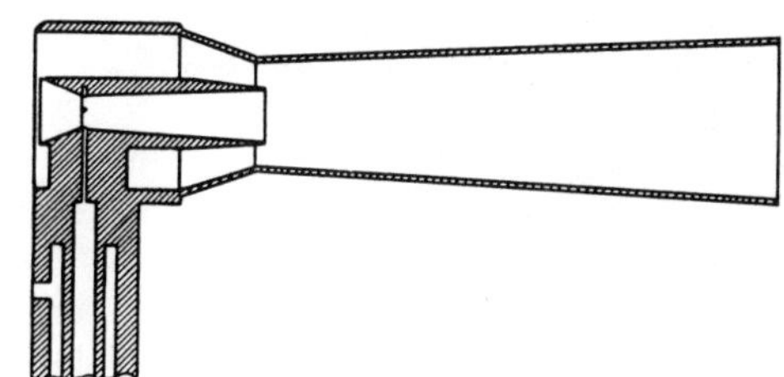

Figure 4.25. The double Venturi also produces a higher differential pressure than the standard Pitot tube. (Courtesy of Foxboro Co.)

stream. They are economical and are easily installed and removed from the line. They are sensitive to upstream disturbances and are not recommended for dirty or sticky fluids. Accuracy may range from ±½ to ±5%. They are useful when pressure drop or power loss cannot be tolerated and where accuracy is not the prime concern.

Annubar Tubes

The Annubar element (Figure 4.26) is a recent addition to the line of primary flow devices. It is a simple, unique design based upon the principles of the fundamental Bernoulli flow equation. It consists of two probes inserted into the line in much the same manner as a Pitot tube, with one probe facing the flow to sense the velocity pressure, and a second probe behind the first with its opening facing downstream sensing the static pressure (see Figure 4.27).

The probe which faces upstream has four sensing ports, each one representing an annular segment of the line. An equalizing line inserted into the plenum of the upstream probe senses the average of the four pressures representing the four line segments, thus providing an average pressure for the entire body of flowing material.

This primary device, with accuracies varying from approximately ±½ to ±1½% over a wide range of pipe sizes, has received good acceptance in a short period of time. Available for pipe sizes from ½ to 150 inches, it is economical to install, particularly in the larger pipe sizes where pipe couplings can replace large, expensive orifice flanges. In the larger sizes, welding couplings are placed on the opposite side of the pipe (from the element entry) to hold the probe and provide a rigid installation.

The Annubar may be installed in any plane desired. It can be installed easily in an existing line. By using "hot taps"; it can be installed and placed in service while the line is under pressure. The upstream probe may be rotated while in service, allowing the sensing ports to be turned downstream, thus eliminating the need for purging in some applications.

Figure 4.26. The Annubar has two elements: one facing upstream measures average velocity; one facing downstream measures static pressure. (Courtesy of Ellison Instrument Division of Dieterich Standard Corp.)

Elbow Taps

Elbow taps provide another economical approach to flow measurement. Differential pressure is developed by centrifugal force as the direction of fluid is changed in a pipe elbow. Taps are located either at 45° (Figure 4.28) or at 22½° (Figure 4.29) from the fluid inlet.

Advantages of this type measurement include the following: (a) the elbows usually already exist in the line and no additional material is necessary; (b) there is no added pressure loss; (c) there are no obstructions in the line.

Accuracy is poor, varying from ±5 to ±10%. Repeatability is good, which is the essential requirement for most flow applications. Installation care is needed; straight runs of pipe should include at least 25 pipe diameters upstream and 10 pipe diameters downstream.

The preferred orientation is to have both runs in the horizontal. The next choice is to have the approach in a vertical pipe with the flow upward and the downstream run on a horizontal plane. The differential developed on most elbow tap measurements is relatively small and has a square root relationship with fluid flow.

Weirs and Flumes

A detailed description of liquid measurement in open channels is beyond the scope of this volume. The purpose here is to acquaint the reader with this type measurement and refer him to another source for more precise information. *Principles and Practice of Flow Meter Engineering* by L.K. Spink is such a source.

A weir is an obstruction in a flowing stream over which liquid is made to pass. Application is primarily in waterworks and waste and sewage systems. Measurements can be made for rates from a few gpm to millions of gallons per day. Flow rate is a function of the head and the width of the opening.

Three common forms of weir notches exist: rectangular, V-notch and Cippoletti or trapezoidal (Figure 4.30). Figure 4.31 shows the flow of fluid over a weir. It becomes obvious that the total flow is a function of the head, H_a, and the crest of the notch (or angle of the notch in the case of the V-notch weir.) Figure 4.32 shows a composite dimension sketch of a weir.

The Parshall flume is a special type of Venturi flume developed by the Colorado Agricultural Experiment Station. Used essentially for the same purposes as other open flow measurements, it is considered in many cases to be more practical than weirs because its loss of head is about one-fourth that of a weir. Its Venturi pattern (Figure 4.33) tends to reduce velocity at the converging section, allowing higher degrees of submergence and lower head losses.

Weirs and flumes have not previously found a great deal of application in the process industries, but their use may increase with the increasing demand for water pollution control. These types of measurements are likely when large volumes of water are handled in waste treatment facilities.

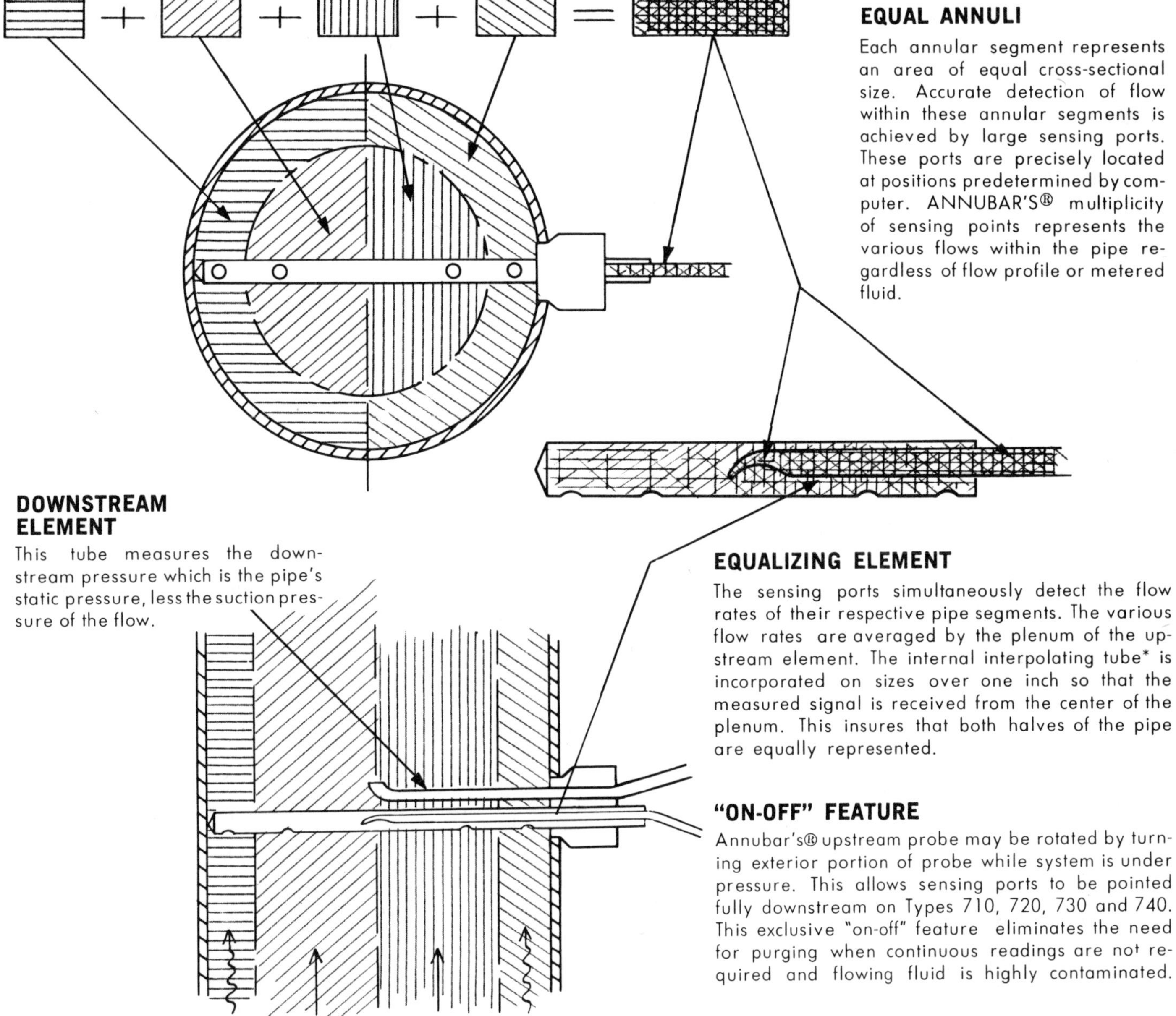

EQUAL ANNULI
Each annular segment represents an area of equal cross-sectional size. Accurate detection of flow within these annular segments is achieved by large sensing ports. These ports are precisely located at positions predetermined by computer. ANNUBAR'S® multiplicity of sensing points represents the various flows within the pipe regardless of flow profile or metered fluid.
DOWNSTREAM ELEMENT
This tube measures the downstream pressure which is the pipe's static pressure, less the suction pressure of the flow.
EQUALIZING ELEMENT
The sensing ports simultaneously detect the flow rates of their respective pipe segments. The various flow rates are averaged by the plenum of the upstream element. The internal interpolating tube* is incorporated on sizes over one inch so that the measured signal is received from the center of the plenum. This insures that both halves of the pipe are equally represented.
"ON-OFF" FEATURE
Annubar's® upstream probe may be rotated by turning exterior portion of probe while system is under pressure. This allows sensing ports to be pointed fully downstream on Types 710, 720, 730 and 740. This exclusive "on-off" feature eliminates the need for purging when continuous readings are not required and flowing fluid is highly contaminated.

Figure 4.27. A description of the function and operation of the Annubar element. (Courtesy of Ellison Instrument Division of Dieterich Standard Corp.)

Laminar Flow Elements

One distinct disadvantage of most primary flow elements for head flow measurement is the square root relationship existing between flow rate and differential pressure. The use of laminar flow elements overcomes that difficulty.

Osborne Reynolds in the late 1800s discovered that many fluid flow phenomena depend on whether the flow is laminar or turbulent and that this characteristic depends primarily on the pipe diameter, fluid viscosity, density and velocity. These characteristics may be combined mathematically to produce the equation:

$$R_d = vD\rho/\mu \tag{4.4}$$

where

R_d = Reynolds number

v = average velocity

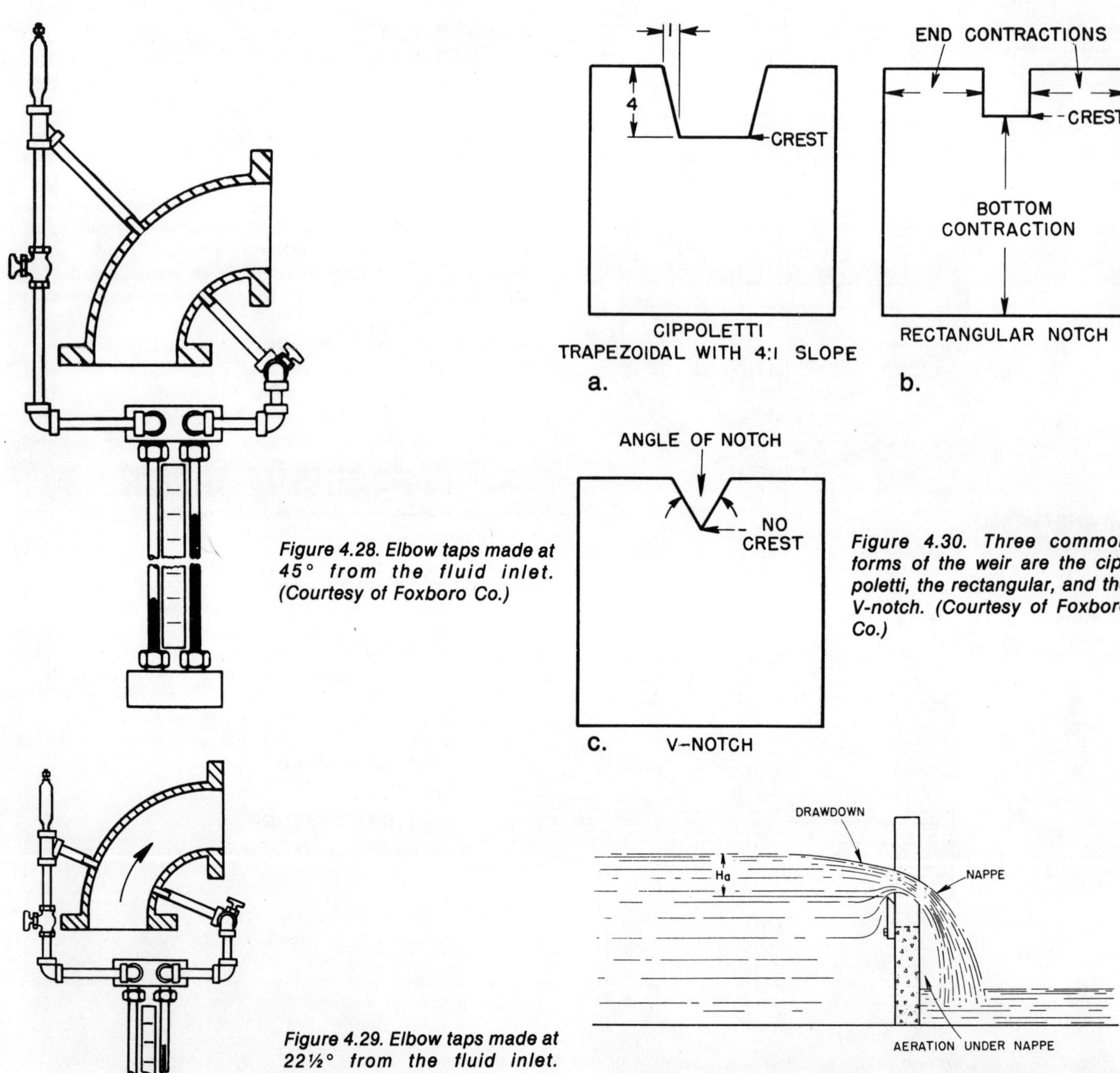

Figure 4.28. Elbow taps made at 45° from the fluid inlet. (Courtesy of Foxboro Co.)

Figure 4.30. Three common forms of the weir are the cippoletti, the rectangular, and the V-notch. (Courtesy of Foxboro Co.)

Figure 4.29. Elbow taps made at 22½° from the fluid inlet. (Courtesy of Foxboro Co.)

Figure 4.31. The free flow of water over the weir demonstrates the function of the height of liquid above the notch. (Courtesy of Foxboro Co.)

D = inside pipe diameter
ρ = density of flowing fluid
μ = absolute viscosity

The characteristic, R_d (Reynolds number), was named in honor of Osborne Reynolds. It is a dimensionless unit when consistent units are used in its determination. Test results show that fluid flow is laminar when the Reynolds number is below 2,100.

Equation 4.4 shows that the Reynolds number varies directly with the pipe diameter, so laminar flow elements are made by reducing the flowing stream effectively to a

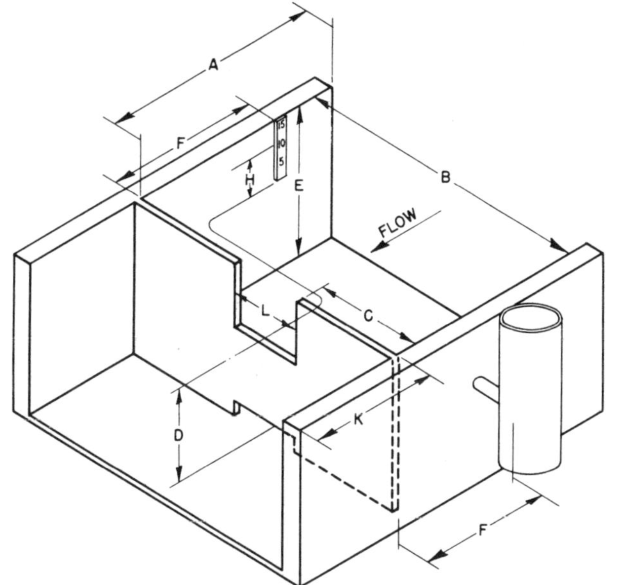

Figure 4.32. Composite dimension sketch of a rectangular weir. (Courtesy of Foxboro Co.)

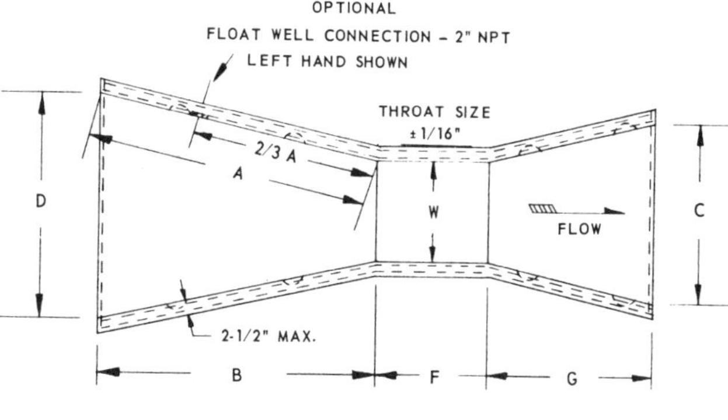

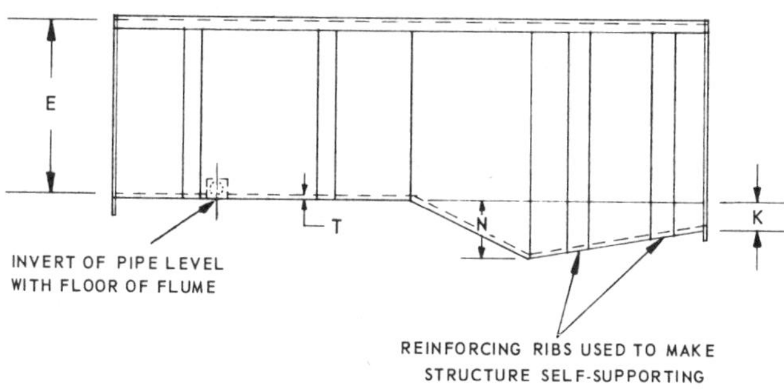

Figure 4.33. The Parshall flume is a special type of Venturi flume used for open flow measurement and characteristically has a low head loss. (Courtesy of Fischer and Porter Co.)

large number of small streams with greatly reduced pipe diameters. This is accomplished with matrices, concentric cylinders and tightly packed spheres placed inside a suitable piece of pipe. Figure 4.34 shows an example of a matrix used in an element pictured in Figure 4.35.

Laminar flow elements are made for both liquid and gas flows, but they are used primarily in gas or vapor service. They are particularly useful for the measurement of low flow rates. They are usually calibrated to accuracies from ± ¼ to ± ½%.

Laminar flow elements are sensitive to viscosity and density changes. They are sensitive to temperature and pressure changes only as they affect fluid density and viscosity. In many situations these changes would be insignificant.

The entire pressure drop incurred by a laminar flow element is permanent. There is no partial recovery as there is with other primary elements. Normally, however, they are designed for relatively low pressure drops because of their use on low flow rate applications. They are used more for laboratory and pilot-plant activities than for industrial size plants. Calibration is necessary at one point only, and there is no compensation for viscosity changes.

Installation of Primary Devices

Piping configurations and disturbances in the line affect the accuracy of all primary elements in head flow measurements. The inlet piping has greater effect than outlet piping. Standards have been adopted for several con-

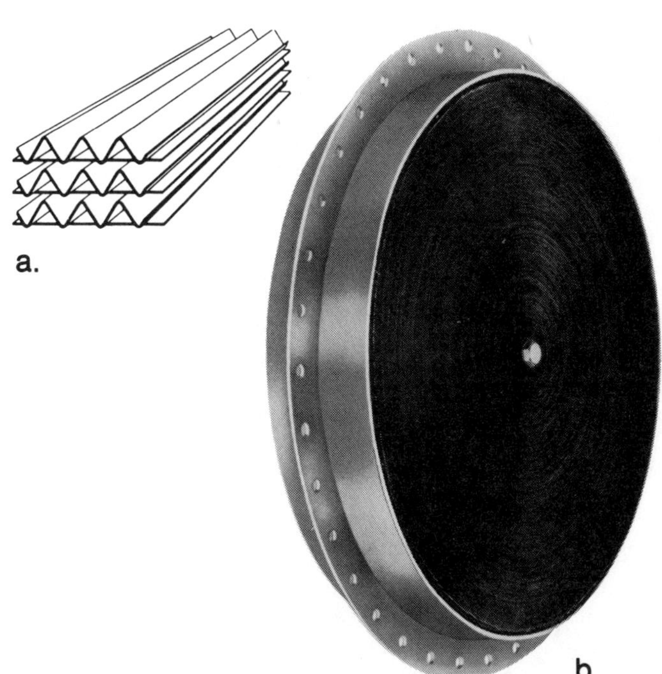

Figure 4.34. (a) A typical matrix for liquid or gas service placed in pipes to make laminar flow elements. (b) An element used for gas service in units similar to that shown in Figure 4.35. (Courtesy of Meriam Instrument Division, Scott and Fetzer Corp.)

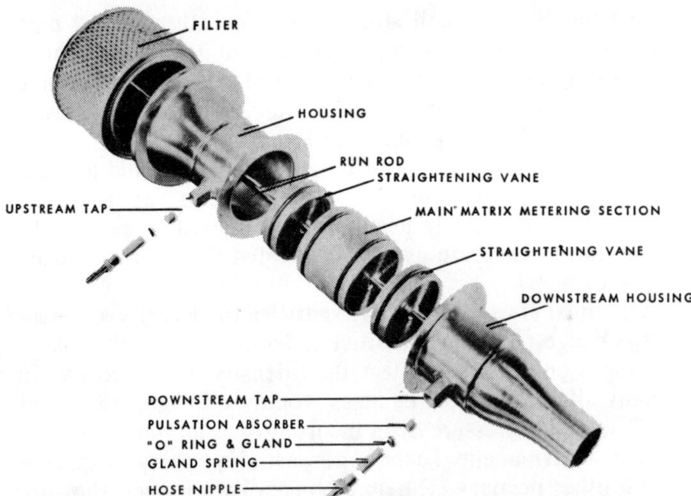

Figure 4.35. An exploded view of a laminar flow element used for gas service. The unit included a filter, a straightening vane and the matrix which assures laminar flow. (Courtesy of Meriam Instrument Division, Scott and Fetzer Corp.)

figurations and disturbances in meter piping which specify minimum distances of elements from piping turns or disturbances in order to maintain measurement accuracy.

Figure 4.36 provides a comparison of two U.S. standards with those of some foreign standards. It is evident that the American standards (A.G.A. and A.S.M.E.) are less conservative than the European standards shown.

In the process industries of the United States, the American Gas Association Standards (A.G.A.) are most widely accepted. Most operating and engineering companies incorporate the A.G.A. Standards into their own standards and specifications. The A.G.A. Standards (for metering piping) has also been incorporated into the Instrument Society of America (I.S.A.) Standards.

Figures 4.37 through 4.41 show the required straight run lengths of piping upstream and downstream of orifice plates for the configuration and/or disturbance indicated in the figure. A plot of length (in terms of pipe diameter) versus beta ratio (between 0.1 and 0.8) is given. The curves shown are for flange taps. When pipe taps (2½ and 8 diameters) are used, all upstream lengths should be increased by two diameters, and downstream lengths should be increased by eight diameters.

Straightening Vanes. Many situations arise where it is impractical to allow the required straight run length of pipe—installations in an existing line, installations involving the use of expensive alloy pipe, the need to fit between two closely located pieces of equipment or other valid reasons. Straightening vanes are used to obtain the flow pattern needed for accurate measurement with reduced straight run lengths.

Two types are commonly used, the tubular (Figure 4.42) and the one-piece radial (Figure 4.43). These devices eliminate or reduce the swirl or irregular patterns produced by pipe turns and valves and reduce the measurement error that would otherwise be incurred.

The use of straightening vanes is discouraged by some knowledgeable people, but evidence exists that they are worthwhile in many instances.

Orifice Unions. Unions for holding orifice plates are usually consistent with the rest of the piping specifications; the flanges may be screwed, slip-on, weld neck, ring-type joint, etc., and at the proper pressure rating. The flange width must be great enough to allow drilling of the pressure taps which are normally ½-inch pipe size. Two sets of pressure taps are drilled 180° apart. Jack screws are provided to jack the flanges apart for easy installation and removal of orifice plates.

When there is a need for quick changeout of orifice plates or for frequent plate inspections, or when flange-spreading is undesirable, special plate holders are provided. One of the simplest is a plate holder (Figure 4.44) designed to allow plate inspection or changeout by depressuring the line, loosening the clamping screws, sliding out the clamping bar and then lifting out the sealing bar-plate carrier ring which contains the orifice plate.

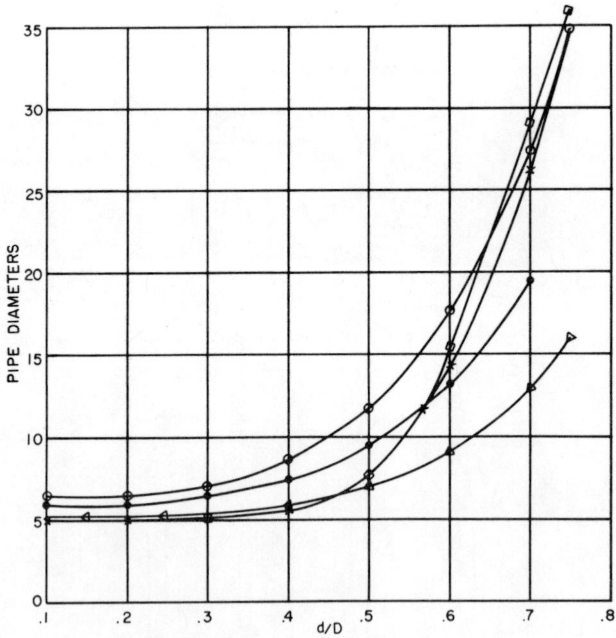

Figure 4.36. The comparison of American and European standards for a 90° elbow ahead of an orifice shows that European standards are more conservative—they require greater lengths of straight run pipe. (Courtesy of Foxboro Co.)

(text continued on page 68)

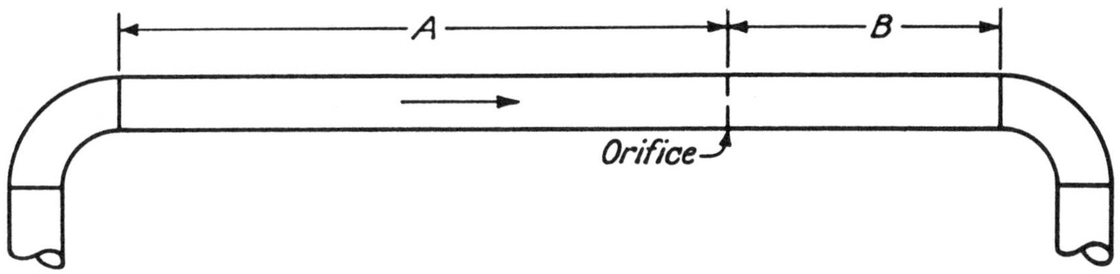

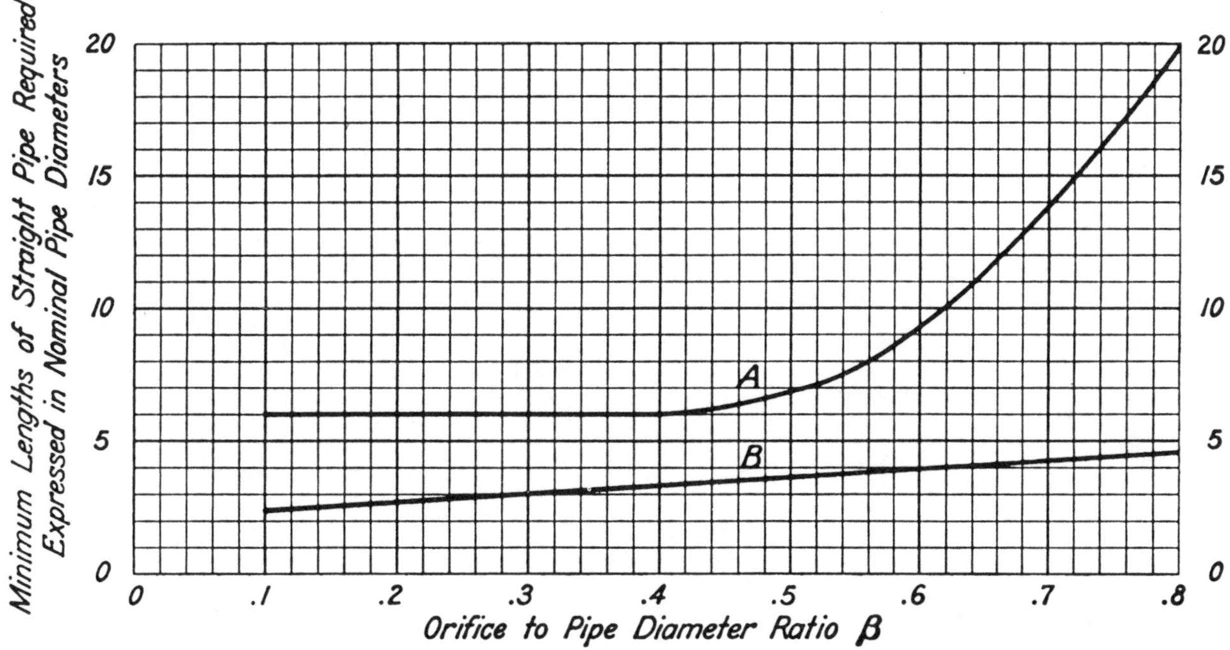

NOTE 1 - When "Pipe Taps" are used, A should be increased by 2 pipe diameters and B by 8 pipe diameters.

NOTE 2 - When the diameter of the orifice may require changing to meet different conditions, the lengths of straight pipe should be those required for the maximum orifice to pipe diameter ratio that may be used.

Figure 4.37. Minimum run of straight length pipe upstream and downstream of an orifice for piping configuration shown as recommended by the AGA. (Courtesy of American Gas Association)

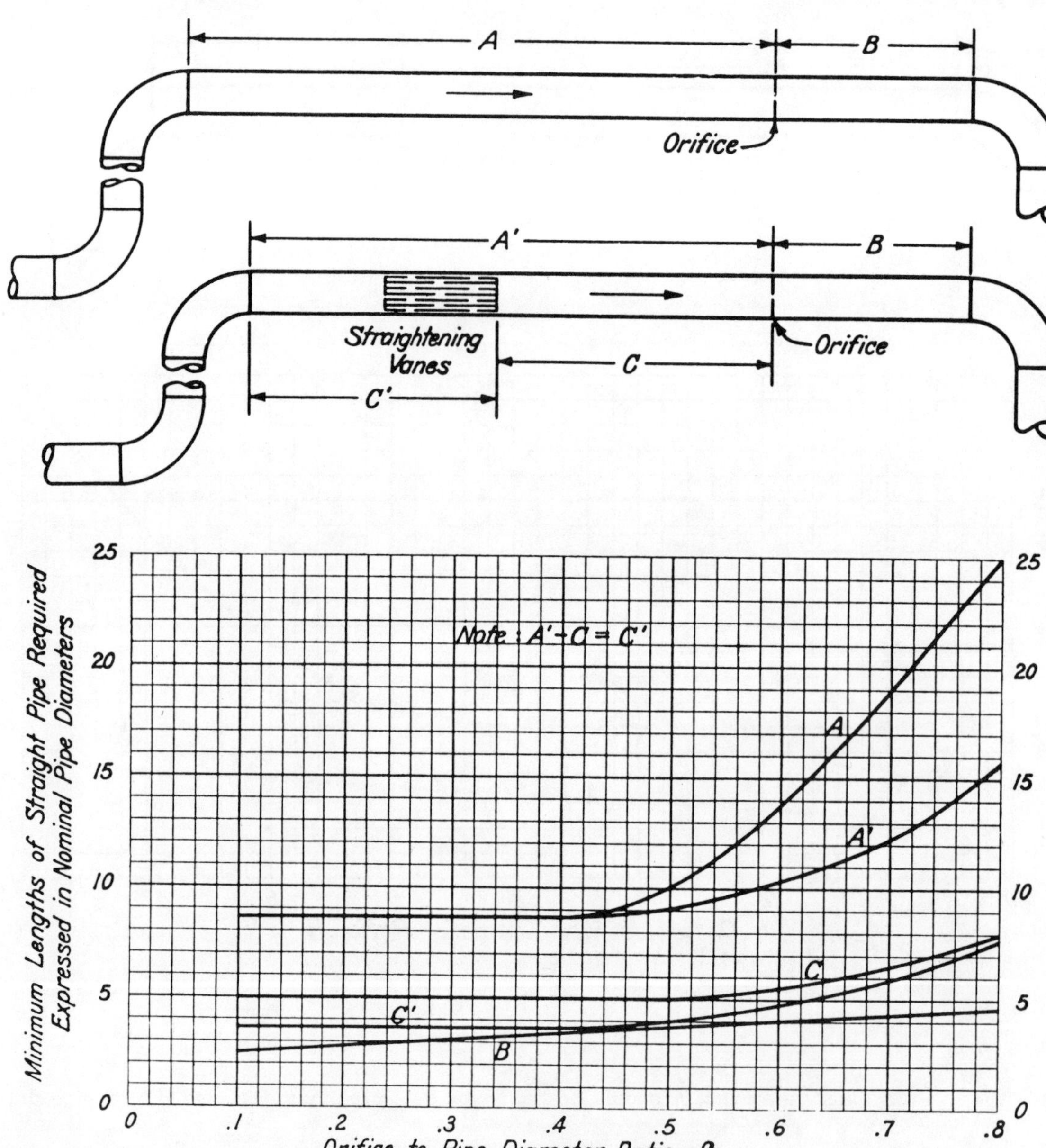

NOTE 1 - When "Pipe Taps" are used, A, A', and C should be increased by 2 pipe diameters, and B by 8 pipe diameters.

NOTE 2 - When the diameter of the orifice may require changing to meet different conditions, the lengths of straight pipe should be those required for the maximum orifice to pipe diameter ratio that may be used.

Figure 4.38. Minimum run of straight length pipe upstream and downstream of an orifice for piping configuration shown as recommended by the AGA. (Courtesy of American Gas Association).

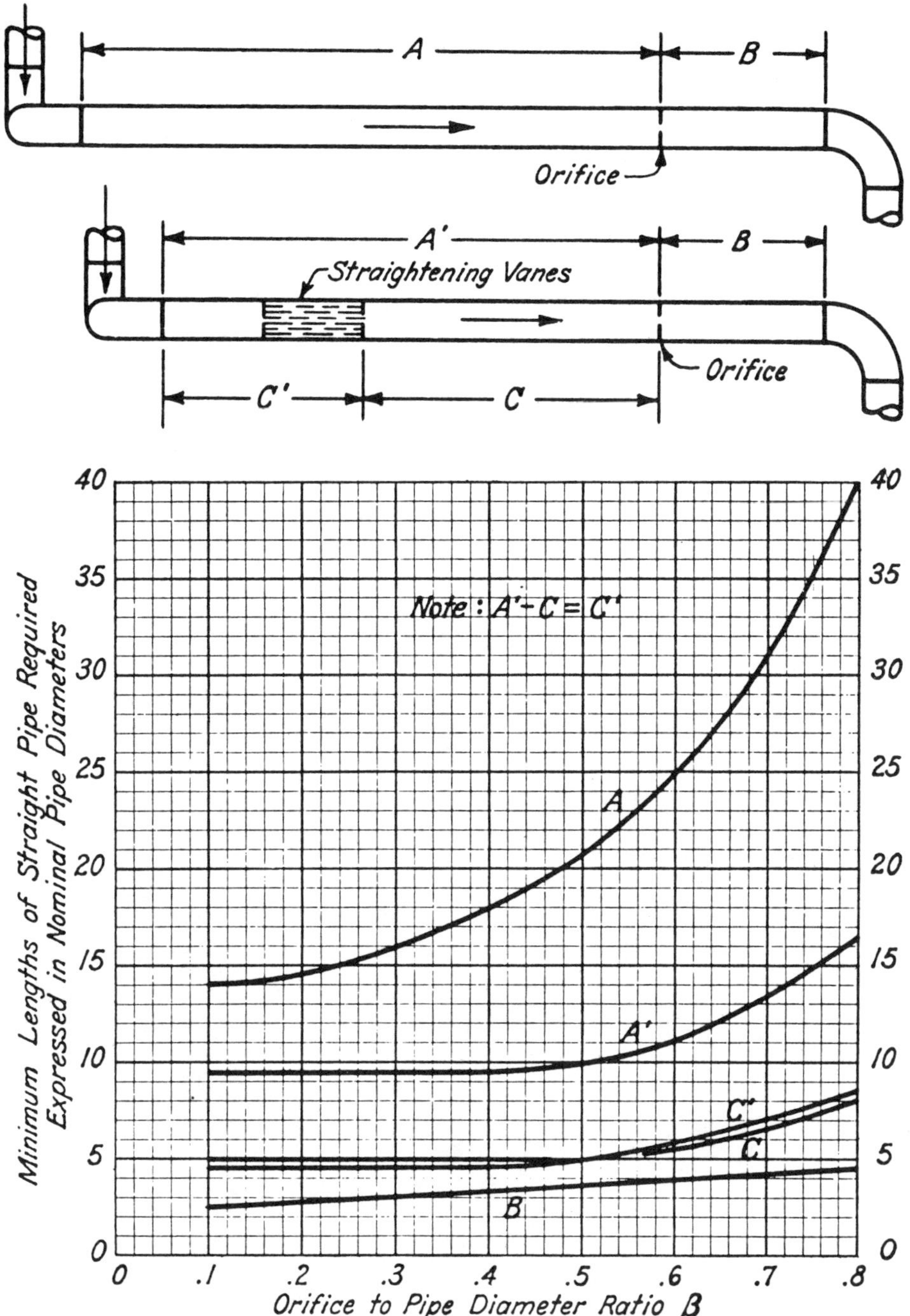

NOTE 1 - When "Pipe Taps" are used, A, A', and C should be increased by 2 pipe diameters and B by 8 pipe diameters.

NOTE 2 - When the 2 ells shown in the above sketches are closely preceded by a third which is not in the same plane as the middle or second ell, the piping requirements shown by A should be doubled.

NOTE 3 - When the diameter of the orifice may require changing to meet different conditions, the lengths of straight pipe should be those required for the maximum orifice to pipe diameter ratio that may be used.

Figure 4.39. Minimum run of straight length pipe upstream and downstream of an orifice for piping configurations shown as recommended by the AGA. (Courtesy of American Gas Association)

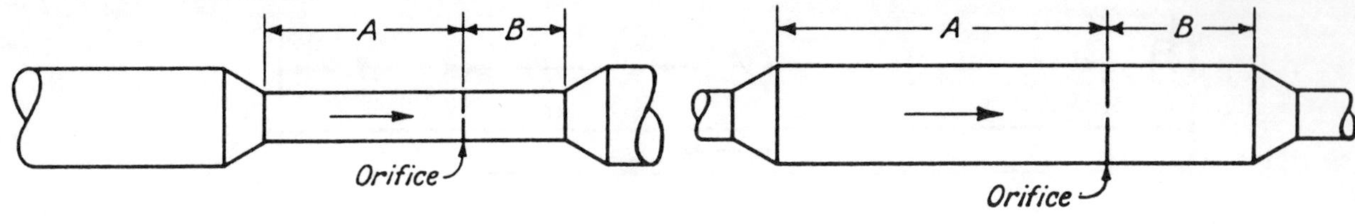

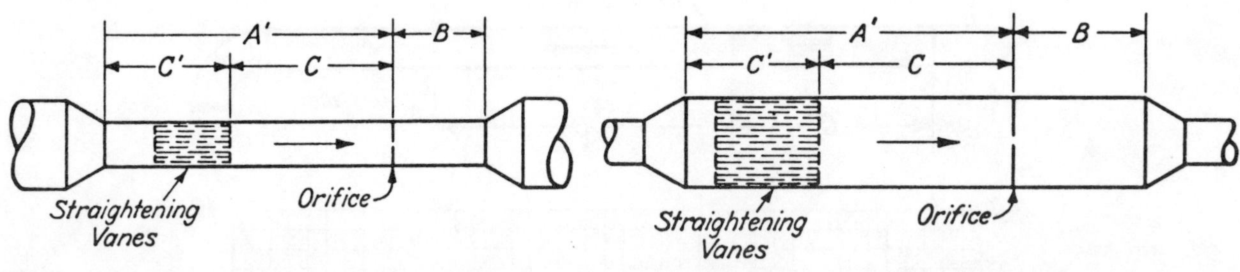

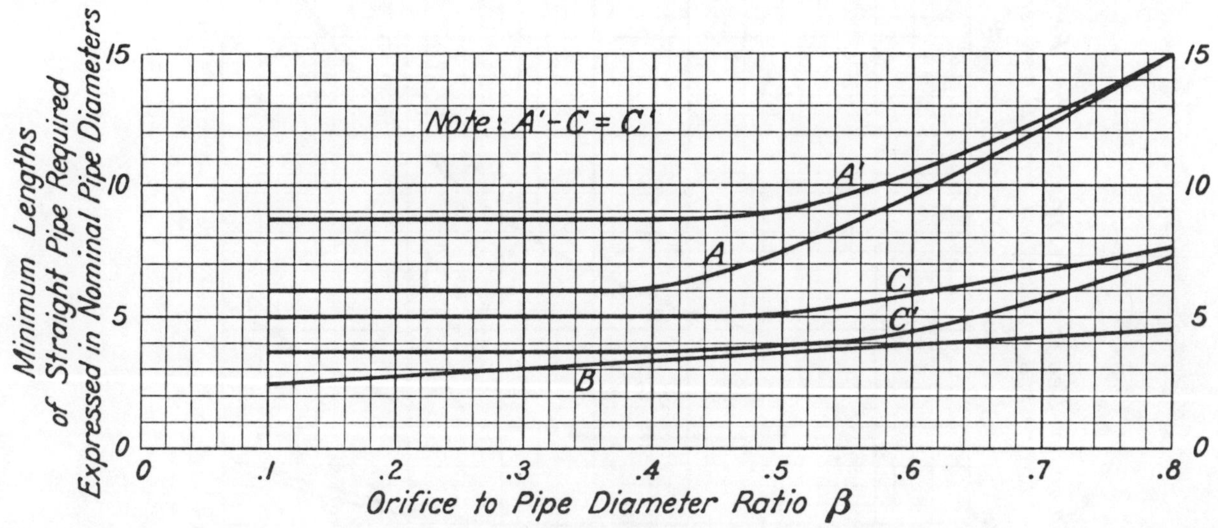

NOTE 1 - When "Pipe Taps" are used, length A, A′, and C should be increased by 2 pipe diameters, and B by 8 pipe diameters.

NOTE 2 - Straightening vanes will not reduce lengths of straight pipe A. Straightening vanes are not required because of the reducers, they are required because of other fittings which precede the reducer. Length A is to be increased by an amount equal to the length of the straightening vanes whenever they are used.

NOTE 3 - When the diameter of the orifice may require changing to meet different conditions, the lengths of straight pipe should be those required for the maximum orifice to pipe diameter ratio that may be used.

Figure 4.40. Minimum run straight length pipe upstream and downstream of an orifice for piping configurations shown as recommended by the AGA. (Courtesy of American Gas Association)

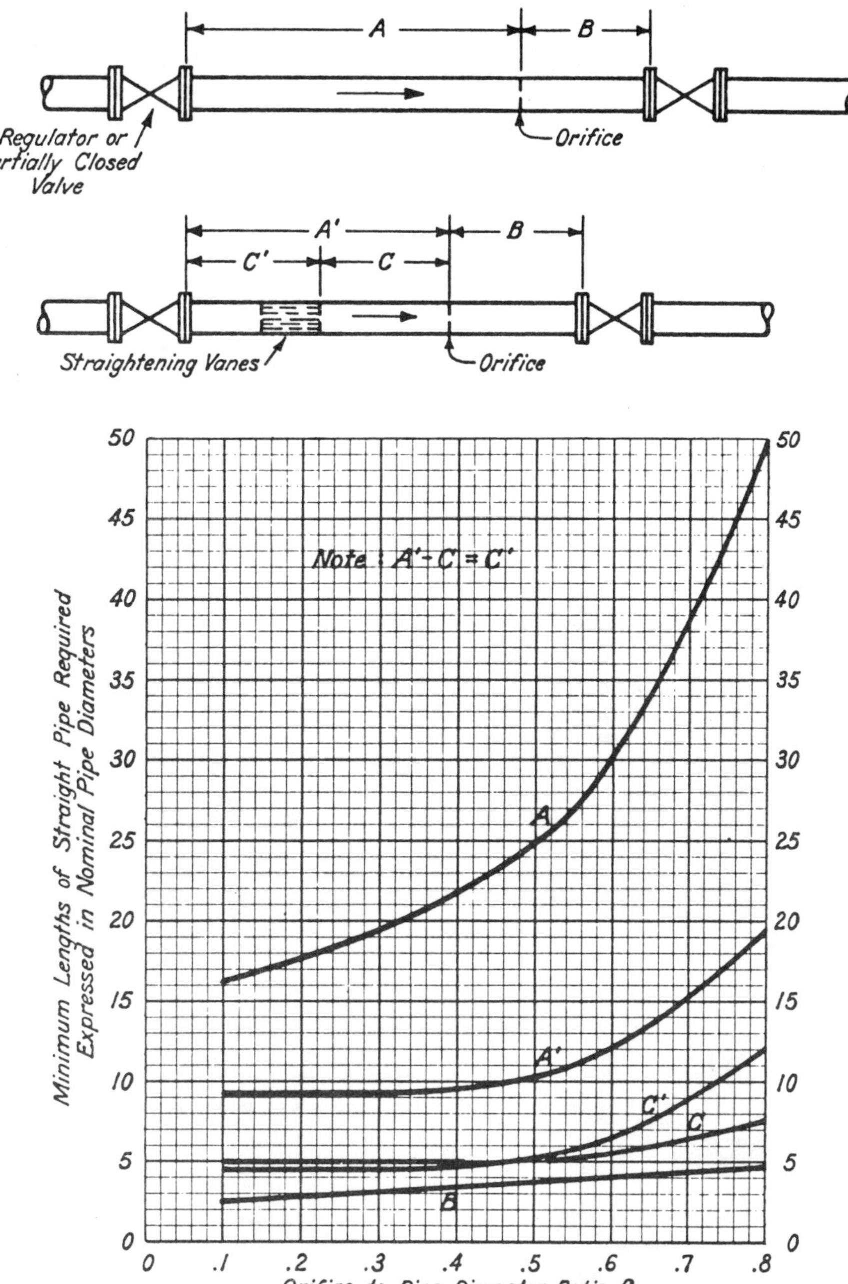

NOTE 1 - When "Pipe Taps" are used, length A, A', and C should be increased by 2 pipe diameters, and B by 8 pipe diameters.

NOTE 2 - Line A, A', C, and C' apply to regulators or gate valves, globe valves, and plug valves which are used for throttling the flow (partially closed).

NOTE 3 - When the diameter of orifice may require changing to meet different conditions, the lengths of straight pipe should be those required for the maximum orifice to pipe diameter ratio that may be used.

Figure 4.41. Minimum run of straight length pipe upstream and downstream of an orifice for piping configurations shown as recommended by the AGA. (Courtesy of American Gas Association)

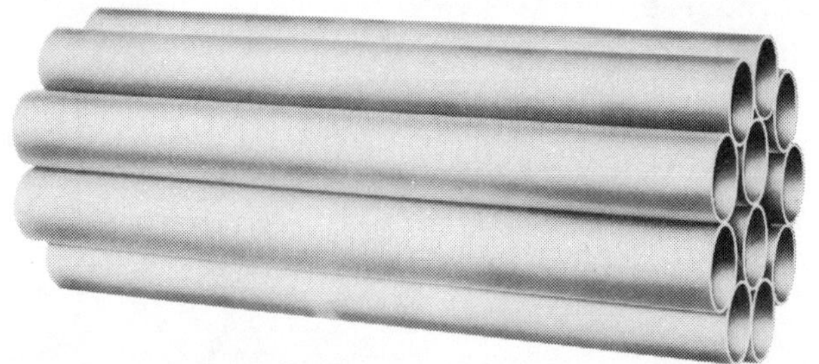

Figure 4.42. Typical tubular type straightening vanes installed ahead of meter runs to reduce the straight run requirements. (Courtesy of Foxboro Co.)

Figure 4.43. One piece radial type straightening vane. (Courtesy of Foxboro Co.)

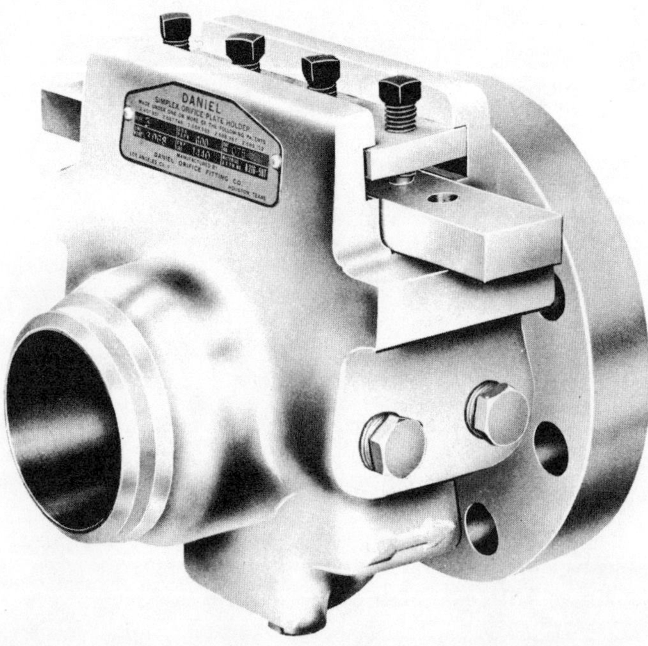

Figure 4.44. Daniel Simplex orifice holder, a special plate holder for quick and easy changeout of orifice plate. Line depressuring is necessary. (Courtesy of Daniel Industries, Inc.)

On larger pipe sizes a pinion gear is added to this type fitting so that the plate can be cranked out quickly after depressuring and removing the clamping.

To meet more exacting requirements, another holder or fitting is available (Figure 4.45) which allows the orifice plate to be removed or inspected without depressuring the line. This is accomplished with a two-compartment fitting (Figure 4.46). During installation or removal, the top compartment can be sealed off from the bottom compartment which holds the orifice plate in the operating position.

This type fitting eliminates costly valves and bypasses required for conventional orifice flange installations in services where frequent inspection and changeout is necessary.

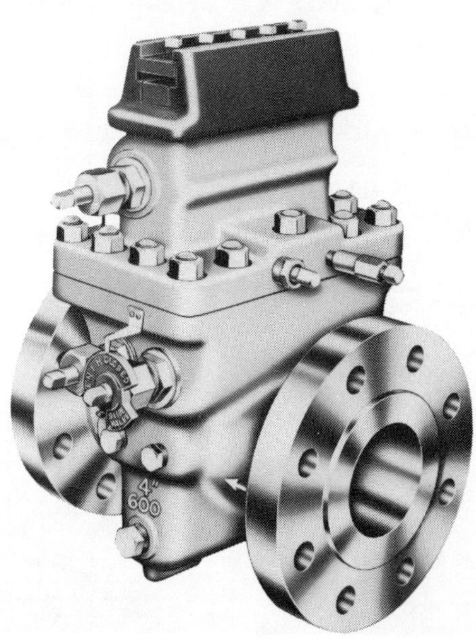

Figure 4.45. This Daniel Senior Orifice Fitting allows orifice change without depressuring the line. It contains two compartments as shown in Figure 4.46. (Courtesy of Daniel Industries, Inc.)

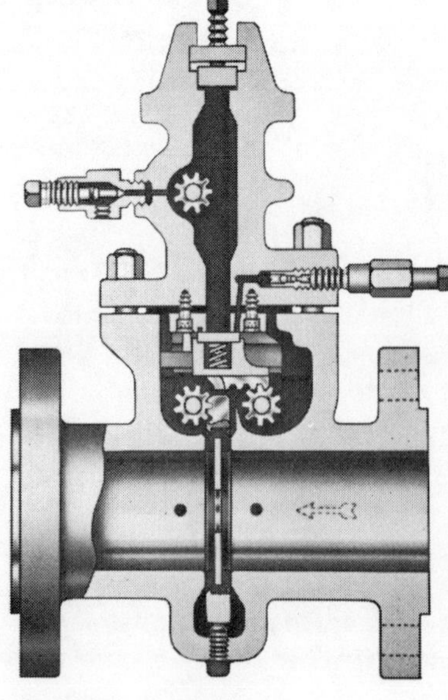

Figure 4.46. Cutaway shows the two compartments of the senior orifice fitting which allow the orifice plate to be isolated and removed without depressuring the line. (Courtesy of Daniel Industries, Inc.)

Prefabricated Meter Tubes. Mention was made previously of the need for straight lengths of pipe upstream and downstream of the orifice plate. Requirements for this straight run include not only specified dimensions on either side of the plate but also that the pipe meet other criteria (specified by the A.G.A. and A.S.M.E.) essential to accurate measurement, including control of material specifications (as to roundness, smoothness and internal diameter), tube preparation (cleaning), fabrication (grinding of welds, freedom from burrs, etc.) and inspection and testing.

Prefabricated meter tubes which meet all these requirements are available from several manufacturers. Pipe is selected to meet the tolerances specified, and preparation, fabrication, and testing are performed by trained crews who are knowledgeable with the total requirements. Use of these prefabricated runs often eliminates costly mistakes by untrained crews and can eliminate some unnecessary inspection costs.

Comparison of Primary Devices

After discussing the many primary devices available, a comparison of their uses and characteristics is in order.

Orifice plates are the most widely used primary device by a large margin. For economy they are difficult to beat until large pipe sizes are reached; the cost of flanges increase significantly with each increase in pipe size. In large pipe sizes Pitot tubes and Annubars are more economical if their accuracy is acceptable for the application. Elbow taps are very economical, but their accuracy is the poorest of all the methods listed.

From an accuracy standpoint there is not much difference among orifice plates (concentric), Venturis and flow nozzles. Much more test data is available on orifice plates, however, than on Venturis and flow nozzles, leading to more uncertainty about their coefficients, particularly for low Reynolds numbers. The accuracy of Annubar elements apparently is close to those just mentioned although their use is not yet supported by much data. In order of decreasing accuracy, Pitot tubes and elbow taps follow in that order. Weirs and flumes are not included in the comparison for they are used for vastly different purposes.

From the pressure loss standpoint, Venturis have a decided advantage over the orifice plate or the flow nozzle (Figure 4.47). Assume for example that all these were calibrated for a 100-inch differential. For a beta ratio of 0.50, the pressure drop at the orifice plate would be 75 inches of water, the flow nozzle 62 inches and lo-loss tube only 5 inches of water. On the other hand, the pressure loss on Pitot tubes and Annubars is almost negligible. There would be no loss at all due to elbow taps.

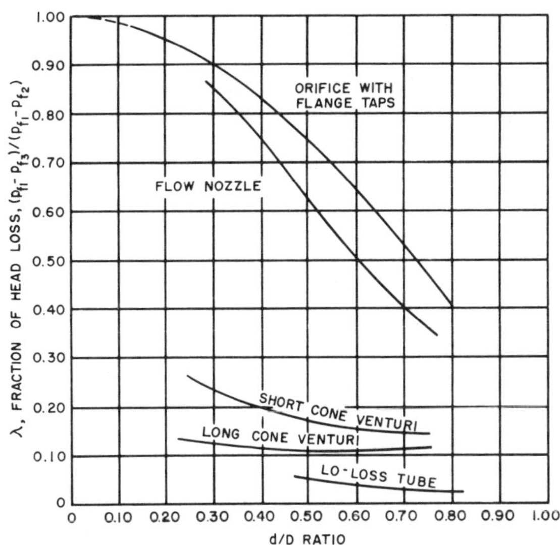

Figure 4.47. Comparison of pressure losses among primary flow elements reveals that Venturis are superior in terms of permanent head loss among the types shown. (Courtesy of Foxboro Co.)

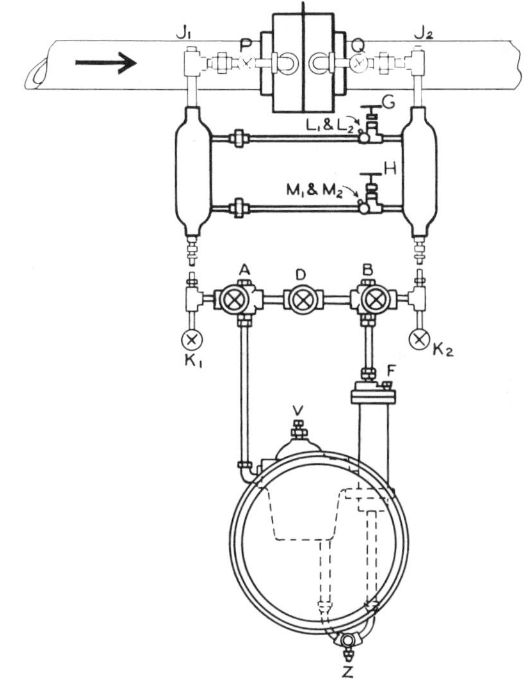

Figure 4.48. Typical installation of a mercury water (wet type measuring device) on a flow application. Liquid seals are used here to isolate the flowing fluid from the mercury. (Courtesy of Foxboro Co.)

Secondary Devices

The devices which respond to differential pressure measurements and convert them to usable forces or signals may be classified in several ways. In the 1950s it was common to refer to them as wet or dry types. Wet types were ones that used liquid heads as a measure of the differential pressure across the primary element (Figure 4.48). The liquid used was usually high purity mercury, which provided an accurate measuring system. Other liquids were also used, particularly when low differential ranges (therefore, lighter fluids) were required and reading accuracy became more important.

Wet type meters are discussed in considerable detail, not because they are widely used now, but because they illustrate so well the principles of head flow measurement. For many years after the introduction of accurate dry type meters, mercury meters continued to be used for accounting measurements. Slowly, however, dry types have gained acceptance as being as accurate and reliable as mercury meters, and mercury meters are now used infrequently.

Wet Meters

Wet meters fall into several classes, among which are manometers, mercury float type and liquid seal meters. In all cases the flowing media makes contact with a manometer or meter fluid which measures the differential produced by the flow.

The liquid manometer is the oldest and simplest type of differential measurement. In its simplest form it consists of a glass tube bent in the shape of a "U" (Figure 4.49). This form is seldom used for field measurements but is used for laboratory work as a calibration standard.

There are various modifications of the basic U-tube manometer. One leg of the manometer may be suppressed by using a chamber of large surface area for one leg (Figure 4.50). The scale can be compressed to correct for the level change in the large chamber, or a regular scale can be used and the zero adjusted for each reading.

Another type of manometer is the float liquid manometer in which the level in a large chamber is sensed by a float. The motion of the float is used for indication, recording or eventual control through a system of linkages and a lever rotating in a low friction pressure seal bearing. Pneumatic or electrical transmission of float motion is sometimes used. The range of this type device is from 20 to 200 inches of water; it is not used below 20 inches because float motion is reduced to a point where accuracy is already affected. For many years this type measurement was the accepted standard for orifice flow measurement.

The liquid seal, or inverted bell, meter is well suited for low differential pressures. The differential pressure to be measured is applied across an inverted bell which produces a force and consequent bell movement (Figure 4.51). This movement, which is opposed by a spring, is linked through a low friction bearing and used to indicate or transmit the differential. These meters are used for ranges below 20 inches of water. A liquid (usually mercury) serves as a seal between the high and low pressure.

An adaptation of the bell meter, the Ledoux Bell (Figure 4.52) uses a shaped floating bell characterized so that its motion is proportional to the square root of the differential pressure and, hence, is linear with flow. It has been used extensively in power plants and other similar applications where a uniform flow scale was needed for a wide range of flows.

An interesting device used rather extensively in Europe but seldom used in the United States is the ring balance meter (Figure 4.53). A ring of steel tubing mounted in a ver-

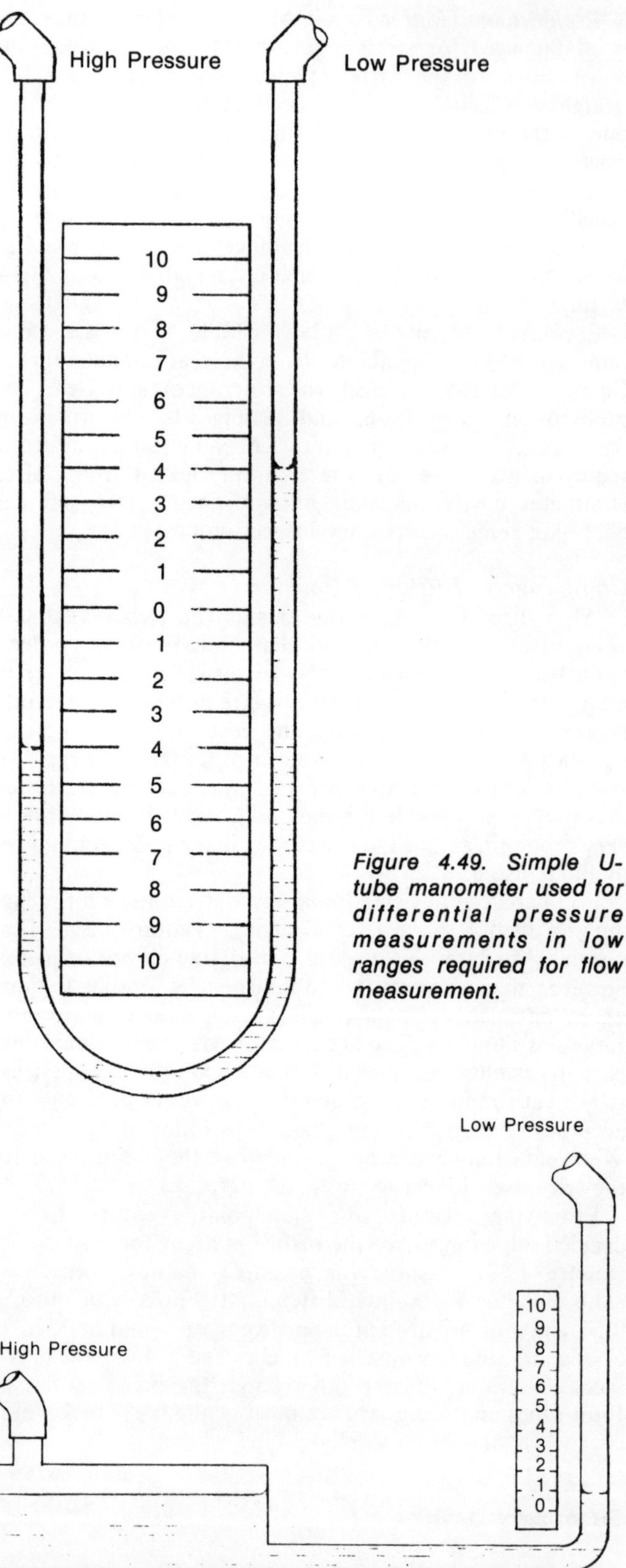

Figure 4.49. Simple U-tube manometer used for differential pressure measurements in low ranges required for flow measurement.

Figure 4.50. The well type manometer allows direct differential reading instead of a difference reading made with U-tube type.

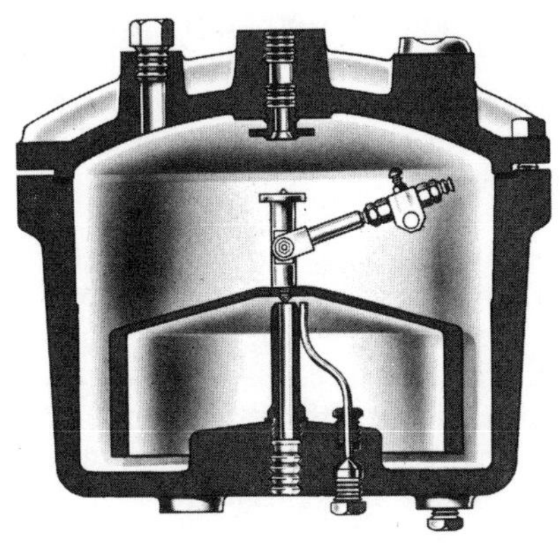

Figure 4.51. The liquid sealed, inverted bell meter is used for low differential range requirements. Ranges below 20 inches H_2O. (Courtesy of Bristol Division of ACCO.)

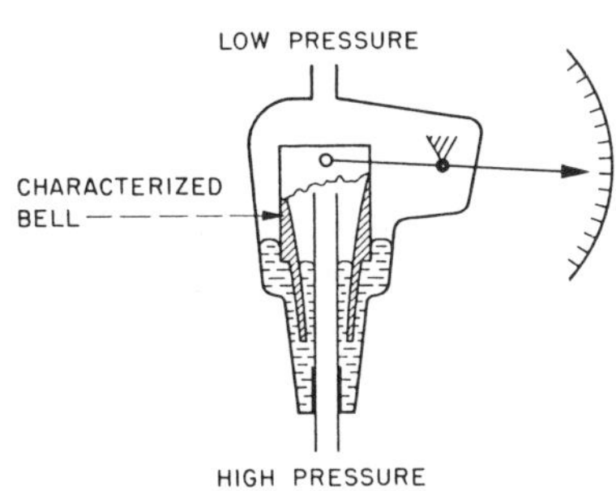

Figure 4.52. The Ledoux Bell is a special adaptation of the bell meter in which the floating bell is characterized to produce a linear motion with respect to flow instead of a square root relationship. (Courtesy of Foxboro Co.)

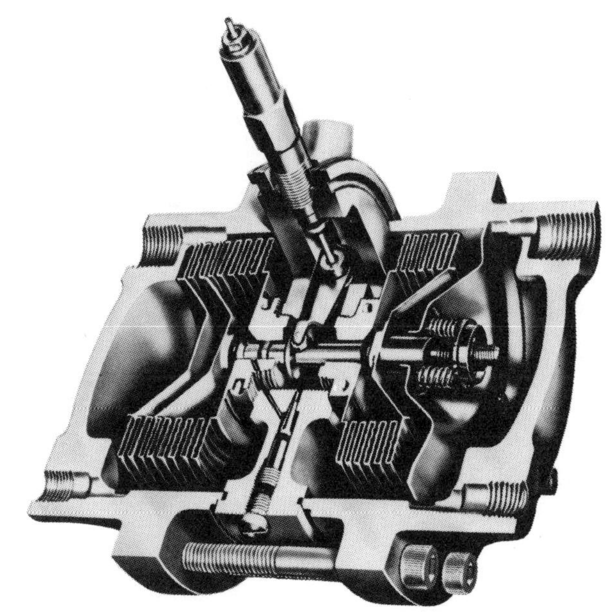

Figure 4.54. A typical bellows type differential pressure unit. The bellows are liquid filled. (Courtesy of ITT Barton)

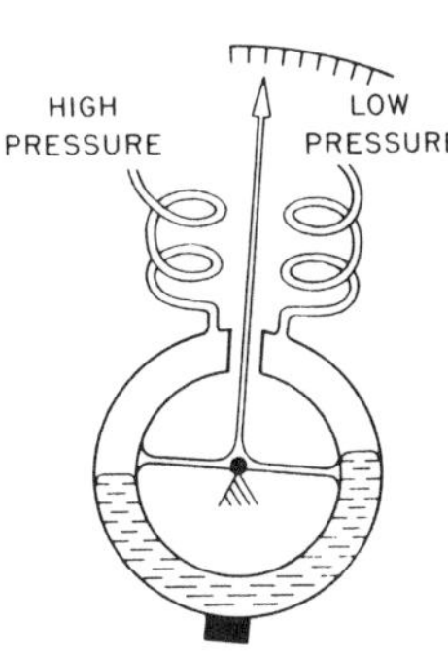

Figure 4.53. Ring balance meters requiring flexible connections pose problems particularly for high-pressure services. (Courtesy of Foxboro Co.)

tical plane is pivoted at its central axis. The differential pressure exerts an unbalanced force on opposite ends of the ring, tending to rotate it. A weight attached to the ring develops an opposing torque as it moves with the ring. Flexible connections required by the moving ring poses problems, particularly at high static pressures.

Dry Meters

Because mercury meters were used so extensively in the early days of process control, other type meters were referred to as dry or mercury-less meters as they came into use. Two types are prominent in dry meters—the bellows type (Figure 4.54), a motion balance design, and the diaphragm capsule (Figure 4.55), a force balance type design introduced in the late 1940s. These meters are more economical, easier to calibrate and maintain and have almost replaced mercury meters altogether.

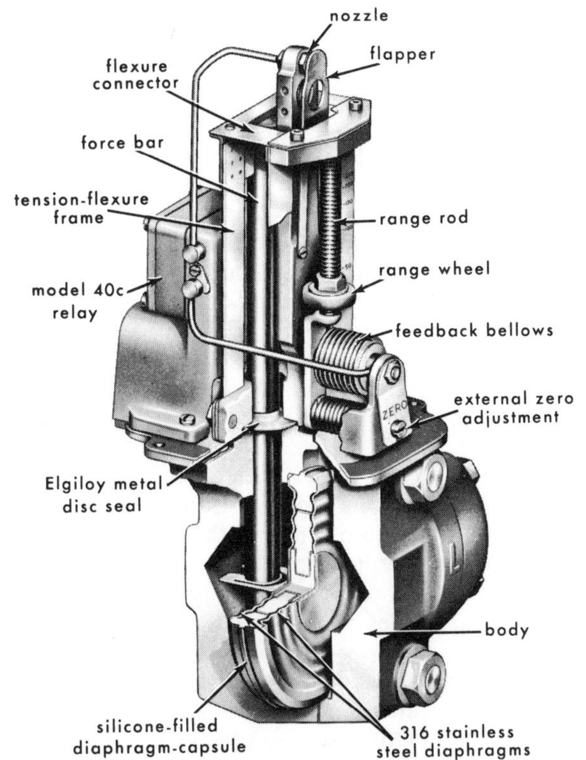

Figure 4.55. A typical force balance type differential pressure unit widely used for flow measurement. (Courtesy of Foxboro Co.)

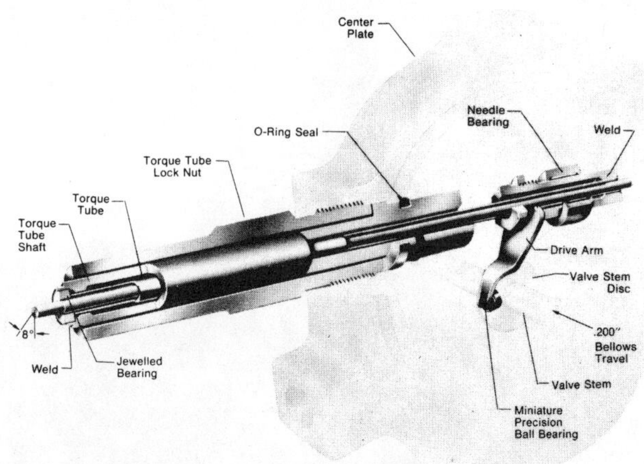

Figure 4.56. A torque tube assembly transmits bellows motion to an external unit for readout or control. (Courtesy of ITT Barton)

Bellows Meters. In the bellows meter (Figure 4.54) differential pressure is applied across two liquid filled opposing bellows. This produces a force which is opposed by a calibrating range spring. The resulting motion is proportional to the differential pressure. In the bellows unit pictured, the opposing bellows are connected to a center plate, their outer ends are sealed and are rigidly connected internally by a stem passing through an annular passage in the center plate. The internal volume of the bellows and center plate is completely filled and sealed with a clean, noncorrosive, low freezing-point liquid. An additional free-floating bellows is attached to the high-pressure side of the bellows unit to allow for expansion and contraction of the fill liquid, thus providing positive temperature compensation through a wide range of ambient temperatures. A torque tube assembly (Figure 4.56) is an integral part of the sealed bellows unit assembly and is employed to transmit motion of the bellows to the exterior of the unit. The torque tube provides a positive, frictionless seal and requires no lubrication or maintenance.

In operation, the bellows moves in proportion to the difference in pressure applied across the bellows unit assembly. The linear motion of the bellows, which is picked up by a drive arm, is mechanically transmitted as a rotary motion through the torque tube assembly. Should the bellows be subjected to a pressure difference greater than the differential pressure range of the unit, they will move through their calibrated travel plus a small amount of "overtravel" until a valve mounted on the center stem seals against its corresponding valve seat. As the valve closes, it "traps" the fill liquid in the bellows, and since the liquid is essentially noncompressible, the bellows are fully supported and are not ruptured by the overpessure applied. Furthermore, since opposed valves are provided, full protection is afforded against an "overrange" in either direction.

Desirable characteristics of these units include the temperature compensating bellows which holds temperature shift to less than 0.4% per 100°F temperature change, linearity between output and differential pressure, overrange capability and the ease of changing ranges. Range spring assemblies are interchangeable between 0 and 20 inches water columns and 0 and 400 inches water column for the unit shown in Figure 4.54.

Force Balance Meters. Figure 4.57 shows a typical pneumatic transmitter of the force balance type that is widely used for flow measurement today—a d/p cell.

In this type the differential pressure is applied across a pair of opposing liquid-filled diaphragms welded on the opposite sides of a capsule. The resulting force is brought out by a force bar to which is applied an opposing force from a pneumatic feedback bellows (see Figure 4.58). A small movement of the force bar repositions a nozzle, causing a change in output pressure. The output pressure is used in the feedback bellows to establish equilibrium of the force bar. The output pressure is therefore proportional to the differential pressure. The direct balance between the two forces contributes to its accuracy and stability.

An electronic version of the force balance d/p cell (Figure 4.59) is similar to the pneumatic type. An electric current flowing in a coil supported in a permanent magnet field develops a force. The difference between this force and that developed by the measured differential pressure tends to produce a motion which is detected by a highly sensitive electrical unit. The output of this unit operates to maintain the coil current at a value to exactly balance the force produced by the differential pressure and is used as the transmission signal.

The force balance transmitter is compact, economical and reliable. It is easy to calibrate and provides continuous wide-range adjustment. D/p cells are available in differential ranges from 1 to 1,000 inches of water.

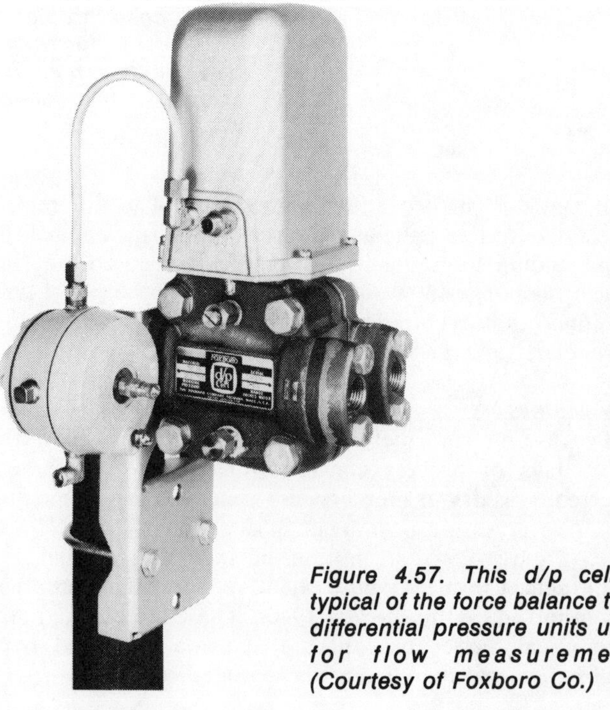

Figure 4.57. This d/p cell is typical of the force balance type differential pressure units used for flow measurement. (Courtesy of Foxboro Co.)

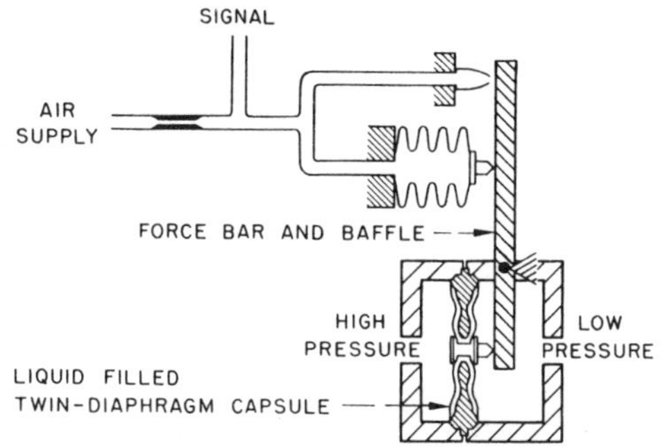

Figure 4.58. A schematic of the force balance unit shown in Figure 4.57. (Courtesy of Foxboro Co.)

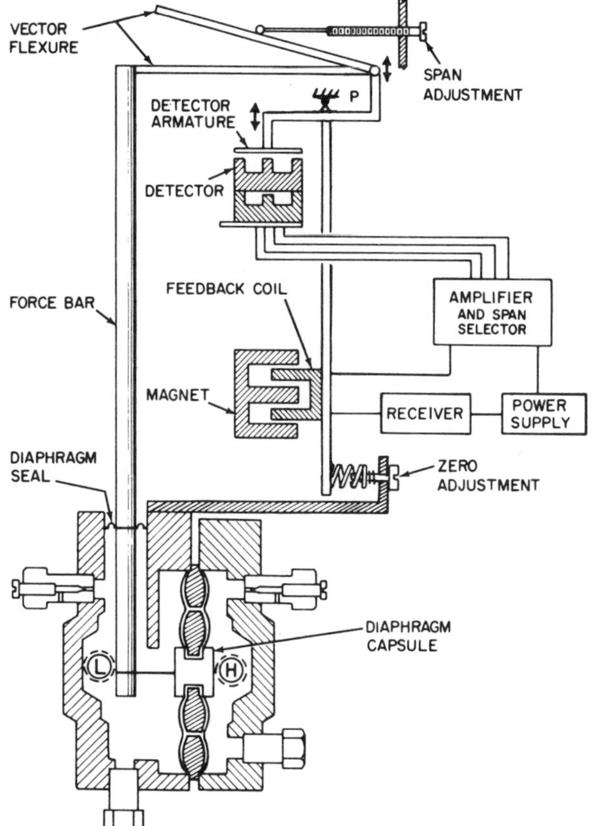

Figure 4.59. The electronic version of the force balance d/p cell shown in Figures 4.57 and 4.58. (Courtesy of Foxboro Co.)

A d/p cell, when equipped with a "zero elevation" kit, can be used to measure bidirectional flow. An explanation of "zero elevation" and "zero suppression" is found in Chapter 5, Level Measurement.

The Integral Orifice

The need to measure small flow rates led to the development of a special meter, the integral orifice d/p cell, so called because the orifice is mounted integrally with the transmitter body.

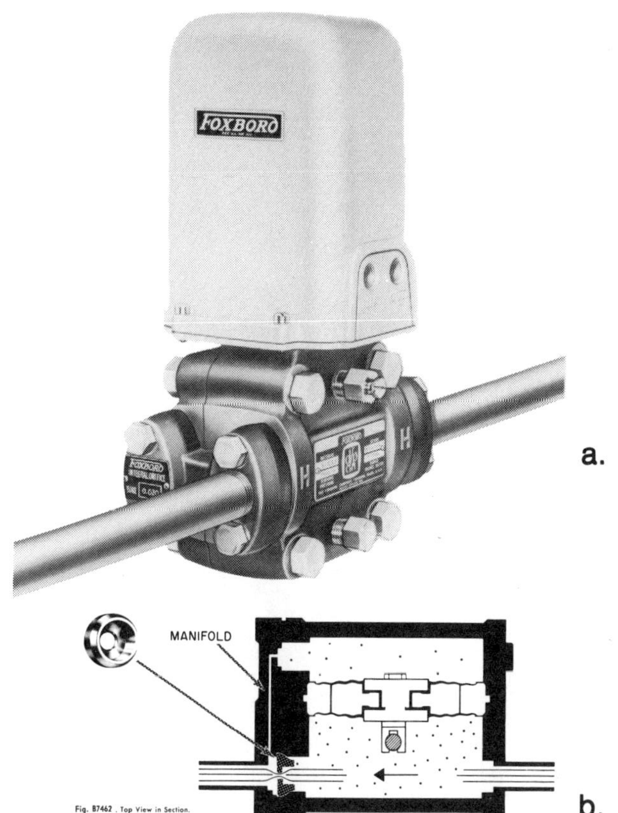

Figure 4.60. The integral orifice d/p cell (a) mounts directly in the line. The downstream side connects to the low side of the diaphragm capsule through a manifold (b). (Courtesy of Foxboro Co.)

The assembly is installed directly in the flow line (Figure 4.60). Measured fluid passes through the orifice in the manifold and continues in the flow line. The downstream side of the orifice is connected to the low-pressure chamber of the transmitter through the manifold. Some "integral orifices" are mounted in a hairpin tubing section piped so that the measured fluid flows through the "high" side of the transmitter body, out and through the orifice in the tubing section, back through the "low" side of the transmitter body and into the process piping. This arrangement is very compatible with bolt-on zeroing manifolds. The difference in pressure across the orifice, applied to the diaphragm capsule, produces the same action as in the regular d/p cell transmitter and is indicated or recorded as flow on a square root scale of a conventional receiver.

Orifices are sized from 0.0020 to 0.350 inch from one manufacturer and can be changed easily without a major dismantling operation. Transmitters are available for both pneumatic and electronic transmission. Seldom are regular orifices installed in line sizes under 1 ½ inches, so the integral orifice d/p cell meets a definite need for metering low flows of clean liquids and gases—in research problems, pilot-plant work and for many applications in commercial size plants.

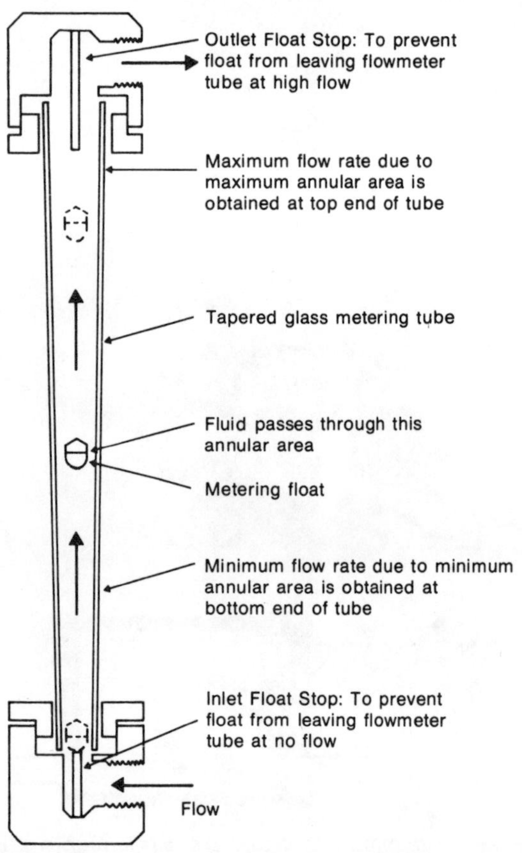

Figure 4.61. The rotameter uses a float in a tapered tube to indicate flow rate, which is linearly proportional to float rise.

Rotameters

The rotameter is a variable area type flowmeter consisting of a vertical tapered tube with a "float" which is free to move up or down within the tube (Figure 4.61). The measured fluid enters the tube from the bottom and passes upward around the float and out at the top. As the flow varies, the float rises or falls, varying the area of the annular passage between the float and tube. The float maintains an equilibrium position in which the upward hydraulic forces acting on it are in balance with its weight less the buoyant force. The tube is tapered so that there is a linear relationship between the flow rate and the position of the float within the tube. A calibration scale printed on the tube or near it provides a direct indication of flow rate.

Rotameters are available in a wide variety of styles. Tube materials, float shapes and materials, end connections, pressure and temperature ratings, and lengths of scales vary to meet a wide range of service conditions.

Glass Tubes

Borosilicate glass metering tubes are commonly used for relatively low pressure and temperature services of non-hazardous fluids, such as water and air. Depending on the meter size, glass tubes can withstand some fairly high pressures. They are not often used at high pressures, however, not because they are likely to rupture from internal pressure but because the glass can be broken so easily. Figure 4.62 shows a typical pressure rating curve for a particular glass tube rotameter. In this curve safe working pressure is plotted against meter size. Note the steep drop in pressure rating with increase in meter size. A maximum temperature rating is given in these cases.

Another method of presenting the same limitations of pressure and temperature versus size is given in a typical curve (Figure 4.63) in which temperautre is plotted against pressure for each meter size. Data furnished by the manufacturer provides the user with service-limiting con-

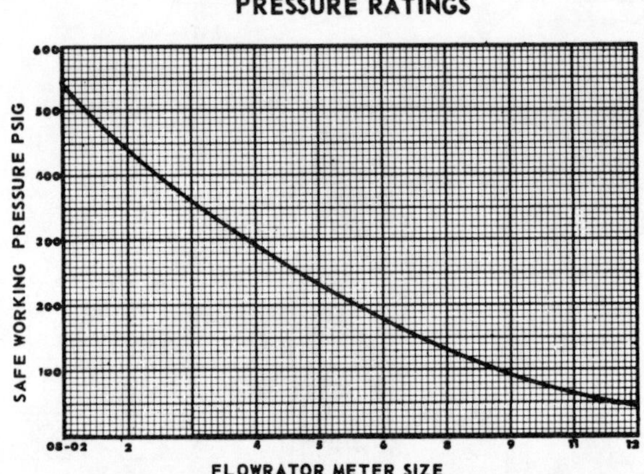

Figure 4.62. Typical pressure rating curve for rotameters. Each model group has its own pressure rating curve. (Courtesy of Fischer and Porter Co.)

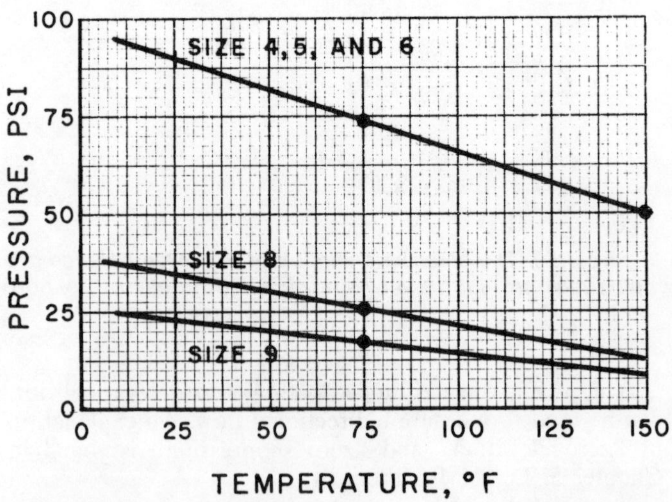

Figure 4.63. Another method of presenting pressure-temperature relationships is to plot pressure versus temperature for the various rotameter sizes offered for meter models. (Courtesy of Fischer and Porter Co.)

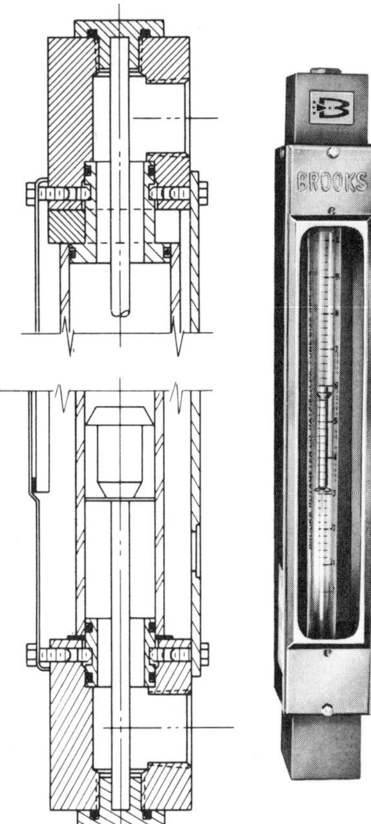

Figure 4.64. Protective shields around glass rotameter tubes reduce hazards when tubes are broken. (Courtesy of Brooks Instruments)

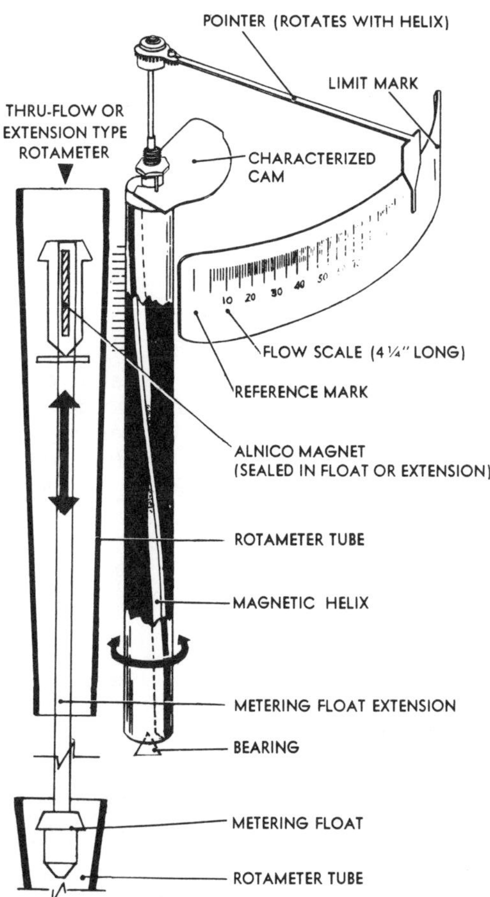

Figure 4.65. A magnetically actuated pointer mechanism is used to provide indication for a metal tube rotameter. (Courtesy of Brooks Instruments)

ditions for benign situations. The user then uses his own discretion in determining whether outside influences should further limit the "safe" temperatures and pressures to be used. Normal practice in industry is to restrict the use of glass tube meters to nontoxic and nonflammable fluids.

One method used to reduce the hazard of glass tubes is the use of protective shields (Figure 4.64) around the tubes. The shield prevents accidental breakage of the tube except for major accidents, and it prevents excessive splashing or dispersion of the fluid if breakage or rupture does occur.

Metal Tubes

Metal metering tubes (often referred to as armored meters) have been developed for applications where glass is not acceptable. A different technique for indication is necessary, of course, since the float cannot be seen inside the tube. Figure 4.65 shows such a scheme. The linear motion of the float is translated into a rotary motion for direct indication or signal transmission. The converter is a magnetic iron helix fixed permanently in an aluminum cylinder and mounted adjacent to and parallel with the rotameter extension tube. A metering float extension, with imbedded permanent magnet, moves within the tube in direct response to changes in flow rate. The leading edge of the helix is continuously attracted to the magnet. As it turns it converts the linear motion of the float into a rotary motion for indication. A characterized cam is attached to the helix cylinder which is sensed pneumatically or electrically to deliver linear transmission signals.

Another method used for indication or transmission when metal tubes are required is shown in Figure 4.66. Extension rods are placed on both ends of the float, the upper rod to provide an indication or transmission device and the lower extension to provide stability and travel guiding service. When indication only is desired, a glass section provides a readout scale. When signal transmission is necessary, an armature is used with a magnetic follower to translate the motion to an appropriate pneumatic or electronic signal.

A feature of this type meter which should be emphasized is that a dead space exists in the upper chamber around the readout or transmission area. It cannot be sealed off because the extension rod must move freely up and down through the adapter section. It is not adaptable to dirty or sticky services unless it is purged (purge connections can be provided).

Slurry Services

Some rotameter designs are adaptable to minor slurry services. Figure 4.67 illustrates its operation. It is a straight

Figure 4.66. In this armored rotameter, indication is obtained by a float extension to an upper chamber containing a visible glass section for readout. (Courtesy of Fischer and Porter Co.)

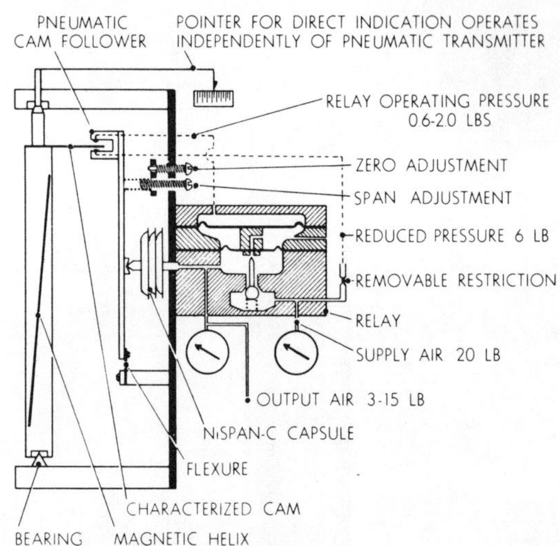

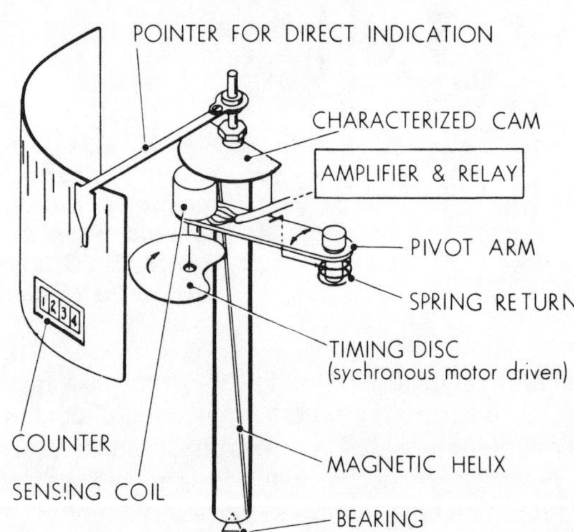

Figure 4.68. Schematic drawings showing transmission techniques used for flow through type rotameter shown in Figure 4.67. Pneumatic operation on top and electronic on bottom. (Courtesy of Brooks Instruments)

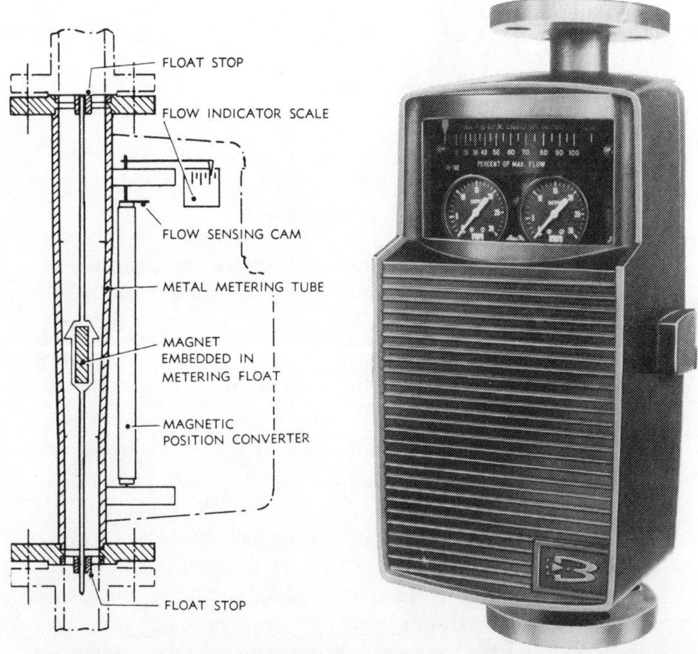

Figure 4.67. The straight through type armored rotameter is adaptable to minor slurry services. There are no crevices in which solids may settle. (Courtesy of Brooks Instruments)

through section using a magnet embedded in the float and a magnetic follower to provide indication. Electronic or pneumatic transmission is accomplished as shown schematically in Figure 4.68.

There are no stagnant pockets or crevices for solids or crystals to settle into, but particles must be small enough to pass around the float with undue disturbance for flow measurement to be effective.

Summary of Operating and Design Characteristics

Accuracy of rotameters varies a great deal depending upon size, type and calibration. Most rotameter accuracies fall into an area of ±2% of full scale; some are rated at ±1%, and, with proper calibration, some are available with accuracies within ±½% of full scale. On the other hand, some short scale types have accuracies no better than ±5% and even ±10% of full-scale flow.

The rangeability of rotameters is considered to be about 10:1. Float movement is normally unstable until 10% of rated flow is attained. From that point on, float motion is stable and linear with flow rate.

Meter ranges are available from a fraction of a cc/minute to over 3,000 gpm. The primary use of rotameters, however,

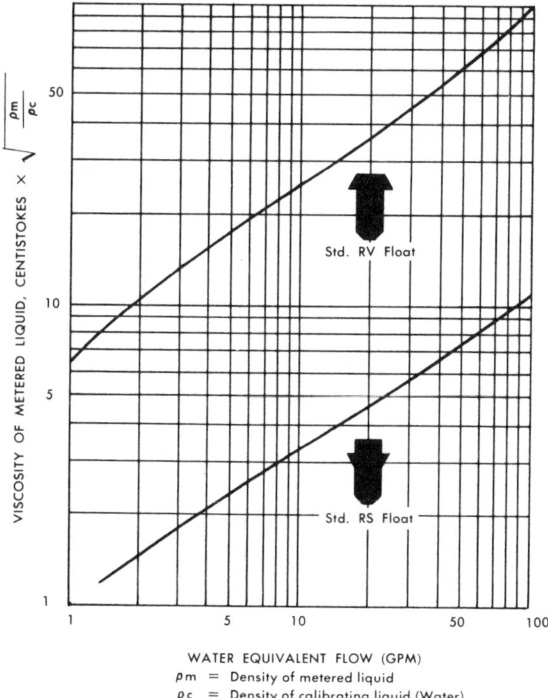

Figure 4.69. Viscosity ceiling curves for various floats are used to determine whether special calibration is needed. (Courtesy of Brooks Instruments)

is in the smaller sizes. They are used extensively on purge applications and for many process flows in pipe sizes ½ to 2 inches.

Rotameters are essentially immune from viscosity effects to a certain point (about 100 centipoise). Small sizes are more sensitive than large sizes. Some manufacturers publish viscosity ceiling curves (Figure 4.69) so that the buyer can determine the viscosity limits to be used. When the limits fall above the viscosity ceiling curve for the float used, water calibration is no longer suitable, and the meter should be calibrated with the actual fluid to be metered or a suitable substitute.

The pressure drop across a meter is essentialy constant from the point the float lifts off the float stop to 100% of rated flow. It ranges from a few inches of water for small meters to a pound or so for larger sizes.

Piping configuration has no effect on rotameter operation. Because they must be mounted vertically and fluid flow is upward, many configurations are used—top and bottom, side and top, bottom and side, side and side. The most convenient piping arrangement may be used. Screwed connections are used in all the small sizes. Flanged connections are usually available in pipe sizes from ½ inch and up in many models.

Sizing

Rotameter sizing is based on the flow of air for vapor or gas services and on water for liquid services. Manufacturers furnish capacity charts for various meter tubes and float shapes (Figure 4.70) for air and water at standard conditions. Actual flows (of the metered fluid) must be converted to gpm water equivalent or scfm air equivalent for selection of the correct meter size. Where low flow rates are involved, capacities are usually given in cc/minute of air and water. The tables are also based on stainless steel floats. Conversion formulas take into account the density of the float material as well as the density of the measured fluid for other float materials and other measured fluids.

The following are typical conversion equations used by one manufacturer.

Liquid Conversion:

$$\text{gpm } H_2O = \text{gpm} \sqrt{\frac{7.02 \times \rho}{\rho_f - \rho}} \tag{4.5}$$

or

$$\text{gpm } H_2O = \frac{\text{lbs/min}}{8.33 \times \rho} \sqrt{\frac{7.02 \times \rho}{\rho_f - \rho}} \tag{4.6}$$

where gpm = desired maximum flow rate in gpm
lbs/min = desired maximum flow rate in pounds per minute
ρ_f = density of the float required for the application and selected from the following list:

316 stainless steel = 8.02
Hastelloy C = 8.94
nickel = 8.91
Monel = 8.84

ρ = fluid density, g/cc at operating conditions
gpm H_2O = equivalent flow rate in gpm H_2O

Gas Conversion

$$\text{scfm air at 14.7 psia and 70°F} = \text{scfm} \sqrt{\frac{8.02 \times \text{SpGr} \times 14.7 \times T_{op}}{\rho_f \times 1.0 \times P_{op} \times 530}} \tag{4.7}$$

or

$$\text{scfm air at 14.7 psia and 70°F} = \text{lbs/min} \times 13.34 \sqrt{\frac{8.02 \times 1.0 \times 14.7 \times T_{op}}{\rho_f \times \text{SpGr} \times P_{op} \times 530}} \tag{4.8}$$

Meter Size	Maximum Flow gpm H_2O equiv.	Maximum Flow scfm air equiv.	Metering Tube 5 or 10 inch	Float Number	Total ΔP (1)	V.I.C. (2)	psia Critical (3)
4	.267	1.10	B4-17	BUSVT-40	.74	2.9	5.5
	.328	1.35	B4-21	BUSVT-40	.77	2.9	3.5
	.442	1.82	B4-27	BUSVT-40	.78	2.0	2.7
	.467	1.92	B4-17	BSVT-45	2.0	5.1	17.9
	.600	2.47	B4-21	BSVT-45	2.3	5.1	11.5
	.650	2.65	B4-17	BSVT-44	3.8	7.1	33.4
	.736	3.04	B4-21	BNSVT-45	3.2	0.8	11.5
	.810	3.35	B4-27	BSVT-45	2.4	5.1	8.4
	.880	3.62	B4-21	GSVT-48	4.5	7.6	24.6
	1.00	4.12	B4-27	BNSVT-45	3.7	0.8	8.4
	1.12	4.60	B4-27	BSVT-44	4.6	7.1	16.2
	1.44	6.00	B4-27	BSVT-43	8.9	9.2	27.5
	1.84	7.50	B4-27	BNSVT-43	10.4	1.3	27.5
	2.04	8.40	B4-21	BL-444	30	1.0	86.0
	3.1	12.8	B4-27	BL-444	40	1.0	66.0
5	1.96	8.00	B5-21	BSVT-54	3.8	10.4	13.9
	2.49	10.2	B5-21	BNSVT-54	4.6	1.6	13.9
	2.70	11.0	B5-27	BSVT-54	4.4	10.4	9.6
	3.15	13.0	B5-21	BSVT-53	11.0	16.8	36.0
	3.55	14.6	B5-27	BNSVT-54	5.4	1.6	9.6
	4.00	16.6	B5-21	BNSVT-53	10.2	2.5	36.0
	4.35	18.0	B5-27	BSVT-53	13.0	16.8	25.0
	5.70	23.6	B5-27	BNSVT-53	15.9	2.5	25.0
6	4.70	19.2	B6-27	BSVT-64	7.6	14.8	11.5
	5.35	22.1	B6-27	GSVT-65	10.0	16.9	15.2
	6.24	25.7	B6-27	BNSVT-64	10.4	2.2	11.5
	6.60	27.0	B6-35	BSVT-64	11.1	14.8	6.8
	6.70	27.8	B6-27	BSVT-63	15.4	20.8	23.7
	7.50	31.0	B6-35	GSVT-65	14.3	16.9	8.9
	8.70	36.0	B6-35	BNSVT-64	16.6	2.2	6.8
	9.40	39.0	B6-35	BSVT-63	24.0	20.8	13.9
	10.0	41.0	B6-35	GNSVT-65	21.4	2.5	8.9
	12.4	51.0	B6-35	BNSVT-63	34.0	2.9	13.9
	14.2 (4)	58.5(4)	B6-27	BL-650	40.0	1.5	38.0
	20.0 (4)	82.0(4)	B6-35	BL-650	60.0	1.5	25.0
8	13.4	55.0	B8-27	BSVT-84	8.8	27.6	15.4
	16.0	66.0	B8-27	BSVT-83	12.4	33.0	22.0
	17.6	72.0	B8-27	BNSVT-84	10.4	4.2	15.4
	21.2	87.0	B8-27	BNSVT-83	14.6	4.9	22.0
	31.0 (4)	128 (4)	B8-27	BL-850	32	2.0	48
	40.0 (4)	165 (4)	B8-27	BL-851	53	2.0	80
9	26.5	110.0	B9-27	BSVT-94	12.4	40.5	16.4
	31.0	128.0	B9-27	BSVT-93	18.7	49.0	24.0
	35.0	144.0	B9-27	BNSVT-94	17.4	6.1	16.4
	42.4	174.0	B9-27	BNSVT-93	24.4	7.3	24.0
	67.0	275.0	B9-27	BL-90	36.0	3.0	35.0
	75.0	310.0	B9-27	BL-950	48.0	3.0	40.0
	100 (4)	412 (4)	B9-27	BL-951	85	3.0	80
11	57.0	236.0	B11-20	BSVT-11-3	20.9	83.0	15.5
	78.0	320.0	B11-20	BNSVT-11-3	26.6	12.6	15.5
12	102.0	420.0	B12-20	BSVT-12-3	23.7	115.0	15.1
	140.0	580.0	B12-20	BNSVT-12-3	29.8	17.4	15.1

Figure 4.70. Capacity chart for Fischer and Porter meter tubes and floats is typical of such charts furnished by rotameter manufacturers. (Courtesy of Fischer and Porter Co.)

where scfm = desired maximum flow rate in scfm

Sp Gr = specific gravity of gas at standard temperature and pressure, referred to air at standard temperature and pressure (14.7 psia and 70°F)

T_{op} = absolute temperature, (460+°F) at operating pressure

P_{op} = absolute pressure in psia at operating conditions

scfm air = equivalent flow rate in scfm of air at 14.7 psia and 70°F

Bypass Rotameters

Bypass rotameter installations are used occasionally to measure flow rates in large pipelines by using low-cost orifice plates (differential producers) while still retaining the linear function of the rotameter (see Figure 4.71).

The principle orifice is located in the main line, sized to take a standard (50 inches H_20, 100 inches H_20, etc.) pressure drop at maximum flow. The arrangement includes the necessary runs of straight pipe upstream and downstream of the orifice and employs the usual pressure taps. The rotameter is piped to these taps as shown in Figure 4.71 and includes an additional range orifice which is sized so that the flow through the meter, at maximum pressure drop across the orifice plate, is equal to the flow rate necessary to lift the float to its maximum position. Flow rates in both the main line and the rotameter bypass line depend upon the pressure drop across the orifice. Because the pressure drop exists across each line, the flow rate through the bypass line is always in fixed proportion to that through the main line, providing an exact, linear indication of flow rate. Rangeabilities of this type installation range from 3.3:1 to 10:1.

Magnetic Meters

Magnetic flow meters utilize the principle of Faraday's Law of Induction which states that an electrical potential is developed by the relative motion at right angles between a conductor and a magnetic field. This is the principle used in electrical machinery where the conductor is a wire; the principle is equally effected in the magnetic flowmeter where the conductor is a moving, electrically conductive liquid.

The meter consists of an electrically insulated tube or pipe with a pair of electrodes mounted opposite each other and flush with the inside walls of its tube. Electrical

Figure 4.71. Rotameters installed across orifice plates are termed bypass rotameters and provide economically a linear indication of line fluid flow. (Courtesy of Wallace and Tiernan, Division of Penwalt Corp.)

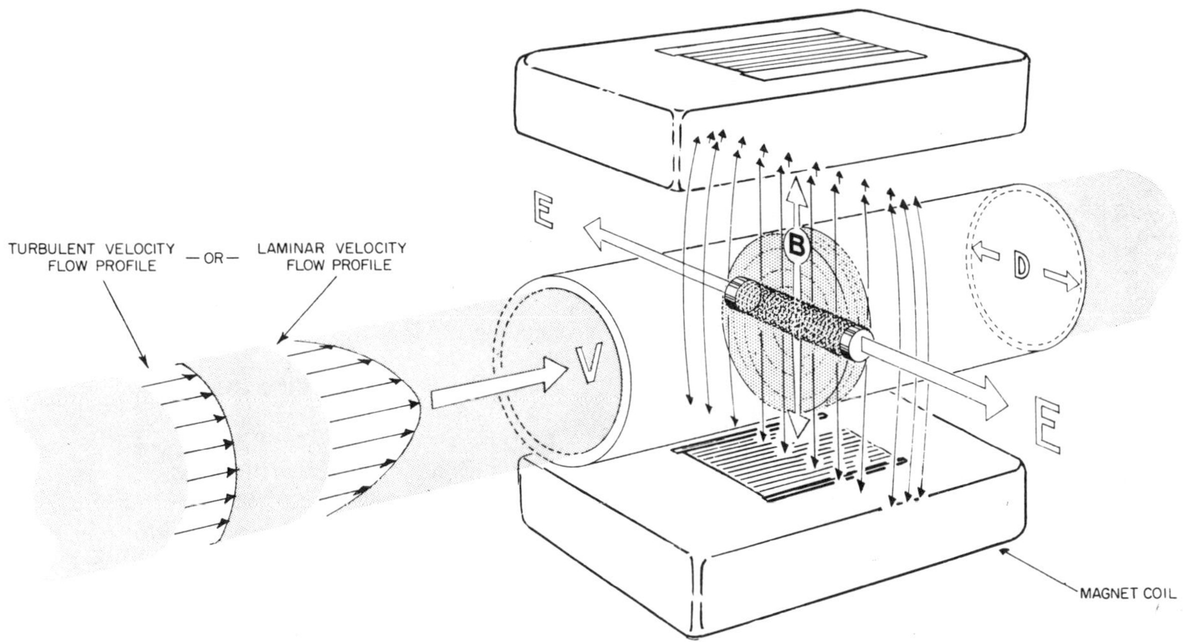

Figure 4.72. The operating principle of the magnetic meter is depicted schematically—a voltage is produced by a moving conductor (fluid) cutting a magnetic field. The conductor, the magnetic field and the produced voltage are mutually perpendicular. (Courtesy of Fischer and Porter Co.)

coils are mounted around the tube so that a magnetic field is generated in a plane mutually perpendicular to the axis of the meter body and to the plane of the electrodes. Figure 4.72 is a schematic representation of the system. The fluid can be conceived as a group of parallel conductors, each at a different velocity because the velocity varies across the entire cross-sectional area. The voltage generated by these moving conductors is the average of the many conductors, however, so that the total instantaneous voltage developed is proportional to the average fluid velocity, producing a true volume measurement of the fluid. Since the output signal (voltage) is proportional to the average velocity, it does not matter whether the flow is laminar or turbulent. The measurement is independent of viscosity, density, temperature and pressure. It is also independent of the conductivity of the fluid once a minimum conductivity value has been reached. The minimum value varies from 20 micromhos down to 0.1 micromho, depending on the capability of the manufacturer's interfacing equipment to handle various signal strengths.

Some of the desirable features of the magnetic flowmeter follow.

1. There are no obstructions to fluid flow.
2. Pressure drop is minimal—no greater than an equivalent length of straight pipe of the same size.
3. Measurement of slurries and of corrosive or abrasive or other "difficult" fluids is easily made.
4. Piping configurations are not critical since the meter measures average velocity.
5. Bidirectional flow can be measured by reversing connections. It can be done manually or accomplished automatically.
6. Meters are unaffected by viscosity, density, temperature, pressure or fluid turbulence.

The following precautions need to be taken:

1. Conductivity must be as high as the minimum required by the particular manufacturer—from 0.1 to 20 micromhos. Stream conditions must be such that this minimum is always maintained or exceeded.
2. The meter must be full at all times because the meter sees velocity as analogous to volume flow rate.
3. Entrained gas bubbles result in measurement error.
4. Care must be exercised in handling the low voltage generated by the meter.
5. Fouling of the electrodes occurs in some fluids, coating them and either reducing or completely eliminating the generated signal. In such cases cleaning methods have been introduced (electrical or mechanical) to keep the electrodes clean and conductive. Piping bypasses are sometimes installed so the meter can be removed from service and cleaned.
6. Versions compatible with the area electrical classification must be provided. Meters are available for Class 1, Group D, Division I.

The early designs of magnetic meters utilized nonmagnetic tubes to carry the flowing fluid and surrounded the

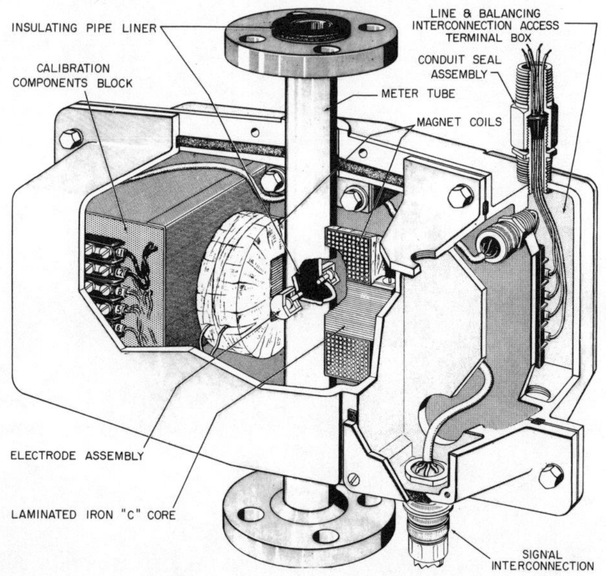

Figure 4.73. Cutaway view of the old style magnetic flowmeter in which laminated iron cores are placed in magnetic coils to help produce the magnetic field needed for flow measurement. (Courtesy of Fischer and Porter Co.)

tube with magnetic coil assemblies and laminated iron cores to produce the magnetic field required (see Figure 4.73). Later designs, which are shorter and lighter, utilize the meter body (which must have magnetic properties) to perform the function of the laminated iron core (see Figure 4.74). The magnet coils are potted, and an insulting liner is used to isolate the coil windings from the conductive process fluid. Magnetic meters are available in sizes from 1/10-inch to over 100-inch diameter, with a wide choice of materials. Accuracy is from ± ½ to ±2%, depending upon calibration and interfacing equipment between the primary device and the receiver.

Mass flow rate can be obtained by making a density measurement of the fluid and multiplying the two signals.

Selection of meter sizes and materials is relatively simple. Manufacturers are able to recommend linings and electrode materials, but the user must evaluate the compatibility between these and the measured fluids. Capacity nomagraphs (Figure 4.75) are available which provide the range of flows accommodated by various size meters. According to the nomagraph, the rangeability for most meters would be 30:1; however, the normally accepted range accommodation is about 20:1.

A recent addition to magmeter designs is one that can be inserted into the line through couplings (Figure 4.76). Built with the electrodes mounted on each side of the probe and with magnetic coils also integral to the probe, it functions as an "inside-out" magmeter. The probe can be mounted on pipes of virtually any size 6 inches and above, can be mounted by conventional hot-tap methods on pipes which remain in service, and can easily be mounted for open-channel flow. Since the meter is essentially a velocity measuring device, the probe must be used in conjunction with a level measurement to determine actual flow rate when used where pipes and channels do not flow full. Level sensors available include bubblers, capacitance probes, and sonic devices.

In pipes the probe is inserted into the line so that the electrodes are ⅛ diameter from the inside pipe wall. At that point, velocity is constant with respect to Reynolds number. Linearities are quoted at ± 1 or ± 2 percent full scale, depending on converter selection.

Turbine Meters

Turbine meters consist of a straight flow tube within which a turbine or fan is free to rotate about its axis which is fixed along the centerline of the tube. The velocity of the flowing stream imparts a force to the turbine blades or rotor which rotate at a speed proportional to flow rate. In most units a magnetic pickup system senses the rotation of the rotor through the tube wall. As each rotor blade passes the magnetic pickup coil, one pulse of AC voltage is induced, each pulse representing a definite flow quantity.

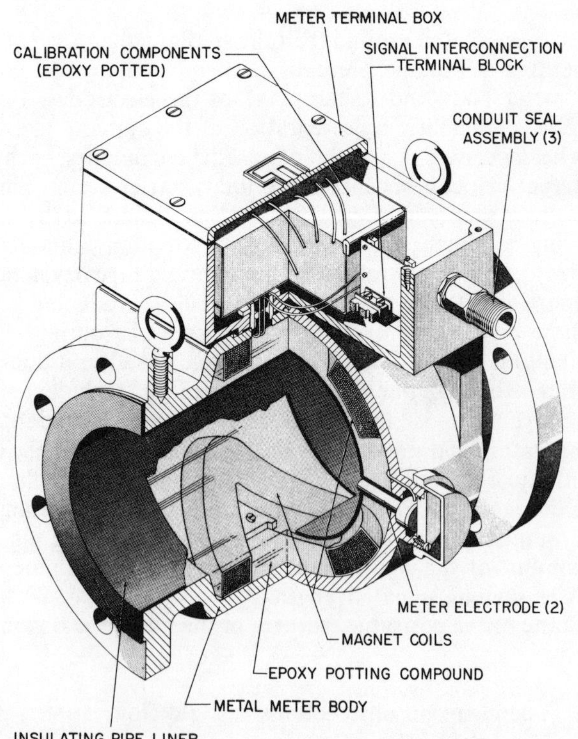

Figure 4.74. Late magnetic meter designs are shorter and lighter, utilizing the meter body to perform the function of the previous laminated iron core and using nonconductive liner to insulate the coil winding from the conductive fluid. (Courtesy of Fischer and Porter Co.)

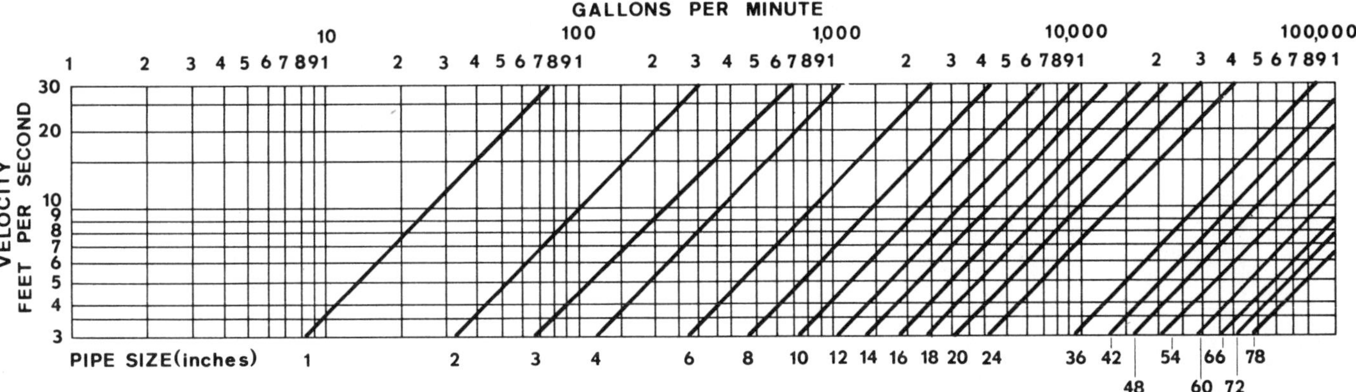

Figure 4.75. Nomagraph shows capacities of various meter sizes for velocity ranges normally encountered. (Courtesy Brooks Instruments)

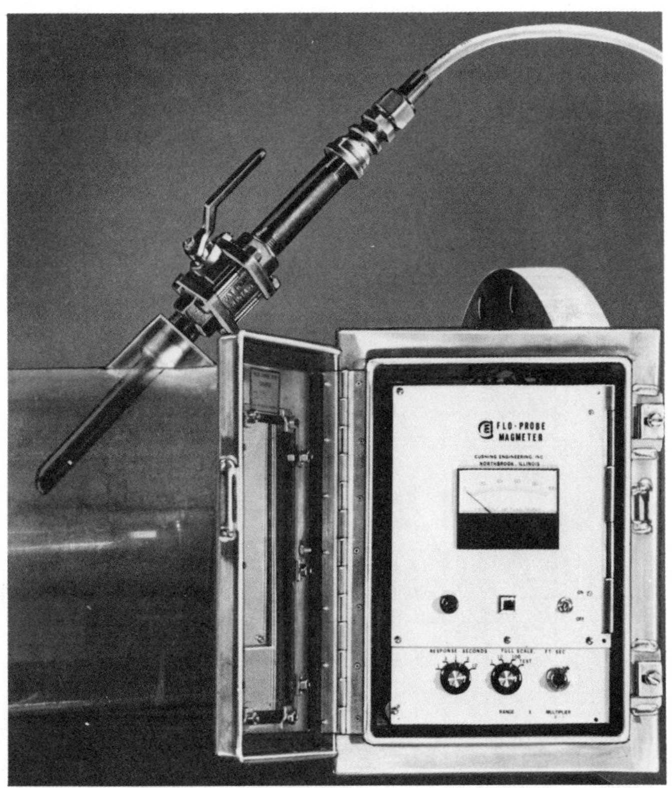

Figure 4.76. An "inside-out" magnetic meter can be inserted into the mine using conventional spot-tap methods. (Courtesy of Monitek, Inc.)

A turbine meter (Figure 4.77) consists essentially of three basic components: the housing, the rotor assembly and the magnetic pickup coil.

One of the major advantages of the turbine meter is its adaptability to flow totalizing. The electrical pulses can be totaled, differenced and manipulated by digital techniques so that a "zero error" characteristic of digital handling is provided from the pulse generator to the final readout. This means that the accuracy of turbine systems is dependent almost entirely on the accuracy of the rotor system in its relationship to flow.

Accuracy is obtained by operating the meter in the linear portion of its range (refer to Figure 4.78). At low fluid velocities (or at low percentages of rated flow), the number of pulses generated per gallon of flow (K-factor) is small and changes appreciably with flow rate. At about 10% of rated flow, however, the K-factor starts to level out and maintains an almost linear relationship with flow rate. The curve shown is called the K-factor curve, and the desired accuracy is chosen by operating the meter in the range where the relationship falls within the desired limits. The limits normally used provide an accuracy range from ± ¼ to ± ½%.

The repeatability of turbine meters is excellent, ranging from ±0.25% to as good as ±0.02% or better according to manufacturers' data.

Figure 4.77. The turbine meter consists essentially of three components—housing, rotor assembly and pickup coil. (Courtesy of Brooks Instruments)

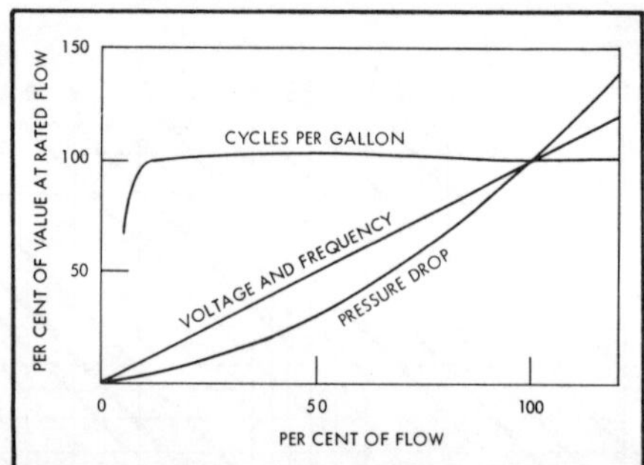

Figure 4.78. From approximately 10% to 100% of rated flow, the turbine meter output maintains a constant relationship with flow. (Courtesy of Brooks Instruments)

The rangeabilities of turbine meters are generally considered to be between 10:1 and 20:1. In low flow ranges, however, it is often less than 10:1. On the other hand, military type meters designed to rigid specifications have achieved rangeabilities greater than 100:1.

The pressure drop across turbine meters varies as the square of the flow rate. Figure 4.79 shows a typical curve of flow rate versus pressure drop for meter sizes ½ through 24 inches. Since the square root relationship prevails between flow and pressure drop, it is often desirable to choose a meter for operation considerably below its rated capacity in order to save pumping costs. This is also desirable from the standpoint of having additional capacity if operating requirements change to higher volumes.

The insertion of a turbine in a flow line poses some obvious problems. The entire mechanism is subjected to dirt, corrosive chemicals, solids that may be in the line and erosive action of the fluid. The rotors are mounted on sleeve bearings, which rely on the lubricating quality of the flowing medium for lubrication; those in nonlubricating services rely on self-lubricated or highly polished surfaces. Sleeve bearings made from nonmetallic materials such as reinforced Teflon, reinforced carbon, carbon graphite and ceramics are often used in nonlubricating and dirty liquid services. Figure 4.80 shows a complete turbine assembly which illustrates the method of rotor mounting on the turbine shaft, the ball bearing providing the low friction contact between the shaft and the rotor. Figure 4.81 shows the rotor of another manufacturer which mounts on a cemented tungsten carbide sleeve.

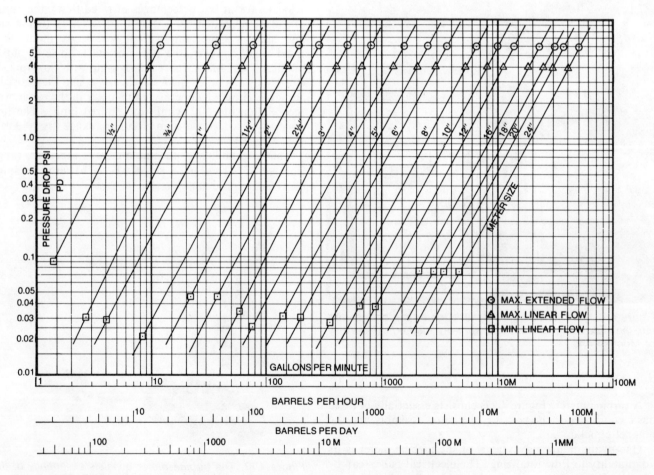

Figure 4.79. Pressure drop characteristics of meter sizes from ½ to 24 inches. (Courtesy of Daniel Industries, Inc.)

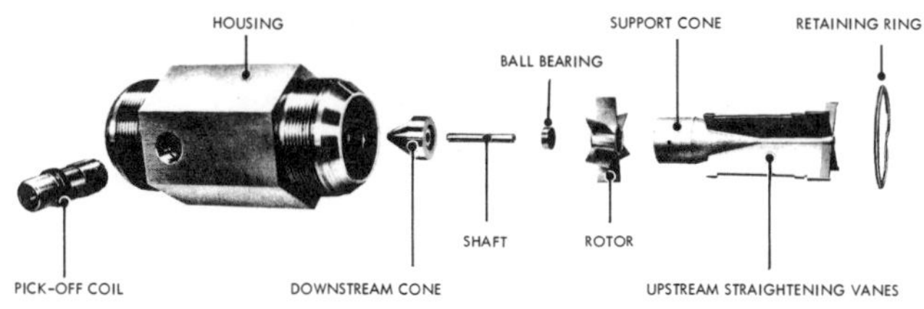

Figure 4.80. The exploded view of the turbine meter assembly illustrates the method of mounting the rotor assembly. (Courtesy of Brooks Instruments)

Figure 4.81. This rotor assembly, shown removed from the body housing, mounts on a cemented tungsten carbide sleeve. (Courtesy of Daniel Industries, Inc.)

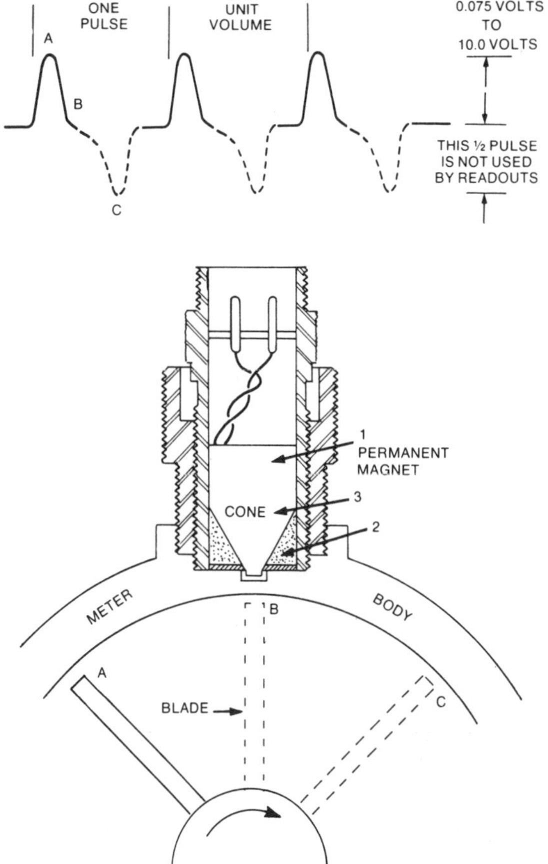

Figure 4.82. Schematic pickup assembly of turbine meter showing major components involved. A typical generated waveform is shown. (Courtesy of Daniel Industries, Inc.)

The output from the magnetic pickup assemblies can be fed directly into digital instruments for indication or totalization, or it can be converted into signals compatible with other analog system devices. Figure 4.82 shows a schematic of a pickoff assembly. When the blade (A) passes the cone point (B), the magnetic field produced by an Alnico magnet (1) is deflected by its presence, generating a voltage in the coil. The voltage builds back in the opposite direction as the leaving blade reaches position C, producing an irregular wave form as indicated. One pulse represents a unit volume so that the turbine output can be rated in pulses per gallon or in any other units desired.

Figure 4.83 illustrates a preferred installation method. Straight runs of pipe upstream and downstream of the meter are recommended for turbine meters as well as head meters. The suggested lengths from various manufacturers vary somewhat; distances shown are fairly typical.

Turbine meters have been used widely for military applications and are finding increasing use in commercial applications. They are used both for liquids and for gas flow measurement. They are particularly useful in blending systems for the petroleum industry and for other industries where clean liquids are mixed or blended in carefully controlled ratios. They are not especially suitable for dirty services although there are some special designs (Figure 4.84) that handle slurries and suspended solids. This type does not have the accuracy of turbine meters in general; as a matter of fact, it is not advertised as a turbine meter. Its principle of operation places it in the turbine meter category, however.

Turbine meters range in size from ¼ to 24 inches and liquid flow ranges from 0.1 to over 50,000 gpm.

They are usually individually calibrated to meet the exacting demands of the services in which they are used.

Target Meters

The target meter measures flow by measuring the force on a disc or "target" centered in the pipe with the plane of the target at right angles to the direction of fluid flow. The flow develops a force on the target or disc which is proportional to the square of the flow.

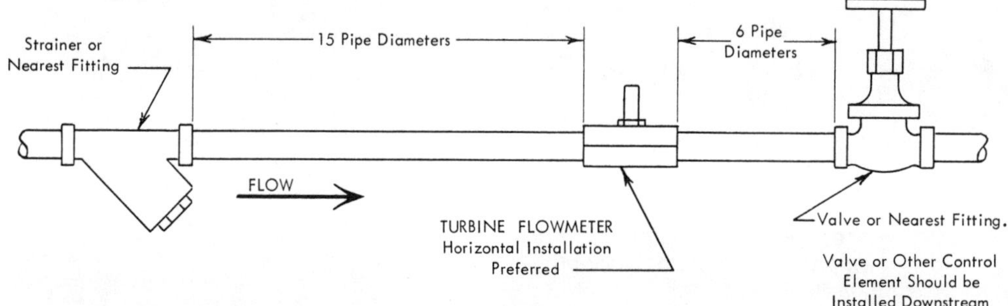

Figure 4.83. Recommendations for turbine meter installations vary; that shown is typical. (Courtesy of Brooks Instruments)

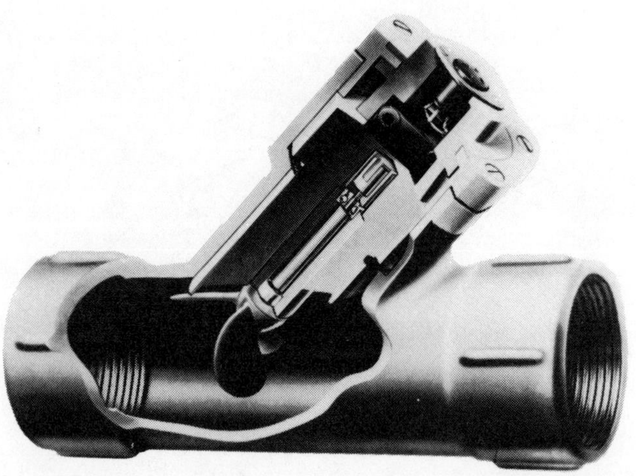

Figure 4.84. The principle on which this meter operates places it in the category of turbine meters. It is called a Y-type transmitter. (Courtesy of Simmonds Precision Instruments)

Figure 4.85. Target meters measure flow by the force the flowing fluid imparts to a disc or target located in the stream. (Courtesy of Foxboro Co.)

Two versions of the principle are available. In one version (Figure 4.85) the disc is mounted on a rod or force bar passing through a flexible seal. The force bar transmits this force directly to a transmitter mechanism (pneumatic or electronic) to deliver consistent, repeatable signals to remote equipment within force limits of 2 to 16 pounds. The relationship between flow rate and force is expressed by

$$Q = K \sqrt{(F)} \tag{4.9}$$

where Q is the flow, K is a known coefficient, and F is a force that can be from 2 to 16 pounds. The accuracy of this device is claimed to be ±½% with proper flow calibration. It is available in 2, 3 and 4-inch sizes, flanged or butt-welded into the line. The ratio of target to pipe size is from 0.5 to 0.8.

The other version of the target meter uses the same principle where the force is expressed as follows:

$$\text{Force} = C_d A \rho \,(V^2/2g) \tag{4.10}$$

where

C_d = drag coefficient
A = sensor area
ρ = fluid density
$V^2/2g$ = velocity head

Bonded strain gauges are used in a four active arm bridge circuit to translate this force into an electrical output which is proportional to the rate of flow squared. The electrical output can be made compatible with any strain gauge signal system.

The accuracy of this device ranges from ±½ to ±3% depending on whether actual or theoretical calibration is used. In-line units are available in tube sizes from ½ to 2 inches and flange mounted units (Figure 4.86) are available for line sizes from 4 to 60 inches (any size can be made).

Target meters are especially useful in measuring heavy viscous, dirty or corrosive fluids. The force balance type can handle pressures to 1,500 psig and temperatures to 750°F. The strain gauge type handles pressures to 5,000 psig and temperatures to 600°F.

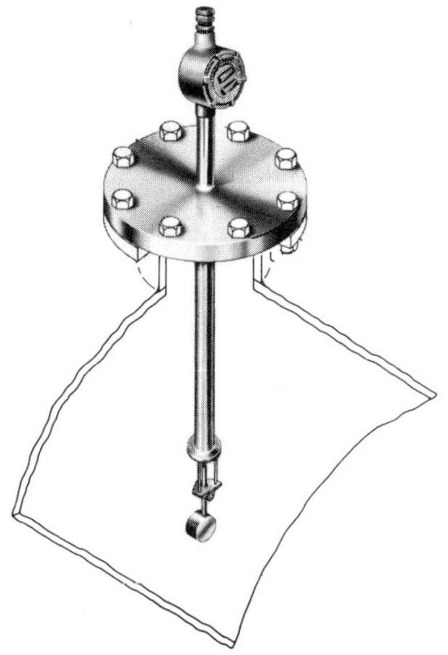

Figure 4.86. Flange-mounted target meter probes are available for line sizes from 4 to 60 inches. (Courtesy of Ramapo Instrument Co., Inc.)

Vortex Meters

Two meters are available that may be classified as vortex meters. One is called a precessing vortex meter or Swirlmeter, and the other uses the term vortex shedding.

Swirlmeter

The principle of vortex precession is used in the Swirlmeter. It is a digital volumetric device which has no moving parts. Its output is in the form of pulses whose frequency is proportional to flow rate. Figure 4.87 shows how the unit is constructed. A swirling motion is imparted to the fluid as it passes the fixed set of swirl blades at the inlet. Downstream of the blades there is a Venturi-like contraction and expansion of the flow passage. In the area where expansion occurs, the swirling flow precesses or oscillates at a frequency proportional to flow rate. This precession of the fluid causes variations in temperature and resistance of a thermistor (sensor) placed in the area, and these variations are converted into voltage pulses which are amplified, filtered and transformed into constant amplitude, high level pulses of square waveform.

The frequency of the pulses are easily measured by an electronic counter or can be transduced for use with other type control instruments.

The deswirl blades at the exit of the meter serve to straighten out the flow leaving the meter. The purpose of the deswirl blades is to isolate the meter from downstream piping effects.

The Swirlmeter has an accuracy of ± 0.75% within its linear operating range of ± 1%. Figure 4.88 shows a typical performance curve for the meter where the linearity is within those limits for Reynolds numbers between 10,000 and 1,000,000. The meter is repeatable to ±0.25% of rate in the linear portion of its operating range, and its rangeability is 100:1.

The meter is currently available in meter sizes from 1 to 6 inches in flanged bodies of 150- and 300-psig ratings. The meters can be used as intrinsically safe devices in hazardous atmospheres of Class 1, Group C, Division I and lower. Fluid temperature limits are from -100° to 500°F. The meter must be preceeded by 10 diameters of straight pipe of the same size as the meter.

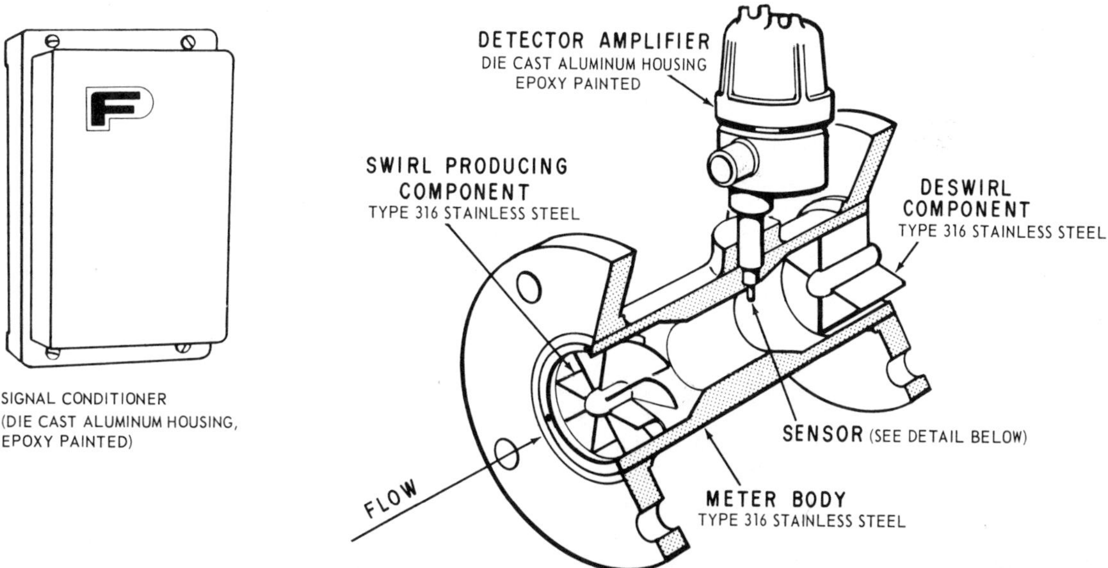

Figure 4.87. The swirlmeter imparts a swirling motion to fluid which causes small temperature variations detected by a sensitive thermistor, the frequency of which is proportional to flow. (Courtesy of Fischer and Porter Co.)

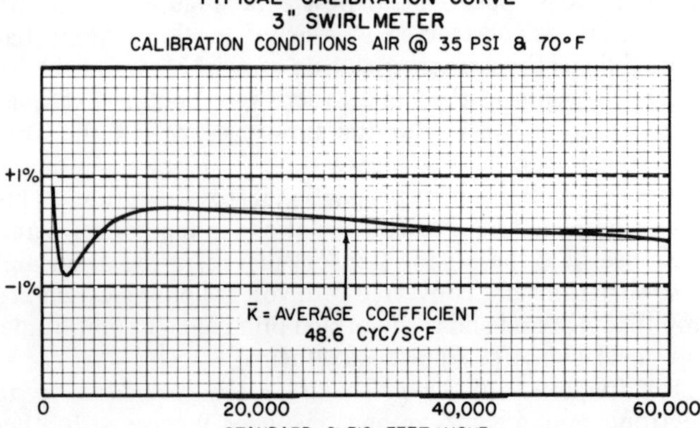

Figure 4.88. A typical swirlmeter performance curve shows a linear response with ±1% for Reynolds numbers between 10,000 and 1,000,000. (Courtesy of Fischer and Porter Co.)

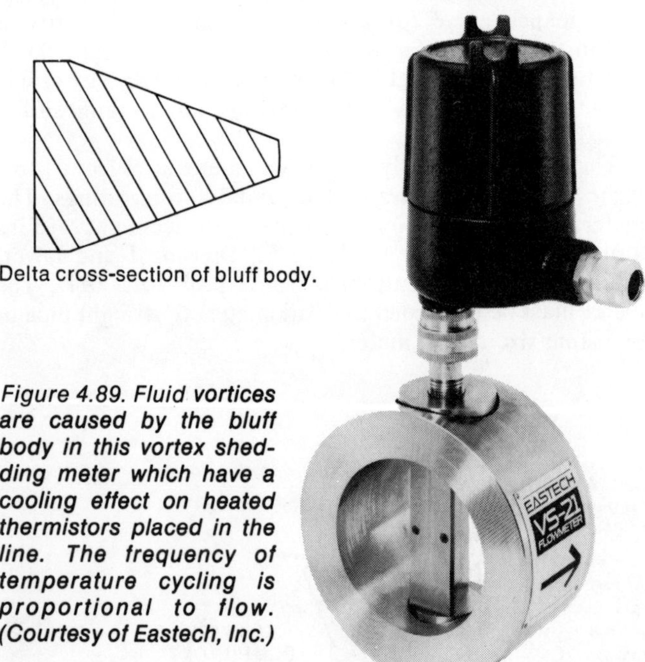

Figure 4.89. Fluid vortices are caused by the bluff body in this vortex shedding meter which have a cooling effect on heated thermistors placed in the line. The frequency of temperature cycling is proportional to flow. (Courtesy of Eastech, Inc.)

Vortex Shedding Meter

Another meter, which operates on the principle of vortex shedding, uses a specially shaped "bluff body" (Figure 4.89) that spans the meter area. It has a delta-shaped cross-sectional area (see insert to Figure 4.89) and is installed with the base facing upstream. Fluid vortices are formed against the bluff body and are shed off its downstream faces in a regularly oscillating fashion. The frequency of oscillation is directly proportional to volumetric flow rate for either liquids or gases.

Vortex shedding frequency is sensed by a pair of thermistors embedded in the upstream face. They are heated electrically to a temperature above that of the flowing stream and sense the cooling effect of the vortex shedding by changing temperature (and resistance) at the vortex shedding frequency which is proportional to flow rate.

Another bluff body and sensor configuration employs a single removable thermistor set in a passage drilled through the flow element, and features easy replacement. For large pipes, the heated thermistor is located externally in a passage connecting sensing taps on opposite sides of the flow element. When vapor transients give measurement problems, a bluff body can be used which has an electromagnetic pickup sensing the motion of a small shuttle ball or cylinder. Obviously, this latter device should be employed in clean fluids *only*. Models are available for insertion into the line by conventional hot tap methods, and, using an oscillating disc design, for measurement of steam through 500 psi.

Readout is available in digital form or may be converted to standard analog signals for rate, totalized flow or control applications. The meter is available in 2, 3, 4 and 6-inch sizes with end connections up to 600-psig ratings. Fluid temperatures may range from 0° to 400°F.

Linearity is within ±½% and rangeability is 100:1. Viscosity effects are negligible for Reynolds numbers above 10,000 on the minimum side; limits on the maximum side are based on fluid velocity, and 20 ft/sec is given as a maximum for liquid flows while 100 ft/sec is maximum for gas flows.

This meter also has no moving parts. Piping configuration requires that upstream and downstream lengths conform to A.G.A. Standards for orifice meters.

Operation of either the Swirlmeter or the Vortex Shedding Meter would be impaired by materials coating the temperature elements and affecting the rate of heat exchange between the elements and the flowing fluid.

Sonic Meters

Many flow measurements can be made by sampling velocity measurements in the pipe. When velocity profile information is obtainable or can be reasonably predicted, an average velocity multiplied by pipe cross-sectional area gives volumetric flow rate. Several types of sonic flow meters are currently in use. They fall primarily into three categories: broadband noise sensors, and invasive (wetted) and non-invasive (externally mounted) Doppler shift devices.

Broadband noise, from audible frequencies to high ultrasonic, is generated whenever fluids or granules move. The sound energy at the low end of the spectrum usually also includes sounds from mechanical pump vibrations, etc., but the upper frequency energy gives a good analog for flow rate. A sensitive microphonic sensor is attached to the pipe, and signals are electronically integrated to provide measurement. Repeatability of values is about 1%, and calibration

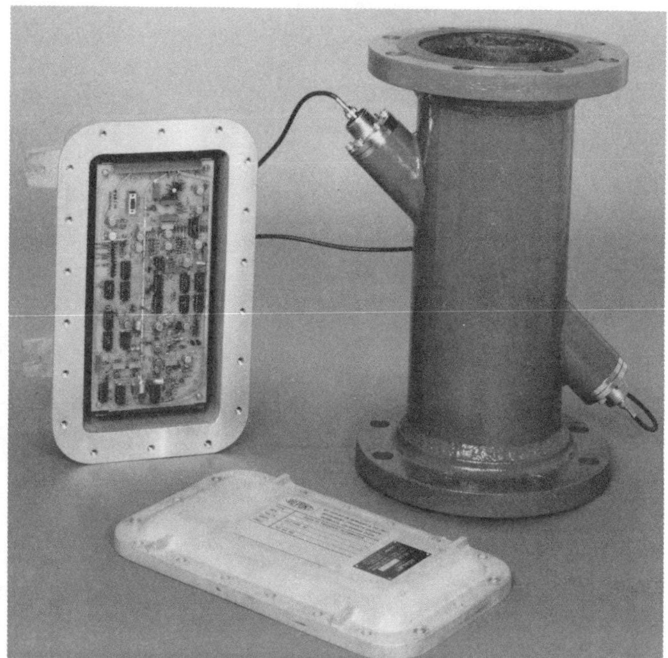

Figure 4.90. The sonic flow meter uses the Doppler shift technique for high accuracy measurements. (Courtesy of E.I. du-Pont de Nemours and Co.)

should be done after installation. Location of sensors should be as far away from pumps and compressors as is feasible; on or near elbows and flanges is often desirable. Both readout and switching functions are available.

The Doppler shift technique allows flow measurement accuracy within ± 0.5% FS. Sensors are installed diagonally along the pipe (Figure 4.90), and signals are sent through the flowing medium to measure the shift in frequency caused by fluid velocity. By measuring with two sets of sensors, or with sensors which both transmit and receive, a differential value can be obtained which is a function of fluid velocity only.

True flow rate measurement is dependent on having a known (and hopefully symmetrical) velocity profile within the pipe. Sonic beams average velocities in their beam envelope, and therefore would give less-accurate readings in non-symmetrical velocity profiles. Multiple-beam meters, while available, are very expensive. In cases where profile is symmetric with respect to pipe axis, single-beam sonic meters can provide ± 1% accuracy or better when measuring a clean stream.

Sensors are generally mounted through angled couplings in the pipe wall. Sensor tip does not extend into the fluid stream, so the pipe can be "pigged" clean without damage to the sensor. Sensors can also be installed in tees at the ends of short runs of pipe (Figure 4.91) for improved measurement of low Reynolds number fluids.

Non-invasive sensors attached to the outside pipe wall can be used; these send sonic pulses through the pipe walls and the flowing medium. This type of sensor can, of course,

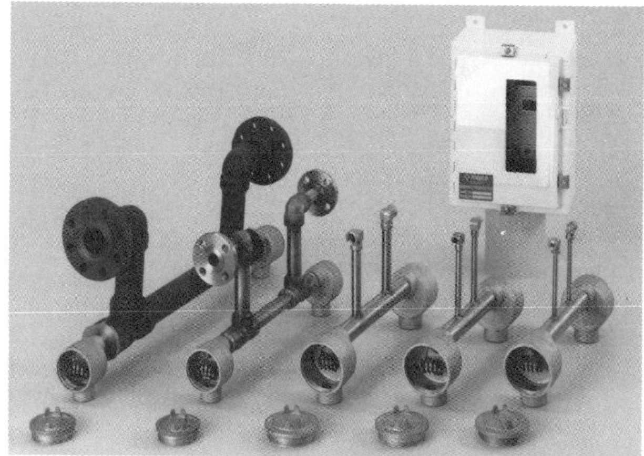

Figure 4.91. Directing the sonic pulse along the axis of flow improves the measurement of fluids with low Reynolds numbers. (Courtesy of Mapco, Inc.)

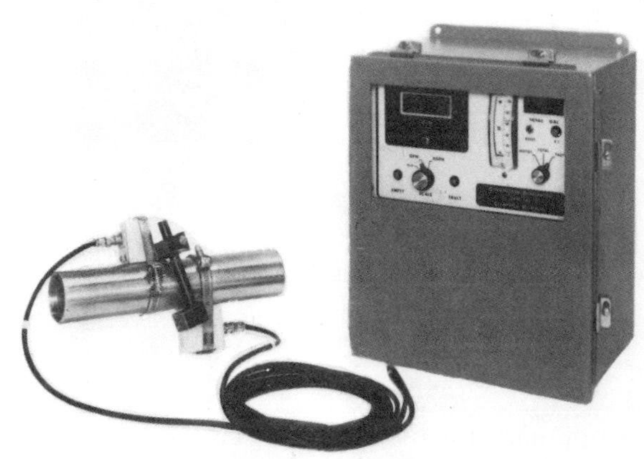

Figure 4.92. Non-invasive sonic meters can be installed without interruption to the process. (Courtesy of Controlotron Corp.)

be installed without interruption of the process (Figure 4.92). Some sensors have both send and receive sensors mounted in the same clamp-on housing. Sound waves must be reflected from air or gas bubbles or particulate matter in the flowing medium, and therefore a clean fluid gives a negligible or "zero" reading.

Sonic meters can also be used to measure specific gravity. Sound travels at a given rate in a fluid (liquid or gaseous) for specific conditions (temperature, pressure, and specific gravity). Sonic propagation is sufficiently different from fluid to fluid to allow detection between grades of gasolines, for example, between leaded and unleaded gasolines. This type of device can be employed to detect the interface between fluids in a pipeline. The probe, which also houses temperature and pressure sensors for compensation, can be installed in the line through a flanged connection (Figure 4.93), can be recessed for cleanout, or inserted and

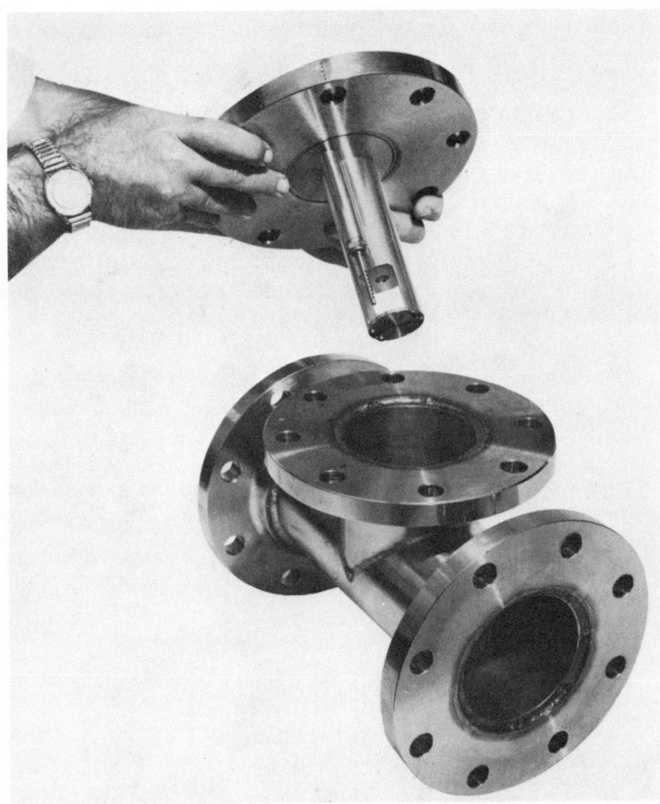

Figure 4.93. Pipeline interface, measuring sonic velocity in the flowing fluids, can be installed for a flanged connection. (Courtesy of Mapco, Inc.)

retracted through usual hot-tap methods. Normal paraffin buildups do not seriously degrade performance. Repeatability is in the order of 0.01%.

Positive Displacement Meters

Positive displacement meters are devices that separate a flowing stream into individual volumetric increments and count these increments. The meters are machined such that each volumetric increment is accurately known, and the summation of the increments give a very accurate measurement of the total volume passed through the meter.

Most positive displacement meters are mechanical types used primarily to measure total quantities of fluids to be transferred. They can be equipped with auxiliary devices for flow indication, on-off control of predetermined quantities, for remote readout, ticket printout, control, computer input, etc.

Positive displacement meters are mainly used for accounting applications or in batch processes. In batch operations fixed quantities are usually fed into the process; when a predetermined amount is reached, the flow automatically shuts off. The accuracy of these meters may be within ±0.10% of full span of the meter. They may be constructed of practically any desired metals. However, use of exotic or difficult-to-machine metals tend to make their cost prohibitive for use in most industrial applications.

Figure 4.94. The nutating disc type positive displacement meter has a disc mounted on a ball bearing which moves a specific volume of fluid with each complete oscillatory motion it makes. (Courtesy of Ametrol Division of Hersey Products Inc.)

Positive displacement meters may be classified by the movement of the metering element. Included are nutating disc, oscillating piston, rotating types and reciprocating piston types.

Nutating Disc

The nutating disc type (shown in Figure 4.94) is used in many industrial services and is practically the only type used for residental water service measurement.

The disc employed in this meter is mounted on a ball bearing. It is slotted radially and the slot is fitted over a partition plate in the body of the meter. The motion of the disc is controlled by a pen from the center of the ball in contact with a roller at the top of the body. The disc fits into the body with very little clearance so that leakage past the disc is negligible.

As liquid passes through the meter, a wobbling or nutating motion is imparted to the disc. For each nutation of the disc, a specific volume of liquid passes through the body. The pen on the top of the disc causes a shaft to turn which is coupled to an external register, or totalizer. Thus, the quantity may be displayed on a counter type indicator, or it may be transmitted to a remote readout.

Oscillating Piston Meter

An oscillating piston meter equipped with an automatic shutoff valve is shown in Figure 4.95. The internal makeup

Figure 4.95. An external view of an oscillating piston meter shown with totalizer and automatic shutoff valve. (Courtesy of Neptune Meter Co.)

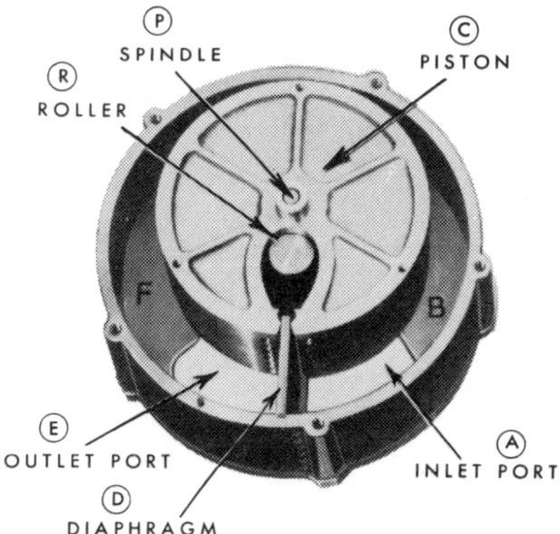

Figure 4.96. Internal makeup of the oscillating piston type meter shown in Figure 4.90. (Courtesy of Neptune Meter Co.)

of this meter is shown in Figure 4.96. It is very similar in operation to the nutating disc type except that the measuring device is a split ring which oscillates in one plane only.

The operational cycle of the meter is shown in Figure 4.97. Liquid enters through inlet port A into chamber *B* (see positions 1 and 2) pushing the piston C in a counterclockwise oscillating movement. This movement forces the liquid in space *F* to be discharged through port *E* as shown in position 2. In moving from position 2 to position 3, liquid enters the space between the inner chamber and the inside wall of the piston to give a continuous smooth motion to the piston.

The piston is guided in its oscillating motion by the diaphragm *D* and by the spindle *P* which rotates in the channel between roller *R* and the inside chamber wall.

Each oscillation delivers a very precisely measured volume. Due to the very small clearances between the rotating and stationary parts, there is practically no leakage. The small amount of leakage which does exist may cause an appreciable error at low flows but not at full rated flow.

Rotating Meters

The rotating positive displacement meter may be constructed in several different arrangements. Figure 4.98 shows the arrangement of a lobed impeller type in which two machined lobes rotate within a chamber. The lobes are kept in alignment by external gears. For each rotation of the lobes, an accurately measured volume passes through the meter. As may be seen in Diagram 1 of Figure 4.99, the pressure differential on the upper gear will cause it to rotate clockwise. The lower gear is balanced but rotates since it is geared to the upper gear. As the gears rotate, the volume trapped between the upper gear and the upper section of the case is passed into the discharge section of the case as shown in Diagram 2. In Diagram 3, the pressure of the incoming liquid causes the lower rotor to rotate. The volume trapped between the rotor and the case is passed on downstream. Thus, two accurately metered volumes have been discharged for each complete revolution of the gears. The application of

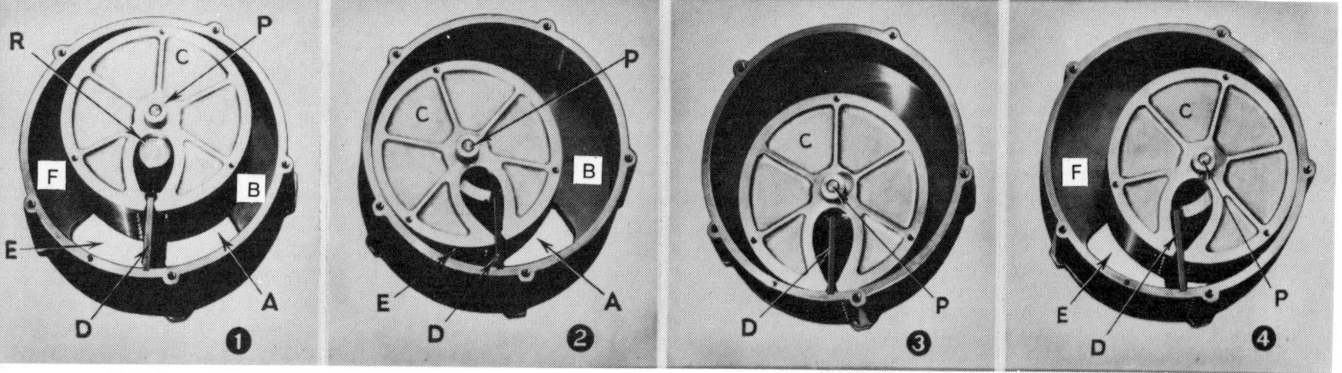

Figure 4.97. Schematic views 1 through 4 show the oscillatory movement of the oscillating piston meter shown in Figures 4.95 and 4.96. (Courtesy of Neptune Meter Co.)

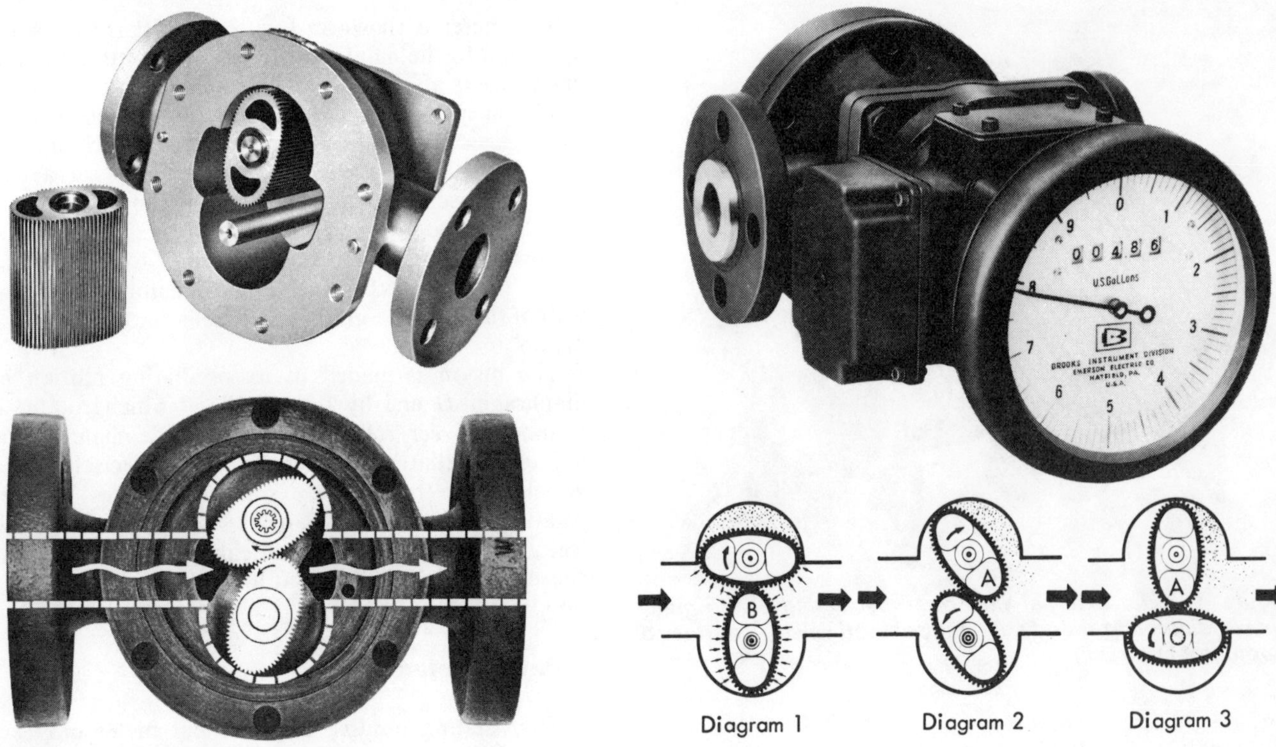

Figure 4.98. Lobed impellar type rotating meter delivers an accurate volume of fluid with each lobe rotation. (Courtesy of Brooks Instruments)

Figure 4.99. Pictorial operation of the lobe oval meter. (Courtesy of Brooks Instruments)

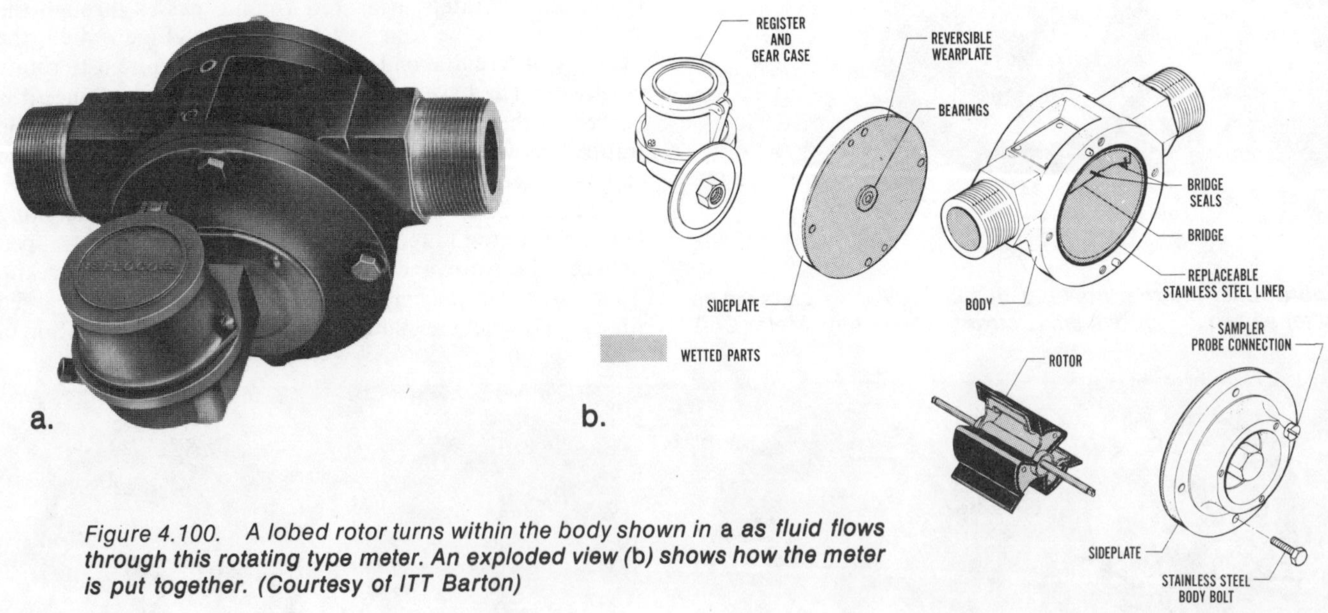

Figure 4.100. A lobed rotor turns within the body shown in a *as fluid flows through this rotating type meter. An exploded view (*b*) shows how the meter is put together. (Courtesy of ITT Barton)*

the proper volume factor multiplied by the number of rotations gives the total quantity metered.

Another type of rotating meter used for water, oil and mildly corrosive liquids is shown in Figure 4.100. Figure 4.101 depicts schematically the fluid path through the meter. As liquid enters from the left, it fills the cavity between the lobes of the rotor. Since the ends of the lobes are at full length at the bottom of the case and are folded back at the top, there is a greater area exposed to fluid pressure at the lower section than at the upper. Therefore, the rotor rotates counterclockwise. As the lobes reach the bridge at the top of the case, they are bent over, thus reducing the volume trapped between each lobe. This volume is forced out of the meter discharge port at the right. Each rotation of the rotor

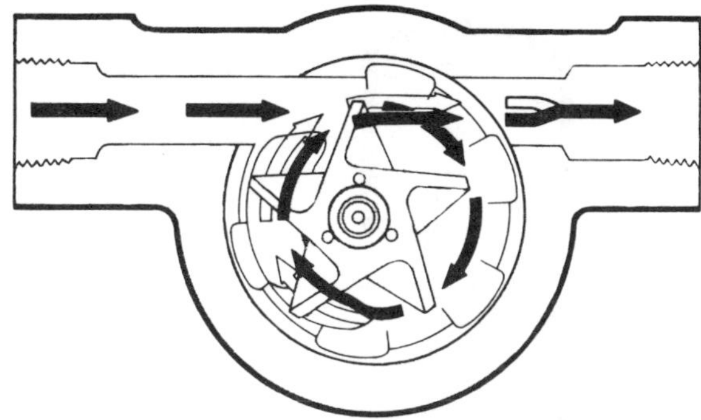

Figure 4.101. Schematic view of the operation of the rotating meter shown in Figure 4.95. (Courtesy of ITT Barton)

gives a real close-fitting, accurate measurement of flow due to the precise, close-fitting of the operating parts of the meter.

Reciprocating Piston

This meter is very similar in construction to a reciprocating steam engine piston and cylinder. The construction of one reciprocating piston type is shown in Figure 4.102. The fluid to be measured enters the right side of the body, forcing the piston to the left until the cylinder is full and the piston is in its extreme left position. At this point, an external leakage causes both slide valves to move. Liquid then enters the left cylinder forcing the piston to its extreme right position. When the cylinder fills, the slide valves again move and the cycle is repeated. The external arm that causes the slide valve to move also drives a counter which provides a total of the fluid quantity that has passed through the meter.

The reciprocating piston meter is not widely used. While it is an accurate meter, it produces a pulsating flow when used for liquid measurement. A variation of this type, using bellows instead of the piston and cylinder, is widely used to measure natural gas to residences and commercial establishements.

Miscellaneous Types of Flow Meters

Several other types of flow meters are mentioned briefly to make the reader aware of the many methods of measurement that can be used. Some have been used rather extensively in military applications and to a much lesser extent in process applications. Some of these find application only in research and development work. A few techniques described have been on the market for only a short time and the full extent of their applicability is still unknown.

Thermal

Thermal meters may be divided into several categories. One category is the hot-wire type. One such system (used for gas flow measurement) places two thermistors in the flowing stream; one is heated by an external source to a

Figure 4.102. A reciprocating piston type meter provides accurate measurement but pulsating characteristic is objectionable in liquid service. (Courtesy of A.O. Smith Corp.)

temperature about 150°F above ambient while the other is unheated. The unheated thermistor serves to compensate for temperature changes in the measured gas. Energy is removed from the externally heated thermistor as a function of mass flow rate. Removal of energy causes cooling, resulting in a changing resistance and voltage across the thermistor. A nonlinear signal voltage is produced which is operated on by a function generator to produce an output voltage which is linear with mass flow.

This particular device has good repeatability, is accurate to ±2% of span from 20 to 100% of full scale and is available for flows ranging from 0.12 to 0.6 scfm to 100 to 1,000 scfm.

Another type of thermal flow meter is one which supplies a constant amount of power (heat) to a gas stream. The temperature of the gas is measured upstream and downstream of this heat source by two resistance thermometers. The temperature rise of the gas stream is a function of the mass flow rate, the amount of heat added and the thermal properties of the gas. The latter two values are known or must be known. The other value—the difference in temperatures—is measured by the two resistance thermometers and is linearly proportional to mass flow.

The unit described is available for pressures to 150 psig at temperatures from 40° to 110°F. Accuracy and linearity is ±2% of full scale. Ranges are from 0 to 5 scfm to 0 to 5,000 scfm.

A thermal meter used for measuring flare gas flows consists of a stainless steel probe constructed with two openings at the probe tip (Figure 4.103). These openings are connected internally by a smaller stainless steel tube. A portion of this tube is heated and thermoelectric sensors measure temperature gradients along the wall, external to the flowing stream.

Purge air is injected into the tubing which forms a pneumatic bridge. At zero line velocity the bridge is

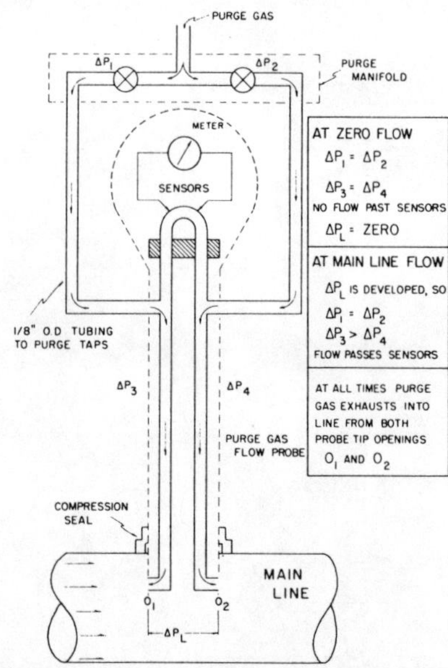

Figure 4.103. A thermal flow measuring device, utilizing a pneumatic purge system whose pneumatic bridge is unbalanced as a function of flow rate through the main line. (Courtesy of Hastings-Raydist, a Teledyne Co.)

balanced so that no flow occurs through the sensing portion of the tube. The purge air exhausts from both openings equally at the probe tip.

As flow across the tip occurs, a differential pressure develops, unbalancing the bridge and causing a small amount of purge air to flow through the sensing section. It exhausts from both openings, but the amount is slightly unequal because of the differential pressure. The thermoelectric sensors measure the shift in temperature gradients along the heated portion of the tube which is a function of the main gas flow creating the differential pressure.

Calibration of this device is nonlinear, a function of the square root of the ratio of densities of the purge gas to the line gas. This is a velocity measurement, and the measurements must be related properly to line size.

Solids Flow Meters

Solids flow measurements are usually made on a weight basis, and these type measurements are discussed in Chapter 8. However, volumetric systems have been developed which offer good accuracy at reasonable cost.

One device measures mass flow rate by measuring the force generated by the weight of the material passing through an accelerator. Material enters the meter inlet (Figure 4.104) and is funneled through the accelerator which rotates at a constant speed. As material flows through the accelerator, its velocity is increased. The weight of the material is determined by the force required to reach the constant exit velocity. This measurement principle is shown by the commonly accepted formula:

$$F = MA \tag{4.11}$$

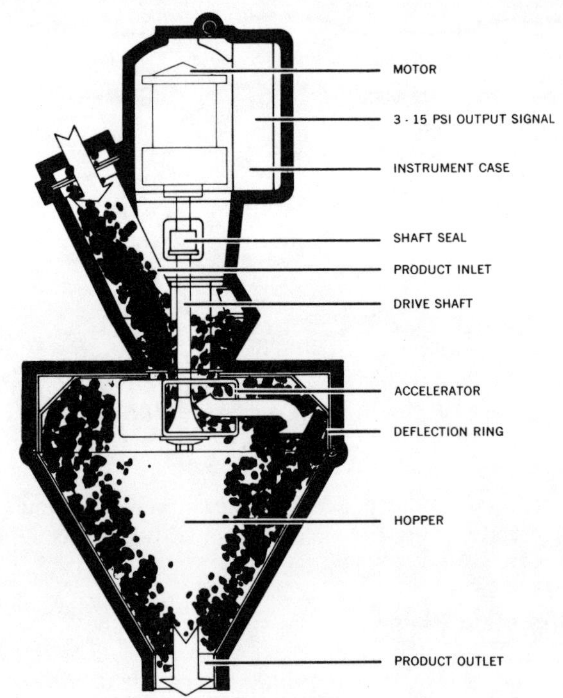

Figure 4.104. This solids flowmeter measures flow by measuring the torque necessary to maintain constant exit velocity of the solids material. (Courtesy of Dezurik Corp.)

where the force (F) generated by increasing the material to a constant exit velocity is equal to material mass (M) times the amount of acceleration (A). Since acceleration is constant, flow rate (or mass) is proportional to force which is represented by the torque at the drive motor.

Torque is measured by a torque transducer which senses movement of a torque arm bolted to the synchronous drive motor. The output of the torque transducer is a standard 3- to 15-psi signal directly proportional to mass flow rate. The signal may be used for rate indication, recording, totalizing or controlling.

The device is accurate to $\pm\frac{1}{2}\%$ and has a rangeability of 25:1, measuring flows from 1.6 to 40 tons per hour.

Mass Flow Calculation

Although a few devices are offered which measure mass flow directly, they are more often made by combining volumetric measurements with density or temperature and pressure measurements. Several manufacturers offer these combinations as complete systems. Sometimes more convenient than specifying individual components to make up a system, this places system responsibility at one source. Because of the many variations in systems offered, no attempt will be made to discuss any particular system. Accuracies claimed for these systems are usually about the same as those for the primary element used.

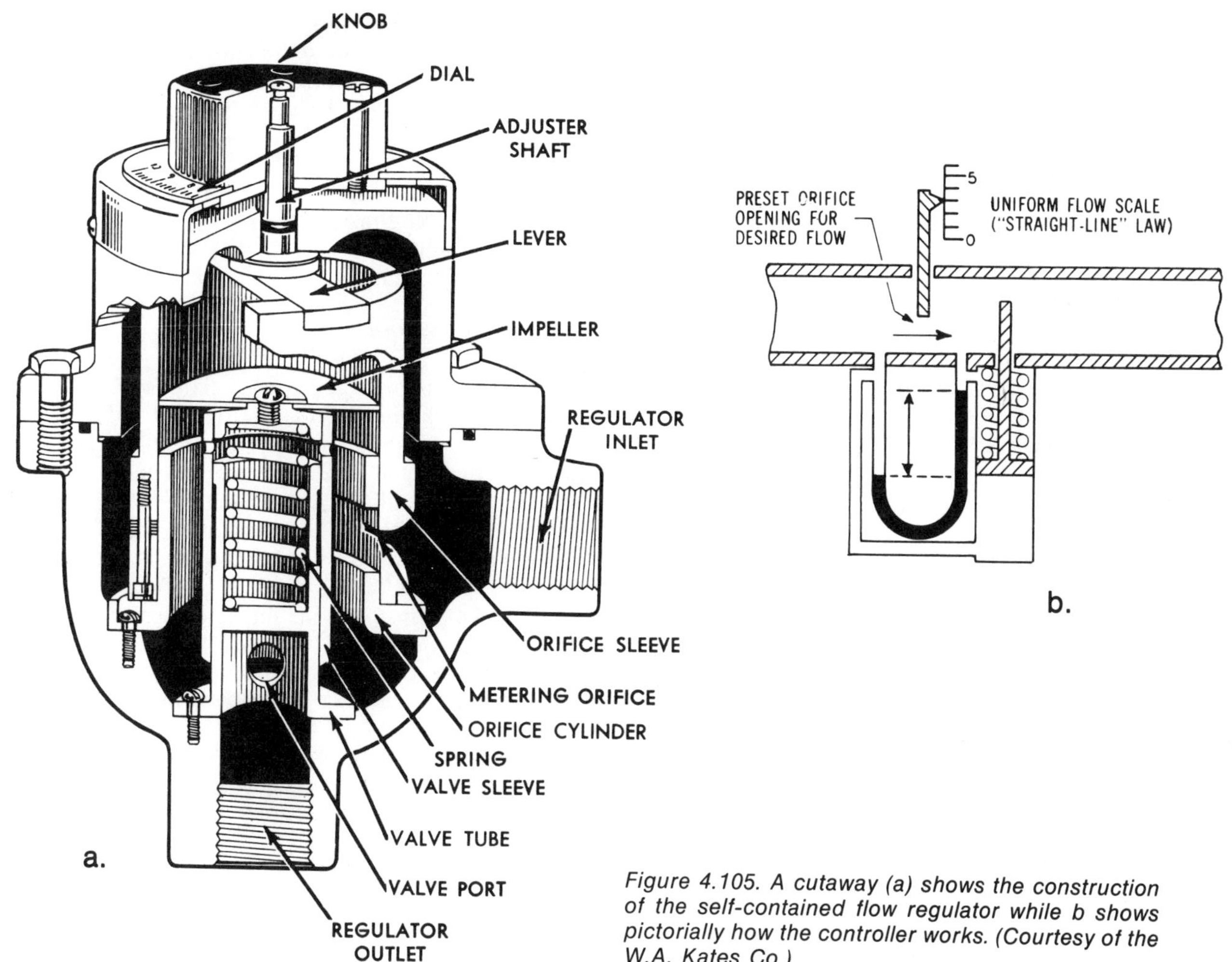

Figure 4.105. A cutaway (a) shows the construction of the self-contained flow regulator while b shows pictorially how the controller works. (Courtesy of the W.A. Kates Co.)

Self-Contained Flow Regulators

The flow devices discussed previously are used only to measure flow. In order to provide a closed loop control system, two other components are usually required in conjunction with the measuring device—a controller and a control valve. There is a device which accomplishes all three purposes—a self-contained flow regulator. It combines an adjustable orifice with an automatic internal regulating valve. Indication of rate is on a calibrated dial which adjusts the orifice or the desired flow rate. The dial is calibrated directly in gpm (used only in liquid services).

Figure 4.105 shows the construction of the regulator and illustrates schematically how the orifice is adjusted. The orifice is rectangular, narrow and long so its area is essentially proportional to its length, thus providing a linear relationship between scale and area—hence, flow rate. The regulating valve is positioned by the differential pressure across the orifice; it maintains a constant differential pressure, so flow rate is determined by the size of the orifice.

The orifice is a slot in a cylindrical sleeve which surrounds a second sleeve similarly slotted and may be rotated around it. When the slots coincide, the metering orifice is at its maximum opening; at any other position, the orifice area is directly proportional to the angle of rotation.

The regulating valve is within the orifice sleeve combination. It includes a valve sleeve which slides on a valve tube to open or close the valve ports. The pressure differential across the orifice drives an impeller disc downward and moves the valve sleeve to close the ports. The force of a spring opposes this closing action. The position of the sleeve results from the direct balance of the two forces, and it sizes the valve port openings to cause the force balance. The force balance holds the orifice pressure differential constant, regardless of the pressures upstream or downstream (for pressure variations as high as 120 psi) or the size of the orifice.

These flow regulating valves are available in sizes from ½ to 4 inches in cast steel or stainless steel. Flow rates are available from 0.02 to 0.3 gpm to 100 to 550 gpm. There is a

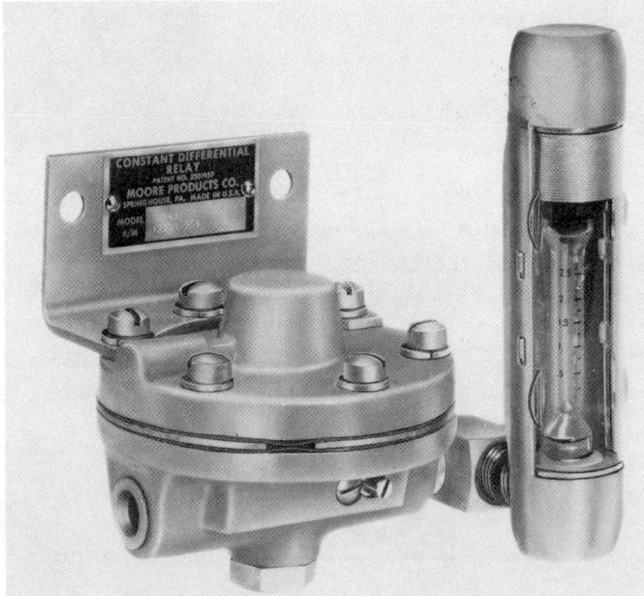

Figure 4.106. A differential regulator combined with a rotameter is used as a constant-flow purge meter. (Courtesy of Moore Products Co.)

minimum threshold pressure at which the valves begin to function as described; this pressure varies from 8 to 22 psig for the sizes listed. The units have an accuracy of ±1½% of set point and have rangeabilities up to 15:1.

Units are also available on which the flow rate can be set or changed remotely, using pneumatic or electric operators.

For purge control, regulators are available which hold flow rate constant through a small rotameter. Often, they are integrally mounted (Figure 4.106). Flow rates are accommodated to about ½ gpm. Units can be selected that will give a constant flow rate, dependent on which varies, upstream or downstream pressure. If upstream pressure varies, one model is selected; if downstream pressure varies, a different model is selected.

Metering Pumps

Metering pumps are used to provide a desired rate of flow to a process or to a blending operation. They operate on a positive displacement principle (somewhat similar to positive displacement meters) and are usually provided with some means of setting the desired flow rate. When the rate setting mechanism is positioned by a controller output, it becomes the final control element of an instrument control loop. Pumps provided with a manual control unit may be set to feed a (constant) desired rate of flow by adjustment of the control.

Any positive displacement pump may be used as a metering pump. However, only pumps which have very little internal leakage are used for metering purposes. Leakages that do occur vary with the differential pressure across the pump, causing the discharges rates to vary accordingly.

The discharge flow rates from metering pumps may be changed by variations of drive motor speeds, adjustable speed transmissions or stroke adjustment.

Metering pumps may be classified by their method of pumping:

1. Rotary gear pump
2. Reciprocating piston pump
3. Diaphragm pump
4. Peristaltic pump

Rotary Gear Pumps

A rotary gear pump using gears with many teeth is shown in Figure 4.107. Liquid enters the left chamber, filling it and the space between the exposed gear teeth. The upper gear rotates clockwise and the lower one counterclockwise. As they rotate, liquid filling the voids between the teeth is confined between them and the outer case of the unit and is forced out the discharge port at the right side of the pump as the teeth mesh together.

Another pump using what might be termed gears with only two teeth is shown in Figure 4.108. It operates as described above, except that the void between the teeth is much larger and a greater quantity of liquid is discharged per rotation.

There is some slippage in gear type pumps since the gear teeth cannot be in actual contact with the case. Clearance is very small, however, and if the pressure differential across the pump is low, slippage will be minimal. By incorporating two pumping units separated by a transfer plate with a poppet valve and bypass system, the pressure is equalized across the second pump and accuracies of ±½% of span are achieved. By using this arrangement, there is no pressure differential across the second pump, thereby, eliminating slippage or back flow.

Figure 4.107. This ge[ar] pump has an arrang[e]ment of multiple-too[th] gears which provide constant discharg[e] rate with little pulsatio[n] effect. (Courtesy [of] Zenith Products Co.)

FRONT VIEW OF METERING SECTION WITHOUT ROTORS
FRONT VIEW OF METERING SECTION WITH ROTORS

Figure 4.108. Gear pump with special two-tooth gears is capable of larger flow capacities than multitooth gear pumps. (Courtesy of Waukesha Foundry Co., Inc.)

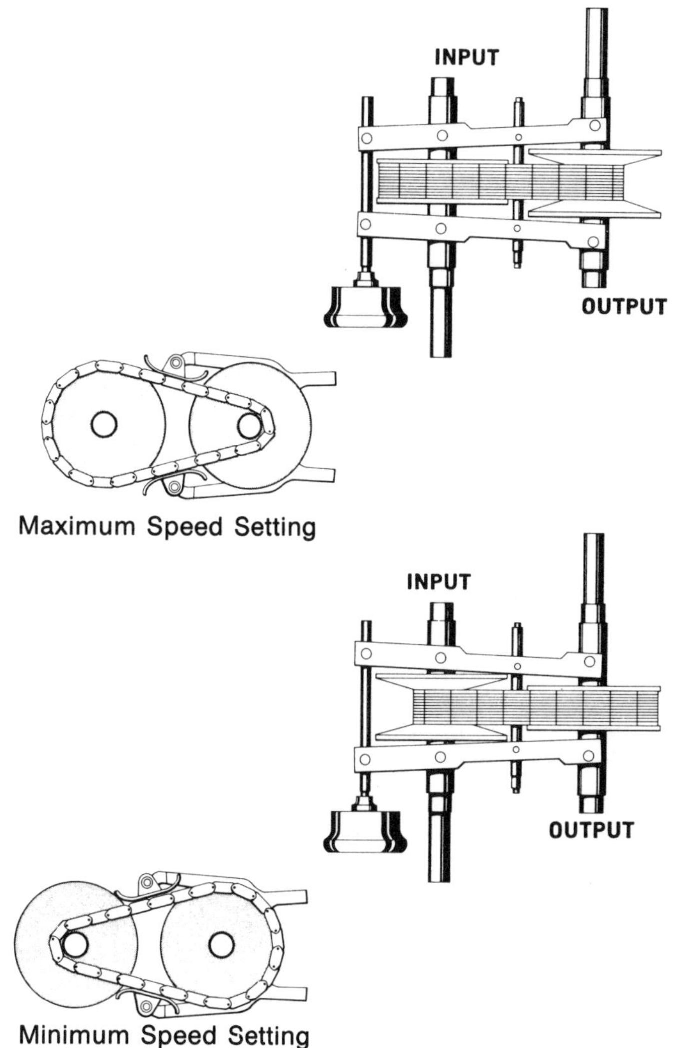

Figure 4.109. Variable speed drives may have speed ranges as high as 6:1. (Courtesy of Fairchild Industries)

The flow rates from gear pumps can be varied only by changing the speed of rotation of the gears. A cutaway view of a variable speed transmission which may be used for this purpose is shown in Figure 4.109. This device consists of two pulleys with grooved, tapered wheel faces connected by a metal chain which rides on the tapered faces. The tapered pulleys are connected by levers which will pull the sides of one pulley together while, at the same time, they spread the sides of the other pulley apart.

At maximum speed setting, the input shaft wheels are close together, forming a larger driving diameter. The output, or variable speed, shaft wheels are automatically spread apart to form a small driven diameter.

Minimum speed setting spaces the input shaft wheels far apart, forming a small driving diameter. Thus, the output shaft wheels are automatically closed to form a large driven diameter.

When a gear pump with very little slippage is driven by this type speed reducer, very repeatable flow rates may be obtained for each setting of the speed adjustment screw. Flow rates must be determined by actual calibration and should be checked frequently. However, gear pumps can be used with satisfactory results for most industrial applications.

Remote control can be obtained on a variable speed transmission by using a reversible gear motor with a slow output speed to drive the adjustment screw.

The drive motor may be electrically, pneumatically or hydraulically driven. The feed rate indication is usually transmitted by a speed transmitter since flow rate is directly proportional to revolutions per unit of time.

Gear pumps are suitable for clean liquid streams. The teeth mesh so closely that particles entering them may cause freezing and damage to the gears. Therefore, filters should be installed in the pump suction lines.

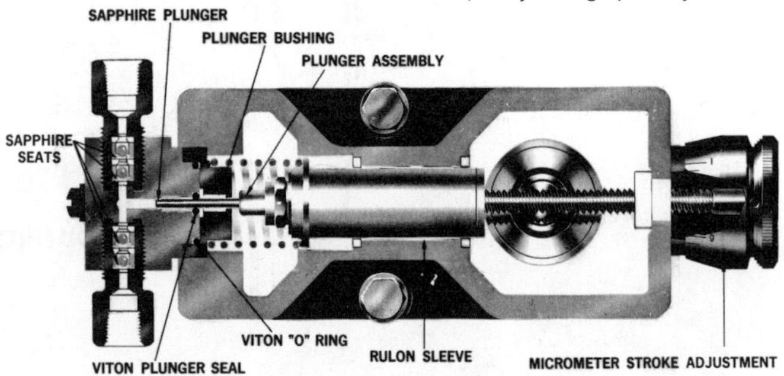

Figure 4.110. A reciprocating piston pump delivers a fixed volume of fluid with each strike. (Courtesy of Milton Roy Co.)

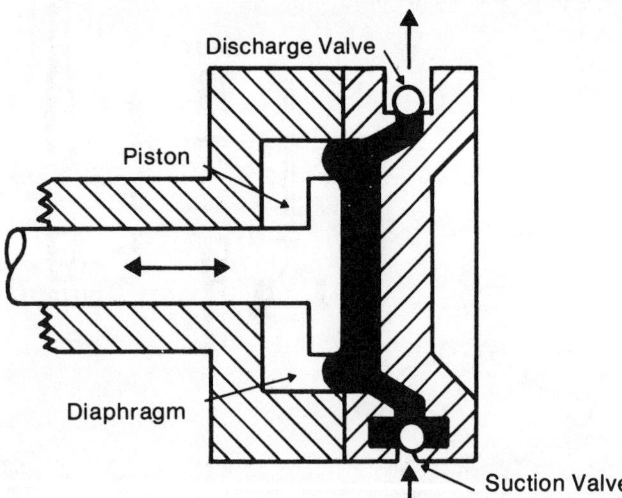

Figure 4.111. Diaphragm pump unit with direct contact between piston and diaphragm is limited to low-pressure and low-flow rate applications.

Reciprocating Piston Pumps

Piston pumps employ a reciprocating piston (or plunger) with inlet and outlet check valves as shown in Figure 4.110. As the plunger retracts from its cylinder, the inlet check valve opens and the cylinder is filled. As the plunger reenters the cylinder, the inlet check valve closes and liquid is forced out the outlet check valve and into the discharge piping.

Diaphragm Pumps

A diaphragm pump is basically the same design as the piston pump except that the process fluid is separated from the piston by a flexible diaphragm. Figure 4.111 shows a type in which the diaphragm is directly flexed by the piston. The diaphragm is subjected to the full pressure of the pump discharge and thus is limited to low-pressure applications. The piston stroke must also be short to prevent excessive flexing of the diaphragm.

Figure 4.112 shows a diaphragm pump in which the hydraulic fluid pressure on one side causes the diaphragm to flex. The hydraulic fluid is pressured by a reciprocating piston with fluid between it and the diaphragm. This causes the diaphragm to flex and thus produce the pumping effect. The diaphragm is hydraulically balanced and is not required to withstand the process pressure. For this reason, it may be made more flexible. A longer stroke may be used and higher pressures handled than with the type where the diaphragm is mechanically flexed. Both piston and diaphragm pumps are usually driven by a constant speed motor drive and transmission. Pumping rate is varied by changing the length of the piston stroke. An illustration of an adjustable stroke pump with a pneumatic positioner is shown in Figure 4.113.

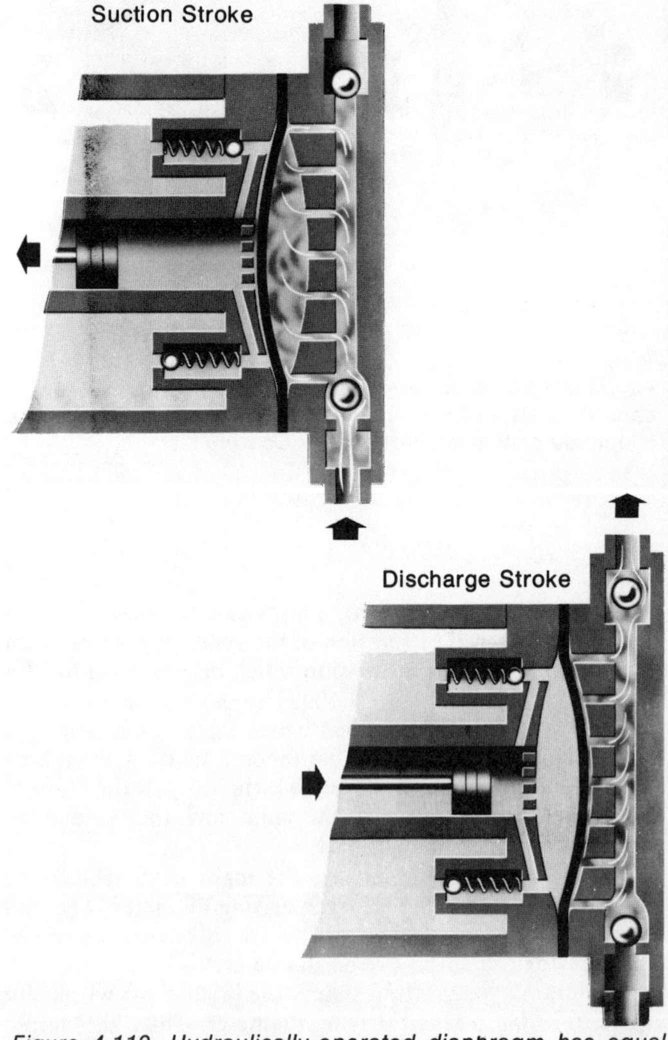

Figure 4.112. Hydraulically-operated diaphragm has equal pressures on each side of diaphragm and can be used for higher pressures and higher flow rates than mechanically flexed diaphragm types. (Courtesy of Interpace Corp./Pulsafeeder Products)

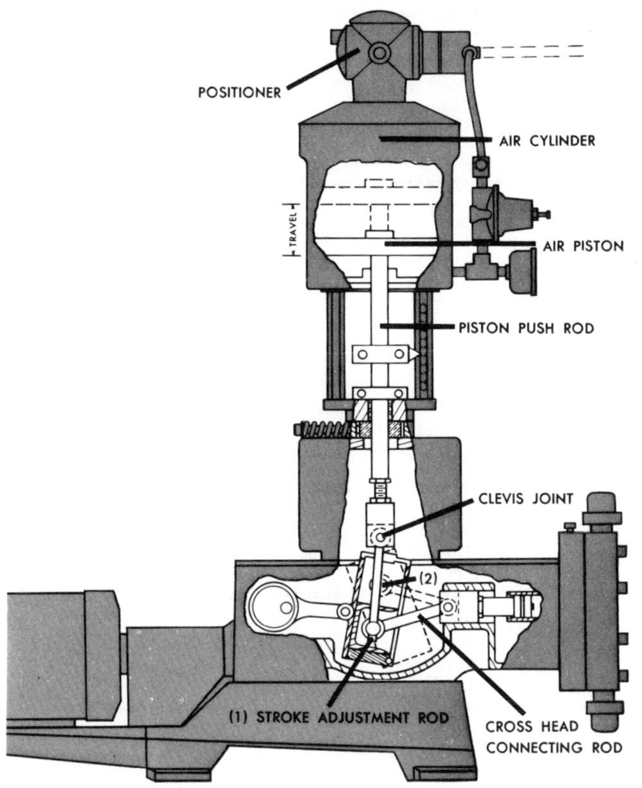

Figure 4.113. Diaphragm pump may have length of stroke adjusted by pneumatically operated positioner. (Courtesy of Interpace Corp./Pulsafeeder Products)

The piston is caused to pump by the crosshead connecting rod which is attached to the stroke adjusting rod. When the stroke adjusting rod is positioned as shown, full stroke is being delivered. When the stroke adjusting rod is withdrawn to the point where its pivot point coincides with the pivot point of the adjustable connection rod (at point 2) no pumping action is present. This type mechanism may be adjusted from zero to full capacity of the pump.

The air piston and positioner shown in Figure 4.113 may receive a signal from a controller output for automatic control. A change in signal input causes the air piston to move, positioning the stroke adjustment rod and thus changing the stroke length of the piston.

These pumps are not suitable for viscous, abrasive or slurry liquids which would prevent the check valves from seating properly.

The discharge of piston and diaphragm pumps have a pulsating characteristic. These pulsations may be reduced by using several pistons arranged such that discharge pulses are staggered. Also dampening chambers and flow restrictors may be installed to smooth out the flow. However, it is practically impossible to eliminate completely all the pulsation effect.

Peristaltic Pumps

Peristaltic pumps produce a wave-like motion or an alternate contraction and refilling of a tubular type vessel to produce a pumping action. In the pump shown in Figure 4.114, liquid enters a chamber formed by a groove in the outer edge of the fixed mandrel and the diaphragm surrounding it. Four spring loaded rollers revolve over the

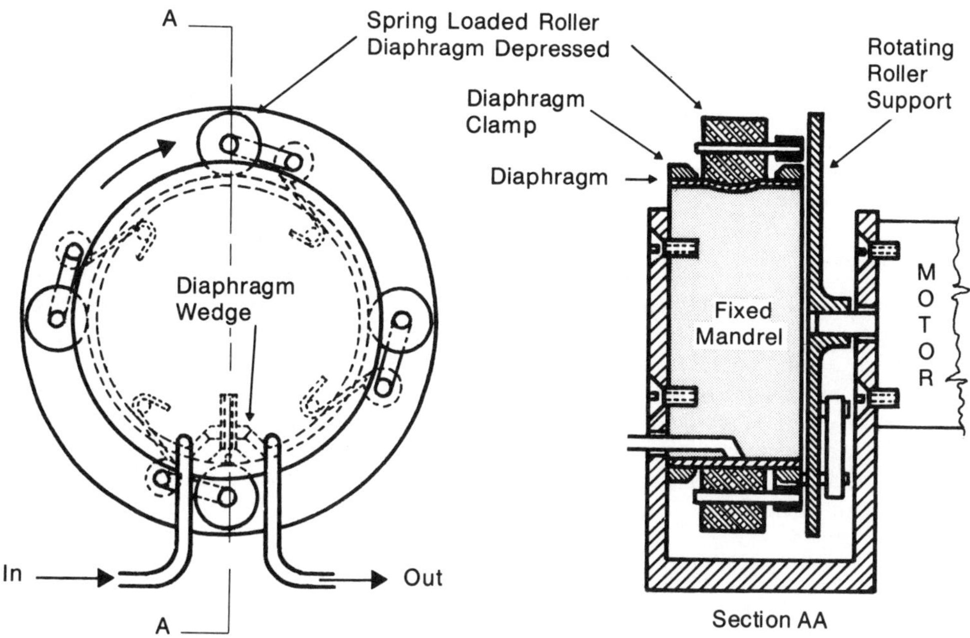

Figure 4.114. Peristaltic pump uses spring loaded rollers to force liquid along a groove and out the discharge port. (Courtesy of Milton Roy Co.)

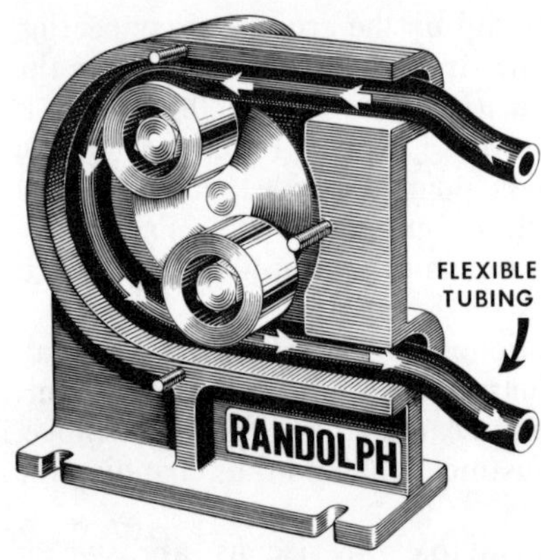

Figure 4.115. Peristaltic pump using flexible tubing which may be easily replaced while pump is in operation. (Courtesy of The Randolph Co.)

diaphragm and mandrel assembly, deflecting the diaphragm into the groove, thus forcing fluid along the groove to the discharge port.

Another type, shown in Figure 4.115, employs a flexible plastic hose on the inner side of a section of a cylinder. Two rollers depress the hose, forcing the liquid through it in the same manner as described above. This type is limited to low-pressure, low-flow applications and is widely used in laboratory and medical work.

Figure 4.116 illustrates yet another peristaltic type which employs a series of cam-operated steel fingers to produce a wave-like motion to force liquids or gases through flexible tubing. Several tubes may be operated by the fingers to produce proportional flows.

Varying the speed of peristaltic pumps, like gear pumps, is usually the method employed to vary flow rates. Variable speed drives are installed between the drive motor and the pump to accomplish the variation needed. Metering pumps are used for small flows and in difficult-to-measure services that have high accuracy requirements. Gear pumps and piston pumps are used for high-pressure services, while peristaltic and diaphragm pumps are used for low pressures. Metering pumps can be damaged easily if their suctions run dry or are partially starved. The resulting cavitation can cause severe damage to pump parts.

Conclusion

Many types of flow meters have been discussed, some briefly and some in great detail. Other types are available which have not been included at all. Some of these have been introduced quite recently and appear to be applicable to many flow requirements. However, they have not been used extensively to date, at least not in the processing industries. They are listed as a matter of information should the reader want to pursue them further.

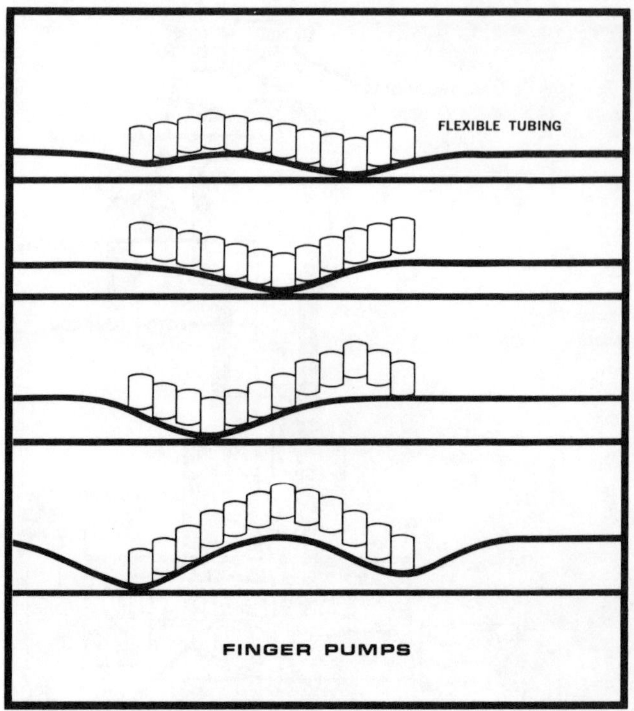

Figure 4.116. A series of cam operated fingers press against flexible tubing in sequence imparting a unidirectional flow of liquid or gas through the tubing in this version of perstaltic pumps. (Courtesy of Sigmamotor, Inc.)

1. Nuclear magnetic resonance flowmeters which operate on the principle that some fluids become slightly magnetized when passed through a magnetic field. Radio frequency energy is then used to "resonate" or amplify the magnetic effect to produce usable signals.
2. Angular momentum flowmeters whose designs impart a force, measurable by torque, which is proportional to mass flow rate. There are several designs that fit this category: impeller-turbine meters, the Coriolis meter and the gyroscopic mass flowmeter.
3. The use and detection of radioactive materials in fluid streams.

The development of new flow measuring techniques and the continuing search for others point out the importance of flow measurement in industry. The methods covered in this chapter provides a wide range of capabilities. The choice of methods is covered in Volume 2.

5 Level Measurement

William G. Andrew, Keith G. Rhea

A wide variety of level measuring devices are available to meet the diverse level requirements of the processing industries. Uses range from the simple and economical mechanical types, such as tapes, boards or gauge glasses, to the more sophisticated types, such as complete weighing systems or nuclear radiation methods. Level measurements are needed for inventory and accounting of product storage, inventory and distribution of raw material for processes and proper operation of fractionating towers and distillation columns, controlling water level in boiler drums and other similar process control applications where accurate level determinations are required.

There are many level applications where accuracy per se is relatively unimportant—one need only know that a level has reached a point where remedial action must be taken. This action may take place automatically, or an operator may be warned or alerted to take the appropriate action. Level switches usually perform these functions.

On the other hand, there are many situations where precise control is necessary or desirable. The level has a significant effect on process quality, controllability and/or cost. Typical examples include:

1. Reactor levels—poorly controlled levels may upset reaction equilibria resulting in poor product quality, damaged equipment or spilled material.
2. Feed control—steady process flow from storage vessels is sometimes dependent on a constant head pressure on a feed line from the vessel.
3. Steam boiler operation—efficient operation of a boiler is dependent on the proper inventory of liquid to maintain feed and on a proper vapor volume space for vapor capacity.
4. Equipment cost—close level control often permits the use of smaller size vessels in the intermediate stages of processes.
5. Accounting applications—level measurements are sometimes used for product or raw material accounting and are directly related to buying and selling. These, of course, have a direct bearing on profit.

These few examples point to the necessity for good level measuring techniques for economy, for product quality, for equipment and personnel safety and for the reduction of losses or spoilage.

Classification of Methods

Level measurements may be classified broadly into two general groups—direct and inferential measurements. Direct level measurements are simple and have been used for a long time. They are used primarily for local indication. Some inferential methods are also long-lived but many have been utilized only in the past few years. Inferential methods are used when remote indication, recording or control is desired.

Direct

Direct level measurements are straightforward, usually visual. Gauge glasses are the most widely accepted method—transparent (or reflex) tubes of glass are connected to a vessel, and the visible fluid in the tube reveals the level inside the vessel.

Other direct methods include calibrated tapes or dip sticks placed in vessels and calibrated to read directly in the desired units—depth, volume or weight of material.

Direct measurement methods are simple and economical but not easily adapted to signal transmission techniques for remote indication or control.

Inferential

Indirect or inferred methods of level measurement depend on the material having a physical property which can be measured and related to level. Many physical and electrical properties have been used for this purpose and are well suited to producing proportional output signals for remote transmission. Included in these methods are:

1. Buoyancy—the force produced by a submerged body which is equal to the weight of the fluid it displaces.
2. Hydrostatic head—the force or weight produced by the height of the liquid.
3. Sonar or ultrasonic—materials to be measured reflect or affect in a detectable manner high frequency sound signals generated at appropriate locations near the measured material.
4. Conductance—at desired points of level detection, the material to be measured conducts (or ceases to conduct) electricity between two fixed probe locations or between a probe and a vessel wall.
5. Capacitance—the material to be measured serves as a variable dielectric between two fixed capacitor plates. In reality, there are two substances which form the dielectric—the material whose measurement is desired and the vapor space above it. The total dielectric value changes as the amount of one material increases while the other decreases.
6. Radiation—the material measured absorbs radiated energy. As in the capacitance method, vapor space above the measured material also has an absorbing characteristic, but the difference in absorption between the two is great enough that the measurement change can be related quite accurately to the measured material.
7. Weight—the force due to weight can be related very closely to level when its density is constant. Variable concentrations of components or temperature variations present difficulties, however.
8. Resistance—pressure of the measured material squeezes two narrowly separated conductors together, reducing overall circuit resistance in an amount proportional to level.

Caution must be used in applying any inferred level measurement to ensure that the measured property has a well-defined (constant, if possible) relationship to level; otherwise, large errors can occur if compensation is not used.

A most unusual inferential method of observing level has been used on a surge drum in a refrigeration unit. The drum operated below 32°F and was heavily insulated. Nipples extended from the drum outward through the insulation. Level was determined by observing the nipples with frozen moisture from the air due to the cold liquid at or below liquid level.

These various methods of level detection and the various electrical and mechanical techniques to convert the measurements into usable signals are discussed in the following sections.

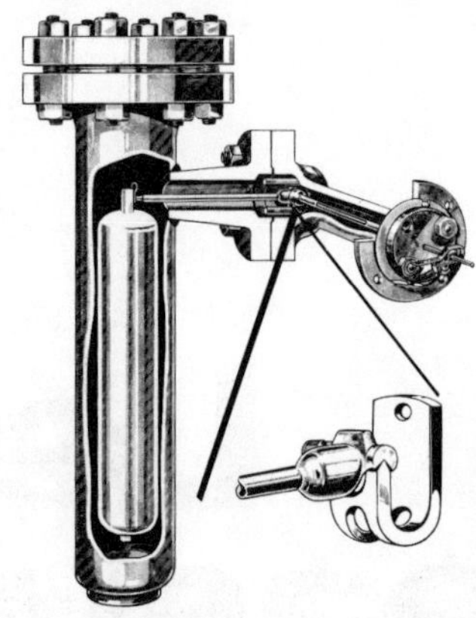

Figure 5.1. Cutaway view of displacement level device showing displacer and torque tube assembly. (Courtesy of Fisher Controls Co.)

Displacement Level Devices

The displacement level device (Figure 5.1) was for many years, and probably still is, the most commonly used of all level measuring techniques. Its operation is based on Archimedes' principle which states that a body immersed in a liquid will be buoyed by a force equal to the weight of the water displaced.

The body immersed in this case is called a displacer. The displacer always weighs more than the upward buoyant force which may be developed by the liquid in which it is submerged. Therefore, it may be made very long or short as required. Displacers are usually made cylindrical in shape (constant cross-sectional area) so that for each equal increment of submersion depth, an equal increment of buoyancy change results, yielding the desired linear or proportional relationship. (This assumes that liquid density remains constant.)

Almost invariably, displacers are hollow, and small amounts of lead shot are used for filling for weighing purposes to take into account variations in specific gravity for various applications.

When a displacer float is completely uncovered, its full weight is brought to bear on whatever force is chosen to contract its weight. As the liquid level rises to cover the displacer, its weight is reduced in proportion to its covered length, and signals responsive to this force change are proportional to level changes.

Several types of displacer units are manufactured. All, of course, make use of the buoyancy principle. The differences in their designs are based primarily on the method of transmitting this force change to a readout device or a transmitting mechanism. Various methods used are discussed below.

Torque Tube

The torque tube is the best known method of using displacers for level measurement. The displacer is attached to a

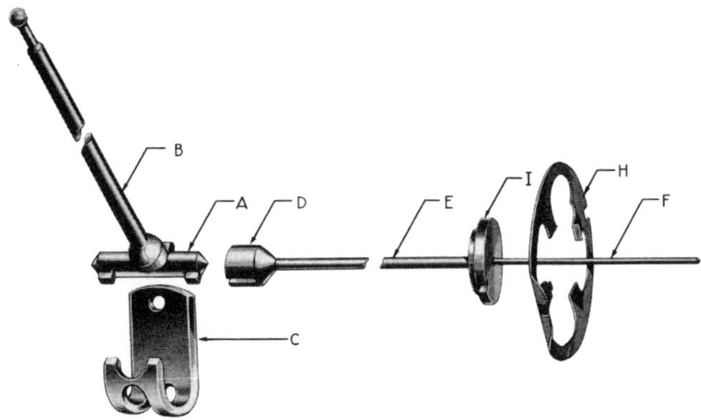

Figure 5.2. Exploded view of torque tube assembly. (Courtesy of Fisher Controls Co.)

torque tube assembly (Figure 5.2) whose rotary motion is used for readout or control.

The displacer is suspended from the torque tube or torsional measuring spring which restricts its movement to prevent contact with any part of the vessel in which it is placed. The torsional spring or torque tube is designed to twist a specific amount for each increment of buoyancy change and to be insensitive to pressure changes in the vessel. The displacer rod which connects the displacer to the torque tube is designed to absorb sideward forces and minimize friction by a knife-edge bearing. Backlash is prevented in the linkage because the displacer always pulls in the same direction on the torsional spring.

The torque tube is tubular so that the small rotary shaft which is fastened to the inner end of the tube may transmit the degree of rotation of the inner end accurately to the outside of the vessel. The outer end of the tube is gasketed and clamped rigidly to the vessel wall. This allows the interior of the torque tube to be at atmospheric pressure; therefore, packing is not required and the disadvantages of packing friction are eliminated. This method of sealing is one of the unique features of this device that makes it so useful in level applications.

With reference to Figures 5.1 and 5.2, the parts of the level device that are wetted by the process material are the cage, the displacer and the components of the torque tube assembly, except items *H* and *F*.

Displacer float cage materials are usually made of iron or steel, although they can be made of other materials. Pressure ratings are available to 600 psig and temperature ratings to 850°F. Connection sizes are usually 1-½ inches screwed or 2 inches flanged. Cages are made in several configurations so that connections may be made from the side, top and bottom, top and side or bottom and side.

The torque tube is made of K Monel as standard, or it may be furnished in inconel, stainless steel or Hastelloy. The use of inconel allows it to be used in service temperatures as high as 850°F.

Standard displacers have a volume of 100 cubic inches and vary from 1 to 3 inches in diameter. Standard lengths vary from 14 to 120 inches in nine sizes. The displacer is cylindrical in shape to permit the highest pressure rating for the amount of material used and to provide a uniform cross-sectional area so that the buoyant force changes proportionally with level.

Displacers are usually made of 304 stainless steel but are furnished in 316 stainless steel Hastelloy, Monel, plastic and special alloys.

In addition to the external cage assemblies, displacer units may also be mounted internally. Figure 5.3 shows a typical internally mounted unit. Several different configurations of internal mounts are available.

The use of displacement type units is often associated with local control loops. Figure 5.4 shows how a typical loop operates. The change in force as level changes is transmitted through the torque tube assembly to change the flapper-nozzle relationship *B* and *A* in the diagram.

In Figure 5.4, air is supplied from the filter-regulator to orifice *J* into relay diaphragm chamber *L* to nozzle *A*. Nozzle *A*, when not restricted by flapper *B*, is large enough to bleed off all the air coming through orifice *J*, and the pressure is zero between the orifice and the nozzle. When the nozzle is restricted by the flapper, due to a rise in liquid level, pressure is built up in the system between *A* and *J*. Thus, any change in liquid level results in a change in pressure in chamber *L*.

The double diaphragm assembly with exhaust ports between diaphragms *M* and *P* in the intermittent bleed relay is free floating and always pressure balanced. If there is an increase in pressure in chamber *L*, the diaphragm assembly is pushed downward, and the inlet valve *O* is pushed open. This allows supply pressure to come into chamber *N* until it pushes the relay diaphragm assembly back into its original

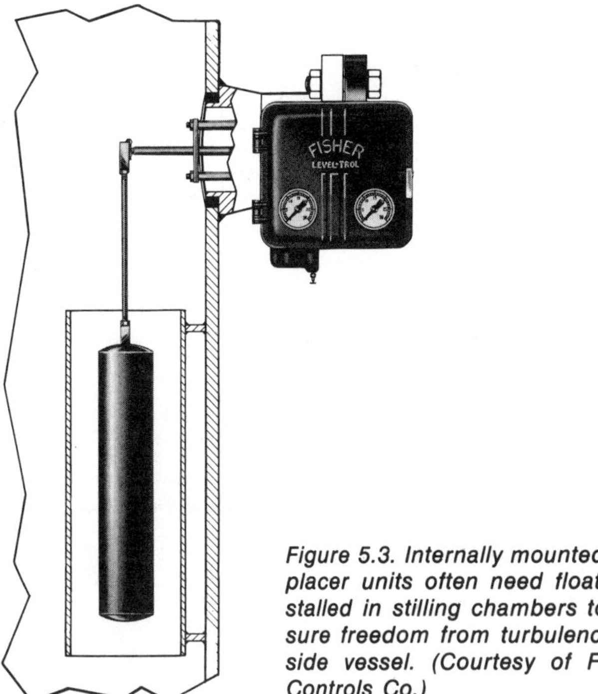

Figure 5.3. Internally mounted displacer units often need floats installed in stilling chambers to ensure freedom from turbulence inside vessel. (Courtesy of Fisher Controls Co.)

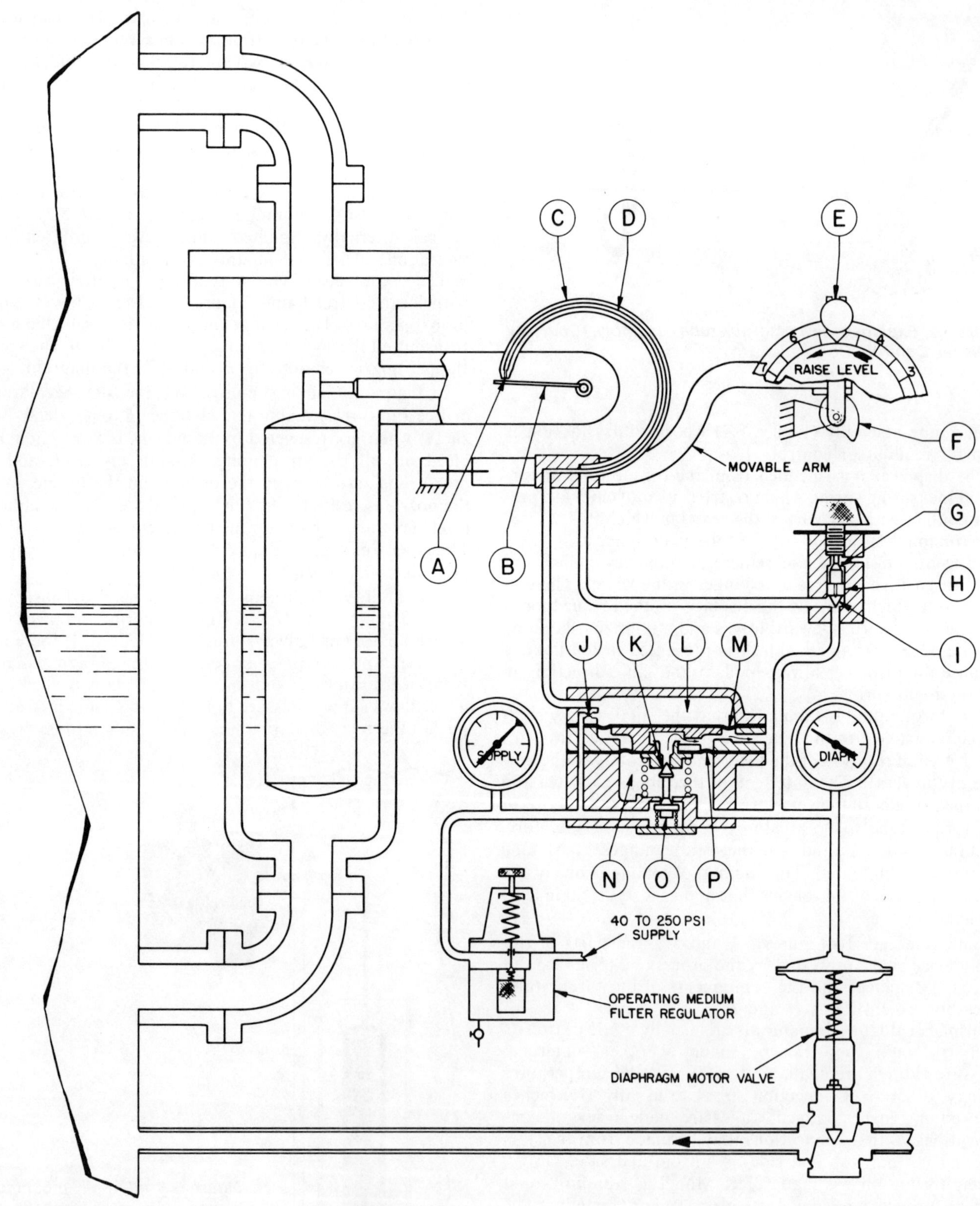

Figure 5.4. Schematic illustration of the operation of the displacement level controller. (Courtesy of Fisher Controls Co.)

position and the inlet valve *O* is closed again. A decrease in pressure in chamber *L* causes the diaphragm assembly to move upward and to open exhaust valve *K*, allowing pressure under small diaphragm *P* to bleed out until the diaphragm assembly again returns to its normal position and exhaust valve *K* is closed.

The ratio of the two diaphragm areas in the relay is 3 to 1 or such that a 4-pound change on large diaphragm *M* results in a 12-pound change in pressure to the diaphragm control valve.

Another part of the controller assembly is the proportional band adjustment mechanism, which consists of a three-way valve assembly *H* in a branch from the control valve supply line to bourdon tube *C*. The three-way valve *H* is manually positioned between the inlet port *I* and the exhaust port *G*. When the valve is seated against exhaust port *G*, all of the diaphragm pressure is transmitted to bourdon tube *C*. This causes the bourdon tube to "back away," and the flapper has to move a relatively large distance to close the nozzle. On the other hand, if the valve is seated against the inlet port *I*, no pressure is transmitted to the bourdon tube, with the result that a very small flapper movement is all that is necessary to close the nozzle. Intermediate positions of the valve, of course, result in intermediate pressure to the bourdon tube.

If one assumes the controller and diaphragm control valve are both direct acting as shown in the schematic drawing, the operation cycle of the complete level controller is explained in the following discussion.

The level in the vessel is at a point midway on the displacer, it is assumed, and the unit is adjusted to give 9 psi on the diaphragm of the control valve. Inlet flow to the vessel equals outlet flow. If there is a decrease in outlet flow, the level in the vessel and in the displacer cage will rise. The displacer then will rise, causing flapper *B* to rise toward nozzle *A*. This will build up pressure in the relay chamber *L* and the relay diaphragm assembly will move downward, opening relay supply valve *O*. Operating medium then flows into chamber *N* until the relay diaphragm assembly is pushed back into its original position, and valve *O* is closed again. The pressure in chamber *N* is transmitted to the diaphragm of the control valve causing it to move toward its seat.

At the same time, the pressure in bourdon tube *C* is being increased through the three-way valve assembly *H* which causes nozzle *A* to move away from the flapper, thus stopping the pressure buildup in chamber *L*. The unit is again in equilibrium with the level at a higher point, and the diaphragm control pressure is increased to close the control valve partially so that inlet flow again equals outlet flow. If an increase in outflow takes place, the reverse of the above cycle will occur, with a decrease in liquid level causing an increase in control valve opening.

The proportional band adjustment assembly furnished with this unit provides a quick and easy method of setting the control band desired. The proportional band of a level controller may be defined as the percentage of the level range that the level must change to actuate the valve through its entire stroke. Adjustment is obtained by turning a calibrated dial whose settings range from 0 to 10. Determination of dial setting is made from a chart (Figure 5.5) furnished inside each controller. The following example illustrates its use.

Assume a service whose fluid specific gravity is 0.8 and where a 60% proportional band setting is desired on a 32-inch displacer. This means there is to be 19.2-inch (32 inch x 0.60) level change to obtain full output pressure change from the controller. From the vertical scale, find specific gravity of 0.8. Follow horizontally to the intersection of the proportional band line of 60%. Then read straight down to the dial setting scale which reads 4.8—the setting needed to obtain the desired result.

When signal transmission only is desired, a modified version of the controller is furnished. For transmission, only 100% proportional band is needed. To this a zero adjustment is added, and the output is linearly proportional to level over the full length of the displacer.

The control and transmitter units described are pneumatic. Electronic versions are available also.

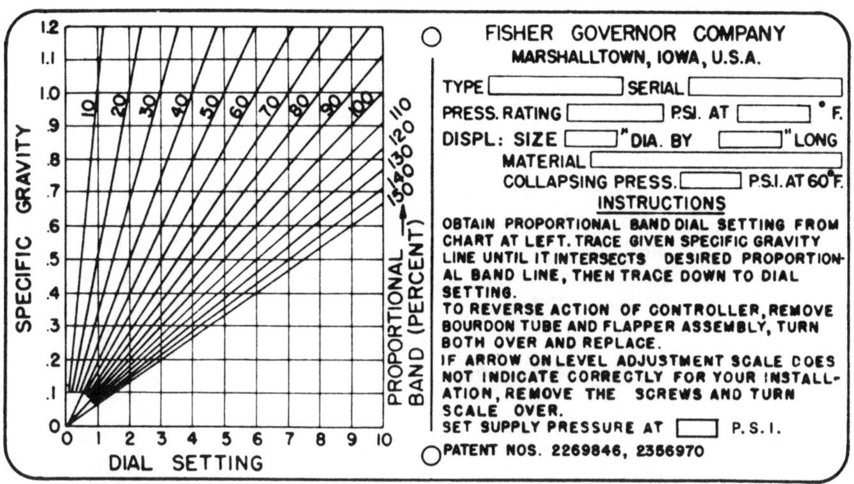

Figure 5.5. Dial settings for various proportional band adjustments are determined from chart when specific gravity is known. (Courtesy of Fisher Controls Co.)

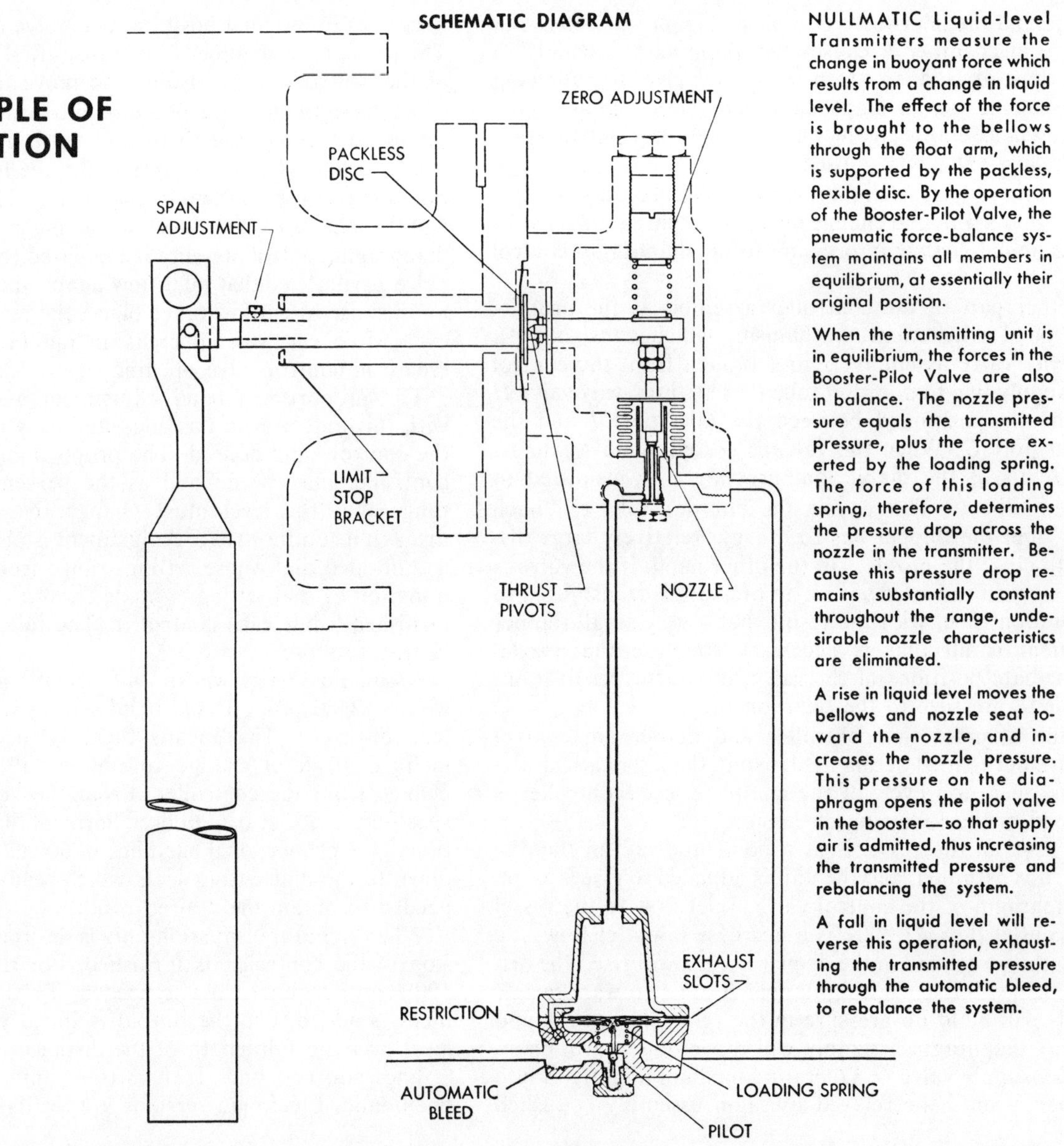

PRINCIPLE OF OPERATION
SCHEMATIC DIAGRAM
ZERO ADJUSTMENT
PACKLESS DISC
SPAN ADJUSTMENT
LIMIT STOP BRACKET
THRUST PIVOTS
NOZZLE
EXHAUST SLOTS
RESTRICTION
AUTOMATIC BLEED
LOADING SPRING
PILOT
NULLMATIC Liquid-level Transmitters measure the change in buoyant force which results from a change in liquid level. The effect of the force is brought to the bellows through the float arm, which is supported by the packless, flexible disc. By the operation of the Booster-Pilot Valve, the pneumatic force-balance system maintains all members in equilibrium, at essentially their original position.
When the transmitting unit is in equilibrium, the forces in the Booster-Pilot Valve are also in balance. The nozzle pressure equals the transmitted pressure, plus the force exerted by the loading spring. The force of this loading spring, therefore, determines the pressure drop across the nozzle in the transmitter. Because this pressure drop remains substantially constant throughout the range, undesirable nozzle characteristics are eliminated.
A rise in liquid level moves the bellows and nozzle seat toward the nozzle, and increases the nozzle pressure. This pressure on the diaphragm opens the pilot valve in the booster—so that supply air is admitted, thus increasing the transmitted pressure and rebalancing the system.
A fall in liquid level will reverse this operation, exhausting the transmitted pressure through the automatic bleed, to rebalance the system.

Figure 5.6. This displacer transmits force changes through a packless flexible disc and a pneumatic force balance system maintains equilibrium in the entire system. (Courtesy of Moore Products)

Flexible Disc

Another version of the displacement principle of level measurement is shown in Figure 5.6. This device is used as a transmitter and operates on a force-balance principle.

A change in liquid level changes the effective weight transmitted to a bellows through the float arm attached to the displacer. By the operation of a booster pilot valve, a pneumatic force balance system maintains all members in equilibrium.

A rise in liquid level moves the bellows and nozzle seat toward the nozzle and increases nozzle pressure. This pressure opens the pilot valve in the booster to admit supply air, thus increasing the transmitted pressure and rebalancing the system.

A fall in liquid level reverses this operation, exhausting the transmitted pressure through an automatic bleed, to rebalance the system.

These units are primarily for internal tank mounting. Standard mounting flanges or tees are cast iron or cast steel in 4- and 6-inch sizes, 150 and 300 psig. Floats, float arms and discs are 304 and 316 stainless steel but may be furnished in Monel, nickel or phosphor bronze.

Spring Balanced

Figure 5.7 illustrates a spring balance displacer unit that utilizes a magnetic coupling to transmit displacer motion to produce a proportional pneumatic signal. It differs from the

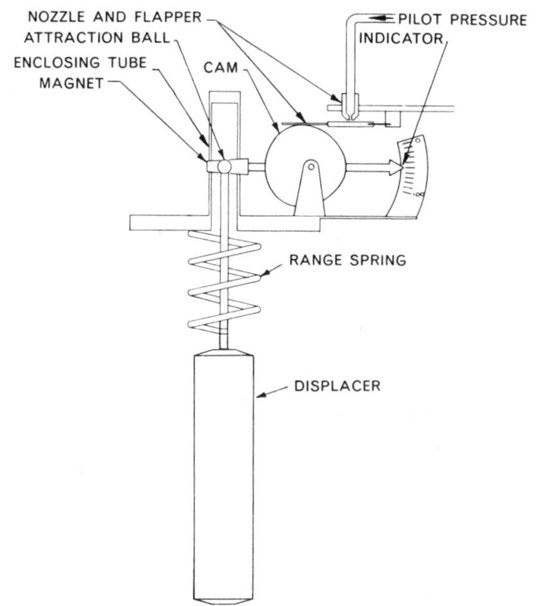

Figure 5.7. The operating schematic of the modulevel controller, a spring balance displacer unit that utilizes a magnetic coupling to produce a readout proportional to displacer motion. (Courtesy of Magnetrol)

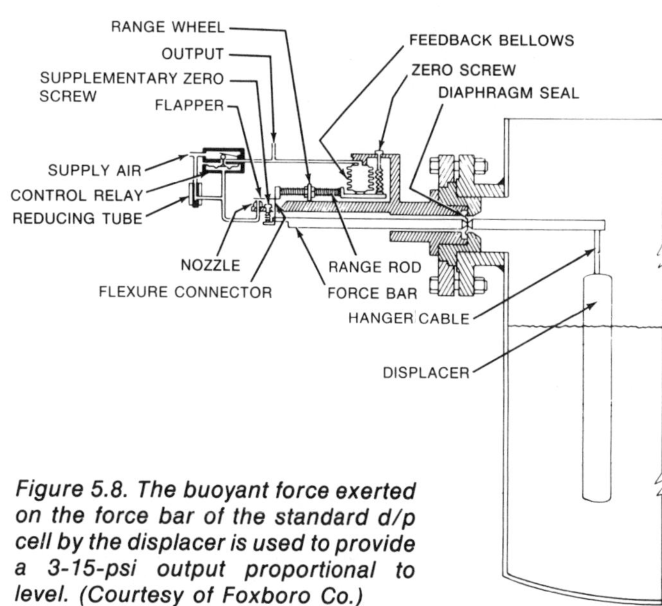

***Figure 5.8.** **The buoyant force exerted on the force bar of the standard d/p cell by the displacer is used to provide a 3-15-psi output proportional to level.** (Courtesy of Foxboro Co.)*

units previously described in that there is an appreciable movement of the displacer. Its upward force, however, is still equal to the weight of the water it displaces. The displacer is suspended from a spring, and its motion varies directly in proportion to the level change imposed upon it. A magnetic coupling is used to transmit motion to the controller.

As the level rises, the buoyant force on the displacer increases and the attraction ball in the displacer-enclosing tube rises. The external magnet follows the attraction ball and the cam rotates, repositioning the pilot flapper which varies the pilot pressure signal. A pilot relay amplifies this pressure to a proportional output signal. A reset function can be added to the controller.

External mounting cages are available in cast iron, steel and stainless steel at pressures up to 2,500 psig. Connection sizes are 1 ½ and 2 inches, threaded or flanged. Internally mounted units are available in 3-, 4-, 6- and 8-inch flange sizes.

Standard displacers are made of 304 stainless steel, and other trim parts in contact with the wetted material are made of 304 or 316 stainless steel. Range springs, displacers and other trim parts can be made of other special alloys such as Hastelloy.

Force Balance

The Foxboro Company uses the same force balance type d/p cell transmitter mechanism used for other fluid characteristic measurements. This device utilizes a flexure seal at the diaphragm face and transmits level changes to a remote readout device. The buoyant force exerted on the displacer is applied to the end of the force bar and is transmitted through the flexure to the range rod with the diaphragm seal acting as a force bar fulcrum as well as a seal against process fluids (refer to Figure 5.8.)

Any movement of the range rod causes a minute change in the position of the flapper in relation to the nozzle, producing a change in the output pressure from the relay to the feedback bellows until the force in the bellows balances the force on the displacer. The output pressure which establishes the force-balance is the transmitted pneumatic signal which is proportional to the buoyant force exerted on the displacer. This signal is transmitted to a pneumatic receiver to record, indicate and/or control.

Units may be internally mounted or cage mounted. Internal units are available in ranges from 14 to 150 inches; cage type units from 14 to 96 inches. Pressures to 600 psig can be accommodated in 4-inch flanges for internal units, while external cage types are furnished with 1½-inch screwed end or 2-inch flange units from 150 to 600 psig.

Displacers are made from 304 stainless steel, 316 stainless steel, Monel or Teflon as standard. Other materials can be used, if needed, for displacers or for other wetted parts.

Interface Measurement

Interface level control is maintaining the level between two immiscible liquids having different specific gravities. Occasionally there is need in a process to control such an interface in a vessel. The displacement level method is used for such applications.

The use of the buoyancy principle for interface measurement is easily understood. The principle states that the displacer exerts an upward force equal to the weight of fluid it displaces. When the specific gravity of a fluid is 1.0, a 100 inch^3 displacer exerts an upward force of 3.6 pounds (100 inches^3 × .036 pounds/inch^3) when it is fully immersed. If the

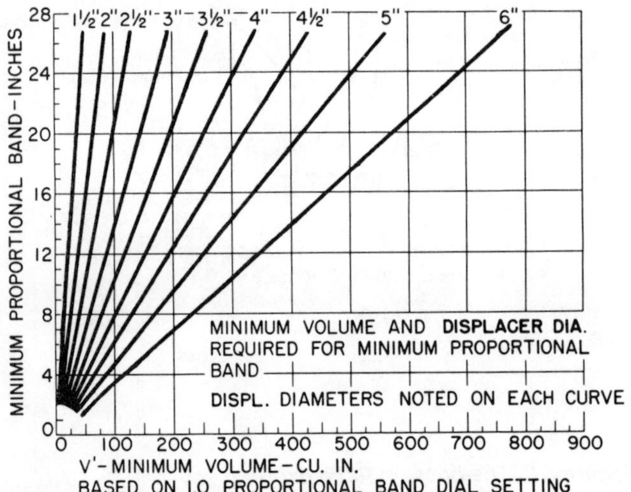

Figure 5.9. Displacer diameter size is selected from curve when displacer volume and minimum level control band are known. (Courtesy of Fisher Controls Co.)

Table 5.1. Displacement Force, F_d, Required for Standard and Thin Wall Torque Tubes

Style of Torque Tube	Displacement Force, F_d, Pounds	
	at 2.0 PBD Setting*	at 1.0 PBD Setting**
Standard	0.72	0.36
Thin Wall	0.36	0.18

*Use these values in sizing.
**Use these values when absolute minimum proportional band is specified.

fluid specific gravity is 0.8, the upward force is only 2.88 pounds. It becomes obvious that the difference in specific gravity of two fluids determines the change in force as the interface level changes through the float range.

When the force is too small for the standard displacer, thin wall displacer tubes are available which have lower torsional rates and, therefore, require less displacement force for the same amount of flapper rotation. The thin walled tube is used only on interface and specific gravity applications where the displacer tube is completely submerged at all times. Its wall thickness is 0.016 inch compared to 0.044 inch for the standard tube.

Another method of getting additional force for interface applications is to use larger diameter displacers. Diameters up to 6 inches are used for total displacer volumes that range from 100 to 1,000 inches³.

Figure 5.9 shows a graph used to determine displacer diameter sizes when displacer volumes and minimum level control bands are known. Table 5.1 provides information on the displacement force, F_d, needed for standard and thin wall torque tubes for proportional band settings of 50 and 100%. The use of a 100% proportional band setting means that if the displacer were 14 inches long, the output from the interface device would be from 3- to 15-psig over the entire 14-inch range.

With F_d and the specific gravities of the two interfacing fluids known, the minimum displacer volume required is calculated from the formula:

$$VI = F_d/[(.036)(SG_H - SG_L)]$$

where

VI = minimum displacer volume required to obtain 3- to 15-psi output

F_d = displacement force in pounds required to give 3- to 15-psi output with a proportional band dial setting of 2.0 (see Table 5.1).

.036 = weight in pounds of 1 cubic inch of water

SG_H = specific gravity of heavier liquid

SG_L = specific gravity of lighter liquid

Displacer diameter size can be selected from the graph in Figure 5.9.

Hydrostatic Head or Differential Pressure Types

Level measurements that fall into the category of hydrostatic head or differential pressure methods comprise a fairly high percentage of all continuous level applications. While displacer units find extensive use in clean liquid services where level ranges fall between 14 and 48 inches, units covered in this section are easily adaptable to services of wide level ranges, high pressures, corrosive conditions, slurries and highly viscous materials.

All the devices described in this section are hydrostatic devices or a combination of hydrostatic head and differential pressure measurement principles. They rely on the pressure of the measured liquid head to provide level indication. When the top surface of the liquid to be measured is at a pressure less or greater than atmospheric, it must be connected to the low side of a differential device to cancel out the pressure effects not attributable to the liquid head. Frequently, a nonvolatile liquid, often an antifreeze compound, is added to this equalizer leg to retard corrosion and to assure that a constant head is present (condensibles might build up and then evaporate periodically, otherwise). This liquid head pressure on the transmitter can be cancelled mechanically; a discussion is included under "Zero Suppression-Elevation" later in this chapter.

Many different types of devices are used for these measurements. Some have been discussed in the previous chapter on flow, and are discussed in the following chapter on pressure. These are mentioned again briefly. Techniques applied to obtain the differential measurements in some cases are unique and are mentioned and discussed in greater detail.

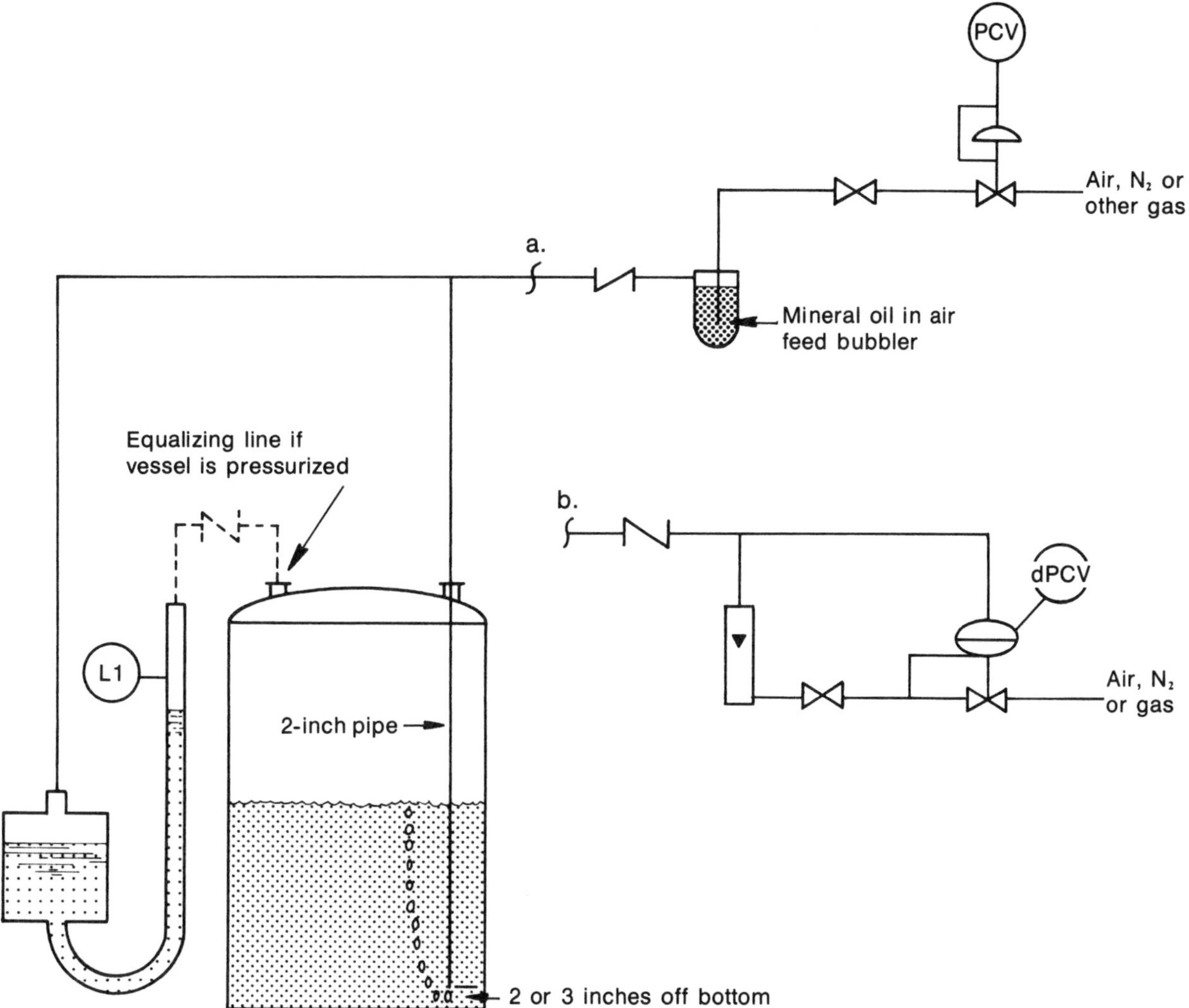

Figure 5.10. Typical air bubbler type level measurement. System a shows simple air feed bubbler; b *shows differential pressure regulator and rotameter used to overcome liquid head.*

Air Bubblers

One of the oldest and simplest methods of level measurement is the air bubbler. It utilizes a dip tube in the measured liquid to which is connected a gas source of sufficient pressure to overcome the liquid head and bubble up through the measured liquid. The level or static head is measured by pressure gauges, manometers or d/p cells. Readout may be local or remote. When transmission distances exceed 50 to 75 feet, differential pressure transmitters are usually used to transmit standard signals to remote locations.

Figure 5.10 illustrates a typical air bubbler and dip tube installation. A regulator is used to reduce gas pressure to a few pounds above the maximum head that it must overcome to bubble through the liquid head. Downstream of the regulator, an air bubbler or a rotameter is installed so that it can be visually determined that a gas flow is established. A combination pressure control valve and rotameter is often used to provide a constant flow rate through the dip tube regardless of the liquid level. The regulator holds a constant differential pressure across a fixed restriction, thus holding a constant flow through the restriction and the rotameter.

Air and nitrogen are the most commonly used gases for bubbler installations. Liquid may be used if there is a reason not to use gas. If process material has a tendency to plug the dip tube, a bypass may be installed around the flow regulator to blow out the line periodically.

Bubbler systems are used rather infrequently now. One drawback is that it is undesirable in many processes to introduce air, nitrogen or other purge material to the process. They do provide economical installations, however, particularly for local readout on clean services.

The accuracy of the bubbler system is about as good as the differential device used for readout. Its accuracy is

Figure 5.11. Level measurements are made with flat or extended diaphragm units in services where cavities or dead ends tend to plug or polymerize. (Courtesy of Foxboro Co.)

dependent on the constancy of the density of the material whose level is measured.

Diaphragm Units

There are several designs of differential or hydrostatic units that may be classified as diaphragm level detectors. These were developed primarily as protection against plugging in slurry services or where dead-end cavities have a tendency to polymerize or solidify material.

Figure 5.11 shows two designs that have been in use for several years. The flat diaphragm type is used on many services where small dead-end cavities present no problems. The extended diaphragms are used where even small cavities would have a tendency to plug or freeze or where undesired concentrations of material might be detrimental to the process.

The diaphragms (or wetted portions) of these units can be furnished in a wide variety of special alloy materials that, despite the higher cost of the diaphragm unit, provide economical installations because other costly piping materials are eliminated. The diaphragms can also be coated with a wide variety of plastic materials that inhibit corrosion and prevent process contamination.

Connection flanges for the flat diaphragm type are usually 3 to 6 inches with pressure ratings standard to 300 psig. The extended diaphragm type is usually furnished with 3- or 4-inch flanges rated at 150 or 300 psig. Extension lengths vary from 2 to 6 inches and can be furnished in odd lengths.

The accuracy of these units range from ±½% of span to ±2% of span depending on the range. From 20 inches H_2O to about 400 inches H_2O, the accuracy is ±½%. In the vicinity of 800 inches H_2O, accuracy is only ±2% of span.

Diaphragm units are furnished with either pneumatic or electronic transmission. Figure 5.12 shows a variation from the designs described above. It, too, uses a force balance d/p cell for pneumatic or electronic transmission. It differs from the ones described in that it uses liquid filled sensing elements connected by capillaries to the differential pressure unit mounted nearby.

The liquid filled sensing elements are flange mounted (3 inches - 150 psig or 3 inches - 300 psig) at both the low- and high-pressure sides, making them readily adaptable to difficult slurry or corrosive services.

The sealed systems also eliminate the problem of using purges or other types of seals to prevent condensation or similar problems in the low-pressure leg.

These sealed systems are available in ranges from 20 inches H_2O to 800 inches H_2O with an accuracy of ±½ to ±1% of span. The wafer elements can withstand process temperatures up to 800°F.

The filling liquids in the sealed systems are carefully selected for low thermal expansion coefficients to reduce ambient temperature effects. The error does exist, however, and is about 1 inch H_2O per 50°F temperature change from the calibrated standard of 80°F.

Diaphragm 1:1 Repeaters

Several manufacturers offer diaphragm elements for liquid level measurement that utilize the force balance measurement principle. Figure 5.13 shows a submersible type that can be used as illustrated. The hydrostatic head of the liquid on one side of the diaphragm is opposed by a flow

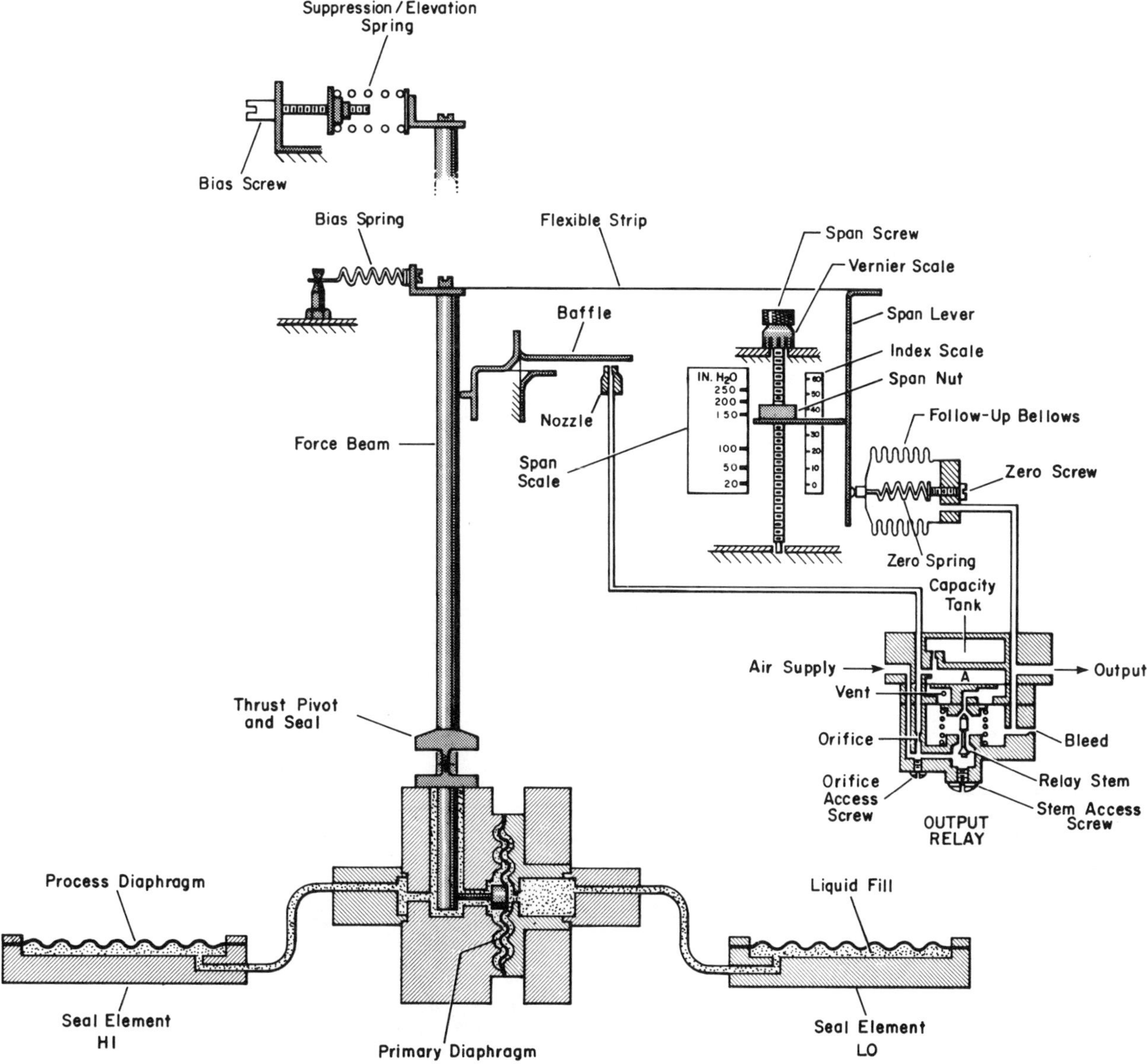

Figure 5.12. Another device used on level measurement of slurries and other difficult-to-measure fluids is a diaphragm type in which filled systems extend from the differential unit to sensing diaphragms mounted between flanges at the measurement points. (Courtesy of Taylor Instrument Process Cont. Div., Sybron Corp.)

controlled air supply on the other side whose developed pressure is equal to the liquid head. Since the output is a direct measurement of the pressure at the diaphragm (1:1 ratio), it is applicable only to open tank measurements unless the output is fed to a differential pressure device whose low side is connected to the vapor space above the liquid level.

Figure 5.14 illustrates a similar flange mounted (3- or 4-inch) device—also a 1:1 repeater. The output of this device can be fed to a controller as shown in Figure 5.15 to produce a standard pneumatic output to a local or remote readout or to a control valve. It operates in the following manner.

Operating air normally is supplied at a pressure of 20 psig to the transmitter through the inlet connection and primary orifice (R). The force exerted on diaphragm (Q) by the liquid head is opposed by air pressure which is termed the transmitted pressure. A bleed orifice (O) is provided in the diaphragm block. The rate of air flow into the transmitter equals the rate of air flow out the bleed when the diaphragm is in equilibrium.

When the liquid level increases, there is a corresponding increase in the measured pressure which moves the diaphragm toward the bleed orifice (O), restricting the flow of air to atmosphere. Supply air continues to enter the unit,

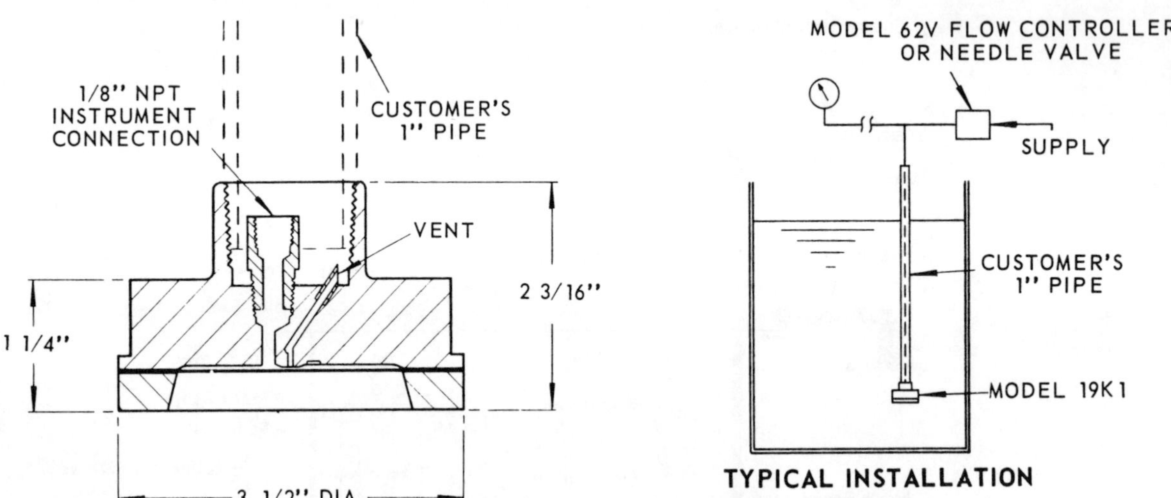

Figure 5.13. Force balance type level device senses hydrostatic head with a diaphragm and produces a 1:1 air signal using a controlled air supply. (Courtesy of Moore Products Co.)

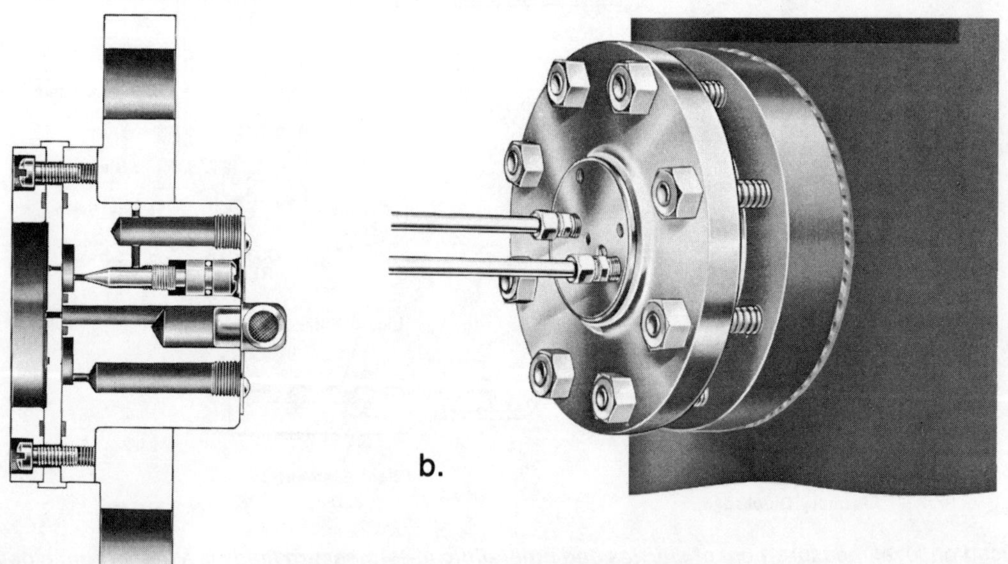

Figure 5.14. A schematic section (a) and a picture (b) of a flange mounted 1:1 repeater used for level measurement in open or closed vessels. (Courtesy of Fisher Controls Co.)

causing a buildup in the transmitted pressure until it is equal to the vessel's head pressure. The diaphragm is forced away from the orifice (*O*) by the increased transmitted pressure, preventing any further buildup, and the unit again goes into equilibrium. The transmitted pressure is piped to the sensing element (*T*) of the controller, expanding the bellows and moving beam (*C*) toward the orifice (*D*). This causes an increase in pressure in the controller relay which reacts to increase the output signal.

When the liquid level decreases, the reverse action takes place. The transmitted pressure acting on diaphragm *Q* exceeds the opposing measured pressure from the tank. The diaphragm moves away from the bleed orifice, exhausting the transmitted pressure until it is equal to the liquid head pressure. The diaphragm is returned to an equilibrium position, and any further decrease in transmitted pressure is halted. The controller senses this change in transmitted pressure and decreases its output pressure.

Proportional feedback is incorporated through the proportional band assembly (*M*) to feedback bellows *E*. Remote set point adjustment can be incorporated, when needed, through bellows *N*.

Because repeaters of this type are fairly accurate at low static pressures, they are useful for open tank level measurements or where static pressures are low. As pressures increase, the measurement error increases also, so that it may be a significant ratio of the level span measured.

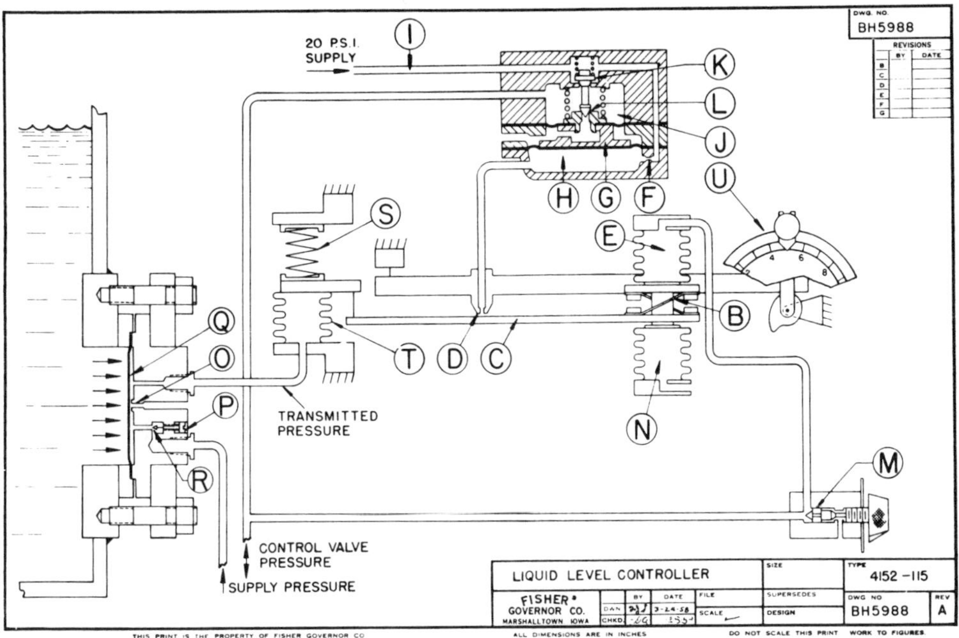

Figure 5.15. Schematic illustration of the 1:1 repeater seen in Figure 5.14 being used with a controller to achieve closed-loop control. (Courtesy of Fisher Controls Co.)

Pressure Bulb Type

Another group of level measuring devices that fits into the category of hydrostatic elements is the pressure bulb type. Figure 5.16 shows typical methods of mounting these bulbs which have different configurations as shown in the inset.

The pressure of the liquid at the bottom of the (open) tank or other body of liquid is sensed by the pressure bulb. This pressure, proportional to the height of liquid above the bulb, is transmitted to the measuring element in the gauge through an air-filled, pressure-tight tube. The resulting deflection of the measuring element provides an indication on a recording pen or indicator pointer.

Pressure bulbs vary from 2 to 10 inches in diameter and connect to the measuring element with ⅛- or 1/16-inch ID tubing. The response time is much faster when the ⅛-inch tubing is used. Figure 5.17 shows a chart used for bulb selection when level range and readout location are known. Maximum tubing lengths are shown for various bulbs and measurement ranges.

Other Differential Types

Differential measuring devices that are applicable to flow measurements using differential sensors are equally adaptable to level measurements. The discussion of manometers, mercury meters, bellows and other differential elements, therefore, will not be repeated. Reference can be made to Chapter 4 where these elements are covered, or to Chapter 6 where reference is also made to differential pressure devices.

One differential device not listed in either Chapter 4 or Chapter 6 is shown in Figure 5.18. It uses a capsular type element which moves in response to changes in differential pressure. This change is then converted to DC and amplified to produce a standard 4 - 20 ma or ±10V DC output. The linear variable differential transformer produces an AC voltage proportional to the change in differential pressure. This change is then converted to DC and amplified to produce a standard 4 - 20 ma or ± 10V DC output.

This transmitter is available in ranges from 0-50 inches H_2O to 0-1,000 inches H_2O at pressures to 6,000 psig. With an accuracy of ±½% of span, the units can be furnished for general purpose or explosion-proof electrical area classifications.

Zero Suppression-Elevation

Most manufacturers of differential pressure transmitters furnish a biasing spring kit which can be installed to compress or shift the span of the transmitter. When the bias acts to *oppose* pressure on the high side of the transmitter, this is designated "zero suppression" by the Scientific Apparatus Makers Association (SAMA). Conversely, when the bias acts to *assist* pressure on the high side, this is called "zero elevation." Some manufacturers use terms (span elevation and span suppression, respectively) that sound as though they conflict with SAMA terminology.

"Zero suppression" is often applied to cancel the effects of the liquid head in the pipe connecting the transmitter to a tank when the transmitter is mounted below the vessel con-

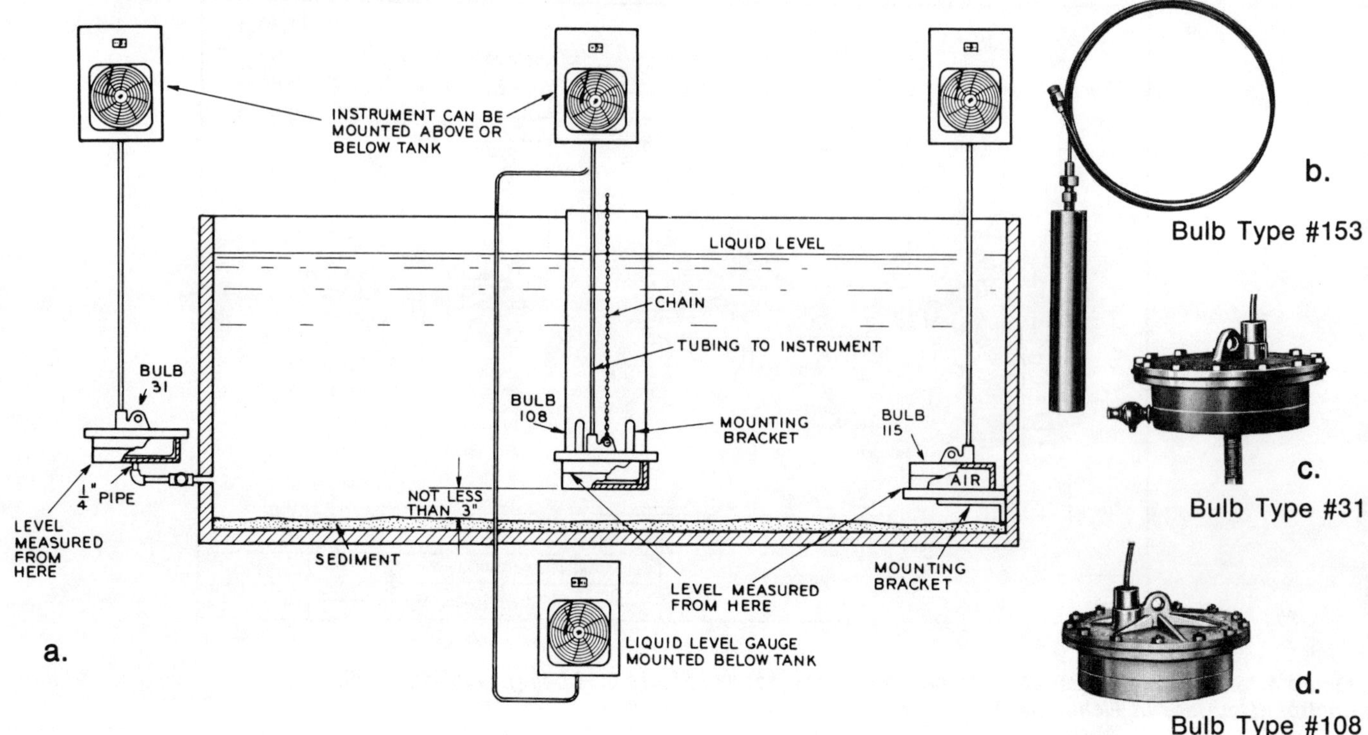

Figure 5.16. Typical installation methods (a) of various types of elements as shown in inserts b, c, *and* d. *(Courtesy of The Bristol Division of ACCO)*

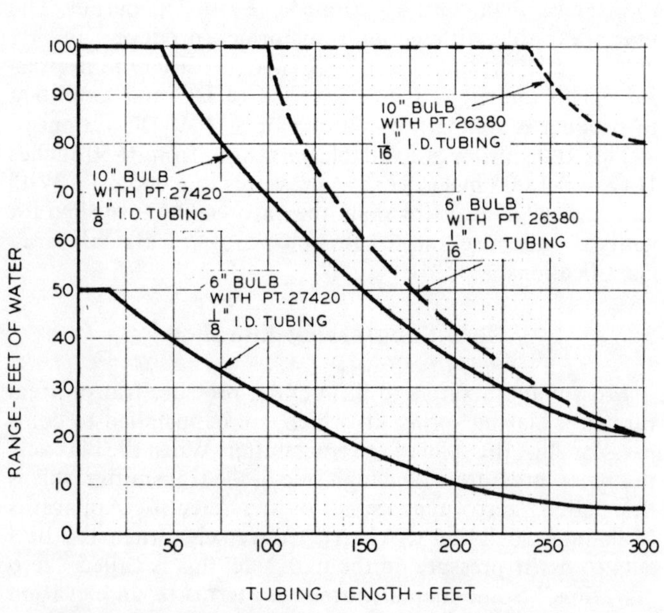

Figure 5.17. When level range and tubing lengths are known, this chart helps select bulb sizes (diameters) of Types 108 and 31 shown in Figure 5.16. (Courtesy of The Bristol Division of ACCO)

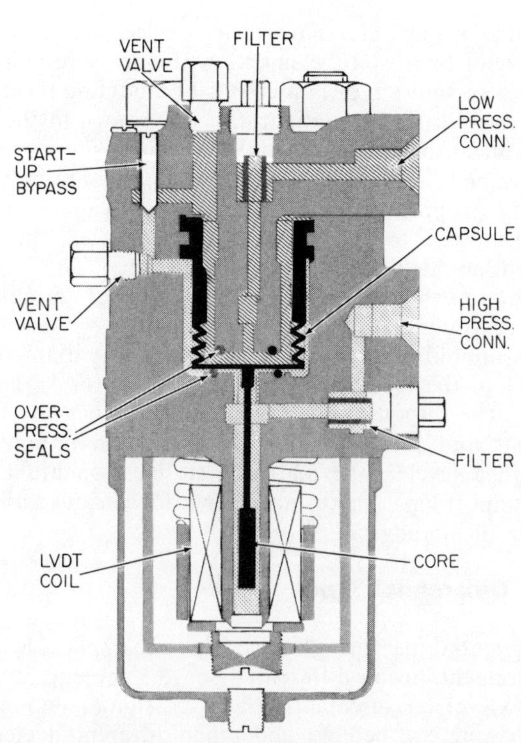

Figure 5.18. Sectional view of electronic LVDT differential pressure transmitter, using a capsular element, that may be used for level or flow service. (Courtesy of Bailey Meter Co.)

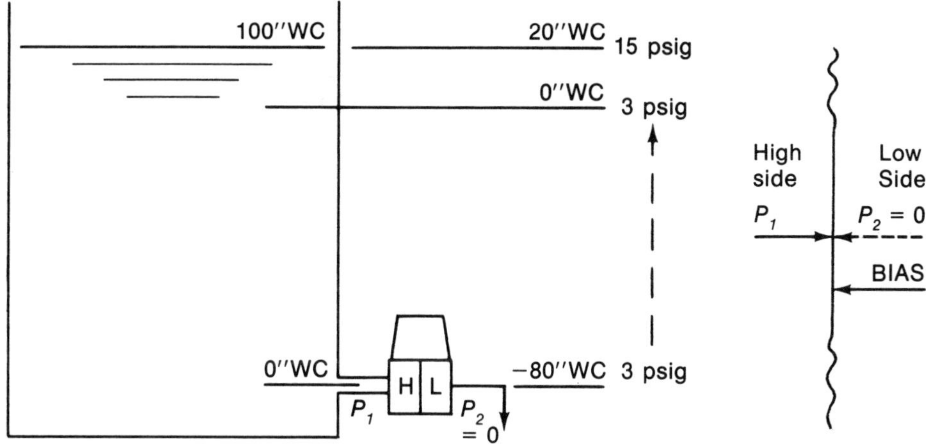

a.

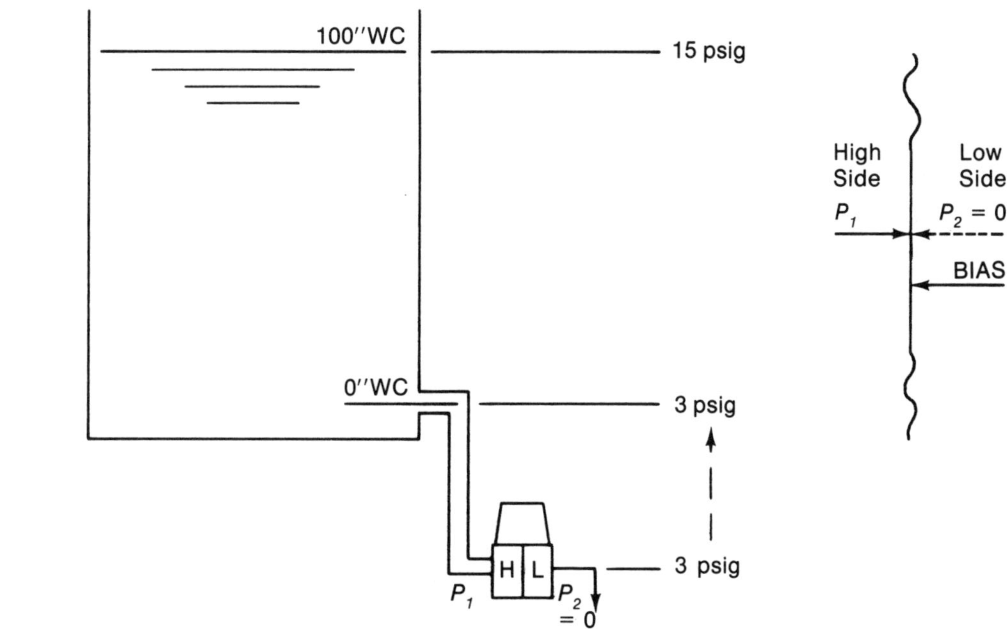

b.

Figure 5.19. Zero suppression is employed to (a) *compress the range of the transmitter or to* (b) *cancel the effects of a liquid head in the pipe connecting the transmitter to the tank.*

nection, or when level is of interest only in the upper portion of the tank (Figure 5.19).

"Zero elevation" is often applied to cancel the effects of the head caused by the seal fluid in the reference leg of a transmitter measuring level in a pressurized vessel (Figure 5.20*a*).

"Zero elevation" is also used to give a center-zero (9 psi, or 12 ma.) signal in bi-directional metering loops (Figure 5.20*b*). An exception is that most integral orifice transmitters use quadrant edge plates, which will not measure reversed flow correctly. Compound pressure ranges can also be measured in this fashion.

Float Types

There are several float type devices that are used for level measurement, indication and control. They differ from the displacement types previously discussed in that they ride or

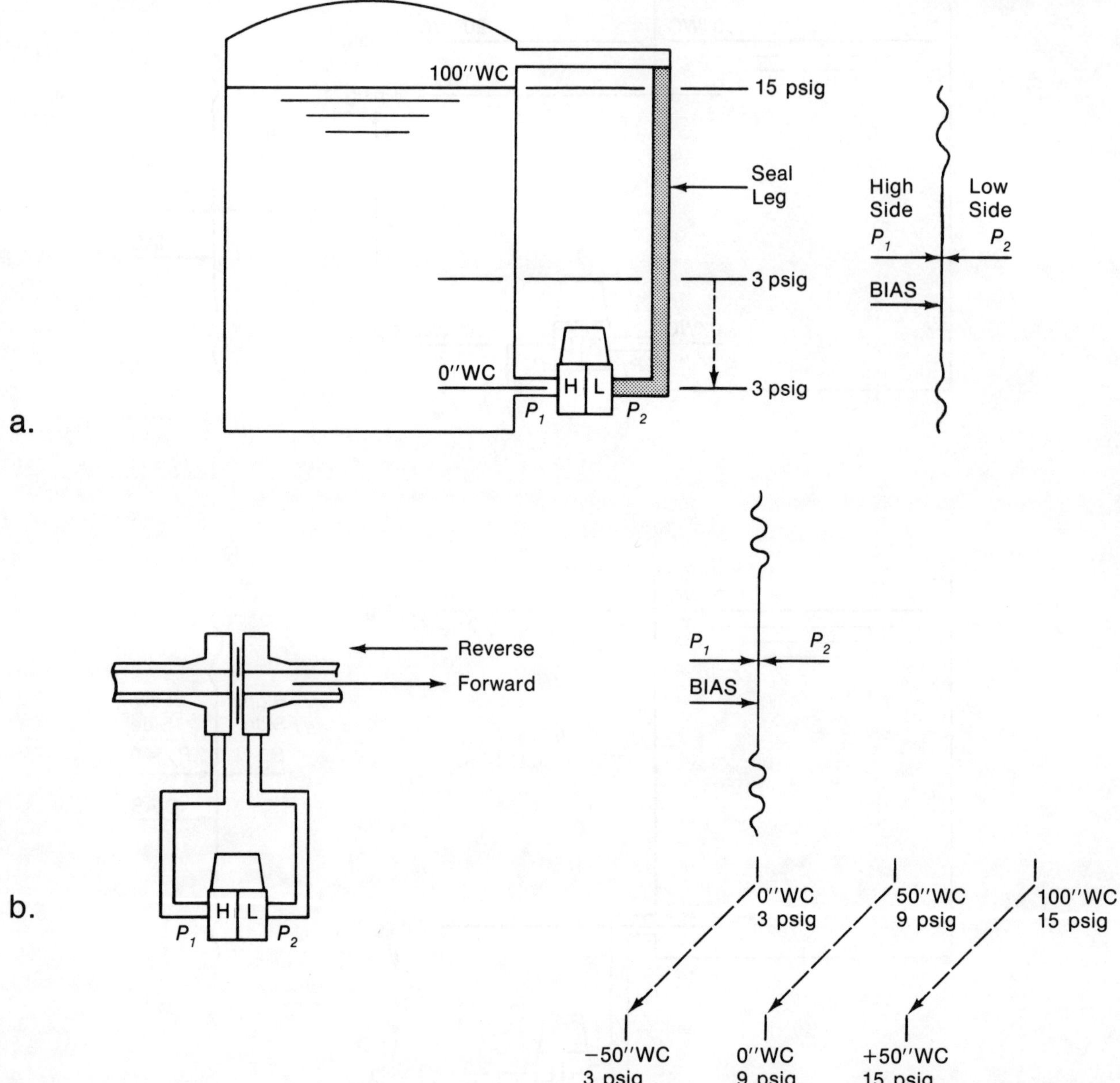

Figure 5.20. Zero elevation is employed to (a) *cancel the effects of the seal fluid and to* (b) *give center-zero signal in bidirectional flow measurements.*

float on the changing liquid level, and their motion is used for signal transmission, indication or control, whereas the displacement float remains essentially at the same elevation, and its buoyant force changes in proportion to the depth of its submerged float to provide a signal proportional to level.

Float type devices are classified broadly into five groups for discussion purposes:

1. Direct float operated transmitters and controllers
2. Pilot-operated devices
3. Direct float operated valves
4. Tape type gauges
5. Float switches.

Float switches are covered in a separate section on level switches, so only the first four groups are discussed below.

The use of float devices for signal transmission, direct control, direct valve operation and switch action is probably not as great as it was a few years ago—at least not percentage-wise. They perform well in clean services, but for slurries, corrosive services and other applications requiring

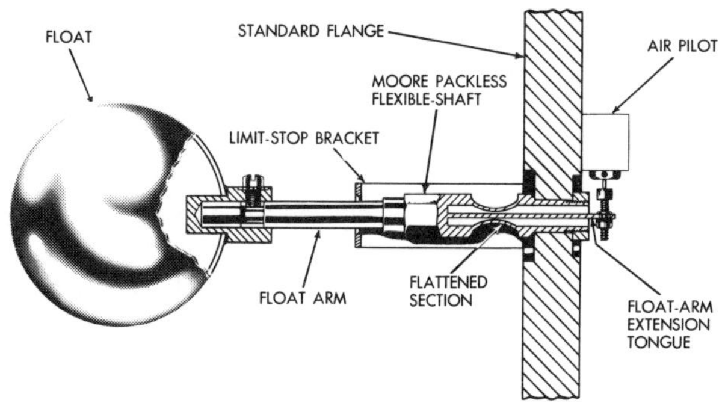

Figure 5.21. Direct float-operated internally-mounted level controller used for pneumatic snap-acting or throttling action (¾ inch span) control. (Courtesy of Moore Products)

special alloy materials, more economical and/or more dependable measuring methods are used.

Direct Float-Operated Level Devices

Figure 5.21 shows an internally mounted float and control assembly that is used for snap acting level control or for throttling applications. It is made by Moore Products and utilizes a packless flexible shaft to transmit float motion to the pneumatic relay.

The tabular shaft has a flattened center section which permits only vertical float motion. It operates on the buoyancy principle rather than on float motion resulting from changing level.

Throttling ranges may vary from 1 inch with 3½-inch diameter balls to 14 inches with larger cylindrical or eliptical floats. Standard pneumatic output is 3- to 15-psig over the throttling range.

Snap-acting pilots may be used instead of throttling relays to provide "on-off" action. The snap-acting pilot actually provides a two-position differential-gap control in which the gap is a fraction of an inch in level change.

These units also may be furnished with microswitches for electrical switching instead of pneumatic operation. Switches are furnished for AC voltages from 110 to 460 and are housed in weatherproof or explosion-proof fittings.

Pneumatic units are available with 4-inch, 150-pound mounting flanges. Electrical units are furnished with 4-inch, 150-pound or 6-inch, 300-pound flanges.

Pilot-Operated Devices

Figure 5.22 illustrates one of several designs of pilot-operated ball float controllers. The pilot valve shown provides a 3- to 15-psig signal for the range of float movement available from the external cage mounted float.

The control system operates in this manner. The pilot relay is adjusted to give a 9-psi (midpoint of range) output to the control valve when the float is at the midpoint of the float cage. When the level rises, the rising float pushes

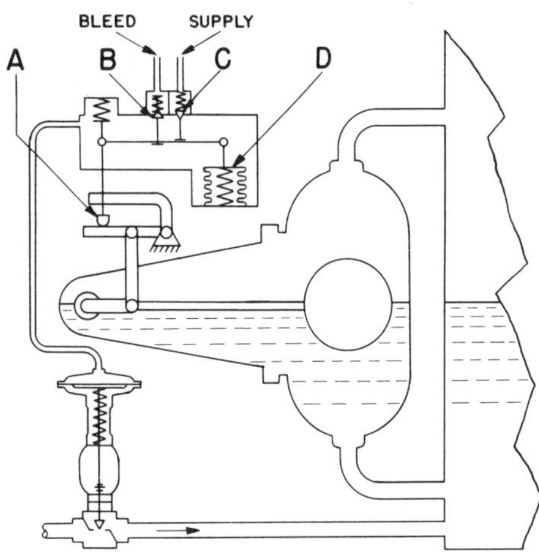

Figure 5.22. The external cage-mounted ball float controller uses pneumatic pilot to provide 3-15-psi throttling range on liquid level control applications. (Courtesy of Fisher Controls Co.)

pusher post *A* upward, tending to close exhaust valve *B* and to open supply valve *C*. Operating medium then flows through supply valve *C*, thus increasing the pressure on the diaphragm of the control valve, causing it to move toward its seat.

At the same time, the pressure is being increased on the bellows *D*, which causes it to move downward and close supply valve *C*. This stops the pressure buildup in the pilot. The unit is again in equilibrium with the level at a higher point. When the level drops, the reverse of the above cycle occurs, with a decrease in liquid level, causing an increase in motor valve opening.

The same type of pilot relay can also be mounted on internally mounted floats to provide similar operation. The relay described is one of several that may be used with the float assemblies shown or described.

Direct Float-Operated Valves

Direct operated float valves have been in use for a long time and still find application in services where narrow level spans are used and where close control is not a prerequisite. They are relatively simple, but because there are several moving parts in a system, a fair amount of maintenance is necessary.

Figure 5.23 shows a typical arrangement of an assembly that is rated at 600 psig and for temperature ranges from -300° to 450°F. The valve is operated by the movement of the float through the levers and turnbuckle assembly. Power generated by the float is dependent on float diameter and density of the process fluid.

The greatest application for float-operated valves is probably on open tank installations requiring level control.

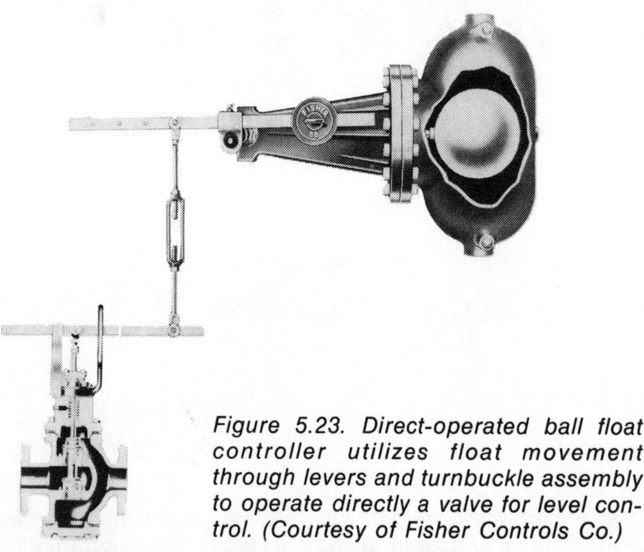

Figure 5.23. Direct-operated ball float controller utilizes float movement through levers and turnbuckle assembly to operate directly a valve for level control. (Courtesy of Fisher Controls Co.)

Figure 5.24. Typical float actuated level valve used on open tanks or vats for level control. (Courtesy of Fisher Controls Co.)

Figure 5.24 shows a typical float-operated level system for open tanks, vats or reservoirs.

The actuator can be positioned to any desired angle by loosening the yoke locknut and rotating the yoke. Valve action can be either rising level to open or rising level to close valve depending on the level arrangement. The position of the float relative to the lever can be adjusted through a 90° arc by means of the adjustable lever segment. By changing the angle, the level of the vessel liquid may be raised or lowered.

For the float system shown, valves are available in sizes from ½ to 4 inches; float sizes are 8 inches (10 inches for 4-inch valves); and maximum pressure drops range from 100 psi for the smallest valve to 25 psi for the 4-inch valve. The float rod is about 30 inches long.

Tape Gauges

Tape level gauges have been used for a long time. One of the simplest designs is the one shown in Figure 5.25. One end of a tape is connected to a float which rises and falls with liquid level. On the other end, a counterweight indicates liquid level as it slides along a calibrated gaugeboard.

The float is guided by two 316 stainless steel guide wires anchored securely at the tank bottom and fastened to a spring-loaded top anchor to keep the wires in tension. The tape runs through sheave elbows to the counterbalance indicator which is also guided by two galvanized steel guidewires. The 15-inch diameter float shown is a 316 stainless steel jacketed foam glass float, but hollow metal floats and solid plastic floats are also used.

The gaugeboard is 1 x 6-inch redwood normally but can be furnished in aluminum, steel or stainless steel on special order. Gaugeboards are usually calibrated in feet and inches but could be calibrated in volumetric units. These types are used primarily on atmospheric services.

In services that operate slightly above or below atmospheric pressure, liquid seals (Figure 5.26) may be used. The seal protects against corrosion, prevents inbreathing and outbreathing through the tape guiding system and avoids vapor condensation in the gauge head assembly when readouts other than gaugeboards are used. The accuracy of this unit is ±1 inch.

Figure 5.27 shows a typical gauge head for local readout when levels of pressurized vessels must be measured. This gauge is located at a convenient level, and a perforated tape moves a dial counter to provide level readings accurate to within ±1/16 inch. A special motor is used to provide a constant tension on the tape, eliminating a cumbersome counterweight.

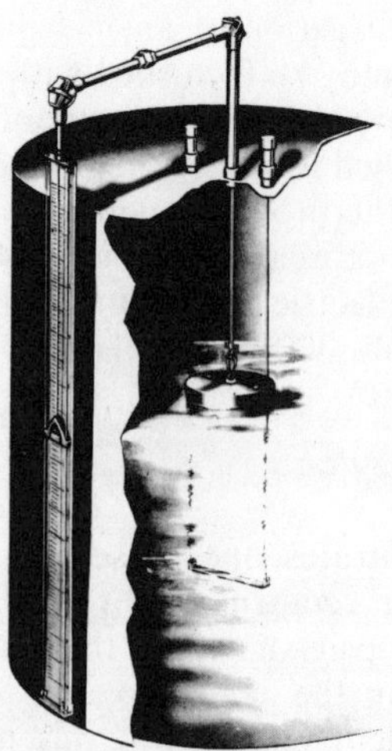

Figure 5.25. A counter-balanced indicator attached to float reads tank level directly on gauge board in whatever units are desired. (Courtesy of Varec, Inc.)

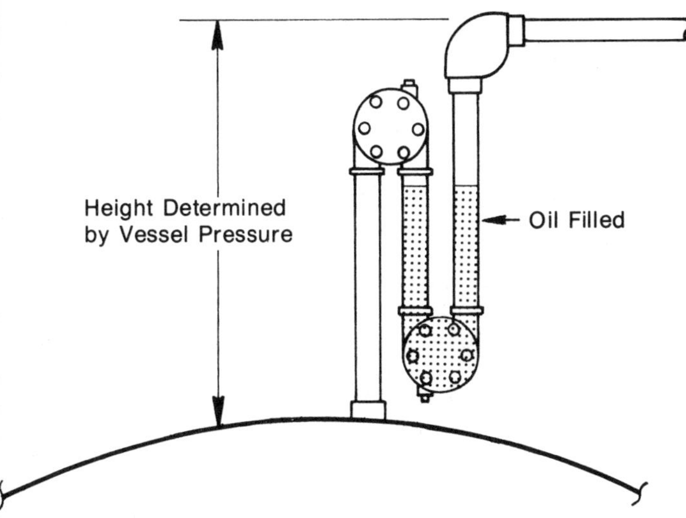

Figure 5.26. On tape type level devices where operation is slightly greater or less than atmospheric pressure, oil seals are often used to protect against corrosion or prevent in-breathing and outbreathing.

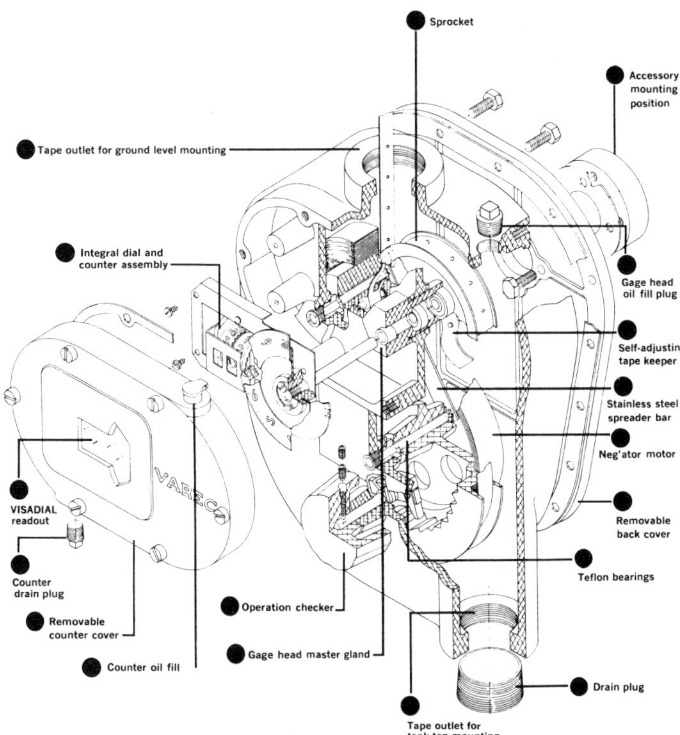

Figure 5.27. Gauge head for local readout of tape type level indicator for pressurized vessels. (Courtesy of Varec, Inc.)

The readout head of the assembly shown can be oil filled to prevent corrosion and rust in the corrosive atmospheres in which these units are often installed.

Limit switches and level gauge printers can be attached to gauge heads to provide control actions and level data printout. Pressures up to 50 psig can be accommodated by this system.

A different concept in tape devices is shown in Figure 5.28. It combines the tape concept with a displacer device and consists of a force measuring mechanism using a torque tube, arm and contacts (Figure 5.29) and a displacer made of a solid-plate material.

The displacer is suspended on the torque arm by a conventional perforated steel tape. The weight of the displacer applies a known load or force on the torque tube. Since the displacer is partially submerged in the material to be measured and thus displaces a certain amount of liquid, this reduces the effective pull of the displacer weight on the torque arm. The difference between the gross weight of the displacer and the net weight of the displacer when partially submerged is simply the weight of the displaced liquid. Since the area of the displacer is constant, any change in the liquid level produces a proportional change in the force on the torque tube.

The displacer is driven up and down (in and out of the liquid) by a reversible two-phase servo motor. The complete system constitutes a force-balance system. If the load (force) on the torque tube is greater than the net displacer load, the motor drives the displacer down so that a greater amount of liquid is displaced, thus reducing the force. If the force on the torque is less than the net displacement load, the motor drives the displacer up so that less liquid is displaced; therefore, the force is increased. By sensing the net force produced by the displacer, the torque tube controls the drive motor so that balance is always restored with the displacer being accurately positioned in relation to the liquid level.

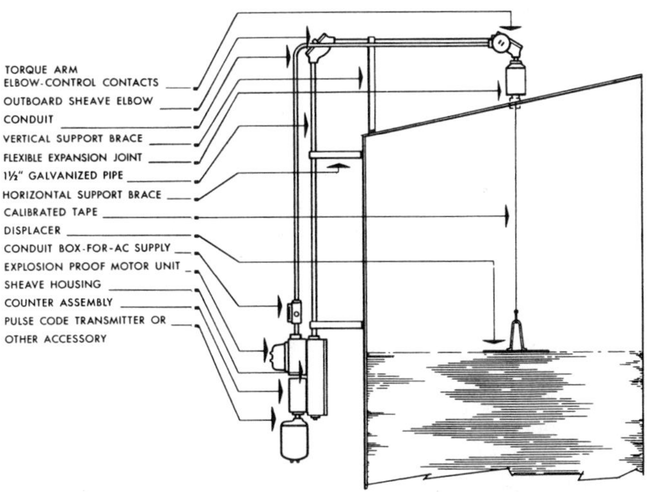

Figure 5.28. A solid-plate displacer unit imparting force to a torque tube assembly provides a varying force to an electrically-driven readout system in this unique level measuring system. (Courtesy of Varec, Inc.)

Net weight of the tape decreases as the liquid level increases. This condition causes the liquid line of the displacer to vary according to the liquid level. Compensation for the

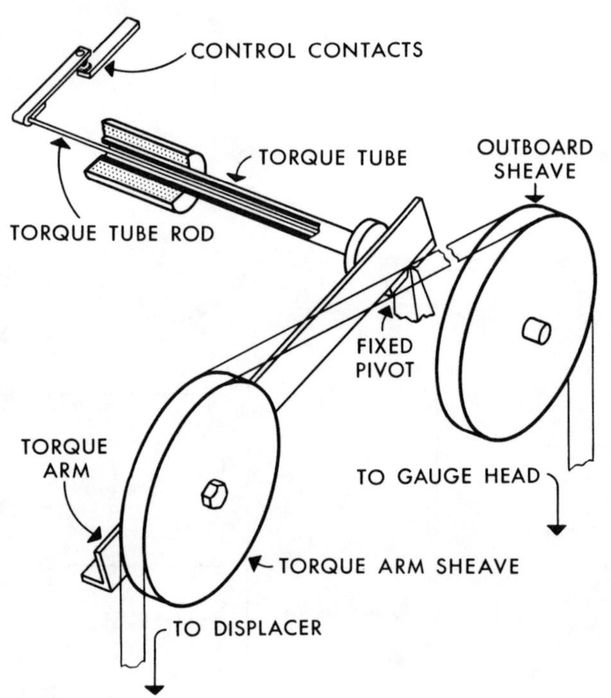

Figure 5.29. The torque tube assembly used in the displacer-tape unit shown in Figure 5.28. (Courtesy of Varec, Inc.)

varying net weight of tapes is accomplished by positioning the storage sheave and sprocket sheave so that tape buildup on the storage sheave shortens the effective length of tape.

This unit is also available for service pressures up to 50 psig. If remote gauging is desired, pulse code or electronic transmitters can be provided along with compatible solid-state receiving equipment. Accuracy of the system is ±1/16 inch.

Capacitance Type

Electrical capacitance sensors are used to detect level changes in liquids and solids. They use one of the fundamental elements of electrical circuits, the capacitor or condenser.

A capacitor consists of two conductors called plates separated by an insulator referred to as the dielectric. Normally one thinks of a capacitor as a pair or set of small parallel plates, separated by air or other special dielectric. In using capacitance to measure level, however, a probe similar to that shown in Figure 5.30 is used as one plate while the vessel serves as the other plate. The material between the probe and the vessel is the dielectric.

The impedance to current flow of a capacitor is determined by its size measured in farads (the farad is a measure of capacitance). The more practical unit of capacitance is the microfarad (10^{-6} farad). Capacitance is a function of the dielectric, the area of the plates and the distance between them. It may be expressed mathematically as

$$C = KA/d$$

where C is capacitance, K is the dielectric constant, A is the area of the plates and d is the distance between the plates.

As one views Figure 5.30, it is evident that the area of the plates and the distance between them are both fixed. The dielectric varies with material level in the vessel, and this variation is used in the measuring bridge circuit to provide a signal proportional to level.

Table 5.2 shows dielectric constants for a variety of materials (note that air has a value of 1 and most gases have values near 1). Water has a dielectric constant of 80 with many common materials somewhere between. The best measurements are obtained when capacity changes are high, and this occurs when the difference between the dielectric constants are relatively high. The capacitance is a direct function of the dielectric constant and increases as the level of the measured material (high K) replaces the air or vapor (low K) in a vessel.

Capacitance devices (usually referred to as capacitance probes) are most frequently used for on-off service as switches for alarm or control functions. For such functions they are usually mounted in a horizontal arrangement (Figure 5.31) so that a larger plate area is effectively used. For liquid services, small round rods are sufficiently large. On some solids measurement applications, larger probe plate areas are made available through use of a wide knife-edge design.

Figure 5.32 shows a block diagram of a typical on-off capacitance probe level switch. The system consists of a detector unit and a display-control unit; both are transistorized solid-state devices.

The detector unit has an oscillator operating at approximately 2 megahertz coupled to an inductance-capacitance (LC) bridge circuit that varies the level of oscillation as a function of the (measuring) bridge unbalance. One leg of the bridge circuit consists of the zero capacitance adjustment, and the other leg consists of the detecting probe capacitance. As the probe capacitance is changed by the process material level, the bridge becomes unbalanced and a voltage output from the bridge is produced, rectified and is fed to a trigger circuit, which converts it to a high level on-off control current for relay operation within the display unit.

The display unit supplies regulated DC power (27 volts) to the detector unit and receives the output current from the detector which actuates its control relay and signal lamps.

The illustration given utilizes separate housings for the detector unit and the display/control unit. Capacitance probes are available with both functions in a single housing, but the alarm lights (display) are usually omitted.

When continuous level measurement is desired, the probe is placed vertically in the vessel (as in Figure 5.30) so that the plate (probe) area exposed to the material dielectric changes with material level.

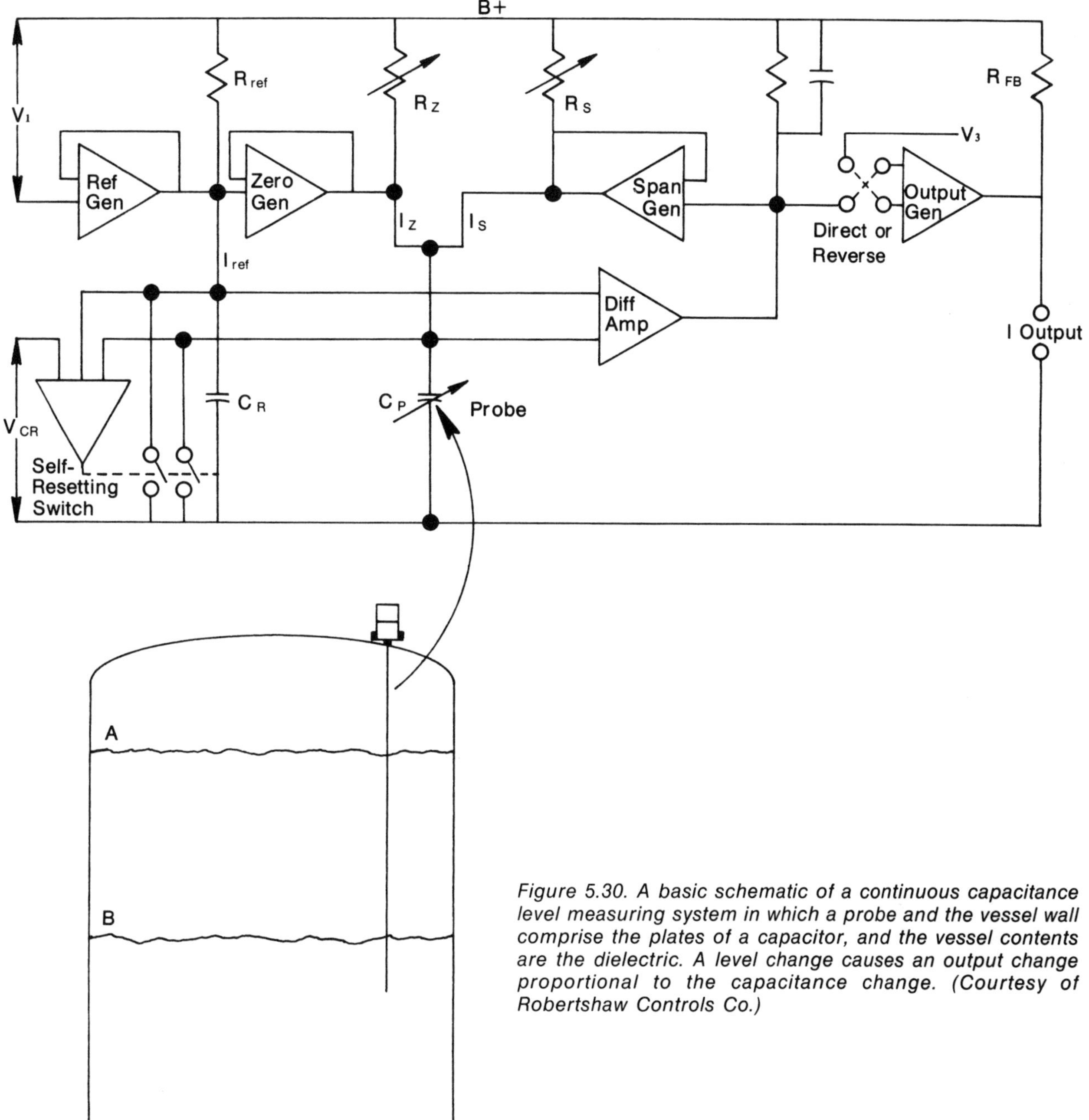

Figure 5.30. A basic schematic of a continuous capacitance level measuring system in which a probe and the vessel wall comprise the plates of a capacitor, and the vessel contents are the dielectric. A level change causes an output change proportional to the capacitance change. (Courtesy of Robertshaw Controls Co.)

One technique for continuous measurement utilizes a design concept (Figure 5.33) which measures the capacitance (or level) by the time required to charge the probe (capacitance). In Figure 5.33, the measurement is accomplished as described below.

The reference generator delivers a constant current to charge capacitor C_R. The voltage across capacitor C_R is charged linearly until it reaches a value V_{CR} equal to the reference voltage. Simultaneously, the probe capacitance C_P is being charged from the zero and span generators. When capacitors C_R and C_P are charged to the reference voltage V_{CR}, the self-resetting electronic switch momentarily closes, causing the discharge of capacitors C_R and C_P. At the same time, the resetting switch automatically opens, and the charging/discharging cycle is again repeated.

A differential amplifier is connected across the capacitors C_R and C_P in order to compare their charging rates, and its output causes the span generator to maintain an equal charging rate to capacitors C_R and C_P. When the probe capacitance C_P increases, the span generator output will

(text continued on page 122)

Table 5.2 Dielectric Constants of Materials

Materials	Dielectric Constant	Materials	Dielectric Constant
Gases			
Air	1.0		
Solids			
Acetic Acid (36° F)	4.1	Paper	45.0
Aluminum Phospate	6.1	Phenol (50° F)	2.0
Asbestos	4.8	Polyethylene	4-5.0
Asphalt	2.7	Polypropylene	1.5
Bakelite	5.0	Porcelain	5-7.0
Barium Sulfate (60° F)	11.4	Potassium Carbonate (60° F)	5.6
Calcium Carbonate	9.1	Quartz	4.3
Cellulose	3.9	Rice	3.5
Cereals	3-5.0	Rubber (hard)	3.0
Ferrous Oxide (60° F)	14.2	Sand (Silicon Dioxide)	3-5.0
Glass	3.7	Sulphur	3.4
Lead Oxide	25.9	Sugar	3.0
Lead Sulfate	14.3	Urea	3.5
Magnesium Oxide	9.7	Teflon	2.0
Mica	7.0	Zinc Sulfide	8.2
Napthalene	2.5		
Nylon	45.0		

Granular and Powdery Material

Material	Dielectric Constant-Loose	Dielectric Constant— Packed
Ash (Fly)	1.7	2.0
Coke	65.3	70.0
Gerber Oatmeal	1.47	Not tested
Linde 5A Molecular Sieve, Dry	1.8	Not tested
20% Moisture	10.1	Not tested
Polyethylene	2.2	Not tested
Polyethylene Powder	1.25	Not tested
Sand-Reclaimed Foundry	4.8	4.8
Cheer	1.7	Not tested
Fab (10.9% Moisture)	1.3+	1.3+
Tide	1.55	
VEL (0.8% Moisture)	1.25	1.25

Table 5.2 continued

Table 5.2 continued

Dielectric Constants		
Liquids		
Material	**Temperature, °F**	**Dielectric Constant**
Acetone	71	21.4
Amonia	-30	22.0
Amonia	68	15.5
Aniline	32	7.8
Aniline	68	7.3
Benezene	68	2.3
Bromine	68	3.1
Butane	30	1.4
Carbon Dioxide	32	1.6
Carbon Tetrachloride	68	2.2
Castor Oil	60	4.7
Chlorine	32	2.0
Chlorocyclohexane	76	7.6
Chloroform	32	5.5
Cumene	68	2.4
Cyclohexane	68	2.0
Dibromobenzene	68	8.8
Dibromohexane	76	5.0
Dowtherm	70	3.36
Ethanol	77	24.3
Ethyl Acetate	68	6.4
Ethylene Chloride	68	10.5
Ethyl Ether	-40	5.7
Ethyl Ether	68	4.3
Formic Acid	60	58.5
Freon-12	70	2.4
Glycerine	68	47.0
Glycol	68	41.2
Heptane	68	1.9
Hexane	68	1.9
Hydrogen Chloride	82	4.6
Hydrogen Sulfide	48	5.8
Isobutyl Alcohol	68	18.7
Kerosene	70	1.8
Methyl Alcohol	32	37.5
Methyl Alcohol	68	33.1
Methyl Ether	78	5.0
Napthalene	68	2.5
Octane	68	1.96
Oil, Transformer	68	2.2
Pentane	68	1.8
Phenol	118	9.9
Phenol	104	15.0
Phosphorus	93	4.1
Propane	32	1.6
Styrene (Phenylethene)	77	2.4
Sulphur	752	3.4
Sulphuric Acid	68	84.0
Tetrachloroethylene	70	2.5

Table 5.2 continued

Table 5.2 continued

Material	Temperature, °F	Dielectric Constant
Toluene	68	2.4
Trichloroethylene	61	3.4
Urea	71	3.5
Vinyl Ether	68	3.9
Water	32	88.0
Water	68	80.0
Water	212	48.0
Xylene	68	2.4

increase in order to maintain the charging rate equal to the reference capacitor C_R. The output generator produces an output signal from the instrument, which is proportional to the change in the charging rate of the probe capacitor C_P. Thus, changes in capacitance at the probe produces an output current which is linearly proportional to the capacitance change at the probe.

The span capacitance range of this unit is from 10 to 1,000 picofarad (pf) ; the output signal can be 1 to 5, 4 to 20, 10 to 50, or 0 to 4 ma DC. Linearity is from ±¼% of span (for spans below 200 pf) to ±1.5% of span (for spans greater than 1,000 pf). Repeatability is very good (±0.1%) and ambient temperature effects are relatively small.

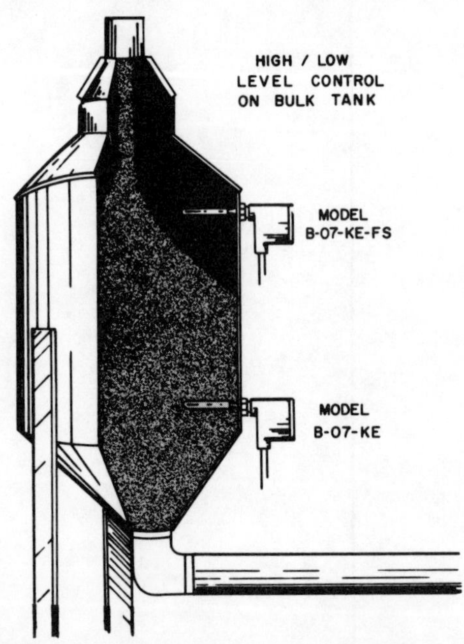

Figure 5.31. In bulk material services, capacitance probes are usually mounted horizontally to provide almost instant use of a large plate area to rising material. (Courtesy of CE IN-VAL-CO.)

Figure 5.34 shows another variation of a continuous measurement. The area of the plate is made large, and the capacitance between the sensing plate and the material surface is compared with a reference capacitor in a control unit. Any variation in the material level changes this variable capacitance and produces a corresponding change in the output current.

This type probe is referred to as a proximity probe, for it is not necessary for the probe to come in contact with the measured material. Materials which coat or build up on capacitance probes sometimes affect their accuracy or render them useless. This is one approach that may avoid that problem.

Another method of minimizing the effects of material build-up on the probe is to precoat the probe with a material whose dielectric constant exceeds that of a reasonable build-up. The build-up then has little effect on the calibration of the probe.

Capacitance probes can be used for many difficult-to-measure level applications, especially for on-off functions. There are several conditions, however, that should be considered prior to their use.

1. Dielectric constants change with temperature about 0.1% per degree Centigrade. Process temperature variations should be checked to determine its effect on accuracy.
2. Chemical changes of the measured material affect the dielectric constant. Moisture content has a pronounced affect.
3. In solids measurement, variations in particle size affect the dielectric constant.
4. Product-coating of the probe may affect the setpoint (relative to actual level), or if the coating is conductive, it may ground the probe. Some probe designs are offered which claim to eliminate this effect.
5. The minimum differential capacitance required for level measurement should be as high as 10pf (smaller ranges are available).

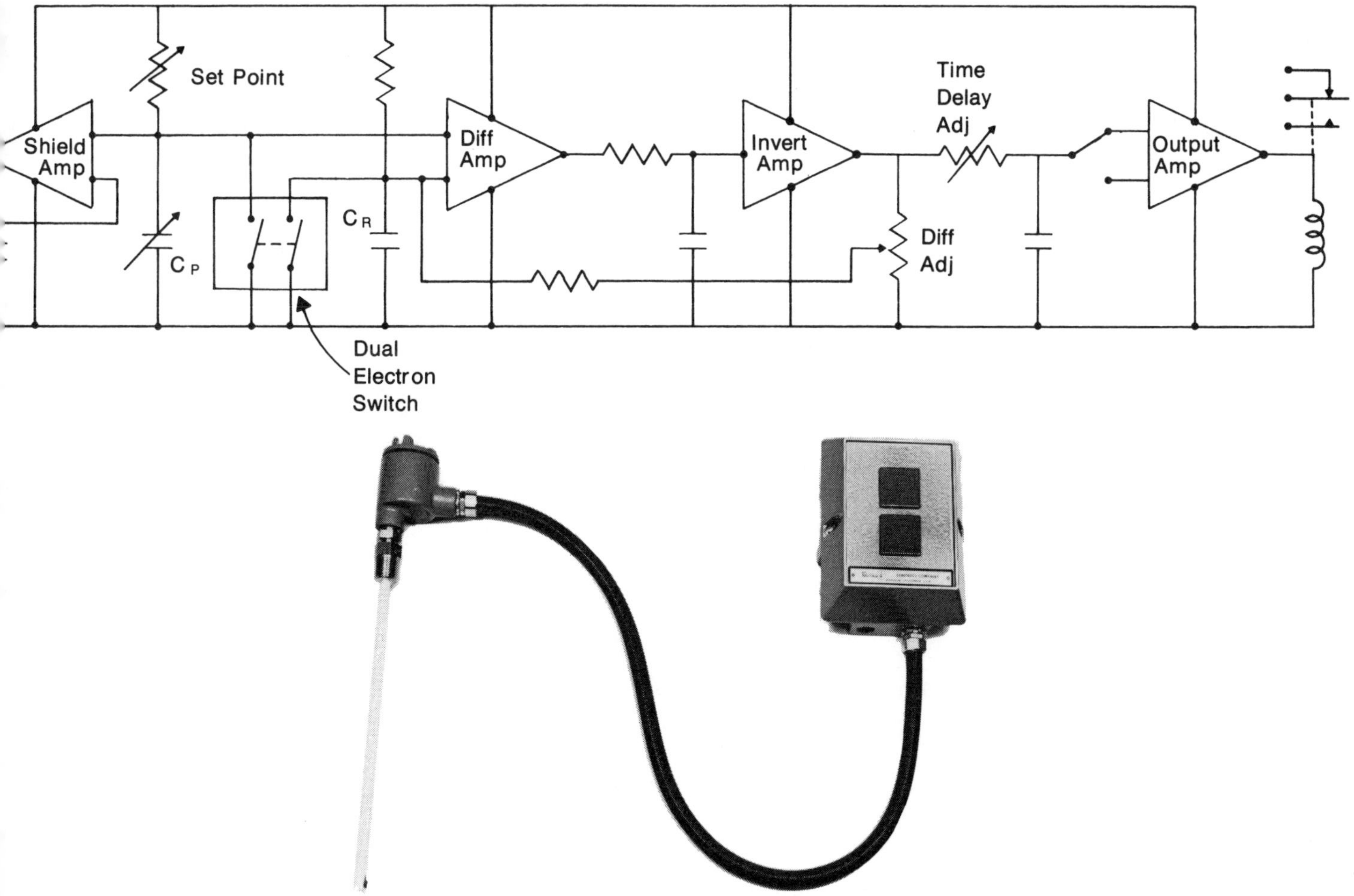

Shield Amp
Set Point
Diff Amp
Invert Amp
Time Delay Adj
Output Amp
C_R
C_P
Diff Adj
Dual Electron Switch

Figure 5.32. Block diagram of Model 310 control unit which is a typical capacitance actuated level switch. Model 310 is housed in explosion-proof or general purpose cases to meet are requirement of classification. (Courtesy of Robertshaw Controls Co.)

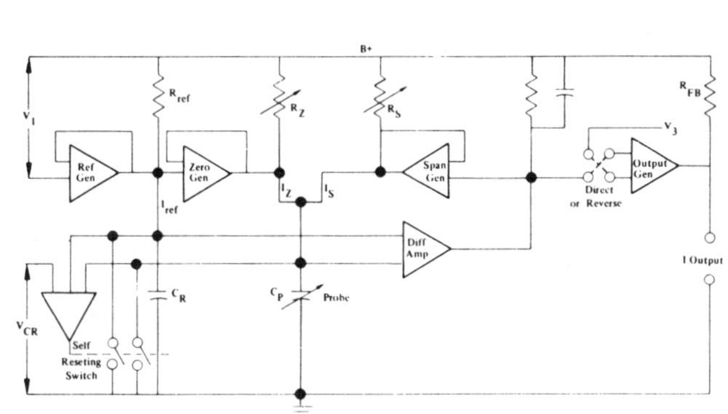

Figure 5.33. Block diagram of a continuous capacitance level measuring system that utilizes the concept of measuring the time required to charge the probe (capacitor) as a measure of level. (Courtesy of Robertshaw Controls Co.)

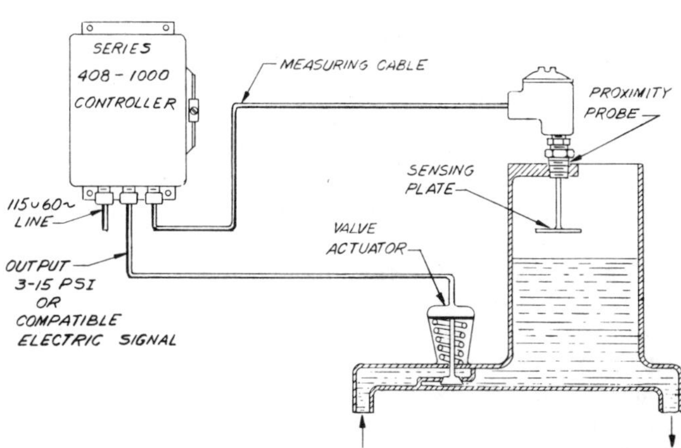

Figure 5.34. A continuous type capacitance system utilizing a large plate area allows enough capacitance change that the plate does not have to come in contact with the measured material. (Courtesy of Drexelbrook Engineering Co.)

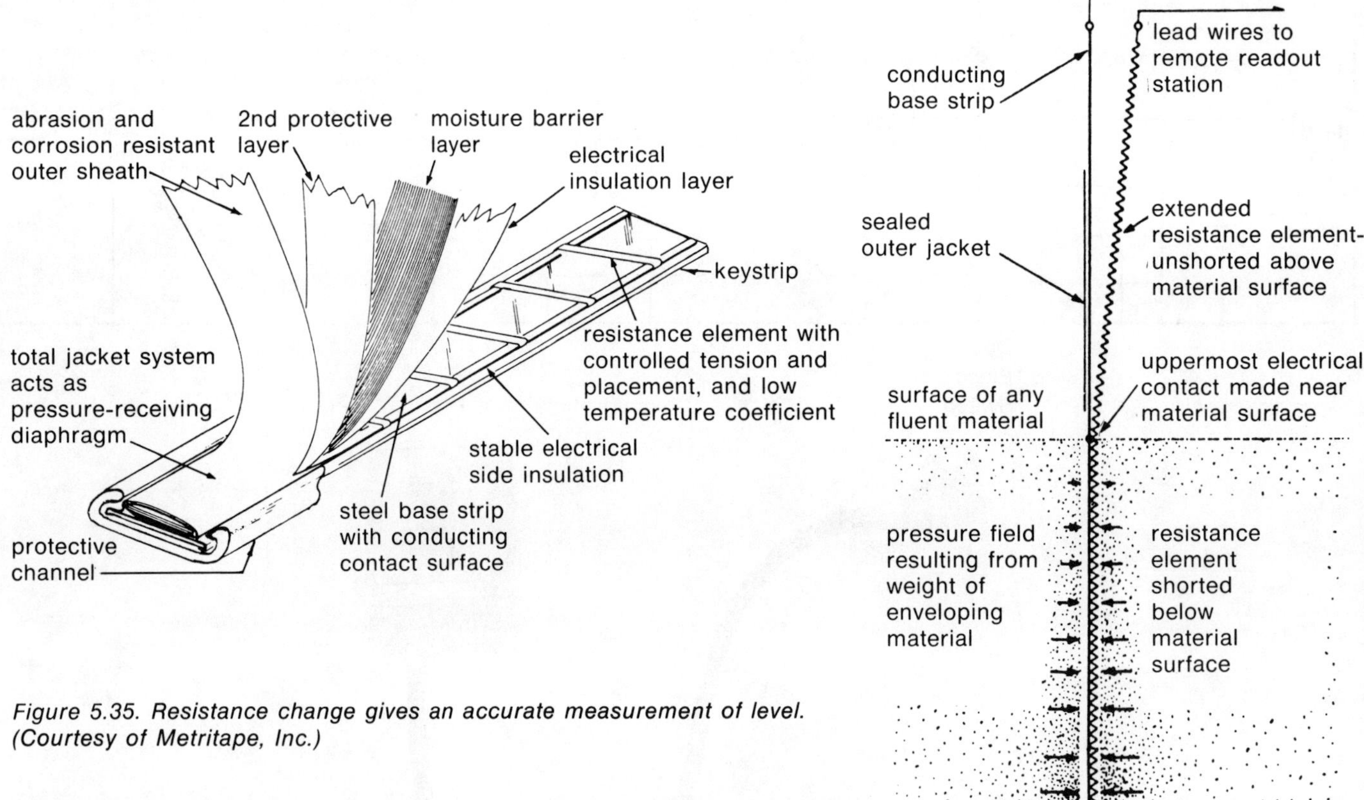

Figure 5.35. Resistance change gives an accurate measurement of level. (Courtesy of Metritape, Inc.)

The advantages of using capacitance probes include their simple design, no moving parts and ease of installation and maintenance.

Resistance Sensor

A recent device measures liquid or dry level when squeezed together. Called Metritape, it operates using resistance as an analog (Figure 5.35). Two elements, one made of stainless steel (301) and one of helically-wound nichrome, are separated by a polyester insulation open on one side. Precision wrapping of the nichrome over the slot gives uniform separation of the conductors. An outside insulating sheath, sealed at the bottom and vented at the top, covers the sensor.

As measured material rises, its weight shorts out part of the helically-wound nichrome, reducing overall sensor resistance. Resistance change is 100 ohms per linear meter (30.5 ohms per foot), and the readout device can be located thousands of feet away from the sensor. Integral RTD temperature sensors can be provided. Resistance of sensors can be measured directly by ohmmeter or by computer, or easily transduced to conventional analog signals. Zener diode barriers are available for installations requiring intrinsic safety.

This type of detector has been used not only for tank and open channel measurements but also for insertion into underground storage caverns. By using a special roller assembly, position signals can be obtained for large moving structures such as cranes, elevators, etc.

Ultrasonic Type

Ultrasonic level devices were introduced on the market about 1960. They are used for both continuous and point level measurement but, like capacitance units, are used primarily as switches for alarms or on-off control actions.

In all designs, a high frequency sound signal is generated (from 20 to 40 Hz normally), and the interruption or detection of the generated signal is the basis for a control action—for the point detectors. On continuous level measurements, the elapsed time is measured from signal emission to reception of a reflected signal, the elapsed time being proportional to level.

Ultrasonic devices are categorized here on the basis of point or continuous measurement. They may be classified further under these two categories.

Continuous Measurement

Single Systems

Figure 5.36 illustrates the principle used for continuous measurement. The concept is embodied in an elapsed time

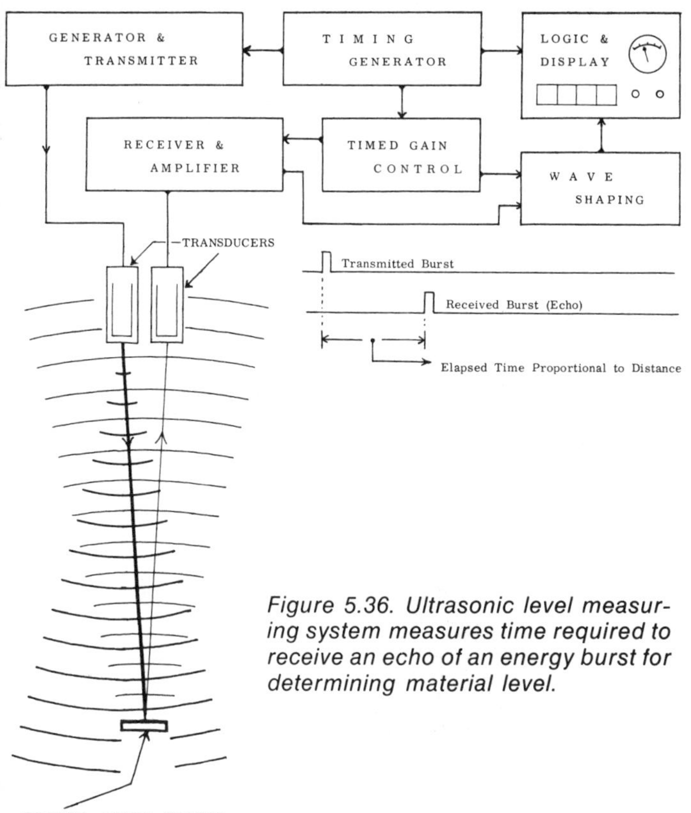

Figure 5.36. Ultrasonic level measuring system measures time required to receive an echo of an energy burst for determining material level.

measurement, an interval which starts at the sending of an ultrasonic signal by the transmitter sensor, and is terminated by the reception of the echo returned by the material being measured. The speed of sound traveling through various media changes appreciably with the media and with temperature and pressure conditions of the media. For example, the velocity of sound through air at 20°C is 1,129 ft/sec while it is 1,266 ft/sec in air at 100°C. Sound velocity through water at 15°C is 4,714 ft/sec and through alcohol at 20.5°C it is 3,890 ft/sec. Obviously, the media and conditions surrounding the media must be known. When they are known, the determination of distance traveled can be related directly to level, time or volume—the readout units are optional.

The sensors must be mounted and aimed to direct the ultrasonic signal toward the material to be sensed to provide the most direct paths of measurement. Measurements can be made for liquids or solids.

Some manufacturers use a single device to transmit the signal and receive its echo. Other manufacturers use separate devices, a transmitting sensor and a receiving sensor.

An interesting application of such a device is the measurement of the product/brine interface in underground salt dome storage wells. A transmitter-receiver is lowered down the well tubing (Figure 5.37). When it impacts the cavern floor, a catch releases a mechanism that allows the transducer to be raised to its upward-pointing operating position. To retrieve the assembly, a spring-clamp is dropped, compressing the mechanism into a size that can be withdrawn up the string. A packing gland at the wellhead seals the winch cable/signal cable against leakage. A pulse is emitted upward toward the interface, which reflects it back to the transducer, where it is converted back to an electrical pulse. The time displacement between the initial pulse and the return pulse is read out as interface level. Knowing the product/brine interface level, the depth of the string and the internal cavern contours, the product volume stored in the cavern can be calculated.

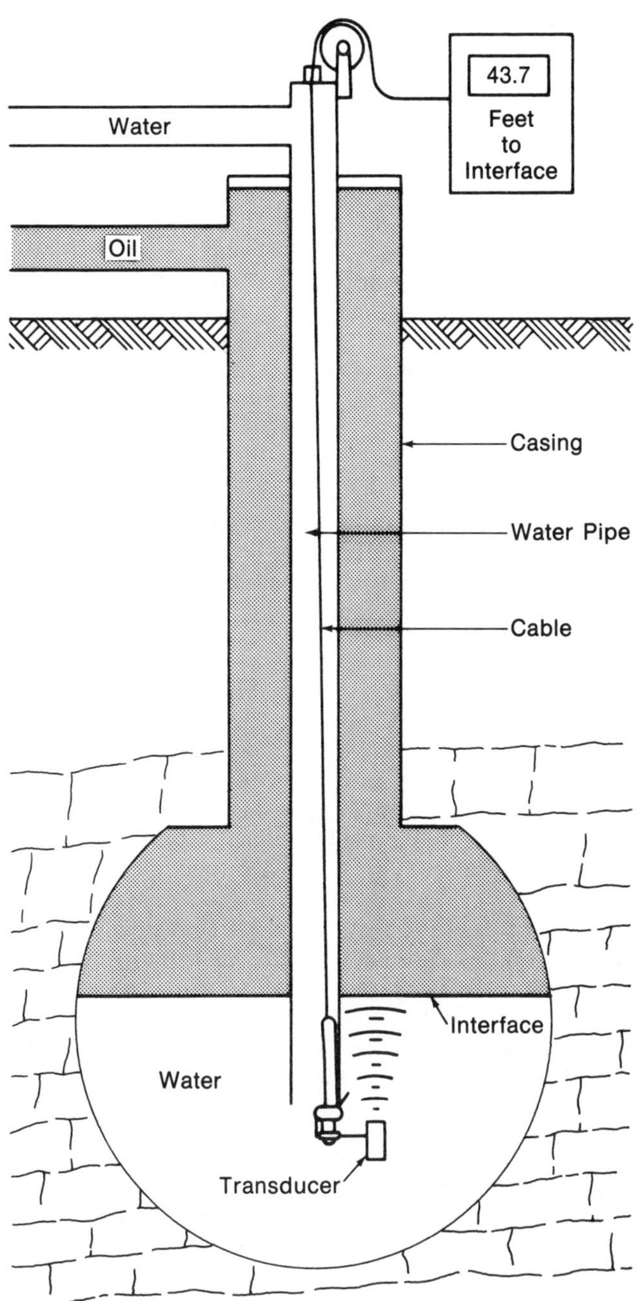

Figure 5.37. An ultrasonic system is used to measure brine-product interface in underground storage caverns. (Courtesy of Interface Detector Co.)

When measuring solids materials, the transmitter-receiver is usually mounted above the material to be measured. For liquid measurements, they may be mounted above or at the bottom of the measured media.

Readout and control devices are housed separately. Readout may be digital, a continuous indicator or a recorder.

The measurement range is from about 6 inches up to 100 feet, although no single manufacturer may offer this wide selection range. Temperature and pressure ranges present no problems. Material temperatures from -320° to 400°F can be accommodated at pressures up to 5,000 psig. The accuracy of these systems is about ±3% of reading.

Scanning Systems

The usefulness of the continuous measuring system is further enhanced by the development of automatic scanning systems for level measurement of multiple units. This provides the opportunity to monitor remotely and conveniently up to 24 levels on a continuing basis.

In this system, a single console houses all the necessary generating, pulsing and receiving equipment. Adjustable low and high level points are available for alarm; and a standard millivolt output signal is available for permanent recordings, if desired.

Point Measurement

An even wider variety of devices are available for point level measurement as switches for alarm or for on-off control functions. They may be classified as either single element or two-element systems, or they might be classified as "damped" or "energized" systems. Several concepts are discussed below.

Two-Sensor Systems

Early models of the two-element system worked as shown in Figure 5.38*a*. One sensor emitted a signal which was detected by a receiver on the opposite side of the vessel until the material level interrupted the signal path. This method is still effective for clean services if the signal path is not too long.

Some drawbacks to that system are eliminated by a two-element probe as shown in Figure 5.38*b*. The distance of the signal path is fixed, regardless of vessel diameter, and it also requires only one vessel entrance. In this system, piezoelectric transducers are used in the sensors separated by a 4-inch gap. When the low-energy ultrasonic beam is interrupted by the presence of material, a control relay is deenergized for control action.

Single Sensor Elements

There are at least two types of single element sensors. One uses an air gap to separate the transmitter and receiver sensors while another presents one face for material contact. Figure 5.39 shows several configurations of single element units that contain piezoelectric crystal transmitters and receivers separated by an air gap. A control unit generates an electric signal which is converted to an ultrasonic signal by the transmitter crystal. When the gap is filled with liquid, the ultrasonic signal reaches the receiver crystal and is amplified to actuate a relay. In the absence of liquid, signal reception is too weak to actuate the control relay. Figure 5.39*a* shows an element used for normal liquid point levels, 5.39*b* shows one used for liquid-liquid interfaces, and 5.39*c* shows how multiple units can be arranged on a single fitting where several measurement points are needed. As many as six points can be accommodated in this fashion.

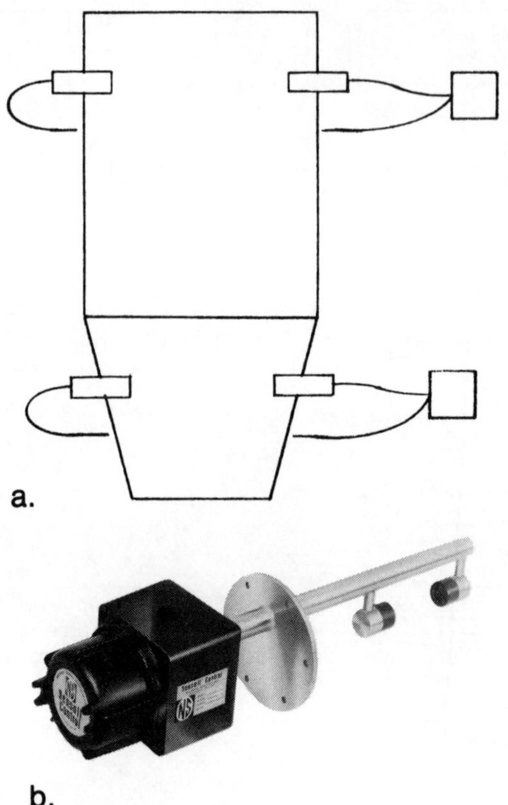

Figure 5.38. Early use of two-element ultrasonic point measuring system (a) *presents more application problems than the type shown in* b *where the distance between elements is better controlled. (Courtesy of National Sonics Corp.)*

Figure 5.40 is typical of devices that use a single face in the measured material. The sensor face is ultrasonically energized and is damped when liquid touches it. When the sensor is dry, its control relay is energized; when it is damped by liquid, the relay deenergizes.

One manufacturer offers a model that mounts on the exterior of the vessel so that tank penetration is not required. It can be mounted on a threaded stud or clamped or epoxied to the vessel wall.

Ultrasonic measurements are successful on many difficult-to-measure applications. Their cost is relatively high, but there are essentially no moving parts, the sensors

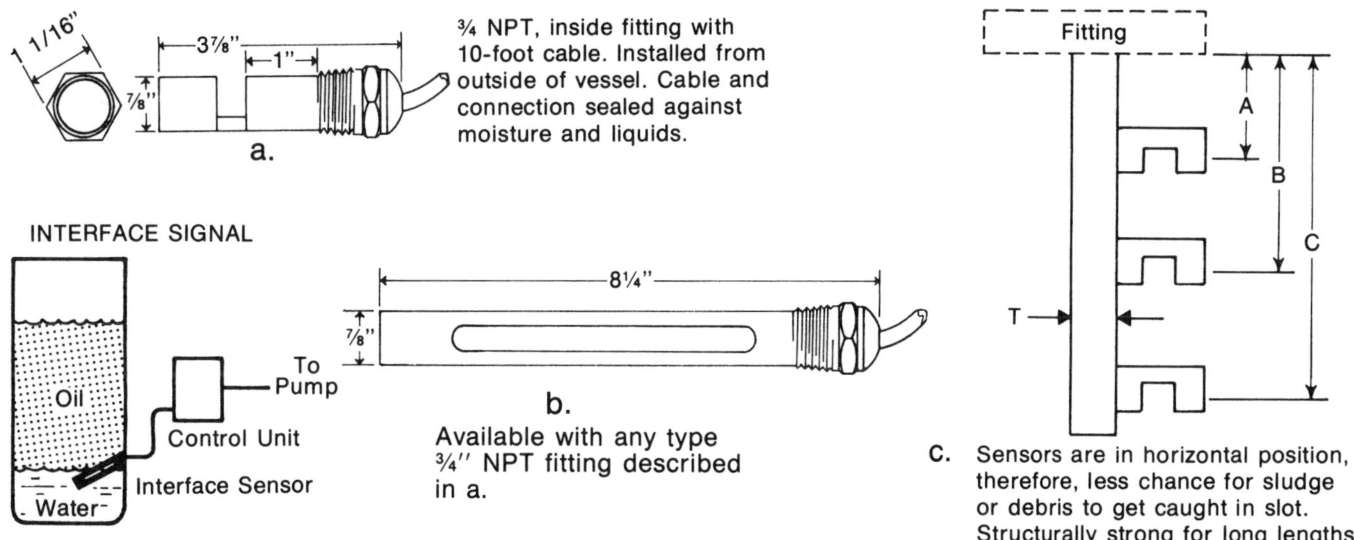

Figure 5.39. Typical one element sensors: (a) liquid level, (b) liquid-liquid interface and (c) multipoint liquid level unit (accommodates up to six points). (Courtesy of National Sonics Corp.)

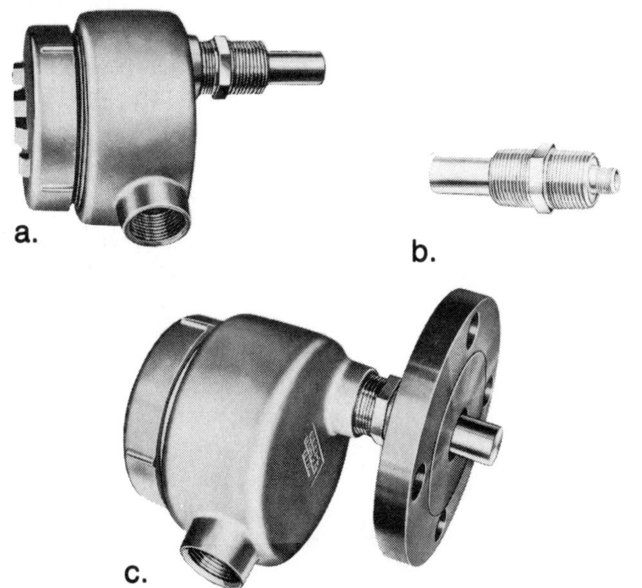

Figure 5.40. Single sensor ultrasonic unit that uses an ultrasonically energized face that is damped by contact with liquid, deenergizing a control relay. (Courtesy of Delavan Electronics, Inc.)

are sealed, and the electronics utilize solid-state circuitry that requires little maintenance.

The accuracy of the point devices is usually within a small fraction of an inch.

Nuclear Radiation Type

Since the early 1950's, nuclear radiation devices have been used for level measurement, usually on applications where standard techniques had little chance for success. Since the nuclear gauge does not come in contact with the product being measured, such adverse conditions as high or low temperature, sticky, abrasive or corrosive materials or clogged or highly pressurized vessels do not affect gauge operation.

To understand nuclear gauging more easily, some fundamental facts and concepts are presented prior to a discussion of measurement methods and devices.

Radiosotopes used for level gauging emit energy in a random fashion but at a rate constant enough to use an average value for calibration. As the radioactive sources disintegrate, alpha, beta and gamma particles are emitted, and gamma radiation is the phenomenon used for level gauging. Alpha particles have a nonpenetrating nature and have found no industrial application. Beta radiation has some penetrating power and is used for thickness measurement of thin materials. Gamma radiation is a high-energy short wave length source of energy, having no substance but great penetrating power and finds wide application not only for level gauging but also for density and weight measurements.

The basic unit of measurement for radiation intensity is the curie. A one-curie source undergoes 3.7×10^{10} disintegrations per second. The practical unit of radiation measurement is the millicurie, one-thousandth of a curie.

There are three radioactive materials normally used for level gauging—radium, Cesium 137 and Cobalt 60. A comparative analysis of their strength is shown by the fact that it requires about 1,000 miligrams of radium to generate one curie of radioactivity, while it requires only 0.88 milligrams of Cobalt 60 and 11.5 milligrams Cesium 137.

Radium is quite expensive but once was used fairly extensively when small sources were required. It occurs naturally and does not require licensing by the Atomic Energy Commission as do Cesium 137 and Cobalt 60. This does not

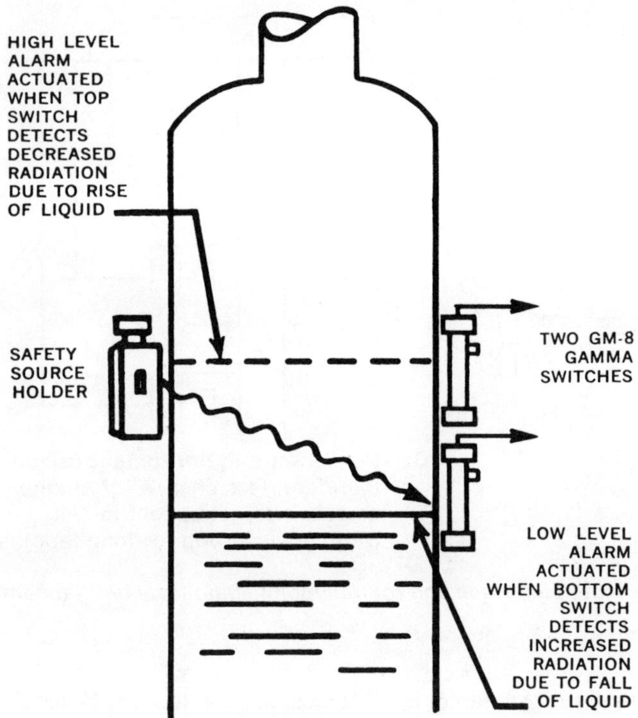
HIGH LEVEL ALARM ACTUATED WHEN TOP SWITCH DETECTS DECREASED RADIATION DUE TO RISE OF LIQUID
SAFETY SOURCE HOLDER
TWO GM-8 GAMMA SWITCHES
LOW LEVEL ALARM ACTUATED WHEN BOTTOM SWITCH DETECTS INCREASED RADIATION DUE TO FALL OF LIQUID

Figure 5.41. For point measurement using radioactive level measurement, the source is usually located on one side of the vessel while the receiver (s) is placed on the opposite side. (Courtesy of The Ohmart Corp.)

mean that radium is not dangerous, for many state regulatory agencies require licensing even though the AEC does not.

Cesium 137 sources are usually the choice for small and medium sources. Cobalt 60 is used primarily when really large sources are needed for penetrating power.

Another yardstick used to determine which source should be used is the comparative half-life of the material. All radioactive sources continue to disintegrate and lose strength with progression of time. Their rate of decay is expressed by their half-life, the time required for them to lose half their strength. The half-life of Cobalt 60 is 5 years; Cesium 137, 33 years; and radium, 1,400 years.

Point Measurement

The principle of point measurement is the same as for continuous measurement, and is discussed later. A level gauge consists of two basic units, a source head and a detector-amplifier. Normally the source is located on one side of a vessel and the detector-amplifier on the opposite side (Figure 5.41). When a single switch action is desired, the detector is located on the opposite side at the same level. When two level points are desired, two detectors may be used with a single source, one located at the source level and one lower, as shown in the illustration. The ability to use two detectors depends on several factors: the source size, the distance between level points, the vessel diameter and the collimating effect at the source.

The source head consists of the radioactive source and the holder. Lead or steel shielding is placed around the source. and a shutter mechanism shields it to prevent radioactive contamination during shipping, installation or maintenance. Sources are encapsulated and sealed to prevent leakage of the radioactive material. Stringent precautions are taken in this regard.

The detector is usually a Gieger-Mueller tube (though scintillation detectors may be used), a transistorized amplifier and a voltage supply.

In operation, gamma rays emitted by the source are absorbed by the material lying between the source and the detector. In a typical application, the source and detector are mounted on the external walls of the vessel in an opposed position. The source emits gamma rays which are collimated and pass through the vessel walls and product material to the detector. A change of level of the material produces a change in the amount of radiation detected, causing a change in the electrical output of the detector. The signal is then amplified and actuates an electrical relay or switch for alarm and/or controls.

Most point sources use small Cesium 137 sources. Although radioacitivity is relatively small, the AEC does require licensing, and precautions must be taken to ensure the safety of operating and maintenance personnel.

Continuous Measurement

Continuous level measurement using radioactive devices is somewhat more sophisticated than point measurement, but it is an extension of the same principle. There are variations in the use of both sources and receivers.

Point sources may be used as shown in Figure 5.42*a*, with detecting cells arranged along the level span to be measured. This method is applicable for short measurement spans and exhibits nonlinearity with increasing spans. One manufacturer uses strip sources or multiple sources and multiple measuring cells along the entire level span as shown in Figure 5.42*b*. The latter method is conducive to greater measurement spans and to greater accuracy and better linearity over the entire span.

As one views the point source installation in Figure 5.42*a*, it is evident that even without interposing material, more radiation is received by the measuring cell directly across from and nearer to the source than is received from the top and bottom cells which are farther away. This means that as material level rises, absorbed radiation increases slowly, builds to a peak as the source is covered and proceeds more slowly to the upper point of the span, thus producing nonlinearity.

In Figure 5.42*b*, each measuring cell receives radiation from each of the sources along the span or from all points of the strip source. Those cells at the center have a shorter mean distance from all source points than those at the end; some nonlinearity still exists. It usually can be reduced to an almost insignificant level.

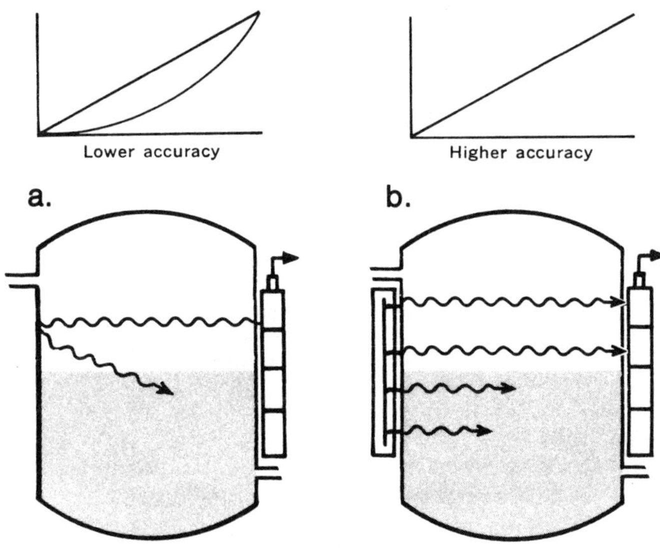

Figure 5.42. Continuous radioactive level measurement is shown in a, *using a point source and stacked receiving cells which give some nonlinearity, and in* b, *using a strip source and stacked cells which provide better linearity and accuracy. (Courtesy of The Ohmart Corp.)*

For increased accuracy and faster response, an electronic measuring cell is used, instead of Geiger-Mueller tubes, to convert radioactivity to electrical current. The current is generated by the ionization of a filling gas in the measuring cell as it is exposed to radiation. Electrodes within the cell attract the ionized electrons, causing a small current flow which is amplified and provides a signal output proportional to level.

In continuous measuring systems, zero suppression is accomplished in two ways: by electrical suppression or by a compensating cell—a small radioactive source hooked up with reverse polarity to alter the output current from the measuring cell. It has an adjustment screw to vary its shielding and provides the needed zero suppression. It is more costly than electrical suppression, but because it is made from the same material as the source, its decay offsets the decay of the main source and provides a stable zero for a long period of time.

For a high degree of accuracy, motor driven sources and measuring cells are available. Figure 5.43 shows a typical installation. The source and cell are mounted on steel tapes that wrap around a motor driven drum. As the drum rotates, the source and measuring cell are moved up and down together, and the position sensor rotates. The position sensor operates through a gear reduction mechanism to two potentiometers that give output voltages that are a function of the positions of the source and cell. One potentiometer reads the full range in feet and the other indicates inches within that range.

A zero-suppression circuit is adjusted so that the meter on the balance indicator unit indicates zero when the level is at the midpoint of the measuring cell and source. As the level changes, the balance indicator reads off-balance, and by means of the motor actuator, the source and cell move until the system returns to balance. The new level is then read on the level indicator. This system has a repeatability of ± ⅛ inch and can cover ranges up to 50 feet.

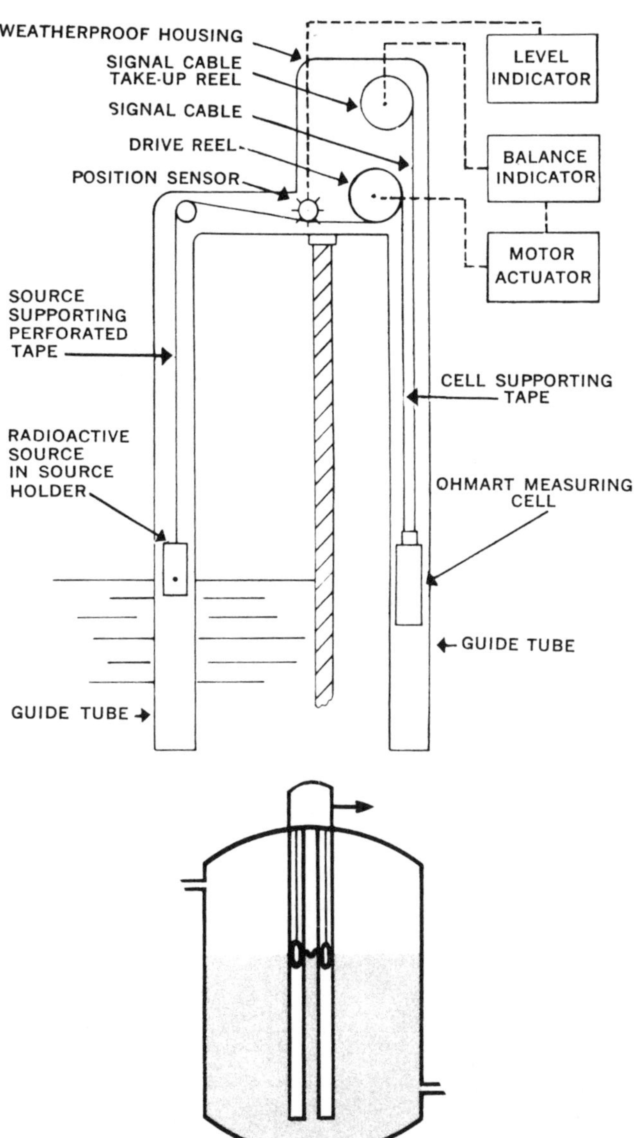

Figure 5.43. A motor-driven unit is available for high accuracy measurement in which the measuring cell and source are located opposite each other at the top of the liquid-vapor interface in a balanced condition. When the level changes, they must move to the new level interface to achieve the balanced condition. (Courtesy of The Ohmart Corp.)

Radiation level gauging is expensive but useful for applications where other methods have questionable value. Pressurized vessels present no difficulty, except that wall thickness must be overcome by large source strengths. Accuracies suffer as the ratio of wall-absorption to material absorption increases.

The safety aspect of handling radioactive materials and having them in an operating area must be considered. Sources are well shielded, and safe containers are used for shipping and can be mounted directly to the source well for transfer to storage while the vessel is serviced.

The source cell strength can often be reduced by installing sources within the vessel whose level is measured. When vessel diameters become excessive, the location of sources and cells can be placed along a chord of the vessel rather than on the diameter. Since absorbed radiation is proportional to mass between the source and measuring cell, density changes have a similar effect to level changes, and compensation is required for services having density variation.

Level Switches

Switch action can be obtained on any type level measuring system previously discussed—buoyancy or displacement types, hydrostatic head, capacitance, ultrasonic, and radiation. In fact, other than the buoyancy and differential level measuring methods, switch applications far outnumber continuous level applications for the methods already mentioned.

This section discusses level switch types that are used, almost without exception, solely as point measurements of level for switch action to actuate alarms or on-off control functions. This does not cover all available types, but it includes the most frequently used types and some that are used rather infrequently. Included are float switches, motorized rotating paddle, conductivity, pressure sensitive (diaphragm), vibrating reed and resistance tape types.

Float Type

As the name implies, float switches utilize a float that follows liquid level as it rises and falls. At some point within its range of travel, it actuates a switch (usually mercury or microswitch) to denote that a certain level point has been reached.

Although the float does rise and fall with changing level, most switches are designed for relatively short travel. Low level switches usually are submerged and actuate as the level recedes pasts its centerline, while high level switches normally hang freely in the vapor space of the vessel and actuate as the level rises to its centerline. When the level rises, or drops as the case may be, its changed position is enough to actuate an electrical switch of some kind to achieve the desired control action.

Floats may be round, oblong, cylindrical or some other configuration (Figure 5.44). Standard round floats vary from 2½ to 7 inches in diameter. The oblong floats vary in length and in diameter, as do cylindrical floats. Most floats are hollow and made of brass or stainless steel. Alloy floats are available as well as solid floats of synthetic materials that are lighter than most liquids.

A very interesting approach to level switching is demonstrated in a device which incorporates a tilt switch directly into its float. As liquid rises and it begins to float, it

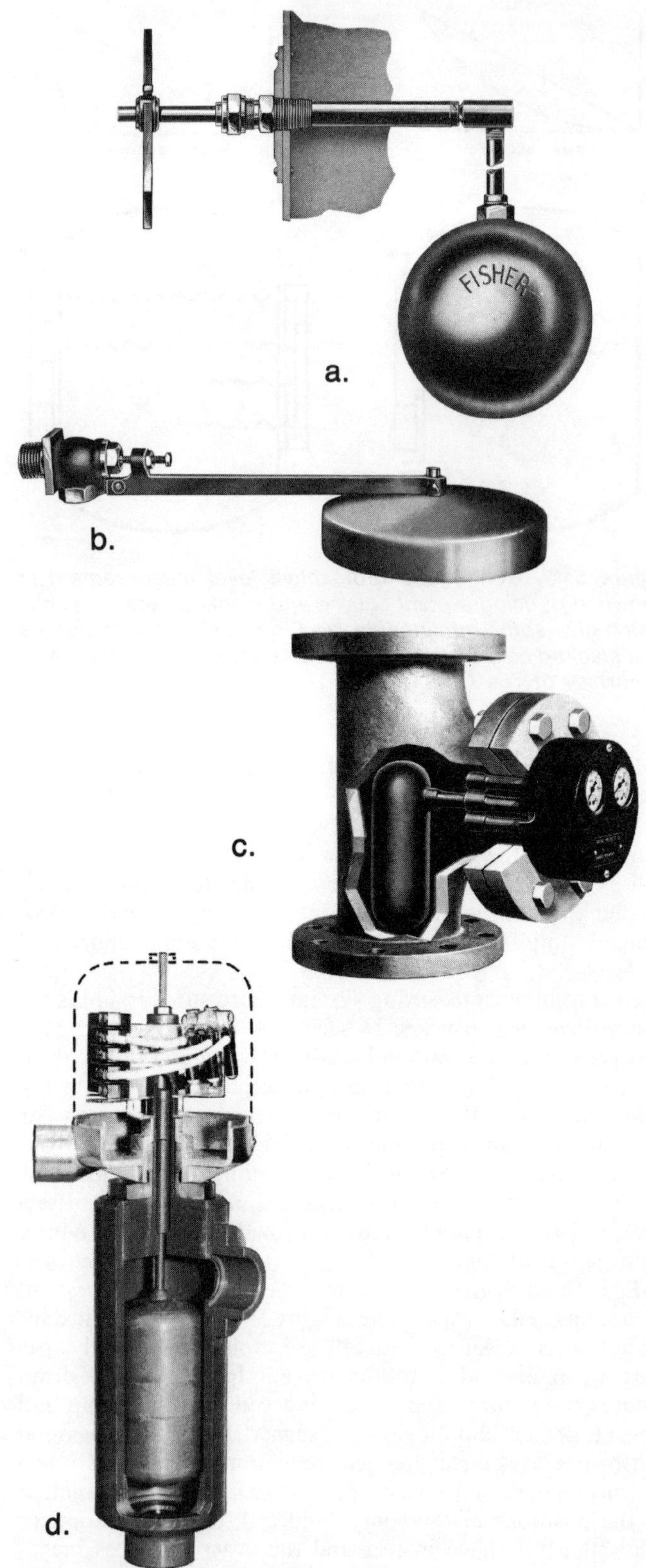

Figure 5.44. Level float configurations include round (a), flat (b), oblong (c) and cylindrical (d). (Courtesy of Fisher Controls Co., Moore Products and Magnetrol)

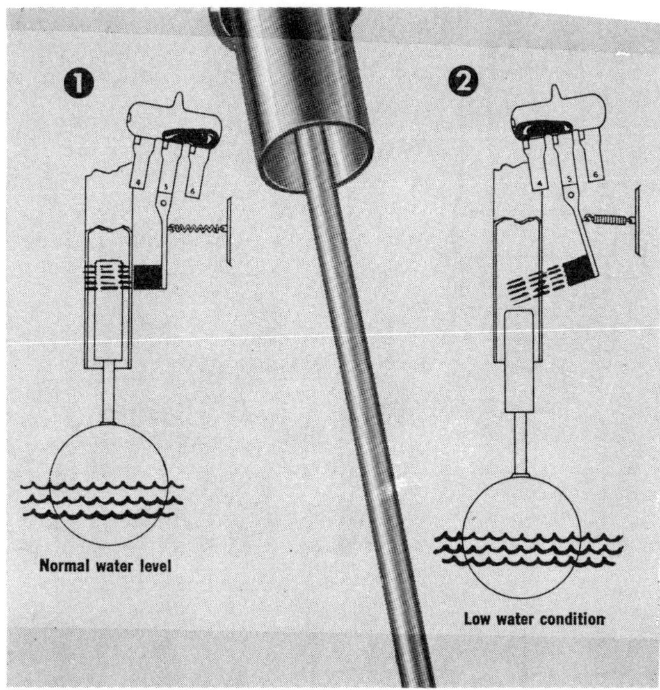

Figure 5.45. A permanent magnet and an attracting sleeve attached to the level float are used to tilt (operate) a mercury switch for single point level control. (Courtesy of Magnetrol)

capsizes causing a damped pendulum to actuate a microswitch. A 15-degree list causes a switching action.

Figure 5.45 shows how the mercury switches work. A permanent magnet is attached to a switching arm which holds the mercury switch. As the float rises with the liquid level, it raises the magnet attracting sleeve into the field of the magnet. This attracts the magnet, which "snaps" in against the enclosing tube and tilts the mercury switch.

When the liquid level falls. the float draws the sleeve below the magnetic field. The magnet swings out by gravity, assisted by a tension spring, and tilts the mercury switch to the reverse position.

Mercury switches are common, but when vibration is excessive or when, for some other reason, dry type switches are preferred, microswitches are used. Switch forms may be SPST, SPDT, DPST or DPDT. Electrical ratings are usually 5a or 10a at 115v, 230v, 440v AC. Switch housings are available for general purpose, weather-proof or explosive atmospheres.

Internally Mounted

Internally mounted switches can be flange mounted or screwed. Flange mounts are made at the top or side, and sizes vary normally from 3- to 6 inch nozzle sizes. Threaded connections are usually 1½ to 3 inches NPT.

Internal float switches are economical, but they have two primary disadvantages: (a) the vessel must be shut down in order to remove the switch for repair or maintenance, and (b) they are subject to false actuation if the vessel contents are turbulent.

Externally Mounted

There are many housing designs for externally mounted floats (Figure 5.46). Entry to housings may be top and bottom, side and side, side and bottom—many variations. Connections may be screwed or flanged. Material and pressure ratings are usualy consistent with piping requirements for the associated piping system.

Externally mounted switches usually cost more than internal ones, but with the installation of isolating valves, they may be removed as needed for repair or maintenance. External piping also serves as an effective stilling chamber for vessels whose contents surge or are turbulent. Higher switch cost and material and labor costs cause them to be appreciably more expensive than internal float switches.

Tandem Switches

Tandem mounted switches may be installed for switch action at more than one point (Figure 5.47). Differential levels up to 132 inches can be accommodated with this type arrangement.

The float assembly consists of a float ball sliding on a rod between two adjustable stops. A toggle clutch assembly is provided to hold the float rod between switch actions. A guard cage, consisting of three guide rods, eliminates any side play in float travel.

In operation, with the adjustable stops on the float rod set at the desired "high" and "low" levels, the float ball is free to travel up and down the rod with the changing liquid level.

When the liquid reaches the desired "high" level, the float rod engages the adjustable stop and lifts the float rod assembly so as to actuate the switch mechanism. The float rod (and mercury switch) is held in the "high" position by the toggle clutch assembly until the float, as it follows the travel of the liquid level, reaches the "low" level. At that point, the weight of the float ball pulls the float rod down to reverse the switch action.

Rotating Paddle

Several manufacturers offer paddle switches for level detection of solids materials. Figure 5.48 shows how one of these units is constructed. A 1/100-hp motor turns a blade paddle at 9 rpm by a torsion spring. The paddle turns continuously as long as no material touches it.

If material makes contact with the paddle, it stops. However, the motor continues to run, expanding the spring until it actuates a limit switch. This switch kicks off the motor and any other equipment controlled by the switch. When the material falls away from the paddle, the torsion spring reactivates the paddle and unwinds from the limit switch, starting the motor and putting the unit in operation again. Any electrical combination can be obtained by the open and closed position of the limit switch.

These types of switches should be mounted so that materials entering the bins or vessels do not flow directly on

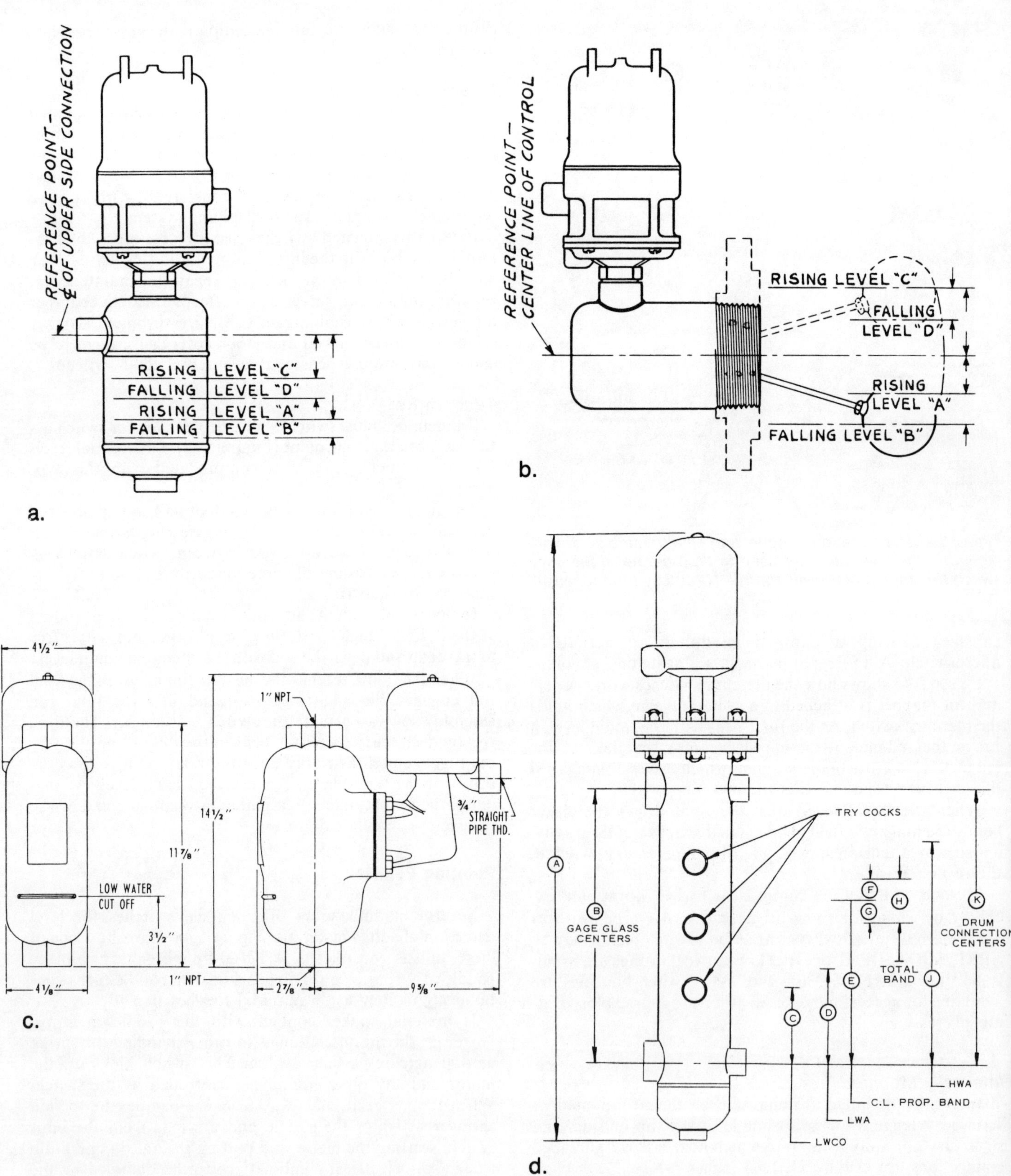

Figure 5.46. External cage floats may be connected in several ways, among which are side and bottom (a), side only (b), top and bottom (c) and side and side (d). (Courtesy of Magnetrol)

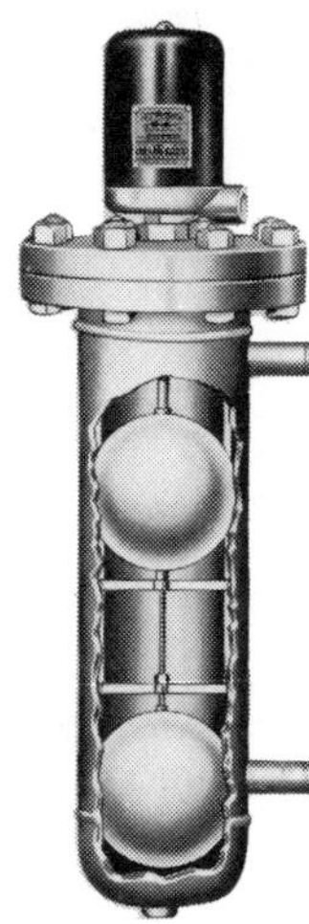

Figure 5.47. Tandem floats are mounted between float rods and are adjustable for settings at various high and low levels. (Courtesy of Magnetrol)

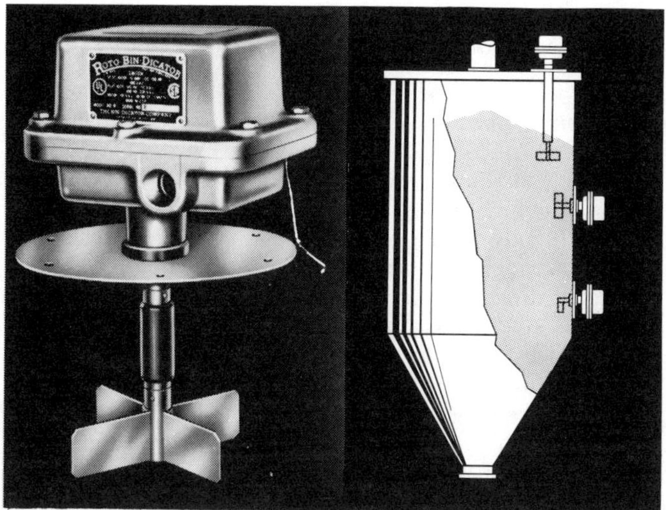

Figure 5.48. Rotating paddle level switch used for bulk materials—paddle powered by 1/100 hp motor—turns continuously until material buildup stops it, activating a switch through a torsion spring. (Courtesy of Bindicator)

them. Baffles placed above the switches provide good protection if falling material cannot otherwise be avoided.

Fine powders have a tendency to enter shaft seals, causing seizure and false alarms; however, dust seals are used to minimize this type failure.

Paddle switches are available in general and explosion-proof models. Units are available for pressures to 15 psig, though applications are usually very near atmospheric pressure. Design temperatures are usually 185°F or under. Switch mounting may be at the top or on the sides of bins.

Conductivity

As the name implies, conductivity level switches operate on the principle of electrical conductivity. Electrodes located within containers at the point of control make or break contact with the conductive material, thus completing an electrical circuit which operates load contacts for pumps and solenoid valves or performs other control or alarm functions.

The schematic of a heavy duty switch of this type is shown in Figure 5.49. When a source of alternating current is connected to the primary coil, a magnetic flux is set up which induces a voltage in the secondary coil. Current flows in the secondary coil;—however, only when the rising liquid completes the circuit between the two electrodes. Completion of the secondary coil circuit and the resultant current flow sets up a bucking action in the lower bar of the transformer core. This tends to divert lines of magnetic flux to the core legs and sets up an attraction that moves the armature, closing or opening load contacts.

One pair of contacts connect the secondary circuit to ground when liquid contacts the upper electrode and acts as a holding circuit to maintain the relay in its closed position until the liquid falls below the lower electrode. This holding circuit provides control of the load circuit through the bottom contacts of the relay over any desired range in the liquid level, depending upon the distance between the upper and lower electrodes.

The flow of current through the low energy secondary circuit is very small and varies with the voltage of the secondary coil. The secondary coil is selected to operate over the resistance of the liquid being controlled.

The two-electrode system shown operates a pump to maintain a desired level in a vessel—from a simple on-off action where one electrode and the container complete a circuit via the conductive material to a multielectrode assembly where several switch actions occur at various container levels.

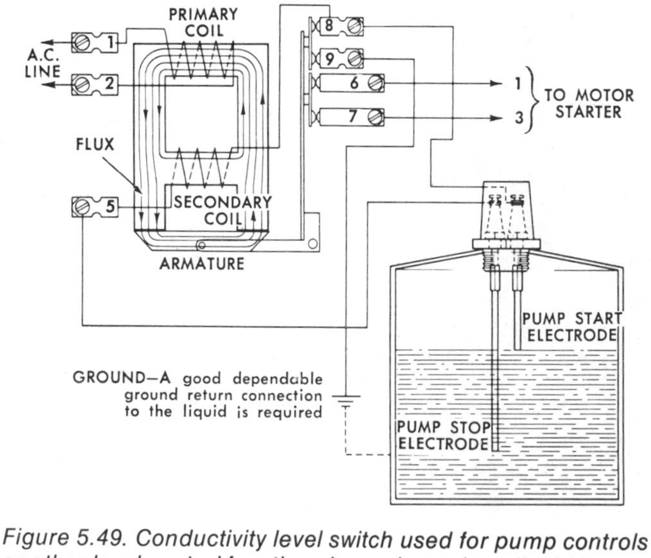

Secondary coil circuit closed; armature up.

PRIMARY COIL
A.C. LINE
FLUX
SECONDARY COIL
ARMATURE
TO MOTOR STARTER
PUMP START ELECTRODE
PUMP STOP ELECTRODE
GROUND—A good dependable ground return connection to the liquid is required

Figure 5.49. Conductivity level switch used for pump controls or other level control functions in services where liquid is conductive. (Courtesy of B/W Controller Corp.)

Conductivity switches are economical and have no moving parts in contact with the measured material. Early, heavy-duty designs were used primarily on water applications because of the high energy levels which could result in sparking to produce fires or explosions. However, manufacturers now offer solid-state designs, operating at low energy levels, which can be considered intrinsically safe.

Pressure Sensitive (Diaphragm) Type

Figure 5.50 shows a pressure sensitive diaphragm level switch which is used primarily on bulk solids level applications.

In this technique the force of two magnets, positioned with like poles repelling one another, is utilized to cause the tripping of a switch, with the magnet fields operating through the wall of the switch housing. The switch compartment may then be completely sealed with no moving parts necessary to cause activation. The unit performs as follows.

1. When the diaphragm is forced inward by product, the "driving" magnet moves inward as does its magnetic field.
2. The field of the "driving" magnet repels the field of the "driven" magnet and the switch is tripped.
3. Deactivation occurs when product is removed and the diaphragm and "driving" magnet return to their original position.

Diaphragm materials are usually of stainless steel, neoprene or plastic fabric. SPST or SPDT microswitches rated at 10a, 125v are standard; housings may be dust-proof or explosion-proof.

Units may be flange mounted on the side of the vessel, or they may be suspended internally on a 1-inch pipe inside the vessel. Operating temperatures for some models may be as high as 400°F and pressures may range to 100 psig.

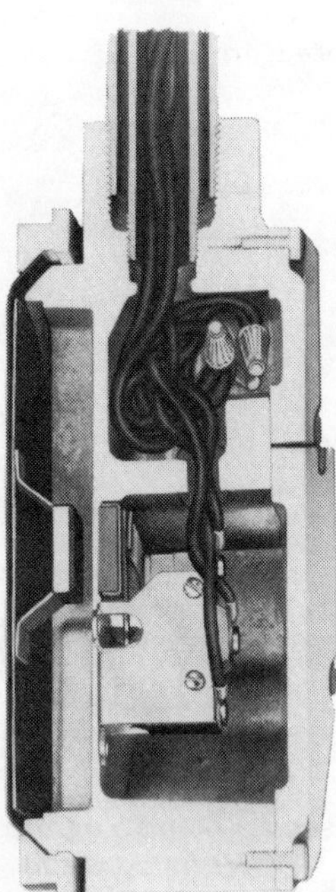

Figure 5.50. Level switch has pressure sensitive diaphragm that, forced inward by head pressure, activates a microswitch by the repelling action of two magnets brought together by diaphragm movement. (Courtesy of Proximity Controls, Inc.)

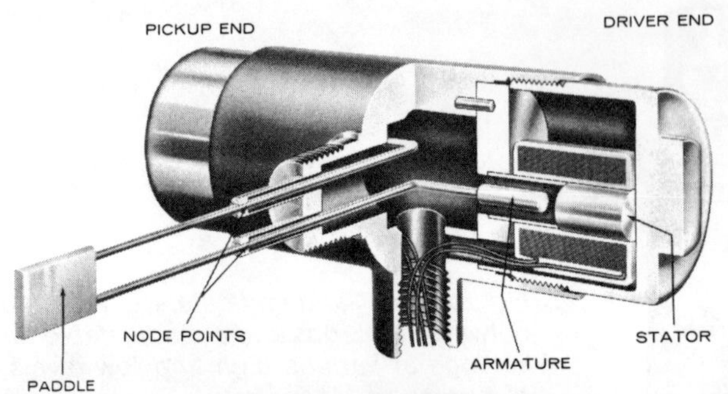

Figure 5.51. Paddle is vibrated at 120 cycles per second and produces a voltage proportional to vibrational amplitude which decreases appreciably when paddle is covered by material. (Courtesy of Automation Products, Inc.)

Vibration Types

Figure 5.51 shows a vibrating paddle that is inserted into a vessel for high or low level detection in liquids or liquid slurries. It consists of a vibrating paddle which is driven into 120-hz mechanical vibration by the driver coil which is connected directly to a 115-volt, 60-hz line. A second coil which has a permanent magnet stator located in the pickup end produces a 120-hz voltage proportional to the paddle vibrational amplitude. When the paddle comes into contact with the process media, its amplitude of vibration decreases and the output voltage drops to a very low value. This change in output signal operates the SPDT contacts of the control relay.

The unit is available in a wide selection of materials and coatings that come in contact with the process material. It is designed for a Class I, Group D, Division I Electrical Area Classification. The pressure rating is 3,000 psig at 100°F. Process temperatures to 300°F are standard and pressure connections are ¾-inch NPT.

For high or low point level detection in bulk solids, a slightly different design is used (Figure 5.52). It consists of a rod which is installed through the wall of the bin or duct at a point of desired level detection. When the probe is uncovered, the drive coil drives the rod into self-sustained mechanical oscillations at the natural resonant frequency of the rod. The pickup coil, located opposite the drive coil, is excited by the mechanical oscillations of the rod and produces an AC signal voltage. The presence of this signal voltage indicates that the rod is uncovered or that a low level exists.

When the process media covers the rod, a dampening of the rod oscillations occurs. The magnitude of the rod oscillations is greatly reduced, and the output from the pickup coil drops to a very low value, indicating that the rod is covered or that a high level exists.

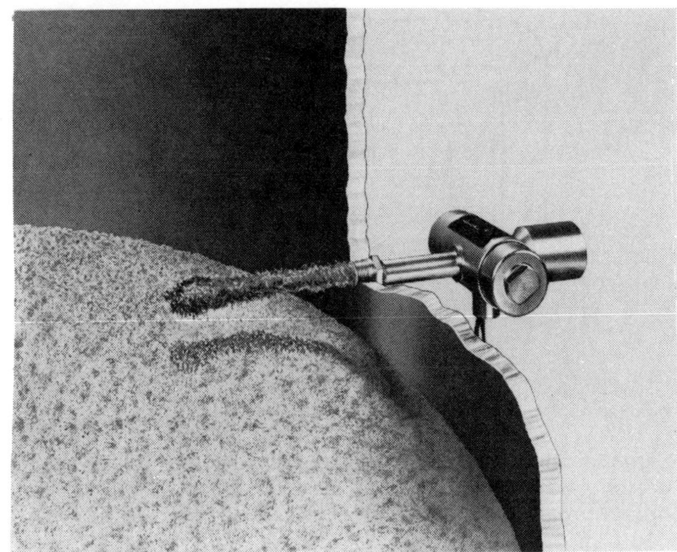

Figure 5.52. For solids level applications, this probe, when uncovered, is driven into self-sustained oscillations producing a voltage picked up by an adjacent pickup coil. Voltage level changes reveal increasing and decreasing levels. (Courtesy of Automation Products, Inc.)

The on-off signal from the detector operates a SPDT relay in the control unit. The contacts of this relay can be used to actuate alarms, indicator lights or process control equipment.

Service and installations are the same as for the unit described above, except that pressures up to 1,000 psig can be accommodated.

The single probe design was made to overcome a deficiency in the paddle type design for some solids services. Some materials tend to pack under the vibrating condition of the paddle type, creating a cavity whose effect is the same as a lowered level, thus providing a false level detection. This type detector unit is also used to detect liquid interfaces and provide density or viscosity measurements.

Level Gauges

Level gauges have been used to observe vessel levels in the process industries almost since the evolution of industry. A high percentage of vessels, columns, reactors, accumulators, etc., are equipped with gauges unless high pressure prohibits or the fluid characteristics make them ineffective. Normally economical and applicable to a wide range of fluids and fluid conditions, they fall into three distinct groups: tubular glass, flat glass and magnetic gauge.

The gauge glass, in effect, gives the operator a window to view the process fluid. For ease of maintenance and from safety considerations, the window is frequently installed outside the vessel in a chamber and connected by piping and shutoff valves to the process vessel (Figure 5.53).

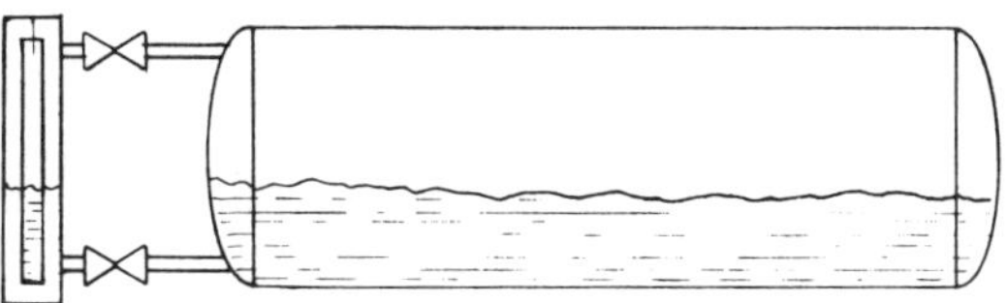

Figure 5.53. Elementary sketch of gauge glass installation.

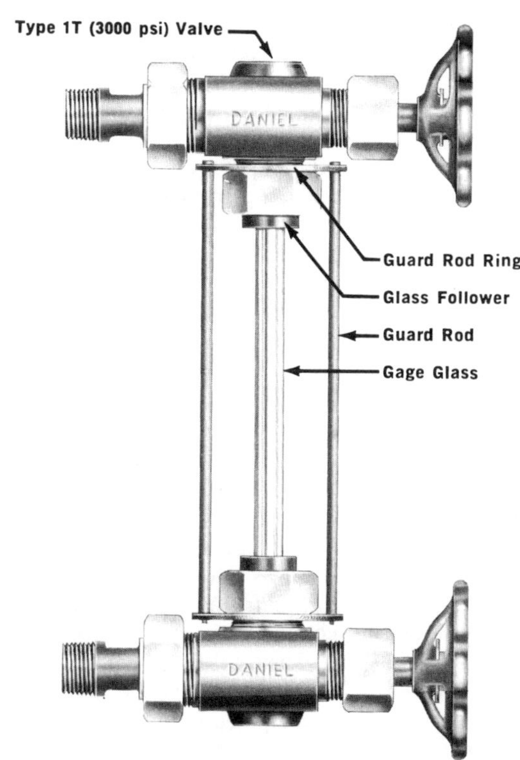

Figure 5.54. Tubular gauge glasses are normally used on low-pressure, nonhazardous fluid applications. Pressures to 600 psig can be accommodated, however. (Courtesy of Daniel Industries, Inc.)

Tubular Glass

Tubular glass gauges (Figure 5.54) were the first type to be used for level indication. They are seldom used now in industrial plants because they are so easily broken and present unnecessary operating and safety hazards. The gauge consists of a glass tube held between two shutoff valves containing packing glands to seal against the tube. The design pressure is dependent on the length and diameter of the glass tube but still falls in the low-pressure ranges.

The tubular gauge can be protected to some degree from accidental breakage by guard rods and/or transparent plastic around the tubes. Their use is difficult to justify against the slightly more costly and much safer flat glass gauges.

Flat Glass Types

Since the introduction of flat glass gauges in the early 1900s, they have almost replaced tubular gauges. There are two types manufactured—the reflex and the transparent.

Reflex Type

The reflex gauge (Figure 5.55) utilizes the optical phenomenon of light refraction changing to reflection at the critical angle of incident rays. Prismatic grooves are molded into the backside of the gauge glass. When the gauge is empty, the incident light is reflected from the prismatic surfaces, causing the glass portion of the gauge to have a silvery appearance. As liquid rises in the gauge, the critical angle changes because the refractive index of the rising liquid is different from that of the vapors above. Visible light is refracted into the fluid, causing the gauge to appear dark where covered by liquid. Thus, a gauge containing liquid shows a sharp demarcation between the dark area of the liquid and the silvery appearing area of the vapor above.

The standard reflex gauge is constructed from a one-piece chamber machined from bar steel. The glass is tempered with borosilicate with a low thermal expansion coefficient and resistant to both thermal and mechanical shock. The steel covers are of forged alloy steel. Gaskets are placed between the glass and chamber for sealing purposes and between the glass and cover to provide a cushion. There is no contact between glass and metal. U-bolts are used to clamp the whole assembly together (Figure 5.56).

Gauge glass sections are available in several standard sizes. Visible portions range from 3¾ to 19⅝ inches in several sizes. Multiple sections are assembled together when longer gauges are needed. The total length should be limited to about 6 feet because of weight and thermal expansion considerations. If gauge length requirements exceed 6 feet, overlapping units may be installed on separate nozzles (Figure 5.57).

These gauges are available from a wide variety of materials, but steels and stainless steels are prevalent. Linings may be inserted in the chamber to decrease cost when an exotic metal normally would be required. Gauge chambers are available with end or side connections which may be screwed or flanged. The solid back chamber of the reflex gauge yields higher pressure and temperature ratings than for an equivalent transparent gauge. Figure 5.58 shows a typical temperature-pressure operating curve for these type gauges. Design pressures may run as high as 4,000 psig, and operating temperatures may be as high as 750°F.

Reflex gauges are used primarily for clean, colorless, nonviscous services. Application difficulties occur when the fluid tends to coat the glass such that the refraction of the prisms give a false level indication. Erroneous indications may also occur if the fluid is near saturation conditions in the gauge glass or when condensation or boiling in the gauge glass affects the refraction of light through the prisms. Reflex gauges cannot be used to show interface levels between two liquids.

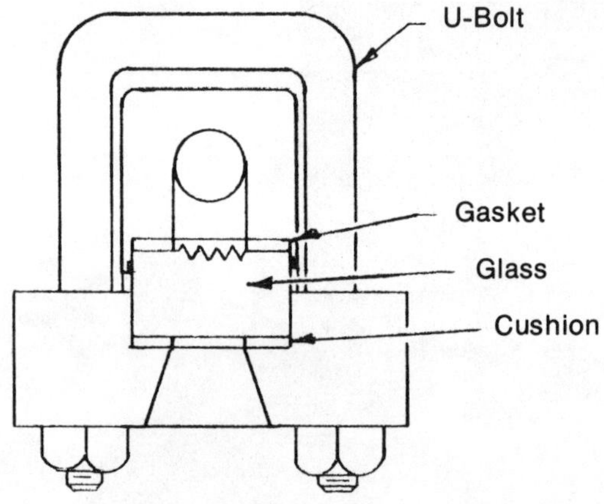

Figure 5.56. Cross section of standard reflex gauge. Note gasket seal between glass and chamber and cushion between glass and cover.

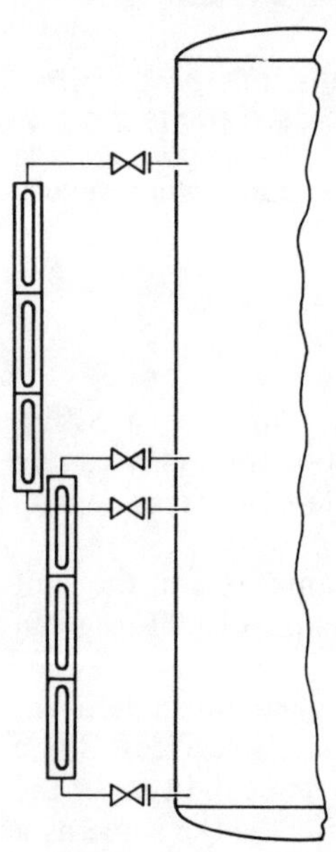

Figure 5.57. When gauge lengths become excessive, separate nozzles are used for staggered guage columns.

Figure 5.55. The reflex gauge depends on light refraction at the liquid prismatic glass interface to reveal liquid level. The liquid area looks dark while the vapor space has a silvery appearance. (Courtesy of Jerguson Gage and Valve Co.)

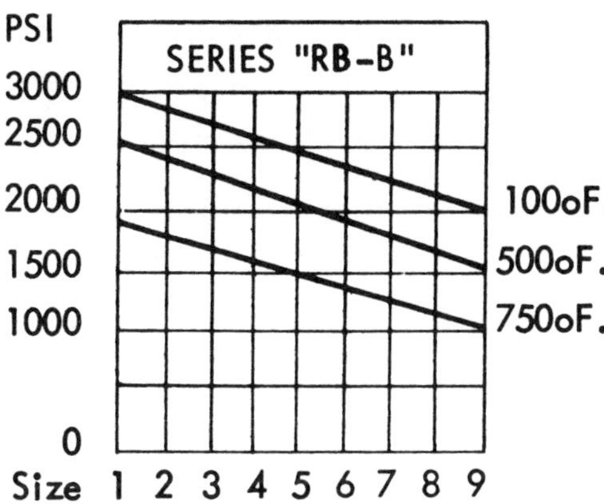

Figure 5.58. Typical temperature-pressure operating curve for reflex liquid level gauge series. (Courtesy of Inferno Mfg. Co.)

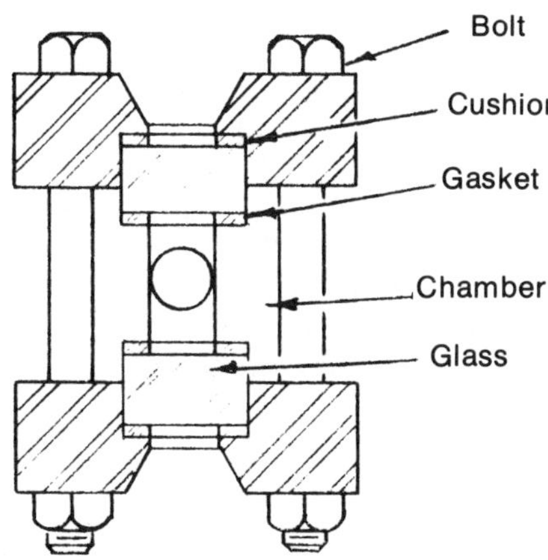

Figure 5.59. Cross section of transparent gauge shows glass on both sides of chamber—seals and cushions used in same fashion as reflex gauges.

Table 5.3. A Comparison of Pressure-Temperature Charts for Same Series of Reflex and Transparent Gauges

Series 20	Size Number of Window								
Test	1	2	3	4	5	6	7	8	9
Pressure	4500	4350	4095	3900	3750	3525	3330	3150	3000
Temp. °F									
100°	3000	2900	2730	2600	2500	2350	2220	2100	2000
200°	2825	2720	2600	2455	2350	2220	2100	2000	1900
300°	2650	2540	2440	2310	2200	2090	1965	1870	1750
400°	2455	2350	2250	2135	2030	1920	1805	1710	1600
500°	2260	2160	2060	1960	1860	1750	1650	1550	1450
600°	2020	1940	1850	1750	1670	1550	1475	1380	1300
750°	1650	1600	1500	1400	1300	1250	1150	1050	1000
Saturated Steam Rating 300 WSP									
Jerguson Transparent Gauges									
Series 20	Size Number of Window								
Test	1	2	3	4	5	6	7	8	9
Pressure	3000	2775	2625	2400	2250	2025	1875	1650	1500
Temp. °F									
100°	2000	1850	1750	1600	1500	1350	1250	1100	1000
200°	1900	1780	1660	1550	1440	1300	1175	1060	950
300°	1770	1660	1550	1450	1330	1220	1100	1000	900
400°	1675	1575	1470	1350	1250	1150	1050	925	850
500°	1530	1450	1350	1250	1150	1050	950	850	750
600°	1350	1275	1180	1100	1010	925	850	750	675
750°	1100	1050	1000	900	850	750	700	600	550
Sat.Steam (using mica)	600	600	600	600	550	500	450	400	350

Other gauges using the refraction principle are available for special applications, such as boiler drums. Generally, these gauges use lights to illuminate the gauge and intensify the indication.

Transparent Gauges

Transparent gauges are used for services where the process material is colored or viscous, for interface detection, or when the fluid is corrosive to glass. Construction is similar to a reflex gauge, except that the glass has no prismatic grooves and is installed both on the front and back of the chamber (Figure 5.59).

Transparent mica or Kel-F can be installed on the inside surface of the flat glass for protective shielding when used in acid or caustic services corrosive to glass.

This type gauge permits direct observation of the fluid—its color, its physical state or its interface with another liquid.

Caution should be used in application of these devices since the double glass construction weakens the gauge assembly, and the pressure and temperature limits are generally much lower than for an equivalent reflex gauge (Table 5.3). Other rating tables apply for different series of gauges. For special applications, heavy designs up to 10,000 psig can be furnished.

Special Gauges

Welding Pad Type

Welding pad gauges (Figure 5.60) are used infrequently but are useful where:

1. Solids are present that would clog gauges and gauge connections

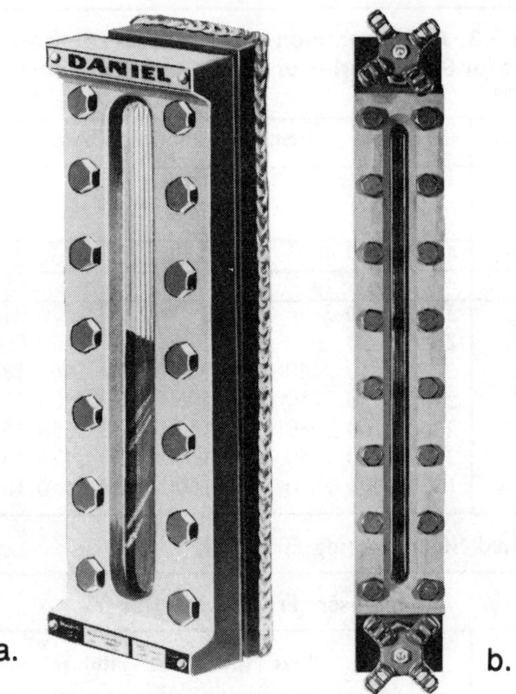

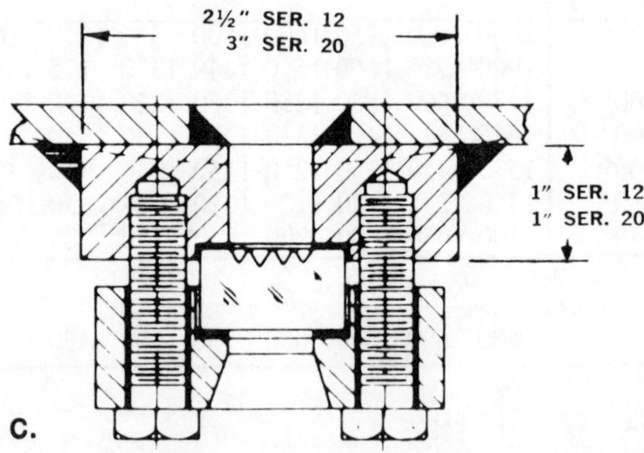

Figure 5.60. Standard welding pad gauges (a) can be welded directly to a vessel (see cross section in c). Welding pad gauge (b) shown with isolating valves to shut off flow if gauge glass breaks. (Courtesy of Daniel Industries, Inc. and Jerguson Gage and Valve Co.)

2. Gauge lengths might require too many external connections
3. Space is not available for external gauges
4. There is a need to minimize connections because contained material poses a hazard
5. Reading errors are caused by liquid specific gravity changes where external chambers have different temperatures from the contents of the vessel

Welding pad gauges are undesirable from several standpoints.

1. They must be reinforced to maintain the vessel strength (pressure rating).
2. A broken gauge glass causes uncontrolled material spillage (however, units can be made with integral isolating valves).
3. Their possible breakage poses dangerous operating conditions for many fluids.

Large Chamber Type

Large chamber gauges (Figure 5.61) are used where vessel fluids contain entrained gases or where fluids are near their boiling point. The entrained gases or boiling fluids cause the levels to surge within small chambers and present reading difficulties. The larger chamber provides a greater area for gas disengagement and reduces or eliminates the surging problems.

Magnetic Type

The magnetic gauge is a unique measuring method used when glass is undesirable because of corrosive, toxic, pyrophoric or otherwise dangerous material. Figure 5.62 illustrates how it works. A float containing a magnet is placed inside a sealed chamber; the float is free to move and rises and falls with liquid level in the adjoining vessel. An indicator consisting of a series of small steel wafers which are free to rotate 18° is mounted outside the chamber. As the float moves, its magnet causes the wafers to rotate. These

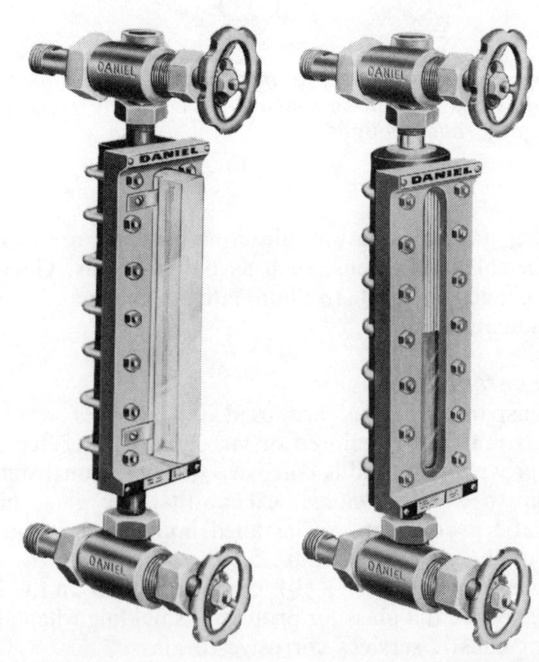

Figure 5.61. Large (2-inch diameter) chamber gauges are used for levels on fluids near boiling point or where entrained gases cause level surges. (Courtesy of Daniel Industries, Inc.)

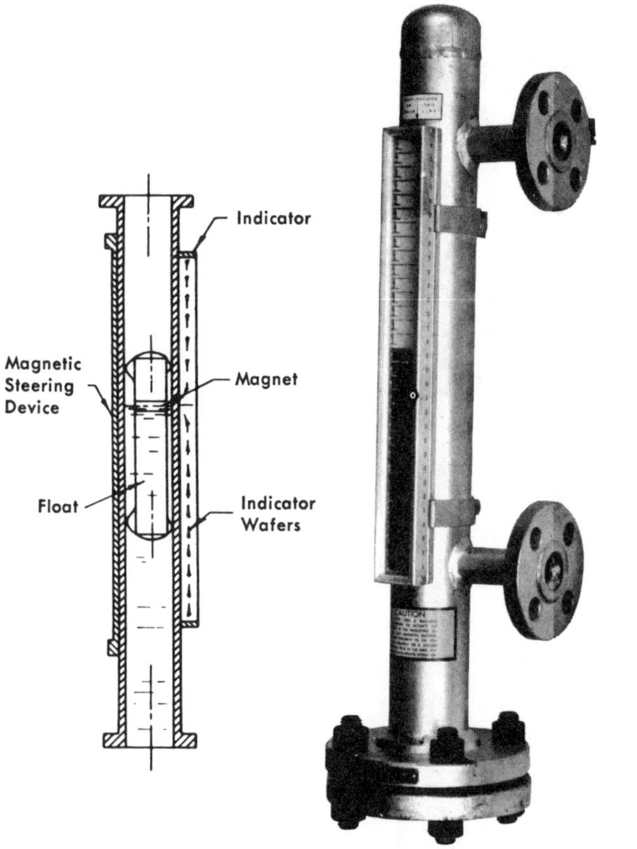

Figure 5.62. The magnetic gauge with a schematic view illustrating the manner in which it operates. (Courtesy of Jerguson Gage and Valve Co.)

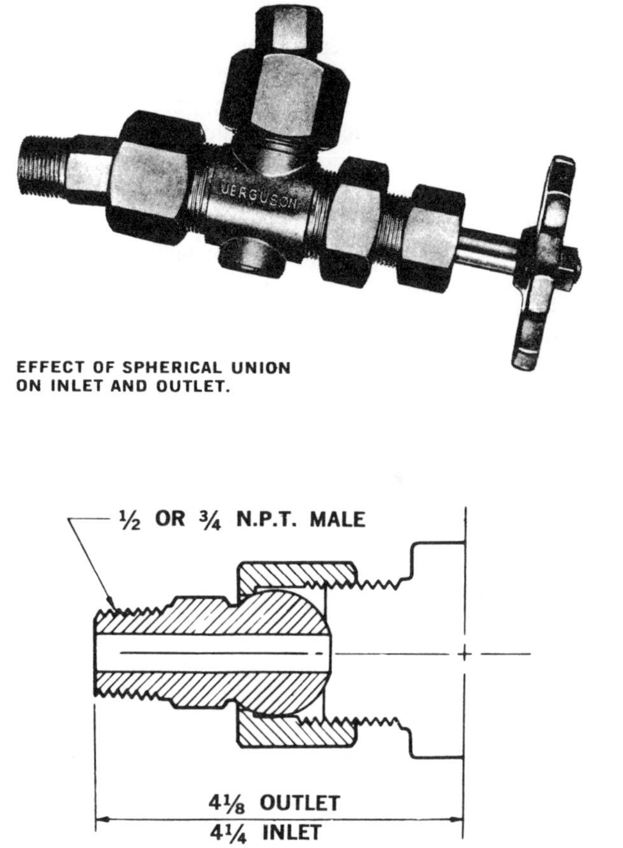

Figure 5.63. Spherical unions are used to correct alignment problems in connecting gauge glasses and valves to vessels. (Courtesy of Jerguson Gage and Valve Co.)

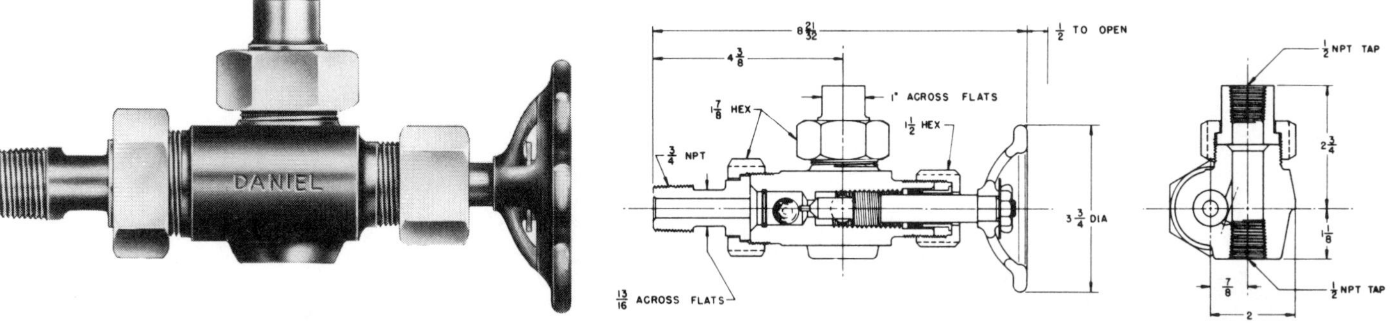

Figure 5.64. Three-way construction of this valve allows tank and gauge connections as well as a vent. Note also the ball-check feature. (Courtesy of Daniel Industries, Inc.)

wafers present a black colored side as the float moves in one direction and show the other side painted yellow when the float moves in the opposite direction. Black denotes liquid and yellow indicates the vapor space.

The accuracy of the gauge is ±¼ inch, about the width of a single wafer. Standard models are available for pressures up to 400 psig and maximum temperatures to 450°F. Special designs can be built for pressures to 10,000 psig and temperatures to 600°F.

Accessories include float extensions for readout at locations above and below the gauge, steam jacketing provisions and level switches for remote alarm or control.

Gauge Accessories

Gauge valves, unions, checks, illuminators and frost preventors are among the accessories available with level gauges.

Valves

A wide variety of valves has evolved out of the peculiar needs for installing and maintaining level gauges. In addition to the ordinary shutoff function, gauge valves are furnished with a wide variety of optional functions such as:

1. Integral union connections for connecting to the vessel and to the gauge glass
2. Spherical unions to allow for connection misalignment (Figure 5.63).
3. Ball checks as a safeguard against a broken glass and a blowout

In some styles of valves, backseating plugs are furnished which seal the stem with the valve in the open position, permitting the packing to be replaced without taking the gauge out of service.

Bonnets for these valves may be integral or outside screw and yoke type construction.

Stems may be threaded for quick closing, permitting the valve to be closed in ¼ turn. This is desirable when operation with a chain operator is needed.

Valves are built in three-way type construction to permit installation of drain and bleed valves with a minimum of fittings (Figure 5.64). A variation of this design is the offset type valve with its stem offset from the centerline of the gauge, presenting a full-bore opening for rodding out and cleaning.

Illuminators

When gauges are installed in poorly lighted areas, illuminators can be attached to provide the needed light (Figure 5.65).

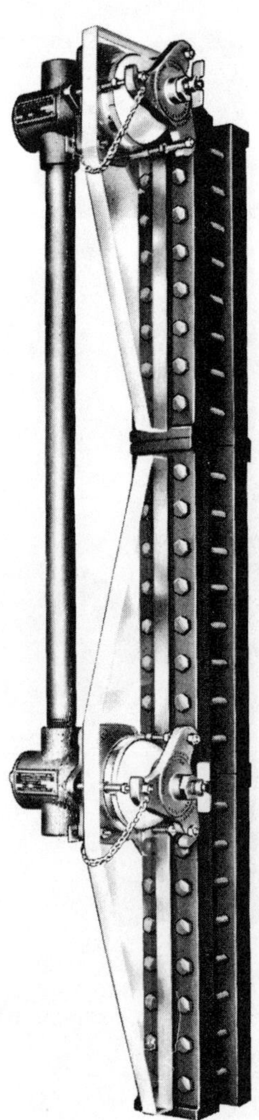

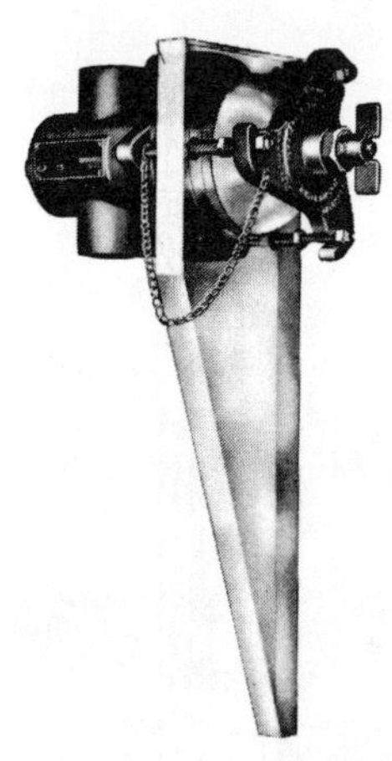

Figure 5.65. Illuminators made of plexiglass mount behind gauge and reflect light through to the gauge. (Courtesy of Jerguson Gage and Valve Co.)

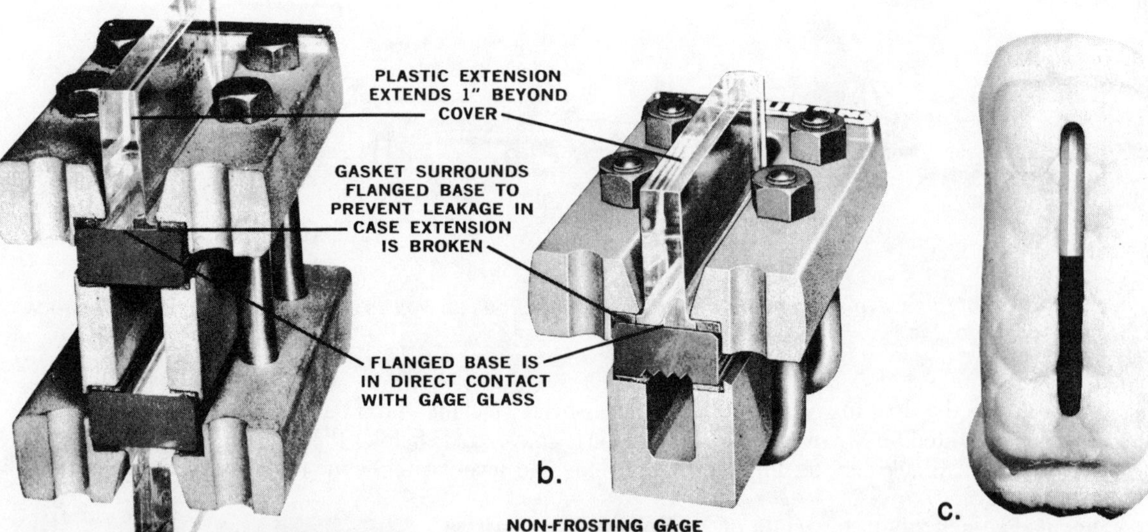
PLASTIC EXTENSION EXTENDS 1" BEYOND COVER
GASKET SURROUNDS FLANGED BASE TO PREVENT LEAKAGE IN CASE EXTENSION IS BROKEN
FLANGED BASE IS IN DIRECT CONTACT WITH GAGE GLASS
a.
NON-FROSTING GAGE TRANSPARENT TYPE (TF-20)
b.
NON-FROSTING GAGE REFLEX TYPE (RF-20)
c.
Non-frosting extension maintains excellent visibility of liquid level despite heavy frost build up on gage.

Figure 5.66. Extensions are placed against gauge glasses to eliminate frost buildup and allow gauge readings in low temperature services. (Courtesy of Jerguson Gage and Valve Co.)

Illuminators attach directly to the gauge and transmit light through the fluid to provide visibility in dark areas as well as to improve the view of some fluids for color, physical state, etc. The illumination is made by a lamp attached to a sheet of plexiglass such that light is transmitted edgewise through the plexiglass to the gauge, giving a uniform distribution of light on the gauge. The illuminators are available for installation in the areas requiring hazardous area ratings.

Frost Preventers

Frost preventer or nonfrosting extensions can be attached to gauges operating below the freezing point (Figure 5.66). The device consists of a thick sheet of translucent material clamped with its edge against the gauge glass. The material must extend outward far enough so that the temperature gradient developed across it will not permit frost to form on the outward edge. They are useful when gauges must be heavily insulated.

6 Pressure Measurement

William G. Andrew, William F. Miller

Pressure may be defined as the action of a force against an opposing force. It has the nature of a thrust evenly distributed over a surface usually within a closed container. Pressure is usually measured as a force per unit area.

The most common units in the English system are the ounce, pound and ton. The practical unit in most cases in the processing industries is the pound, and the unit area is the square inch. Therefore, pressure is normally expressed in pounds per square inch (psi).

Other units used for expressing pressure include inches and feet of water and inches of mercury—in the English system; grams or kilograms per square centimeter, centimeters of water or millimeters of mercury—in the metric system; and in atmospheres. Table 6.1 gives conversion factors for changing any of these units to any other unit.

Two reference points for pressure measurement exist. The most logical one is absolute zero—a condition existing only in a perfect vacuum. Pressures measured from this reference point are called absolute pressures (psia).

The other reference point used is atmospheric pressure. The difficulty with this reference point is that it changes with altitude (referenced to sea level) and to some extent with weather conditions. At or near sea level, this pressure is about 14.7 psia or 29.9 inches (760 mm) of mercury absolute.

When using the latter reference point, pressures above atmospheric are referred to as gauge pressures (psig) or positive pressures. Pressures below atmospheric are referred to as vacuum or negative pressures. A vacuum is merely a reduction from atmospheric pressure.

A term often used in pressure measurement, particularly in low-pressure ranges near atmospheric, is hydrostatic head, or simply head. It represents the pressure at a point below a liquid surface due to the height of liquid above it. This head expressed in force per unit area is dependent on the specific gravity of the liquid exerting the hydrostatic head and on the pressure exerted by any gas or vapor above the liquid.

Pressure Elements

Pressure elements or pressure devices may be categorized in several different ways—on the basis of their design principle, their operating range, their application or their usage factor. The listing here is primarily by design principle.

Most of the discussion is on element types (several are illustrated in Figure 6.1). Very little space is given to methods of translating the various element movements into transmission signals when transmitters are required. It is readily recognized that the principles of signal transmission are essentially the same for flows, temperatures, pressures or levels. Discrete movements (preferably linear) of the various sensing elements are translated into proportional, usable signals for remote readout and/or control.

Elements to be discussed in some detail include manometers, bourdon tubes, bellows, diaphragms and strain gauges. Some other well-known, but infrequently used, devices are mentioned briefly.

Manometers

Manometers may be classified as a gravity-balance pressure device. They measure unknown pressures by balancing against the gravitational force of liquid heads. Although they are often thought of as laboratory devices, they do find application in plant systems, primarily as differential pressure devices or as level devices sensing liquid heads.

Table 6.1 Conversion Factors for Changing from One Pressure Unit to Another

Pressure units	psi	in. H_2O	ft H_2O	in. Hg	atm	g/cm²	Kg/cm²	cm H_2O	mm Hg
1 psi	1.000	27.68	2.307	2.036	0.06805	70.31	0.07031	70.31	51.72
1 in. water (39°F)	0.03613	1.000	0.08333	0.07355	0.002458	2.540	0.002540	2.540	1.868
1 ft water (39°F)	0.4335	12.000	1.000	0.8826	0.02950	30.48	0.03048	30.48	22.42
1 in. mercury (32°F)	0.4912	13.60	1.133	1.000	0.03342	34.53	0.03453	34.53	25.40
1 normal atmosphere	14.7	406.79	33.90	29.92	1.000	1.033	1.033	1,033.0	760.0
1 g/sq cm	0.01422	0.3937	0.03281	0.02896	0.0009678	1.000	0.0010	1.000	0.7356
1 kg/sq cm	14.22	393.7	32.81	28.96	0.9678	1,000	1.000	1,000	735.6
1 cm water at 4°C	0.01422	0.3937	0.03281	0.02896	0.0009678	1.000	0.0010	1.000	0.7355
1 mm Hg at 0°C	0.01934	0.5353	0.04461	0.03937	0.001316	1.360	0.001360	0.001360	1.000

Open Frame Bellows Element (PF)

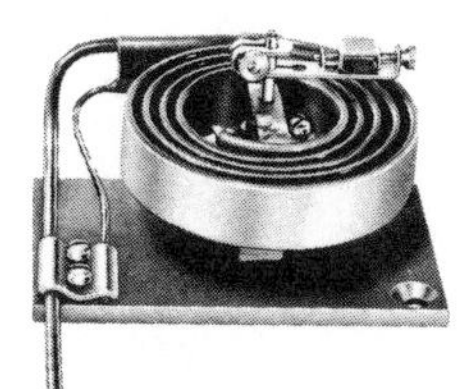

Stacked Capsule Element (PC)

Helical Element (PH)
Liquid Filled Helical Element (PJ)

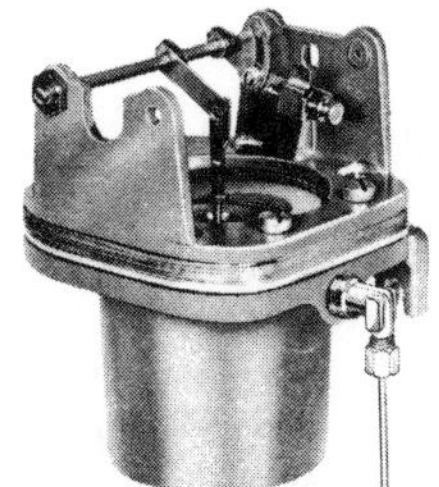

Opposed Helical Element (PW)

Opposed Spiral Element (PV)

Spiral Element (PS)
Liquid Filled Spiral Element (PT)

Opposed Bellows Element (PE)

Enclosed Bellows Element (PR)

Figure 6.1. Various types of pressure elements are in use to cover a wide range of applications. (Courtesy of Fischer and Porter Co.)

Liquid heads (manometers) have been used for hundreds of years to measure gas pressures. In the early years of the processing industries, they were widely used for process applications. They are still considered a primary standard for pressure measurements from the low vacuum range to approximately 1,000 mm of mercury.

A wide variety of fluids are used for manometers. Their specific gravities range from well below 1.0 to 13.5707—the specific gravity of mercury. The main requirements for the fluids are that they should be noncorrosive, stable, nontoxic, not subject to freezing (where ambient temperatures may be subfreezing) and compatible with the measured fluid in contact with the manometer fluid.

Advantages of manometers include simple and time-proven construction, high accuracy, good repeatability, wide range of filling fluids and use as a primary standard or

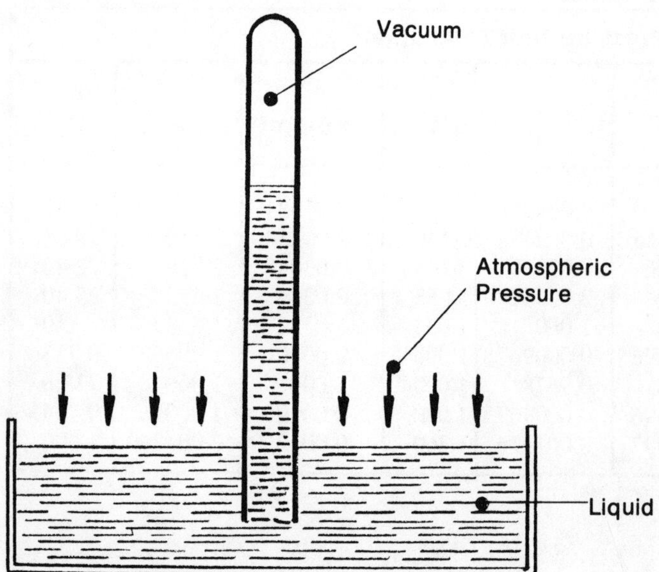

Figure 6.2. Changes in atmospheric pressure are indicated on this simple barometer as they cause liquid level to vary in the evacuated vertical tube.

as a working device. Disadvantages include lack of portability, need of leveling, the hazardous condition existing when mercury is used as the filling fluid and exposed to the atmosphere and the reading error due to the meniscus on small diameter tubes.

There are two basic manometer designs to be discussed: liquid and liquid-sealed. As a sensing device, manometers are referred to as "wet" sensors in contrast to other elements that have no fluids in contact with the measured stream.

Liquid Manometers

Liquid manometers are simple pressure sensors, which are economical, reliable and accurate. There are two types of liquid manometers, the visual and the float. When local indication only is required and static pressure is not too high, glass tube manometers may be used. However, in high-pressure service or when remote readouts are desired, float manometers are needed.

Visual Type. Visual manometers may be classified into four different groups: liquid barometer, U-tube, well type and inclined manometers.

The liquid mercury barometer is a basic instrument for detecting atmospheric pressure as a standard for calibration of other instruments. With a full vacuum in the closed end, it provides a true absolute pressure. Figure 6.2 shows how the filling fluid in a barometer is acted on by changes in atmospheric or process pressures. The accuracy is affected by the visibility of the liquid height, by capillary action in the tube and by ambient temperature variations.

The U-tube, well and inclined manometers operate basically in the same manner as barometers. Figure 6.3 illustrates the U-tube manometer design, which is used fre-

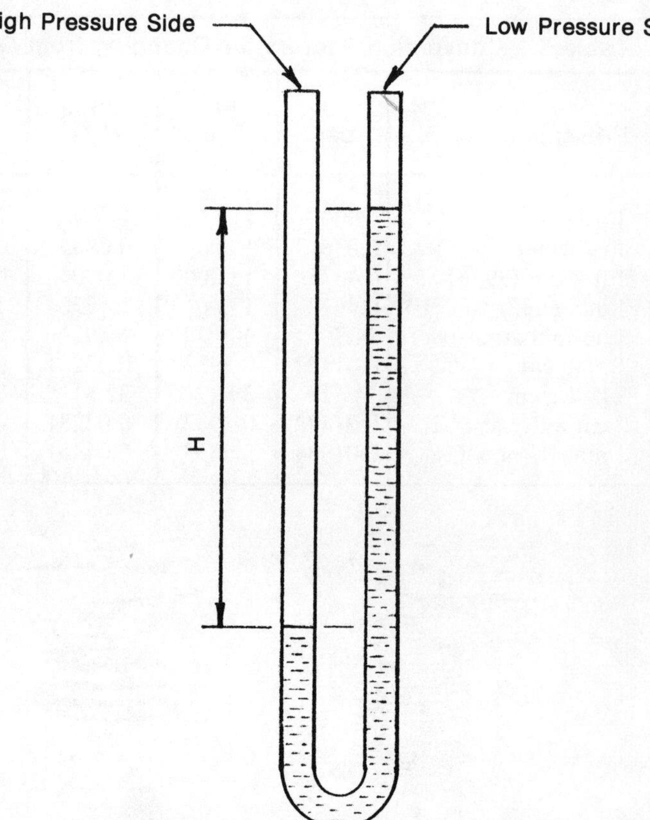

Figure 6.3. U-Tube manometers are often used to measure low differential pressures or low gauge pressures. On low gauge pressures, low side is open to the atmosphere.

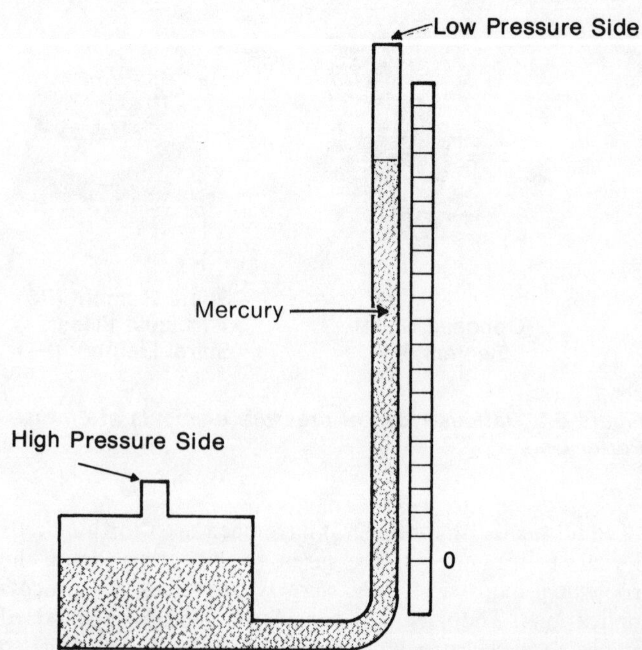

Figure 6.4. This direct reading well-type manometer is used as a matter of convenience. The reservoir level changes are compensated for by correcting the graduation scales.

quently as a differential pressure sensing device. U-tube manometers are available with ranges up to 150 inches H_2O and maximum operating pressures up to 400 psig.

The well manometer, shown in Figure 6.4, is the same as the U-tube, except for a reservoir on the high-pressure side. The well type is used as a matter of convenience and versatility. It provides a single direct reading column rather than a difference between two columns as required by the U-tube. There is also a change in elevation on the well side, but its cross-sectional area is made so large in comparison to the reading column that the error is small. Even that error is compensated for by correcting the graduation scales for measurement readout.

Well manometers are often built in a multitube arrangement to conserve space and reduce mounting difficulties (Figure 6.5). Raised well manometers may be designed with the well located in a raised position (Figure 6.6) so that zero can be located anywhere along the column, allowing vacuum measurements as well as pressure measurements. The extent of vacuum possible is limited, but it does serve a useful purpose. The location of zero (and the well) is optional, but it is usually lower than the midpoint of the scale. Manometers can be furnished with movable wells, either manually adjusted or power driven.

Inclined manometers (Figure 6.7) provide greater reading accuracy. It is evident that a small change in vertical height provides an amplified liquid movement along the slope where the scale is located. Inclined manometers are available in ranges from 0 to 0.5 inches H_2O to 0 to 4 inches H_2O using water and up to 0-50 inches H_2O using mercury.

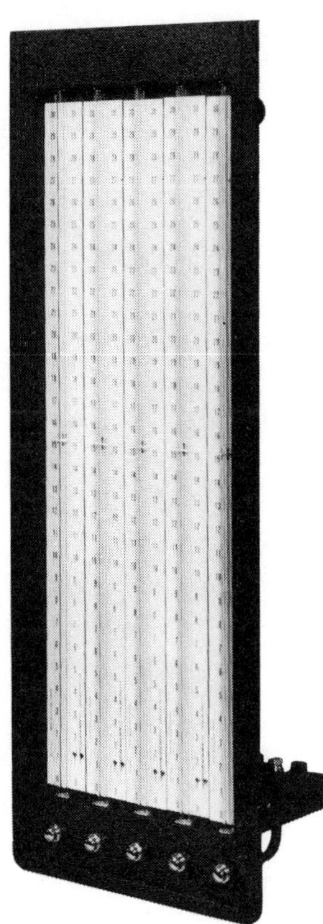

Figure 6.5. Multiple tube manometers conserve space and are easily mounted. (Courtesy of Meriam Instrument Division, Scott and Fetzer Corp.)

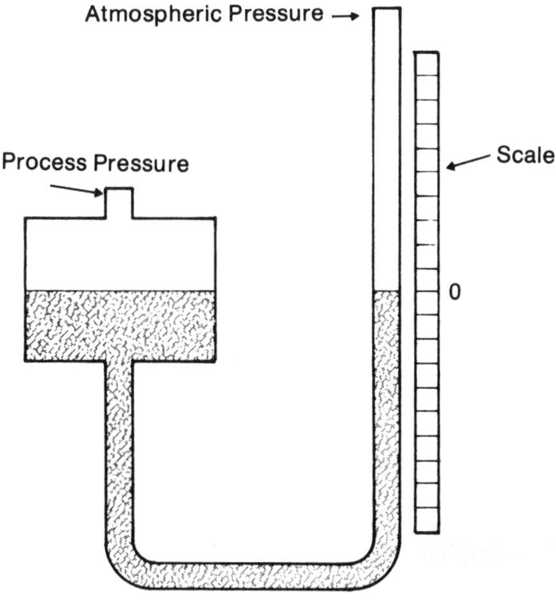

Figure 6.6. A raised well manometer permits optional location of zero. It is usually located below mid-scale, however.

Float Type. Float manometers (Figure 6.8) generally have been replaced with other pressure sensors. In the past they were used extensively where remote signals were required or where the process material was either hazardous or under high pressure. When they are used for process measurement, the process pressure is connected to one side and the reference pressure to the other. The float movement is proportional to the pressure differential between the process and the reference, whether the reference is atmospheric pressure or vacuum. Ranges are available from 10 to 400 inches H_2O, and housings can withstand pressures up to 4,000 psig with measurement accuracies of $\pm 1/2\%$ of span.

The float motion produces sufficient power to drive local indicators or transmitting devices for remote readout.

Liquid-Sealed Manometers

Like some other manometers, the use of the liquid-sealed type is declining. They have been used primarily where low differential pressures were to be measured. Their advantages include their self-powered design and their strong out-

Figure 6.7. Inclined manometers provide an amplified liquid movement along the scale for a small change in vertical height. (Courtesy of Meriam Instrument Division, Scott and Fetzer Corp.)

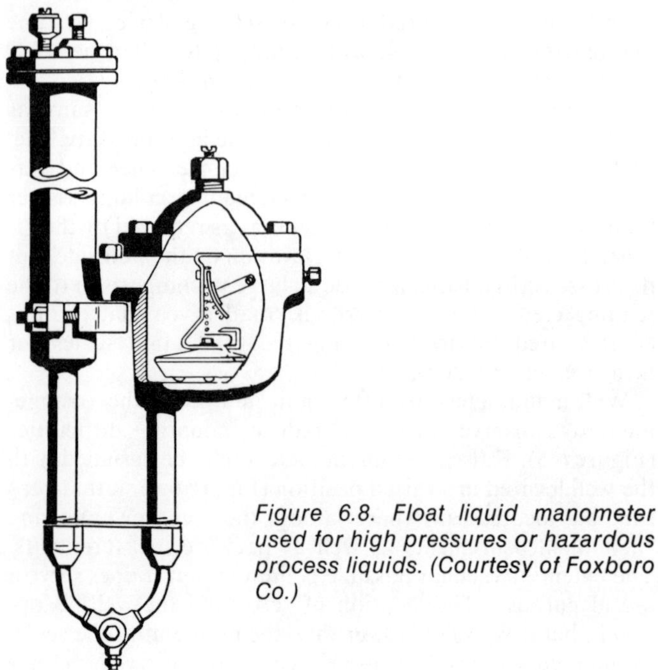

Figure 6.8. Float liquid manometer used for high pressures or hazardous process liquids. (Courtesy of Foxboro Co.)

puts. Their disadvantages include too many moving parts exposed to the process, high initial installation costs and incompatibility with processes where corrosion, temperature, chemical reaction or plugging is a problem. Generally, force and motion balance sensors cost less and require less maintenance than manometer sensors.

A common type of liquid-sealed manometer is the inverted bell design. The operation of this sensor, as illustrated in Figure 6.9, depends on the force developed by the differential pressure acting on the bell being opposed by a spring. The movement of the bell is directly related to the pressure differential. Since the bell is a force amplifier, a small pressure differential can generate a force great enough to drive the readout mechanism. Depending on the sealing fluid and the particular design, these sensors are available in spans from less than 1 inch H_2O to about 15 inches H_2O.

Manometer Materials and Ranges

Manometer bodies are usually made of brass, steel, aluminum or stainless steel. Tubes are made of Pyrex. Scales are usually furnished to read in inches of H_2O or inches of mercury. They can be provided to read in psi, in feet of H_2O or in feet of the liquid measured (when used in level applications).

Ranges are limited primarily on the low side by the ability to compare a liquid head to a scale reading. On the high side, the limitation is simply the physical size required to accommodate the liquid head.

Bourdon Tubes

Bourdon tubes are among the group of pressure sensors that are known as elastic deformation elements. They have been used in industry for over 100 years. Because of their simple design and low cost, they are more widely used than any other pressure element type.

There are three types of bourdon elements, the C, the spiral and the helical. They are widely used both for local indication and for signal transmission to remote locations.

Bourdon elements have several distinct advantages: low cost, simple construction, many years use in various

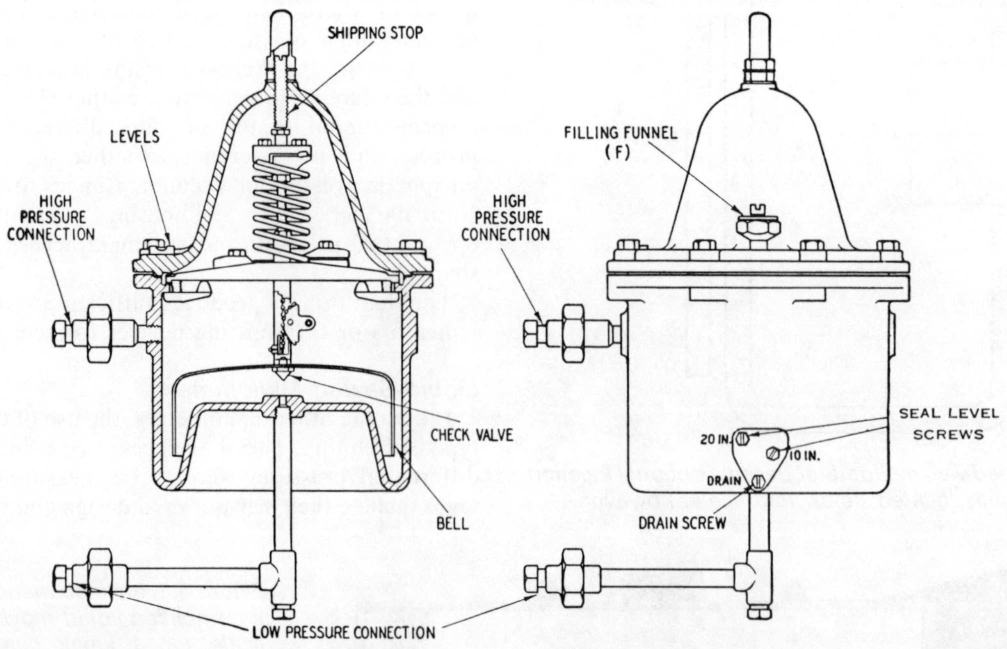

Figure 6.9. The inverted bell design is a liquid-sealed manometer which amplifies the force exerted by a small differential pressure to a force great enough to drive a readout mechanism. (Courtesy of Foxboro Co.)

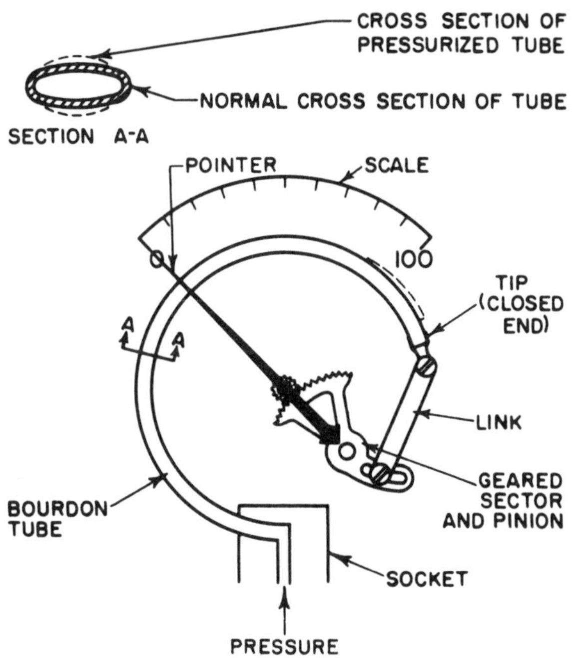

Figure 6.10. C-type bourdon tube pressure element with linkages, gears and pointer as used in most pressure gauges. (Courtesy of Ametek/Calmec)

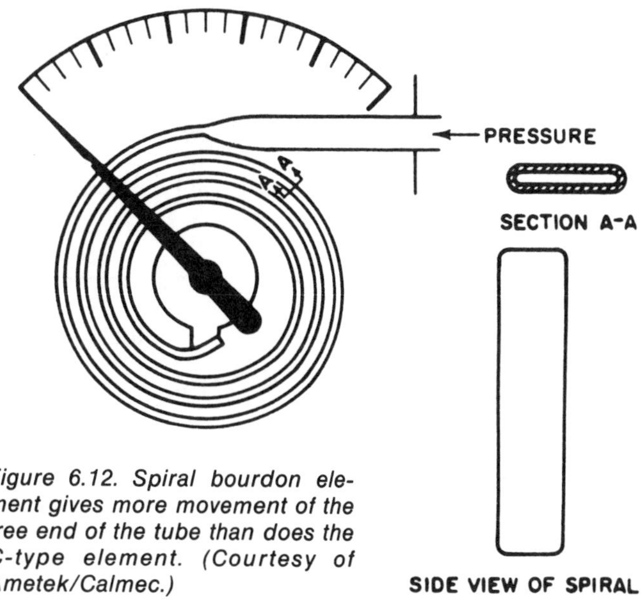

Figure 6.12. Spiral bourdon element gives more movement of the free end of the tube than does the C-type element. (Courtesy of Ametek/Calmec.)

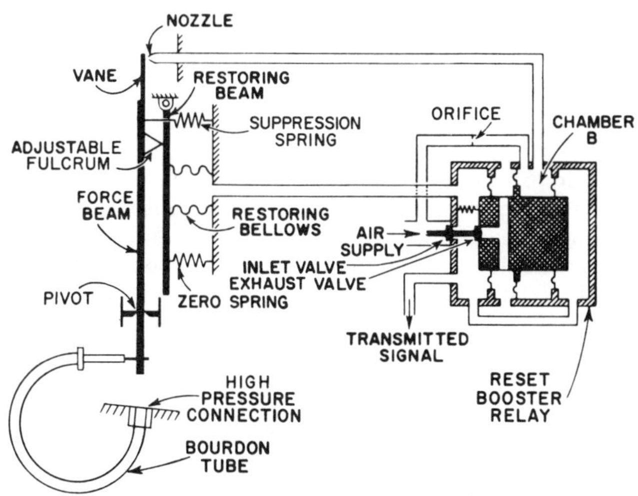

Figure 6.11. The bourdon C-tube is restrained by the force beam whose vane is balanced against the pneumatic nozzle to provide a force balanced system of pressure transmission. (Courtesy of Bailey Meter Co.)

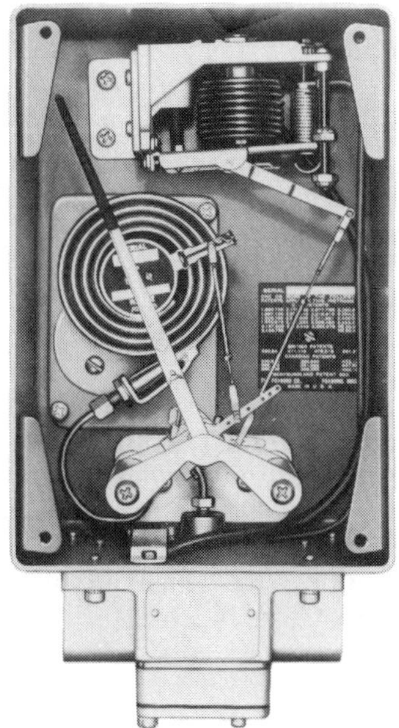

Figure 6.13. Spiral element pressure transmitter-motion balance unit has sufficient movement to operate indicator directly. (Courtesy of Foxboro Co.)

applications, high pressure ranges, good accuracy versus cost except at low ranges, improved designs at high pressures for maximum safety. They are also easily adapted to transducer designs for obtaining electrical outputs. However, some of their limitations should be noted: (a) their spring gradient is low, and precision measurement is limited at pressures below 50 psi; (b) they are susceptible to shock and vibration due to their large overhang; (c) they are subject to hysteresis—the failure of the element to return to its zero position over the same path (force versus position) on signal reversals.

C-Bourdon

The C-bourdon element (Figure 6.10) is used most frequently for local indication but also finds much use in pressure transmission and control applications. The tube is usually formed into an arc of 250°, thus deriving the term *C* because of its configuration. It is secured at one end in a

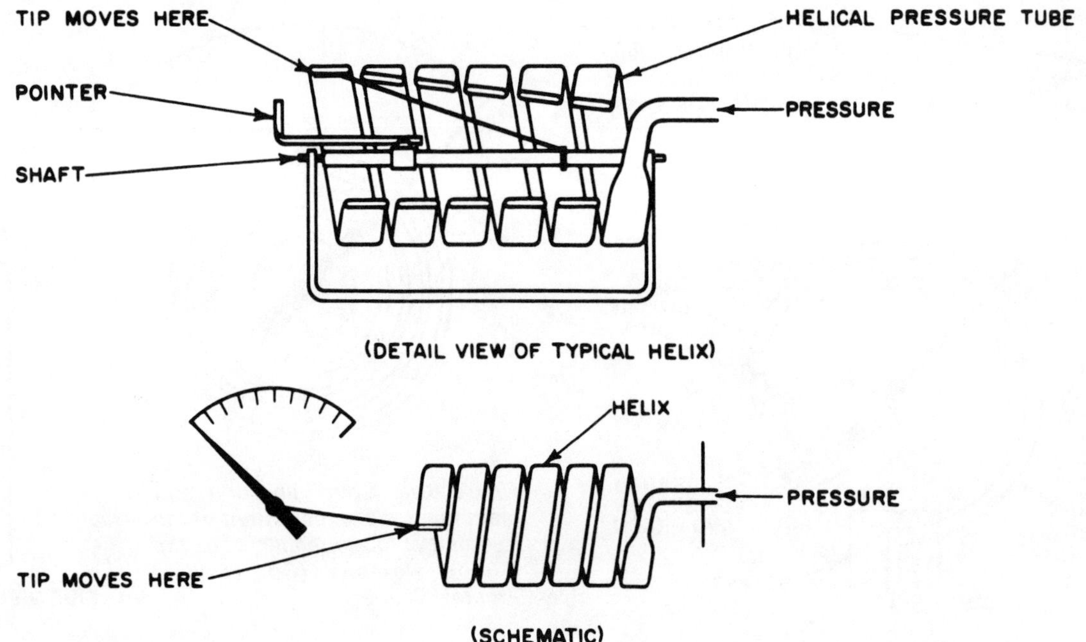

Figure 6.14. Helical bourdon elements provide even more power than spiral elements. They are stable and have high overrange capabilities. (Courtesy of Ametek/Calmec)

fixed socket where the sensed pressure is applied and the tip end is sealed. When internal pressure is applied, the bourdon tube tends to straighten. This movement is balanced against the stiffness or spring of the tube. The motion of the tip is nonlinear—each pressure increment does not produce a corresponding tip movement—but linear response is obtained through a geared sector and pinion movement or other mechanical means.

Figure 6.11 illustrates the use of a bourdon C-element in transmission service. As pressure is applied, the tube tries to straighten but is restricted by the restraining flexure and force beam. The force is transmitted to the force beam whose vane is balanced against a signal output which is always proportional to the sensed pressure. This is a typical force balance transmitter. The C-tube is used for both pneumatic and electronic transmission systems.

The accuracy of C-elements varies from ±0.5 to ±2% or poorer. Elements can be furnished, however, with accuracies as good as ±0.1%. Normal accuracy is about ±1%.

Spiral

The spiral bourdon element (Figure 6.12) is used when the free-end movement of the C-type is not great enough to provide the needed motion. Since greater movement of the free end is attained with the spiral element, it is not necessary, in most cases, for mechanical amplification; thus, better accuracy is obtained. Spiral tubes are made by winding the tube with its flattened cross section in a spiral form of several turns instead of the usual 250° arc made by the C-type design.

As pressure is applied to the spiral, it tends to uncoil, producing the relatively long movement of the tip end whose motion can be used for indication or transmission.

Figure 6.13 is a typical application of the use of a spiral element for a motion-balance pressure transmitter. In the motion balance type, the motion of the element alters the flapper-nozzle relationship to provide an output proportional to the applied pressure.

The normal accuracy of spiral elements is about ±0.5%.

Helical

The helical bourdon element (Figure 6.14) is similar to the spiral element, except it is wound in the form of a helix. It increases the tip travel considerably, producing even greater amplification than the spiral element. Usually a central shaft is installed within the helical element, and the pointer is driven from this shaft by connecting links. This system transmits only the circular motion of the tip to the pointer and, hence, is directly proportional to the changes in pressure. Advantages of the helix element include the high overrange capabilities (the ratio may be as high as 10:1), its stability in fluctuating pressure services and its adaptability for high-pressure service.

The number of coils used for helix elements varies, depending on the pressure range. Low pressure spans may have as few as three coils while high spans may have as many as 16 or more.

The accuracies of helical elements vary from ±½ to ±1.0% of span, depending on the manufacturer.

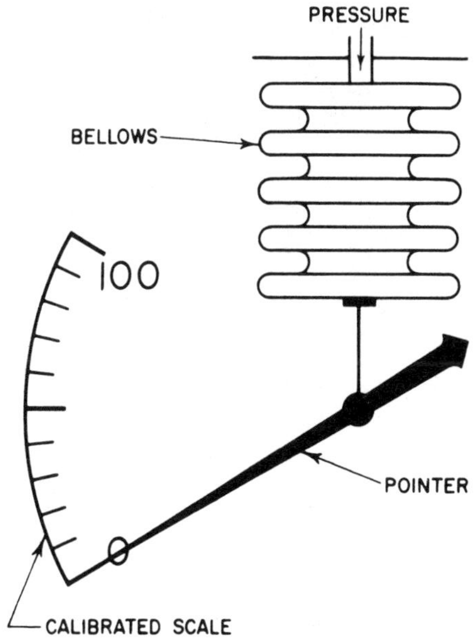

Figure 6.15. Simple bellows element expands when pressure is applied and contracts when pressure decreases. (Courtesy of Ametek/Calmec)

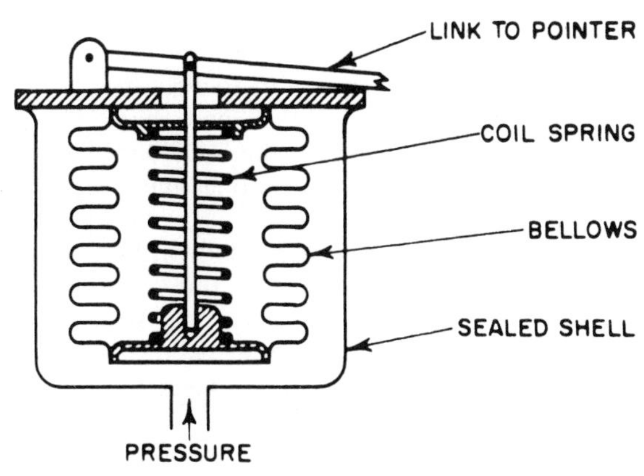

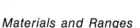

Figure 6.16. Spring loading the bellows reduces its stroke, making longer bellows life. (Courtesy of Ametek/Calmec)

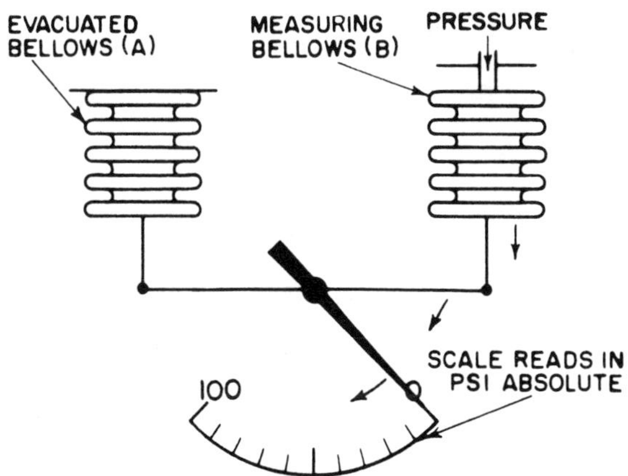

Figure 6.17. A double bellows assembly is used to measure absolute pressure, with the evacuated bellows serving as a reference to vacuum. (Courtesy of Ametek/Calmec)

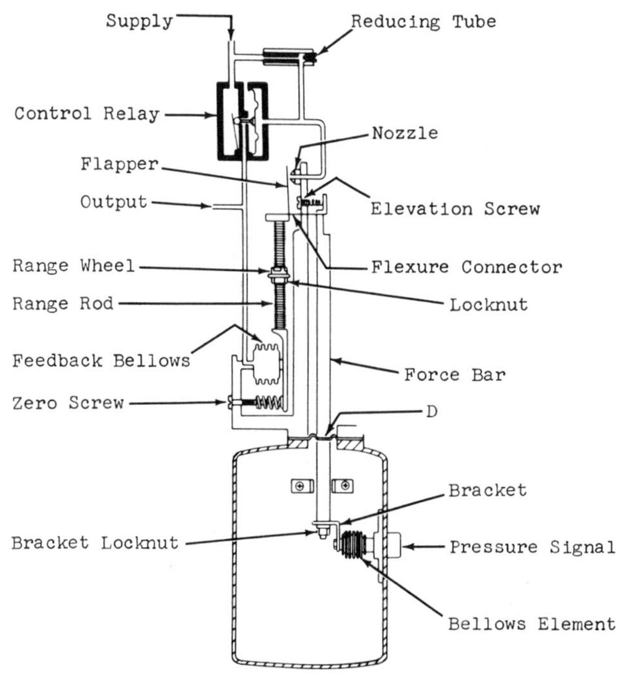

Figure 6.18. Bellows are used for transmitters both as the sensing element and as the feedback element to provide the force balance. (Courtesy of Foxboro Co.)

Materials and Ranges

Bourdon tubes are made of various materials, including phosphor bronze, beryllium-copper, 4130 alloy steel, 316 and 403 stainless steel, K-Monel, Monel and Ni-span C. The choice of the proper tube material depends on the process medium, pressure range, temperature and corrosiveness. For low-pressure and noncorrosive service, phosphor bronze is preferred. When corrosion and/or high pressure is a problem, stainless steel or Monel can be used. On some applications where the commonly used materials do not meet the process requirements, sealed systems may be used. A more detailed account of the use of seals is given in a following section.

Pressure instruments using bourdon elements are made with ranges from 30 inches Hg vacuum to 100,000 psi or higher for special applications. The minimum span is about 10 psi. There is quite an overlap of ranges at which the various bourdon elements operate, both in regard to material and type. One manufacturer, for example, lists spiral elements from 0-10 to 0-200 psi in bronze, beryllium-

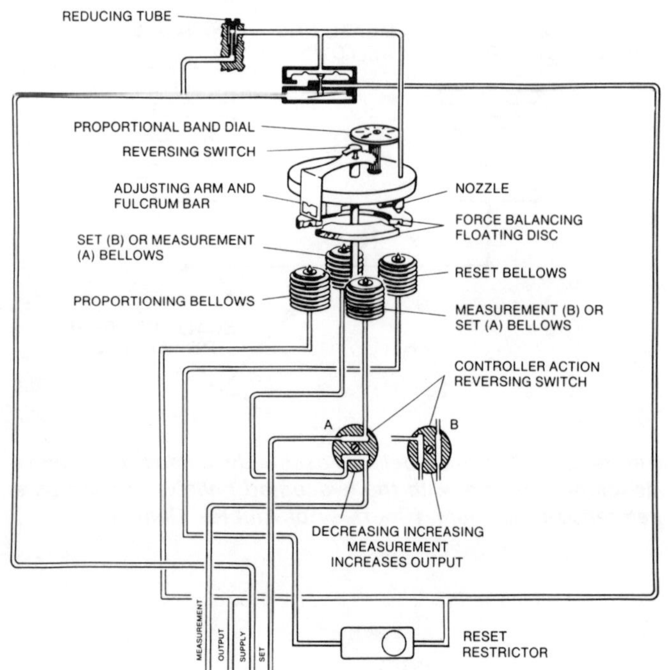

Figure 6.19. Schematic of a pneumatic controller which uses four bellows elements for functions indicated. (Courtesy of Foxboro Co.)

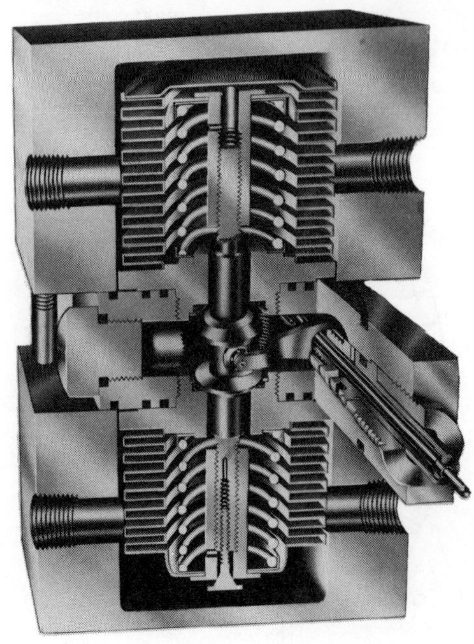

Figure 6.20. Double bellows unit used for differential pressure measurements as high as 0-400 psi. (Courtesy of ITT Barton)

copper, Ni-span C and 316 stainless steel (except 0-12 psi on the low end) and helical elements from 0-200 to 0-80,000 psi. For bronze, the upper span limit is 0-400 psi; for beryllium-copper and Ni-span C, 0-6,000 psi; and for 316 stainless steel, 0-80,000 psi.

Another manufacturer lists bronze bourdon C-elements to 1,000 psi and steel and stainless steel to 20,000 psi. Still another offers C-elements to 100,000 psi. It may be noted generally that C-elements cover the entire range from vacuum measurement to 100,000 psi and higher, spiral elements from vacuum to as high as 4,000 psi and helical elements from 200 to 80,000 psi.

Bellows

The bellows element is an elastic member, usually formed from a thin seamless tube. A simplified diagram of the element is shown in Figure 6.15. In order to give the bellows maximum life and better accuracy, its movement is generally opposed by a calibrated spring so that only part of the maximum stroke is used. This system, as shown in Figure 6.16, is called a "spring-loaded" bellows.

Bellows elements are used to measure absolute, gauge or differential pressures. When measuring absolute pressure, two bellows are used, as shown in Figure 6.17. The evacuated bellows is used for atmospheric pressure compensation while the measuring bellows senses the process pressure. Since the evacuated bellows has negligible internal pressure and is surrounded by atmospheric pressure, any change in atmospheric pressure either adds to or subtracts from the movement of the sensing bellows.

Bellows elements are sometimes used in pressure transmitters as shown in Figure 6.18. The process pressure is applied to the inside of the bellows which in turn exerts a force on the lower end of the force bar. The metal diaphragm acts as a fulcrum for the force bar. The force is balanced and converted to either a pneumatic or electronic signal in the transmitter.

The greatest use for bellows elements probably is that of receiver elements for receivers and controllers for pneumatic instruments. Figure 6.19 is a schematic of a pneumatic controller which uses four bellows, one each for measurement and set point and one each for the proportional and reset control functions.

In this configuration, all four bellows act with the "floating disc" to form a force balance arrangement for control, the floating disc acting also as the flapper of a conventional flapper-nozzle system. Each of the bellows exerts an upward force on the disc, performing its intended function toward a net effect of all forces acting simultaneously to establish a disc position which provides the desired output pressure.

Like many other elements, bellows elements have been greatly improved over the past few years. The reduction of drift and hysteresis allows their use in functions requiring ±½% of full span accuracy. They deliver relatively high forces and are well adapted to vacuum and low-pressure ranges. They are used to some extent in medium pressure ranges. Figure 6.20 shows a double bellows arrangement used for differential pressures as high as 0-400 psi in 6,000 psig static pressure service.

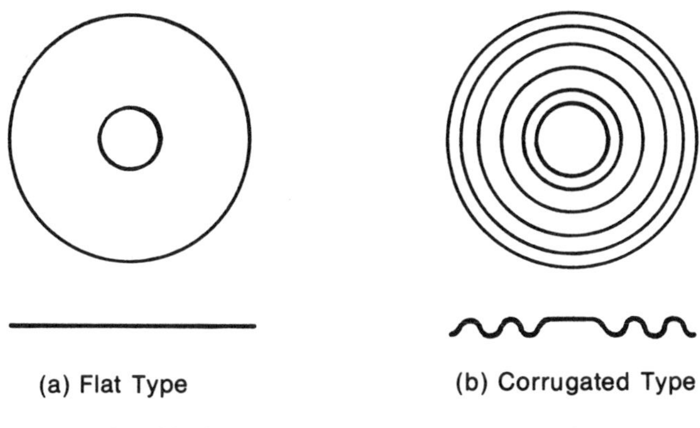

Figure 6.21. Single diaphragm elements may utilize flat or corrugated surfaces.

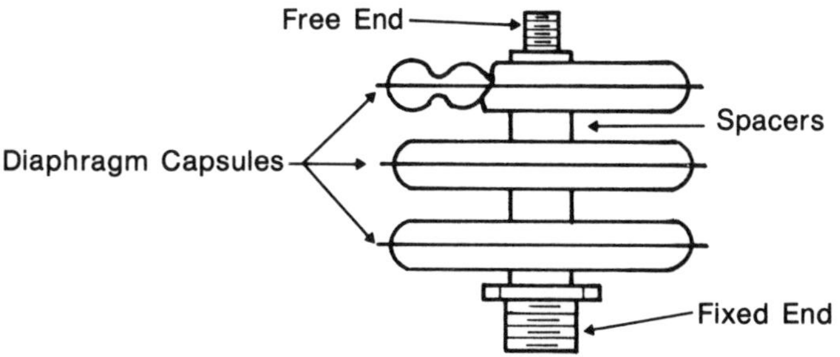

Figure 6.22. Two or more capsules are often used together to form a single multiple-capsule element.

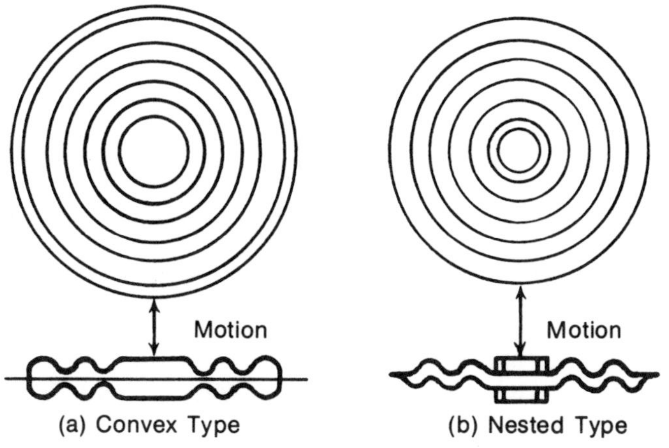

Figure 6.23. Two types of diaphragm capsules—convex and nested—are in common use.

Materials of construction for bellows elements depend on the medium to which they are exposed. Usually they are made from brass, phosphor bronze, beryllium-copper, stainless steel or Monel.

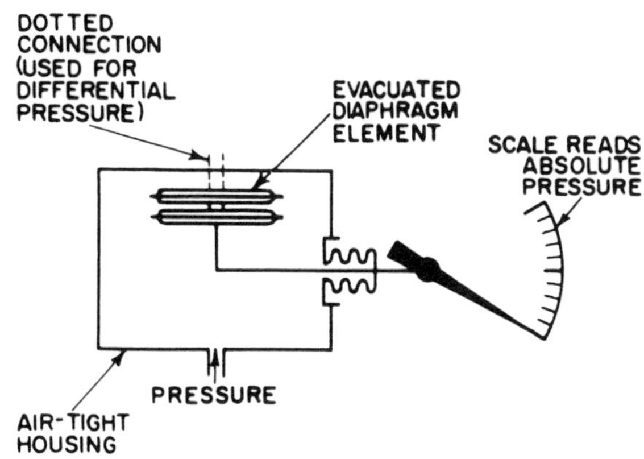

Figure 6.24. An evacuated diaphragm element used in a motion balance instrument for absolute pressures, with an alternate connection for use when differential pressure measurements are made. (Courtesy of Ametek/Calmec)

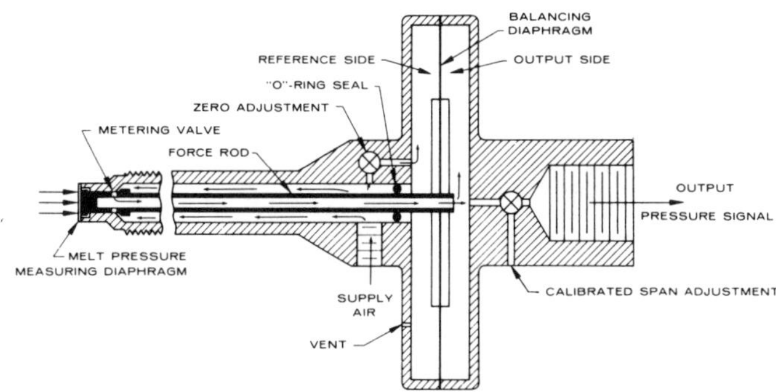

Figure 6.25. Force balance transmitter using a button type diaphragm element. (Courtesy of Rosemount, Inc.)

Diaphragm

The operating principle of the diaphragm element is similar to that of the bellows. Pressure is applied to the element, causing it to expand in direct proportion to the applied pressure. The movement of the diaphragm is small, depending on its thickness and the diameter of the metal. Hence, no calibrated springs are required as in the case of the bellows element.

The diaphragm element is essentially a flexible disc with either a flat or corrugated surface as illustrated in Figure 6.21. In some cases, the element may consist of a single disc; in others, two diaphragms are bonded together at their circumference by soldering or pressure welding to form a capsule. Two or more capsules may be used together to make one sensing element (Figure 6.22). Two different configurations are used to form capsular elements, the convex and the nested types (Figure 6.23).

When measuring absolute pressure, evacuated capsules are required. Figure 6.24 illustrates an evacuated capsule being used in a motion balance instrument with an alternate connection for differential pressure service.

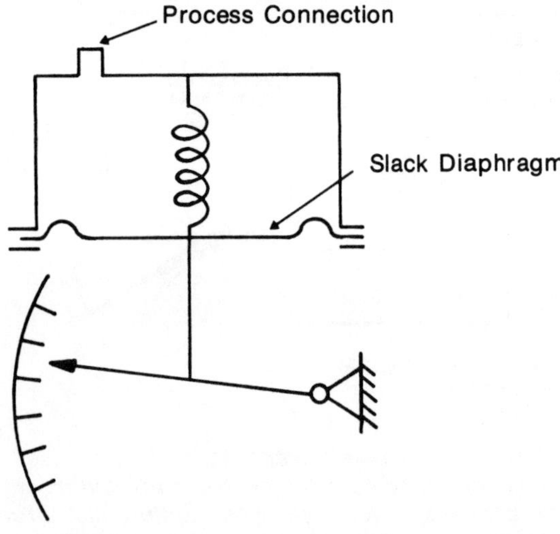

Figure 6.26. Typical pressure sensor using a slack diaphragm as the primary element.

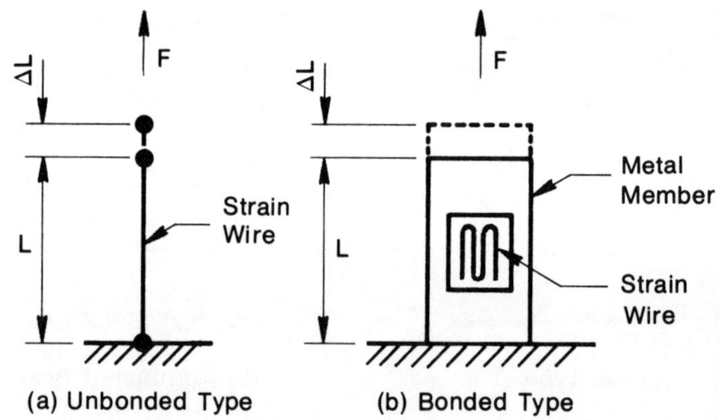

Figure 6.27. Two types of strain gauges—bonded and unbonded. The bonded type adheres to the elongating member while the unbonded type is fixed only at one end and stretches as the member elongates.

Diaphragm elements are used not only in motion balance instruments but also in force balance units. Figure 6.25 shows the operation of a button diaphragm transmitter in which a small (0.307-inch diameter) diaphragm is used to sense process pressure. The process pressure acting on the diaphragm is supported by regulated air pressure on a balancing diaphragm located at the other end of the force balance transmitter.

Supply air entering the transmitter passes through the regulating valve located in the insertion tip of the unit. The valve position is controlled by movement of a force rod; therefore, the system seeks an equilibrium or force balance position.

When the system is at force balance, the air pressure on the balancing diaphragm is proportional to the process pressure by the inverse ratio of the two diaphragm areas. The process pressure is then determined by measuring the air pressure with an indicator, recorder or transducer calibrated to the range of the transmitter.

This device and others similar to it have been developed primarily for the plastic extrusion and synthetic fiber industries. Their pressure ranges extend to 10,000 psig with temperature ratings to 800°F.

Some devices, similar to the one described in Figure 6.25, utilize filled systems. Capillaries connect the sensing button diaphragm to an indicator, recorder or transmitter. It is available for pressure ranges to 15,000 psig. Temperature limits of this device are as high as 650°F, depending on the fluid used for filling. Several fluids are used, including water, water-glycerin and mercury.

A diaphragm unit design (Figure 6.26) referred to as a slack-diaphragm element utilizes a spring action to balance the process pressure. Very sensitive, it is used to measure pressures near atmospheric, such as draft pressures. The elements provide sufficient power for direct readout, or they may be used as transmitters for remote readout. They are available in spans as low as 0.5 inch H_2O and can be used on the vacuum or positive pressure side or for compound ranges. They are also used for differential pressure services.

Diaphragm sensors have been in use for many years. Refinements in design in recent years have provided many improvements as they have for many other sensing elements. Materials of construction include phosphor-bronze, stainless steels, beryllium-copper, Ni-span C, Inconel, Monel, Hastelloy and nickel. Nonmetals have been used in some applications. Buna N rubber, nylon, Teflon and Kel-F are adaptable to services that are corrosive to metals.

Accuracies range from ±½ to ±1¼% of full span. Applications are usually in low-pressure ranges, except for the button diaphragms which are used for pressures to 15,000 psig.

Electronic Types

Since the introduction of electronic instruments over two decades ago, there have been some tremendous advances in the design of electronic pressure sensors and/or transducers. These advances have come not as a result of the use of electronic process instruments per se, although electronic systems have hastened their use in the process industries. The real impetus has been their use in the aerospace field. Another valid reason, of course, has been the development and use of semiconductors, improved electronic circuitry, thin-film technology, etc.

Most electronic pressure sensors incorporate one of the previously discussed elements as the primary pressure detector, and it is used to vary a measurable electrical quantity to produce a proportionately variable electronic signal. Because the energy form is transferred from a mechanical to an electrical nature, these devices are often classified as transducers.

Among the electronic devices used to measure pressure are strain gauges; capacitive, inductive and variable reluctance transducers; linear variable differential transformers; and piezoelectric transducers.

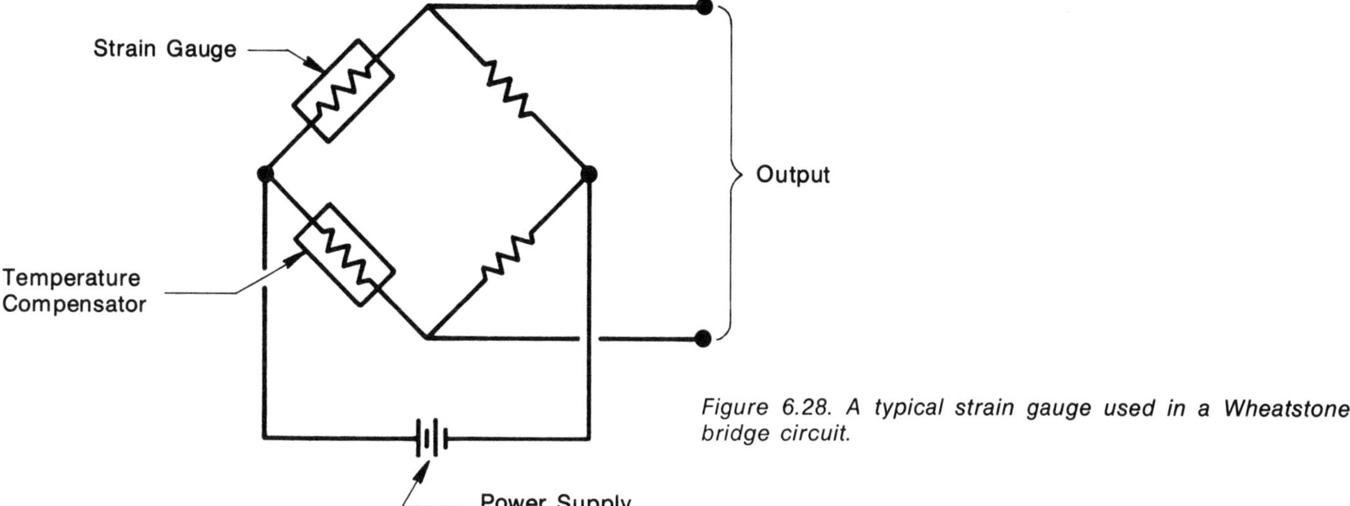

Figure 6.28. A typical strain gauge used in a Wheatstone bridge circuit.

One of the major limitations to these sensors is the high cost of the units and their accessories. However, under certain operating conditions, these elements are preferable, if not necessary. Generally electrical pressure detectors are more accurate and have much faster response times than the previously discussed elements. This is due in part to the accuracy of their electronic circuitry and in part to the extremely small movement required of the elastic elements in order to obtain the needed electrical change. The reduced movement very nearly eliminates drift, friction and hysteresis common to bellows, diaphragms and bourdon elements that require relatively large movements.

Strain Gauges

Strain is defined as a deformation or change in the shape of a material as a consequence of applied forces. A strain gauge is a device which uses the change of electrical resistance of a wire under strain to measure pressure. The strain gauge changes a mechanical motion into an electrical signal when a wire length is changed by tension or compression, altering the wire diameter and, hence, changing the electrical resistance. The change in resistance is a measure of the pressure producing the mechanical distortion.

The complete pressure measuring device includes a sensing element (bourdon tube, bellows or diaphragm), a strain gauge attached to the element, a stable power source and a readout device.

Figure 6.27 shows the two general types of strain gauges that have been developed since the invention of the strain gauge—the bonded and the unbonded.

The unbonded gauge is a strain-sensitive wire which has one end fixed, the other end movable. Since the force requirement of the unbonded strain gauge is low, small size and low-range sensors are readily available. Overload protection and damping are easily combined in this type sensor. By its very nature, unbonded strain gauges are adaptable to tension measurements only.

The bonded gauge also is a strain-sensitive wire or foil and is attached by an adhesive (bonding agent) to the member whose strain is to be measured. Electrical insulation is provided by the adhesive and/or insulating backing material on the strain gauge. The force required to produce a measurable displacement is larger than that required for an unbonded gauge because of the additional stiffness of the member. The bonded strain member functions in tension, compression or bending mode. Because of the higher force requirement, the bonded strain gauge has limitations in small size and low-range applications. The measurement is obtained by transferring strain from the metal member through the bonding adhesive and backing material to the strain gauge. The accuracy of the measurement is limited by the characteristics of the adhesive and backing material, since they are in the force path. The adhesives and backing materials commonly used are of the epoxy type.

Metal and metal alloys in wire and foil forms are used for standard strain gauges. Although all electrical conductors exhibit a "strain-gauge" effect only a few have the necessary properties to be useful as strain gauges. The major properties of concern are gauge factor, resistance, thermal coefficient of gauge factor, thermal coefficient of resistance and stability. For common metallic strain gauge materials, the gauge factor is generally from 2.0 to 5.0. High gauge factor materials are usually more sensitive to temperature and not as stable as lower factor materials. Strain gauge factor is defined as $10^6 (\Delta R/R)$ where R is the resistance of the strain gauge element and ΔR is the change in resistance due to the applied strain.

A new development in the strain gauge field was the introduction of the semiconductor (silicon) strain gauge. The properties of semiconductors as a group are controlled and optimized for a specific application by controlling the type and amount of dopant in the silicon. In general, silicon strain gauges provide a higher gauge factor than the metallic strain gauges, but with higher temperature coefficients. Typical gauge factors range from 50 to 200.

A recent advance in strain gauge technology was the development of the "thin-film" strain gauges. These gauges, like other types, can be made of various resistor materials.

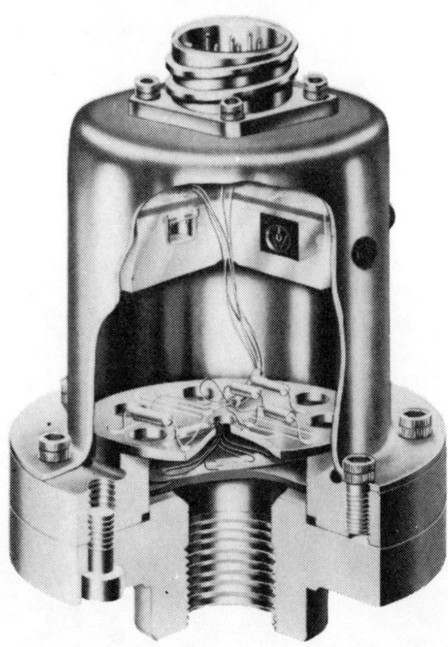

Figure 6.29. A typical pressure transducer using a metal diaphragm with four bonded strain gauges attached. (Courtesy of Viatran Corp.)

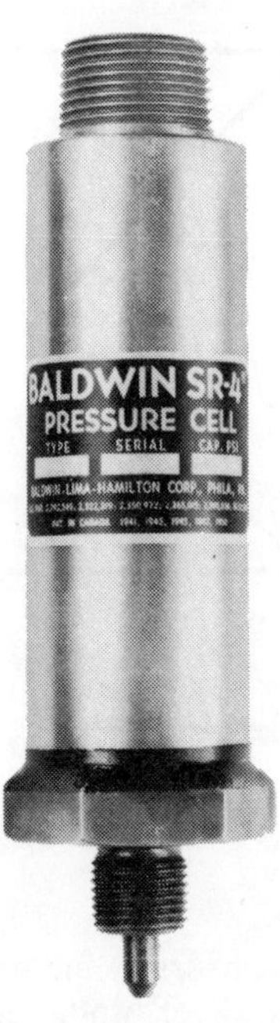

Figure 6.30. This type strain gauge pressure transducer is available for pressure ranges to 50,000 psig at accuracies of ±0.25% of span. (Courtesy of BLH Electronics, Inc.)

They are made by vacuum deposition, using techniques similar to electronic microcircuitry. A metal substrate is used to provide the needed mechanical properties, and a ceramic film is deposited on the metal to provide electrical insulation. The strain gauges are then deposited on the insulator, and they are electrically connected into a bridge circuit (a four-arm bridge is normally used) by vacuum-deposited interconnecting leads. Lead wire is attached to the film by techniques such as welding and thermocompression bonding of small noble-metal wire.

The sensing elements may have any configuration, and they may be deposited on diaphragms, beams, columns or other sensing elements. The best properties are obtained by controlling the strain gauge material and deposition conditions, such as pressure, temperature, rate of evaporation and time. The gauge factors obtained from thin-film devices are as high as those reached by the silicon type. The resistance of the gauge is determined by the film thickness and physical size, which can be varied to give the proper value. Resistances of 350Ω to 5,000Ω are easily produced. The higher the resistance value, the higher is the excitation voltage which can be used with the same power input.

The ceramic insulating material of thin-film gauges provides excellent insulating qualities over a wide temperature range and exhibits no hysteresis up to the yield point of the metallic sensing element, permitting high accuracy.

Regardless of the particular type strain gauge employed, it is almost always used in a Wheatstone bridge circuit similar to Figure 6.28. When process pressure is applied, the gauge resistance changes with the deformation of the sensing element (wire, foil or film), and the output voltage of the bridge changes. This output cannot be measured by a conventional voltmeter, so amplification is required to drive a receiver instrument.

When strain gauges are used, it is often necessary to compensate for process temperature changes because both the strain gauge and the bonding material expand and contract with the temperature changes. The most frequent method of compensation is the use of compensating resistors in the Wheatstone bridge measuring citcuit. Dual elements and other techniques are also employed.

Strain gauge transducers (not thin-film type) normally have bridge resistances of 100 to 500 ohms. AC or DC power is used for excitation voltage. The output voltage generated is usually 1 to 3 millivolts per volt input. Output voltages are readily converted to standard signals for the readout devices which are needed.

Strain gauge transmitters are now available with the bridge electronics built into the same head assembly as the measuring device. The transmitter puts out a standard 4-20 ma signal, with bridge excitation derived from the power available below the live signal level.

Generally, strain gauge accuracies range from ±0.1 to ±2% of full scale, depending on the materials and design. They are available in ranges from a few inches of water to 200,000 psig.

Strain gauges are used for measurements other than pressure—torque, weight, velocity and acceleration. Some typical pressure measuring devices are described below.

Figure 6.29 shows a typical pressure transducer that uses a metal diaphragm as the sensing element. The diaphragm element transmits movement to the strain sensing element which consists of four bonded strain components connected in a balanced bridge circuit. The transducer output is proportional to the applied pressure.

On high-pressure applications, pressure ports are made to accept high-pressure fittings. Proof pressures are from 1½

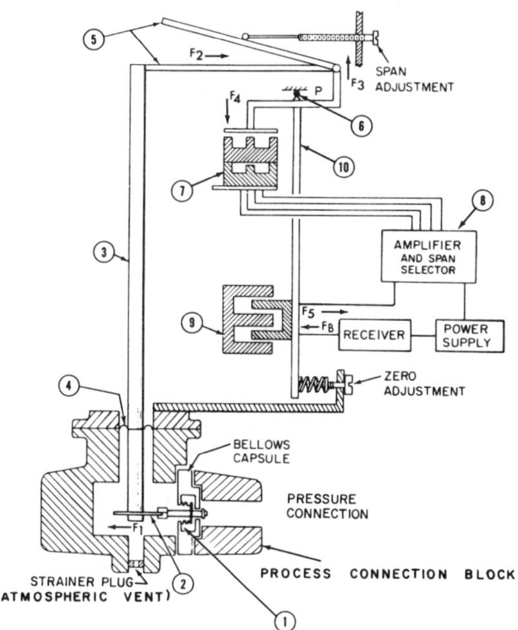

Figure 6.31. A typical schematic of a linear variable differential transformer (LVDT) system that is actuated by a pressure element acting on the force bar. (Courtesy of Foxboro Co.)

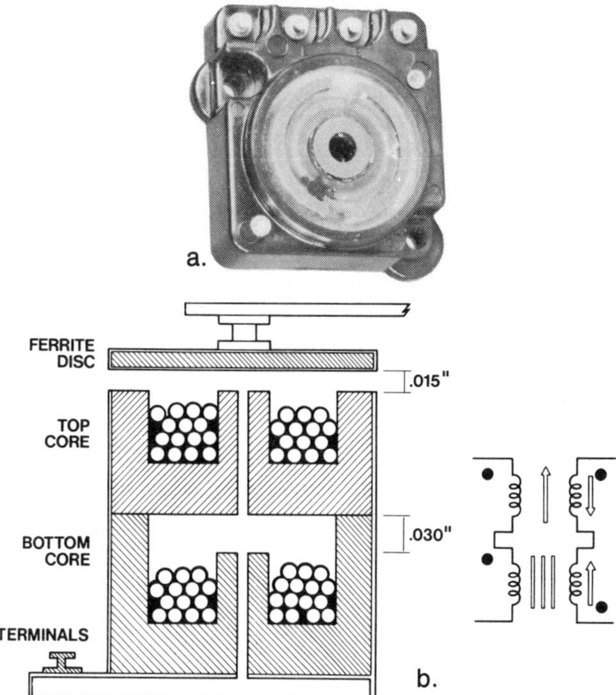

Figure 6.32. The differential transformer detector consisting of two ferrite coils on top of one another. (Courtesy of Foxboro Co.)

to 10 times the rated range, depending on the particular design.

The ranges of these small and compact devices are from 0 to 5 psig through 0 to 50,000 psig. Accuracies range from ±0.2 to ±1.0% of span.

Figure 6.30 pictures another pressure transducer used for industrial and military applications. Four-arm bonded strain gauges are mounted on a tube-type sensing element to provide pressure measurements for ranges from 0-5,000 to 0-50,000 psig standard and on special order to 150,000 psig.

The calibration accuracy of this device is ±0.25% of span. Temperature compensation is provided from 15° to 115°F. The unit may be operated safely at temperatures to 150°F.

A high-pressure version of this transducer is available for pressures from 0-100,000 to 0-200,000 psig. Its accuracy is ±0.5% of span.

Other Electrical Types

There are several other electrical pressure devices used for electronic transmission. These electrical transducers are not limited solely to pressure and differential pressure elements but also function from other motion sources. All of these used as pressure sensors (and many from other motion sources) utilize one of the primary sensing elements already described and cause its movement to vary an electrical characteristic that converts to a measurable proportional output signal. Among the electrical transducers discussed are linear variable differential transformers (LVDTs), variable reluctance type, variable impedance type, capacitive transducers and piezoelectric transducers. Other types sometimes used but not discussed include photoelectric, potentiometric, resonant frequency, etc.

Linear Variable Differential Transformers. The differential transformer is one of the most widely used methods of converting mechanical motion to proportional electronic signals. The LVDT is an inductive device that contains a primary coil and two series-opposed secondary coils, as well as a magnetic core, connected to the sensing device. Core movement which is a function of sensor movement produces voltage variations that are measured directly. The required core displacement is within a range that provides a high degree of linearity. Sensitivity is also good so that stiff sensors with very little movement can be used to reduce environmental effects. Frequency response is also good, with commercial units able to respond to pulses greater than 100,000 Hz.

Figure 6.31 shows the schematic of a differential transformer system that operates on the force-balance principle. In operation the pressure applied to the capsule is sensed by the bellows element inside the capsule, where the process fluid is external to the bellows (1). A force equal to the pressure times the effective area of the bellows exists. The resultant force is transferred through the flexure (2) to the lower end of the force bar (3). Attached to the force bar is a cobalt-nickel alloy diaphragm which serves as a fulcrum point for the force bar and also a seal to any atmosphere. As a result of the force generated, the force bar pivots about the cobalt-nickel alloy seal (4), transferring a force to the vector mechanism (5).

The force transmitted by the vector mechanism to the lever system (10) is dependent on the adjustable angle.

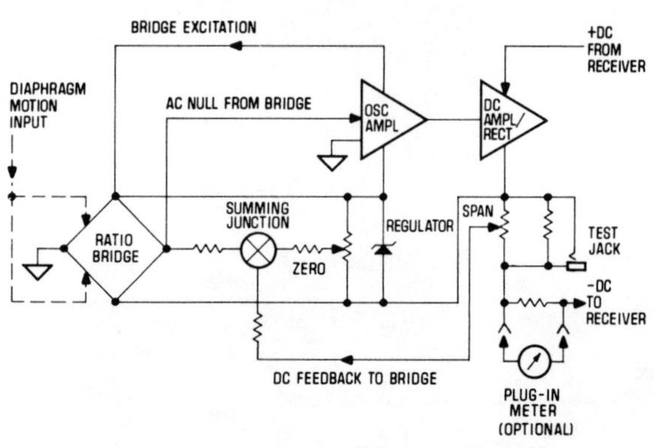

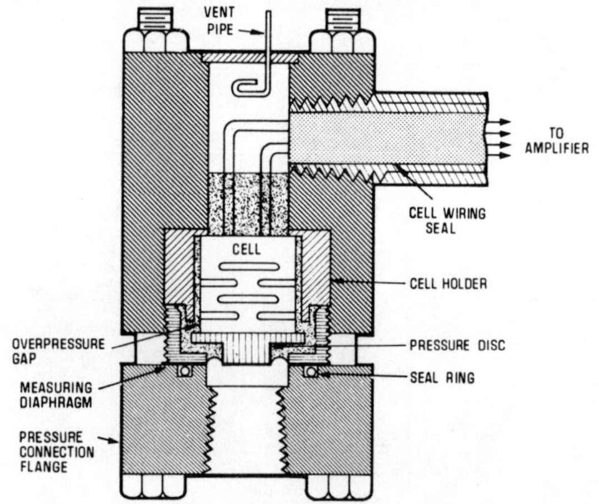

Figure 6.33. A reluctance pressure cell is shown along with its electrical wiring schematic. (Courtesy of Westinghouse Electric Corp.)

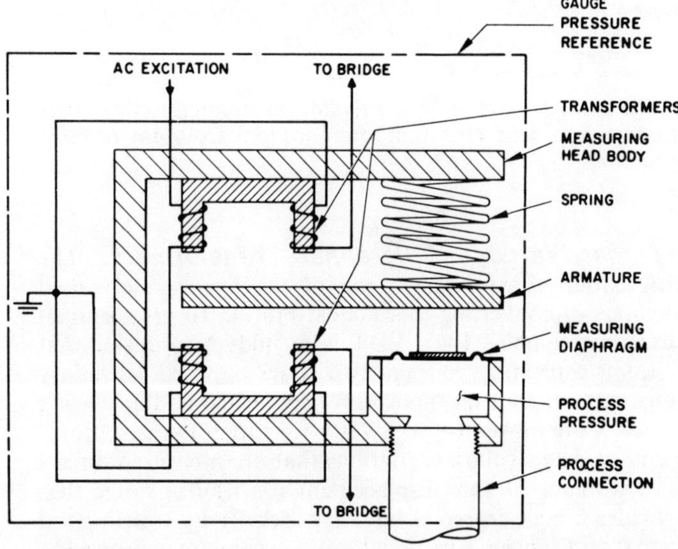

Figure 6.34. A schematic view of the reluctance pressure head showing the measuring diaphragm, the armature and the transformer coils. (Courtesy of Westinghouse Electric Corp.)

Changing this angle adjusts the span of the instrument. At point 6, the lever system pivots and moves a ferrite disc, part of the differential transformer (7) which serves as a detector. Any position change of the ferrite disc changes the output of the differential transformer determining the output to an oscillator (8). The oscillator output is rectified to a DC signal and amplified, resulting in a 4-20 ma DC transmitter output signal. A feedback motor (9) in series with the output signal exerts a force proportional to the error signal generated by the differential transformer. This force rebalances the lever system. Accordingly, the output signal of the transmitter is directly proportional to the applied pressure with respect to atmosphere at the capsule.

For any given applied gauge pressure, within the calibrated measurement range, the ferrite disc of the detector is continuously throttling, maintaining an output signal from the amplifier proportional to the measurement and retaining the force-balance system in equilibrium.

The pressure sensing element in this case is the bellows capsule. In other devices, it might be a bourdon tube or other pressure element.

The differential transformer detector (Figure 6.32) consists of two ferrite coils on top of one another. The air gap between the middle of the bottom core and the bottom of the upper core is fixed. A ferrite disc on top of the top core, when in a certain position with respect to the top core, forms an effective air gap equal to the one in the bottom core. Whenever the disc departs from this position, an error signal is produced. As the ferrite disc shortens the air gap, the inductive coupling to the secondary transformer increases, increasing the output amplitude of the oscillator. This results from an increase in measurement.

Variable Reluctance Type. Figure 6.33 shows a pressure sensor that operates on a variable reluctance principle. The deflection of a measuring diaphragm moves a ferrite armature between two ferrite cup cores which, in turn, change the inductance ratio in a bridge network whose unbalance is detected and amplified, producing a signal proportional to pressure variations to the sensor. Schematically, the pressure head—pressure diaphragm, armature, coils and coil cores—is shown in Figure 6.34.

Physically, the cell is a slotted cylindrical spring that surrounds a variable reluctance pickup. The pickup consists of a ferrite armature positioned between two ferrite cup core transformers. The windings are attached to the stationary bottom end of the spring, while the armature is attached to the movable top end of the spring. Force on the measuring diaphragm compresses the spring and moves the armature,

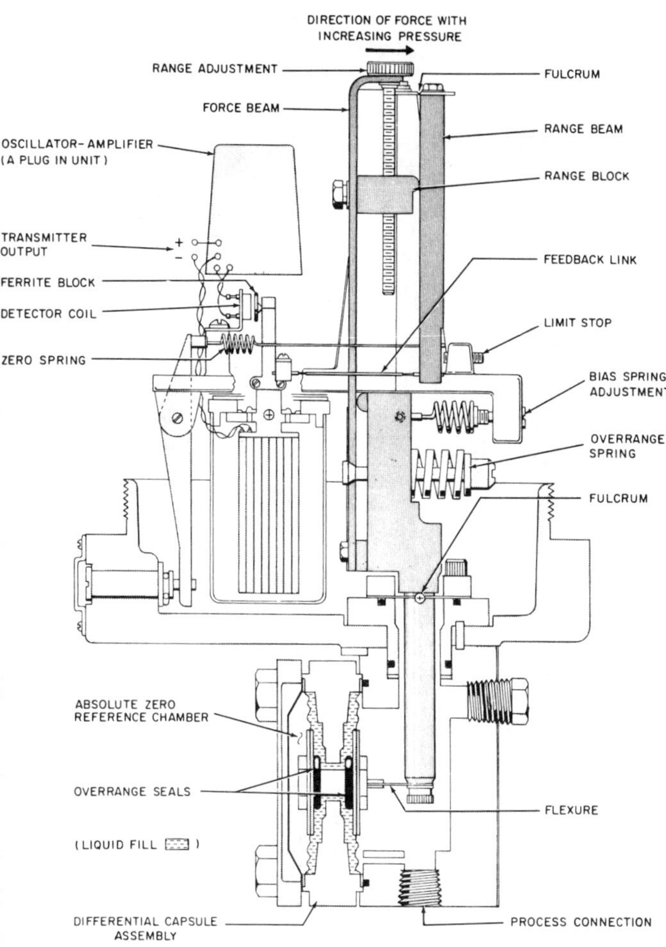

Figure 6.35. Schematic of an electronic pressure transmitter using the variable impedance principle. (Courtesy of Fischer and Porter Co.)

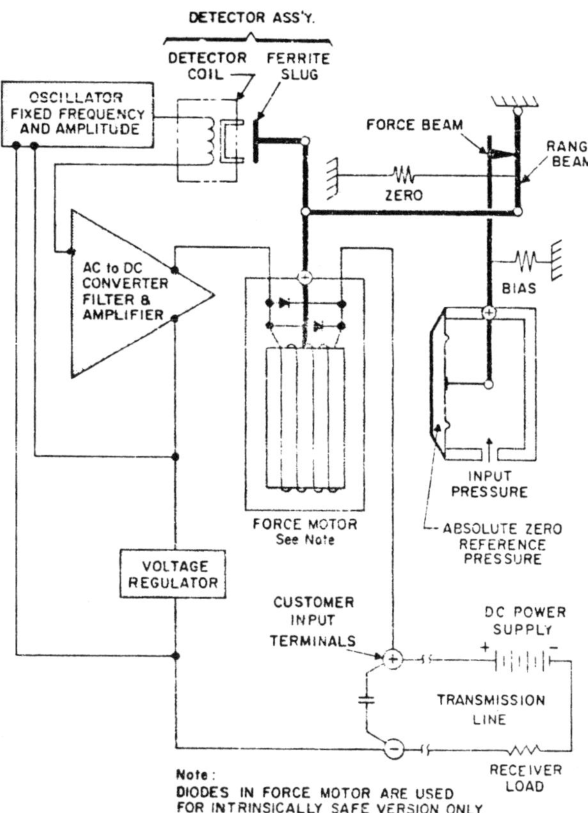

Figure 6.36. Functional diagram of the variable impedance transmitter shown in Figure 6.35. (Courtesy of Fischer and Porter Co.)

which decreases the air-gap of one transformer and increases the air-gap of the other. The air-gap change causes a predetermined change in the inductance ratio of each transformer. For full-range pressure, the diaphragm movement is approximately 0.0025 inch. This small movement assures stability of effective area and results in minimum linearity error.

The measuring cell is fastened into the center of the unit, with four leads from the pickup brought out through a high-pressure seal header. The measuring diaphragm is in contact with the top of the measuring cell. The sensor cavity is partially filled with silicone fluid to prevent corrosive atmosphere from entering the interior of the pressure cell. A vent for the reference side of the measuring diaphragm is brought out through the main body to compensate for changes in atmospheric pressure.

Variable Impedance Type. Figure 6.35 shows another electronic system that operates on what is here termed a variable impedance system. It consists of a measuring element that senses pressure and converts it into a force, a force beam that at one end receives the force from the measuring element and a rebalancing force from a feedback network (variable impedance system) on the other end of the force beam.

The force beam pivots about a fulcrum located at the small diameter horizontal sealing diaphragm. A force proportional to the measured pressure is applied to the lower end of the force beam at a fixed distance from the fulcrum point. A feedback force proportional to the output signal (developed by a force motor) is applied to the force beam by way of the range beam and range block.

The force beam is in balance when the torques produced by the two forces are equal. The slightest change from the balanced condition deflects the force beam and actuates the electronic detector circuits of the oscillator-amplifier. The distance from the fulcrum, at which the feedback force is applied to the force beam, may be adjusted by moving the range block along the force beam. The position of the range block determines the differential pressure necessary to produce full-scale output.

Figure 6.36 is a functional diagram of the electronic system. The detector assembly senses any deviation from a balanced condition between the forces produced by the measuring element and the force motor. The detector coil is attached to the instrument frame and consists of a coil wound on a ferrite core. This coil is excited at a fixed frequency and amplitude by the oscillator. The ferrite slug at-

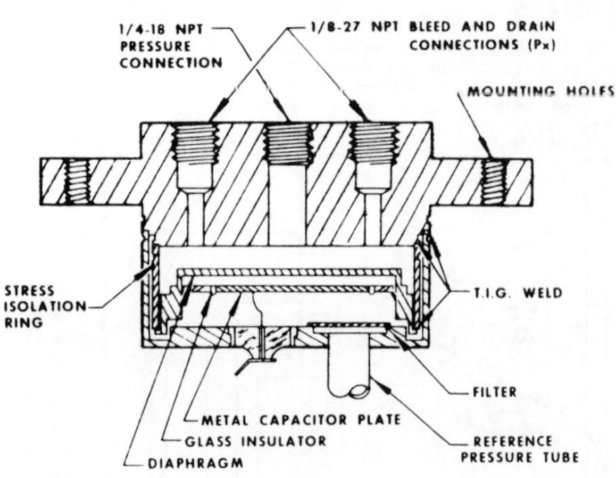

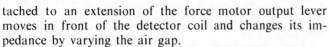

Figure 6.37. Typical capacitance pressure sensor that uses a single metal capacitor plate. (Courtesy of Rosemount, Inc.)

Figure 6.38. This piezoelectric pressure transducer uses a synthetic crystal having a high voltage sensitivity. (Courtesy of Metrix Instrument Co.)

tached to an extension of the force motor output lever moves in front of the detector coil and changes its impedance by varying the air gap.

The change in impedance modifies the coupling between the oscillator and the amplifier to vary the output current of the amplifier. The current is applied to the force motor and to the receiving equipment connected to the terminals.

These units are applicable for differential pressure devices as well as pressure units, and the accuracy is ± 0.5% of span.

Capacitance Type. In all capacitive pressure detectors, the basic operating principle is that a change in capacitance occurs due to the movement of an elastic element. The movement physically changes the distance between two capacitor plates. They can be used to measure absolute gauge or differential pressures.

The sensor shown in Figure 6.37 uses one capacitor plate, while other designs use two. Changes in the process pressure deflect the diaphragm, and the resulting change in capacitance is detected by a bridge circuit. A high-frequency, high-voltage oscillator energizes the sensing element. The capacitance is converted through the bridge circuit to a proportional DC signal. This signal is amplified to a standard milliampere output signal range.

The accuracy of most capacitive sensors is from ±0.1 to ±0.2% of span, and they operate in the range of 3 inches W.C. to 5,000 psig with proper selection of the primary element. The response time is good, their temperature sensitivity and hysteresis are low and their output is linear.

Piezoelectric Type. Piezoelectricity is defined as the production of an electric potential due to pressure on certain crystalline substances such as quartz, Rochelle salt, tourmaline, barium titanate, ammonium dehydrogen phosphate and other ceramic crystals. This piezoelectric effect is used for measurement of pressure, force or acceleration. The primary interest here is in its use as a pressure sensor.

Quartz is the most commonly used crystal that produces the piezoelectric effect. Synthetic crystals have been developed that produce the same effect and they generally have higher sensitivities than natural crystals.

The nature of the piezoelectric device is the production of electric potential as it is deformed or stressed. In a static condition, its potential drops off, producing an error. This characteristic limits its use somewhat. As a pressure device, it is most useful where pressure variations occur frequently. It is particularly suited for measurement of pressure transients in ballistics, in internal-combustion engines or in reaction processes where pressures change quickly.

Major advantages of the piezoelectric devices are the linear relationship between pressure variation and output voltage and their high frequency response (as high as 10^6 Hz for quartz).

A decided disadvantage of the piezoelectric device is its sensitivity to temperature variations. Reproducible results are not obtained unless temperatures are kept within close limits.

Figure 6.38 shows a typical pressure transducer using a synthetic crystal. It has a high-voltage sensitivity of 20 mv/psi, a pressure range of 0-3,000 psi, linearity of ±1%, and it operates at temperatures from −40°F to 300°F. Its temperature coefficient is −.035% per °F from 70° to 300°F.

Miscellaneous Pressure Measurement Methods

Several methods of pressure measurement not included in those described in the preceding pages are discussed briefly. Few of these find extensive use in the process industries. Some are used only for very special applications. They are mentioned briefly to make the reader aware of the principle involved and to meet the possible need of fulfilling special application requirements.

Deadweight Piston Gauge

Deadweight piston gauges are used principally as primary standards to calibrate other pressure sensors. When used with a controlled pressure source, such as a pump or

Table 6.2. Conversion Table for Vacuum Measurement Units

Units	Atm	lb./in.²	in. Hg (32°F)	mm Hg (Torr) (32°F)	Microns (32°F)
atmospheres	1.	14.696	29.92	760	760,000
bars	0.9869	14.50	29.53	750	750,000
kg/cm²	0.9676	14.22	28.96	736	736,000
lb/in.²	.0680	1.	2.036	51.72	51,720
in. mercury	.0334	.491	1.	25.38	25,380
ft.water	.0295	.434	.8826	22.42	22,420
in. water	.00246	.0361	.0736	1.870	1,870
torr (mm Hg)	.00132	.0193	.0395	1.	1,000
millibars	.000987	.0145	.0295	.750	750
gm/cm²	.000968	.0142	.02896	.7356	735.6
microns	1.32×10^{-6}	1.93×10^{-5}	3.95×10^{-5}	10^{-3}	1.

pressure booster, a deadweight gauge is called a deadweight tester.

Construction of piston gauges varies as to methods of loading weights, rotating or oscillating the piston to reduce friction and the design of the piston and cyclinder. In principle, however, a piston of known area is inserted into a close fitting cylinder. Weights loaded on one end of the piston are supported by fluid pressure applied to the other end.

The piston areas are precision built, and the weights are checked against weights certified by the National Bureau of Standards.

In many deadweight test sets, a number of interchangeable piston assemblies are available to provide different ranges of pressure measurement with the same cyclinder and set of weights. Piston areas vary considerably—areas from 0.1 to 0.01 inch are common.

The accuracy of deadweight testers has improved over the past few decades. Standard accuracies now are ±0.1%, and an accuracy of ±0.01% is possible if the tester is in good working order. Standard units are available for ranges from 0-5 to 0-100,000 psig.

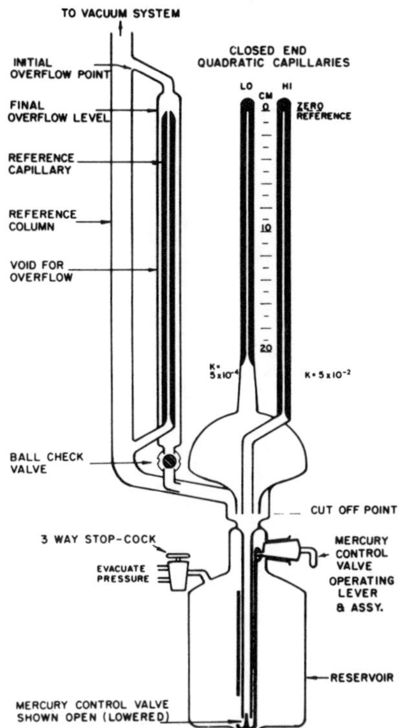

Figure 6.39. Typical McLeod gauge for vacuum measurement. (Courtesy of Hastings-Raydist, Teledyne Co.)

High Vacuum Measurement Techniques

Some of the pressure sensors previously described operate satisfactorily in the vacuum (or negative pressure) range as well as the positive pressure range for practically all commercial chemical processes. From the standpoint of research and for some special processes, high vacuum measurement (not obtainable by these devices) is necessary. High vacuum measurement, for instance, is significant in the production of thin films by evaporation, in the production of television tubes, in cryogenics and in space simulation—just to mention a few application areas.

Table 6.2 shows the relationship of the various units used in vacuum measurement. The units are arranged in descending or ascending order of values, depending on whether one reads vertically or horizontally and up or down. The arrangement makes the relative size of units easy to understand.

The McLeod Gauge

Until the middle 1960s, the accepted standard for absolute vacuum measurement was the McLeod gauge (Figure 6.39). It operates by compressing a known volume of gas into a much smaller volume, whose final value provides an indication of pressure.

For the measurements to be valid, the gas involved must obey Boyle's law over the required range of compression. There are several variations of the McLeod principle, but a sequence of manipulations are required so that it is applicable only to intermittent readings. Accuracy is about ±1% down to 10^{-3} mm of mercury. Capillary effects tend to

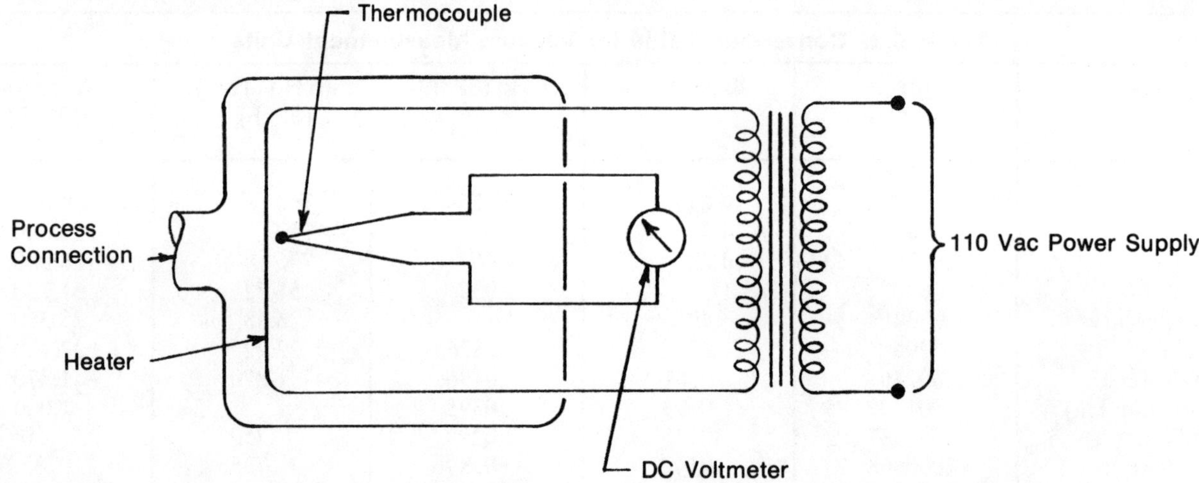

Figure 6.40. A typical thermocouple sensor used in high vacuum measurement.

limit accuracy severely at pressures below 10^{-4} mm of mercury so that between 10^{-3} and 10^{-6} mm of mercury, the accuracy at best may be a few percent.

Ionization Vacuum Sensors

There are several types of ionization vacuum sensors including the hot cathode, cold cathode and the radiation alphatron. The technique has been used since 1916, and measureable vacuums have been lowered from 10^{-7} to 10^{-13} torr since 1916.

The principle of measurement is the detection of current flow in vacuum tubes containing low-pressure gas. Molecules of gas are ionized by electron flow from the filament to the plate where they are discharged, yielding plate current flows proportional to the gas pressure within the tube.

The accuracy of these devices is low compared to most instrument measurements because the physical quantities measured are so small.

Thermal Vacuum Sensors

Thermal vacuum detectors are less sensitive than ionization types. They are capable of measurements as low as 10^{-7} torr. The principle of operation is that thermal conductivity of a gas is a function of gas pressure. Sensors have two basic elements—a heater and a temperature sensor. If the pressure to be measured exhibits process temperature variations, compensating elements are used to eliminate that error.

Two basic designs distinguish this type measurement, the thermocouple and the resistance wire detectors.

Figure 6.40 illustrates a thermocouple type. A constant source of AC or DC power heats an element (wire) to which a thermocouple is welded to provide a temperature measurement. The thermocouple output is a function of the thermal conductivity of the gas which is proportional to gas pressure. The whole device, of course, is connected to the vacuum system to be measured.

Thermopiles (several thermocouples) are often used to detect the heater temperature in order to increase sensitivity. Unheated thermocouples are placed in the system to compensate for process temperature variations.

The Pirani gauge is similar in principle to thermocouple types. A resistance wire, forming part of a Wheatstone bridge, is placed in the vacuum to be measured. Voltage is applied to provide a constant electric current. Heat is conducted away from the element in proportion to the gas pressure surrounding it, and its changing resistance becomes a measure of gas pressure. A second resistance wire is placed in a reference vacuum to provide temperature compensation.

Accuracy is about ±10% over the measurement range of resistance sensors.

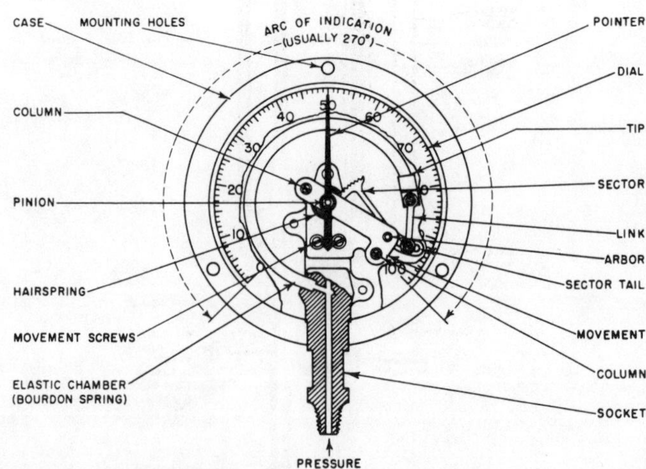

Figure 6.41. Pressure gauges comprise a high percentage of sensing elements used in a processing plant. (Courtesy of Ametek/Calmec)

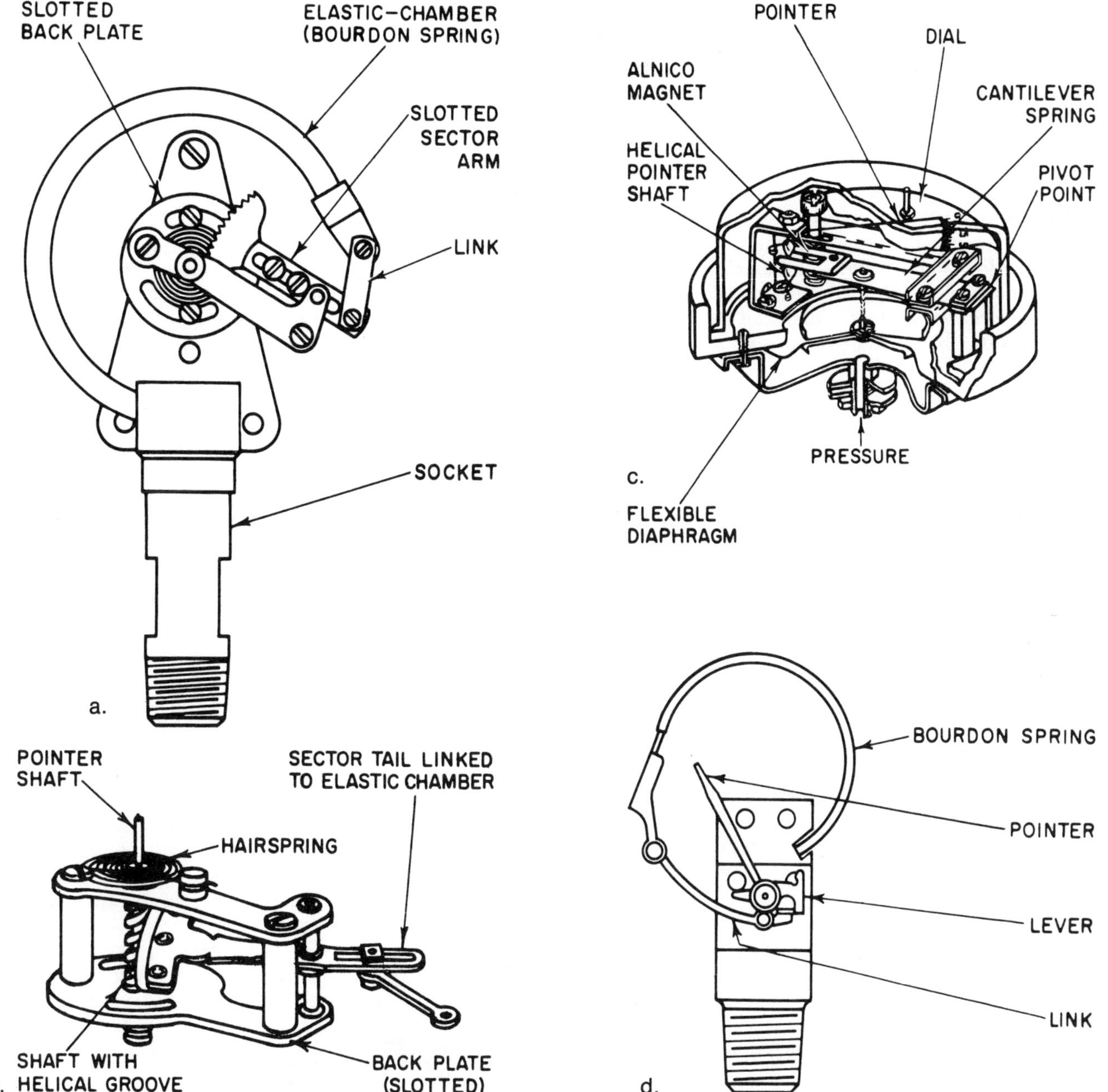

Figure 6.42. (a) Geared sector and pinion shown in the four basic movements used to produce pressure gauge movements from the sensing elements; (b) helical movement; (c) magnetic movement; (d) gearless movement. (Courtesy of Ametek/Calmec)

Pressure Gauges

Pressure gauges are simple devices used primarily as local pressure indicators. Because they comprise such a large percentage of the use of pressure measuring elements, they deserve special attention.

Figure 6.41 shows a typical pressure gauge. The gauge consists essentially of the following components: socket, sensing element, tip, movement, case, dial and pointer, lens, bezel and optional features and accessories.

The socket serves two purposes: it connects the pressure source to the sensing element, and it supports all other components of the gauge. Sockets extend from the gauge case in four basic positions: center back, lower back, bottom or top and right or left. Connection sizes are usually ¼ or ½-inch NPT male. Sockets are made from various metals and alloys, since they come in contact with the process fluid and must be compatible with it. The most commonly used materials are brass, alloy steels, stainless steel and Monel.

There are three different types of sockets: bar stock, forged and cast. The bar stock is the most widely used since it is economical to produce in large quantities. The forged socket is used when a high-quality, special design shape is required. The cast socket is used for intricate shapes and is suitable for low or medium pressure sensing.

The three basic types of sensing elements used in pressure gauges are bourdon tubes, bellows and diaphragms. Bourdon tubes are most commonly used. Elements have been discussed in earlier sections of this chapter, and the reader may refer back for a discussion and comparison of the element types.

The tip of a pressure gauge is attached to the free end of the bourdon tube sensing element. It serves a two-fold purpose: to seal the tubing end and to provide an attachment point for the pointer connecting linkage. There are two types of tips, the male and the female. The male type, generally not as strong as the female, is usually recommended.

The movements of a pressure gauge have two basic purposes: to magnify the small motion of the sensing element and to convert this motion into rotary motion so as to produce a full 270° travel of the pointer. The movement also should function repeatedly with accuracies as high as ±0.25% of full scale. In order to maintain this accuracy, movements are manufactured to close tolerances. Figure 6.42 shows four available types of movements: geared sector and pinion, helical, magnetic and gearless.

The case encloses all the internal parts—socket, measuring element, movement, dial and pointer. The lens and bezel complete the enclosure of the gauge internal parts. The case protects the working parts of the gauge from mechanical damage, dust, dirt and corrosion. It also provides a means for mounting the gauge.

There are five different gauge case styles available: the beaded, back flange, front flange, turret and "solid front" safety case.

The beaded or plain drawn cases have no flange, either back or front, and are generally stem mounted on equipment.

The front and back flange cases have three holes 120° apart in the flange for wall or panel mounting.

The most frequently used case in the processing industries is the turret style. Intended for stem mounting or wall mounting, this type is usually made of a phenolic resin or high-impact plastic. The case strength is greatly enhanced by a tapering wall construction.

For severe or dangerous services, the "solid front" case is recommended. This design has a solid wall which separates the dial and lens from the measuring element, and a safety blowout disc covers the entire back of the case. The disc is gasketed and spring mounted so that it releases when the inside pressure exceeds 1 psig.

Gauge lens can be furnished in plain glass, beveled glass, tempered glass, nonshatterable glass and transparent plastic materials. The plain glass type is the most commonly used, but the others have useful applications. For example, tempered glass is useful in applications subject to severe shock. Plastic lens can be used for rough service.

The dial or scale is generally graduated over a 270° arc which gives a rather long length for the dial diameter, thus improving readability. The number of graduations depends to some extent on the accuracy of the gauge. Three standard dial types are the nonilluminated dial, the illuminated dial and the luminous dial. Scales are graduated in simple pressure units. Where extreme precision of reading on a high accuracy gauge is desired, a mirrored strip around the edge of the scale is used to provide a mirror image of the pointer to reduce reading errors due to parallax.

The indicating pointer of a gauge varies in design with the type of gauge. The weight and shape affect the basic accuracy and readability of the gauge. Pointers are made from aluminum or brass; aluminum is preferable because the load on the pinion bearings is small. "Zero adjustment" is provided for some pointers for calibration compensation for preloading, liquid heads or calibration shift.

The bezel, sometimes called the "ring," secures the lens to the case. In some designs the bezel and lens must be removed for calibration of the gauge or for maintenance of internal parts. Less expensive designs do not have a removable bezel; it is more economical to replace the entire gauge. Bezels are classified by the method in which they fasten to the case—a press fit, a slip fit or a threaded fit.

Optional features can be furnished on pressure gauges such as movements and scales for 360° or 720° rotation of pinion. Suppressed scales (Figure 6.43) are available for improved readability where close reference to 0 is not needed.

When the highest pressure of a process is desired, a "maximum" pointer can be supplied to indicate the highest pressure attained for a given period. The pointer does not fall back automatically but is pushed forward when new highs are reached.

Duplex gauges—two sensing elements in a single case and using a single scale—are available for applications where two related pressures need to be displayed together. If the

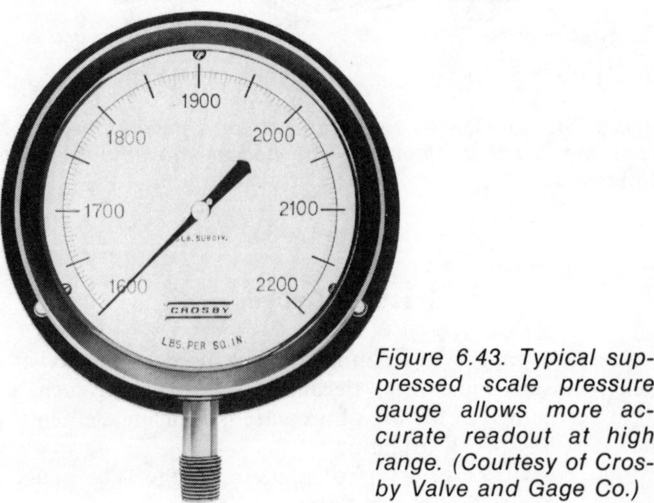

Figure 6.43. Typical suppressed scale pressure gauge allows more accurate readout at high range. (Courtesy of Crosby Valve and Gage Co.)

differential between two pressures is desired, a single pointer is used..

Accessories used with pressure and vacuum gauges for pulsating, corrosive or otherwise adversive services include chemical seals, pulsation dampeners, pigtail siphons, bleeders, blowout discs and heaters. Sealed glycerin-filled cases are also available to reduce shock in rough or pulsating services. The fluid serves to dampen effectively the pointer and movement oscillations as well as to lubricate the movements.

The sizes of pressure gauges vary normally from 2½ to 8-inch diameter. Smaller or larger gauges can be purchased on special order. Two-and-one-half-inch gauges are used primarily for indication of instrument air pressures at pneumatic devices with ranges of 0-15 to 0-100 psig. Three-and-one-half-inch diameter gauges are commonly used to indicate process signals at field transmitters. These are often referred to as receiver gauges, and although they contain 0-15-psi elements, their scale is calibrated to the range of the process measurement (i.e., 0-100 psig, etc.) In such service, the gauge is calibrated to read "0" at the 3-psig level and the process maximum at the 15-psig level.

Four-and-one-half-inch diameter gauges are the standard for local process pressure indicators. Six-inch and larger gauges are sometimes used at remote stations for indication of utility service pressures.

Pressure element materials have been listed in other sections. Phosphor bronze, beryllium-copper, steel, stainless steels and Monel are common.

Ranges vary from a few inches of H_2O to 100,000 psig and from 0.1 to 760 mm of mercury absolute.

Accuracies vary from ±0.1 to ±5%. Precision test gauges of ±0.1% accuracy are available for high accuracy measurements. Process pressure gauges are usually ±0.5 or ±1% of span; and many 2½ and 3½-inch gauges are accurate only to ±2% of span.

Pressure Switches

Pressure switches may be defined as gauges with electrical contacts and no indicators. They are used to energize or deenergize electrical circuits for alarm purposes or to perform control functions at predetermined pressure settings. They are subject to the same design considerations as their counterparts for pressure indication or transmission. They are used to sense absolute, compound, gauge and differential pressures. The primary elements used have been discussed in the earlier parts of this chapter. The diaphragm type is probably the most frequently used element. Some switches use sealed pistons or Bellville discs in their design. The electrical switching mechanisms are either snap-acting microswitches or mercury switches. The latter type should be mounted level and be free from vibration.

The terminology used in switch application and selection should be understood.

The *adjustable range* is the pressure range within which the actuation point (set point) can be adjusted.

The *set point* is that pressure which actuates the switch to open or close the electrical circuit (depending on how the switch is wired). It may actuate either on increasing or decreasing measurement at that point. Set point accuracy defines the ability of a pressure switch to operate repetitively at its set point.

The *dead band* is the difference between the set point and the reactuation point.

Tolerance is usually referred to as the repeatable accuracy of the reactuation point.

Factors to consider in selecting pressure switches include service life, proof pressure, function and electrical rating for the switch and housing.

For most applications, the standard pressure sensors will respond quickly enough and will operate the expected number of cycles. Most elastic elements have an expected service life close to a million cycles.

Pressure switches are often used for set-point values far below the expected maximum design pressure of a system, but the switch must have a proof pressure as high or higher than that expected for the system.

The function of the switch determines whether a single setting switch, a dual switch or a difference switch should be used. Economy often dictates the use of a single element to operate two switches.

Recent packaging designs include plug-in switch modules that can be densely mounted to conserve internal panel space (Figure 6.44). Pneumatic seals prevent signal leakage when the switch module is in place, and spring-loaded valves seal the manifold when the module is removed.

A new design provides separate sensor units and adjustable switching units (Figure 6.45). Nearly a hundred different compatible sensor modules provide pressure and temperature actuations ranging from 12 inches of water to 3000 psi and from −60°F to 800°F, respectively. The actuator-switch assembly functions through rotations of a fulcrum plate about two axes. Individually adjustable springs control the forces at which rotations occur, giving snap-action switching during both actuation and resetting of the switch mechanism. Two-stage (dual) pressure switching and manual reset features are optional.

Figure 6.44. Densely mounted plug-in switches conserve internal panel space. (Courtesy of United Electric Control Co.)

Figure 6.45. Separable sensor units and switching units provide nearly a hundred different compatible modules. (Courtesy of Automatic Switch Company.)

Some switches have external set points while others are factory set and are not adjustable. Differential switches may be adjustable or fixed. All these factors must be related to the function to be served.

The electrical classification of the area must be considered in pressure switch selection. Switches are usually furnished as general purpose, weatherproof (conform to NEMA 3) or explosionproof (NEMA 7 or 9).

The electrical contact rating of the switches must be consistent with its use. Contacts should be as large as possible without overloading the switch and impairing its accuracy. Greater current-carrying capabilities result in less need for auxiliary switching relays.

Standard switch arrangements include single pole-single throw (SPST), single pole-double throw (SPDT), double pole-single throw (DPST) and double pole-double throw (DPDT). Units are available with more poles, however.

Table 6.3 shows one manufacturer's load ratings for AC and DC operated pressure switches. Most pressure switches have accuracies of ±½% of span.

Table 6.3. Typical Electrical Current Ratings for Snap-Acting Microswitch Contacts

AC Ratings							
Inductive Load—50% Power Factor Maximum Continuous Current*							
Class of Switch	A,H,J	B,K	C	D	E	F,G	M
Volts. 125	10	10	10	5	20	10	10
AC. 250	10	10	10	5	20	10	10
. 460	3	10	10		20	10	3
DC Ratings							
Inductive Load—L/R = .026 Maximum Continuous Current*							
Class of Switch	A,H,J	B,K	C	D	E	F,G	M
Volts. 6-8	.5	15	15	10	15	15	8.0
DC 12-14	.5	10	15	10	15	15	15
. . . . 24-28	.5	5	10	5	10	1.0	
. . . . 110-115		.05	.1	.6	.05	.4	.5
. . . . 220-230		.05	.4	.03	.2	.25	

*Current given in amperes.

(Courtesy of Delaval, Barksdale Controls Div.)

Accessories for Pressure Measurement

For economical reasons or as an operating expediency, several accessories are made for use with pressure elements. Their purpose is to offset difficulties due to pressure pulsations, corrosive fluids, slurries or other similar service or stream characteristics. Accessories described include chemical seals, pulsation dampeners and pigtail siphons.

Chemical Seals

Chemicals seals (or diaphragm seals) are designed as protective attachments for pressure instruments. They are used for three basic application services: (a) where the process fluid being measured would normally clog the sensing element, (b) where materials capable of withstanding corrosive effects of certain fluids are not available or are too expensive as element materials and (c) where the process fluid in the element might freeze due to changes in ambient or process temperatures.

A chemical seal is constructed of a thin flexible diaphragm as a separating member between the process fluid and the pressure element. The space above the diaphragm and all connections to the element are completely filled with a selected temperature stable liquid. The flexing of the diaphragm caused by a process pressure change is transmitted through this liquid system directly to the diaphragm seal or remotely connected with a length of capillary tubing.

Chemical seal assemblies, when properly filled, attached and calibrated to operating and ambient temperature conditions, should have the same accuracy as the instrument to which it is attached. However, on vacuum applications, most manufacturers do not guarantee the accuracy beyond about 20 inches Hg. Accuracies for higher vacuum ranges depend on temperature variations, the type and vapor pressure of the fill fluid and the design, type and material of the diaphragm.

Chemical seals operate satisfactorily with nearly all elastic deformation elements. The only limitation is the volumetric capacity of the seal itself. This is kept to a minimum to reduce error caused by temperature changes, yet it must be sufficiently large to activate the element through the required position or movement. The operating

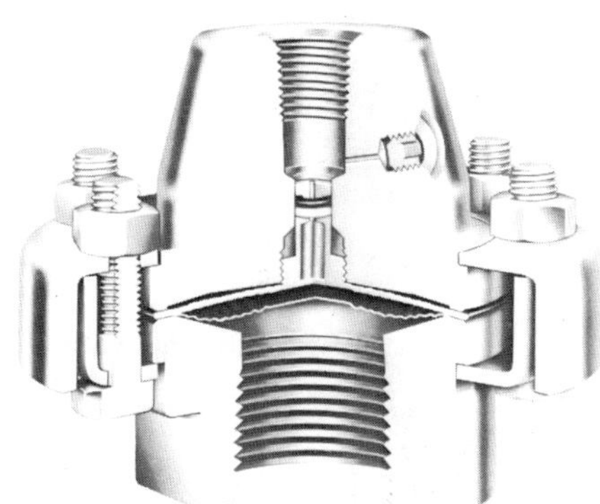

Figure 6.46. Typical standard chemical seal consists of the diaphragm assembly and the upper and bottom housings. (Courtesy of Dresser Industries, Inc.)

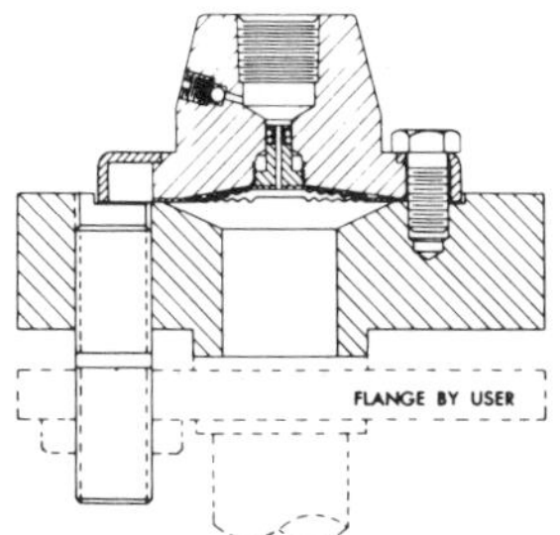

Figure 6.47. Typical flanged diaphragm seal assembly. (Courtesy of Dresser Industries, Inc.)

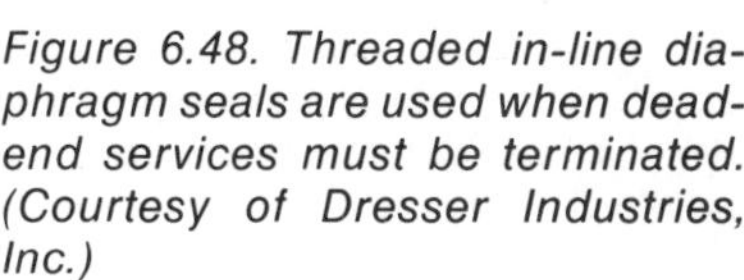

Figure 6.48. Threaded in-line diaphragm seals are used when dead-end services must be terminated. (Courtesy of Dresser Industries, Inc.)

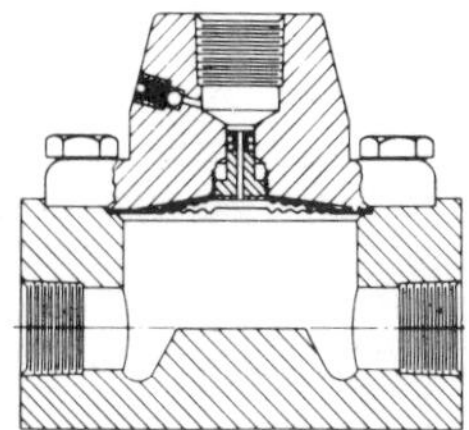

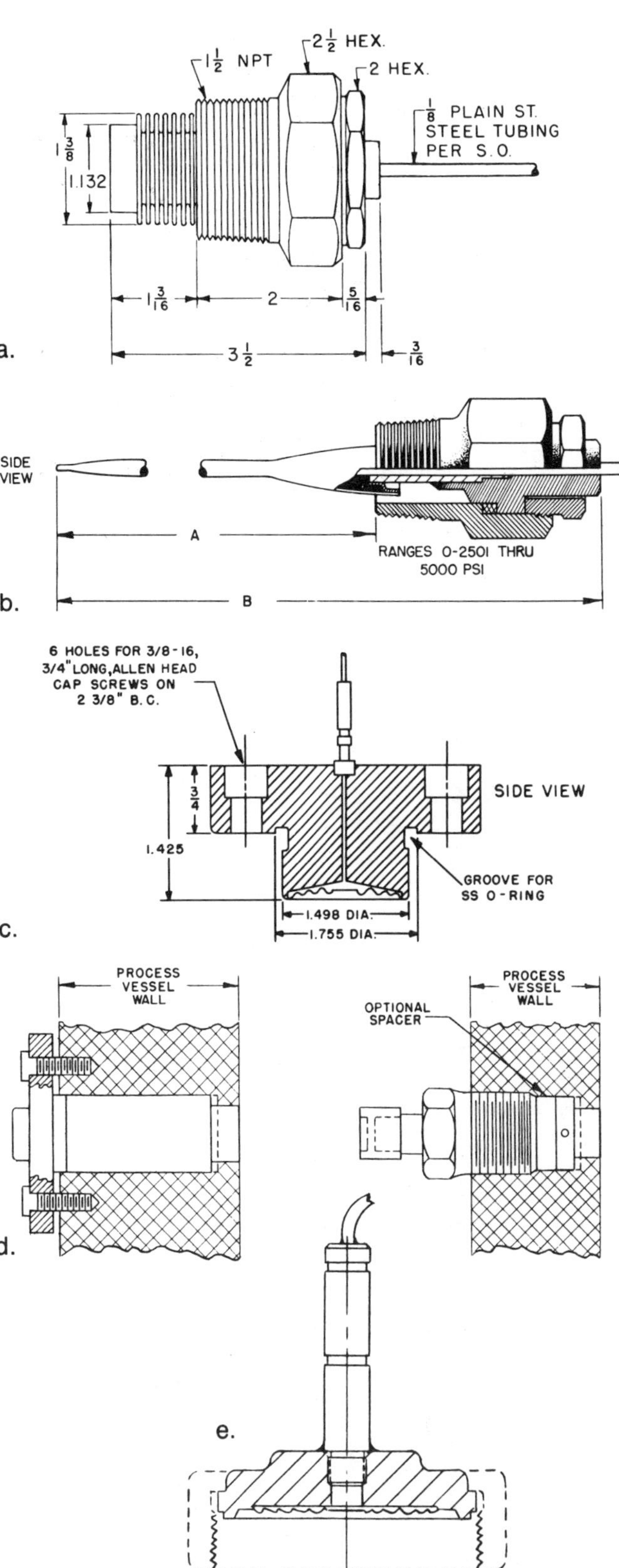

Figure 6.49. (a) Typical bellows volumetric seal; (b) typical bourdon volumetric seal; (c) typical diaphragm volumetric seal; (d) typical flanged and screwed button diaphragm volumetric seals; (e) typical screwed sanitary diaphragm volumetric seal. (Courtesy of Foxboro Co.)

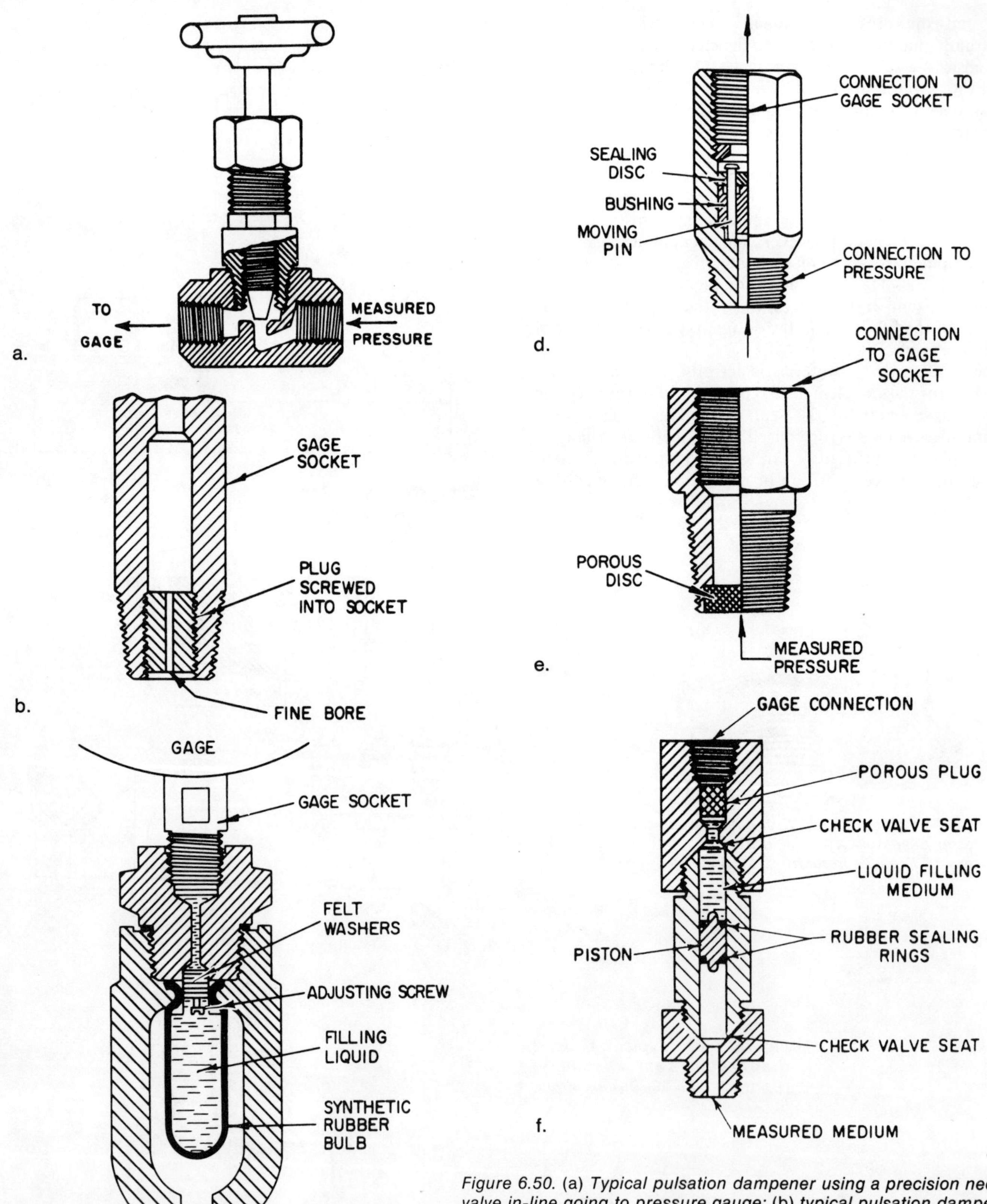

Figure 6.50. (a) *Typical pulsation dampener using a precision needle valve in-line going to pressure gauge;* (b) *typical pulsation dampener using a fine bore plug screwed into gauge socket;* (c) *typical filled bulb pulsation dampener;* (d) *typical moving pin pulsation dampener;* (e) *typical porous metal pulsation dampener;* (f) *typical flow check pulsation dampener. (Courtesy Ametek/Calmec)*

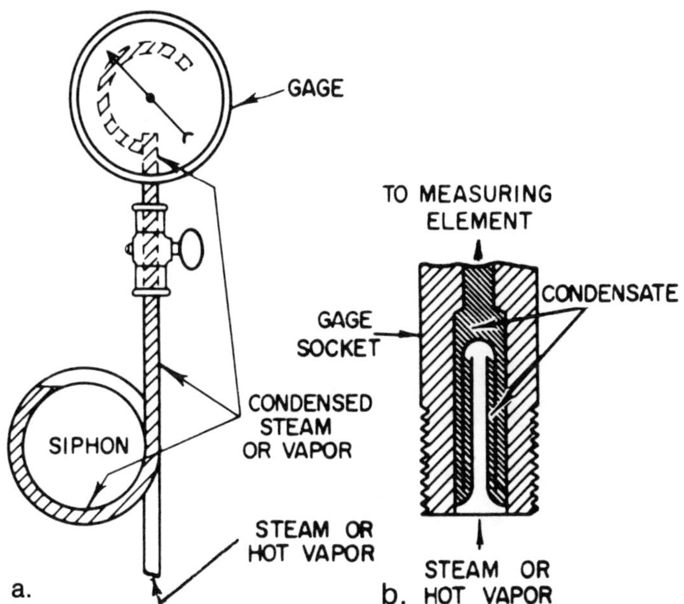

Figure 6.51. (a) *Typical single coil or "pigtail" siphon;* (b) *typical internal siphon. (Courtesy Ametek/Calmec)*

temperatures of diaphragm seals depend primarily on the stability of the fill fluids. Standard fluids are available for temperature ranges from −40°F to +600°F. For temperatures above 600°F, special volumetric seals with sodium-potassium fill fluid can be used.

Standard and volumetric are two types of seals to consider.

Standard Seals

The construction of a standard chemical seal (Figure 6.46) consists of the top and bottom housings and the diaphragm assembly. The upper housing is in contact with the filling fluid only and can be made of carbon steel or 304 stainless steel. The filling screw allows the seal and the instrument to be filled under vacuum as a unit.

The diaphragm assembly shown in Figure 6.46 is the nested diaphragm capsule; however, some manufacturers use a single diaphragm element welded to the upper housing. If leakage occurs above the nested diaphragm capsule, the diaphragm presses against the upper backup plate. The corrugations in the backup plate prevent diaphragm distortion and seals off the process pressure.

Most diaphragms are rupture-proof to about 2,500 psig, the full rated pressure of the seal. The diaphragm is in contact with the process fluid and is available in various metallic materials or with Teflon or Kel-F lining. The upper housing and the diaphragm assembly can be removed easily with the instrument for maintenance.

The lower housing is also in contact with the process fluid and is available in a variety of corrosion-resistant materials. On some seal types, a flushing connection is provided to allow purging of any material which might accumulate below the diaphragm or in the connecting piping.

Figure 6.47 shows the same basic seal as shown in Figure 6.46, except the process connection is flanged. Also, there are some applications where an inline (or flow-through seal) is required to eliminate the dead-end cavity as on other seals. Figure 6.48 shows the threaded flow-through type. Flanged and welded saddle inline seals are also available. The major limitation of these seals is that their removal requires the process line to be out of service.

Volumetric Seals

Volumetric seals operate on the same principle as the standard types just described. The basic difference is that the seal element is connected to the pressure element by capillary tubing filled with a suitable sealing fluid. Figure 6.49 shows several types of these seals. Each of these minimizes or eliminates the dead-end cavity in which the process fluid could collect. The extended diaphragm and wafer elements have operating range limits of 0-50 to 0-1,000 psig.

The button diaphragm seals have minimum spans of about 100 psig and can go as high as 10,000 psig in some designs. The bellows seal, however, is much more sensitive than a comparable size button diaphragm and operates at spans of 100 to 1,000 psig. Since this element has flexibility in both directions, it is more applicable to compound pressure applications than the other types. The bourdon seal has range limits of 0-1,000 to 0-5,000 psig.

The materials of construction for volumetric seals are more limited than for the standard types. Generally, the seal elements are made of stainless steel, but more corrosion-resistant materials can be used.

Pulsation Dampeners

Pulsation dampeners or snubbers are needed when pressure cycling is periodic enough to damage measuring sensors, or they may be used when pressure surges are great enough to produce damage or unnecessary wear on pressure elements. Several techniques are used to dampen the surging or cycling. The common requirement is a restricting device which produces a sufficient resistance to flow that pressure at the element changes slowly enough for the element to respond with relatively slow movement. Figure 6.50 reveals several methods used to accomplish the snubbing or dampening effect.

Pigtail Syphons

Pigtails (Figure 6.51) are used normally for two purposes: (a) to form a liquid seal in condensible vapor services to prevent instability due to mixed-phase conditions and (b) to isolate hot vapors from a measuring element by removing it far enough from the hot stream to allow the vapors to condense. Pressure measurement in steam service is the prime example of its use, though it is equally useful on other hot vapor services.

7 Temperature Measurement

Baxter Williams, William G. Andrew

Earlier chapters have discussed the measurements of flow, level and pressure, all of which can be measured in definite units (flow, in quantity per unit of time; level, in units of linear distance; and pressure, in force per unit of area). Temperature, however, must be measured in terms of the indirect effects it has on the physical properties (such as pressure) of materials or the changes it produces in electrical circuits (voltage or resistance changes). Such changes in temperature must be related to laboratory reproducible phenomena, such as the boiling and freezing points of water. Laboratory calibration points are often based on the temperatures at which equilibrium exists between the liquid and gaseous phases of common pure substances, such as oxygen, water, sulfur, silver and gold.

Over a period of years, at least five different temperature scales have evolved that are used in temperature measurement. The two most commonly used scales, Fahrenheit and Celsius (or Centigrade) use arbitrary spans of 180°F and 100°C, respectively, as the span between the boiling and freezing points of water. The French physicist Réaumur developed a scale (seldom used except in the Teutonic countries) having an 80° span. These scales plus two—Rankine and Kelvin—which are referenced to absolute zero (the temperature at which a perfect gas would exert no pressure on its container) are shown in Figure 7.1. The Rankine scale makes arbitrary use of the Fahrenheit divisions and places its zero at −459.69°F, while the Kelvin scale uses Centigrade divisions and absolute zero has been determined to be −273.16°C.

The most common methods of temperature measurement include thermocouples, filled systems and bimetallic elements. Resistance elements have found increasing use in the past few years. Radiation, optical and infrared pyrometers are used in certain fields and are discussed in sufficient detail to make the reader aware of the principle involved in the measurement. Other miscellaneous types are mentioned only briefly as a matter of information.

The selection of the proper temperature sensor and its related system depends primarily on four factors: system cost, accuracy, dependability and adaptability to the process.

The first factor is obvious. The second one is readily understood as one considers that while one process must be controlled within a degree or two (and sometimes within a fraction of a degree), other processes may vary several degrees with no loss in efficiency or quality.

The dependability factor can be illustrated by an example. An iron-constantan thermocouple can be used in a furnace measurement for temperatures around 1,400°F. However, oxidation causes early failures, and iron-constantan thermocouples would require frequent replacement. A more dependable, longer-lasting thermocouple is desirable, if not necessary. Chromel-alumel thermocouples are normally used for furnace temperatures to 2,300°F.

The factor of adaptability to the process is simply the selection of a sensor which does not upset or interfere with the process functions—such as placing a thermocouple in the path of an agitator or placing a thermowell in a stream where process buildup prevents heat transfer with consequent false temperature indication.

From a historical point of view, the first practical temperature device was the glass bulb thermometer. The earliest reliable variety of these utilized the increase in volume of a liquid (alcohol) in a column of glass to show the effect of an increase in temperature. The mercury, or Fahrenheit, thermometer was developed in the early 1700s

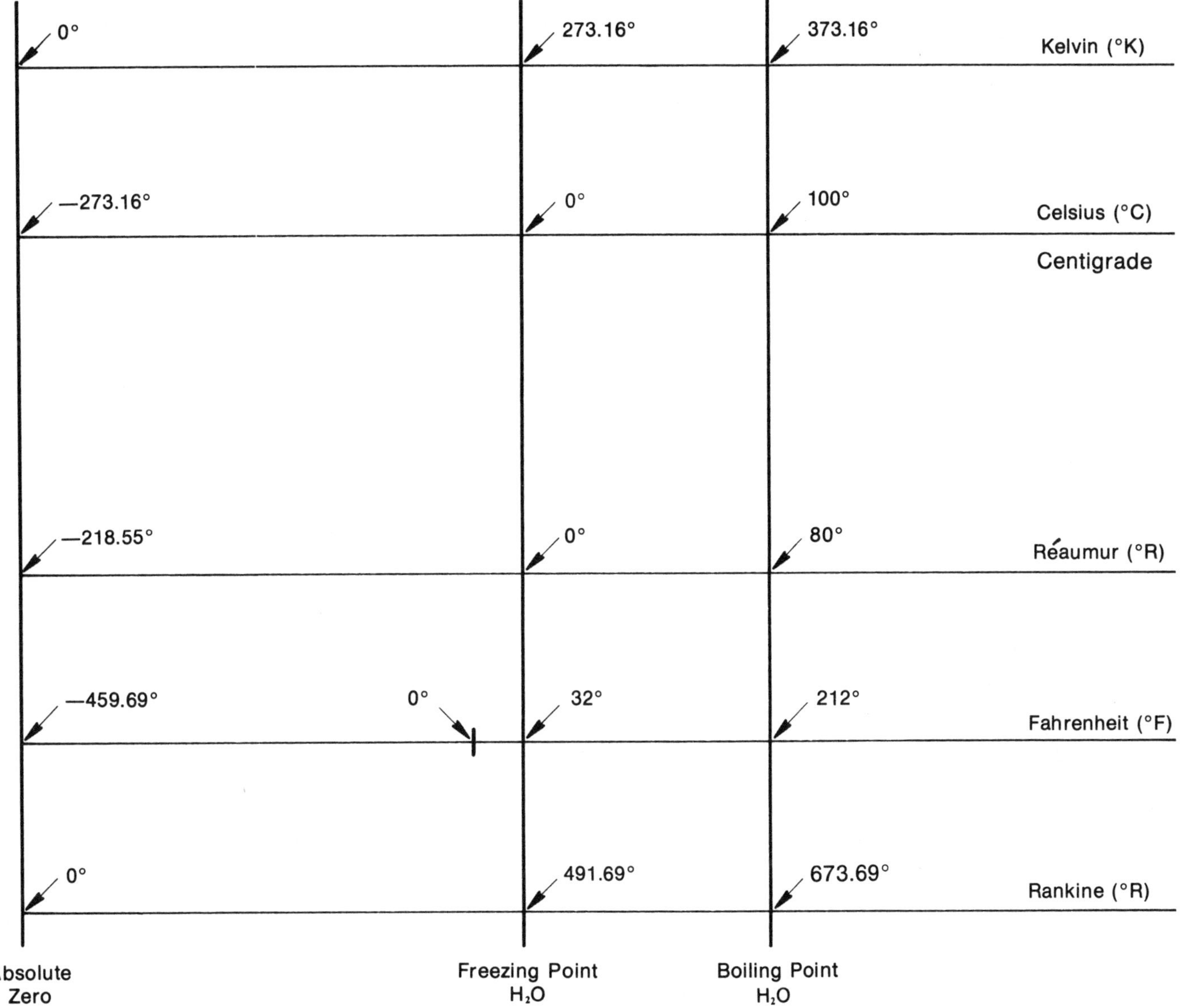

Figure 7.1. Five arbitrary scales are used to measure temperature.

and is still used world wide as an inexpensive standard. Physical dimensions of the glass capillary are carefully checked to prevent inaccuracies due to volume errors. Typical accuracies presently range from ±0.1 to ±0.01°K for narrow ranges between 125 and 875°K.

Thermocouples

Theory

The thermocouple is one of the simplest and most commonly used methods of determining process temperatures. When remote indication is required and periodic readings are adequate and when multiple points can be displayed on a single readout device, no other temperature measuring method can compete cost-wise.

In 1821 T.J. Seebeck discovered that when heat is applied to a junction of two dissimilar metals, an electromotive force is generated which can be measured at the other (cold) junction of these two metals (conductors). These conductors form an electrical circuit, and current flows as a result of the generated emf. A current will continue to flow in such a circuit (shown in Figure 7.2) as long as $T1 \neq T2$. Conductor *B* is described as negative with respect to conductor *A* when current flows into it at the cold junction. This negative conductor is always color-coded a shade of red (reference may be made to ISA and ANSI Standard C96.1-1964) as a matter of convenience for determining the correct polarity when connecting thermocouples.

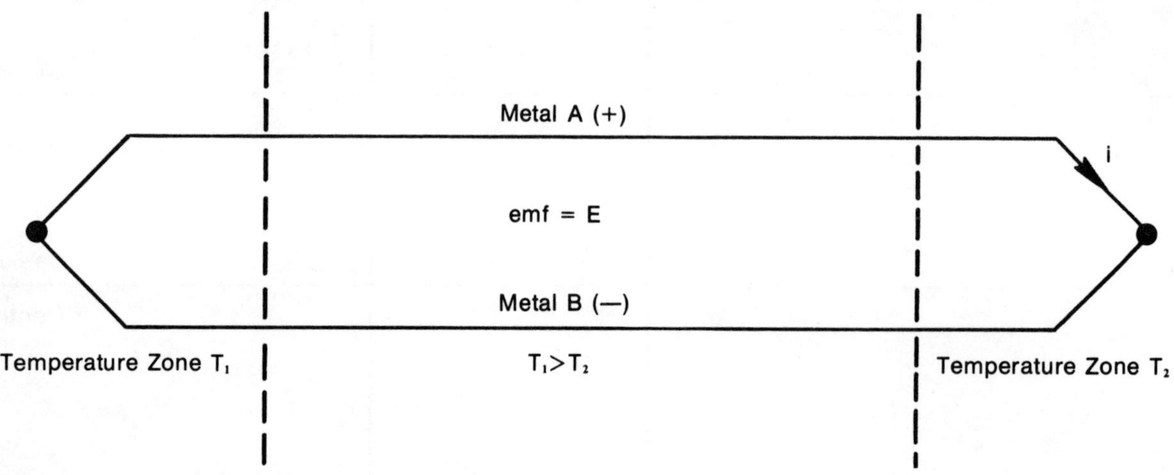

Figure 7.2. Generation of an electromotive force by the application of heat to one of the junctions of two dissimilar metals is known as the Seebeck effect.

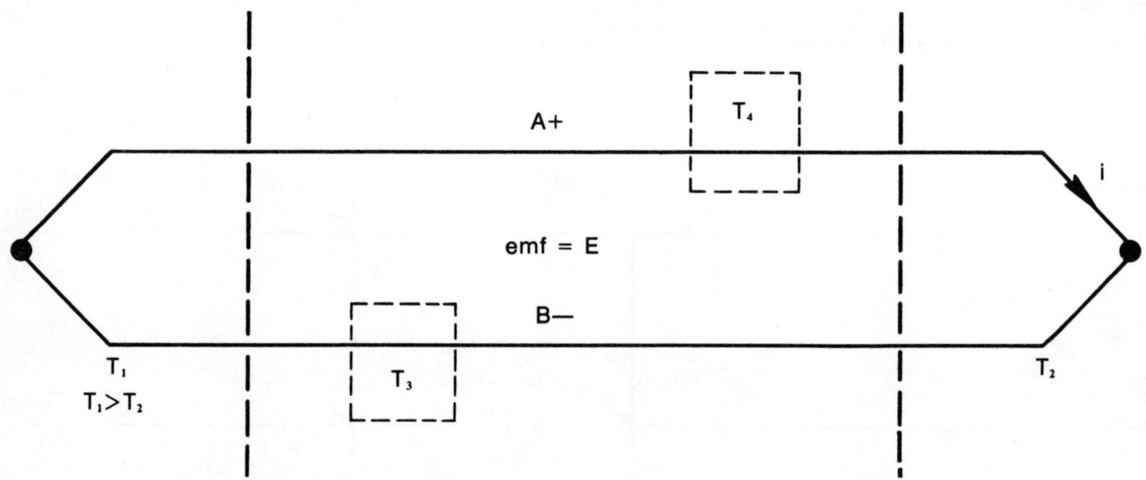

Figure 7.3. Temperature gradients do not generate emf's in a homogeneous conductor.

Several phenomena have been discovered which are accepted as thermoelectric laws.

1. The application of heat to a single homogeneous metal is in itself not capable of producing or sustaining an electric current therein.
2. The thermal emf (electromotive force) developed when the junctions of two dissimilar homogeneous metals are kept at different temperatures is not affected by temperature gradients along the conductors (shown in Figure 7.3).
3. In a circuit consisting of two dissimilar homogeneous metals having the two junctions at different temperatures, the emf developed will not be affected when a third homogeneous metal is made a part of the circuit, provided the temperatures of its two junctions are the same (as in Figure 7.4).
4. The thermal emf of any two homogeneous metals with respect to one another is the algebraic sum of their individual emfs with respect to a third homogeneous metal (see Figure 7.5).
5. The thermal emf produced when a circuit of two homogeneous metals exists between a first temperature and a second and the thermal emf produced when that same circuit exists between the second temperature and a third are algebraically equal to the thermal emf produced when the circuit exists between the first and third temperatures (shown in Figure 7.6).
6. The algebraic sum of the emfs produced in a circuit containing two or more thermocouples all at the same temperature is zero.
7. The overall emf of a circuit containing two thermocouples is unaffected by the addition of more thermocouples at the same temperature as any of the others (see Figure 7.7).

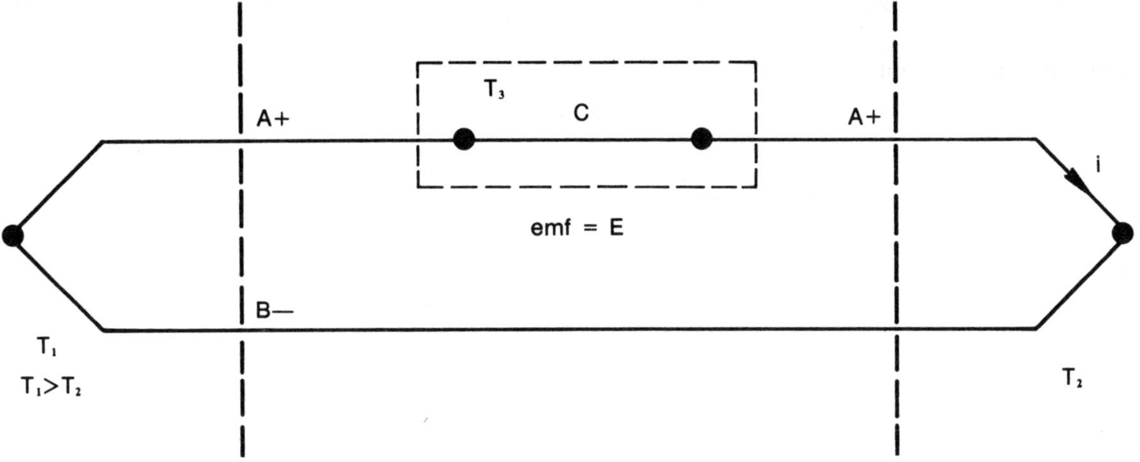

Figure 7.4. A third metal does not affect the overall circuit emf as long as its junctions are at a common temperature.

A+

i

emf = E_{AB}

B—

B+

i

emf = E_{BC}

C—

T_1

T_2

$T_1 > T_2$

A+

$E_{AC} = E_{AB} + E_{BC}$

C—

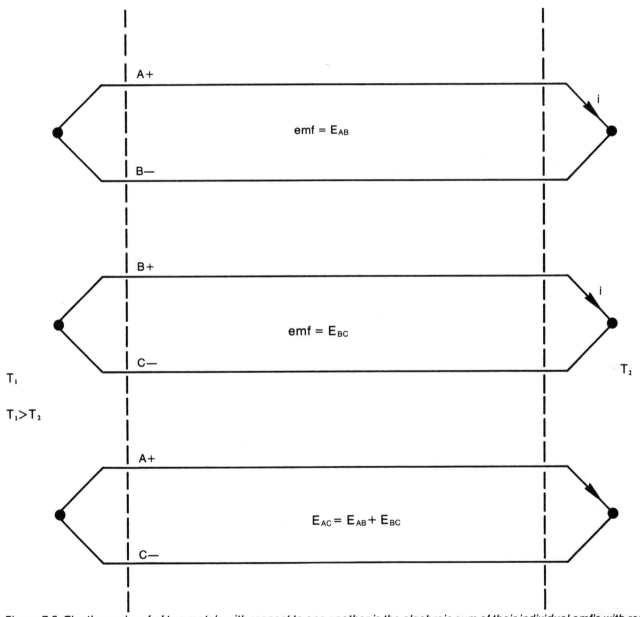

Figure 7.5. The thermal emf of two metals with respect to one another is the algebraic sum of their individual emf's with respect to a third metal.

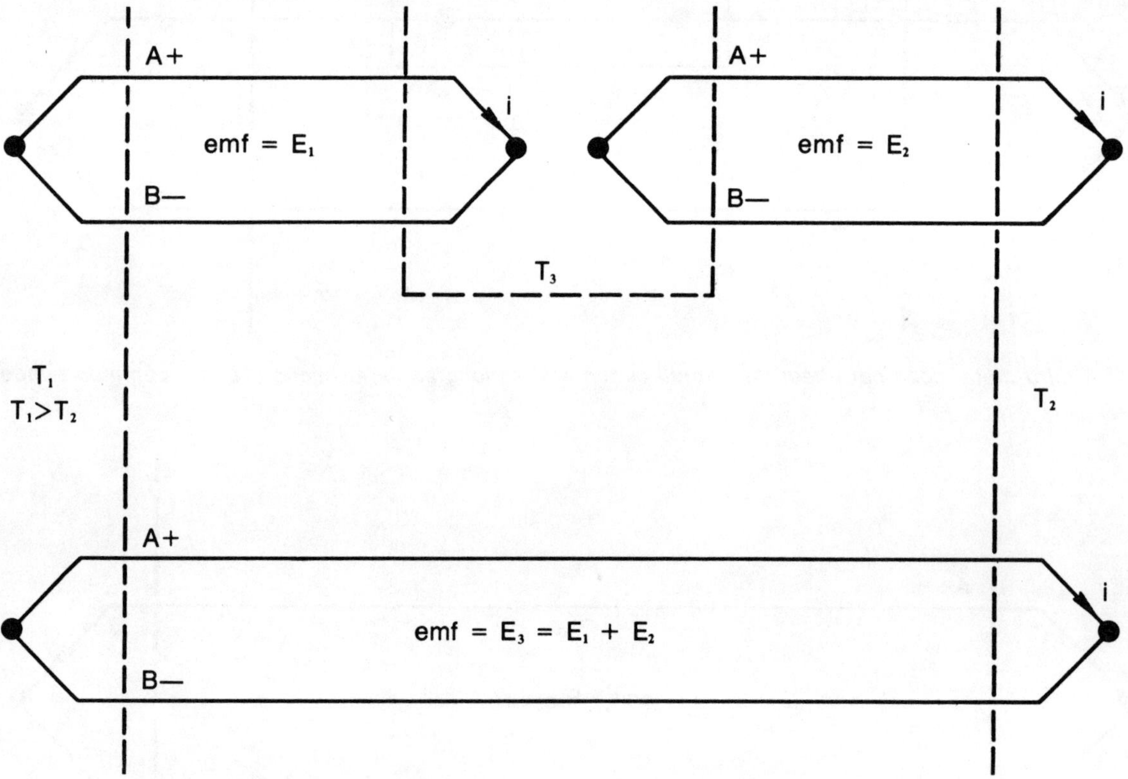

Figure 7.6. The thermal emf produced in circuits between T_1 and T_2 and between T_2 and T_3 are algebraically equal to the emf created in a similar circuit between T_1 and T_3.

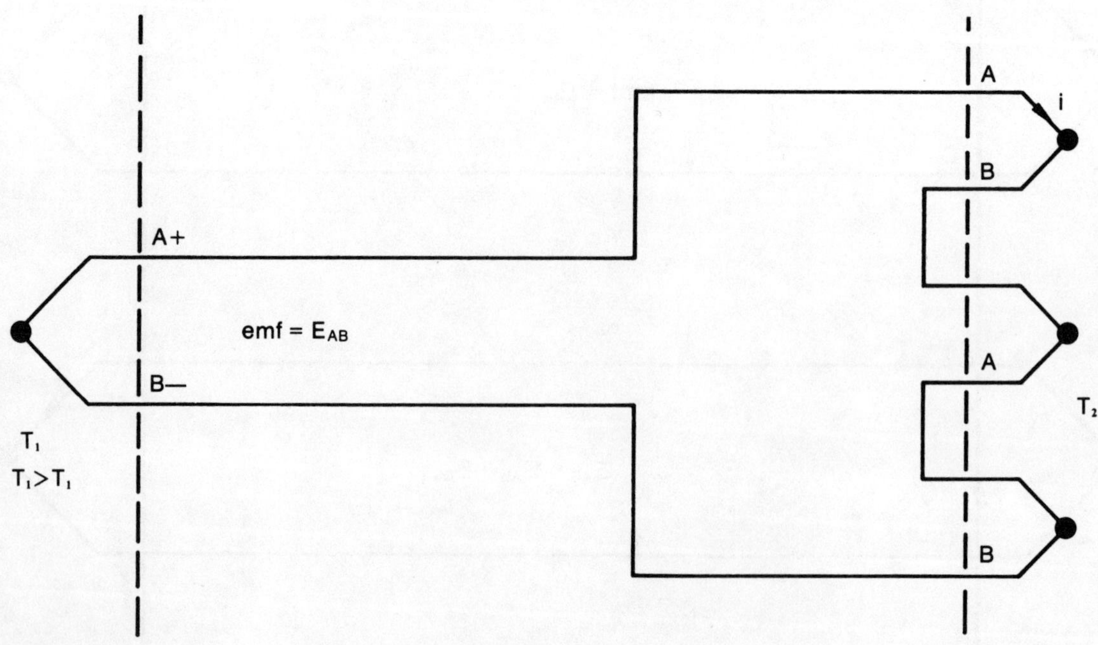

Figure 7.7. Additional thermocouples at T_1 or T_2 do not affect the overall emf.

In industry these thermocouple effects are utilized to measure temperature accurately for indication, recording or process control. A single junction of two dissimilar metals (a thermocouple) may be subjected to a temperature to be measured, and its thermal emf can be compared to a reference chart to obtain a temperature analog. Since the temperature of the second junction of the metals must be known, an ice bath is often used in the calibration laboratory to provide a reference point. Commonly used reference points are 32°F (ice bath), 75°F and 25°C.

Obviously, ice baths must have an available source of ice; they need to be stirred to give homogeneity of temperature; they must be allowed a "warm-up" period; and they must be maintained to prevent contamination. Their initial cost, size, and maintenance are not points in their favor. Ovens or heated baths are subject to similar disadvantages in measuring process temperatures, and usually consume much more energy.

Miniaturized portable reference junctions provide a millivoltage output which cancels those that would be developed by the thermocouples formed when incoming leads would reach the measuring instrument. Solid state circuitry and advances in battery technology have made possible miniaturized "self-powered" reference junctions (Figure 7.8). An integral mercury battery powers a bridge circuit consisting of an RTD, precision resistors and a cold junction block with imbedded reference thermocouples. Often an on-off switch, balancing potentiometer, and quick-disconnect jacks are provided. Copper output terminals eliminate further errors and allow inexpensive copper lead-in wires when the reference junction is installed in the field near the measuring thermocouple.

Reference junctions powered by 115 VAC, 60 HZ are also available. Isolation transformers minimize common mode noise. Multi-channel models packaged for mounting in a standard 19-inch rack are available as well. Reference temperatures are usually 0°C, or 32°F.

Other reference junction devices employ methods such as heated ovens and thermoelectric refrigeration (Figure 7.9). Two heated ovens can be used to give the effect of a single reference junction at 32°F. Oven temperatures are generally 150° and 265.5°F for Chromel-Alumel thermocouples.

Using the Peltier Effect, which is the phenomenon of heating or cooling when current is passed through the junction of two dissimilar conductors, refrigeration can be produced to freeze water in a small chamber. When a layer of ice is formed, a bellows element senses the expansion and turns off the current producing the refrigeration. This thermostatic action maintains a 0°C to +0.1°C temperature in the reference wells.

Converting Millivoltage to Temperature

Since the emf levels developed in most temperature measurements are generally in the range of −10 to +50 millivolts, sensitive measuring instruments are required to convert these low voltages to temperature equivalents. A high impedance millivoltmeter may be used to compare the measured emf to a standard cell (electrolytic) or a very stable millivolt source. More recently, self-balancing potentiometers employing Zener diode regulated power supplies and highly accurate, temperature sensitive copper or nickel compensating resistors have made accurate temperature measurement possible without the need of an ice bath or manual balancing of the bridge circuit.

The third listed phenomenon among the thermoelectric laws stated that a third metal added to the circuit has no effect on the generated emf provided the temperatures of its two junctions are the same. This allows the use of less expensive lead wire when expensive metals are used to make thermocouples. Lead wire cost is often high since so many long runs are normally used to bring multiple measuring points into a central area. Copper wire has been used for thermocouple extension wire. It must be stressed, though,

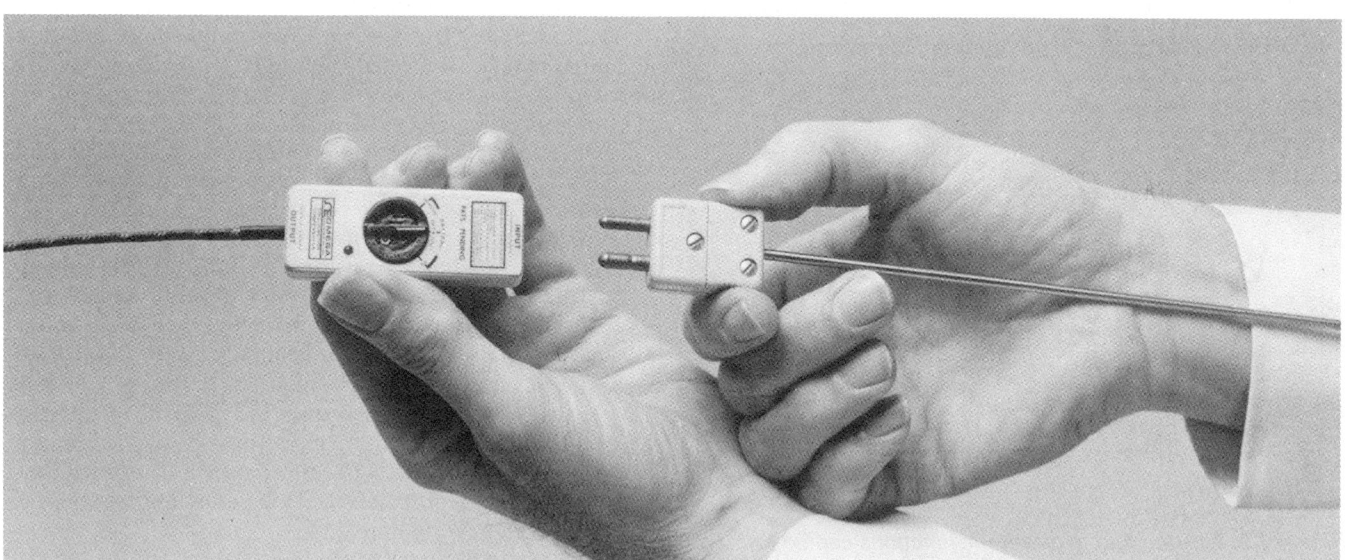

Figure 7.8 Miniaturized thermocouple reference junctions cancel local effects using a battery-powered bridge circuit. (Courtesy of Omega Engineering, Inc.)

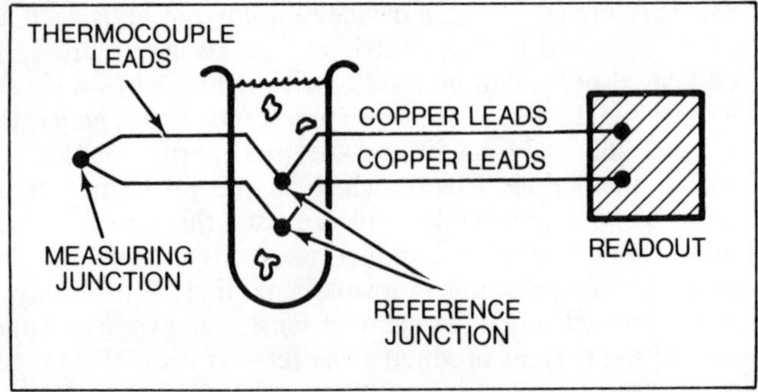

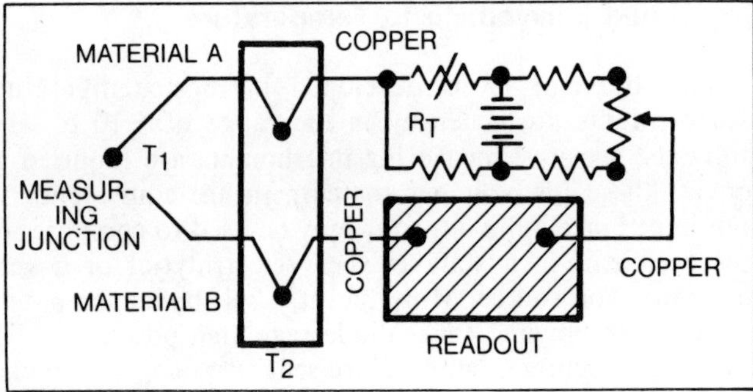

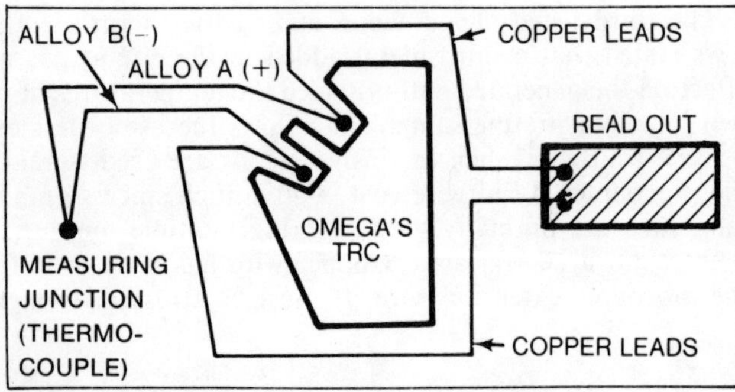

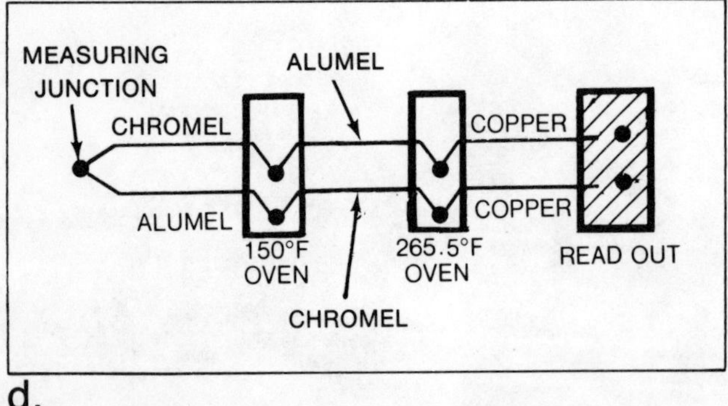

Figure 7.9. Thermocouple reference junctions can be formed using (a) *an ice bath,* (b) *a bridge circuit,* (c) *heated ovens, or* (d) *thermoelectric refrigeration. (Courtesy of Omega Engineering, Inc.)*

that the temperatures of the two ends of the extension wires will cause errors if they are not the same or if the difference is not compensated. Seldom does it occur that junction box temperatures in the field and in the control room are the same.

Two of the most commonly used thermocouples, iron constantan and copper constantan, however, normally use extension lead wire of the same material. Extension wires are color coded to minimize mistakes in connecting thermocouples and lead wires. Table 7.1 shows the accepted thermocouple extension wire color code recommended by the Instrument Society of America (ANSI C96.1-1964) and accepted throughout the industry. The negative lead is always red.

Thermocouple lead wire sizes vary from #14 to #20 gauge normally. Occasionally it may be furnished in sizes as small as #26 gauge. In many instances the primary factor in size selection is the strength required to withstand the tension force of pulling the wire through the protecting conduit. When millivoltmeter readout devices are used, the lead wire resistance is important and affects appreciably the instrument calibration. In the use of the late model, high impedance potentiometric readout devices, however, lead wire resistance is usually insignificant.

Because of the low level emfs generated by thermocouples, precautions must be taken to avoid stray currents that may result from power, lighting or control wiring in the same area with thermocouple lead wires. Thermocouple lead wire should be isolated by installation in separate conduit or raceway systems.

Table 7.2 lists the most commonly used thermocouples in the processing industries. Their recommended useful range is given in the right-hand column. This is the range over which a near-linear relationship exists between temperature and generated emf.

Table 7.3 contains the thermocouple tables for several thermocouple types based on a reference junction temperature of 32°F (0°C).

Thermocouples are not limited to single point measurement but may be connected in parallel to provide a measurement of average temperature (Figure 7.10 on page 177). To minimize the effects of unlike resistances in individual thermocouples and their lead wires to the point of parallel connection at the measuring instrument, a "swamping" resistor can be put in series with each thermocouple. This resistor prevents current flow between the thermocouples, which would induce measurement errors. This resistor should be large in value compared to overall circuit resistance and to resistance changes in the thermocouples caused by temperature variations. It is generally 500 to 2,000 ohms.

A circuit of several thermocouples in series may be used also to measure average temperature (Figure 7.11), although a special reference junction source and special calibration must be used. Extension wires must extend from each thermocouple to the instrument. This series arrangement is often called a "thermopile."

Temperature differences can be measured readily by connecting thermocouples in series but with reversed polarities such that their values subtract (Figure 7.12).

Table 7.1. Thermocouple Extension Wire Color Coding

Single Conductor Insulated Thermocouple Extension Wire

Extension Wire Type			Color of Insulation	
Type	**Positive**	**Negative**	**Positive**	**Negative***
E	EPX	ENX	Purple	Red-Purple Trace
J	JPX	JNX	White	Red-White Trace
K	KPX	KNX	Yellow	Red-Yellow Trace
K	WPX	WNX	Green	Red-Green Trace
R or S	SPX	SNX	Black	Red-Black Trace
T	TPX	TNX	Blue	Red-Blue Trace

Note of Caution: In the procurement of random lengths of single conductor insulated extension wire, it must be recognized that such wire is commercially combined in matching pairs to conform to established calibration curves. Therefore, it is imperative that all single conductor insulated extension wire be procured in pairs, at the same time, and from the same source.

*The color identified as a trace may be applied as a tracer, braid, or by any other readily identifiable means.

Duplex Insulated Thermocouple Extension Wire

Extension Wire Type			Color of Insulation		
Type	**Positive**	**Negative**	**Overall**	**Positive**	**Negative***
E	EPX	ENX	Purple	Purple	Red
J	JPX	JNX	Black	White	Red
K	KPX	KNX	Yellow	Yellow	Red
K	WPX	WNX	White	Green	Red
R or S	SPX	SNX	Green	Black	Red
T	TPX	TNX	Blue	Blue	Red

*A tracer having the color corresponding to the positive wire code color may be used on the negative wire color code.

Both parallel and series multiple thermocouple circuits should be designed using thermocouples whose characteristics are linear within the operating range. This range can be determined by the chart shown in Table 7.2.

Installation of Thermocouples and Thermocouple Extension Wire

Early thermocouple measurements using millivoltmeters and relatively low impedance potentiometers were not seriously affected by noise problems. As receiving instruments have become more sophisticated, however, higher impedance measuring circuits are used. Basically, the lower the voltage level and the higher the impedance of a circuit, the greater is its sensitivity to noise of all types. One needs to be aware of the noise sources and to avoid them if possible.

Thermocouple circuits are subject to three basic noise types:

1. Static
2. Magnetic
3. Common Mode

Static noise is caused by the electric field radiated by a voltage source being coupled capacitively into the thermocouple circuit. The most effective way of fighting static noise is to place the circuit inside a total coverage shield which isolates the pair of thermocouple lead wires from outside influence. The grounded shield intercepts static interference and carries it off to ground. The shield must be grounded; a floating (ungrounded) shield will not reduce noise.

Magnetic noise is produced by currents flowing through conductors and pieces of electrical equipment such as

Table 7.2. Thermocouple Types and Ranges

ISA Type	Metals		Range °F
	Positive	Negative	
E	Chromel	Constantan	—300 to 1600
J	Iron	Constantan	—300 to 1400
K	Chromel	Alumel	—300 to 2300
R	Platinum and 10% Rhodium	Platinum	32 to 2700
S	Platinum and 13% Rhodium	Platinum	32 to 2700
T	Copper	Constantan	—300 to 650

Table 7.3. Typical Thermocouple Tables

Temperature °F	Millivoltage (32°F Reference Junction)		
	Type J	Type K	Type T
—300	—7.52		—5.284
—200	—5.76		—4.111
—100	—3.49		—2.559
0	—0.89	—0.6	—0.670
00	1.94	1.52	1.517
200	4.91	3.82	3.967
300	7.94	6.09	6.647
400	11.03	8.31	9.525
500	14.12	10.57	12.575
600	17.18	12.86	15.733
700	20.26	15.18	19.100
800	23.32	17.53	
900	26.40	19.89	
1000	29.52	22.26	
1100	32.72	24.63	
1200	36.01	26.98	
1300	39.43	29.32	
1400	42.96	31.65	
1500	46.53	33.93	
1600		36.19	
1700		38.43	
1800		40.62	
1900		42.78	
2000		44.91	
2100		47.00	
2200		49.05	
2300		51.05	
2400		53.01	

motors, generators, etc. As the current flows through equipment, a magnetic field is radiated around the conductor. As this field passes through the space between the conductors in an instrument circuit, a current is set up in the instrument circuit to oppose the magnetic field (transformer action). This current causes a noise to be superimposed on the signal in the thermocouple circuit. The best way of reducing this type of noise is to use twisted lead wire. Twisted leads cause the noise to be cancelled in adjacent sections of the wire. This is the least expensive, most effective way of combating magnetic noise.

Common mode interference is a problem presented by having two different grounds in a circuit with current flow between them. Occurring when a high quality receiver is used, it is a particular problem in thermocouple extension wire circuits. Most thermocouples used are the "grounded" type; that is, the couple is connected physically and electrically to the well in which it is installed. When a thermocouple circuit shield (or any nearby metallic object, such as conduit, tray, building frames, etc.) is at a different potential from the couple, charging currents flow in the extension wire, causing interference to be superimposed on the thermocouple signal. Grounding the shield circuit at the couple is the accepted way to eliminate noise problems from common mode. Multipair cables used with thermocouples should be the individually shielded, isolated pair shield type so that the shield circuit may be maintained at the individual couple ground potential all the way back to the readout location.

To summarize grounding procedures for thermocouples:

1. Ground all shields—an ungrounded shield does not provide noise protection.
2. Ground a shield at one point only. Grounding at two or more points causes currents to flow in the shield circuits (due to ground potential differences) producing noise rather than preventing it.
3. For common mode noise rejection when using a grounded couple, ground the shield on the extension wire at the couple. As the shield circuit is carried back to the control room through a junction box and a multipair cable, connect the pair shield in the cable to the single pair which leads to the couple, without grounding the shield in the junction box or connecting it to any other shield (on other pairs). The shield should not be grounded in the control room.

Thermocouple Materials

Many metals may be paired to produce thermocouples; however, certain pairs have become more or less standard in the United States. These are listed in Table 7.2. Thermocouple range limits shown can be exceeded slightly, provided errors are within acceptable limits and the couple is not subjected to temperatures that would destroy it. Errors due to junction manufacturing techniques and metal impurities range from ±5% at the low end of the range of types R and S to ±½% at the upper end of their range. When closer attention is paid to these factors, errors of ±¼% can be ap-

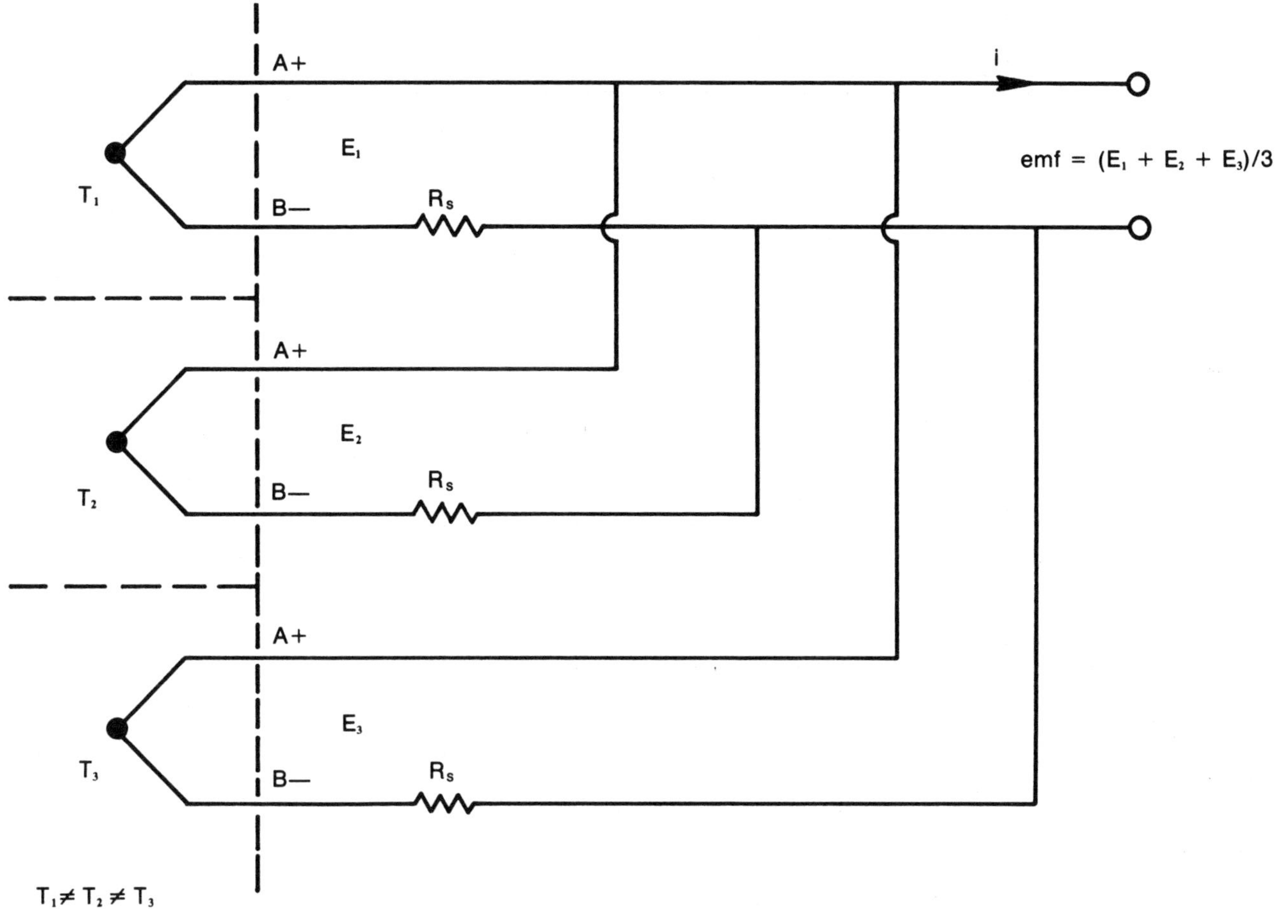

Figure 7.10. Parallel thermocouples are used to measure temperature averages.

proached. Other types of thermocouples exhibit errors within these limits, the smallest errors generally being found near the upper end of their ranges also. Errors are in comparison to a reference junction at 32°F. Significant errors (up to ±5°F) can be introduced into the measuring circuit by inhomogeneities in the thermocouple extension wires.

Fabrication and Insulation Techniques

Thermocouples are easily formed. Any pair of dissimilar metals that are in physical contact constitutes a source of emf, a fact well remembered by those who may have worked with other low-level DC circuits. Twisted, screwed, peened, clamped or welded constructions have all been used, any of which may be satisfactory for a particular application (see Figure 7.13). The most common method of fabrication is that of welding the two wires together. The insulation of the thermocouple and extension leads from other conductors is necessary to prevent unwanted emfs arising from the couples they might form accidentally or from unwanted ground currents.

Bare thermocouples are sometimes used where atmospheric process conditions permit. Generally, though, industrial thermocouples have some protective sheathing surrounding the junction and a portion of the leads. The leads and the junction are internally insulated from the sheath, using various potting compounds, ceramic beads or oxides (such as magnesium oxide or berrylium oxide) as process considerations dictate (Figure 7.14).

Protecting Wells for Thermocouples

Thermocouples are seldom placed in a process stream (pipeline, vessel or other piece of equipment) directly. Instead they are placed inside protecting wells (Figure 7.15) so that they may be removed or replaced without shutting down the process. This not only makes replacement easier, but it simplifies fabrication and insulation techniques for various types of couples, thus reducing their cost.

Protecting wells are normally furnished in sizes from ½ - to 1-inch pipe size, screwed, at pressure ratings compatible with the process. Flanged or other type fittings are available when needed. Wells are normally made of 304 or 316

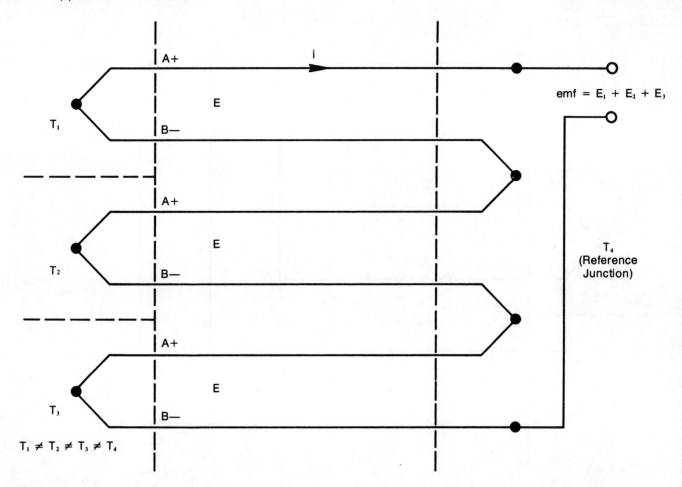

Figure 7.11. Series thermocouples can also be used to measure temperature averages, providing a special reference junction source and calibration is used.

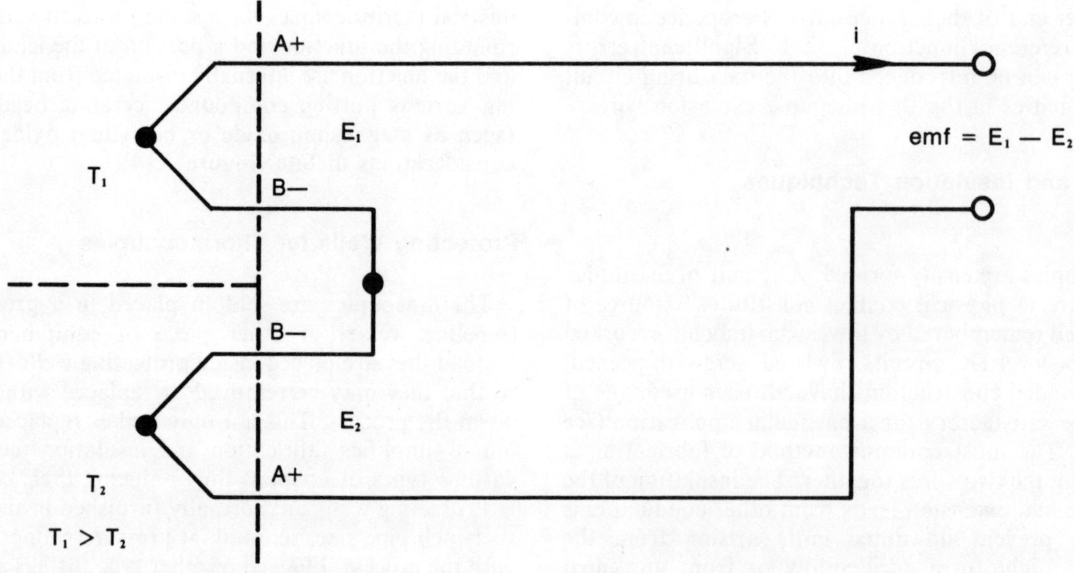

Figure 7.12. Temperature differences are measured using thermocouples with reversed polarities.

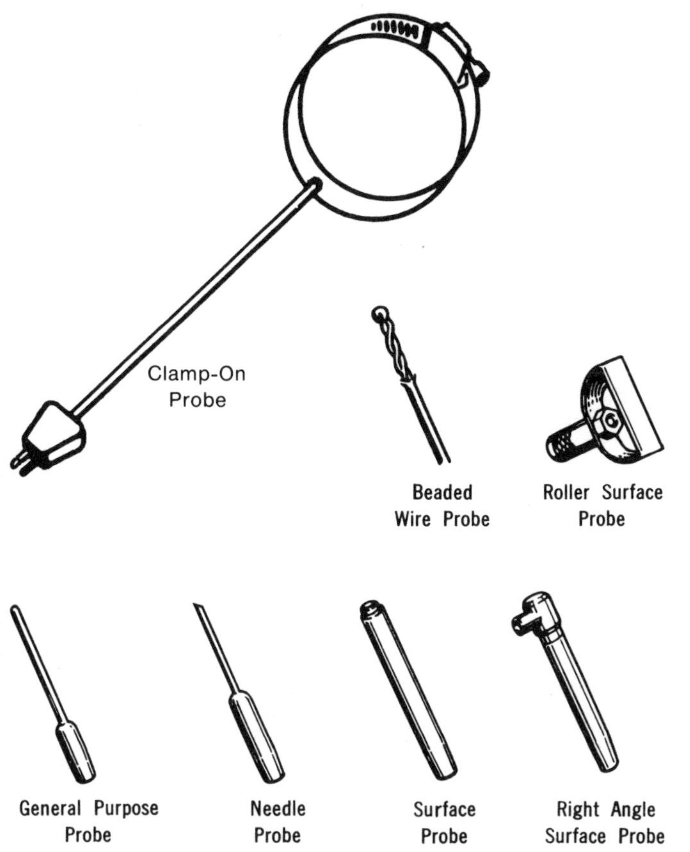

Figure 7.13. Thermocouple joints can be made in a variety of configurations. (Courtesy of Omega Engineering, Inc.)

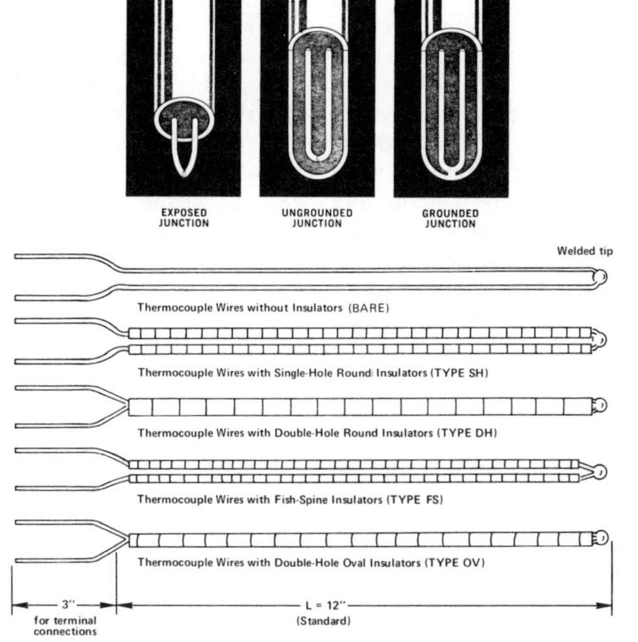

Figure 7.14. Thermocouples are generally insulated away from their protective sheaths. (Courtesy of Omega Engineering, Inc.)

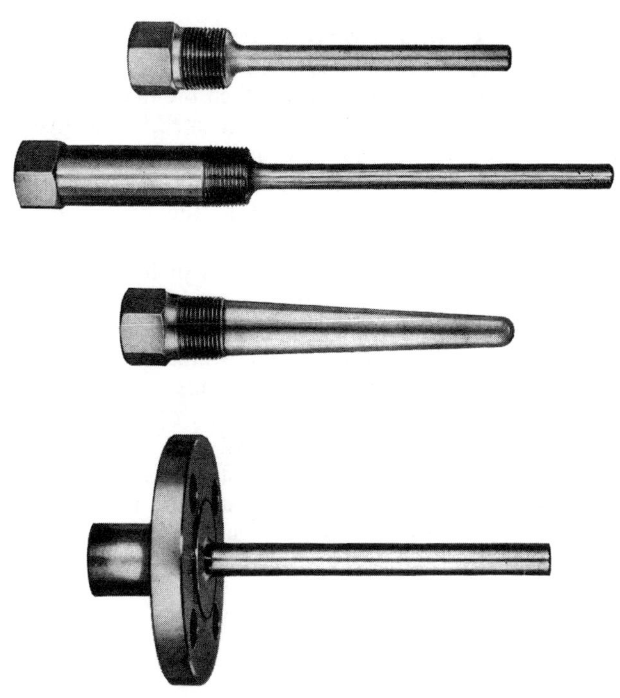

Figure 7.15. Thermowells allow temperature element removal without process interruption. (Courtesy of Foxboro Co.)

stainless steel but are also available in special metals or alloy materials if required.

Protecting wells slow response times appreciably since response is a function of the mass to be heated. When response time is critical, bare or thin-sheathed couples can be furnished that exhibit extremely fast response characteristics. Because of the pressure sealing problems to be overcome, their cost increases appreciably but not prohibitively.

Resistance Temperature Detectors (RTDs)

Resistance thermometers or resistance temperature elements have been used for many years. Their greatest use has primarily been in services requiring high accuracy and where narrow temperature spans were needed. They have served quite well for temperature ranges too narrow for thermocouple devices. Despite the availability of thermocouple devices for even narrower ranges in recent years, the use of RTDs seems to be on the increase.

Theory

The electrical resistance of a conductor changes as its temperature varies. The magnitude of this change with respect to a 1° change in temperature is known as the "temperature coefficient of resistance" of the conducting material. For most pure metals, this coefficient is constant over some range of temperatures; for example, the temperature coefficient of resistance of platinum is 0.00392 ohm/(ohm)/(°C) over a range of 0° to 100°C. For most conductors, this coefficient is positive; i.e., resistance increases as temperature increases. When the measuring and readout instruments are remotely located, transmission wires from the RTDs should be strategically routed over the shortest distance so that the element resistance change will be as large as possible compared to the overall circuit resistance.

Commonly used resistance materials include platinum, nickel, copper, nickel-iron and tungsten. Most frequently used are platinum, which has a linear characteristic over much of its range, and nickel, which exhibits a rather large temperature coefficient of resistance, although its characteristic is nonlinear. The resistivity of platinum is 10 ohm-cm at 20°C, while the resistivity of nickel is 6.844 ohm-cm at 20°C. The change in resistance is a function of the temperature coefficient of resistance, designated α, which represents the slope of the curve expressed by the equation:

$$R_t = R_o(1 + \alpha T) \text{ or} \tag{7.1}$$

$$dR_t/dT = \alpha R_o \tag{7.2}$$

where R_t = resistance, in ohms, at temperature T
R_o = resistance, in ohms, at a reference temperature (often 0°C)
α = temperature coefficient of resistance

This equation expresses a linear function, and since platinum has a linear characteristic, it provides a fairly accurate calculation of resistance at its working temperature. Since nickel has a nonlinear characteristic, the coefficients of the succeeding terms of the general equation,

$$R = R_o(1 + \alpha T + \beta T^2 + \gamma T^3 + \ldots) \tag{7.3}$$

must be calculated and used to project approximate values for temperatures greater than 100°C or less than −50°C. The α values of platinum and nickel at 0°C are 0.00392 ohm/(ohm)/(°C) and 0.0063 ohm/(ohm)/(°C), respectively. β in Equation 7.2 is an empirical quantity obtained from the manufacturer. The γ coefficient can be ignored. Only on wide spans where extreme accuracy is needed should it be considered. As a matter of fact, the β coefficient is seldom used or needed.

Since the actual resistance of a finite length of wire is indirectly proportional to its cross-sectional area and since size should be kept to a minimum to limit thermal mass, wire size is also minimized. For platinum and nickel, the smallest practical wire diameter is 0.002 inch. Typical resistance values range from 10 to 500 ohms, with selected manufacturers producing RTDs of resistance as high as 1,200 ohms. A piece of alloy wire having a zero-temperature coefficient can be series- or shunt-connected internally to raise or lower overall resistance to standardize RTDs for interchangeability.

Construction

Although certain laboratory RTDs are constructed having the resistive element exposed, such as in anemometry experiments, most are constructed so that the fine wire element is coiled and loosely supported on a mica form. The coil is annealed (heated until stresses brought about by the coiling procedure are relieved) and is then placed in a protective sheath. Industrial RTDs are formed in much the same manner. Extra care is taken to fabricate the element so that the effects of mechanical shock are minimized (Figure 7.16).

Typical sheathed RTDs contain coils wound around the support in such a manner as to distribute the resistance evenly while maintaining good thermal contact and electrical insulation from the sheath. The element is annealed and stretched lightly over the support to give firm seating. After its end wires are connected, it is fixed in place with a varnish or other encapsulating material. The entire sheath is often hermetically sealed, a fill material such as magnesium oxide or aluminum oxide being included. A mechanical fitting is usually attached to the sheath. This is often accomplished with an attached electric terminal block or sometimes with a loading spring which gives the tip of the sheath positive contact with its thermowell. This also prevents vibration.

Flat semiflexible RTDs, similar in size and shape to some strain gauge elements, can be surface mounted on irregular surfaces. The low thermal mass of these RTDs provides excellent response.

Readout Devices

The basic bridge circuit in Figure 7.17 can be adapted to give an indication analogous to temperature. As the resistance, X, representing an RTD, changes, the galvanometer, G (a sensitive DC current meter with "zero" at center scale), can be calibrated to deflect accordingly. However, in normal practice, the RTD is apt to be located physically some distance away from the galvanometer. Lead resistance, L, shown in Figure 7.18 becomes a part of the "X-leg" of the bridge circuit.

Since $E = V_{21} + V_{14} = V_{23} + V_{34}$ when the voltage at node 1 equals the voltage at node 2, galvanometer G will experience a null (zero) indication. The ratio of the bridge components,

$$A/(X + 2L) = B/S \text{ or}$$
$$X = S\,(A/B) - 2L \tag{7.4}$$

now holds true.

Deflection Reading

By the addition of one more lead-in wire and a slight circuit change, the effects of lead-in resistance can be eliminated by adding the resistance, L, to both legs of the

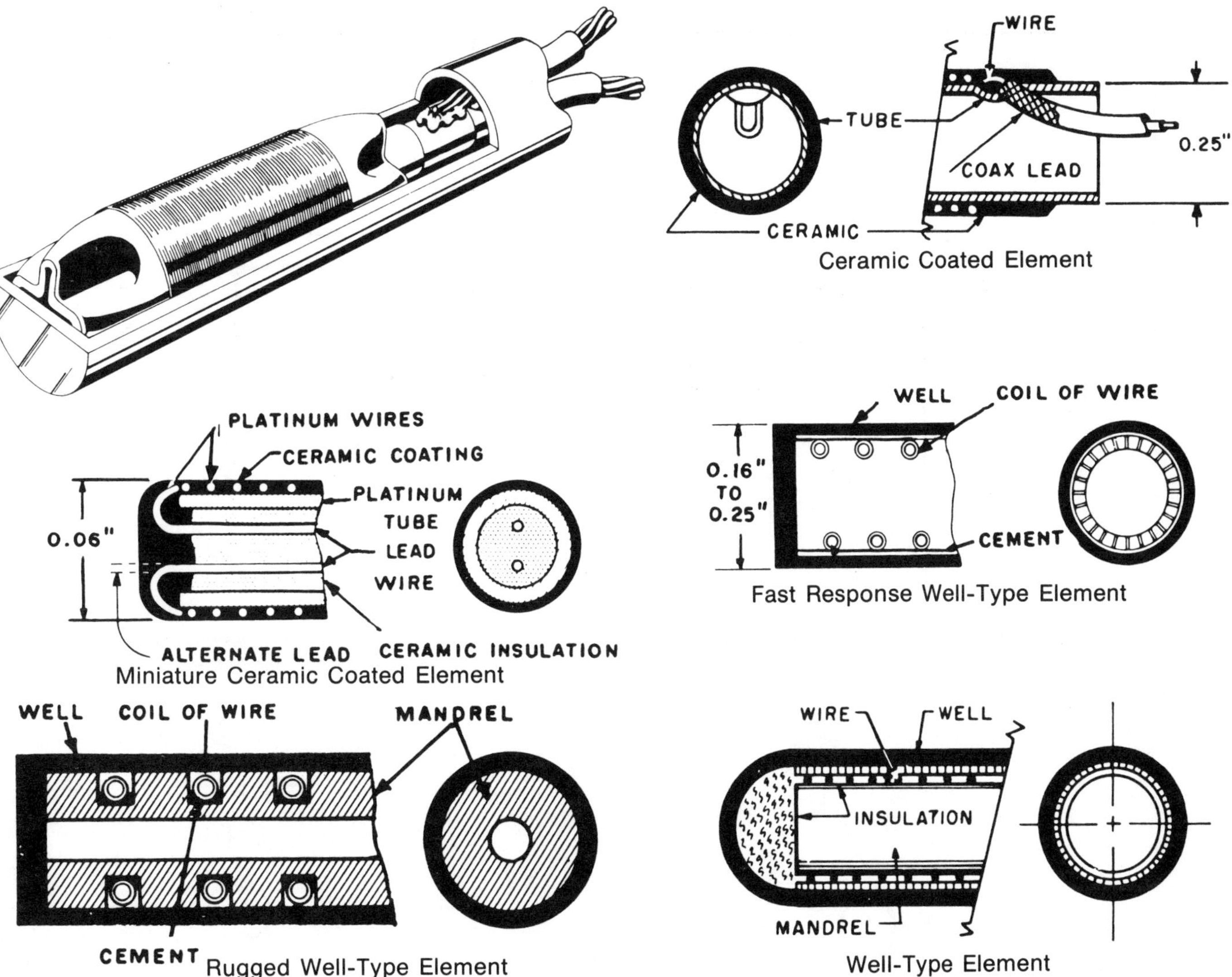

Figure 7.16. RTD elements are fabricated to minimize the effects of mechanical shock. (Courtesy of Rosemont, Inc.)

bridge circuit, as shown in Figure 7.19. Now, for meter null, the following ratio applies:

$$(A + L)/(X + L) = B/S \quad \text{or}$$

$$X = S\,[(A + L)/B] - L \tag{7.5}$$

The addition of the variable resistance, C, provides the adjustment of the galvanometer, G, to some convenient known calibration point. For new values of X, G can be calibrated to read temperature directly.

Null-Direct Reading

Resistance A (Figure 7.20) may be replaced by a highly accurate adjustable resistance, which is calibrated to correspond to the temperatures which X measures. For each new reading, galvanometer G is set to null, manually or electrically, by means of resistance A. Even though highly accurate, the slider in A would be affected by contact resistance. By placing sliding contacts in both the galvanometer and battery circuit loops, such as in Figure 7.21, the effects of contact resistance are eliminated. Resistances R_1 and R_2 are ganged so that percentage of spans, k, is equal. With galvanometer at null, this equation is applicable:

$$\frac{X}{A + kE} = \frac{S + D(1 - k)}{B + kD + E(1 - k)}$$

or

$$X = \frac{A(S + D) + (ED + ES - A)k - Dk^2}{(B + E) + (D - E)k} \tag{7.6}$$

The voltage source, E, can be replaced by an alternating current source (normally, $f - 1{,}000$ Hz), and resistances A

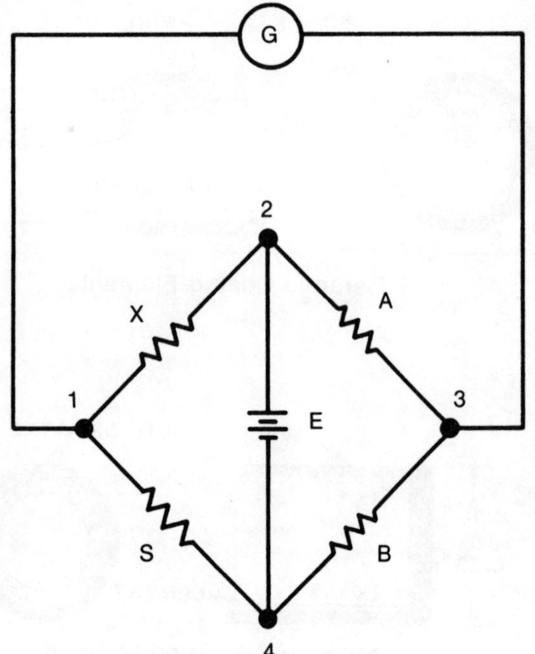

Figure 7.17. A Wheatstone bridge is the basic circuit for most readout devices.

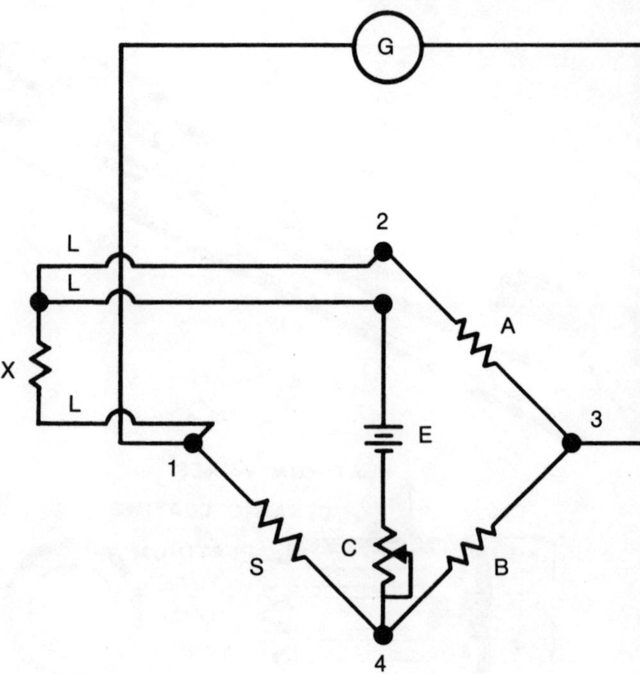

Figure 7.19. The effects of lead resistance can be eliminated by adding resistance in another leg of the bridge circuit, using a three-wire system to cancel out the lead resistance.

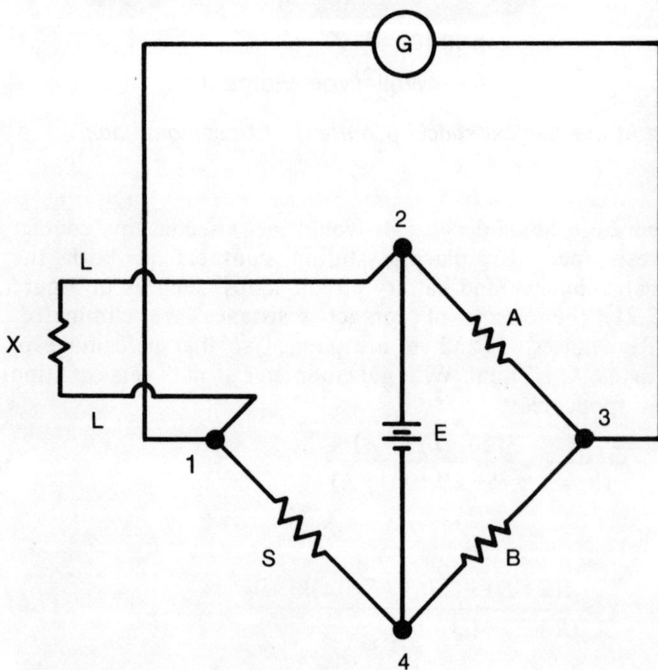

Figure 7.18. Lead resistance becomes a part of the measurement in a two-wire circuit.

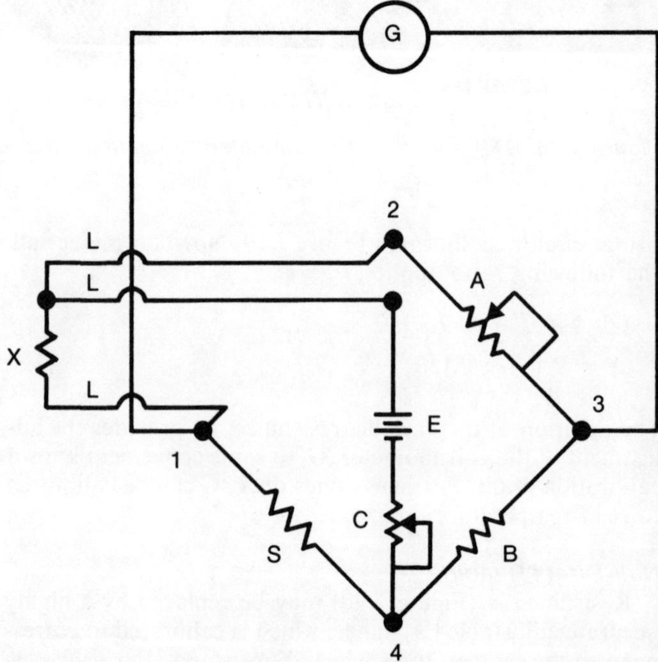

Figure 7.20. Temperature can be made analogous to a potentiometer setting which nulls the meter in a null-direct reading bridge.

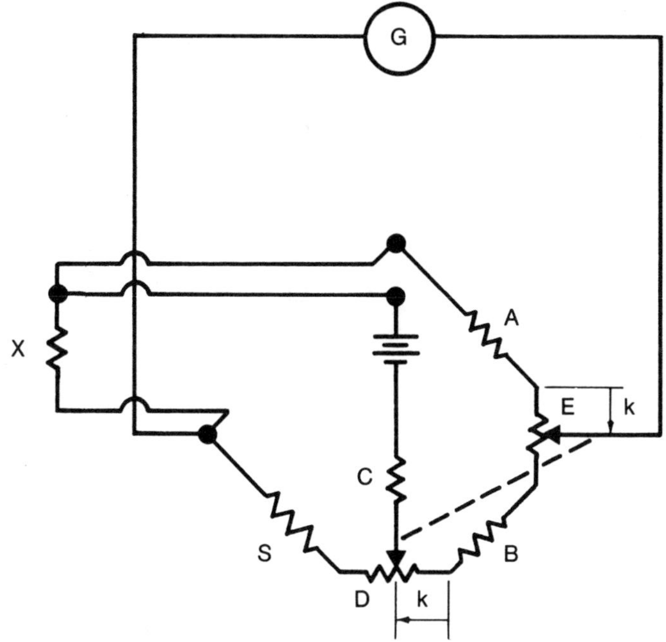

Figure 7.21. A double-slidewire bridge is not affected by contract resistance.

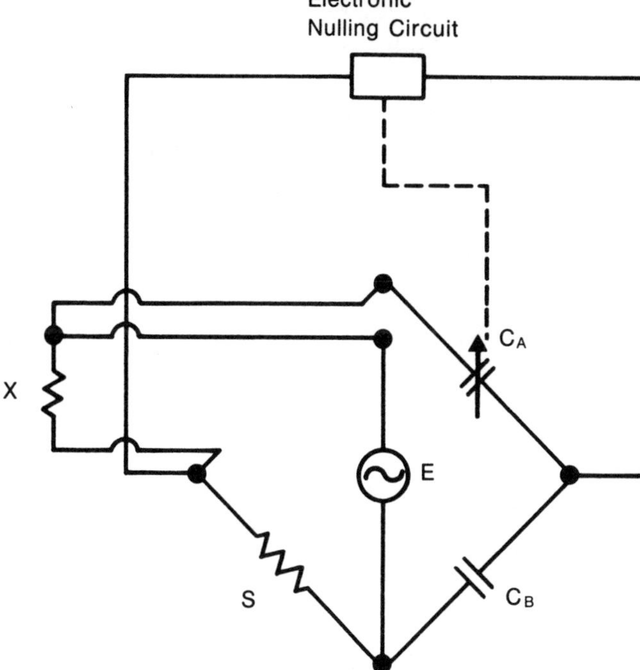

Figure 7.22. AC bridges are also available.

and *B* can be replaced by capacitors, creating an AC bridge (Figure 7.22). In this case,

$$\frac{Z_X}{Z_A} = \frac{Z_S}{Z_B}$$

$$\frac{X}{\dfrac{1}{2\pi f C_A}} = \frac{S}{\dfrac{1}{2\pi f C_B}} \quad \text{or} \quad X = \frac{C_B}{C_A} S \tag{7.7}$$

For individual measurement for indication or control, a blind resistance-to-current (R/I) converter is used to retransmit the temperature analog to other instruments within the loop. One such transmitter currently manufactured is close-coupled to the resistance element, i.e., mounted within the conduit head with circuit power coming from the two-wire external output circuit. For multiple measurements, where only an indication or recording is required, a number of RTD circuit leads are brought to a selector mechanism. The selected RTD is then connected to a linearizing circuit (when necessary) and then to the readout.

Portable battery-operated bridges are used for field checks. Precision Mueller bridges can give calibration accuracies of ±0.0001 ohm down to an external measurement of 0.5 ohm.

Filled Systems

One of the early methods of measuring temperature was the use of pressure-actuated thermometers which utilized "filled systems" (enclosed systems filled with liquid, gas or vapor) responsive to temperature variations.

Almost all fluids, whether liquids, vapors or gases, expand when heated and contract when cooled. This phenomenon is utilized to expand a pressure element (Figure 7.23)—usually some form of bourdon element—which in turn positions an indicator or a transmitter coil or flapper or actuates a switch mechanism. Filled systems such as these have been used extensively in the past for local indicators, recorders and controllers for temperature loops and are still widely used today for temperature transmitters when pneumatic control systems are used. They consist basically of the following:

1. A sensitive pressure bulb immersed in the medium to be measured
2. A capillary tube connecting the bulb to a readout or transmitting device
3. A pressure element actuating device to respond to pressure changes caused by temperature
4. A linkage movement to effect indication, recording or transmission

One of the disadvantages of filled systems is the error caused by variations in ambient temperature. Errors are due to two sources:

1. The capillary volume between the sensing bulb and the responding element in the instrument case
2. The volume of the responding element itself

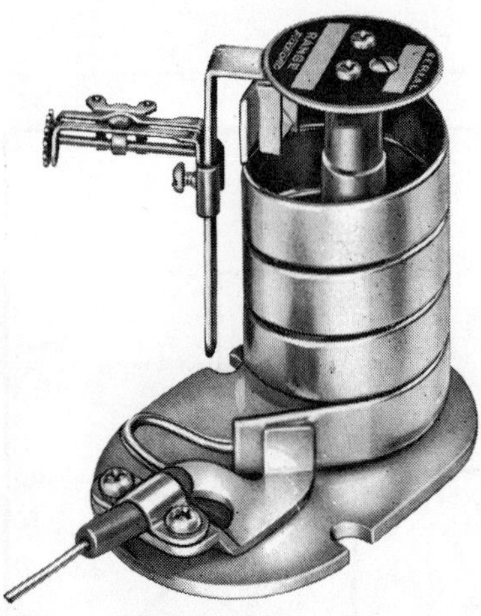

Figure 7.23. The basic filled system uses the expansion of a bourdon element to give position analog of temperature. (Courtesy of Foxboro Co.)

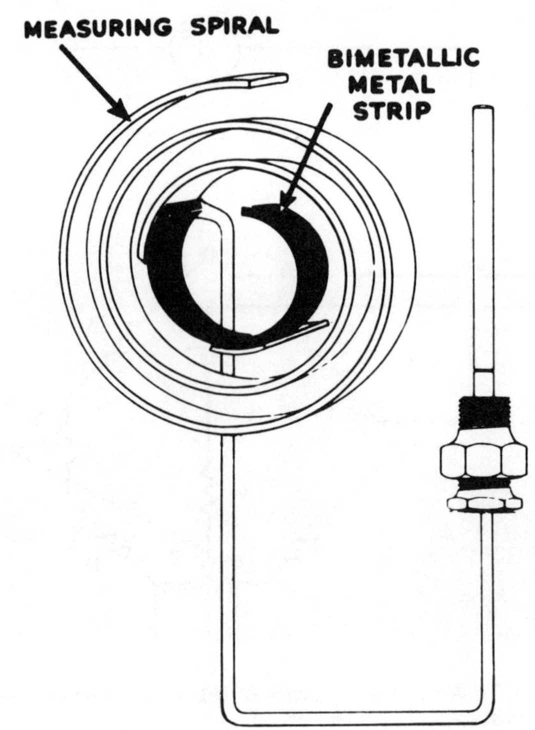

Figure 7.24. A bimetallic compensator is often used to minimize errors caused by ambient effects on the bourdon. (Courtesy of Honeywell, Inc.)

These elements have no way of differentiating between ambient caused pressure changes and process caused pressure changes in the filled system.

In situations where the pressure element is not mounted close to the bulb, compensation is often needed to correct for ambient temperature variations along the capillary. Compensation for temperature effects on the element inside the instrument, known as case compensation, is often desirable.

Case compensation is generally accomplished by incorporating a bimetallic strip into the mechanism, which acts to negate ambient errors (see Figure 7.24). Case plus capillary compensation, known as full compensation, is accomplished by having another full-length capillary run to a second element mounted in the case and linking it to the primary mechanism so as to cancel ambient effects, as in Figure 7.25. A small auxiliary capillary may also be placed inside the case and connected to the other element to provide case compensation. Both methods are used. Systems using mercury as a fill liquid are sometimes compensated by a small wire inserted into the capillary. This wire has an expansion rate which matches that of mercury, thereby eliminating temperature errors induced along the capillary.

Filled systems have been classified into four groups by the Scientific Apparatus Makers Association (SAMA). These classes (I, II, III, and V—there is no Class IV) are differentiated on the basis of fill fluid and range in relation to ambient temperature. Classes I and V, filled with liquid and mercury, respectively, work because of volumetric liquid expansion and contraction. Classes II and III, filled with vapor and gas, respectively, work because of internal pressure changes.

Table 7.4 lists the four classes and provides some general guidelines for application of each group.

Bulb elevation must be considered when mounting the local instrument on liquid and vapor systems. The weight of the fill liquid creates a static head which adds to or subtracts from the pressure which actuates the bourdon (element) mechanism. An error of ±½% of span is possible for an elevation difference of 100 feet for an instrument (Classes I, IIA and IID) having a 100°F span. Class III instruments are not affected by bulb/case elevation differences.

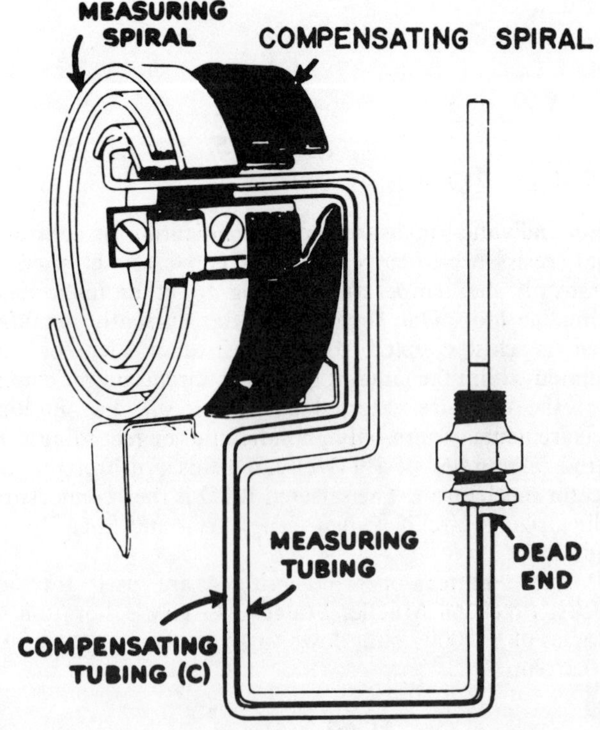

Figure 7.25. Compensation of ambient effects on the capillary, and the case-mounted bourdon is known as "full" compensation. (Courtesy of Honeywell, Inc.)

Table 7.4. Guideline Data for Filled Systems

Filled System Classification (SAMA)	Fill Fluid	Compensation	Scale	Range Limits, °F Lower	Range Limits, °F Upper	Overrange	Bulb Elevation Errors
I	Liquid	Uncompensated	Linear Above —100°F	—125	600	100%	Minor below 100 ft
IA		Full		—125	600	0-100%*	
IB		Case		—125	600	100%	
IIA	Vapor	Not required	Scale divisions increase with temp. increase	Amb.	550	Almost always less than 100%	Minor below 100 ft
IIB				—430	Amb.		None
IIC				—430to Amb.	Amb. to 550		Not allowable**
IID				—400	500		Minor below 100 ft
IIIA	Gas	Full	Linear above —400	—400	1500	100-300%†	None
IIIB		Case		—400	1500		None
(SAMA has no Class IV)							
VA	Mercury	Full	Linear	—40	1000	100%	Minor below 25 ft
VB		Case		—40	1000		

* Depends on bulb length.
† Depends on range.
**Bulb and measuring element must be at the same elevation.

Barometrically induced bourdon errors can amount to ±⅓% or less of the pressure span.

Instrument scales for Class I are linear, except for low temperatures. Class II scales are nonlinear with divisions spreading as temperatures increase. Class III scales are linear, except at very low temperatures. Class V scales are linear over their entire range.

Class I—Liquid Filled

This liquid-filled (mercury-excluded) classification is subdivided into three groups:

Class I—uncompensated
Class IA—fully compensated (capillary and case)
Class IB—case compensated only

Fill liquids chosen by the manufacturer are usually selected to give large expansions per degree change in temperature, thereby holding bulb size to a minimum. Pressure inside the system must exceed the vapor pressure of the fill liquid to prevent vaporization, and minimum temperature of the system should be kept above the freezing point of the fill liquid to prevent plastic deformation of the pressure element.

Bulb size varies inversely with the expansivity of the fill liquid and inversely with span. Bulb size for a commonly used fill liquid, xylene (C_8H_{10}), is approximately 3½ inches long by ¾-inch OD for a span of 100°F, or an approximate capacity of 4 cc. Minimum range spans, determined by practical bulb size, are around 25°F. Maximum span can be as high as 250° to 450°F with range limits near −125° and 600°F, depending on the characteristics of the fill liquid. System reaction is linear, except at low temperatures. Most major instrument manufacturers do not market Class I instruments; they do market, however, Class IA and Class IB instruments. If the capillary length is short (10 feet or less), Class IB systems can be successfully used, since the bimetallic compensator normally used compensates for all or most of the capillary.

Overrange limits for Class IA systems range from about 200% for 5-foot long capillaries to practically 0% for capillaries whose length approaches 200 feet.

Speed of system response, generally around 5 to 10 seconds, is slow compared to other classes because of thermal mass. Response speed can be increased by decreasing the bulb outside diameter. A long capillary with a small diameter is sometimes coiled to provide bulb capacity requirements within minimum space (see Figure 7.26), a practice used frequently for measuring temperatures in large air handling systems, such as building air conditioning systems. Capillary type bulbs used with Classes I, III and V instruments can be left uncoiled to average the temperature over the entire bulb length.

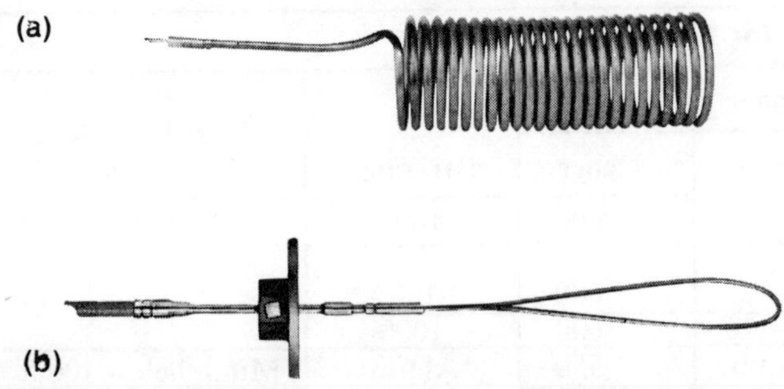

Figure 7.26. System response time is lessened in air handling systems by using preformed capillary-type bulbs. (Courtesy of Foxboro Co.)

Class II—Liquid Vapor Systems

Class II systems utilize the vapor pressure of a volatile fill liquid as a pressure source to actuate the element. The system pressure will always be that occurring at the liquid-vapor interface. No compensation is necessary in vapor pressure type systems. Care, however, must be taken during manufacture to have the fill liquid, bulb, capillary and case at the same temperature.

Overrange capabilities are often less than 100% because as temperature increases, vapor pressure increases at a more rapid rate.

There are four subdivisions of this class:

Class IIA: designed for bulb placement in a process whose temperature always exceeds the case ambient temperature.

Class IIB: designed for bulb placement in a process whose temperature is always below the case ambient temperature.

Class IIC: designed for bulb placement in a process whose temperature can exist periodically both above and below the case ambient, but whose temperature when near case ambient is unimportant.

Class IID: designed for bulb placement in a process whose temperature measurement, when at case ambient, is important, and which temperature can exist periodically both above and below ambient temperature.

The fill fluid in a Class IIA system is in a vapor state in the bulb and in a liquid state throughout the rest of the system (Figure 7.27). The bulb must be sized to contain changes in volumes of the fill fluid in the bourdon, capillary and bulb due to ambient and process temperature fluctuations while still maintaining the liquid-vapor interface within the bulb. Bulb length is therefore directly proportional to capillary length.

Long capillaries slow the response times of Class II systems. The time constant (time, in seconds, necessary for the system to register 63.2% of a step change) of a Class IIA system is approximately 1 second for a 10-foot capillary but is about 10 seconds for long capillaries. Spans vary from 40° to 300°F generally, with range limits between ambient and 550°F.

Class IIB systems have all of the fill liquid contained within the bulb and, therefore, are not affected by ambient conditions (Figure 7.28). Bulb volume is generally less than those for Class IIA systems.

Response times of Class IIB systems are essentially the same as for Class I systems.

Class IIB system spans vary from 40°F to 300°F generally, with range limits from ambient to about −430°F.

Class IIC systems, which can measure temperature above and below ambient but not at ambient, are constructed and filled so that the liquid part of the fill fluid exists in the capillary and bourdon when the bulb temperature is greater than ambient (Figure 7.29*a*). When bulb temperature drops below ambient, the fill liquid transfers into the bulb (Figure 7.29*b*). During the transfer the location of the liquid-vapor interface is undefined; therefore, system pressure does not

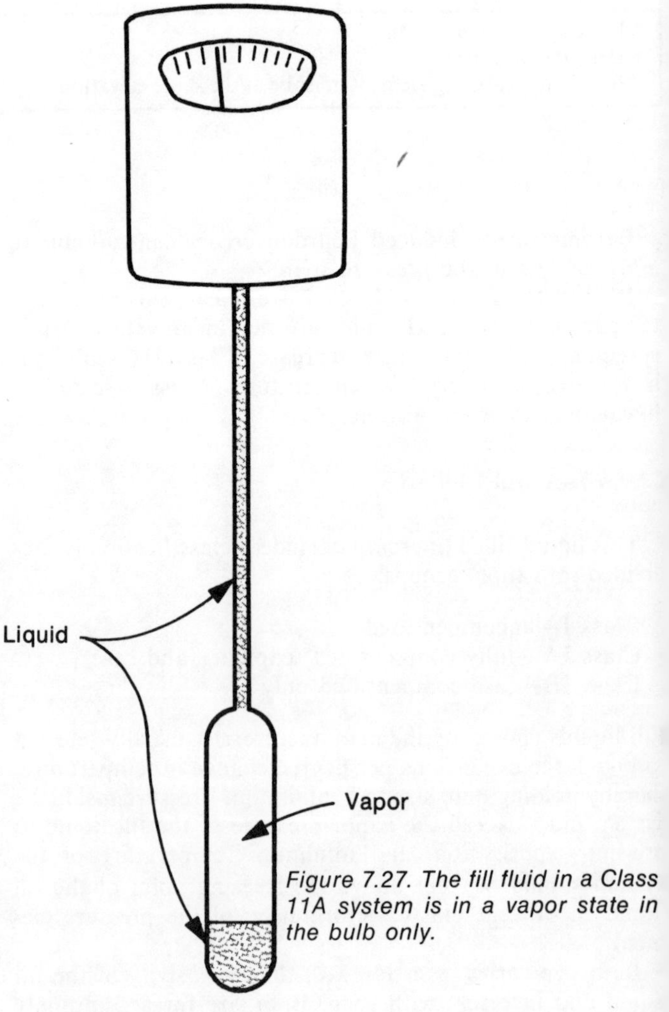

Figure 7.27. The fill fluid in a Class 11A system is in a vapor state in the bulb only.

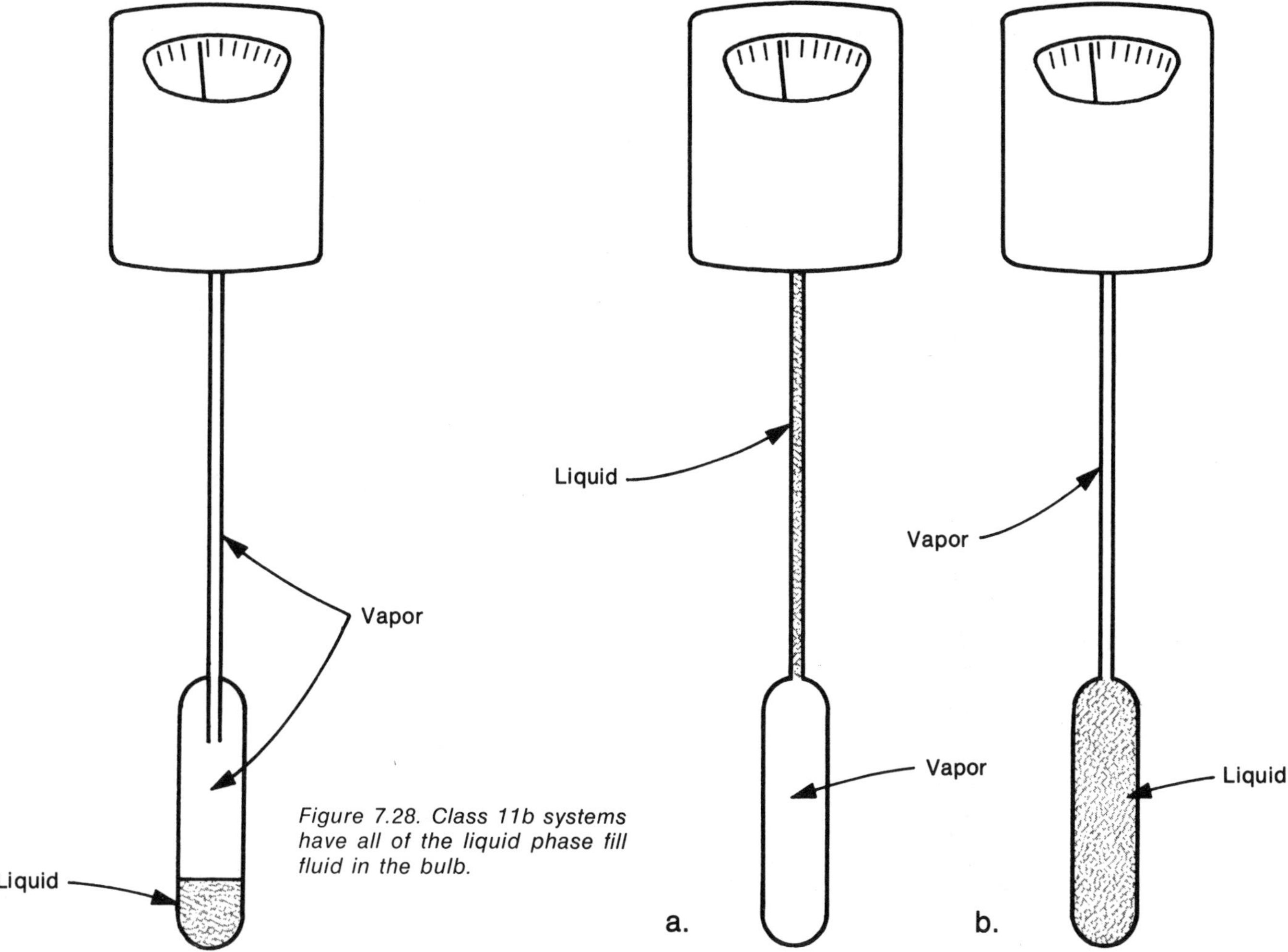

Figure 7.28. Class 11b systems have all of the liquid phase fill fluid in the bulb.

Figure 7.29. (a) *In a Class 11C system, the liquid phase of the fill fluid exists only in the bourdon and capillary when the measured temperature exceeds ambient.* (b) *When the bulb temperature drops below ambient, the liquid transfers into the bulb.*

represent an analog of bulb temperature. Span and range limits essentially encompass those of Class IIA and Class IIB systems.

Bulb and instrument elevations must be virtually the same because of the fluid-vapor balance at the ambient crossover point. No compensation is available for elevation difference. Bulb volume is such that it will contain the volume of both capillary and bourdon and changes that occur in all portions of the system. It is usually larger than volumes for Class IIA and Class IIB systems.

Response times are quite similar to Classes IIA and IIB. Transfer time when bulb temperature passes through ambient ranges from 5 minutes to an hour.

Class IID systems are constructed such that the volatile fill fluid is sealed in the bulb by a nonvolatile liquid which also acts as a transmitting medium between vapor pressure in the bulb and the bourdon mechanism (Figure 7.30).

Bulb size is usually the largest of all Class II systems due to the trap used to prevent the volatile fluid from ever reaching the capillary. The thermal mass of the trap itself and the use of the more viscous sealing fluid result in response times in the 5- to 10-second range.

Span and range limits essentially encompass those of Class IIA and Class IIB systems.

Class III—Gas Filled

Classes IIIA and IIIB thermal systems use gas as the fill fluid. Charles's Law, which indirectly states that the absolute pressure of a mass (constant volume) of a perfect gas is directly proportional to its absolute temperature within moderate temperature ranges (−200° to +600°F), is the basic principle utilized. Capillary-to-bulb size ratios are kept small to reduce capillary-induced errors. While commercially available fill gases do not behave exactly as perfect gases, they do permit ±½ to ±1% accuracies from −400°F to +1,000°F for spans of about 300°F. The span is set by the system pressure at the time of filling and generally

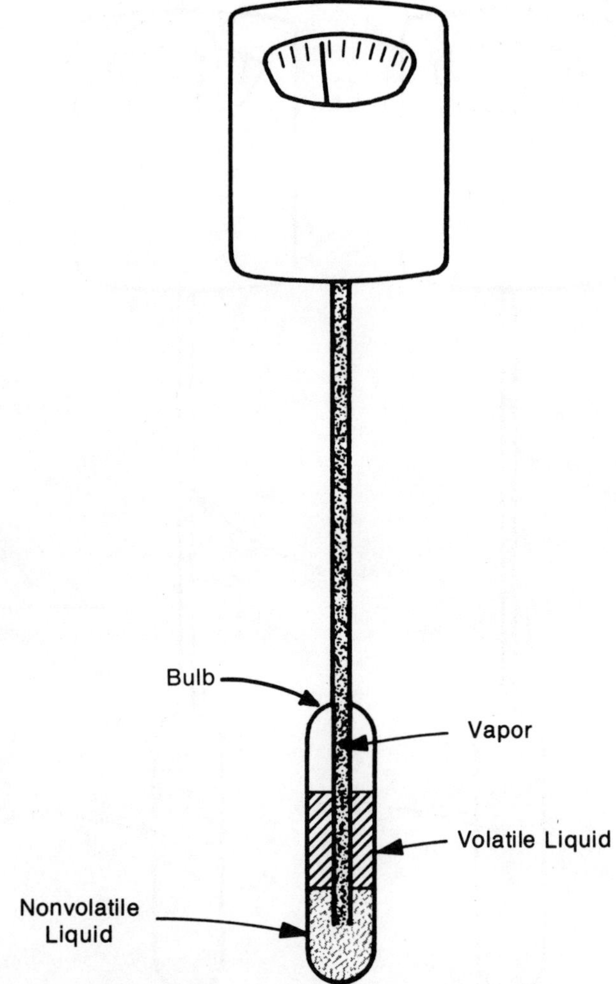

Figure 7.30. A Class 11D system has two fill fluids, one of which is nonvolatile and which seals the actuating fluid in the bulb.

varies from 100° to 1,000°F with range limits near−400° and 1,500°F. Nitrogen, which is inexpensive and readily available, is a good fill fluid for the middle and upper ranges, while helium is preferred for ranges with near-zero absolute temperature lower limit. Minimum system temperature must always be above the critical temperature of the fill fluid, which for nitrogen is −232.7°F and for helium is −450.3°F.

Class IIIA fully-compensated systems are compensated with a bulbless capillary and its bourdon and with a short capillary section oı a bimetallic strip internal to the case Class IIIB case-compensated systems have the usual bimetallic strip compensator.

Errors introduced by barometric pressure or elevation difference are generally less than ±½% of range.

Response times vary between 1 and 5 seconds. Preformed coiled-capillary bulbs are used to reduce response time.

Overrange is dependent on fill fluid and fill pressure and varies from a very few percent for low temperature ranges to perhaps 300% at 1,200°F.

Class V—Mercury Filled

Class VA and V.B systems, completely filled with mercury or mercury-thallium eutectic amalgam, work as a result of volumetric expansion. Class VA fully-compensated systems use standard case and capillary compensation techniques, although some manufacturers use the Invar wire compensation method mentioned earlier, which has no second capillary and bourdon assembly.

The thermal expansion of mercury is less than that of Class I systems for a given temperature change; therefore, a comparatively larger bulb size is inherent. This characteristic also gives rise to inexpensive measurement using Class VB case compensated systems.

Barometric errors are negligible, and elevation difference errors are minor if the difference does not exceed 25 feet.

One hundred percent overrange is a common minimum up to 1,000°F limit. Spans of Class VA and VB systems vary from about 40° to 1,000°F, and range limits are generally −40° to 1,000°F.

Radiation Pyrometers

"Pyros" is the Greek work for fire. Pyrometer is the name generally applied to devices which measure temperatures that are above the average range capabilities of thermocouples, although certain alloy thermocouples have been used for brief periods to 3,000°C. Some pyrometers can be used to measure temperatures as low as 0°C. Radiation pyrometry uses the property of thermal radiation that is emitted from all matter, except inert gases, at absolute zero temperature. Radiation pyrometry is particularly interesting in that it does not require direct contact with the material whose temperature is to be measured.

For temperature measurements, such as those common to furnace interiors, it is convenient to calculate temperature from the amount of energy radiating from the measured body. The Stefan-Boltzmann Law of radiation relates radiant energy, E, body temperature, T_b, and the temperature, T_r, of the receiver of the energy by the equation:

$$E = \sigma(T_b^4 - T_r^4) \qquad (7.8)$$

where the Stefan-Boltzmann constant, σ, has a value of $0.1713(10^{-8})$ btu/(square feet)/(hour)(°R). This law is relative to a perfect radiating body (black body) which absorbs all incident radiation and radiates, both qualitatively and quantitatively, only in proportion to its temperature. However, no surface can be treated as a perfect radiator because of reflection or transparency. The constant relating the radiated energy of a black body and the amount of radiated energy of an imperfect radiating body is called emissivity, E. It is based on both bodies having equal size, shape and temperature and energies which radiate at right angles to plane, polished, opaque portions of the body surfaces. Since most measurements are taken from untreated surfaces, this relationship is modified to take such factors into account and is sometimes renamed emittance, *e*.

All radiant energy is not radiated at the same frequency (wavelength). For each temperature a maximum radiant intensity occurs near a particular frequency, which varies proportionately with temperature.

Radiation pyrometers (Figure 7.31) give a temperature analog readout of a hot body by sensing the radiation intensity of a band of frequencies. Some types of pyrometers have wide-band sensitivity; some, single narrow-band sensitivity; and some have dual narrow-band sensitivity. The latter have band frequencies which are close together to minimize emittance errors, since dividing the output at one frequency by the output at the other frequency tends to cancel emittance altogether.

Since temperatures of a hot body and a receiver body are related through the Stefan-Boltzmann Law, the receiver body temperature must be known, controlled or compensated. Compensation of the sensor housing (often accomplished through use of a nickel resistor), optical reflections and other characteristics of various sensors cause the outputs of industrial pyrometers to follow equations different from the Stefan-Boltzmann equation. Therefore, calibrations are empirical.

Sensors using thermopiles are prevalent in industry, primarily because of stability and response characteristics. Photocells, having a multiplying effect, provide high sensitivity and response times of a few milliseconds, although sensitive to frequencies only in the near-infrared and visible ranges. Transistor-like photovoltaic cells are also response-limited in range to temperature measurements above 1,500°F and must be kept at case operating temperatures below 120°F. Bolometer sensors, which change resistance due to temperature change (which is in proportion to incident radiant energy), are sometimes used where cost and mechanical fragility permit.

When the field viewed by a pyrometer is completely filled by the target, the distance between the two has no effect on the sensor output. If target size is expressed in terms of a circle having an area equal to that of the target, then the ratio of the distance from the target to the diameter of the minimum circle which would completely fill the field of view of the optical system is known as the "distance factor." Distance factors for most industrial pyrometers are about 20:1, although, for measurements of large targets or low temperatures, factors of 2.5:1 are not uncommon.

Several types of errors can affect sensor output. Included are radiation attenuation due to absorption of certain frequencies by intervening substances (such as gases, vapors and particulate matter in fumes and on the lens) and attenuation by the lens itself. Gases, vapors and any body within the field of view can also be radiators themselves at certain temperatures. Reflections and transparencies of targets having emittances ≤ 0.5, such as most nonferrous metals, will cause large errors.

To decrease the effects of intervening gases, vapors and fumes, a "sighting tube" can be extended to within an inch or so of the target (Figure 7.32). A slight purge of about 20 cubic feet per hour of clean, dry air can be put on the tube. Tube material of Inconel is good for use in temperature services to 2,000°F, while ceramic tubes can be used in services up to 4,500°F. A "target tube" having a closed target end can be employed where no reasonable target exists. The field of view is confined to the end of the tube, which is often made of a ceramic material (Figure 7.33).

To prevent damage to the sensor, air and/or water cooled spool pieces can be inserted between the sensor head and the boiler housing to dissipate heat conducted through the mounting bracket or target tube. Heating due to radiation from the boiler housing is easily reduced with shielding.

Optical Pyrometers

Optical pyrometers may be classified into two categories: (a) those which operate within the visible spectrum where the human eye is used as a detector to compare the measured source with a known standard and (b) those which use an electrical detector to compare the measured source with a known standard.

The first mentioned might be classified as a manual type for it requires manual balance by the operator. The second is listed as the infrared type because it senses wavelengths beyond the human spectrum into the infrared region.

Manual Type

Pyrometers sensitive to radiation in the visible spectrum are classified as manual optical pyrometers. They generally have single narrow-band sensitivity but are sometimes manufactured with dual band capabilities. Monochromatic (single-color) brightness of the target to be measured is compared to the brightness of a standard calibrated filament by means of the human eye (Figure 7.34). Although the response of the eye may vary from one observer to the next, brightness comparisons are usually within ± 1%. With the standard filament viewed superimposed on the target, as a background, both being seen through a monochromatic filter (usually red), the filament is varied by means of a calibrated rheostat until it apparently disappears or merges into the background, or an optical attenuator (wedge) can be varied until merging occurs.

In either case the variable adjustment can be calibrated and read in degrees of temperature. Filters can be used to reduce target brilliance optically when it becomes too bright, usually above 2,400°F.

Range limits of the manual optical pyrometer are about 1,400°F on the low end (there is an insufficient emission of visible light for an accurate comparison below this figure) and about 6,300°F on the high side.

Optical pyrometers are hand held but contain their own power supply for operating current (Figure 7.35). They are used as NBS primary and transfer standards above the gold point (1,945.4°F).

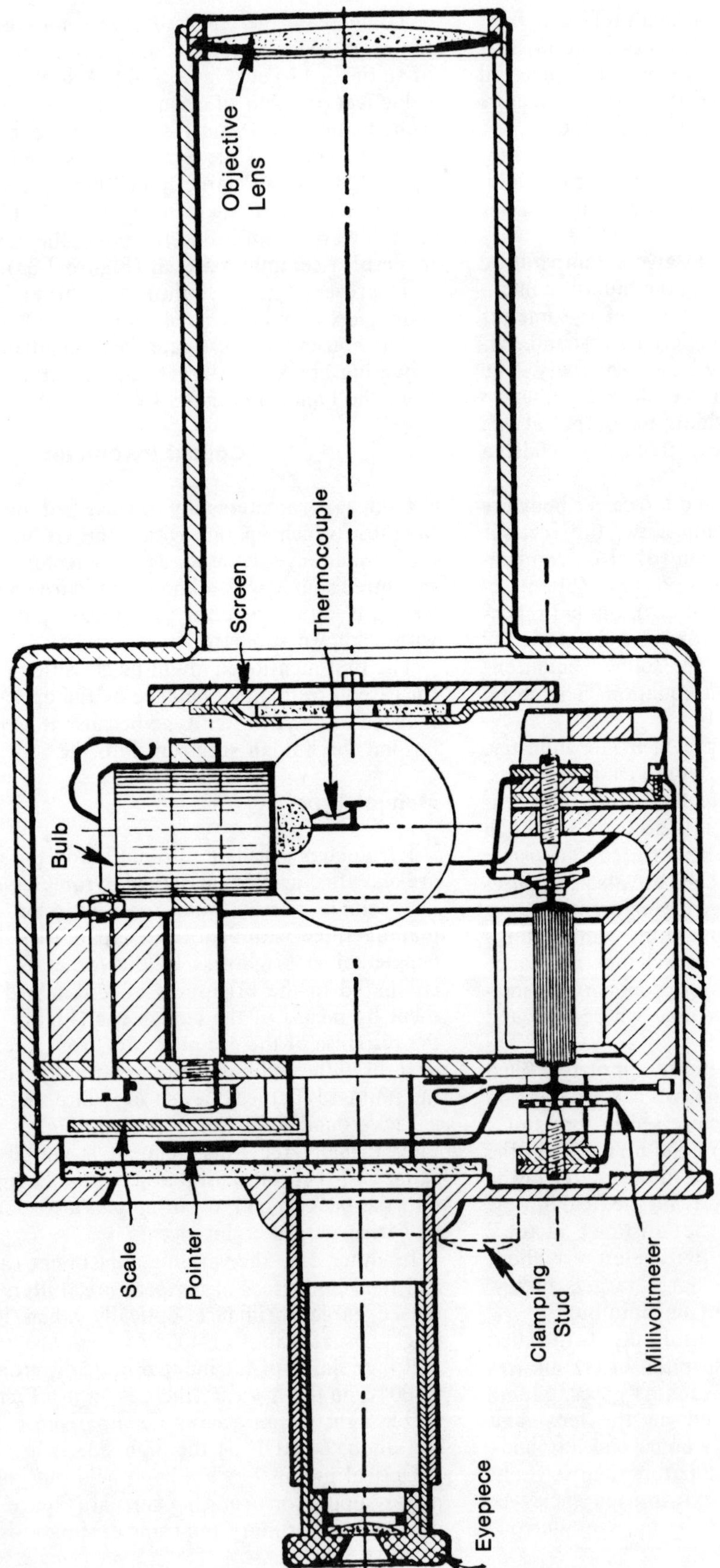

Figure 7.31. Radiation pyrometers give a temperature analog readout by sensing the intensity of radiation emitted because of heat. This cross-sectional view of the PYRO Radiation Pyrometer illustrates the principal working parts. Note particularly the vacuum thermocouple bulb mounted in the optical axis of the instrument and the exclusive PYRO pointer brake device with clamping stud. (Courtesy of Pyrometer Instrument Co.)

Figure 7.32. A sighting tube can be used to decrease the effects of intervening gases, vapors, and fumes. (Courtesy of Honeywell, Inc.)

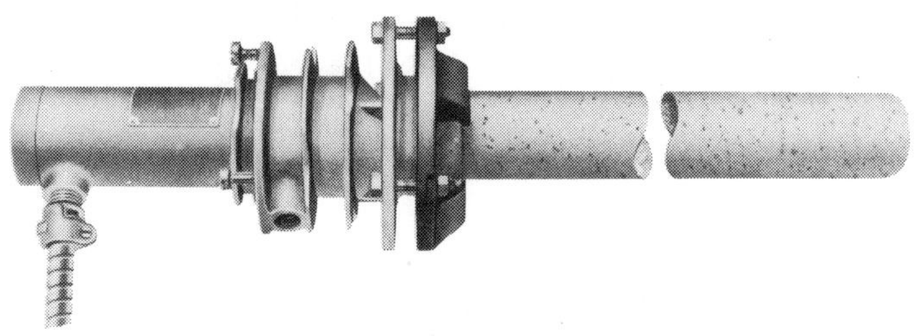

Figure 7.33. Ceramic sighting tubes extend temperature ranges up to approximately 4500°F. (Courtesy of Honeywell, Inc.)

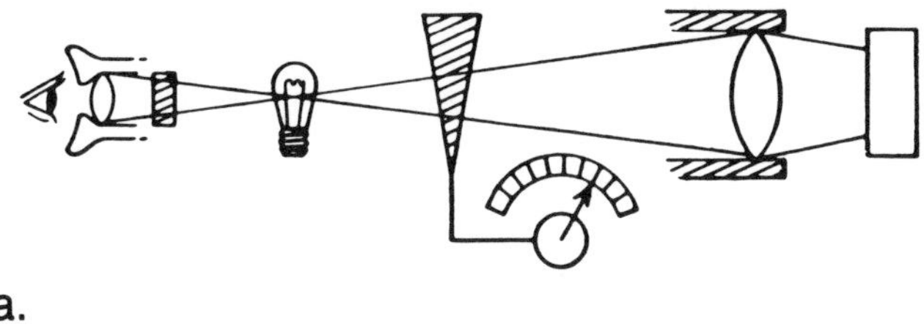

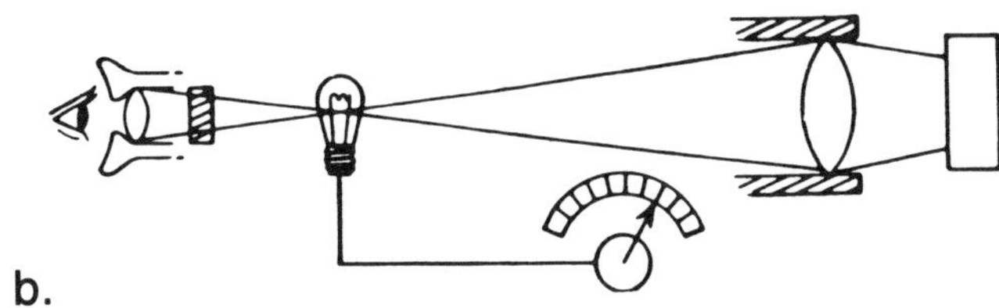

Figure 7.34. (a) Brightness of the source light is varied by an optical wedge, lamp filament temperature remains constant, and unit is self-contained. PYRO Optical Pyrometer operates on this principle. (b) In disappearing filament type, lamp filament temperature is varied by separate potentiometer box. (Courtesy of Pyrometer Instrument Co.)

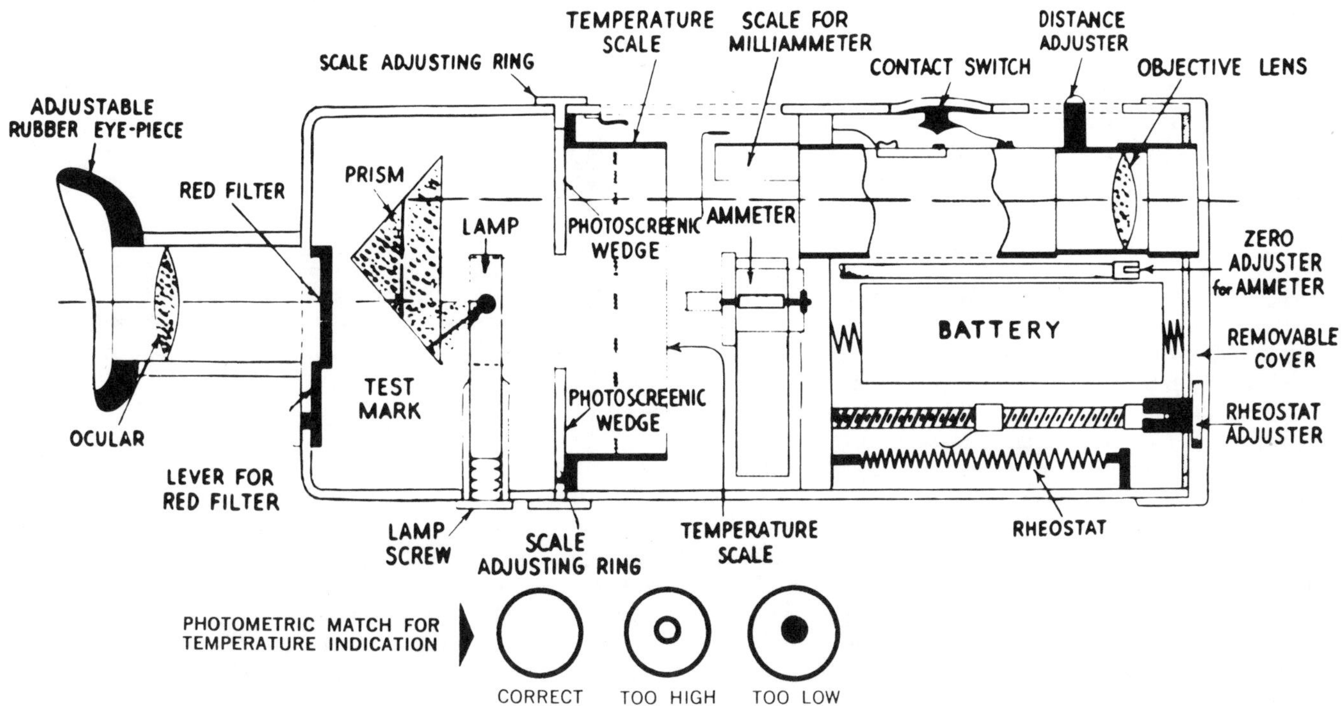

Figure 7.35. Hand-held optical pyrometers are completely self-contained. (Courtesy of Pyrometer Instrument Co.)

Infrared Type

Optical pyrometers can be constructed to utilize an electronic means of comparison rather than using the human eye. These electro-optical devices can be made to sense frequencies both within and outside of the visible spectrum. With the use of optical filters, the sensitive span can be narrowed and placed within the sensor range at a convenient band. When the frequency of the radiation being sensed is such that wavelength is between 8,000 Angstrom units (Å) and 10,000,000 Å (375×10^{12} Hz and 0.3×10^{12} Hz, respectively), the sensor is said to be operating in the infrared range, which is below the range of the human eye (4,000 to 7,000Å). Frequency and wavelength are related by the following equations:

$$\text{Wavelength (meters)} = 300{,}000{,}000 \,/\, \text{Frequency (Hz)} \tag{7.9}$$

and

$$\text{Wavelength (Å)} = 3\,(10^{18}) \,/\, \text{Frequency (Hz)} \tag{7.10}$$

The sensor is alternately exposed first to a calibrated source and then to the target by a rotating slotted disc known as an optical chopper. The difference of the two signals is amplified and read out in temperature units. Some types of infrared detectors are visually sighted through a dichroic mirror, which reflects incoming infrared onto the sensor while passing visible radiation to the eyepiece.

Infrared detection allows measurement of temperatures up to about 8,000°F, with accuracies to ±½%. Detected readout is nonlinear, varying with the fourth power in most units.

Equipment portability is a definite benefit, with several manufacturers making hand-held, battery-operated models (Figure 7.36). These portable models can be sighted on distant targets, and used for checking hot spots. They are very easily adapted for checking the temperature of moving objects.

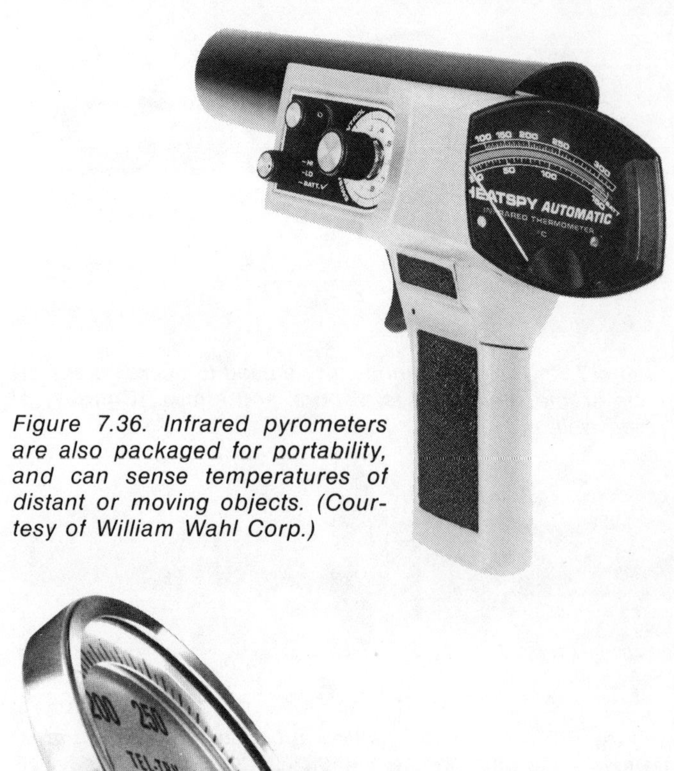

Figure 7.36. Infrared pyrometers are also packaged for portability, and can sense temperatures of distant or moving objects. (Courtesy of William Wahl Corp.)

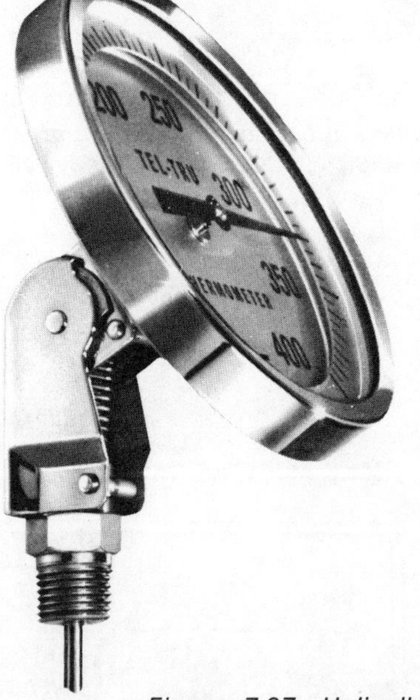

Figure 7.37. Helically formed bimetallic strips allow rugged construction of small stem diameter thermometers. (Courtesy of Tel-Tru Manufacturing Co.)

Bimetallic Thermometers

Bimetal thermometers are used extensively in the processing industries for local temperature indication. They are not as economical from an original cost standpoint as glass stem thermometers. However, because glass stem thermometers are so easily broken, bimetal thermometers are easily justified economically.

Bimetallic thermometers use two fundamental principles: (a) metals expand or contract with temperature changes, and (b) their coefficient of change is not the same for all metals. This difference in thermal expansion rate is used to produce deflections proportional to temperature changes.

The dissimilar metals are bonded together, usually being configured in a spiral like a watch spring or in a twisted helical shape. Because of its ability to be manufactured into a small diameter sheath while maintaining ruggedness, the twisted helical arrangement is used for most process measurements (Figure 7.37). The spiral spring element is often used in ambient temperature measurement devices and air conditioning thermostats (Figure 7.38).

When two dissimilar metals are bonded together, they tend, with increasing temperature, to bend toward the side whose metal has the lower expansion rate. Deflection is proportional to (a) the deflection constant (a function of the

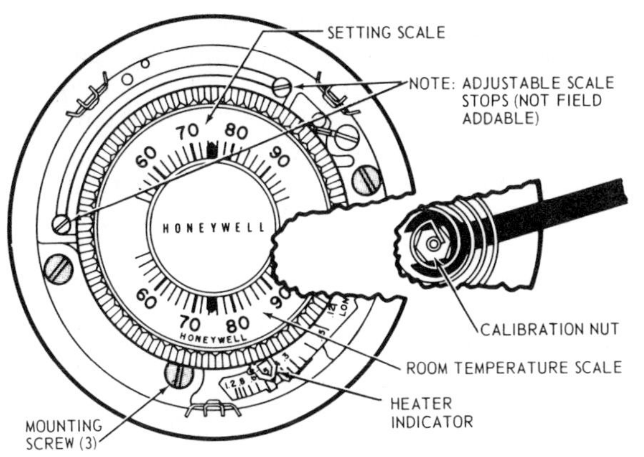

Figure 7.38. A spiral spring-type element is often used in air conditioning thermostats. (Courtesy of Honeywell, Inc.)

two metals), (b) the change in temperature and (c) the square of the length. It is inversely proportional to the metal thickness.

In making bimetallic thermometers, metals are chosen which have widely different thermal expansion coefficients. Invar (an alloy containing 36% nickel and 64% iron) is a favorite low-expansion metal. Its expansion coefficient remains stable over a wide temperature range. Nickel-iron alloys with chromium or manganese added are often used for high expansion coefficient metal.

Bimetallic temperature gauges are furnished in various stem lengths for different immersion depths. Readout dials vary from 2 to 5 inches in diameter. Gauges may be adjustable or fixed. Those that can be adjusted are generally adjustable only with regard to scale position under the pointer. Twisting a stem with a nonadjustable fixed pointer to cause it to align with a calibration point is often sufficient to deform the bimetallic helix to the point where calibration will not be possible at the other end of the dial. Since laboratory bimetallic thermometers are generally accurate to only ±½% and process thermometers to about ±2%, any permanent deformation of the element could make them unusable.

Initial manufacture and calibration usually begins by cutting the bimetallic strip to a previously determined length and attaching it to the stem and to the pointer shaft. Pointer span is noted on a scale having only calibration numbers as the probe is subjected to temperature baths near each end of the desired range. A scale which most nearly matches the determined span is then selected from several of like ranges with different spans. That scale is installed on the gauge and is rotated until the pointer is set at a calibration point and is then secured in place.

Bimetallic thermometers, like thermocouples, are usually mounted in protecting wells. The same consideration concerning wells as noted previously in the thermocouple section applies to thermometers also. Although wells slow response times appreciably, that is usually more desirable than having to shut down and drain a system in order to check or replace a thermometer.

Miscellaneous Methods of Temperature Measurement

Many different methods of temperature measurement have evolved throughout the years as new requirements have presented incompatibilities with older methods. While many are beyond the scope of a text on basic methods for chemical and petroleum industries, a brief discussion of several methods will serve to inform the reader of some of the more applicable techniques used from time to time.

Thermistors

Thermistors are semiconductors made from carbon, germanium, silicon and mixtures of certain metallic oxides that exhibit high temperature coefficients, usually negative.

Characteristically, they are nonlinear and exhibit the greatest changes in cryogenic temperature ranges below 100°K. Their resistance is a function of absolute temperature.

Thermistors were recognized in the early 1800s but their use was first introduced by Bell Laboratories about 1940. Only in recent years have they found general industry-wide acceptance, after one manufacturer patented a process enabling their production on an interchangeable basis. Previously, their production techniques required individual calibration when used in a circuit.

Calibration curves are generally available from the manufacturer (Figure 7.39). Accuracies are dependent on temperature range. One manufacturer guarantees calibration of germanium resistance sensors to ±0.005°K over the cryogenic range of 1.5° to 5.0°K, and to ±0.1°K over the range of 40° to 100°K. Calibration should be checked periodically for sensor aging.

Carbon thermistors are the least expensive ones made, although stability factors sometimes necessitate the selection of other types of sensors.

Readout devices for highest accuracy are generally Wheatstone or other bridge types. More accurate (less than ±0.05°K) readings can be obtained by a potentiometric measurement across a precision resistor in series with the thermistor to form a voltage divider circuit. Large changes in resistance allow long lead wires. Small size (0.1-inch diameter, typically) is often advantageous. One patented thermistor is available with a 0.06°K interchangeability with others over small spans. As for resistance temperature detectors, current through the sensor must be limited so that self-heating will not affect the reading.

In addition to their use as temperature devices, thermistors are used for voltage regulation, power level controls and analyzer detectors.

For temperature measurement their resistance relationship to temperature is very nonlinear, but they are very good for narrow measurement spans because of their high temperature coefficient. They are particularly suitable for measurements in the cryogenic range.

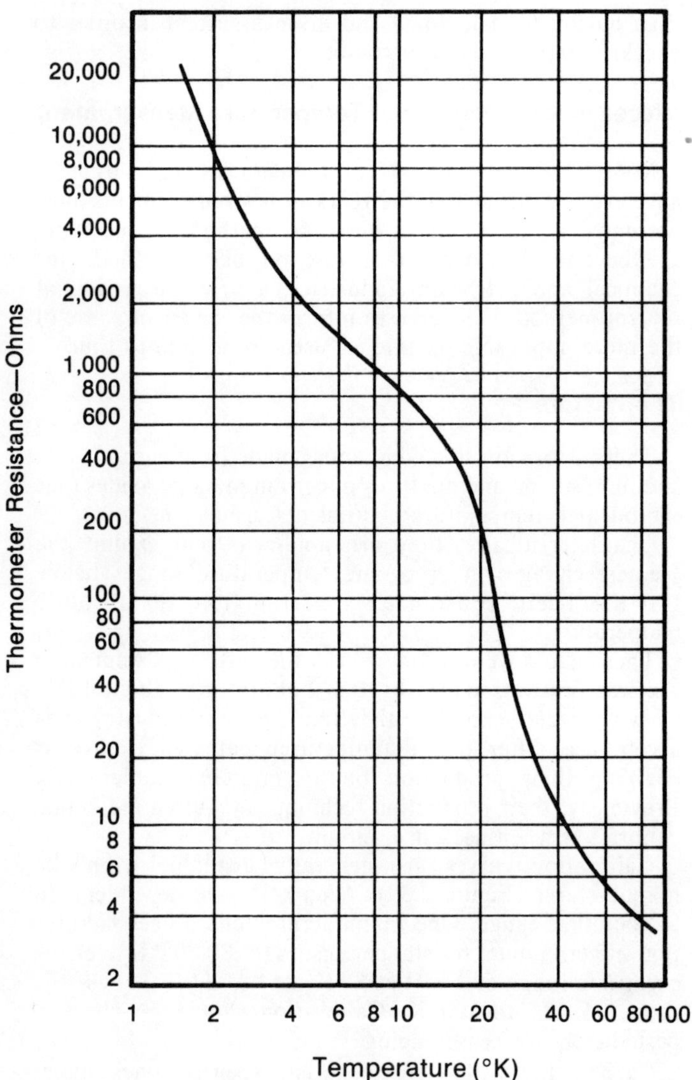

Figure 7.39. A calibration curve of a typical thermistor shows its non-linear negative coefficient.

Color Change

Used as a one-shot, highest-achieved temperature indication, chemicals can be applied to a surface whose temperature is to be checked. As the chemicals reach their temperature of sensitivity, their color changes, generally turning black.

The methods of applying these chemicals vary. A dot, or dots, of the chemicals printed onto adhesive-backed paper can be affixed to the surface, or the chemicals can be applied as paints, crayons or tablets (Figure 7.40). One equipment manufacturer uses an adhesive-backed dot placed in an inconspicuous place as a temperature warranty validation technique. Such chemicals can be made to respond to almost any temperature between 100° and 1,100°F, and 10°F increments are possible. Accuracies are about ±1%.

These type chemicals are not made to show continuous temperatures, but rather to indicate that they have reached some maximum temperature. They are a very inexpensive means of checking maximum bearing housing temperature and other equipment failures due to high temperatures.

Thermographic

While most thermographic temperature measurements have been used in medical applications to date for locations of possible cancerous conditions, industry may well profit eventually from this method. A thermally sensitive chemical is painted over an area to be checked. When temperature stabilizes, the area is photographed under ultraviolet light, showing temperature profiles as bands of colors.

Infrared photography has likewise been used to make records of hot components on circuit boards and hot outfalls into lakes and rivers for pollution detection and to show healthy vegetation for farmers and ecologists.

Acoustical Thermometry

One of the most interesting principles used for temperature determination is its relationship to velocity of sound in a medium to be measured. Acoustical techniques have been applied to measurement of temperatures from 30,000°F plasma arcs to near 0°K cryogenic processes, as

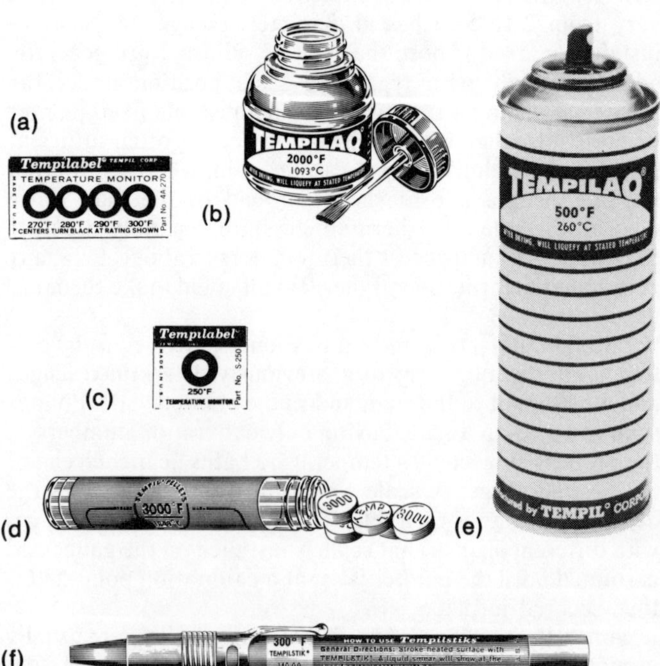

Figure 7.40. Temperature sensitive chemicals for one-shot highest-achieved temperature checks are available in many forms. (Courtesy of Tempil Division, Big Three Industries)

well as nuclear reactor beds, molten glasses and metals and oceanographic and high-altitude studies. Acoustic waves can be sent through a material in thermal equilibrium with the medium or directly through the medium itself (such as through the ocean).

While this method of temperature measurement is expensive and complicated, it does yield other physical properties and measurements simultaneously as byproducts of temperature measurement. They include pressure, viscosity, flow velocity, laminar or turbulent flow mode, boundary layer thickness and several other properties.

An allied type of device measures the resonant frequency of waves within a cavity. Resonant frequency varies with temperature. Called a fluidic oscillator, the device was originally designed for the supersonic X-15 aircraft. Temperature measurement ranges from 0° to 4,500°F with ±2% accuracy.

Pyrometric Cones

Ceramic industries use geometrically shaped, slender cones of certain ceramic compositions to determine kiln temperatures. As temperature rises, the cones soften and deflect. They may be used singly or in groups such that each responds or softens at a different temperature. The numbered cone whose tip last bent all the way down to touch its resting surface indicates kiln temperature to within a few degrees.

This type indicator is not used as an exact temperature measurement but rather to determine the end point for a particular application.

Quartz Crystal Thermometers

Quartz crystals have been used in frequency generating circuits for many years. Their frequency of resonance is, for the most part, a function of the geometry of the crystals. It also varies quite linearly with temperature. For that very reason, commercial broadcasters are required to maintain their crystals in closely controlled temperature ovens. This latter feature can be and is used for temperature measurement. Cutting the crystal to a "Y" pattern yields a high degree of linearity.

Quartz has a positive temperature-frequency coefficient, and linearity is on the order of ±0.05% over a range of −440° to +500°F. Resolution of measurement can be as close as 0.00018°F. Above 1,000°F, quartz does not self-oscillate. Within its operable range, repeatability, stability, freedom from self-heating and interchangeability are distinct advantages over common RTDs and thermocouples. Quartz crystal thermometers are accurate and rugged. But they are expensive, and their use is usually confined to laboratory environments.

Laboratory Experimental Techniques

Some temperature measuring devices, while ideal for certain uses, are confined to laboratory usage because of cost, complexity or fragility. Some of these are listed below.

1. A device using a paramagnetic salt, whose electromagnetic susceptability changes below 2°K, is a very complex device. This thermometer is one of the few types available for use at this low cryogenic temperature.
2. With a noise thermometer, the thermal noise voltage generated in a thin (0.01-mm diameter) platinum wire is measured as a temperature analog. Range is about 0° to 1,000°C.
3. Nuclear quadrupole resonance is used in the 20° to 300°K range. This device actually measures the frequency of a potassium chlorate sample placed within the tank circuit of an oscillator.
4. Other acoustical techniques, not mentioned in the previous paragraphs, are those which involve sonar-like pulses down a wire sensor, ultrasonic thickness gauging equipment, boiling and melting point changes, refractions, pneumatic probes, Poisson's ratio and combinations of these and other principles.

8 

Fred E. Edmondson, Gareth Eugene Ivey, William G. Andrew

To "weigh" an object means to measure the force of gravity on its mass at a particular location. The relationship that exists among weight, force and mass is expressed by the equation:

$$W = mg$$

where:

W = weight
m = mass
g = acceleration due to gravity

This is an expression recognizing Newton's fundamental law of gravitation which points out that all material objects attract all other material objects. Thus, everything on or near the earth's surface is attracted toward the earth's center. This attractive force of gravity has been termed "weight."

When dealing with quantity measurements in the process industries, mass is usually the real value being considered even when the measurement is in pounds, gallons, cubic feet or other volume measurements. In the case of weight measurement the interest is in mass rather than the force by which that mass is attracted to the earth. However, the acceleration due to gravity varies by such a small amount over the earth's surface under usual conditions that it is considered constant and weight values are used instead of mass. It may be noted that the terms *weight* and *force* are used interchangeably.

In many manufacturing operations and in most processing systems, flow measurements are used to define and control the transfer of mass, especially in gas, vapor or liquid form. This is not an arbitrary selection, but one that has evolved because of convenience and economics. The measurement of solids flow by conventional flow techniques is not easy; therefore, "Weight" methods are used to determine mass transfers of solids, as well as other difficult-to-measure liquid and slurry flows.

Weighing techniques that are available and that are used by industry provide very accurate methods of control. Generally they are more accurate than the commonly used flow devices. They possess one fundamental advantage over volume measurements—a freedom from temperature effects.

Weigh devices which are discussed in this section include the mechanical lever type which dates back thousands of years, the spring balance type which dates back hundreds of years and, more recently, hydraulic load cells, pneumatic load cells and electric strain gauges.

Mechanical Lever Scales

By far the largest number of weighing installations are based on mechanical lever systems. They are used for almost all types of weighing applications. They are very accurate, reliable and relatively economical to install. Maintenance, however, can be rather expensive.

Most mechanical lever scales are used for indication or printout of weight at a location on the scale. They can be equipped with transmitting devices, but other methods should be considered first when signal transmission is needed because their outputs are more easily adapted to standard signals.

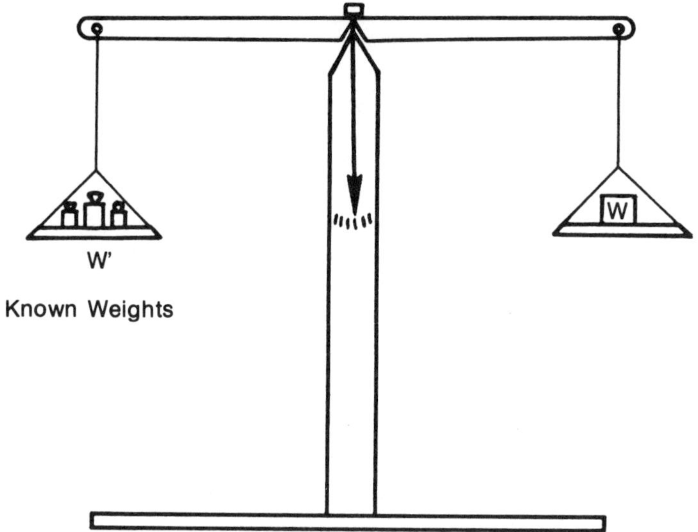

Figure 8.1. The even arm balance is the oldest known weighing method.

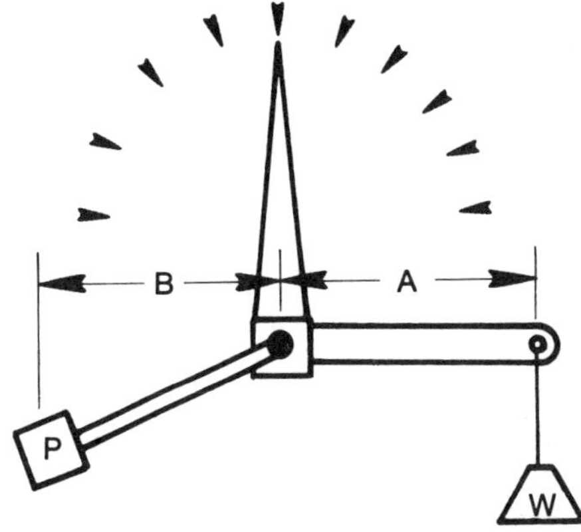

Figure 8.3. A pendulum scale is another example of the mechanical lever scales where weight times lever arms are balanced.

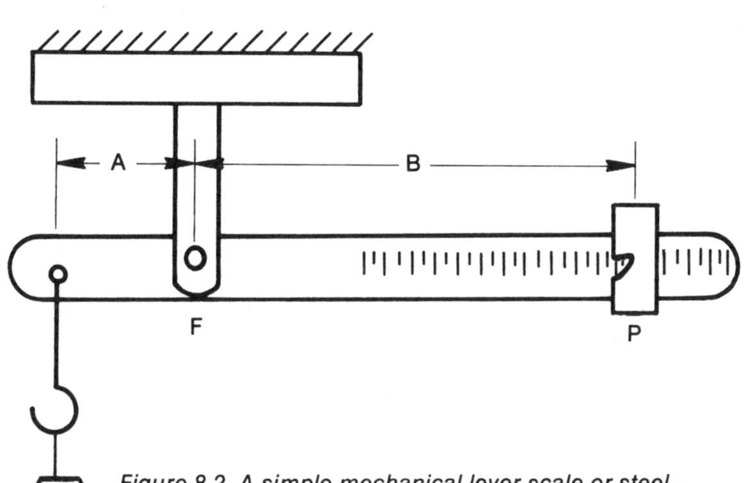

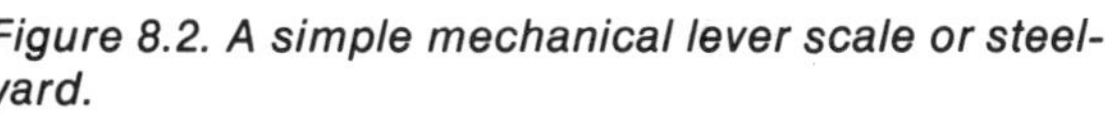

Figure 8.2. A simple mechanical lever scale or steelyard.

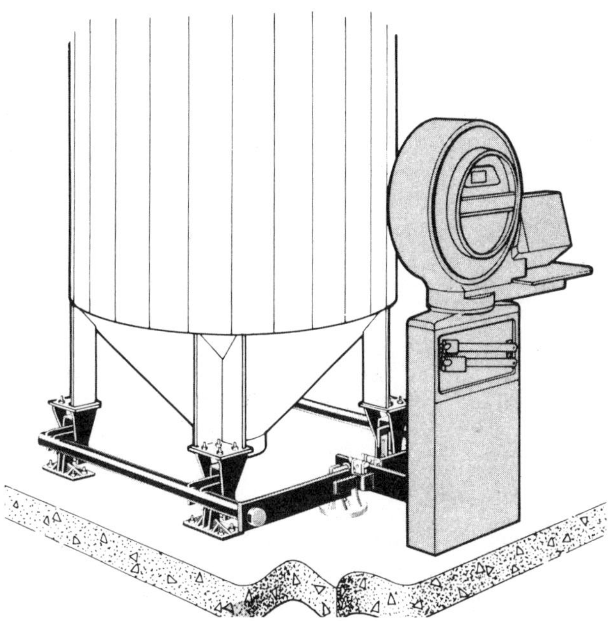

Figure 8.4. A platform or mechanical lever scale using simple levers with pendulum counterbalance operating an indicator. (Courtesy of Toledo Scales, Division of Reliance Co.)

The oldest known weighing device, and still the most accurate, is the even arm balance as shown in Figure 8.1. The mechanism is designed such that the two pans are level with no weight on either pan. An unknown weight (W) is placed on one pan and is balanced using very accurate weights (W^1) on the other pan. The total of the known weight (W^1) is equal to the unknown weight (W). This scale is used for weighing small quantities very accurately.

The simplest mechanical lever scale is the steelyard (Figure 8.2). An unknown weight (W) positioned at a known distance (A) from the fulcrum point (F) of the lever is balanced by a known weight (P) at an adjustable distance (B) from point F. For the unit to be in balance. $WA = PB$. Since the distance (B) is the only variable factor in the equation, it is directly proportional to weight (W). In scales of this type, B is calibrated directly in weight units.

A pendulum scale (Figure 8.3) may be employed to indicate weight on a graduated scale. Here, the same relationship, $WA = PB$, exists as described above. However, B increases as the unit rotates until it comes into balance. The distance (B) varies proportionately to the sine of the angle of rotation of the scale beam. The dial may be calibrated to read weight directly.

In actual practice, most mechanical lever scales are a combination of simple levers, with the final indication utilizing the adjustable poise as does a simple steelyard or an indicating dial as for the pendulum scale.

Figure 8.4 illustrates a platform scale which uses simple levers for transmission of the weight to the pendulum

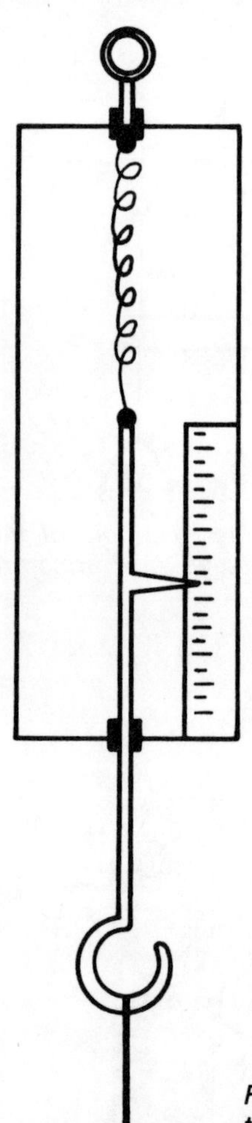

Figure 8.5. Spring scale demonstrates the principle of the elongation of an elastic element which is proportional to the attached weight.

counterbalance unit which, in turn, operates the weight indicating pointer. The poise weights are used to null out any tare weight, such as the weight of the container which holds the material to be weighed.

Mechanical lever platform scales are available with manually operated steelyard beam balances or with direct indicating dial heads. They can be furnished with various combinations of card and tape printers. Some printers require outside power sources. They are usually mounted in a scale pit whose depth varies with scale size. The scale pit should be well drained. If spillage of hazardous materials might occur, adequate natural or forced ventilation may be needed.

When heavy vehicles may enter the scale platform, the entry area should be prepared to avoid heavy shock or impact loading of the platform. The useful life of the scales is seriously affected by high impact loads on the lever knife edges and pivots. Heavy angle iron or other suitable reinforcement placed on the same level as the platform is effective in reducing shock loading.

Mechanical lever scales find application in processing plants for such uses as hopper scales and tank scales as well as other applications. Although they utilize the basic lever principles described previously, years of development by various manufacturers have resulted in many combinations of load support arrangements.

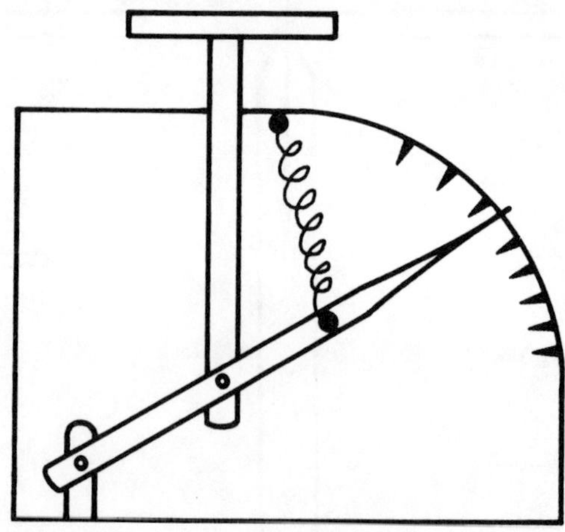

Figure 8.6. The combination of levers and springs is used to demonstrate the spring balance scale.

Although mechanical lever scales are high maintenance items, maintenance is easy to obtain because of their basic simplicity. The principle upon which they operate includes the use of knife edges and pivot bearings (or their equivalent) for lever support. They must be designed for low frictional losses to ensure accuracy. They must be designed for high impact loading and rough usage. However, they must still withstand rust, corrosion and functional wear while maintaining accuracy and dependability. They may be subjected to vibration. Since mechanical scales are characterized by low natural frequency, vibration or dynamic loading conditions may cause a resonance condition destructive to the scale mechanism.

Spring Balance Scales

Spring scales are based on the principle that the deflection of an elastic member is directly proportional to the force applied to the member. Thus, a simple scale may be constructed as shown in Figure 8.5. When weight (W) is suspended from the spring, the spring elongates by an amount proportional to W. The indicator scale may be graduated in weights units and the weight read directly at the pointer.

By combining simple levers with the spring, a simple weight indicator may be constructed as shown in Figure 8.6. By combining with a rack and pinion, a circular dial indicator can be added to the unit.

As the temperature of a spring is increased, the length of the spring increases and the modulus of elasticity is decreased. This causes a zero shift and an error in the deflection per weight unit. In commercial weighing devices, this error is compensated by using special spring alloys and compensating springs.

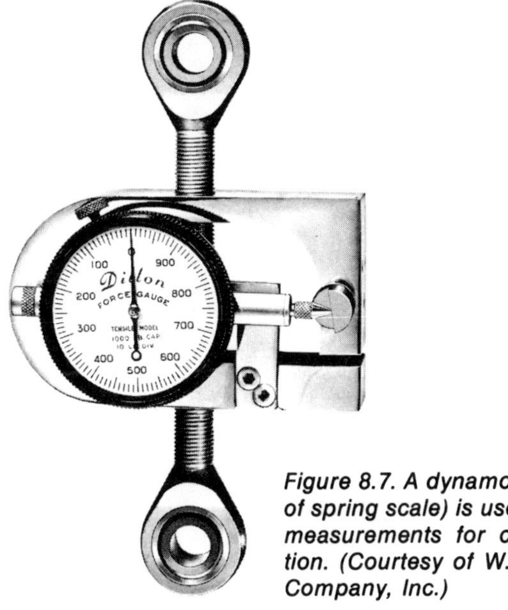

Figure 8.7. A dynamometer (a type of spring scale) is used for tension measurements for cable installation. (Courtesy of W.C. Dillon and Company, Inc.)

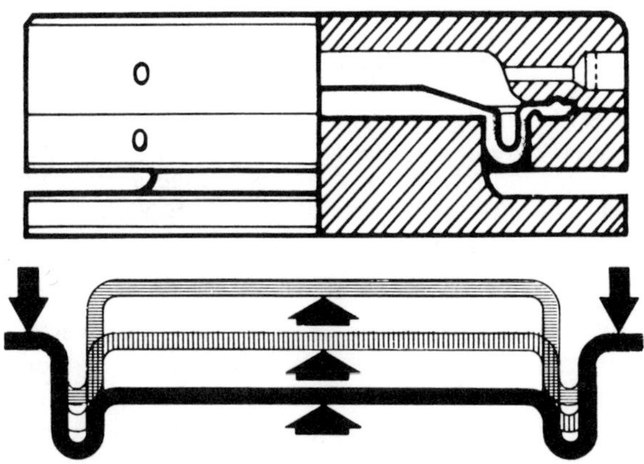

Figure 8.8. This hydraulic load cell uses a rolling diaphragm that eliminates friction present in a conventional piston cell. The lower drawing shows the diaphragm in three positions. (Courtesy of Martin-Decker Corp.)

All spring weighing devices are accurate only to the proportional limit of the spring. This means that when a spring scale is overranged, it does not return to its original zero, and a permanent error is exhibited by the scale.

Scales of the types illustrated in Figures 8.5 and 8.6 are used for quick approximate weighing applications where an expensive scale is not necessary.

Another type of spring scale known as a dynamometer, used primarily for measurement of tension in a cable, is illustrated in Figure 8.7.

Strain gauge load cells use spring mechanisms with the deflection indicated by the change in electrical resistance of a wire grid attached to the spring. The deflection measured may be the elongation or compression of a column or the elongation of a side of a beam due to bending. These are

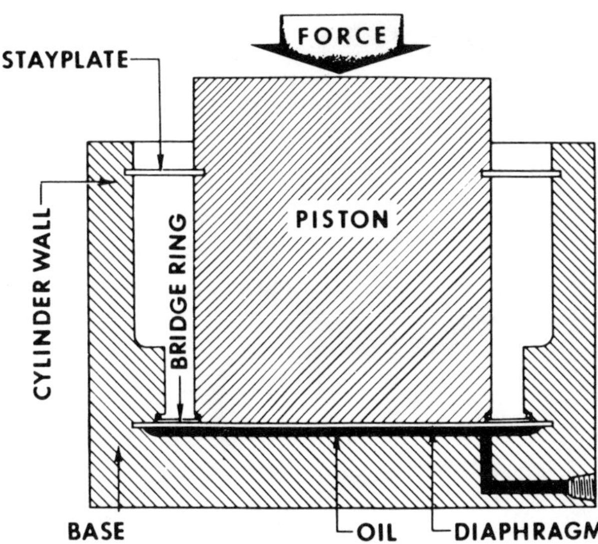

Figure 8.9. This hydraulic load cell consists of a piston positioned centrally within a cylinder by means of a stay plate at the top and a bridge ring at the bottom. A fluid approximately .030 inch thick is sealed beneath the piston by a stainless steel diaphragm. (Courtesy of A.H. Emery Co.)

both variations of the spring balance principle. They are described later in this chapter.

Spring balance scales are best suited for relatively light loads and are an economical approach to accurate weighing.

Hydraulic Load Cells

Hydraulic load cells operate on the principle of a force counterbalance. They are constructed such that the weight to be measured is supported by a piston, which in turn is supported by hydraulic fluid confined within a sealed chamber. Figures 8.8 and 8.9 illustrate two types currently being used. Figure 8.8 shows a type that utilizes a rolling diaphragm that eliminates friction. The nature of its design provides self-aligning features. Figure 8.9 shows a load cell with a centrally positioned cylinder by means of a stay plate at the top and a bridge ring at the bottom to provide lateral stability and resist torque movements. Very little vertical movement (0.030 inch) is required for full-scale deflection. In both types the pressure exerted in the hydraulic fluid chamber is proportional to the force exerted on the weight-supporting piston.

$$P = F/A$$

where

P = pressure, pounds per square inch
F = pounds
A = area, square inches

Since the area of the piston is fixed, the hydrostatic pressure of the fluid is directly proportional to the force.

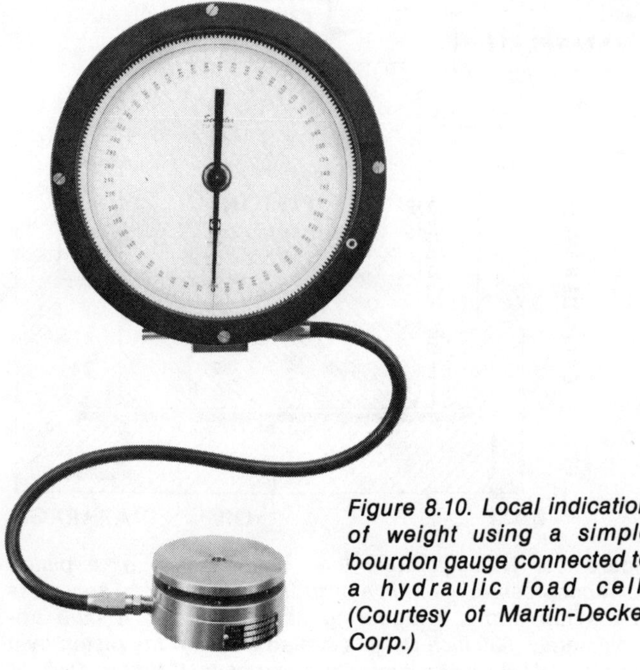

Figure 8.10. Local indication of weight using a simple bourdon gauge connected to a hydraulic load cell. (Courtesy of Martin-Decker Corp.)

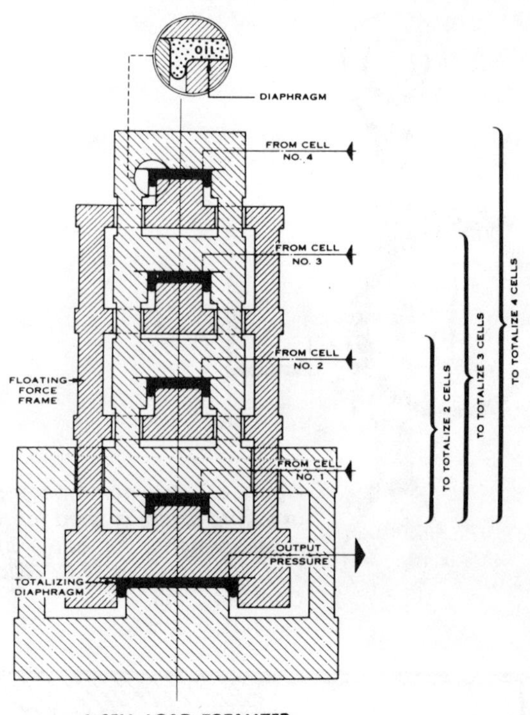

Figure 8.11. Totalizing loads from several points is accomplished by applying the force from each individual load cell to a floating frame which exerts pressure on the single totalizing chamber. (Courtesy of A. H. Emery Co.)

However, to obtain the weight of the product, the system must be zeroed with no product being measured. This compensates for the weight of the supporting structural members, tanks, load transmitting members, etc., and any hydrostatic pressure of the system due to the weight of the hydraulic fluid. Any alteration or movement of any component of a hydraulic weighing system necessitates rezeroing of the system.

The hydraulic pressure may be indicated by connecting to a conventional bourdon tube pressure gauge which has a readout scale showing weight units. It may also be connected to a conventional pressure transmitter which allows the value to be converted to any desired signal range for input to a recorder, controller, computer, etc.

Hydraulic load cells are relatively trouble-free over extended periods of use. They require no outside source of power for indication only, respond quickly to load changes, are relatively insensitive to temperature changes and, within limits, are unaffected by the amount of hydraulic filling. This permits use of flexible hoses in transmission, if they are desired. It should also be noted that these systems are well suited for use in hazardous areas since they require no electrical power.

Any leak in a hydraulic system quickly causes the unit to be inoperative. Leaks must also be considered as possible contamination sources of the product being weighed.

Figure 8.10 shows a simple hydraulic load cell application with the hydraulic system connected to a pressure indicator graduated in weight units.

When it is necessary to use several weigh cells to obtain the total weight of a structure or tank being supported at several points, the outputs of the cells may be added by using a hydraulic totalizer as shown in Figure 8.11. The outputs from four cells are connected to the totalizer input connections where the pressure of each input is converted to a force on the floating force frame. This force is then converted to a hydraulic pressure in the totalizing chamber and may be indicated or transmitted as previously described. When using a totalizer, indication from each individual cell is desirable. Otherwise, a cell may develop a leak which would be difficult to detect.

Hydraulic load cells are well suited for high-impact loads. They can withstand high overloads without loss of accuracy or zero shift. Temperature change effects are small, although well-designed load cells do provide temperature compensation for both span and zero effect. Most manufacturers specify standard operating limits of 0° to 120°F as a basis for performance guarantees. Temperature correction charts are available for reference when operating outside these limits.

Hydraulic load cells have a high natural frequency and exhibit a fast response rate. Therefore, they are adaptable to torque measurement systems requiring fast dynamic responses.

Pneumatic Load Cells

Pneumatic load cells are used primarily on continuous feeders but may be used for practically any weighing application where the maximum load per cell is no greater than 9,800 pounds.

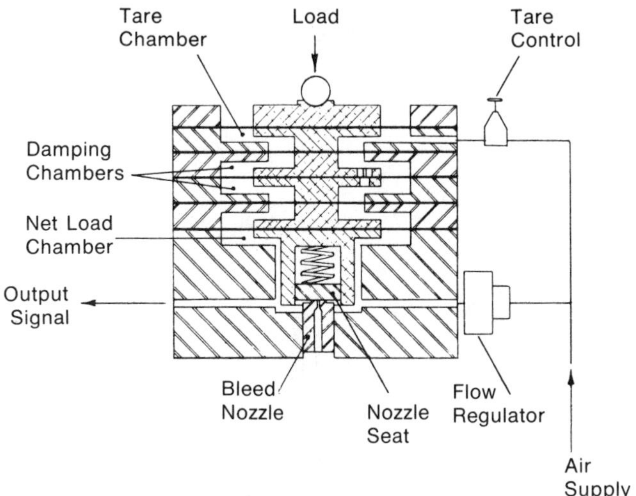

Figure 8.12. Pneumatic load cell works on force balance principle: increasing load restricts flow through bleed nozzle, builds up signal output pressure in proportion to loading. (Courtesy of K-Tron Corp.)

Figure 8.12 illustrates the pneumatic load cell, or force transmitter, used in many weight measurement applications. This cell operates on the force balance principle. Pressure under the net load diaphragm is regulated automatically to oppose the force placed on the unit. An increase of force on the "load" point moves the nozzle seat toward the bleed nozzle, restricting the air flow through the nozzle. The air pressure in the net load chamber increases until the air pressure times the area of the diaphragm equals the force applied. The nozzle seat is repositioned relative to the nozzle such that the normal air bleed is established and the unit is in balance. If the force is decreased, the seat is moved away from the nozzle, the pressure is bled from the unit weight chamber until the nozzle seat is repositioned and the unit is again in balance. The flow regulator controls the flow through the bleed nozzle by maintaining a constant flow regardless of the differential pressure.

The dampening chambers eliminate pulsation and average out rapid variations in load.

The tare weight chamber is loaded by an external regulator to zero out static weights or forces not desired in the measurement.

The pneumatic output signal is the same pressure as is required to balance the force on the cell. It may be indicated by a conventional pressure gauge or may be used as the input to a pressure transmitter for conversion to any type of standard signal. The scales of indicators may be marked to read directly in weight or force units as required.

Cells of this type are available in ranges of 0-7 to 0-2,450 pounds with a standard 3 to 15-psig output signal. By using nonstandard signals, they are manufactured with a maximum load rating of 9,800 pounds with a 51-psig output signal.

When greater capacities are desired, or several cells are required, the outputs of all the cells may be totalized in a hydraulic totalizer as described in the previous section. If the signals are maintained in the 3 to 15-psig range, they may be averaged using an averaging-scaling relay.

Pneumatic load cells are suitable for use in hazardous or explosive areas, require no special sealed transmitting systems, and cannot cause product contamination due to leakage of hydraulic fluid. They are relatively free from temperature related errors. The air or other gases used as the operating medium must be dry enough to prevent condensation of water vapor and subsequent freezing at temperatures below 32°F. They respond slowly to sudden load changes and therefore are not generally used for test work. However, this characteristic is desirable for weighing moving materials, such as material on a belt conveyor, or in weighing a vessel which is being filled by falling materials.

Electric Load Cells

Strain Gauge Cells

Strain gauge load cells used in industry are constructed of wire grids known as strain gauges, bonded to precisely machined supporting columns as shown in Figure 8.13. These grids are connected electrically in a Wheatstone bridge circuit as shown in Figure 8.14.

When a load (Figure 8.13), represented by *F*, is applied to the supporting column, the column is compressed, causing the wires in the grids bonded to sides *X1* and *X2* to decrease in length and increase in cross-sectional area, decreasing their electrical resistance. The grids bonded to sides *Y1* and *Y2* are virtually unaffected by the compression of the column. These grids are attached to the column to minimize errors due to temperature variations.

When the column is stressed in tension, the measuring grid wires are lengthened and their resistance is increased.

The unbalance in the bridge circuit produced by the change in resistance is directly proportional to the load *F* applied to the column, providing a linear output signal with respect to load.

Strain gauge load cells are made for either compression, tension, or universal loading of either type. Universal load cells are manufactured such that mounting hardware for tension measurements can be threaded into the ends, in place of compression loading caps. They are also available with two sets of strain gauge elements for "double bridges," supplying output signals to two separate readout systems.

Output signals of strain gauges are relatively small and are proportional to the excitation. Common values range from 1 to 3 millivolts per volt of excitation. Excitation voltage can be AC or DC and may range from 5 to 25 volts. The design of the excitation power supply, amplifiers and other backup equipment is important to the successful application of strain cells to weighing systems.

Strain gauge cells should be protected from angular or nonaxial loads as they have no way to discriminate between bending and axial loads. Special mounting methods can be used when nonaxial loading is a problem. Overloads greatly

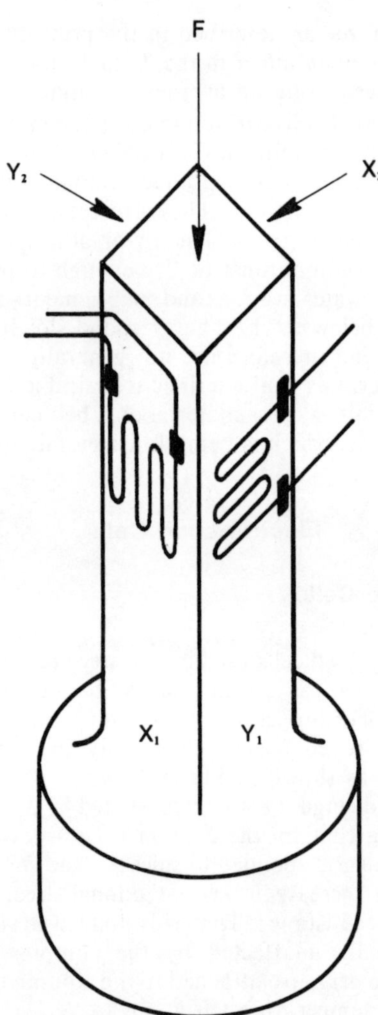

Figure 8.13. Strain gauge cell support column with bonded strain gauges X1 and X2 for weight measurement and Y1 and Y2 for temperature compensation.

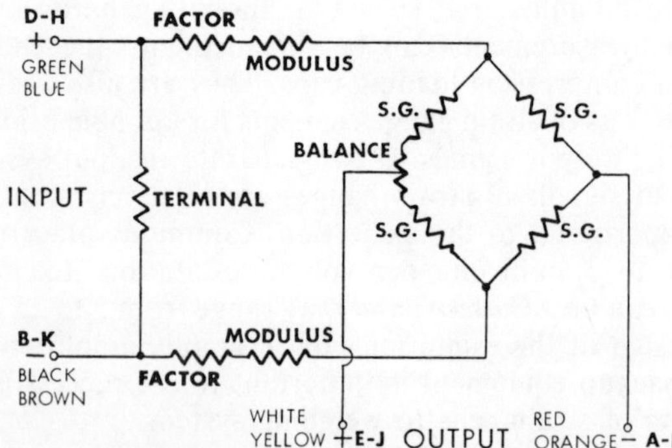

Figure 8.14. Load cell wiring schematic uses familiar Wheatstone bridge measuring circuit with strain gauges as arms in bridge. (Courtesy of Toroid Corp.)

Figure 8.15. Strain gauge load cells are compact, accurate and ideally suited for electronic measuring systems. (Courtesy of Toroid Corp.)

in excess of their rating should be avoided. Generally it should not exceed 125% of rated load. In special models, however, they may be overloaded from 150% to as much as 500%.

Strain gauge load cells are particularly well suited for measurements where an electrical output signal is desired. They are small in size, compact and measure up to rated capacity with deflections in the range of .005 to .01 inches. They respond rapidly to load variations. They are relatively maintenance free and, if hermetically sealed, may be installed in practically any environment.

Although strain gauge load cells themselves are relatively inexpensive, the related electrical equipment required for remote readout may be rather costly. Their operational temperature limits range between approximately 15° to 115°F without special compensation circuitry. They are limited to maximum temperatures of approximately 275° to 300°F.

Strain gauge load cells are provided with temperature compensation for zero shift and span. It is accomplished by making the strain wires out of temperature-insensitive alloys and using compensating resistors in the measuring bridge network. The output of several cells may be added electrically to obtain the total weight or force exerted when it becomes necessary to support the load with multiple load cells.

The compact design of strain gauge cells is shown in Figure 8.15. This unit with a capacity of 3,000 pounds is approximately 4½ inches high and 2¾ inches in diameter. A 250,000-pound capacity unit is approximately 22 inches high and 10⅛ inches in diameter.

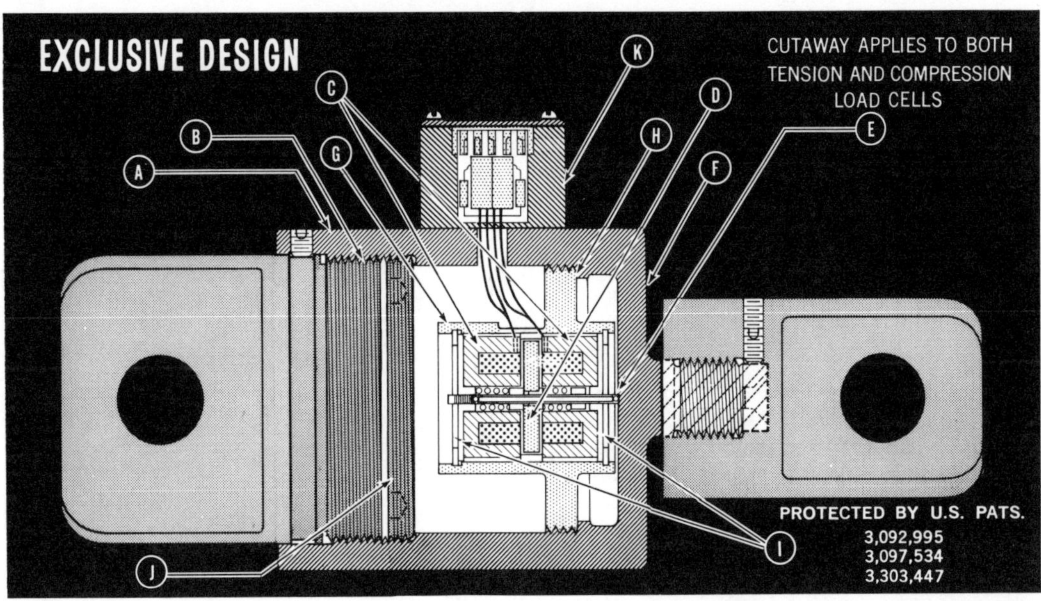

Figure 8.16. Induction electrical load cell utilizes changed inductance as load deforms cell dome, moving armature away from stationary inductor coils. The cell consists of a stainless steel case (a)*, internal threads* (b)*, two ferrite inductors* (c)*, armature* (d)*, armature shaft* (e)*, diaphragm* (f)*, sleeve* (g)*, shoulder* (h)*, end plates* (i)*, threaded plug* (j) *and saddle* (k)*. (Courtesy of W.C. Dillon and Company, Inc.)*

Induction Cell

A load cell constructed to use ferrite inductors and a movable armature for a motion sensing element is illustrated in Figure 8.16. The case (*A*) of the cell is the load bearing member of the unit. It is so constructed that the amount of deformation of the cell dome (*F*) is proportional to the load applied in either tension or compression.

Figure 8.17. Induction load cells range in size from 1⅝ inch diameter by 2½ inches long for 600-pound capacity to 6¾ inch diameter by 10¾ inches long for 300,000-pound capacity. Cell shown is for tension measurement. (Courtesy of W.C. Dillon and Company, Inc.)

The movement sensing mechanism of the cell consists of two ferrite inductors *(C)*, each containing a single coil, and an armature and shaft assembly *(D)*. The end of the armature shaft is in contact with the dome at point *(E)*. As a load is applied, the dome is deformed slightly and the armature position relative to the inductors is changed. When the inductor coils are excited by a low voltage, high frequency electrical input, the output voltage of the cell varies directly with the change in the position of the armature. The output, therefore, is directly proportional to the load when allowances for tare weight are considered. The input voltage may range from 5 to 150 millivolts or from 0.105 to 1.5 volts. The deflection of the dome is 0.003 inch at full load.

These cells were developed in an effort to produce load cells which would be free from drift due to temperature changes and from creep due to stress changes in the electrical sensing unit and which would be unaffected by humidity changes.

Because they are designed to be intrinsically safe, they are suitable for use in hazardous areas. The electrical signal may be converted into any standard signal desired for transmission or control.

A tension cell is shown in Figure 8.17. It is compact, small in size, ranging from approximately 1⅝ inch diameter and 1½ inches long for a 600-pound capacity cell to 6¾ inch diameter and 10¾ inches long for a 300,000-pound cell, exclusive of end connection fittings.

Belt Conveyor Weighing Systems

The weigh devices and weigh systems described previously are used to weigh both liquids and solids. Either of them

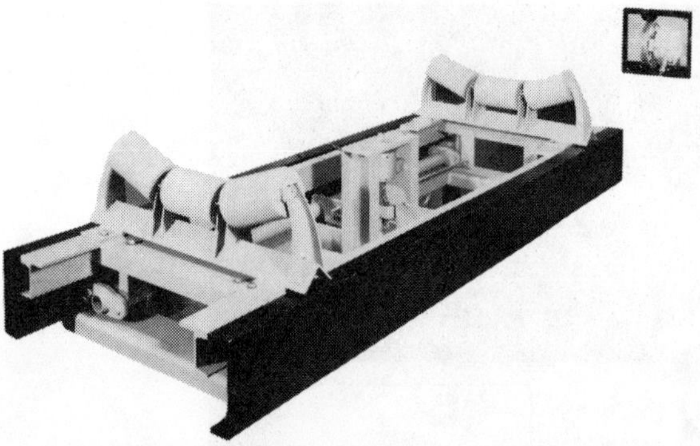

Figure 8.18. A continuous weigh device is obtained by weighing a known section with a load cell as it travels by and integrating the output. When speed variation exists, speed transmitters are used to modify the output accordingly. (Courtesy of Toledo Scales, Division of Reliance Electric Co.)

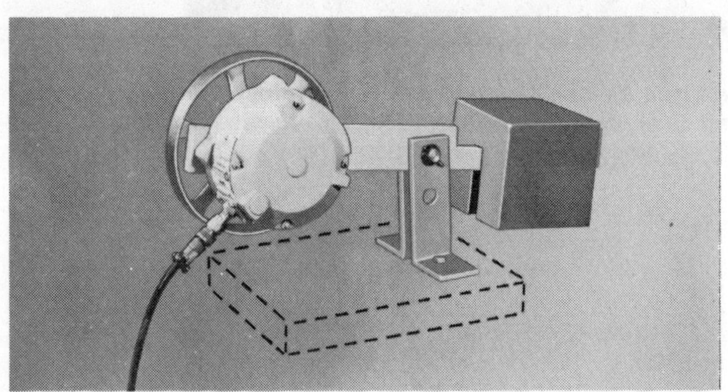

Figure 8.19. A belt speed transmitter for use in continuous weighing systems when speed variations occur. The roller is held in constant contact with the belt to ensure accuracy. (Courtesy of Toledo Scales, Division of Reliance Electric Co.)

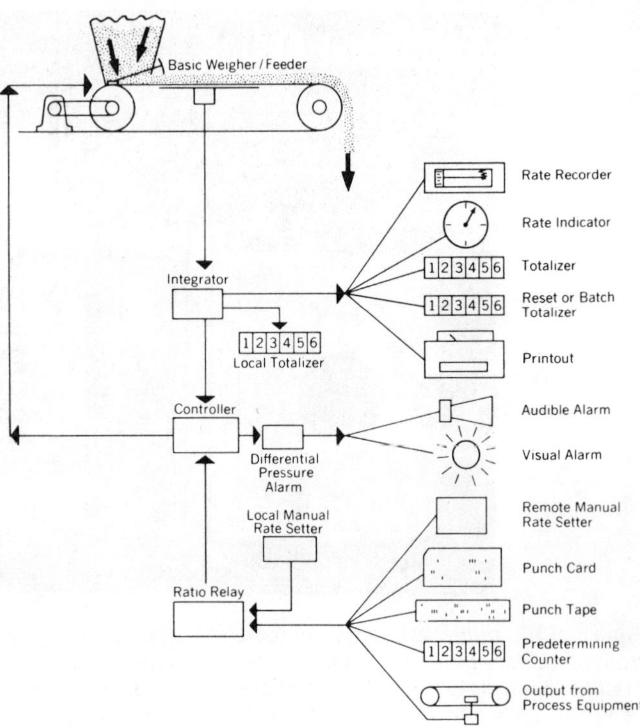

Figure 8.20. Gravimetric feeder that employs belt load control to maintain constant feed rate. Various types of auxiliary functions are shown. (Courtesy of BIF, a unit of General Signal Corp.)

can be weighed easily on a batch basis by the methods described. Liquids can be weighed easily on a continuous basis with some of the systems presented. Continuous liquid weighing measurements are usually accomplished with load cells—pneumatic, hydraulic or electric. The continuous weighing of most solid streams, however, requires a different approach.The most common method is one that weighs a section of a conveyor belt and utilizes that measurement together with the belt speed to obtain a total weight measurement. Several variations of this basic concept are used, and some of these are discussed in the succeeding paragraphs. The systems described here are also commonly referred to as gravimetric feeders.

Continuous weighing systems for solids are usually needed primarily for flow control rather than a knowledge of the weight per se. There are some instances, though, where mass or weight is the primary objective, such as in blending or mixing operations. But even these applications are closely related to flow control.

In its simplest form, a continuous belt feeder uses a load cell similar to that shown in Figure 8.18 to weigh a section of the load. The weight of the belt and the material it contains is transmitted to the load cell. If the belt is running at a continuous speed, the total amount of weight passing over the scale is easily weighed by integration. If the belt speed is variable, or if close accuracy is required, a belt speed transmitter can be used. Total weight is obtained by multiplying the two signals. A belt speed transmitter is shown in Figure 8.19. It is usually mounted under the belt with the roller in contact with the belt.

Methods of Control

Three basic methods are used to determine total weight or to maintain flow control when using belt feeders: (a) varying the belt speed, (b) varying the belt load or (c) varying both belt speed and load.

Figure 8.20 shows a diagram of a basic pneumatic weigh feeder system. It consists of the basic weigh feeder, integrator, controller, ratio relay, local manual rate setter and local totalizer for the control components. Several auxiliary components are also shown to indicate the ease with which many accessories may be added to the system.

The basic weighing unit may be used simply as a primary sensing element to determine the mass flow of the dry material, or it may be used as a highly sensitive and responsive gate-controlled feeder by use of a control gate where the feed dumps onto the conveyor.

In operation, the materials enter the weigh feeder and are continuously fed onto a positively driven, nonslip weigh belt

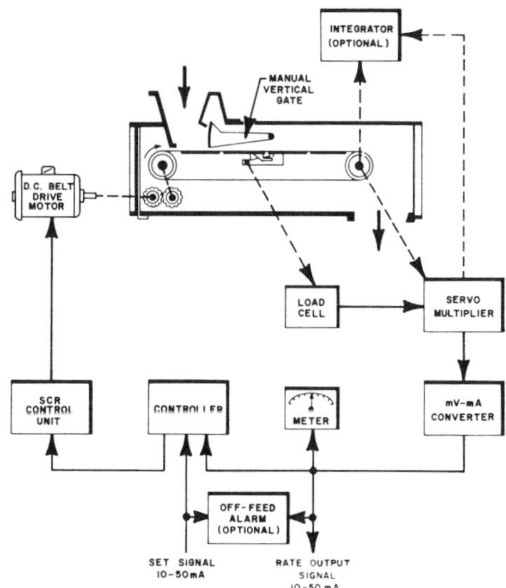

Figure 8.21. Gravimetric feeder that employs belt speed control to maintain a constant weight feed rate. (Courtesy of Wallace and Tiernan, Division of Pennwalt Corp.)

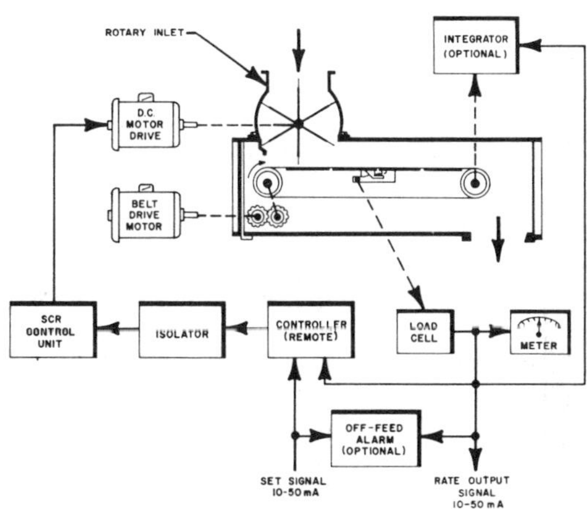

Figure 8.22. Belt feed rate is achieved by varying the speed of a rotary vane at the discharge of a belt feed section. (Courtesy of Wallace and Tiernan, Division of Pennwalt Corp.)

which conveys the material over a weigh sensing section near the discharge point. The weight of the load over a flexure mounted weighing section is sensed by a load cell whose signal output is directly proportional to belt load. This pneumatic signal may be used for control and/or for auxiliary instruments for indication, recording, totalizing or other uses.

A continuous integrator is used for making accurate accounting of the amount of material fed to the unit. Actual belt load, sensed by the load cell, is integrated with respect to the belt travel to give true weight totalization.

The system shown in Figure 8.20 is described as a pneumatic system. The signal system used can just as easily be furnished with standard electronic signal ranges.

Figure 8.23. Flexible metallic connections (or rubber hose) minimize load carrying forces when weighing tank contents. (Courtesy of Anaconda American Brass Company)

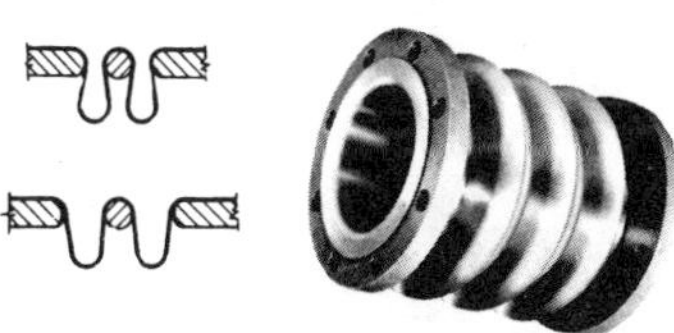

Figure 8.24. Bellows-type expansion joints minimize load carrying capabilities of piping sections. (Courtesy of Adsco Division of Yuba Industries)

Figure 8.21 illustrates a system that uses speed control to maintain steady material flow. A flexure mounted section of the weigh deck supports a short section of the belt and material directly on it. This load is transmitted via a lever system to an electronic load cell. Belt speed is obtained from a tachometer-generator driven off the idler roll to provide a true belt speed. It furnishes excitation voltage to a potentiometer whose output is the product of belt speed and belt loading (via the load cell). This signal is converted to a standard electronic signal representing feed rate.

The feed-rate signal is compared to the set point in the controller, and any difference changes the controller output. This output to a SCR control unit varies belt speed until the feed-rate signal matches the process set point. An optional integrator and counter may be added to the feeder arrangement.

Systems may be designed where both belt load and speed are variable. One method used is to make step changes in speed and maintain automatic control of belt loading using one of the conventional belt load control methods.

Belt Speed versus Belt Load Control

The primary advantage of belt load control in belt conveyor weighing systems is that they are less complex and less costly than variable speed control systems or combined speed and load control systems. Variable speed systems have few inherent disadvantages aside from complexity and cost. Instantaneous feed rate must be calculated, but this can be done easily.

The principle disadvantage of belt load control is the transport lag time between the load control gate and the point of weight measurement. From a few inches to a few feet of travel may take place between the gate and weigh section, depending on the configuration of the feeder. The lag error may be significant where blending ratios are closely controlled.

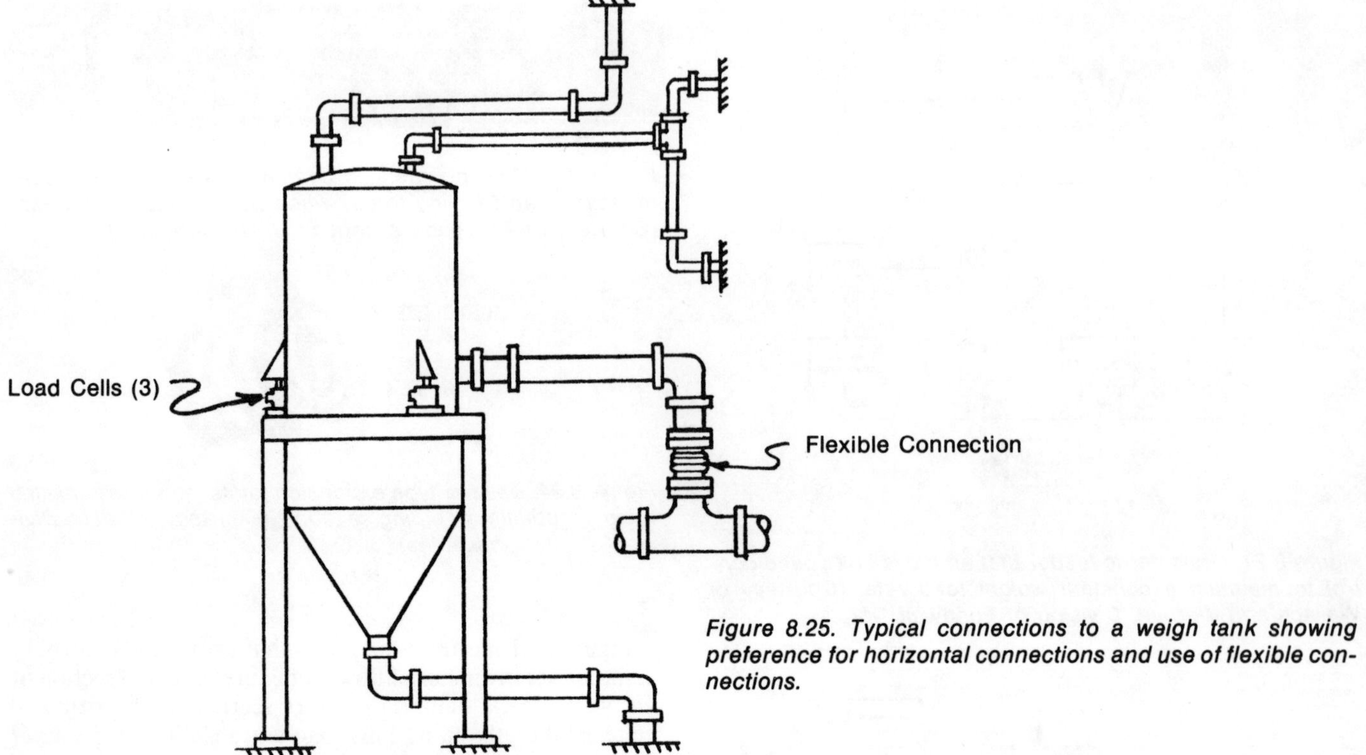

Figure 8.25. Typical connections to a weigh tank showing preference for horizontal connections and use of flexible connections.

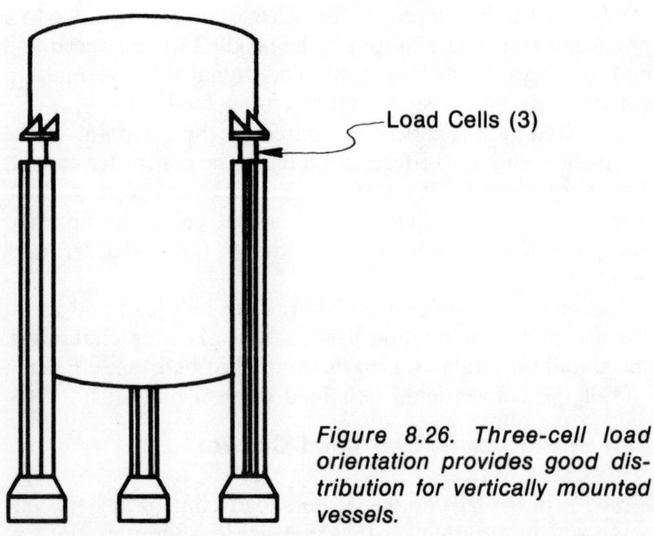

Figure 8.26. Three-cell load orientation provides good distribution for vertically mounted vessels.

There are essentially four different methods of controlling belt loading: (a) vertical gate, (b) rotary vane feeder, (c) screw feeder and (d) vibratory feeder.

The vertical gate feeder is depicted schematically in Figure 8.21 and is simply a controlled opening from the feed section to the conveyor belt capable of varying the material feed rate. This is probably the most commonly used belt loading method. A disadvantage is the difficulty of handling solids of large and/or irregular particle size or in handling aerated materials. Large particles do not flow freely through restricted openings. Equipment manufacturers should be consulted when particle size is appreciably large or when sizes vary considerably.

Vertical gate control is also questionable if the flowing material tends to form lumps or sticks to container walls.

Figure 8.22 shows a weigh system that varies feed loading by controlling the speed of a rotary vane in the feed section. Rotary vanes are used for powders and for other free-flowing materials of small particle size. They are also useful for aerated materials and low density products that are not easily controlled by vertical gates.

Variable speed screw feeders are also used to control low density or aerated materials. They can also be used for fibrous materials and for sticky or lumpy materials that could not be easily controlled by vertical gates or rotary valves. The design of screw sections can easily be custom-made to suit material requirements.

Vibratory feeders are quite flexible in the range of feed materials that they can handle successfully. They can control flows not easily controlled by other methods. Feed rate is varied by vibration amplitude imparted by an electromagnetic vibrator attached to the feed section. Power is regulated to the vibrator by rheostat control of its voltage supply. Manual adjustment of feed rate is made by adjusting the opening between the feed hopper and the trough.

Size Considerations

The size or the amount of material handled by gravimetric feeders varies widely. Material flow is governed primarily by belt width and belt speed. To some degree, it depends on load depth also, but this factor depends

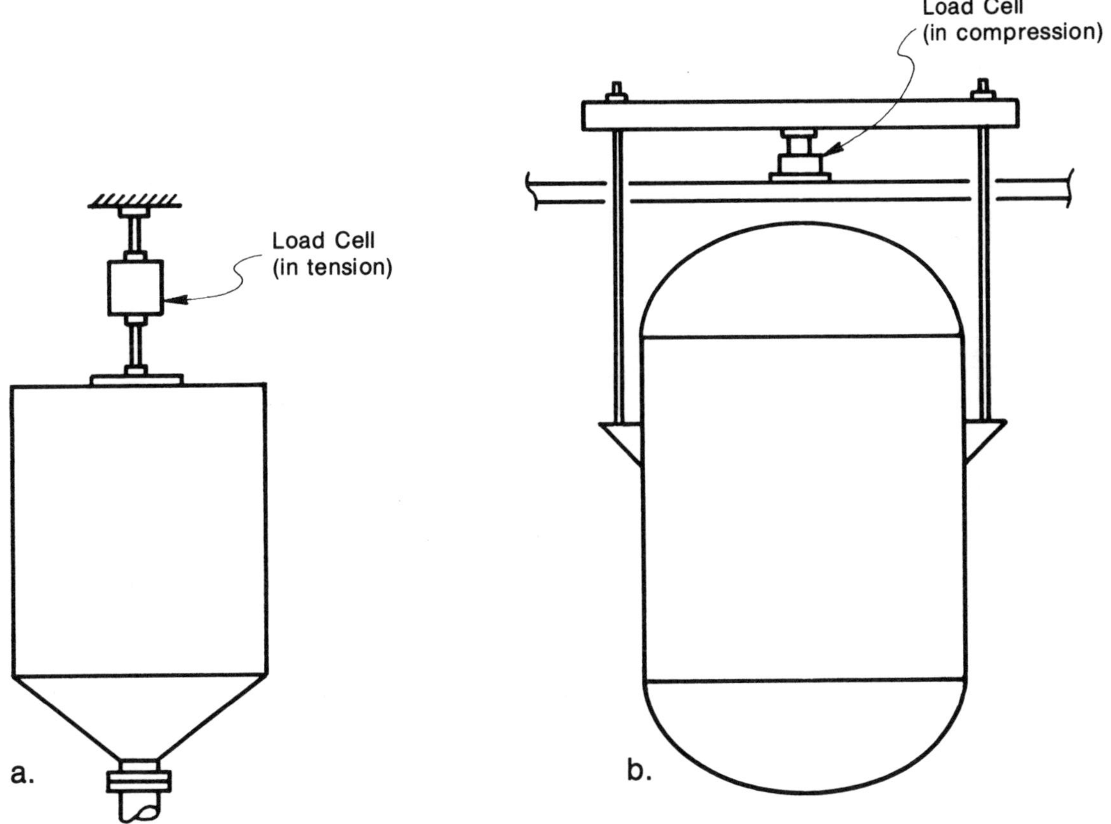

Figure 8.27. Single-load cells may be used for small loads in both tension (a) and compression (b) measurements.

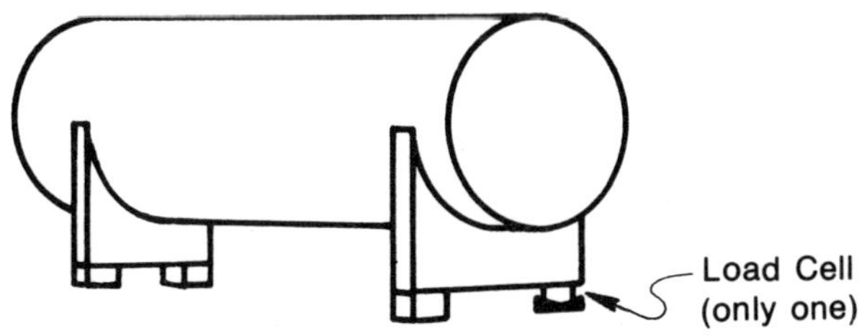

Figure 8.28. One cell may be used for multisupport weigh system with some sacrifice of accuracy.

somewhat on the nature of the material itself. Typical belt widths may range from 12 to 36 inches. Widths as great as 54 inches are available on the high side, and much lower widths are available for special purposes on the low side.

Belt speeds also vary widely. One manufacturer advertises speeds for one model that varies from 0.3 to 180 feet per minute. This provides great flexibility in the amount of material handled.

Application Considerations for Load Cells

The average engineer accustomed to control systems engineering probably has specified fewer weighing systems than most any other major measurement category. Several important factors are given below which should be considered in applying load cells to weighing problems.

Vessel Connections. Probably the most important factor affecting the weighing system's performance is the connection of the vessel to the other parts of the process system. Ordinarily pipe is used to convey incoming and outflowing materials. Usually there are at least two connections, perhaps more.

Two solutions that minimize connection problems are (a) use of flexible connectors such as rubber or metal hose (Figure 8.23) or bellows sections (Figure 8.24) and (b) use of horizontal connections (rather than vertical) to minimize load carrying capabilities of the connecting pieces (Figure 8.25). When electrical strain cells are used, nonaxial loading is particularly harmful. The manufacturer or his literature should be consulted for determining the maximum spring rate or stiffness allowed for such loading.

When designing the connecting lines to weigh vessels, the supports for connecting pipes must also be borne in mind, because support locations affect the spring rate. Ambient temperature variations and temperature changes of the piped fluid also cause expansion or contraction of the connecting pipe and impose changing forces on the vessel, which add to or subtract from the true vessel weight. It is apparent that the checkout of such a system must be thorough. Zeroing (for tare weight) of such a system should be done under operating conditions if temperature changes are appreciable.

Tare-to-Net Ratio. Small tare-to-net ratios are desirable; less than one is certainly perferable. High ratios (10:1 and greater) can be accommodated, but with some sacrifice of accuracy. Manufacturers should be contacted when high ratios are needed unless the user is experienced in this area of use.

High Impact Loading. Hydraulic load cells are more suitable for shock loading than other types. The same is true for systems subject to vibration.

Ambient Temperature Changes. The effect of temperature changes has already been mentioned as it relates to vessel connections. The effect on hydraulic fluids and on electrical components can be appreciable in systems requiring high accuracy or when excessive temperature variations occur.

Quantity of Cells Needed. The choice of the number of load cells needed depends on many factors: (a) the load size, (b) contents, (c) vessel configurations, (d) location, (e) wind loading, (f) available structural supports, (g) agitation and (h) rigidity. A three-cell orientation (Figure 8.26) is generally considered to offer the best load distribution. Care must be used in the application of four-cell systems, since rigid structures can give rise to removal of compression loading from one of the cells.

One-cell systems (Figure 8.27) are sufficient in many instances for small loads. As an economical measure, one cell may be used for two-, three- and four-support systems (Figure 8.28), but the indicated weight will be calibrated, using factors of two, three or four, whichever is applicable. This arrangement assumes an even load distribution—a good assumption if vessel connections and supports are carefully engineered.

Other factors that bear on the type load cell used, the number of cells needed and the supporting method to be used include the accuracy desired, the comparative economic factors, the vessel elevations, available supporting structural members or foundations and operating conditions.

9 Analytical Instruments

Roy L. McCullough, William G. Andrew

A process analyzer may be defined as an unattended instrument that continuously or semicontinuously monitors a process stream for one or more chemical components.

Most process analyzers operate on the same basic principles as their laboratory counterparts but with the addition of mechanisms and circuitry to perform the required analysis unattended and to present resulting data as desired. In addition, process analyzers must be housed both to comply with electrical standards and for protection from weather and physical abuse.

Process analyzers may be classified in various ways depending upon the purpose of the classification. Some classifications are made (a) by operating principle (infrared, ultraviolet, chromatographic, etc.), (b) by type of analysis (oxygen, carbon dioxide, etc.) and (c) by selection—selective or nonselective (infrared may be sensitized to monitor only one component while a chromatograph may monitor several components).

In application, process analyzers are almost always calibrated empirically on standard samples prepared and analyzed by the laboratory. For this reason, process calibration can be no better than the laboratory analysis since the errors of both units are compounded. Repeatability, however, is superior for process units since errors due to human variances and ambient conditions are virtually eliminated.

Following are discussions of the most frequently encountered analyzers in the processing industries. A glossary of analyzer terminology is presented at the end of the chapter.

Gas Chromatographs

Gas chromatography may be defined as a method of physically separating and quantitatively identifying two or more components of a mixture by distributing them between two phases, one phase being a stationary bed of adsorbent surface and the other phase a carrier gas which percolates through and along the stationary bed.

Operating Principle

In Figure 9.1, the carrier gas flows continously through the column at a constant rate. With the injection of a vapor sample into the carrier stream, the components and the sample are mixed with the carrier and adsorbed into the stationary bed.

The bed substrate has an affinity for each component in the sample, and since this affinity varies with each component, generally the lighter components are swept away (eluted) faster than the heavier. As a result of this action down the entire length of the column, the lighter components reach the end of the column before the heavier ones. By use of the proper column length, the components are separated from each other, and the individual elution time serves to identify the particular component (Figure 9.2).

Chromatographs consist of two basic sections: analyzing and control. Normally the analyzer section is located in the field near the sample point (or points), and the control section is located remotely, usually in a central control room. For discussion purposes, the analyzing section may be sub-

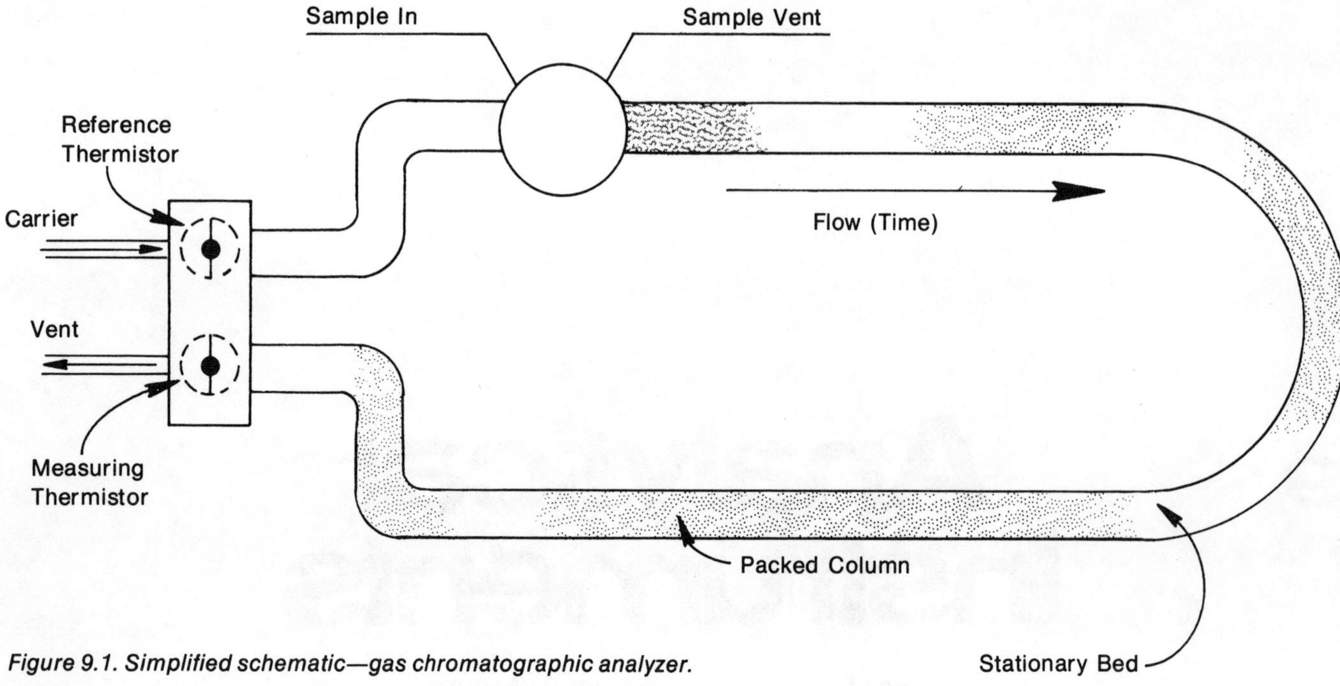

Figure 9.1. Simplified schematic—gas chromatographic analyzer.

divided into smaller units: valves, columns and detectors. The control section consists of programmer, recorder and auxiliary units, including stream selector and peak-picker memory unit.

Analyzing Sections

Sample Valves

Chromatographic valves serve two basic functions: (a) to introduce a fixed volume of sample into the carrier stream (sample valve) and (b) to switch segments of the column out of and into the carrier stream (column valve). Several types of valves are available. Figures 9.3 and 9.4 depict the varied construction of the four most popular types—namely, slide, spool, rotary and flexible-diaphragm.

The slide valve has a rectangular shaped Teflon block which slides between two stainless steel plates. Three holes through the slides permit the flow of sample and carrier gas. The diameter of the center hole is calculated to furnish a predetermined volume of sample. When the valve is actuated by air pressure on the diaphragm, the slider moves the sample volume hole from the sample stream to the carrier stream, thereby injecting the sample into the column.

The spool valve has a round shaft which slides when pushed by the diaphragm. Three grooves are machined around the shaft to permit flow between side fittings. O-rings isolate each groove to form a compartment. When the valve is actuated, the sample compartment (volume) is moved into the carrier stream and injected into the column.

The rotary valve has the sample volume machined into the base section of the valve. When actuated, the base section rotates 90° and places the sample volume in the carrier stream. Models of this valve are available with either air or electrical drive.

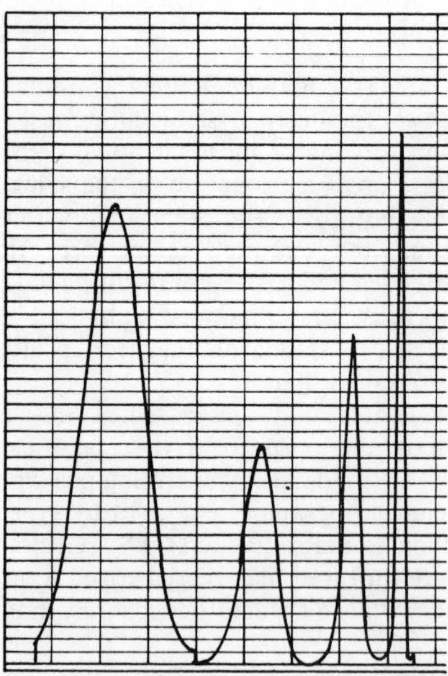

Figure 9.2. Typical chromatogram showing the elution of four components.

The diaphragm valve (Figure 9.4) is constructed in two sections. The top section has holes drilled from the six fittings to the lower surface of the block. Between each hole on

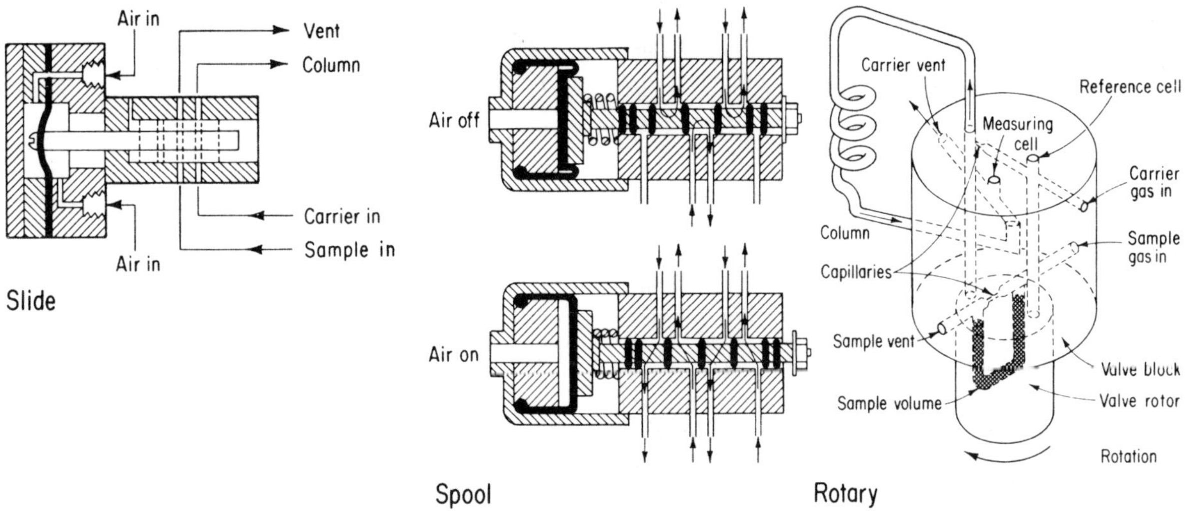

Figure 9.3. Typical operational diagram of the slide, spool and rotary valves. (Courtesy of Instrumentation Technology)

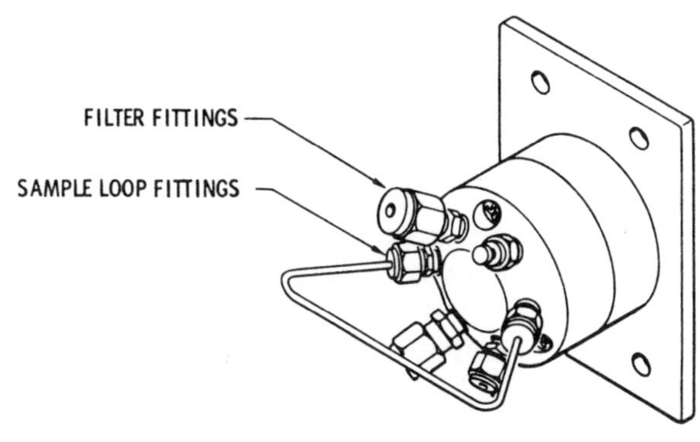

Figure 9.4. A six-port arrangement of the flexible diaphragm type sample valve.

the lower surface is a concave "dimple" to permit flow between the holes. The lower block has six holes drilled to its top surface, and these are spaced to match the "dimples" in the upper block. A flexible diaphragm is placed between the two sections. When air pressure is placed on a hole in the bottom section, the flexible diaphragm moves upward at that point and closes the "dimple," thus stopping flow between the two ports. In the valve "on" (sample) position, three alternate "dimples" are blocked with a resultant flow pattern as shown in Figures 9.5*a* and 9.5*b*. In the valve "off" (normal) position, the remaining three "dimples" are blocked. Figure 9.5*c* shows this same action being accomplished with plungers as the actuating force.

High quality is essential in a sample valve because it must be leak-proof, both to the outside and between valve ports, while transporting liquid or gas streams which are

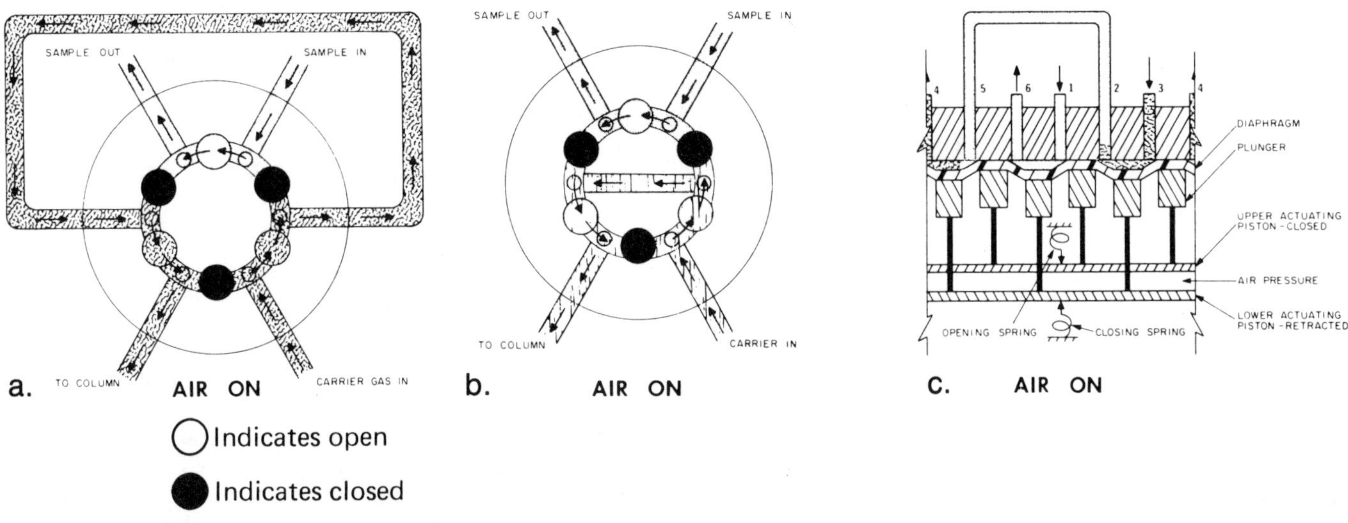

Figure 9.5. Schematic operation—six-port diaphragm valve. (Courtesy of Seiscor, Division of Seismographic Service Corp.)

sometimes in a high-pressure and high-temperature state. The valve must operate thousands of times without failure and be so constructed as to permit field maintenance and repair.

Column Switching Valves

In making some analyses, columns are sometimes used that are affected by other stream components (moisture or oils for example) in which there is no interest. These other components can damage the column by plating or washing the column material. To avoid this, a precut column is used to separate the contaminating components from those of interest, allowing the desired components to flow into the second or analysis column. The undesirable components are then backflushed or purged into the vent system by an auxiliary carrier supply. Column switching valves are used in this manner to accomplish the separation desired.

The column switching valve differs from the sample valve only in function. Figure 9.6 depicts a single column configuration where only a four-port sample valve is required. When the sample valve is energized, the slider moves one position to the left, placing the sample volume in the carrier stream. The carrier gas sweeps the sample through the column, where separation occurs on the way to the detector.

In the event that only light components in a sample are of interest, the heavy components may be discarded, saving the time required to sweep them through the column. Figure 9.7 shows how this is done using a sample valve, a column switching valve, a precut (or stripper) column and an analysis column. After injection of the sample into the stripper column, the lighter components are separated from the heavier and move on into the analysis column. At this time, the switching valve is actuated, reversing the flow in the stripper column and purging the heavy components out to vent while the lighter components continue on through the analysis column and to the detector.

By proper application of valves and columns, many separations may be achieved which would otherwise be impossible or perhaps would require much longer time cycles.

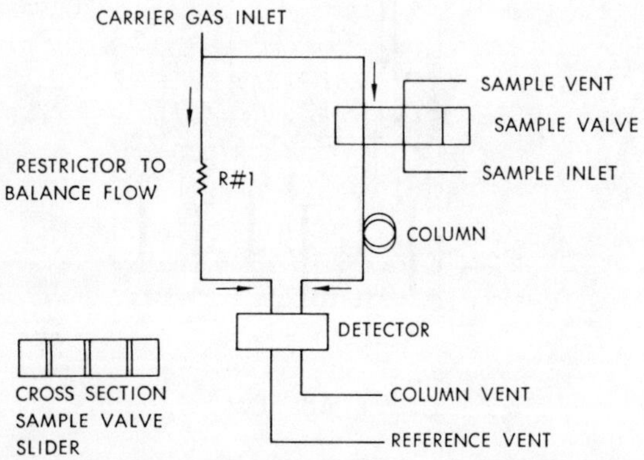

Figure 9.6. Schematic representation of an analyzer with single column and a four-point sample valve. (Courtesy of Beckman Instruments, Inc.)

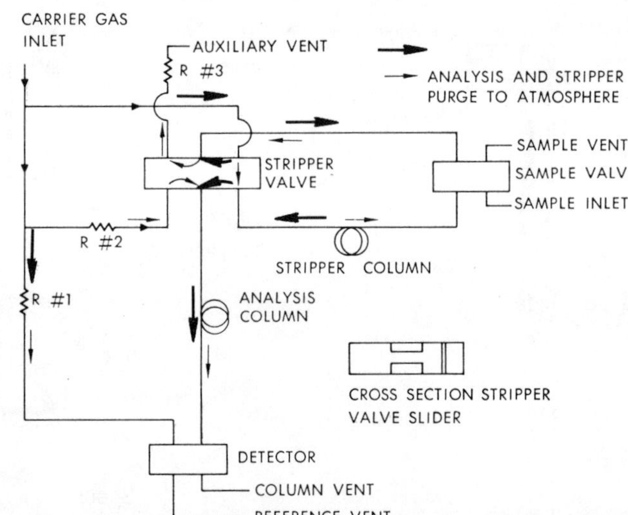

Figure 9.7. Analyzer with dual columns, sample valve and column switching valve. (Courtesy of Beckman Instruments, Inc.)

Figure 9.8 illustrates the use of 10-port, flexible-diaphragm valves in a configuration using two different carrier gases to perform analysis in two loops, each containing a precut (stripper) column and an analytical column. This switching arrangement permits the use of two different types of columns with separate and different carrier gases for separations which could not be achieved otherwise.

Columns

The working element of a chromatograph is the column, the device that effects the separation of the component or components to be measured. It consists of small diameter tubing packed with a bed of material which offers varying degrees of resistance to the stream components. Beds are selected that have the ability to retard (retain) some components while passing (eluting) others quickly. Column lengths, as well as material, affect component separation. A schematic of a column is shown in Figure 9.1.

In gas-solids chromatography the stationary bed is made of fine solids such as silica gel, molecular sieve, charcoal, etc. The sample in the carrier gas is adsorbed onto the porous surfaces of these solids. In gas-liquid chromatography the stationary bed is made of inert solids such as firebrick, diatomaceous earth, glass beads or other materials on the surface of which a liquid adsorbing agent (substrate) is deposited. In some instances the substrate may even be deposited on the inside wall of extremely small diameter tubing. In this type column the sample in the carrier mixture is adsorbed by the substrate instead of the solids material. The solids, in either case, are packed into ⅛- or ¼-inch OD tubing, and the column tubing is usually coiled for ease of mounting in the analyzer enclosure.

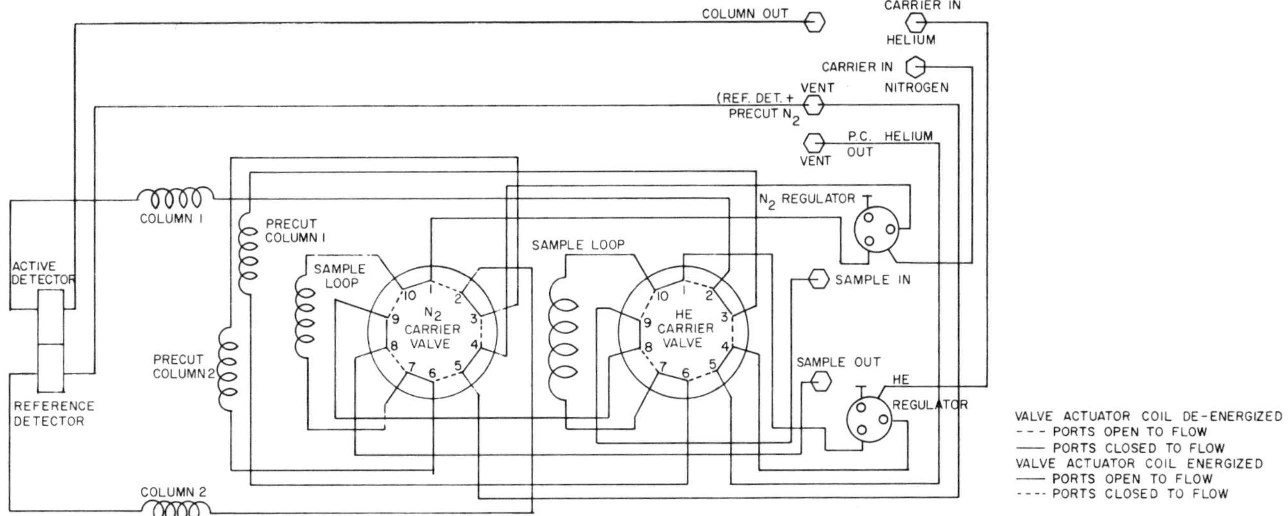

Figure 9.8. Two 10-port flexible diaphragm valves using two different carrier gases to analyze two loops. Each loop contains a stripper and an analytical column. (Courtesy of Consolidated Electrodynamics)

The following are the most important factors affecting the separation efficiency of a column:

1. Bed particle size: smaller granules offer a larger surface area for vapor contact but pack tighter and require higher carrier pressure. The increased surface area improves resolution, but this advantage is offset by the tendency of the smaller particles to channel with flow.
2. Liquid substrate: the amount of substrate deposited on the bed affects separation within limits; no one substrate is suitable for all mixtures.
3. Column length: increasing length increases separation and also increases elution time and requires higher carrier pressure.
4. Column temperature: increased temperature increases separation but decreases resolution (or peak shape).

Detectors

As the sample and carrier mixture elutes from the column, it passes through a detector where the components are measured in their separated states. By far the most common detectors in use today are the thermal conductivity and hydrogen flame ionization. These two types are discussed in some detail below. Other types of detectors include the beta ionization, electro capture, ultrasonic whistle, gas density and photoionization.

Thermal Conductivity

Figure 9.1 depicts thermister elements in a thermal conductivity detector, and Figure 9.9 shows the elementary electrical schematic of a Wheatstone bridge in which they are used. As a sample component passes the measuring thermistor, the element is heated due to the lower thermal conductivity of the component compared to that of the pure carrier gas. This temperature rise causes a change in thermistor resistance and a subsequent change in bridge balance.

In the hot-wire thermal conductivity detector, wire filaments replace the thermistors as resistance elements and the operation is the same. To achieve higher sensitivity, the standard resistors shown in Figure 9.9 are also replaced with wire filaments, and the detector is modified so that all four elements are in the carrier gas and effluent streams.

Hydrogen Flame

The hydrogen flame ionization detector is most frequently used to monitor components in small concentrations. Capable of analyzing in the range of 0 to 1 ppm, the detector uses sophisticated circuitry to amplify signals in the range of 10^{-14} amperes. Figure 9.10 depicts a schematic diagram of a commercial detector cell. A DC voltage of 100 to 1,500 volts is connected in series with the electrodes. The thermal energy in the hydrogen flame plasma is sufficient to induce emission of electrons from hydrocarbon molecules in the column effluent stream. The ions are collected at the electrodes and amplified in a high quality electrometer-amplifier. When the unit is properly calibrated, the sensitivity of the detector is proportional to the carbon content of the mixture.

Analyzer Housing

The design of analyzer hardware housing is dictated by three requirements: (a) to comply with electrical codes for hazardous areas, (b) to provide a temperature controlled

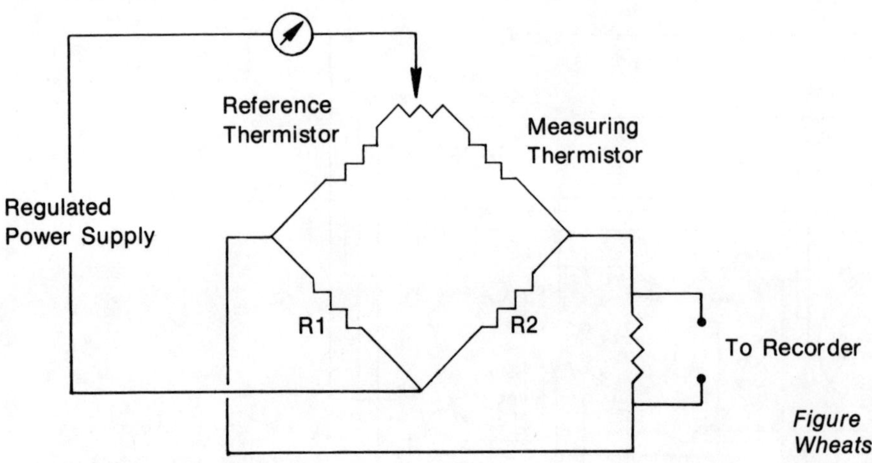

Figure 9.9. Thermal conductivity detector using a Wheatstone bridge measuring circuit.

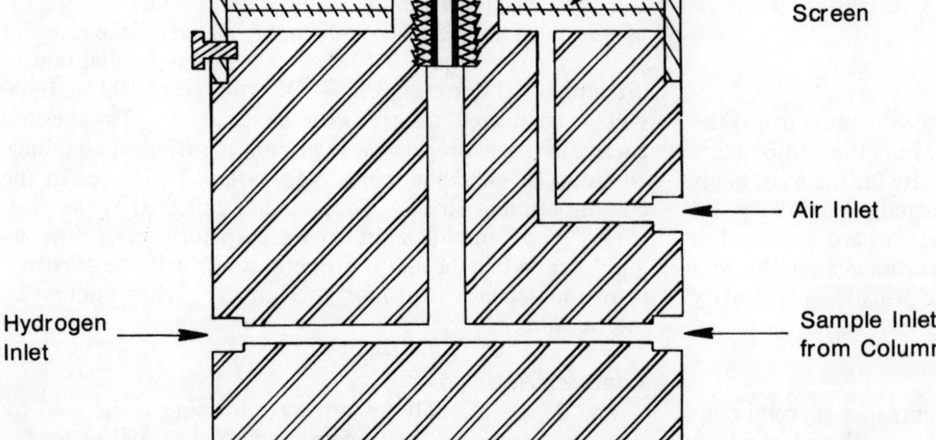

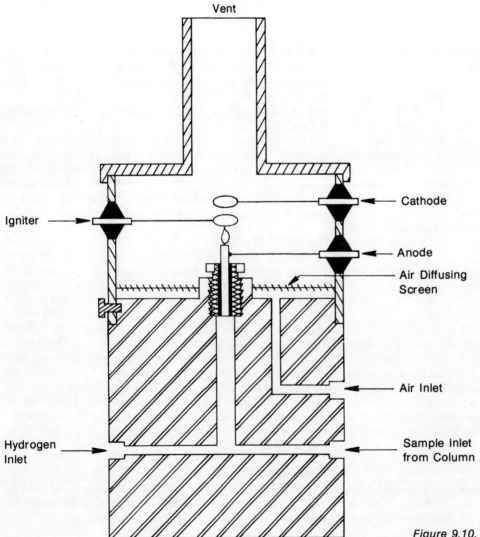

Figure 9.10. Hydrogen-flame ionization detector.

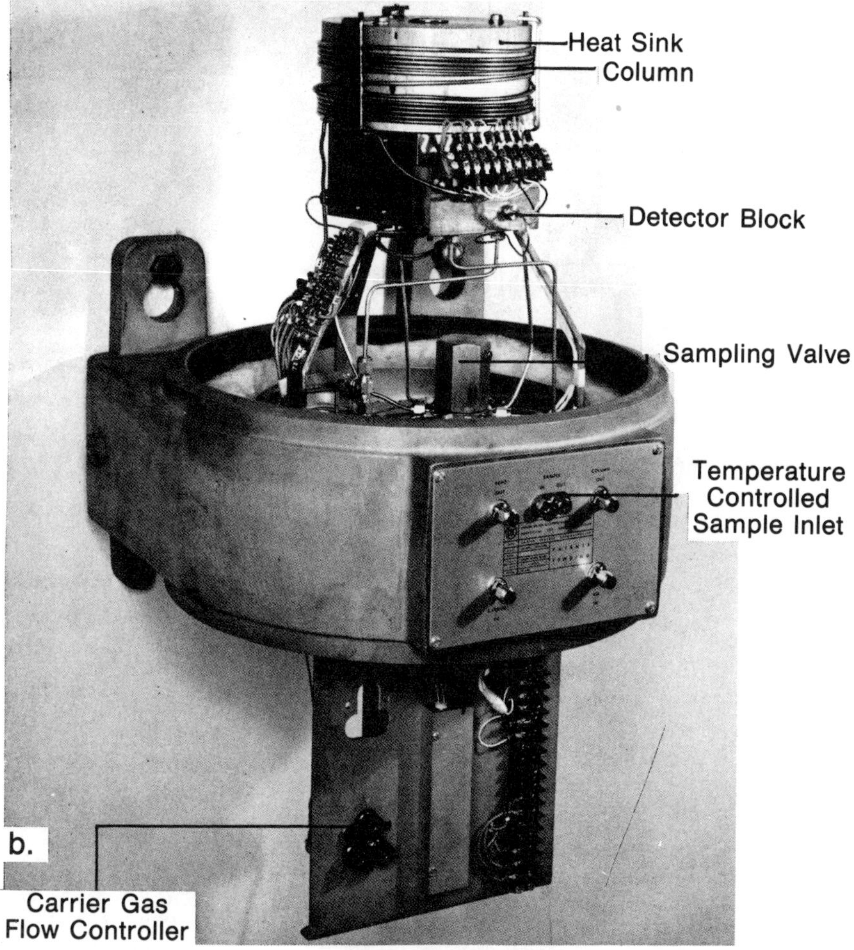

Figure 9.11. (a) A chromatograph mounted in an explosionproof housing; (b) dome removed from explosionproof housing. (Courtesy of Process Analyzers, Inc.)

enclosure and (c) to protect from physical abuse. Two styles of housing comply with these requirements, the explosionproof dome (Figure 9.11*a*) and the cabinet housing (Figures 9.12*a* and 9.13*a*). The domed housing is a compact assembly which provides quick access to all hardware components (Figure 9.11*b*). Temperature control is accomplished with cartridge heaters inserted in the heat sink block around which the column is coiled. Maximum access from three sides makes maintenance and repair easier than on enclosed housings. The cabinet housings shown in Figures 9.12 and 9.13 use explosionproof fittings only for the electrical circuitry. Items requiring temperature control are housed separately. The heater in Figures 9.12*a* and 9.12*b* is a ring-fin coil encircling a blower fan. In Figures 9.13*a* and 9.13*b*, a hot-air blower is used for heat control. Figure 9.12*b* shows a programmed temperature analyzer provided with enclosed temperature zones. Figure 9.14*a* is a single-stream vapor sample conditioning cabinet. Figure 9.14*b* shows a four-stream sample cabinet which may be temperature controlled by adding a heater in the lower compartment.

Control Section

The control section of a process chromatograph includes all units of the analyzer which are related to the power supply or the readout and command circuitry. A discussion of these covers the programmer, recorder and other auxiliary devices.

Programmer

The programmer is the control unit of the process analyzer. Mounted in a sheet steel cabinet, it is not explosionproof but may be air purged for field installations (Figure 9.15) to meet hazardous electrical classification requirements. It contains the timer, power supply for the detector bridge circuit, the automatic zero mechanism, attenuation pots for each component measured and memory amplifier, if required. Its primary function is the amplification and transmittal of the detector signal to the recorder and/or controller. A *function-selector* permits the selection of a chromatogram mode, a calibration mode or automatic

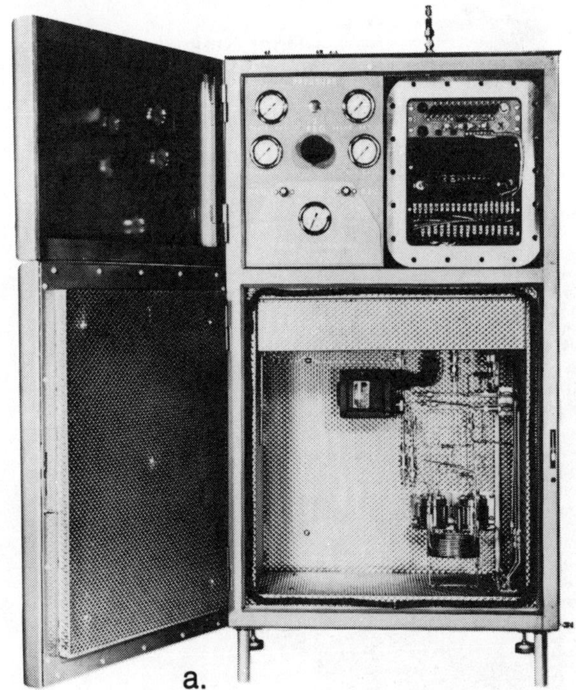

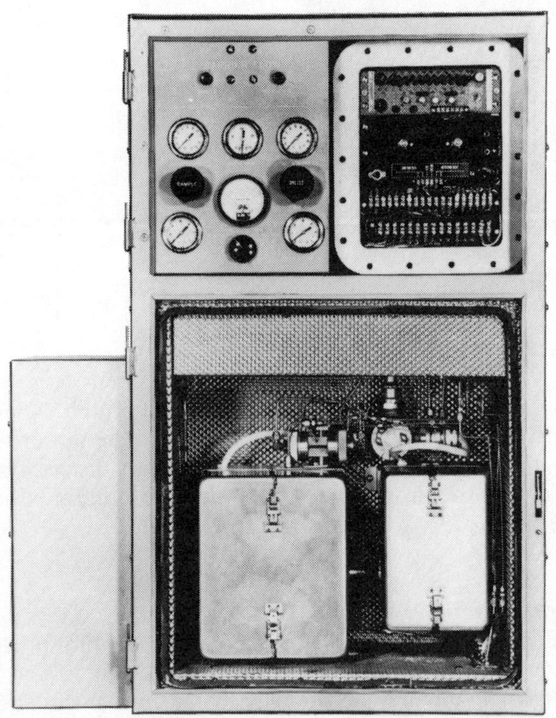

Figure 9.12. (a) Cabinet housing for single column chromatograph; explosionproof fittings enclose electrical devices. (b) Cabinet housing for Multistream chromatograph; unit contains a programmed temperature controller; electrical devices are in E.P. Fittings. (Courtesy of MSA Instruments Div.)

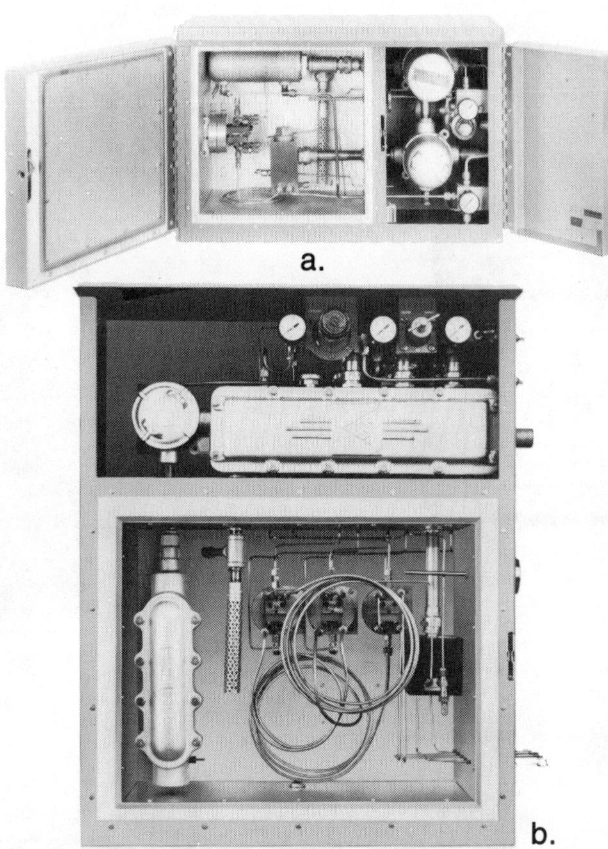

Figure 9.13. (a) Compact arrangement of single column analyzer in cabinet case with temperature zone on left; (b) cabinet housing for Multicolumn chromatograph with controlled temperature zone below. (Courtesy of Beckman Instruments, Inc.)

analysis. Additional functions of the programmer include command of the sample and column valves in the field analyzer unit, stream switching circuitry (although this function is often housed in a separate chassis) and programmed temperature control.

Computer-Based Controls

The use of chromatographic component analysis has increased dramatically in recent years. Emphasis on process efficiency, reduction of impurities, and minimization of fuel costs has provided strong economic incentive for on-stream analysis and control. And reservations concerning the abilities of equipment to adequately control the process has given way, in the face of advances in the state of the art of equipment design and application, to a gradual dependence on analytical systems by both operations and management.

This confidence in analytical control has been strongly influenced by the ease with which complex process control equations can now be solved. Computing equipment used a decade ago was custom-built for each application, both hardware- and software-wise, from systems originally only a

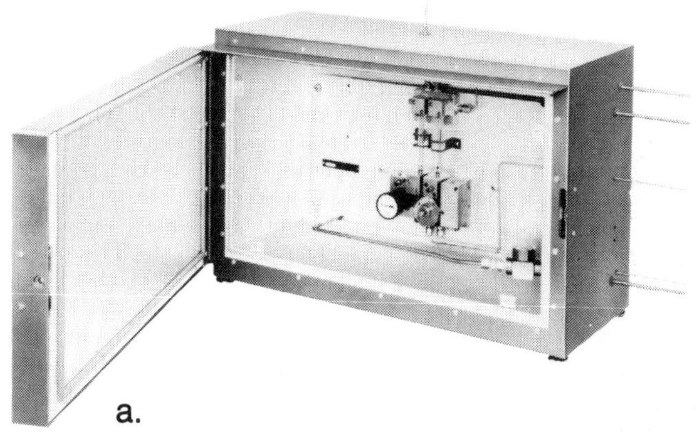

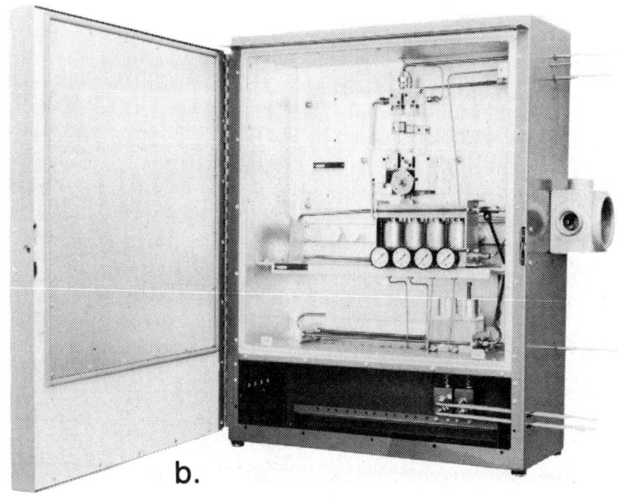

Figure 9.14. (a) Vapor Conditioning cabinet for single stream unit; (b) four-stream vapor conditioning cabinet—a temperature control unit could be added in the lower compartment. (Courtesy of Beckman Instruments, Inc.)

Figure 9.15. The programmer or control unit for the G. C. is housed in a steel cabinet. It is not explosionproof and must be air purged if used in a hazardous atmosphere. (Courtesy of Beckman Instruments, Inc.)

step or two removed from office data processing. Current analytical measurement and control systems are a blend of common control modes (set point, feed forward, predictive, data acquisition, etc.) and powerful data functions (hierarchies, bulk storage, data manipulation, data communication, man-machine interfaces, system diagnostics, etc.).

Computer-based technology has brought the engineer and system designer an enormously flexible tool, and we are face to face with a situation in which we are limited only by our own imaginations. A look at a current system design will point out design features now available.

The Optichrom 2100 Process Chromatograph System combines proven sensor components with integrated circuit analyzer electronics and a microprocessor based programmer (Figure 9.16). Analog-to-digital conversion is done at the analyzer in the field, and data is then transmitted serially over a twisted pair of wires. Signals are optically isolated at the analyzer. The programmer is capable of operating four analyzers, with up to 16 streams per analyzer and up to 50 components per stream. Programmer outputs can consist of up to 40 4-20 ma continuous signals, also bar graphs, trend recording, digital printing, or computer link. Scaling and data reduction can be accomplished within the programmer, as well as programs which are manually loaded through its keyboard or downloaded from a cassette tape.

Pneumatic Composition Transmitter

Of interest is a recent analyzer which uses all pneumatic circuitry (Figure 9.17). Called a "Pneumatic Composition Transmitter" by its manufacturer, Foxboro, it is inherently safe, since it has no electrical inputs or spark-producing circuits. Internal components are mounted on pneumatic circuit boards, which facilitate replacement to minimize downtime. The analyzer housing is less than 2 cubic feet in size and is rated NEMA 3 (weather-proof). Instrument air, 30 psi steam, helium (carrier gas), and samples are supplied to the transmitter, and outputs are sample vent and a 3-15 psi signal.

The analyzer is based upon known chromatographic principles. A small amount of sample is injected into the carrier gas stream, which is already flowing at about 60 cc per minute. The sample, bracketed by carrier gas, flows through a small orifice (a 0.5-mm jewel orifice) and then a capillary tube, and is then vented. A pneumatic force balance differential pressure detector is connected across the orifice. The differential pressure developed is a function of the density of the flowing fluid and its flow rate, and the orifice downstream side pressure is a function of the viscosity of the flowing fluid and its flow rate and is caused by the capillary tube. Compared to parameters related to the helium carrier gas, the sample produces changes in all three factors—flowrate, density, and viscosity. Density changes have a positive effect, and viscosity changes generally have a negative effect, the algebraic sum of which leads to a stronger measurement effect. By judicious choice of dif-

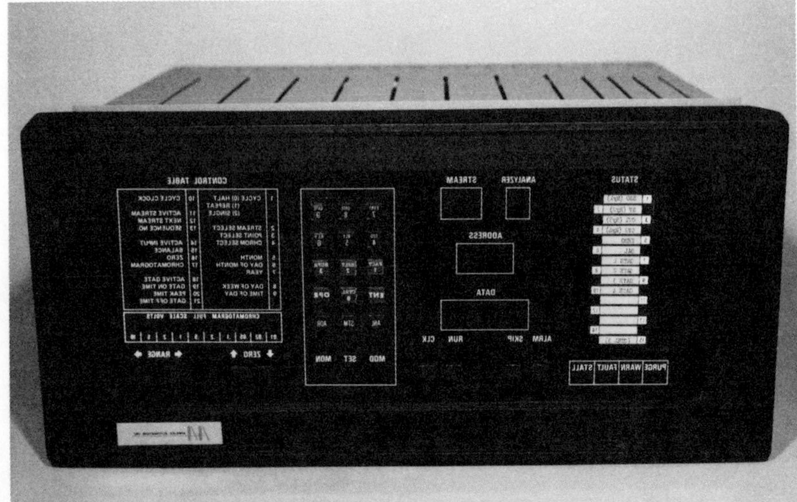

Figure 9.16. The Optichrom 2100 Process Chromatograph System combines proven sensor components with integrated circuit components and a microprocessor-based programmer. (Courtesy of Applied Automation, Inc.)

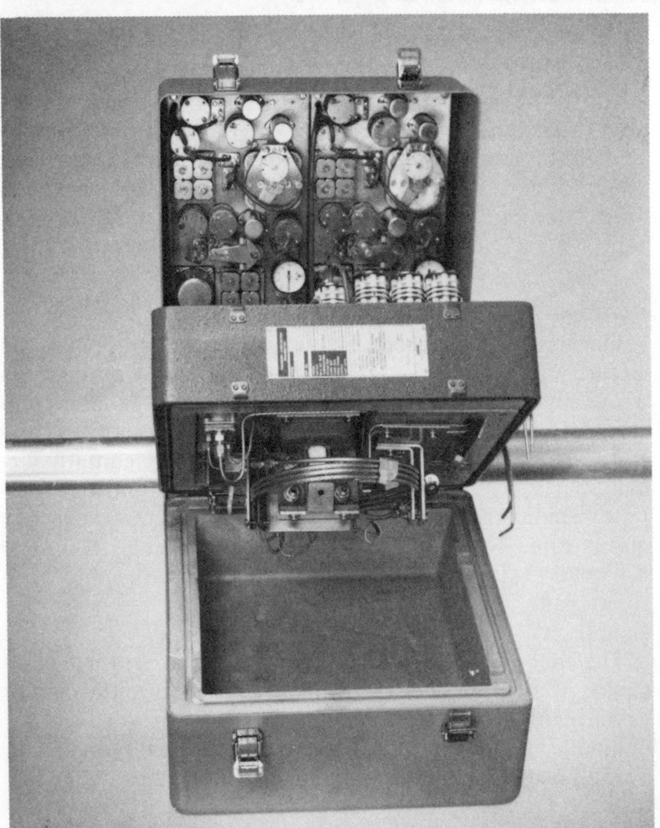

Figure 9.17. The Pneumatic Composition Transmitter is a chromatograph using all pneumatic components. (Courtesy of Foxboro Co.)

ferential pressure bellows sizes, and thereby a convenient "gain ratio," the slight flow rate variation effects cancel, giving a differential pressure which is due to component properties only.

Temperature effects of this type of detector are much less than those of thermal conductivity detectors, and are usually about 0.3-0.6% per degree centigrade. A steam supply to a concentric-tube heat exchanger furnishes heat to the analyzer section. Using a novel venturi-airjet arrangement and a bimetallic flapper nozzle assembly, temperature is controlled to within ±1°C.

Pneumatic trip switches program events within the processor section, being set for values occurring on a pressure ramp signal generated in a pneumatic "clock" module.

The Pneumatic Composition Transmitter is designed for component concentrations exceeding 0.1 mol percent and for measuring one or two components in a single process stream. It is ideally suited for location in hazardous areas and is a low-cost means for developing a control signal from a product concentration.

Programmed Temperature Gas Chromatography (PTGC)

Gas chromatograms are usually run with the column maintained at a constant temperature, just above ambient, to prevent errors due to temperature changes. Operating at a low temperature sometimes results in two undesirable consequences:

1. Early peaks are sharp and close together while later peaks are broad and widely spaced (Figure 9.18*b*). This means that resolution is poor in the early stages of the chromatogram and excessive in the latter stages.
2. High boiling compounds may go undetected. This is particularly true of compounds having wide boiling ranges.

The percentage concentration of a component in a sample is calculated from the area of that peak times a thermal conductivity factor, divided by the total area of all peaks.

$$\% = A(\text{component peak}) \times K/A\ (\text{total all peaks}) \times 100$$

Comparing the areas of peaks 5, 6, 7 in Figure 9.18*b* indicates that the three are approximately equal in concentration, yet peak 5 is twice as high as peak 6 and six times

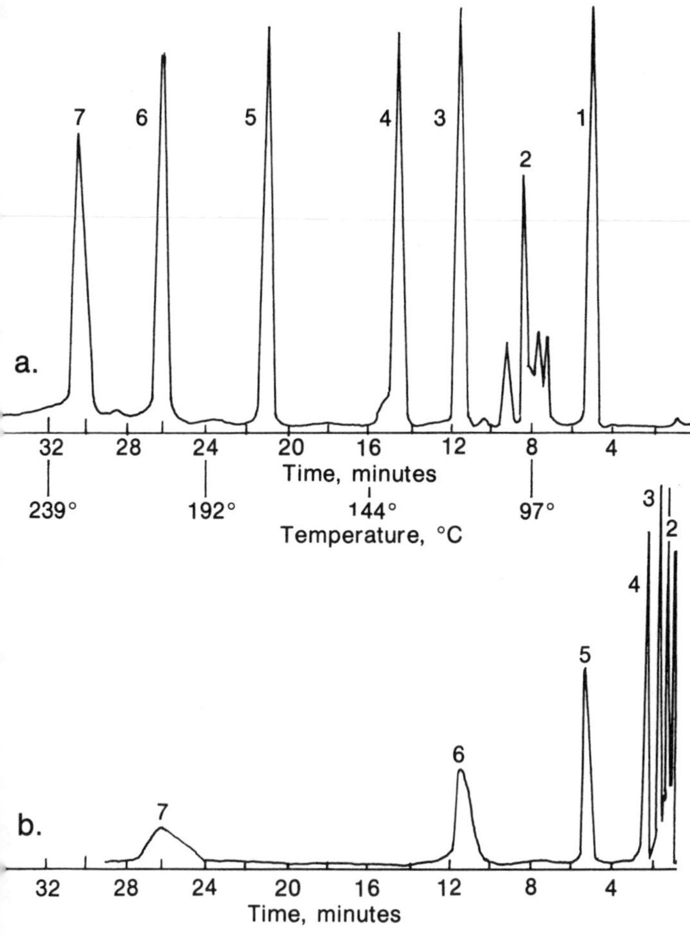

Figure 9.18. Chromatogram a demonstrates the effectiveness of programmed temperature control. It is the same sample as shown in b, but temperature control has separated and emphasized peaks which usually are used for process control.

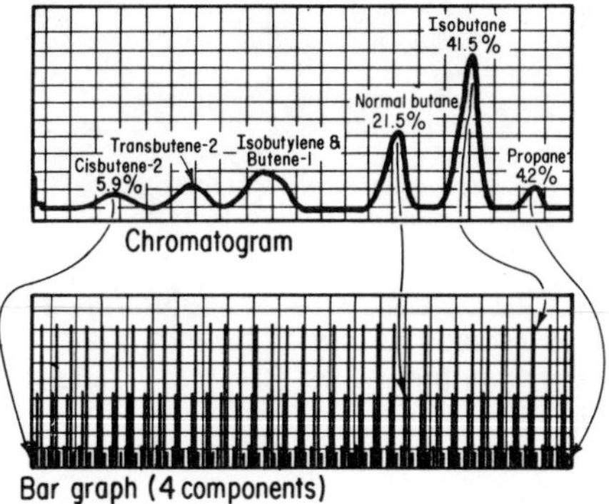

Figure 9.19. The two primary methods of recording chromatographic readouts are bargraphs (bottom) and chromatograms (top). The same analysis is shown for both methods. (Courtesy of Instrumentation Technology)

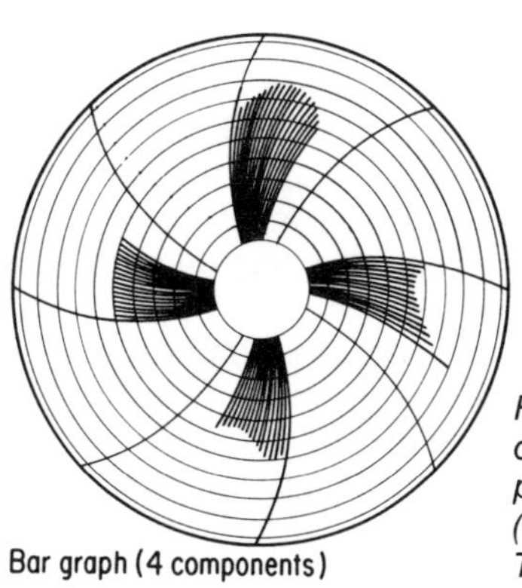

Figure 9.20. Bargraph records are sometimes displayed on circular charts. (Courtesy of Instrumentation Technology)

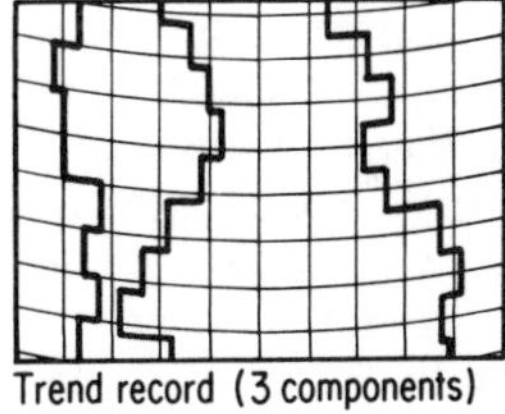

Figure 9.21. Trend records are shown for three components. Each one records the output of a memory amplifier which holds a peak until a succeeding analysis is made. (Courtesy of Instrumentation Technology)

that of peak 7. Linearity presents a problem in this instance since process chromatographs are calibrated on peak height. Figure 9.18*a* is a chromatograph of the same sample analyzed with programmed temperature control, in which the temperature was increased 6°C/minute. Peaks 5, 6, 7 are now of similar height and should be calibrated on peak height with no difficulty. The early peaks in Figure 9.18*a* are also spread out and clearly defined. In addition, peak 2 has been split into four peaks, three of which were not discernable in the lower temperature analysis. Process analyzers frequently use the programmed temperature method of analysis, or separate columns in two heat zones may be used to accomplish the desired separations.

Recorder

Standard strip-chart recorders are generally used for chromatographic readout. Figure 9.19 illustrates two methods of recording the same analysis—the chromatogram and the bargraph. The chromatogram is used during calibration to identify the peaks and to determine individual peak elution time. Bargraph recording is used during normal operation. Figure 9.20 shows a bargraph recorded on a circular chart, though strip charts are much more common. In the chromatogram mode, the detector signal is recorded continuously as the chart is driven. In the bargraph mode, the chart drive is off, and the pen is alive only while a specified peak is eluting. The pen is dead between component presentations, and the chart is advanced a short distance between component peaks to separate the bargraph presentations. The circular chart bargraph (Figure 9.20) is recorded with four pens, and the chart is advanced at the end of each analysis. The trend record in Figure 9.21 uses

Figure 9.22. Automatic stream selection (10 streams) mechanism is housed in metal cabinet. Selection is made through use of solenoid valves. (Courtesy of Beckman Instruments, Inc.)

three pens recording the output of three memory amplifiers (one on each component); this signal is adjusted during each cycle by an amount equal to the change of the component analysis.

Auxiliary Units

The automatic stream selector mechanism (Figure 9.22) is for selecting, either manually or automatically, the stream to be analyzed. The selection is made by actuating solenoid valves in the sample system. The rotating mechanism of the selector is "stepped" each cycle of the programmer and moves to the next stream. To skip a stream, the operator merely flips the appropriate toggle switch on the front panel of the selector. Small panel lights glow dimly when the stream is in the program and glow brightly while the stream is being analyzed. Most manufacturers furnish stream selectors in units from two to ten streams.

The peak-picker memory unit for an analyzer is an electronic circuit to furnish a steady, analog signal for trend readout or for use in computer or control loops. It is switched into the bridge output circuit while a particular peak is eluting. As the peak reaches its maximum reading, a charge is accumulated across a capacitance circuit, and this potential is held until the peak is presented on the next cycle. The unit is mounted as a circuit in the programmer in large case units and in a separate chassis when panel space is scarce.

Calibration

Standard samples are necessary to calibrate the chromatograph. These may be purchased, or actual stream samples may be used by checking the stream sample in the control laboratory. A multistream analyzer requires several standard samples if the full-scale range of any of the components is different.

Infrared

Operating Principle

The infrared process analyzer is used to monitor selectively components which exhibit absorptive characteristics in the infrared region of the wave length spectrum. When a specific molecule of gas (or liquid) is subjected to electromagnetic radiation, it is caused to rotate or vibrate by absorption of energy from an infrared beam of light. The amount of energy transmitted through a sample cell (that is, not absorbed) is monitored and used to measure specifically the stream concentration. By plotting transmittance against wavelength, a spectrogram (or spectrum) is produced which is unique for that molecule (Figure 9.23). Spectra of four samples are shown in the illustration. Ethane is indicated in three and is more pronounced in the lower spectrum where transmittance drops to about 5. The spectra reveals clearly that the ethane component absorbs IR (infrared) radiation energy in the wavelength between 3 and 4 microns. The other components shown absorb IR energy at other wavelengths. Only the elemental gases, hydrogen, nitrogen, oxygen, chlorine and the rare gases, do not absorb energy in the infrared region.

An infrared analyzer consists of the light source, the sensitization (or optical) section and the detector.

Sources

The spectrum in Figure 9.23 indicates that each compound absorbs strongly at some characteristic wavelength. A process analyzer does not scan the infrared band but is designed with a source which emits light spanning this wavelength of strong absorbance. The spectrum of most molecules falls in the range of 1 to 25 microns. The three most widely used sources are the Globar (silicon carbide), Nernst glower (a filament of oxides) and the Nichrome (wire) coil. The latter is the most widely used in process analyzers because it is inexpensive and longer lived and emits the longer wavelengths frequently required.

Sensitizing

One of the simplest schemes for an infrared spectrophotometer is shown diagrammatically in Figure 9.24. The light transmitted through a sample is dispersed by prisms, and the undesirable wavelengths are eliminated by the use of slits. This system is called dispersive spectrometry. The instrument records very accurately the wavelengths at which the sample absorbs, but it cannot measure the percentage of such absorption due to the variation of source light intensity at different wavelengths. This difficulty is overcome by using two beams of light from the same source. By using a rotating chopper, the beam through the sample is compared to the beam bypassing the sample. Because of difficulty in keeping the prisms, slits and mirrors aligned properly, dispersive analyzers are used primarily for laboratory analysis.

Nondispersive analyzers, used predominately in plant applications, measure all the transmitted light from a source except that wavelength absorbed by the sample gas or sensitizing gases. Figure 9.25 shows that light from the sources is limited to the desired wavelength by the material of the windows. The beam chopper alternately blocks the beam to the sample cell and to the comparison cell. When the beams are equal, equal amounts of radiation enter the detector from each beam. When the gas to be analyzed flows into the sample cell, it absorbs and reduces the radiation reaching

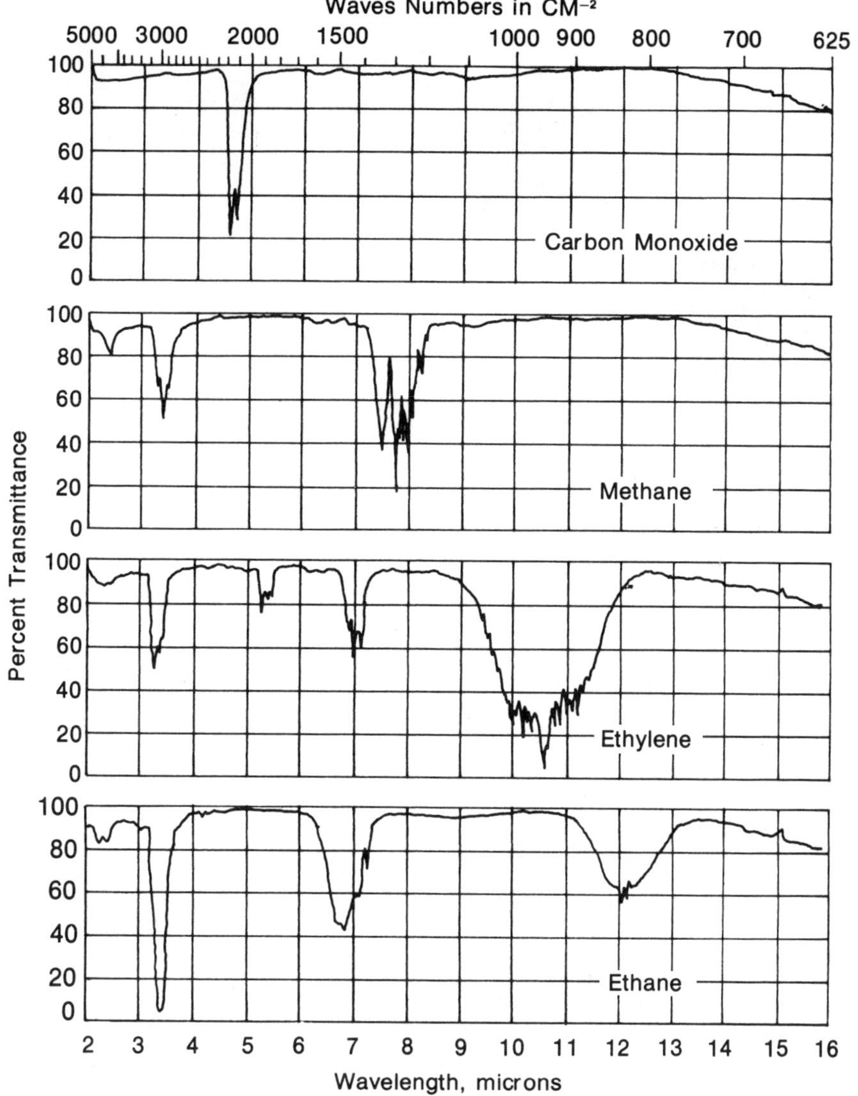

Figure 9.23. Typical infrared spectra revealed by a laboratory test. Sharp drops from near 100% transmittance show wavelengths where listed components absorb IR energy. (From "Process Instruments and Controls" by D.M. Considine, published by McGraw-Hill)

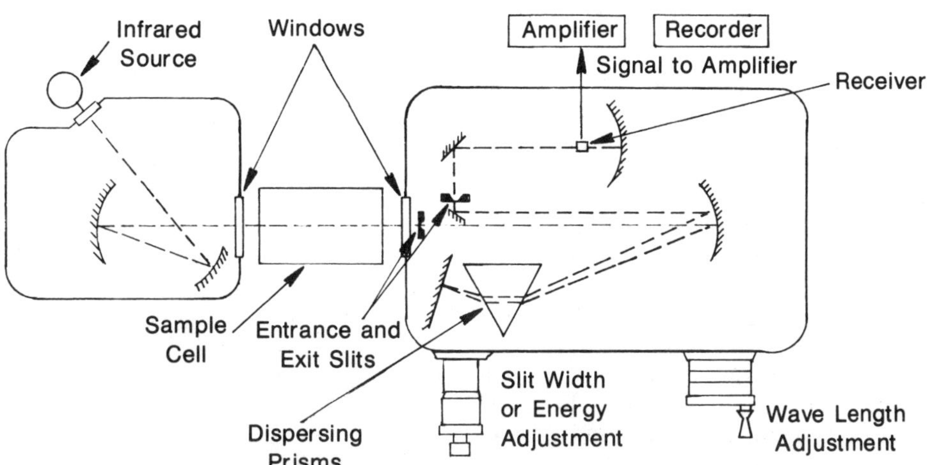

Figure 9.24. A single-beam dispersive spectrophotometer. The use of prisms, slits and mirrors eliminates all but the narrow IR wavelength desired for a particular component. (From "Process Instruments and Controls" by D.M. Considine, published by McGraw-Hill)

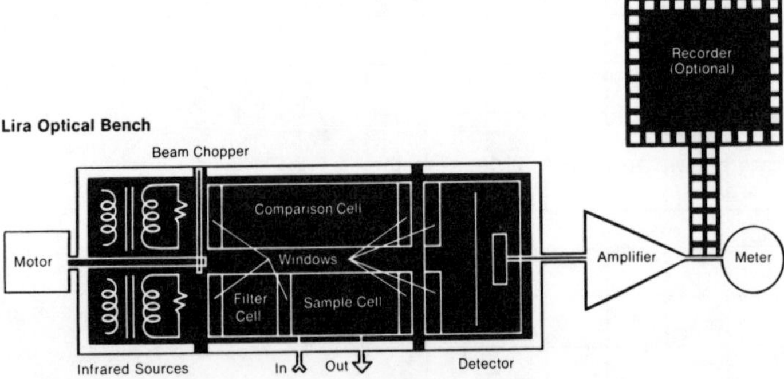

Figure 9.25. Schematic diagram of a nondispersive infrared analyzer. Undesired wavelengths are eliminated by the windows. (Courtesy of Mine Safety Appliance Co.)

the detector from the sample beam. The beam becomes unequal and the radiation entering the detector varies in intensity as the beams are alternated by the chopper.

Nondispersive analyzers are classified as negative-filtering (nonselective detector) or positive-filtering (selective detector). In Figure 9.25, the detector may be a phototube or bolometer, both of which respond to all wavelengths of radiation; hence, it is a nonselective detector. On the other hand, the detector may be a condenser type, filled with pure gas of the component to be measured. The detector becomes sensitive to this component gas; hence, it is classed as selective.

Infrared absorption is logarithmic with concentration; therefore, if a linear receiver is used, there is a lack of sensitivity in applications where the range approaches 100. The logarithmic response is desirable for trace analysis (narrow ranges), but for higher values, the range must be greater. Filter cell sensitization and special cell length modifications may be used to alter the curve toward linear response.

Detectors

The pneumatic (capacitance) detector is one of the types now available in infrared analyzers. Also known as a microphone detector, it is used in two basic forms, single chamber (combined beam), and dual chamber (opposed beam). Both systems compare the amount of energy absorbed from a reference cell and from the measured source of infrared radiation.

In the dual chamber method, IR energy emerging from both the optical systems (the reference side and the measuring side) are compared by their effect on a flexible metallic diaphragm (Figure 9.26 bottom). The detector chambers are filled with a gas which absorbs energy at the wavelength of interest. The gas expands as energy is radiated. Unequal amounts of energy cause the flexible metallic membrane to deflect and change the capacitance between it and a fixed metal plate. The capacitance change is proportional to the material component being measured.

In the single chamber method, both the reference and measured sides affect the single gas-filled chamber (Figure 9.26 top). The detector system utilizes a "chopping" action which alternately compares the IR energy radiation from each side of the optical system. When unequal amounts of energy come from this detection system, unequal pulses are generated from the electrical measuring systems of which the capacitor is a part. The difference in pulse height from the two systems is proportional to the measured component.

Another detector—the bolometer type—is made of resistance elements whose resistance changes as a result of incident radiation directed upon it. For high-speed response, the element can be platinum ribbon or a sensitive thermistor. Single or dual elements (for single or dual beam analyzers) are used in the arms of a Wheatstone bridge circuit. Single elements measure the change in incident light, and dual elements measure the difference in light between two beams.

The thermopile detector is made of several thermocouples connected in series. Although slower in response than other methods, the thermal mass can be made small enough to respond to "chopped" radiation (15 to 30 Hz). Like the bolometer detector described earlier, a thermopile may be used in either one of two arms of a Wheatstone bridge to measure changes or differences in incident radiation.

Calibration

Process infrared analyzers usually are equipped with auto-zero capability to compensate periodically for drift or changes in light intensity.

A minimum of three standard gases are normally required for calibration, one zero gas and two span gases. Span gases are prepared (or purchased) with varying percentages of the gas to be analyzed. By flowing these gases through the analyzer, points may be plotted for minimum and maximum span, thereby establishing a calibration curve.

pH

The pH analysis is one of the most widely used analytical measurements in the manufacturing industry. In the refining and petrochemical industry, pH measurement is used extensively for the following purposes:

1. Quality control: where pH control determines product uniformity.
2. Corrosion inhibition: pH monitoring is used in conjunction with oxygen control to minimize corrosion in high-pressure boilers and other water streams.
3. Effluent pH control: neutralization of liquid discharges into public streams.

The measure of the acidity or alkalinity of a solution, pH, is defined as the negative logarithm of the hydrogen ion activity. The pH range covers both acid and alkaline solutions. In water, the equilibrium product of the hydrogen ($H+$) and hydroxyl-ion ($OH-$) concentration is a constant, 10^{-14}, at 22°C. Since the $H+$ and $OH-$ concentrattions are thus

Single Chamber Pneumatic Detector

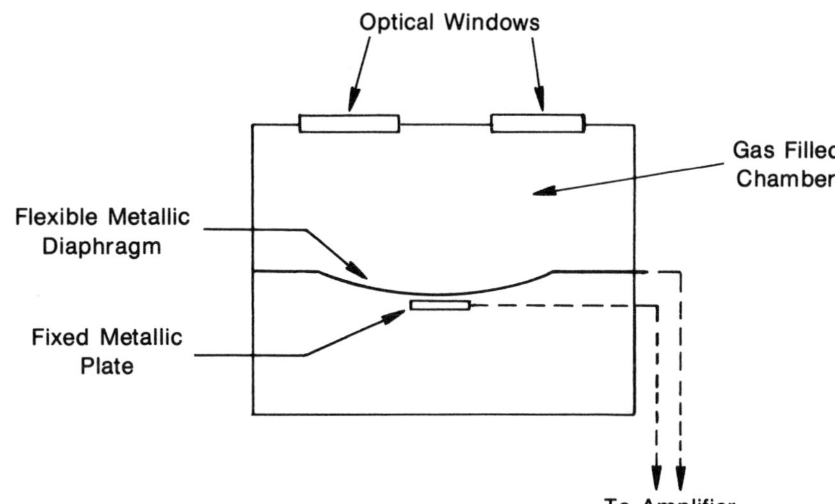

Dual Chamber Pneumatic Detector

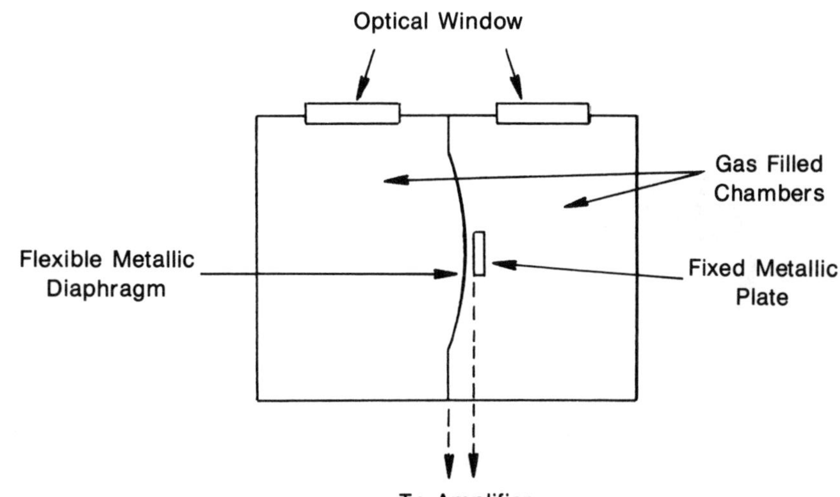

Figure 9.26. Schematic diagrams of single and dual pneumatic detectors for infrared analyzers.

equal, the H+ ion concentration must be 10^{-7} and the pH= log $(1/10^{-7})=7.0$. Acid solutions increase in strength as the pH value decreases below 7, and alkaline solutions increase as the pH rises above 7. The scale is not linear since a change of one unit of pH represents a change by a factor of 10 in the effective strength of the acid or base. Only the number of hydrogen ions actually dissociated in a solution is measured by pH and not the total H+ or OH− ions present.

Figure 9.27 shows how errors due to temperature changes can occur in pH measurements. As the temperature rises, more hydrogen ions are dissociated; hence, a change in the millivolt output occurs to either lower the apparent pH in the case of acid solutions or raise the apparent pH in the case of base solutions. The magnitude of such changes are evident in Figure 9.27. Temperature errors are inherent in uncorrected electrode output; a method of temperature compensation is necessary.

A basic process pH measurement system consists of a pH sensitive electrode, a reference electrode, a temperature compensating electrode and a potential measuring device.

Sensing Electrodes

The sensing electrode consists of a pH sensitive glass membrane joined to an insulating envelope or tube (Figure 9.28). When this electrode is immersed in an aqueous solution, a potential is developed across the membrane which is proportional to the concentration of hydrogen ions in the

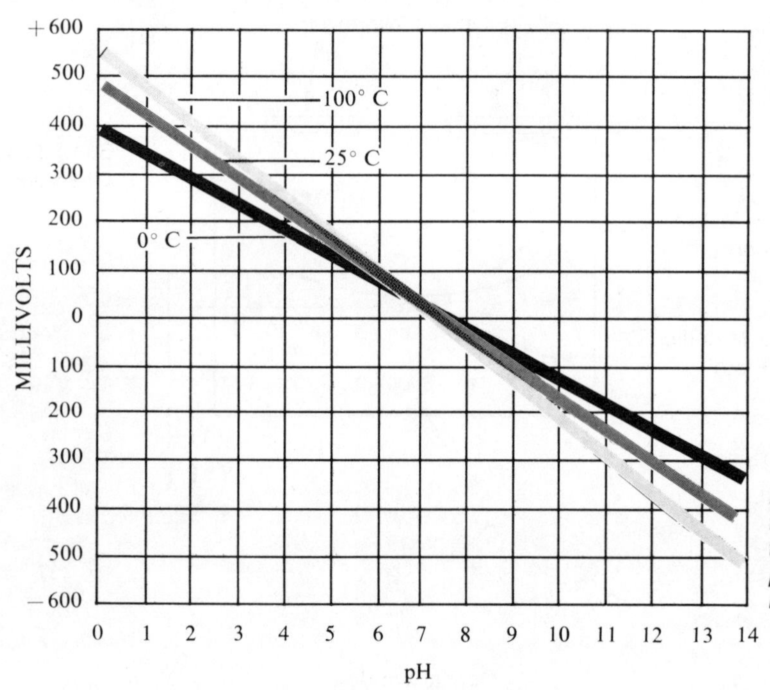

Figure 9.27. pH-millivolt relationship shown at three different temperatures reveals the necessity of stable temperature control or temperature compensation in pH measurements. (Courtesy of Beckman Instruments, Inc.)

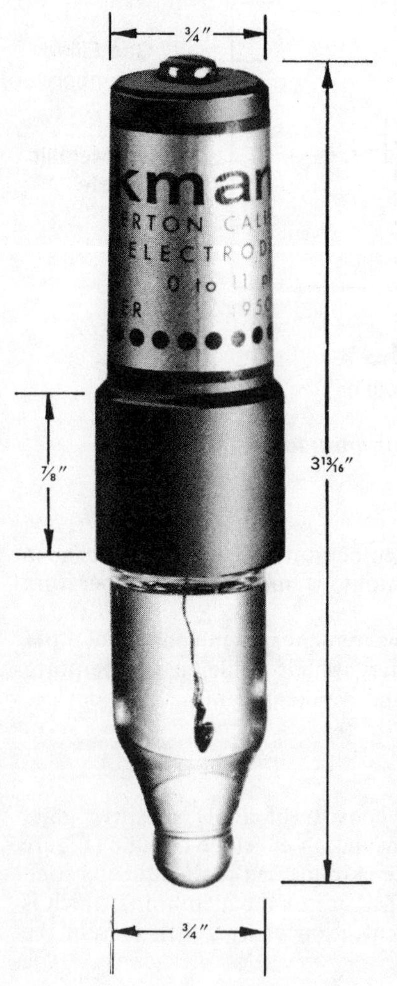

Figure 9.28. pH sensing electrode. A potential is developed across the glass membrane when immersed in an aqueous solution. The inner electrode is visible. (Courtesy of Beckman Instruments, Inc.)

solution. Since the glass membrane offers a high resistance to ion flow (100 megohms), this potential must be measured from a point inside the envelope to a point in the solution outside the envelope. A silver/silver chloride element is immersed in a buffer solution to act as the inner contact. An adjacent reference electrode contacts the outer surface through the measured solution. An electrical measuring circuit then measures the potential across the glass membrane.

Reference Electrodes

Voltage measurements must be made with reference to some stable electrode which does not change with pH. This is accomplished with a reference electrode (Figure 9.29), one type of which is the calomel electrode. This electrode uses a platinum wire immersed in mercury which in turn is in contact with KCl-mercurous chloride paste. Another popular reference electrode is the silver/silver chloride liquid junction type, made of a silver wire coated with silver chloride in contact with a reservoir of electrolyte. Electrical continuity is achieved through a liquid junction which permits electrolyte to flow outward from the electrode into the sample solution. Four types of liquid junctions are used to accomplish this: (a) the palladium junction which consists of a length of crimped palladium wire imbedded in the glass with a microscopic opening between the wire and the glass, (b) the fiber junction which uses a long asbestos fiber inserted through the glass stem, permitting the flow of electrolyte by capillary action, (c) the frit junction, a pressed-sintered Carborundum which forms a permeable passage for the flow of electrolyte, and (d) the ground glass sleeve junction which is made up of a ground glass taper on the electrode body and a matched ground sleeve. Two holes in the matched taper permit flow of the electrolyte. Each of the

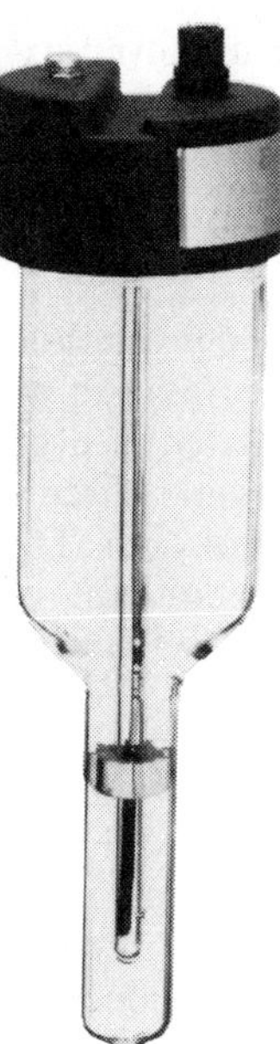

Figure 9.29. A typical calomel reference electrode, used as a reference voltage source in pH measurement. (Courtesy of Beckman Instruments, Inc.)

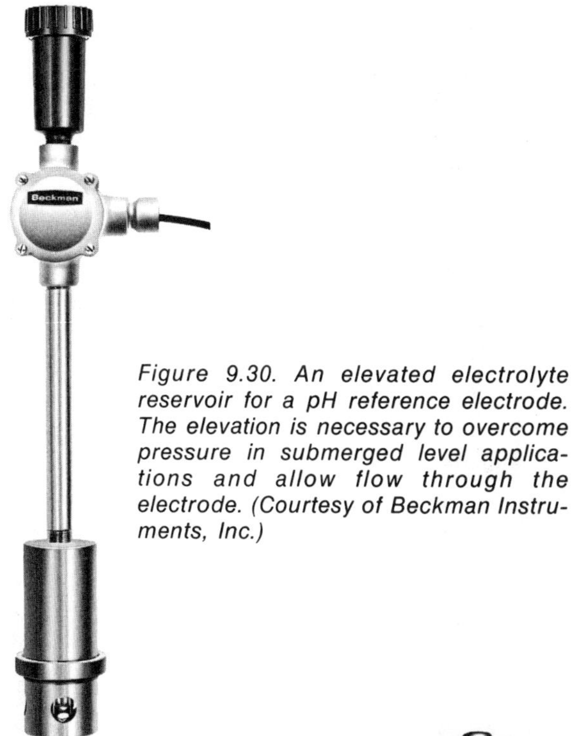

Figure 9.30. An elevated electrolyte reservoir for a pH reference electrode. The elevation is necessary to overcome pressure in submerged level applications and allow flow through the electrode. (Courtesy of Beckman Instruments, Inc.)

four types is best suited for a particular type liquid or a particular set of operating conditions.

In pressure or submerged-level applications, the process pressure must be overcome in order to have a flow through the liquid junction of a reference electrode. This is accomplished by (a) installation of an electrolyte reservoir at an elevated height such that its weight is sufficient to overcome the pressure at the electrode (Figure 9.30), (b) use of a positive air pressure on the reference electrode as shown in Figure 9.31 or (c) use of a self-pressurizing electrode as shown in Figure 9.32. When the electrode is installed in a flow chamber, the process solution rises in the lower section of the electrode, compressing the air to a pressure equal to that of the process stream. This pressure is transmitted onto the electrolyte in the upper chamber. Pressure at the liquid junction is greater than the process pressure by an amount equal to the weight of the electrolyte column. This type electrode is limited to applications where the pressure differential does not exceed 15 psig and the temperature is no higher than 175°F.

Errors

Temperature fluctuations affect the calibration of a pH analyzer. Small errors occur in the reference electrode voltage and the internal voltage of the sensing electrode. These errors, however, are minimized by using electrodes of similar material composition. Solubility of chemicals in the cells varies with temperature, resulting in a period of signal drift until a new state of equilibrium is reached following the change in temperature (Figure 9.33).

Modern pH process analyzers are mostly the direct-reading type because changes are more easily read over the wide range of 0 to 14. Accuracy is good (±0.03 pH) especially with the dual or "folded" scales of 0 to 8 and 6 to 15 pH units. The electronic measuring circuit includes a feedback amplifier where the amplifier output is applied across a precision resistor. The resulting current, measured by an ammeter, may be used as an analog output. This type circuit minimizes amplifier and capacitance errors in the sensing electrode leads. The unit zeroes itself automatically each second for more stable operation.

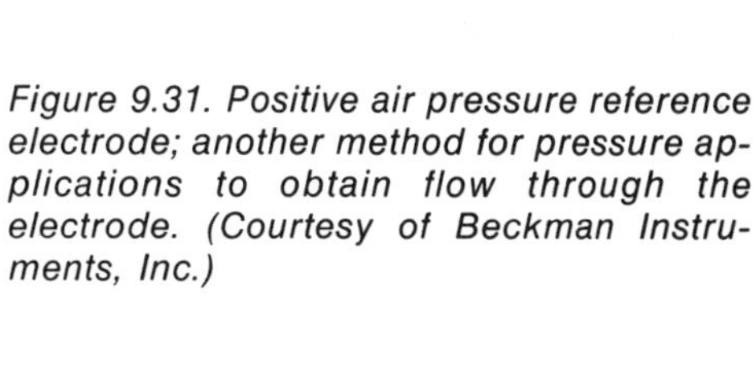

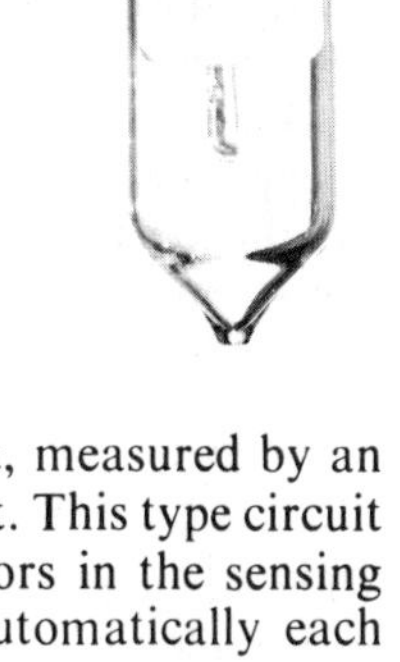

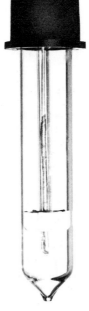

Figure 9.31. Positive air pressure reference electrode; another method for pressure applications to obtain flow through the electrode. (Courtesy of Beckman Instruments, Inc.)

Calibration

Calibration of the pH analyzer is basically simple although tedious. Zero adjustment is automatic since the analyzer incorporates feedback circuitry. Reference elec-

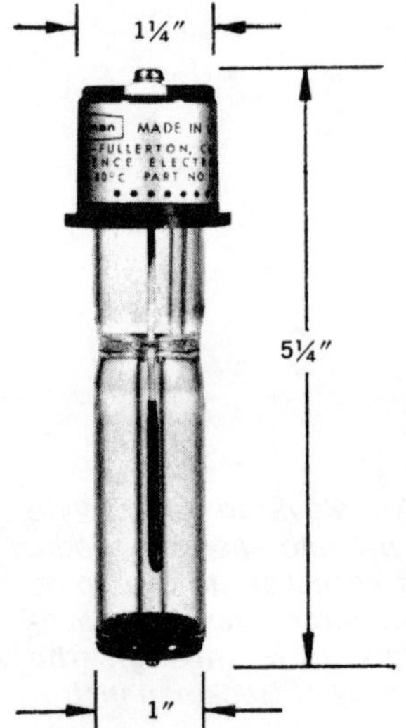

Figure 9.32. Self-pressurizing reference electrode. Used when process pressure does not exceed 15 psig. (Courtesy of Beckman Instruments, Inc.)

Figure 9.33. Thermocompensator cell is used to compensate for errors due to temperature fluctuations. The voltage to pH ratio changes with temperature. (Courtesy of Beckman Instruments, Inc.)

trodes designed for continuous operation provide for sufficient solution, and positive flow of this solution maintains reliable liquid junctions. Sensing electrodes require buffer solutions in the approximate range at which the analyzer will operate. Standard buffer solutions are available from manufacturers to cover the entire pH range. Calibration with standard buffer solutions should be at two points, one below and one above the expected normal pH value.

Oxygen

The universal demand for oxygen analysis due to its essential role in oxidation, combustion and industrial processing applications has led to a large number of varied techniques applied to process analyzers. The recent intense interest in the ecological field probably will lead to further advances, especially in the analysis of dissolved oxygen. The most widely used methods of oxygen analysis are the deflection paramagnetic, thermal paramagnetic, catalytic combustion, microfuel cell and galvanic cell. For dissolved oxygen analysis, polarographic, galvanic and thallium cells are used.

Deflection Paramagnetic

As the name implies, the deflection paramagnetic method of analysis is based on the strong attraction of oxygen by magnetic fields while other gases are less attracted or even repelled. Figure 9.34 shows the relative magnetic susceptibilities of a few common gases compared to oxygen. The only other gases that exhibit an appreciable amount of susceptibility are NO and NO_2, and they are rarely found as contaminates in process streams.

Paramagnetic susceptibility is measured directly by determining the change of the magnetic force acting on a test body suspended in a nonuniform magnetic field while the test body is surrounded by test gas. As indicated in Figure 9.35, if the sample gas is more paramagnetic than the test body, the test body is repelled from the field of maximum flux density. If the sample gas is less paramagnetic, the test

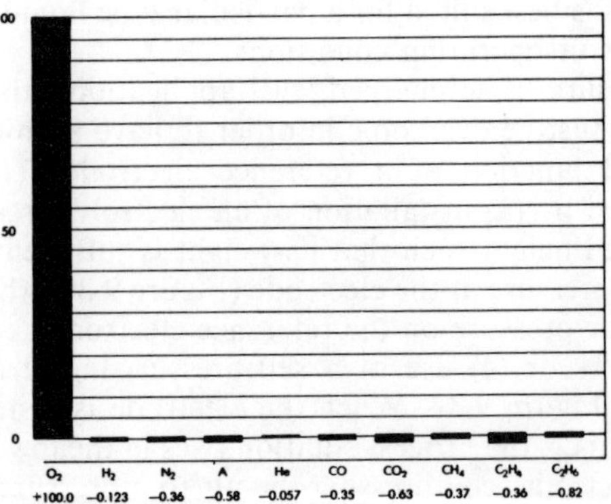

Figure 9.34. Relative magnetic susceptibilities of common gases. (Courtesy of Beckman Instruments, Inc.)

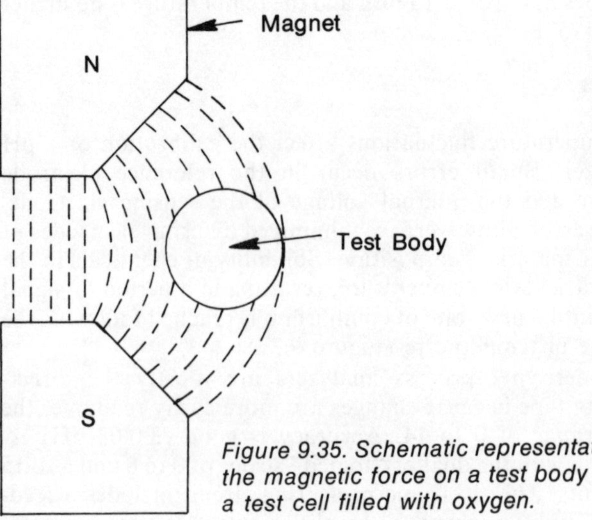

Figure 9.35. Schematic representation of the magnetic force on a test body within a test cell filled with oxygen.

body is drawn toward the field. Figure 9.36 depicts the test body made up of two hollow glass spheres on a bar which is supported on a torsion fiber. A small mirror is mounted at the center of the bar to reflect light from a source to a light-dividing mirror. The mirror reflects the light to two phototubes whose outputs are proportional to the reflected light incident on their surfaces. With no oxygen present, the reflected light is equal on each phototube, and their outputs are equal. In the presence of oxygen, however, the test body is rotated on its fiber suspension, the amount of light reflected to the phototubes is unequal, and their outputs are unequal. Since the difference in the phototube outputs is proportional to the difference between the magnetic susceptibilities of the test body and the gas which the test body displaces, the imbalance is then a linear function of oxygen concentration. The phototube imbalance is amplified and fed back to the test body to control its potential. Figure 9.36 shows a null balance circuit with a motor driven potentiometer which supplies a variable electrostatic potential to the body, and the body is held in null position. Any change in oxygen concentration is followed by subsequent change in electrostatic force, and null conditions are achieved. Recorded output is taken from the DC potential required to maintain null position.

Close temperature control is required for high accuracy, but this is achieved by using low sample flows and allowing the gas to reach the temperature of the analyzer. Sample gas flow above 100°F, however, must be cooled before entering the analyzer.

The sensing unit is designed with internal shock mounting to protect the test body suspension system and the optical mirrors.

Automatic self-standardization is available. Periodic calibration checks consist of checking zero with an oxygen-free standard and using a standard sample of known concentration to check one point of the linear curve.

Thermal Paramagnetic

The thermal paramagnetic analyzer combines the principles of the paramagnetic susceptibility of oxygen with the thermal conductivity principle of TC cells. This method avoids the use of light beams, reflective lenses, torsional fibers, diaphragms and moving parts that must be kept in critical adjustment.

Figure 9.37 depicts a thermomagnetic cell which is merely a thermal conductivity cell (with Wheatstone bridge circuitry) equipped with a magnet to provide magnetic flux in the area of one filament. When sample gas flows through two cells as shown, the filaments are cooled equally, and there is no measurable difference in their resistance. However, when a magnet is moved into position so that one filament is located in a region of high magnetic flux, oxygen, which is paramagnetic, is concentrated in that region, displacing the other gases. As the oxygen is heated by the element, it loses its magnetic susceptibility inversely as the square of the temperature. The heated oxygen is displaced by cooler oxygen with higher magnetic susceptibility. The induced flow of oxygen gas past the filament cools and

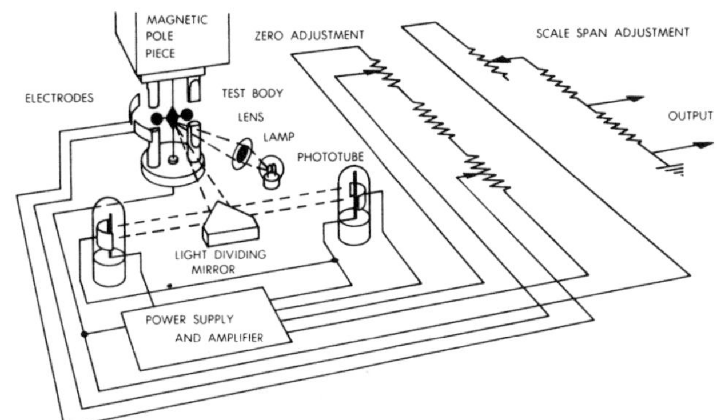

Figure 9.36. Schematic diagram of paramagnetic oxygen analyzer. (Courtesy of Beckman Instruments, Inc.)

changes its resistance, thereby upsetting the electrical balance of the bridge measuring circuit. This imbalance is proportional to the effective magnetic susceptibility of the gas and, hence, to the amount of oxygen present.

Calibration of this analyzer is dependent upon constant temperature and pressure conditions. Temperature changes result in resistance variations of the thermal conductivity element which, in turn, changes the output of the bridge measuring circuit. This error is prevented by case temperature control.

Changes in the sample stream or atmospheric pressure affect the flow rate through the cell and therefore increase the conductivity of heat away from the thermal conductivity element. One method of compensating for changes in atmospheric pressure, shown in Figure 9.37, makes use of two compensating cells of different diameters. Since both cells are exposed to atmospheric pressure through a common port, the larger cell contains a larger volume of air and is more strongly influenced by convection from pressure change than is the smaller cell. By precise sizing of the two cells, this difference is used in the bridge measuring circuit to offset the simultaneous change in the sample cell due to the atmospheric pressure change.

Calibration of this type analyzer is extremely easy. Zero is checked by swinging the magnet away from the measuring cell. The calibration is completed by checking one upscale point. If the range spans the concentration of oxygen in air (20.7%), air is passed through the cell and the analyzer is adjusted to read 20.7%. If the range is lower than 21%, air may still be used by switching in a calibrating resistor.

Catalytic Combustion

The catalytic-combustion analyzer measures oxygen indirectly by measuring the heat content of an oxidizing fuel. This is accomplished by adding hydrogen to the oxygen-containing gas sample and passing the mixture over a heated noble-metal filament which causes combustion to take place. The filament assembly in Figure 9.38 consists of two noble-metal thermal conductivity filaments mounted in separate compartments of a sample cell. The measuring fila-

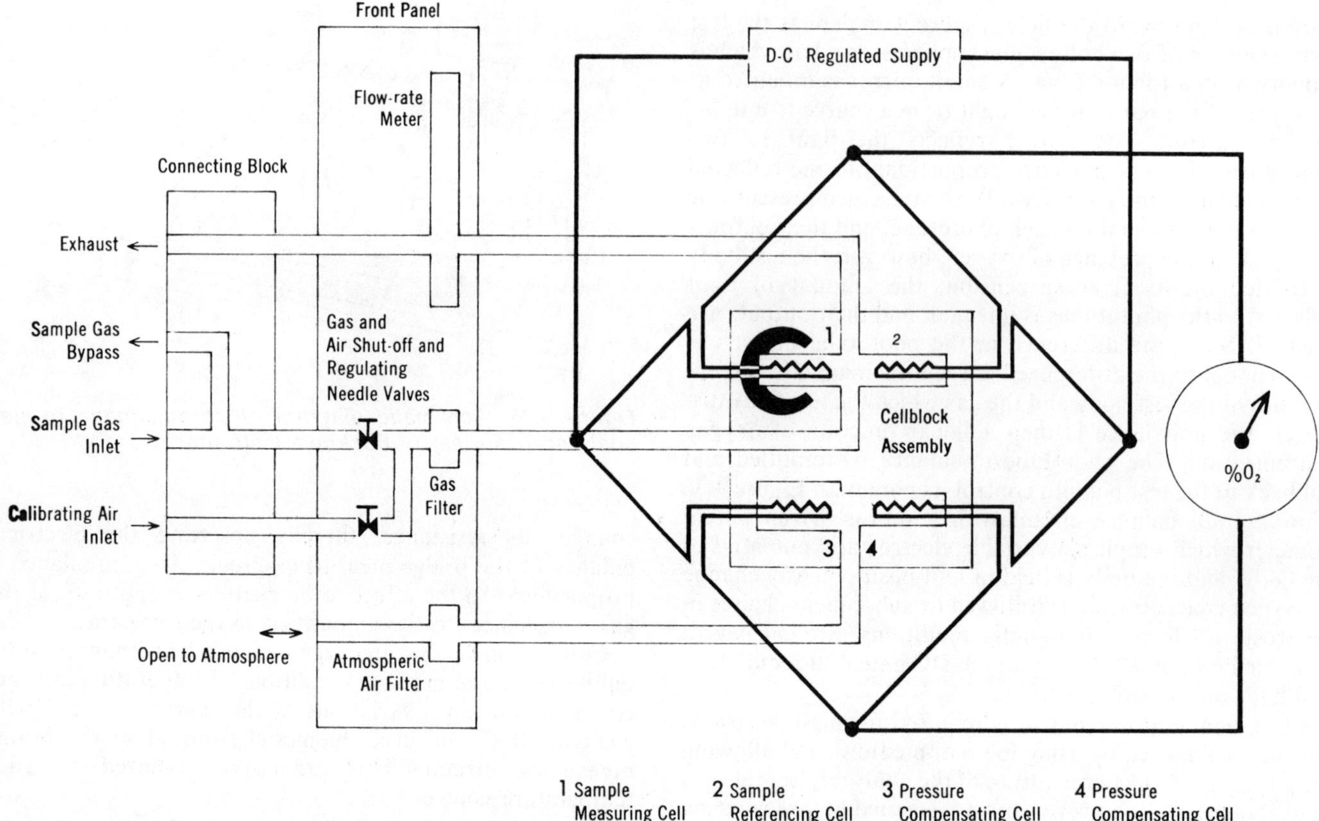

Figure 9.37. A schematic arrangement of a thermomagnetic O_2 Analyzer. (Courtesy of Leeds and Northrup)

ment is covered with a screen so that it is fully exposed to the hydrogen-oxygen gas mixture. The compensating filament compartment is closed except for a small hole which permits a small amount of the mixed gas to diffuse in, compensating for temperature and normal conductivity variations. Combustion takes place in each chamber due to the temperatures of the filaments. The rise in temperature increases the resistance of the measuring filament proportional to the amount of oxygen present in the mixed gas stream and causes an imbalance in the Wheatstone bridge signal circuit. Calibration check is made simply by flowing samples of known oxygen content through the analyzer and making adjustments to the recorder readings.

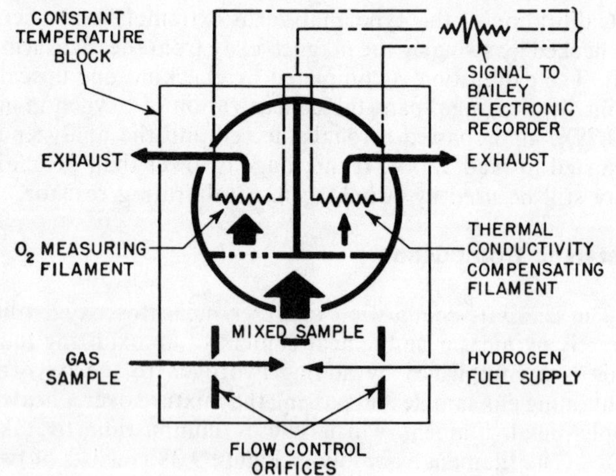

Figure 9.38. Catalytic combustion detector cell for oxygen analyzer. (Courtesy of Bailey Meter Co.)

Few process streams are free of combustible gases, and in a combustible analyzer, these gases burn at the same time with the oxygen. They must be measured and accounted for in the 0_2 analysis. The same cell configuration used for the hydrogen sample gas mixture is also used to burn the combustibles in the sample. In this cell, however, air is mixed with the sample gas instead of hydrogen. In fact, both cells may be built into the same block with one reading oxygen content and the other reading combustibles content.

Microfuel Cell

One analyzer is available which operates on a patented electrochemical transducer with unique features. The cell is a sealed disposable disc, specific to oxygen, which generates current flow in the presence of oxygen. In operation, oxygen diffuses through a Teflon membrane and is reduced on the surface of the cathode. A corresponding oxidation occurs at the anode internally, and a current is produced which is proportional to the concentration of oxygen. In the absence of oxygen, no current is produced; therefore, no zero calibration is required. The cell output is very stable, and the useful life of the cell depends upon the length of time and

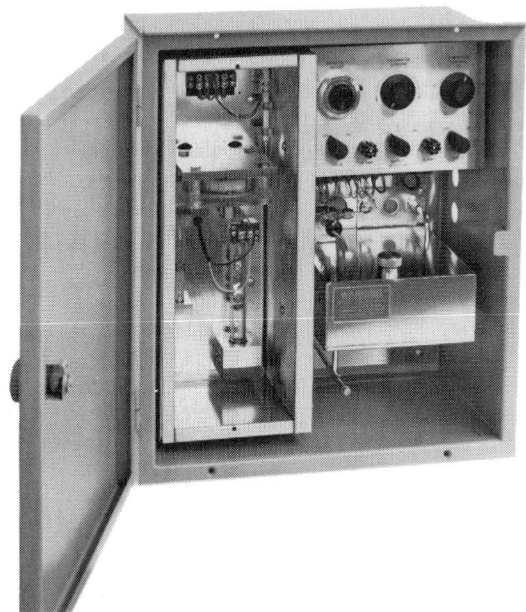

Figure 9.39. Galvanic cell O_2 analyzer operates on an electrochemical principle. (Courtesy of Teledyne Analytical Instruments)

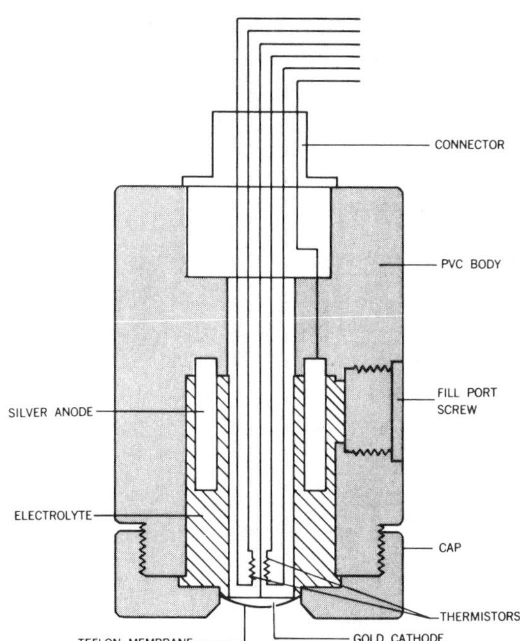

Figure 9.40. Polarographic dissolved oxygen cell. Oxygen diffusion through Teflon membrane causes electrical current proportional to oxygen content. (Courtesy of Beckman Instruments, Inc.)

magnitude of oxygen exposure. One of the attractive features of this type unit is its low maintenance cost. The cell is merely replaced periodically by inserting a new one.

Galvanic Cell

Another method of monitoring trace amounts of oxygen is the galvanic cell analyzer. Recent developments in paramagnetic and fuel cell types have caused a decrease in its use. There are some applications, however, where it is still superior. The cell (Figure 9.39) consists of a noble-metal cathode such as silver, a lead anode and a suitable electrolyte such as potassium hydroxide. The electrodes are polarized by an applied voltage causing an electrochemical reaction when oxygen contacts the electrodes. The reaction releases electrons, producing a current flow through the electrolyte, the magnitude of which is proportional to the oxygen content in the electrolyte.

The electrolyte must be selected for compatibility with the process stream. With proper selection of the components, sensitivities in the range of 1 ppm are possible. Maintenance and downtime are higher than for other type O_2 analyzers because of the need for frequent electrolyte replacement.

Dissolved Oxygen

Polarographic Cell

The polarographic cell operates on the electrolytic principle as shown in Figure 9.40. A silver anode and gold cathode are both protected from the sample by a thin membrane of Teflon. An aqueous solution of KCl serves as an electrolytic agent. The Teflon membrane is permeable to gases; therefore, oxygen diffuses to the cathode where it is reduced, causing a flow of ions to the anode. This current, proportional to the amount of oxygen present, is amplified and displayed in parts per million by weight of dissolved oxygen. The polarographic cell requires temperature compensation for accuracies of ± 1 to $\pm 2\%$. Two thermistors in the sensor are used for this purpose.

The sensor may be submerged to depths up to 100 feet or used in sample streams up to 50 pounds pressure. The only maintenance required on the sensor is the infrequent replenishing of the electrolyte. Calibration is done electronically with the sensor exposed to air.

Galvanic Cell

Galvanic action is the basis of operation for the dissolved oxygen analyzer for water shown in Figure 9.41. This unit is a simple galvanic cell consisting of a noble-metal electrode and an anode of less noble metal. With a flow of sample through the cell, O_2 is ionized, and the drift of these ions creates a measureable current. Minimum conductivity of the water is assured by the constant addition of calcium carbonate before the sample reaches the cell. Current flow is proportional to the amount of oxygen dissolved in the water. Sample temperature and pressure must be controlled. The unit contains a self calibrator, enabling all calibrations to be done at the analyzer.

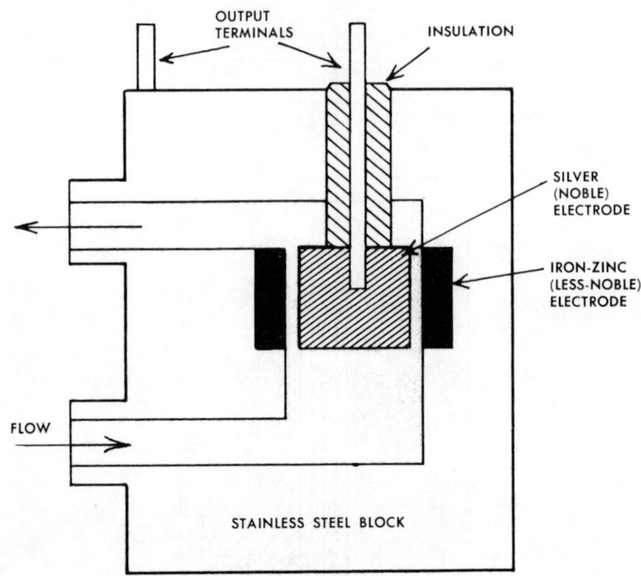

Figure 9.41. Galvanic dissolved oxygen cell. (Courtesy of Milton Roy Co., Hays Div.)

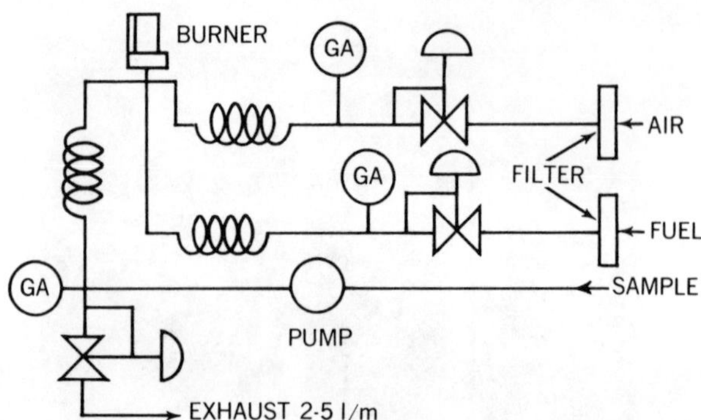

Figure 9.42. Typical sample system for hydrocarbon analyzer using hydrogen flame ionization detector in a vapor system. (Courtesy of Beckman Instruments, Inc.)

Thallium Electrode

Requiring no membrane or electrolytic gel, the thallium electrode detector measures oxygen potentiometrically in an aqueous solution, using the reactions of oxygen with an exposed thallium electrode. This device utilizes a ring of thallium metal and a temperature-sensitive resistance element molded into a standard flanged probe body. A specially designed reference electrode fits into the hollow center of the probe and is secured with a gland nut. When oxygen in water contacts the thallium electrode, a reaction takes place at the surface, producing a thallous ion concentration proportional to the dissolved oxygen. Electronic circuitry converts this output to a linear function. Temperature compensation is furnished for temperature variation from 0° to 50°C.

Total Hydrocarbon

Vapor Phase

The analysis of vapors for total hydrocarbon content is performed with a modified version of the hydrogen flame ionization detector as described in the section on gas chromatography. In the chromatographic analyzer, the column effluent is mixed with the hydrogen fuel and burned in an air atmosphere. In total hydrocarbon applications, nitrogen is premixed with hydrogen in order to control flame temperature, and this fuel is mixed in a sample system with sample gas and air to form a combustible mixture (Figure 9.42).

The thermal energy in the hydrogen flame plasma induces an emission of electrons from the hydrocarbon molecules in the gas mixture. The ion current collected at the electrodes is amplified by an electrometer. The magnitude of this signal is a function of the number of carbon atoms passing through the flame at any given time. This means that the unit is not only flow sensitive but that the signal from C_6 molecules is twice as great as that from C_3 molecules.

Response of the total hydrocarbon analyzer is linear, and sensitivity is 0 to 1 ppm with ranges from parts per million to percent, depending upon the application.

Liquid Phase

The analysis unit consists of sample preparation equipment and an infrared analyzer (Figure 9.43). A microliter liquid sample is injected into an air carrier stream. The carrier flow transfers the sample into a catalytic combustion tube controlled at 950°C. Here the water is vaporized to steam, and the carbonaceous material is oxidized to carbon dioxide (CO_2). Leaving the combustion tube, the steam is condensed and the CO_2 enters the infrared cell which is sensitized to measure CO_2 as described in a previous section on infrared analyzers. Figure 9.44 depicts a typical recorder chart readout. This data must then be correlated with BOD (biological oxygen demand) and COD (chemical oxygen demand) analysis.

Analysis time for this analyzer is 2 to 5 minutes, compared to several days for standard BOD and COD analysis. Standard solutions are used to calibrate the analyzer in milligrams of carbon per liter of sample. Calibration ranges are available from 0 to 50 to 0 to 4,000 milligrams per liter carbon.

Total Organic Carbon

The total organic carbon (TOC) analyzer replaces the standard laboratory BOD test with a continuous flow process analyzer. The analyzer described below may be divided into three sections for the purpose of discussion: (a) on infrared analyzers. Figure 9.44 depicts a typical recorder chart readout. This data must then be correlated with BOD

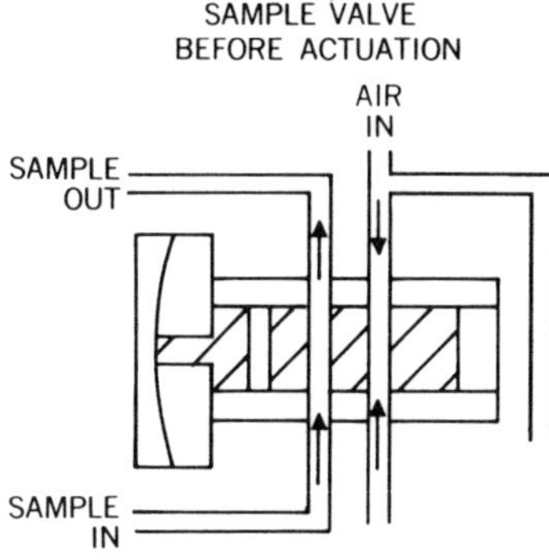

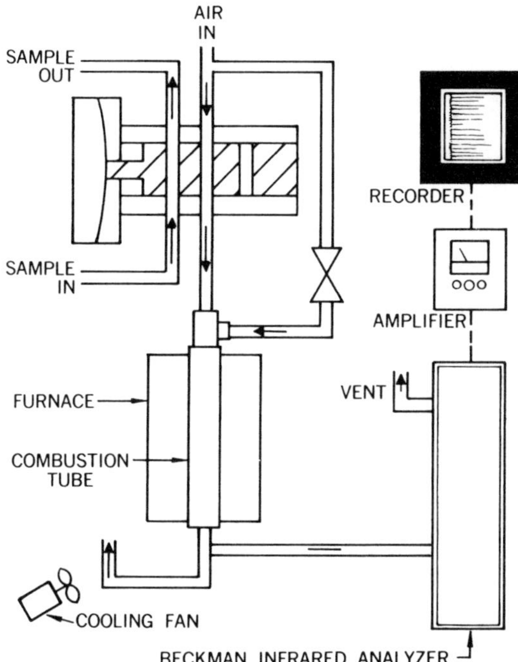

Figure 9.43. Total hydrocarbon analyzer for liquids. (Courtesy of Beckman Instruments, Inc.)

tion system which converts the organic carbon to CO_2 and (c) an infrared analyzer to measure quantitatively the CO_2 which is directly proportional to the TOC in the stream.

In Figure 9.45, the sample from a constant-head tank is treated with hydrochloric acid as it enters the sparge column to shift the carbonate equilibria. In the sparge column, which is a countercurrent multistage gas stripper, CO_2 is removed from the water stream, resulting in an effluent free of inorganic carbon. The carbonate-free sample is metered into the fluidized bed reactor by a positive displacement pump. The reactor is a gas-fluidized bed of 180 mesh aluminum oxide particles at 850°C. Oxygen is supplied to the reactor at 10 times the stoichiometric requirement to oxidize the maximum amount of TOC expected. The oxygen and the vaporization of the sample provide the gas requirements for fluidization of the reactor bed. In the reactor, organic carbon in the sample is oxidized to CO_2. The reactor effluent is a gas stream of water, CO_2 and O_2, plus oxidation products of other noncarbonaceous compounds. A two-stage condenser cools the effluent vapor stream to 50°C, and condensation is removed. The CO_2 in the effluent gas is measured by a nondispersive infrared analyzer. The measured CO_2 is proportional to TOC in the sample stream with ranges from 0-50 to 0-4,000 milligrams per liter.

Moisture Analysis

Of the numerous methods developed to measure moisture, the following types are probably used most often in the refining/petrochemical industry:

1. Electrolytic hygrometer
2. Sorption hygrometer
3. Microwave adsorption
4. Heat of adsorption

Electrolytic Hygrometer

The primary element of the electrolytic hygrometer is the phosphorus pentoxide (P_2O_5) cell (Figure 9.46). Rhodium electrodes of the cell are wound in a helix on the inner wall of a cylinder potted in plastic, and the electrodes and walls are coated with a thin film of P_2O_5. A flow regulated vapor sample is passed through the cell, and moisture from the sample is absorbed by the P_2O_5. The moisture renders the P_2O_5 electrically conductive, causing a DC current to flow between the electrodes. The current flow electrolyzes the absorbed moisture into oxygen and hydrogen molecules. A continuing moisture content and consequent electrolysis maintains a current flow between the electrodes that is proportional to moisture content. Since electron flow is a result of the number of water molecules present, the unit is flow sensitive and must be calibrated at an established flow.

The P_2O_5 cell is subject to error from the recombination effect at low moisture levels in hydrogen-rich or oxygen-rich gases. This phenomenon is the reverting back to water of the electrolysis products, hydrogen and oxygen. The use of rhodium electrodes minimizes recombination. This error can be corrected, however, by using two cells, one with twice the flow of the other. Since the measurement is one of total moisture mass and since the recombination error is constant, their difference is a true reading.

The electrolytic hygrometer is suitable for vapor analysis only. Liquid samples may be prepared by vaporization or moisture stripping. This second technique involves the stripping of moisture from a liquid sample in a stripper column. Dry nitrogen is used to remove the moisture from the liquid and serves as a carrier to transport the moisture to the analysis cell.

Sorption Hygrometer

The sorption hygrometer compares the changes in frequency of two hygroscopically-coated quartz crystal oscillators which vibrate at a frequency of 9 million Hz. Water vapor is alternately absorbed and desorbed rapidly and

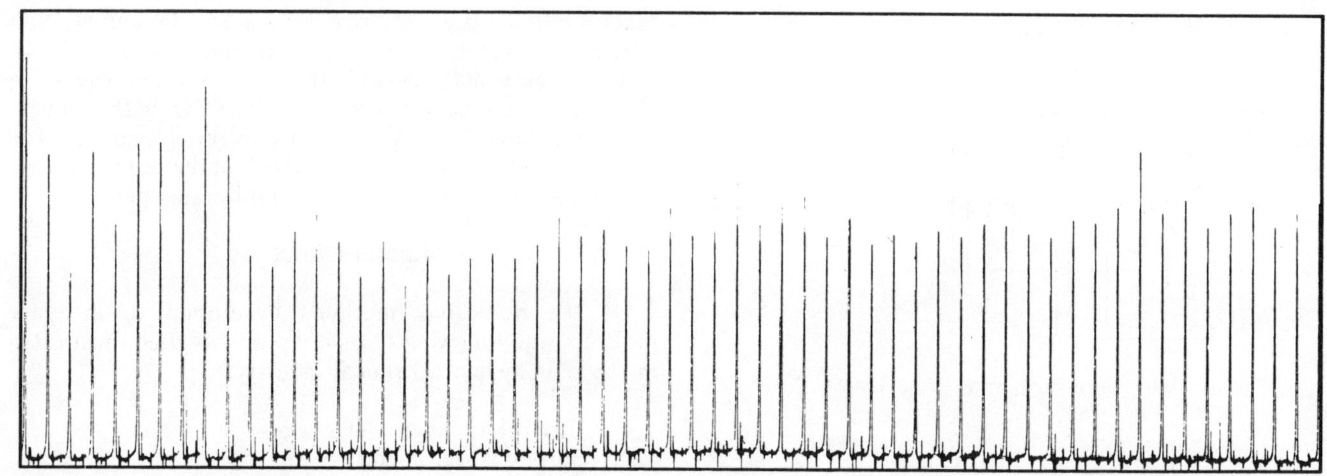

Figure 9.44. Typical carbonaceous analyzer chart readout showing the individual peaks of each analysis in a continuing series of analyses. (Courtesy of Beckman Instruments, Inc.)

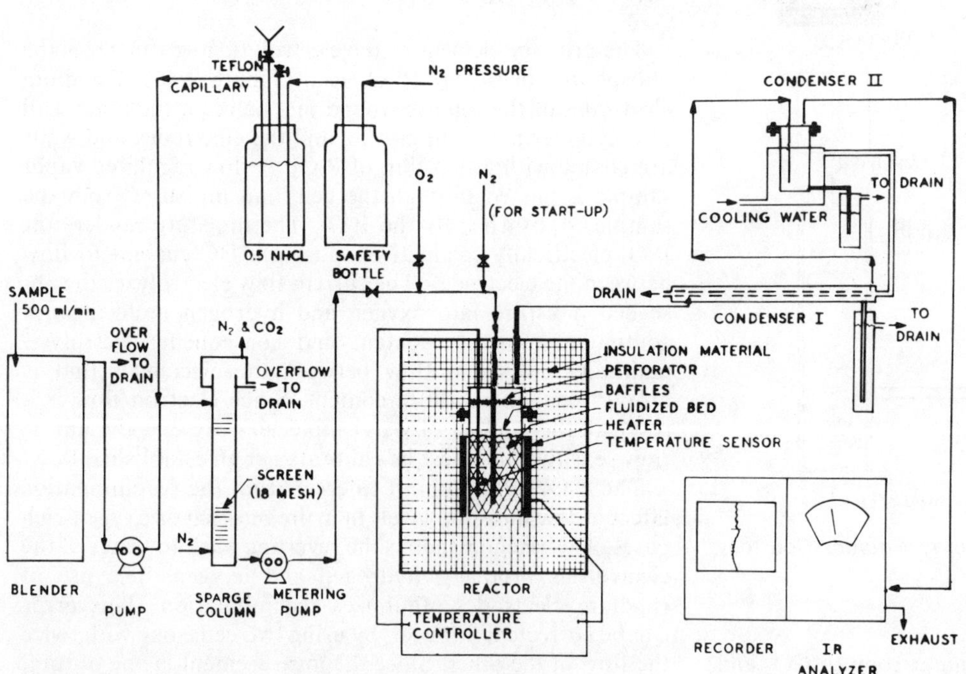

Figure 9.45. Continuous TOC monitor—Flow Diagram.

Figure 9.46. Phosphorus pentoxide cell with rhodium electrodes—primary element of an electrolytic hygrometer used for moisture analysis. (Courtesy of Beckman Instruments, Inc.)

reversibly on each crystal, resulting in mass changes of the crystals. These mass changes produce corresponding changes in frequency. The frequency changes are compared electronically and are proportional to moisture content which is indicated on the analyzer scale in parts per million water vapor by volume. To overcome the problem of measuring dynamic changes of decreasing moisture concentration, each crystal is exposed alternately to the moist sample gas and then to a dry reference gas for 30 seconds (Figure 9.47). While one crystal is absorbing moisture, the other is drying.

The quartz crystals are extremely sensitive to moisture in the atmosphere, petroleum and refrigerant gases, alcohols and low moisture levels in some corrosive gases. Low maintenance and long life are characteristic of their operation.

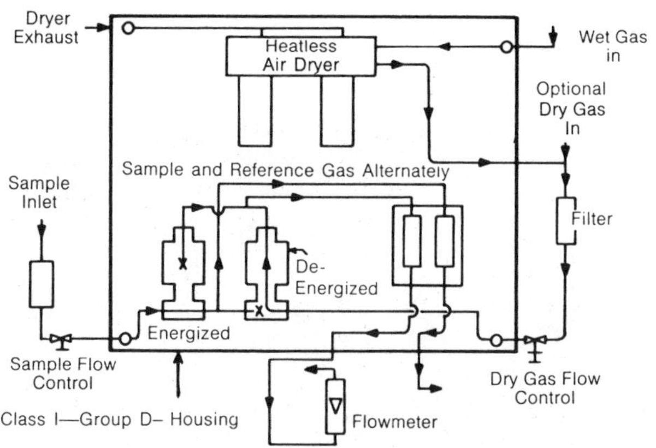

Figure 9.47. Flow diagram of a sorption hygrometer moisture analyzer. Uses moisture sensitive quartz crystals to measure moisture in low ppm range. (Courtesy of E.I. du Pont de Nemours and Co.)

Figure 9.48. Microwave probe for moisture analysis in solids services. (Courtesy of Microwave Instruments Co.)

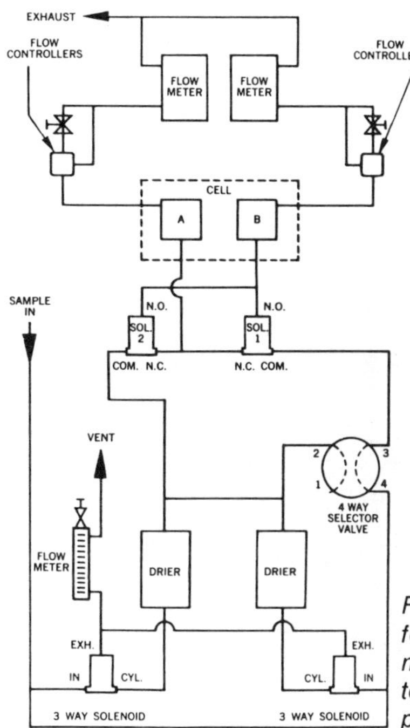

Figure 9.49. Flow scheme for heat of adsorption moisture analyzer. (Courtesy of Mine Safety Appliances Co.)

Microwave Absorption

Microwaves are electromagnetic radiation with frequencies between those of radio and infrared waves (1 to 100 GHz). The microwave analyzer measures moisture by the infrared technique. An infrared optical system similar to those discussed in an earlier section on infrared analysis provides a microwave beam which is transmitted through or reflected by the sample.

The water molecule, a very good absorber of microwaves at certain frequencies, will absorb several thousand times more microwave energy than a similar volume of almost any perfectly dry substance. Hence, the presence of even a few tenths of 1% moisture may double the microwave absorption factor of a substance, and, in the case of percentage increases, this factor may be several thousand. The analyzer senses mass of moisture in the beam path, so readout is in terms of mass moisture per unit volume. A calibration curve is produced by plotting mass moisture per unit volume against percentage. No calibration curve is needed to show only plus or minus deviations from a normal moisture condition. This is frequently used for control applications. Calibration ranges from 0-.5 to 0-100% are available. Temperature compensation should be provided for best accuracy.

The microwave analyzer consists of the main instrument cabinet containing the electronics and the sensing head. The instrument cabinet contains the power supply, electronic circuitry and a meter. The sensing head is available in various configurations depending upon the application. Figure 9.48 shows transmitter and receiver units housed for solids analysis in bins or chutes. Some heads operate on the reflection principle and incorporate both the transmitter and receiver in one housing.

Heat of Adsorption

A unique scheme of moisture measurement, shown in Figure 9.49, is based upon the principle that heat is relased when a vapor is condensed and is adsorbed when a liquid is evaporated. The stream to be measured is split; one stream flows through a heat exchanger while the other is dried in a regenerative dryer and directed through a second heat exchanger. For 2 minutes the dry gas adsorbs moisture from the heat exchanger desiccant, cooling the desiccant. During this same period, the wet sample passes through the other half of the detector-desiccant cell, condensing the water vapor, giving up heat and raising the temperature of that side of the detector cell (Figure 9.50). This condition causes a temperature difference to be developed across the two halves of the detector, which consists of a pair of thermistors in each chamber arranged in a Wheatstone bridge to measure the temperature difference. It produces an output whose magnitude is a measure of the water vapor present. At the end of 2 minutes, each of the two flows is switched to the other heat exchanger. During each half cycle, the temperature differential rises rapidly, passes thorugh a max-

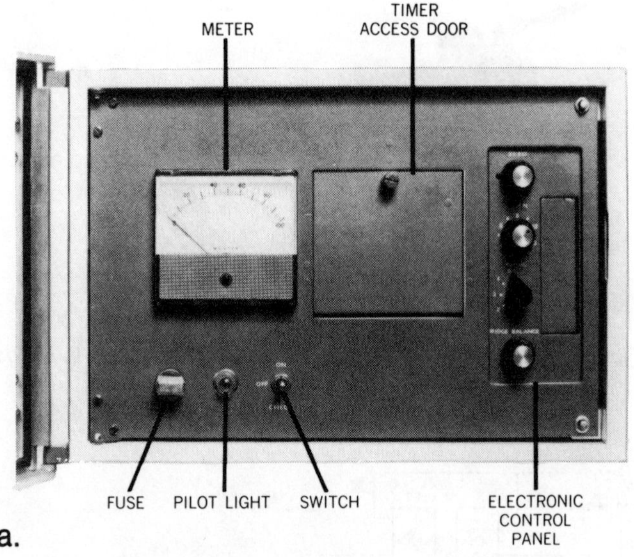

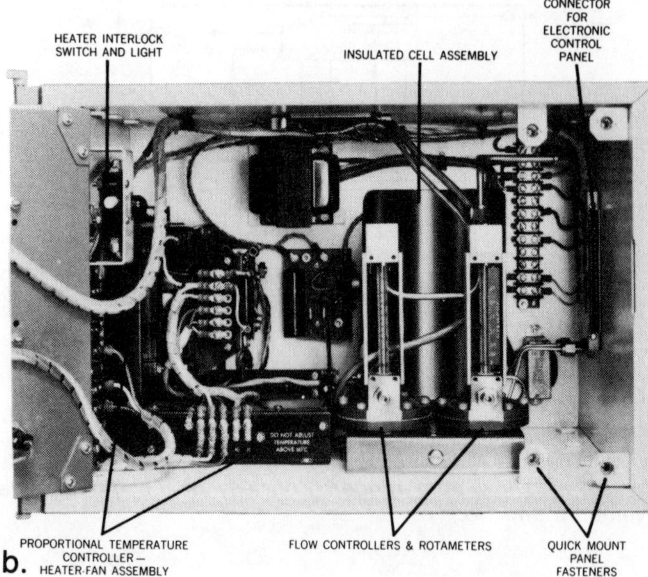

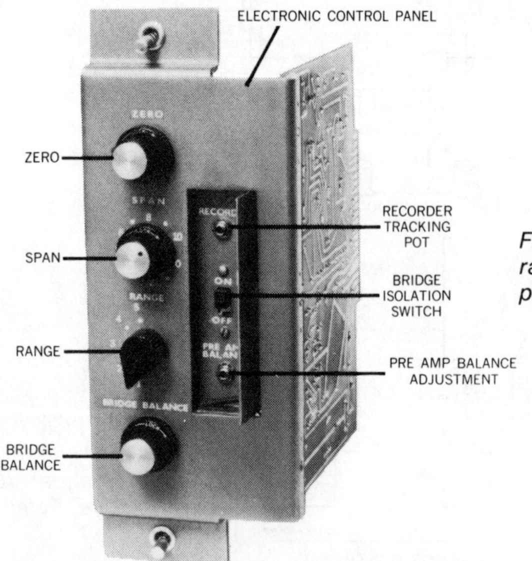

Figure 9.50. Heat of adsorption detector cell showing thermopile arrangement for temperature detection. (Courtesy of Mine Safety Appliances Co.)

imum and diminishes slowly. In each half cycle, the desiccant which has lost heat in the first half cycle gains heat and vice versa. The emf generated is equal to that in the previous half cycle but opposite in polarity; therefore, the thermopile output is reversed each half cycle as well.

The two streams must have constant and equal flows to obtain accurate measurements, and the cycle time must be long enough to establish equilibrium between the desiccants and the streams. The required calibration time may be controlled within limits by the type and size of desiccant material and by gas flow.

The heat of adsorption detector is suitable for vapor service only. Liquid streams, however, may be analyzed by one of two methods: (a) by vaporization of the sample or (b) by using nitrogen to strip the moisture from the stream and passing this vapor through the detector.

Application requirements determine the type desiccant applicable, flow conditions and calibration range. Generally, ranges are possible from 0-10 to 0-5000 ppm.

Air Quality Monitors

Recent environmental considerations necessitate stringent monitoring of stack gases. With fuel costs rising, economic factors also dictate combustion optimization. Excellent devices are now available to aid the power plant operator in optimal steam and power production. Measurements usually include one or more of the following pollutants: carbon monoxide (CO), carbon dioxide (CO_2), methane (CH_4), total hydrocarbons, ozone (O_3), oxides of nitrogen (NO, NO_2, NO_x), and sulfur dioxide (SO_2).

Carbon Dioxide

For precise measurement of carbon dioxide, the analyzers usually selected are the infrared and gas chromatograph.

Infrared

Figure 9.51 shows an infrared absorption spectrum of a mixture containing CO_2 and CO gases. The primary CO_2 infrared absorption band is shown to be at 4.2 microns and the CO band is 4.6 microns. The discussion on infrared analyzers in a previous section of this chapter explained that

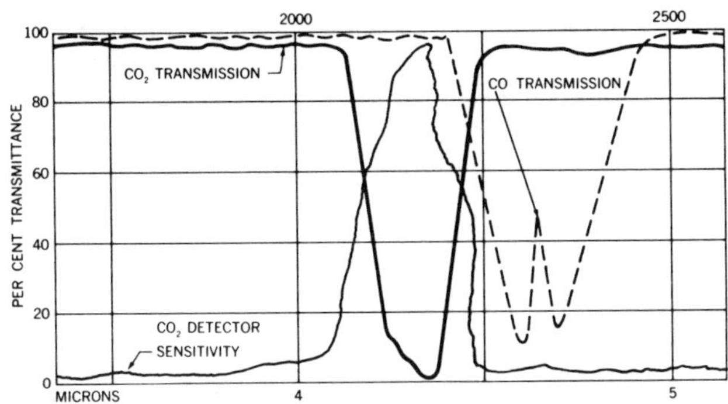

Figure 9.51. An infrared absorption spectrum of a mixture containing CO_2 and CO. The IR absorption band for CO_2 shows to be 4.2 microns, and for CO, 4.6 microns. (Courtesy of Beckman Instruments, Inc.)

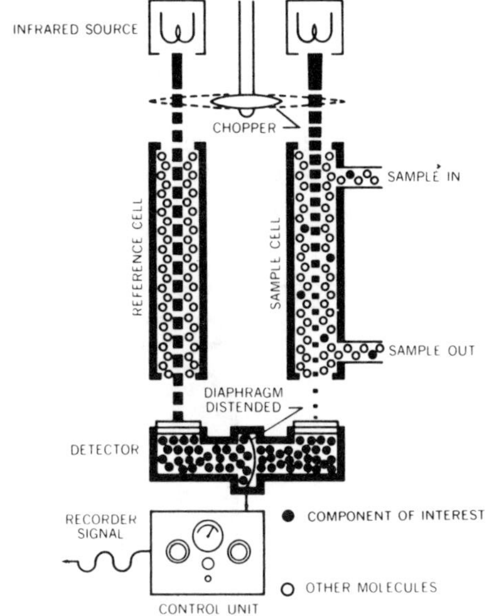

Figure 9.52. Simplified diagram of an air process IR analyzer which may be sensitized for CO_2 analysis. (Courtesy of Beckman Instruments, Inc.)

in process analyzers a source is selected to emit light at a specific wavelength only. Figure 9.52 is a simplified diagram of an air process infrared analyzer which may be sensitized for CO_2 analysis. In this instance, both sides of the capacitance detector are filled with CO_2. The reference and sample cells are filled initially with a gas which is nonabsorbing in the region of 4.6 microns. At these conditions, the light intensity incident on both sides of the detector is equal, both sides are heated an equal amount, and their pressures are equal. With the admission of a sample containing CO_2 into the sample cell, part of the light beam is absorbed, decreasing the intensity of light reaching one side of the detector. The reference side of the detector absorbs more heat from the light beam than does the sample side, the pressure in the reference side increases more, and the metal diaphragm is moved by this pressure to change the capacitance of the oscillator. The signal is demodulated and transmitted to the amplifier section where it is amplified, indicated on a meter and then used for a recorder or control signal.

Sensitivity of this type analyzer is 0.5% of full scale with calibration ranges available from ppm to 100% depending on the application. Calibration procedures are those normally used for infrared analyzers.

Gas Chromatograph

When using the gas chromatograph for CO_2 analysis, the CO_2 is usually one of several components to be monitored, such as methane and total hydrocarbons. Since each chromatographic analysis is unique, the reader is referred to the previous section on gas chromatography.

Oxides of Nitrogen

Oxides of nitrogen can be measured by the amount of light emitted when they react with ozone. This chemiluminescence exists when nitrous oxide (NO) is allowed to react with ozone (O_3), forming oxygen (O_2) and charged molecules of nitrogen dioxide (NO_2*). When the charged molecule gives up its charge, it does so by emitting light. A photomultiplier measures the intensity, which is proportional to the NO concentration.

Measurement ranges include 0-0.25 to 0-25 ppm, with a minimum detectable limit of 0.005 ppm. Precision is ±1% of full scale.

Ozone

Ozone (O_3) measurement is by chemi-luminescent method similar to that for oxides of nitrogen. Ozone reacts with ethylene, producing measurable light. A photomultiplier tube detects the light, the amount of which is proportional to O_3 concentration. A dilute solution of 10% ethylene/90% carbon dioxide is used as a nonexplosive reactant solution.

Measurement ranges include 0-0.025 to 0-2.5 ppm, with a minimum detectable limit of 0.001 ppm. Precision is ±1% of full scale.

Sulfur Dioxide

Sulfur dioxide (SO_2) is measured by first reacting the sample with iodine. A buffered solution of potassium iodide is oxidized to iodine at the anode of the detector cell, and the sample is then bubbled through it. Anode and cathode currents are measured and compared to a reference electrode. Unbalance current is a measure of the concentration of the SO_2 in the sample. Unreacted iodine, reduced back to iodide at the at the cathode, is circulated back to the anode by the injection-pump action of the rising bubbles of the sample. A thermistor senses electrolyte level in the detector, and distilled water is automatically added through a solenoid valve when required.

Ranges of measurement include 0-0.25 to 0-2.0 ppm by volume, with 0.01 ppm being a minimum detectable amount. Accuracy is ±5% of full scale.

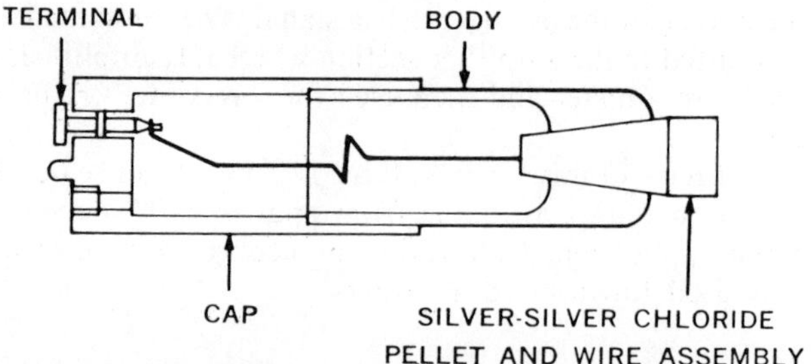

Figure 9.53. Chloride ion electrode used for chloride ion analysis—similar in operation to pH electrodes. (Courtesy of Beckman Instruments, Inc.)

Chloride Ion

The chloride ion analyzer works on the same principle as the pH analyzer, except the sensing cell uses a silver-silver chloride electrode. The potential developed across the electrodes is proportional to the log of chloride ion concentration in the solution. Figure 9.53 depicts a popular chloride ion electrode. As in pH measurement, the potential is measured from this electrode to a reference electrode with a third thermocompensator cell for temperature compensation. Another device uses a similar silver-silver chloride electrode system with all three electrodes in one assembly.

Calibration is by means of standard solutions or by grab sample techniques. Analysis ranges are available from 1-1,000 to 10-10,000 ppm. Accuracy is generally within ±5% of reading.

Sodium Ion

The sodium ion analyzer works on the same principle as the pH analyzer, except the sensing cell uses an electrode that responds to changes in sodium ion concentration. The potential developed across the electrodes is proportional to the log of sodium ion concentration in the solution. To complete the circuit and provide a constant reference potential, a conventional reference electrode similar to the one used in pH systems is used. The sodium ion electrode is also sensitive to hydrogen ions. To permit sodium measurement regardless of the pH of the sample, anhydrous ammonia is added continuously to control the pH (Figure 9.54).

Calibration is achieved by either standard reference samples or by grab sample technique. Sensitivity is extremely good, and ranges are available from 1-1,000 ppb to 1-10,000 ppm. Accuracy is within ±5% of reading.

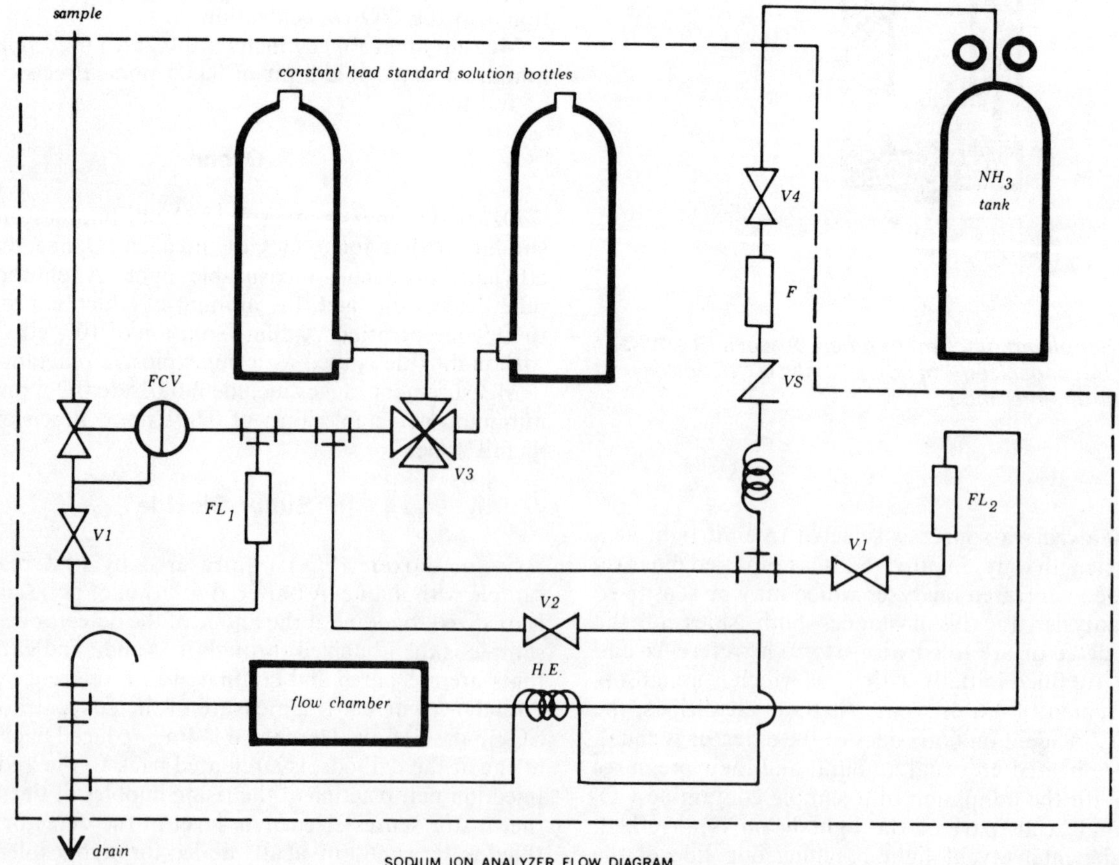

Figure 9.54. Sodium ion analyzer flow diagram shows addition of ammonia to control pH of sample to permit correct analysis regardless of pH variations in sample. (Courtesy of Beckman Instruments, Inc.)

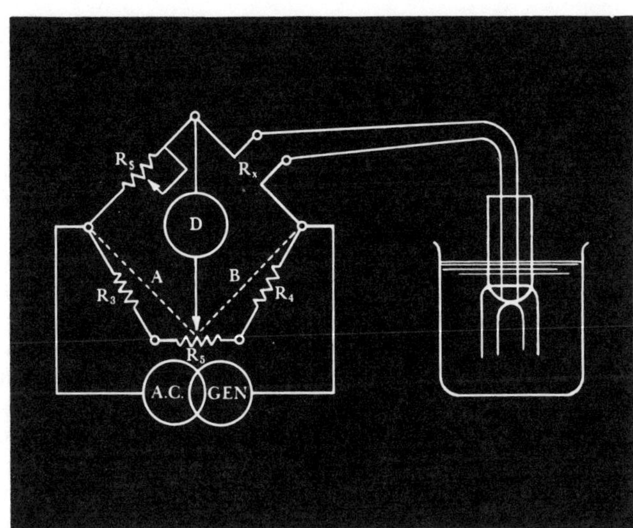

Figure 9.55. Simplified conductivity analyzer schematic showing simplified Wheatstone bridge measuring circuit. (Courtesy of Beckman Instruments, Inc.)

Figure 9.56. Dip-type conductivity cell. (Courtesy of Beckman Instruments, Inc.)

Figure 9.57. Gate-valve mounted conductivity cell. (Courtesy of Beckman Instruments, Inc.)

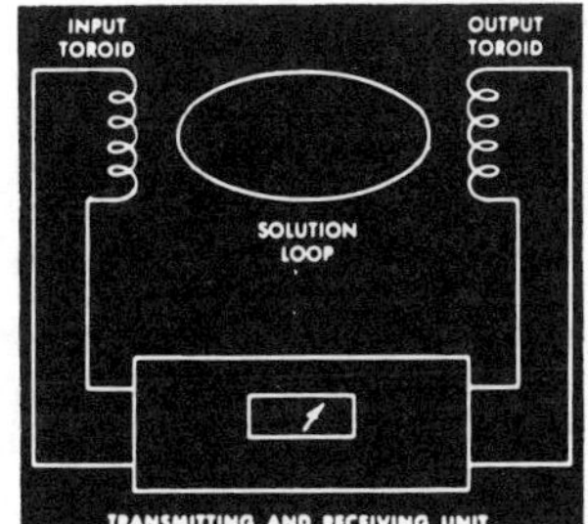

Figure 9.58. Electrodeless conductivity unit. (Courtesy of Beckman Instruments, Inc.)

Conductivity Meters

Electrolytic conductivity is a measure of the ability of a solution to carry an electrical current. Also called specific conductance, it is defined as the reciprocal of the resistance, in ohms, of a 1 centimeter cube of the liquid at a specific temperature. The unit of electrolytic conductivity is the mho/cm which is the reciprocal of ohm/cm, a resistance measurement. One millionth of this value, a micro-mho/cm, is the practical unit widely used. The flow of electricity in solutions is different from that in metals. In solutions, conduction is by the drift of positive and negative ions compared to the movement of free electrons in metallic conductors.

The conductivity of a process stream is usually measured to determine the stream purity. Cooling water in an exchanger, for instance, with a normal conductance of 2 micro-mho/cm, may have its conductance doubled by the addition of 1 ppm of a typical salt. Any sudden change of the conductance of a process stream is often construed as the appearance of an undesirable contaminant.

The determination of conductivity involves the measurement of the electrical resistance of a volume of solution. Since the use of direct current leads to "polarization" in the electrode area, alternating current is more frequently used, utilizing an AC Whetstone bridge as shown in Figure 9.55 The AC voltage level varies with different applications, from the millivolt range to as high as 50 volts, while the frequency varies from 60 Hz to as high as 1,000 Hz. The lower frequencies are used in applications of high resistance, such as pure water.

Cells

Conductivity cells are simple in structure, consisting essentially of metal plates (or discs) mounted in an insulated body structure. The dip-type cell in Figure 9.56 has platinized nickel discs. The gate-valve mounted cell (Figure 9.57) is designed in several models for pipe service. Also available is the flange and packing gland mount type.

Electrodeless Conductivity

Although much has been accomplished in modifying and applying the basic laboratory conductance cell to process applications, problem areas remain, such as solutions containing abrasive or fibrous solids and very corrosive solutions. Servicing the cell and maintaining the electrodes in good working order in these services created a demand for an electrodeless system. Figure 9.58 illustrates schematically an electrodeless conductivity system which measures the resistance of a closed loop of solution by the extent to which the loop couples two transformer coils. In Figure 9.59, two toroidal coils are mounted on nonconductive or nonconductive-lined metallic piping. One coil is connected to a transmitter which supplies a voltage in the high audio frequency range and is stable and free from drift in

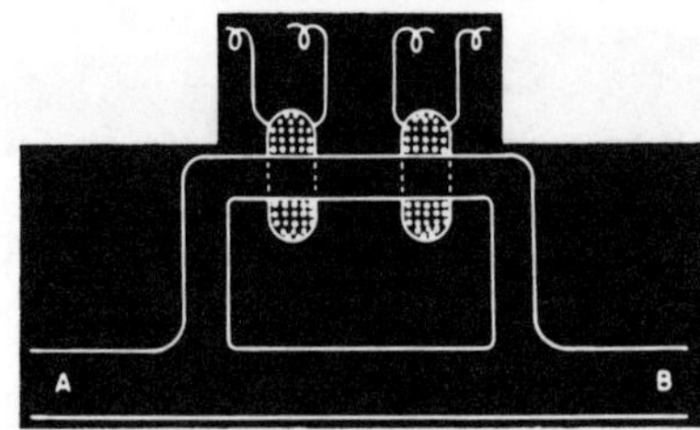

Figure 9.59. Electrodeless conductivity unit with toroidal coils. (Courtesy of Beckman Instruments, Inc.)

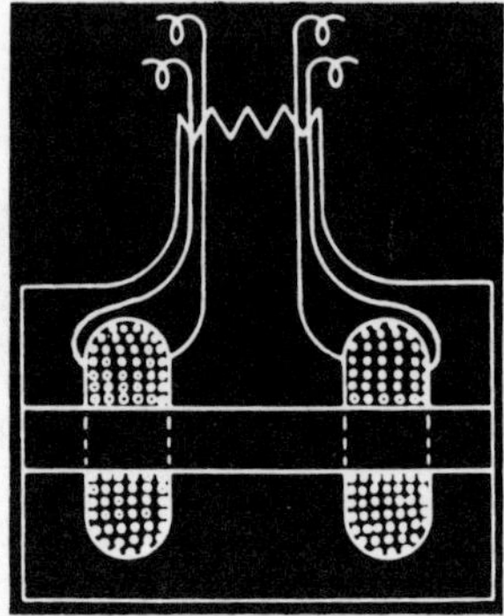

Figure 9.60. Electrodeless conductivity unit with potted coils for corrosive liquid applications. (Courtesy of Beckman Instruments, Inc.)

both frequency and amplitude. The second coil is connected to a receiver which measures the output voltage from this winding. With a constant input voltage, the output of the system is proportional to the conductivity of the solution. Several different configurations of the electrodeless "cell" are possible. For immersion in corrosive liquids, two toroids are "potted" on an axis with a cylindrical bore (Figure 9.60). The solution in the bore and the external solution constitute the loop linking the transformers.

Selection of Cells

In the selection of conductivity cells, three service requirements must be considered with care: electrical, mechanical and chemical.

Electrical

There are practical limits of measured electrolytic resistance for any desired accuracy and sensitivity. For process requirements where ±½% error can be tolerated, experience indicates the span of resistance should be about 50 to 100,000 ohms (20K to 10 micro-mhos).

When measuring high electrolytic resistances, the effect of capacitive impedances in series is insignificant, whereas shunt capacitance (as in long lengths of lead wire) may affect the bridge signal significantly. This is particularly true at high frequencies, dictating the use of low bridge frequencies. Conversely, when measuring low electrolytic resistances, the effect of parallel capacitive impedances is low while the effect of series capacitance impedance is significant; hence, high bridge frequencies are required.

Mechanical

Design factors include temperature, pressure, flow and entrained solids. Temperature must be held within the range of cell design. The conductivity of most solutions increases as much as 30% for each degree centigrade rise in temperature; therefore, accuracy is poor with fluctuating temperatures.

Pressure effects are negligible on conductivity readings, but cell structure normally restricts the pressure range to a low value. In high-pressure applications, it is sometimes necessary to install the cell in a bypass stream whose pressure may be reduced.

Flow rate has no significant effect on the conductivity reading but should be sufficient to assure circulation between the plates. In high flow applications, the cell should be installed so that the plates are recessed out of the main stream velocity. Undissolved solids in the solution increases the scouring effect and strips the coating from the electrodes. The cell should be installed to prevent direct impingement of flow on the electrodes. In addition, the cell should be mounted with the open end pointing downward to avoid accumulation of solids. In cases of high solids content, consideration should be given to the use of an electrodeless system.

Chemical

Chemical action on the electrodes poses a frequently encountered problem in the selection of cell materials. Metals which are normally considered sufficient in a particular solution may be slowly dissolved due to the AC voltage impressed across the electrodes. Most manufacturers maintain a file of materials that have proven successful in difficult applications. The only recourse in some instances is the design of an electrical bridge circuit providing voltages in the low millivolt range.

Temperature Compensation

Solution conductivity varies with temperature as well as concentration, with the coefficient varying from 0.5 to 3% per degree centigrade, depending upon the type and concentration of ions. To compensate for temperature changes, the bridge signal may be corrected by a manual rheostat calibrated in temperature units, a reference conductivity cell or a thermistor and resistor network in contact with the solution. Automatic temperature compensation is normally recommended.

Density Analyzer

Density may be defined as mass per unit volume and is most commonly expressed in grams per cubic centimeter, pounds per cubic feet or pounds per gallon. Specific gravity is a term often used interchangeably with density and is expressed as the ratio of the density of the fluid to the density

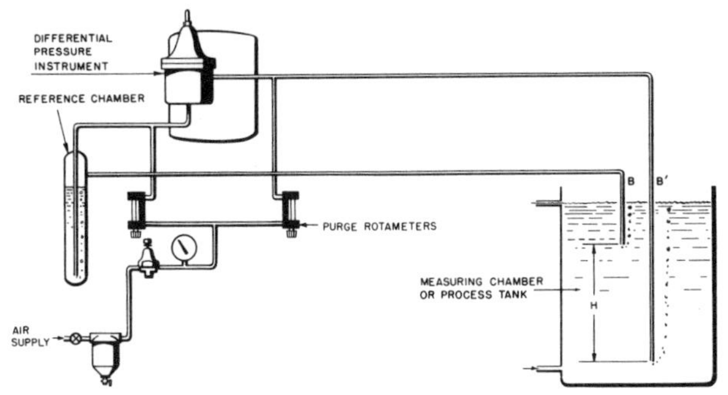

Figure 9.61. Liquid density measuring system. Specific gravity changes as small as 0.001 can be read clearly and accurately using dip tubes. (Courtesy of Foxboro Co.)

of water at a specified temperature. In engineering practice, the reference for density or specific gravity is generally water at 60°F.

Because liquids are essentially noncompressible, their densities are unaffected by operating pressure. With temperature compensation, specific gravity and density of liquids are interchangeable. The same does not hold true for gases since they are compressible, and specific gravity (or molecular weight) is no longer equal to the density at operating conditions. Hence, gases are generally referenced to their specific gravity relative to air or to their direct density at operating conditions for mass flow measurement.

Air Bubble

The simplest and perhaps most widely used process density measurement is the installation of two bubbler tubes in a vessel of fluid at different levels (Figure 9.61). The pressure required to bubble air (or any suitable vapor) into the fluid is a measure of the pressure at the level of the tube outlet. The difference between these two pressures is measured by a differential pressure device and is equal to the weight of a constant-height column of the liquid. The changes in differential pressure as density changes is proportional to the fluid denisty. When the differential device is a transmitter, the signal from the d/p transmitter may be transmitted as desired.

Displacement Type

Displacement density instruments operate on the buoyancy principle of a completely submerged body (Figure 9.62). The force acting on the balance (or torque) arm is directly related to the density of the fluid displaced by the float. In a bypass unit as shown, a continuous flow is required to purge the unit with fresh liquid; low flows are necessary, and temperature must remain constant. The liquid must be clean to prevent settling of solids in the chamber and buildup of material on the float.

U-Tube (Balanced Beam)

A system to measure density from mass weight is shown in Figure 9.63. Process fluid flows through a U-tube. The weight of the U-tube and its constant volume of minimum density liquid is balanced on a weigh beam. Any increase in density of the liquid results in an additional force on the weigh beam. The weigh beam is mechanically linked to the nozzle-flapper assembly of a force balance transmitter. Increased density decreases the flapper-to-nozzle distance, increases the back pressure on the nozzle and increases the transmitter output signal. In Figure 9.63, the process liquid flows through the tube loop (*A*) pivoted on flexures (*B*) about an axis which passes through the flexible connectors (*C*). The weight of the tube loop and contents is transferred to a weight-beam (*D*) and counterpoised by the balance weights (*E*) which are adjustable along the beam. A change in density of the process liquid produces a directly proportional change in force on the weigh beam. Density changes can be measured in liquids with a nominal density between 0.4 and 2.5 g/ml.

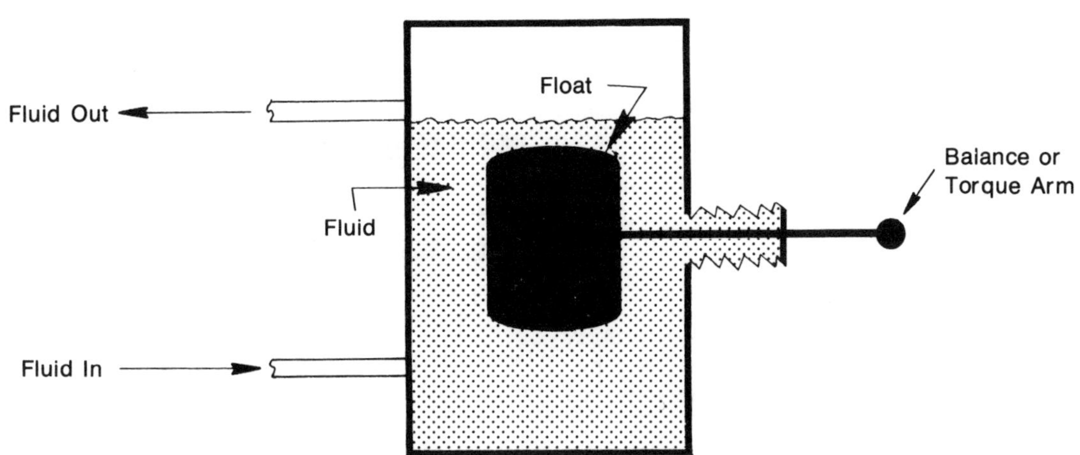

Figure 9.62. Displacement density unit. Buoyant force on submerged body is proportional to density of fluid.

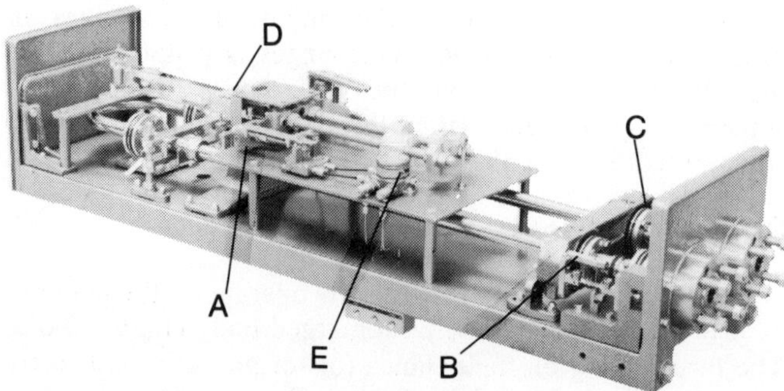

Figure 9.63. U-tube (balanced beam) density device. Force on a balanced beam arrangement varies with density. (Courtesy of Hallikainen Instruments)

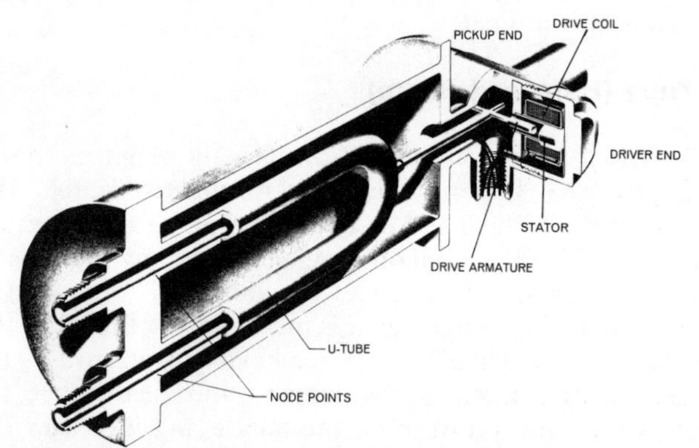

Figure 9.64. Vibrating U-tube density detector in which vibration amplitude is a function of mass. (Courtesy of Automation Products, Inc.)

Vibrating U-Tube

The vibrating U-tube operates on the principle that a body may be caused to vibrate by a pulsating force, and the amplitude of this vibration is proportional to its mass. In Figure 9.64, the total mass of the U-tube includes the mass of the flowing fluid and, therefore, changes with a change of fluid density. An increase in density increases the total mass of the U-tube, decreasing the vibration amplitude. An armature and coil assembly form a "pickup" circuit in which an AC voltage is induced by the vibration of the armature. This voltage is proportional to the amplitude of the vibration and hence the density of the fluid. The AC voltage is converted to a DC millivolt signal as required for readout or control.

These sensors are not affected by variations in process pressures, flow rate, viscosity, etc. Ambient temperatures have no effect, and process temperature changes may be electronically compensated. Flows in dirty streams must be high enough to prevent solids buildup.

With rapid response of this sensor to changes in density, accuracies to 0.0005 specific gravity units are claimed as typical. Calibrated spans are from 0.05 to 0.5 specific gravity units as standard with narrower or wider spans available on special order. High-viscosity streams should be checked to assure that plugging will not be a problem.

Radiation Type

Radiation sources generate three forms of energy—alpha and beta particles and gamma rays. Alpha particles (helium nuclei) have no penetrating power and are not used in industrial applications. Beta particles (electrons) have a penetrating power sufficient for thickness gauge applications for thin or low-density materials. Gamma rays, having no mass substance, have the highest penetrating power and have been used extensively in industrial application.

The "curie" is the unit rate at which a radioactive material generates radiation and is equivalent to the disintegration by a radioactive material of 3.7×10^{10} atoms per second. One curie is the rate of disintegration of 1 gram of radium. The more practical unit of rate is the millicurie.

The measure of radiation intensity at a given location is the roentgen; the practical unit is the milliroentgen (mr). In actual installations (Figure 9.65), radiation must penetrate two thicknesses of the fluid container plus the thickness of the fluid itself before reaching the detector. As a rule of thumb, it is desirable to have a radiation intensity of approximately 2 to 5 mr/hr at the detector with no fluid present. The lower value is desirable for proper sensitivity, and the higher figure limit is maintained for personnel safety.

Cesium 137 is most often used as the gamma ray source in density analyzers although radium and cobalt 60 sources are also used. Radium is most desirable as a source because of its extended half-life of 1,400 years compared to 33 years half-life for cesium and 5 years half-life for cobalt 60. Because of its high cost, however, radium use is usually restricted to small source applications. When a larger source is required, cesium 137 is used rather than cobalt 60,

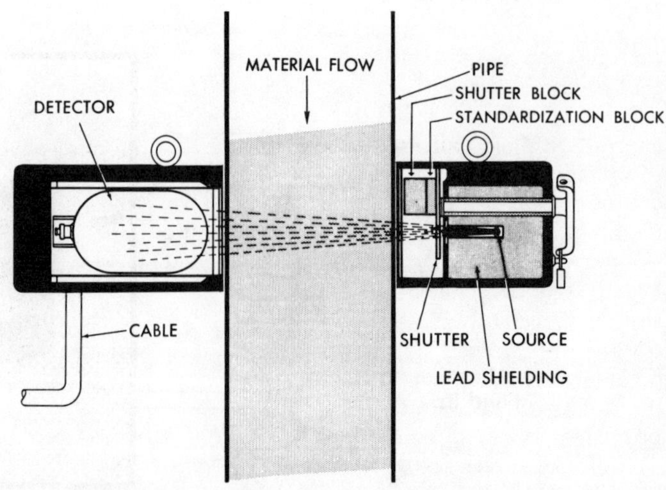

Figure 9.65. Radiation density analyzer. Radiation must penetrate air spaces, two pipe walls and the flowing material. (Courtesy of Nuclear-Chicago)

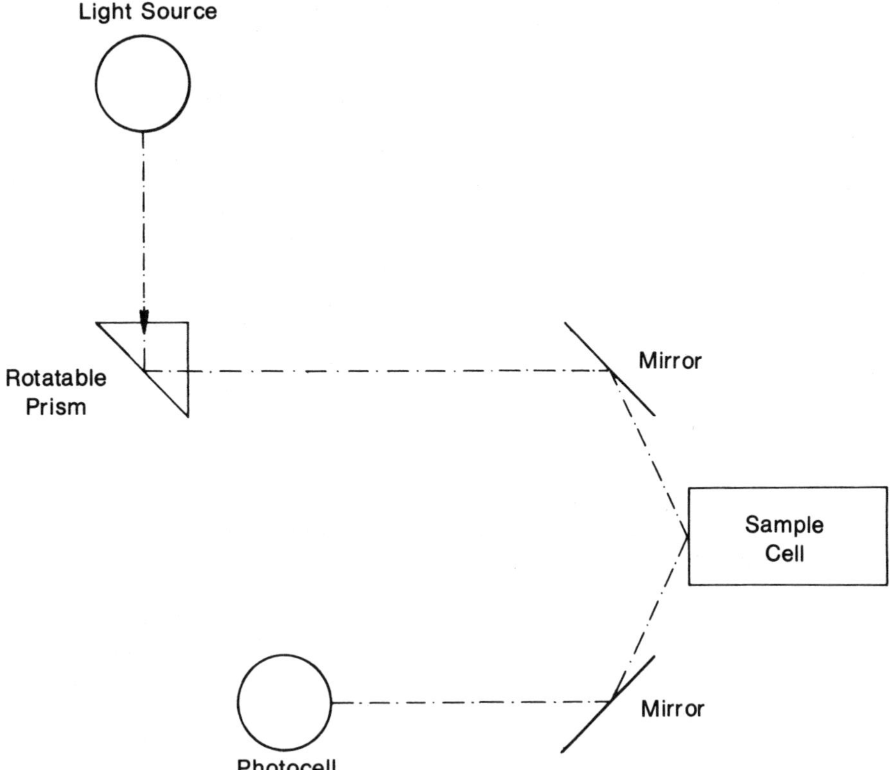

Figure 9.66. Critical-angle refractometer. A small change in reflectance causes a large change in incident light on the photocell.

although cobalt 60 may be selected in applications where high penetration characteristics are required. The shorter half-life sources require more frequent calibration.

The ionization chamber detector is commonly used in industrial applications because of its stability, accuracy and sensitivity. Geiger tubes are less expensive, but have low accuracy. Scintillation detectors are more sensitive than ionization chambers but are less stable and more sensitive to temperature fluctuations.

Gamma rays from the source pass through the measured material and are absorbed in proportion to material density; that is, the number of ions passing through the fluid is inversely proportional to the density. The gamma rays energize an ionization detector which then produces a proportional electrical signal. An amplifier-indicator unit can be remotely mounted to present the signal as required. Collimated beam geometry restricts radiation in all directions, except to the detector. The source is enclosed in a heavily shielded housing which is equipped with a fail safe shutter that closes and locks the radiation port in the event of a power failure or during maintenance periods. The detector chamber should be thermostatically controlled to prevent temperature variations and moisture condensation problems.

Calibration checks for both zero and span are performed by the insertion of absorber plates between the source and detector and with no fluid present in the vessel path. Accuracy is ±1% of span, and range is from 0.05 to 1.0 + specific gravity.

Refractometers

The refractometer, as the name implies, is an optical analyzer to measure the refractive index (RI) of a liquid sample. It is used to measure changes of component concentration in binary streams. Refractive index is the ratio of the velocity of light in a vacuum to its velocity in the substance to be measured. Refractive index is usually expressed in terms of air as a standard. RI is limited in that it is nonselective; that is, the stream must be binary or a binary type. Two types of refractometers are used: the critical-angle refractometer, employed where the stream is turbid or carries suspended material, and the differential refractometer, used for clean, clear streams.

Critical Angle

In any liquid under constant process conditions, there is a specific angle at which a beam of light is transmitted into and perhaps through the liquid. Furthermore, at this angle, a very small change in composition or concentration causes a large change in the amount of light which is transmitted or reflected at the surface. In Figure 9.66, a light beam is transmitted through a prism to a mirror, which reflects the beam to a glass window containing the liquid. At the correct angle the beam is deflected to another mirror and to the photocell. By proper location of the photocell and rotation of the prism, the system can be calibrated such that a small

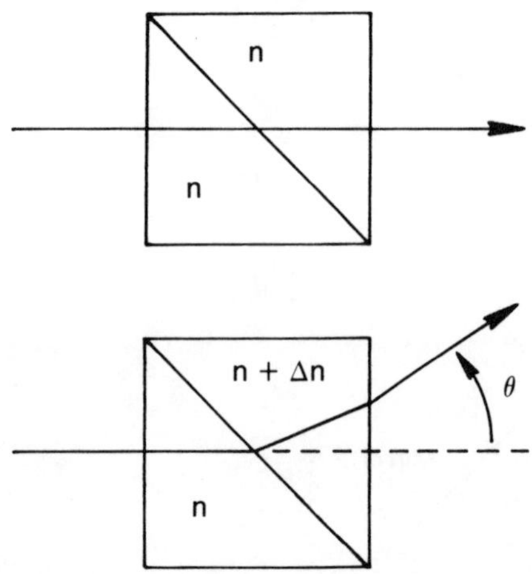

Figure 9.67. A differential refractometer measures the deflection of a light ray passing through two solutions. When the solutions are the same, no deflection occurs, but if they are different, there is deflection.

change in reflected angle causes a large change in the incident light on the photocell. This change in reflectance angle is correlated to changes in concentration of one component in a binary fluid.

Differential

The use of the differential refractometer is limited to streams of clear liquids because of its principle of operation. It measures the angle of refraction of a light beam as it passes through two solution cells—a reference solution and the process solution whose measurement is desired.

When the two solutions are identical in composition, a beam of light passes through both with no refraction. When the process solution differs from the reference solution, however, the refractive index (RI) of the light beam is different and refraction occurs by an amount Δn as shown in Figure 9.67. The resulting deflection of the light beam is a measure of the change.

Temperature affects the RI of solutions, but differential refractometers are constructed in a way to minimize the effects. Ambient changes affect both reference and process solutions equally with no resultant shift in the angle of refraction.

A high degree of accuracy (approximately 1×10^{-7} RI) is attained when good mechanical construction techniques are employed in building differential refractometers—i.e., use of small sample cells and good heat sinking between the cells.

Turbidity

Turbidity analysis—the measurement of clarity in a liquid sample—involves the detection of particulate matter, immiscible droplets, bubbles, etc., that affect the quality or condition of a fluid stream. In its simplest form, turbidity is the measurement of light that has passed through a volume of sample without being absorbed.

Transmission Type

Figure 9.68 illustrates the transmission turbidity analyzer, which contains a light source, a sample cell and a photocell. A fraction of collimated light passes through the cell to the phototube while the balance of the light is either absorbed or reflected. The amount of light reaching the photocell varies inversely with the concentration of solids in the stream. The light that reaches the photocell is converted to electrical energy, amplified in the control unit of the analyzer and transmitted to a readout device. Turbidity, measured in percent of light transmitted, is calibrated by checking zero and then 100% transmission with a clear sample.

Reflection Type

Another similar analyzer measures the amount of light reflected by the suspended particles. Figure 9.69 pictures the sample cell of a turbidimeter which has a shield between the light source and photocell to prevent the transmission of light directly to the cell. Particles in the stream reflect light around the shield to the photocell. The quantity of light reaching the photocell is directly proportional to the concen-

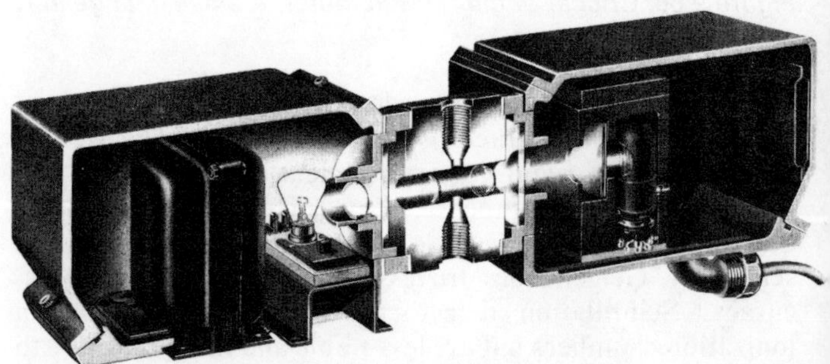

Figure 9.68. A transmission turbidity analyzer in which turbidity is measured by the percent light transmitted through a stream to the photocell. (Courtesy of Beckman Instruments, Inc.)

Figure 9.69. Reflection turbidity analyzer in which a shield is placed between the light and the photocell, and turbidity is measured by the light reflected by the fluid around the shield. (Courtesy of Fischer and Porter Co.)

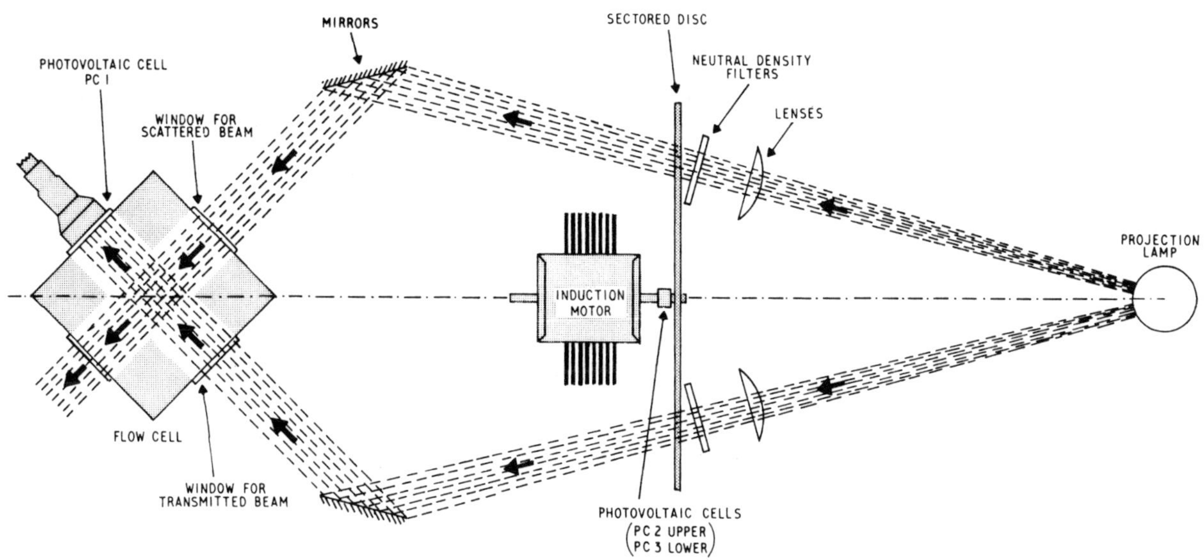

Figure 9.70. Turbidity ratio analyzer uses both the transmission and reflection principles. Readout is in ppm.

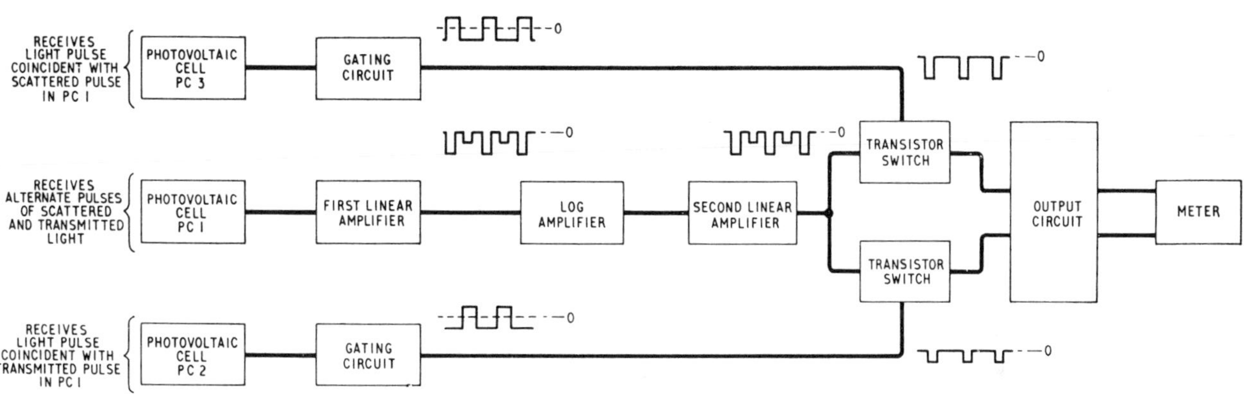

Figure 9.71. Block diagram of the ratio analyzer electronic unit.

tration of particles present. The electrical signal from the cell is received by either an indicating or recording potentiometer calibrated in parts per million turbidity. The potentiometer is equipped with a four-position range switch for 0-1, 0-3, 0-10 or 0-30 ppm turbidity.

The turbidity sensor is equipped with a self-contained calibration feature consisting of a push button which actuates a slide plate, exposing a precisely machined opening in the light shield. This permits an amount of light exactly equivalent to a turbidity of 5 ppm to be transmitted directly to the photoelectric cell. Actuation of this calibration button changes the output of the sensing cell by an amount equal to a turbidity of 5 ppm. This signal change permits the system to be quickly and accurately calibrated.

Ratio Type

A ratio turbidity analyzer uses a more sophisticated approach which measures both the incident and reflected light with the configuration shown in Figure 9.70. Light from a common source passes through a rotating chopper which alternately blanks out one beam at a time. The transmitted beam passes through the sample to the photocell followed by the scattered beam at the right angle. The photocell sees the incident beam during one half of the chopper cycle and the reflected beam during the last half of the cycle. Figure 9.71 illustrates the chopped signal output of PC-1. Photocells PC-2 and PC-3 generate pulses in synchronism with the transmitted and scattered beams and are used only as gating pulses in the electronic unit. The two light beam signals are amplified, switched to the output circuit and transmitted to the readout device as the logarithm of the ratio of the scattered and transmitted beams. Readout is in parts per million.

One type of analyzer which may be referred to as a turbidimeter or a density analyzer, is designed to monitor free flowing gases (Figure 9.72) and to scan stacks or ducts up to 20 feet in diameter for smoke or dust densities. Instead of

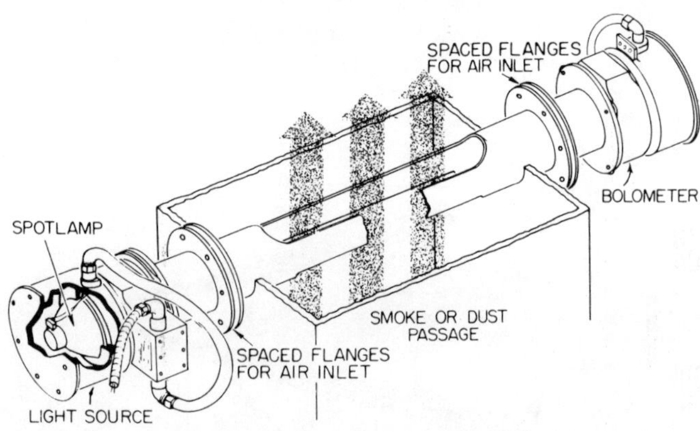

Figure 9.72. Smoke-dust density analyzer uses a bolometer to measure the amount of transmitted light. (Courtesy of Bailey Meter Co.)

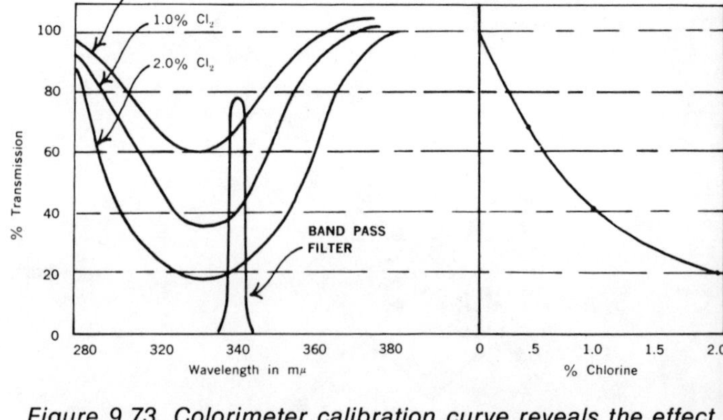

Figure 9.73. Colorimeter calibration curve reveals the effect of chlorine concentration on lightwave transmission at the 340-millimicron bandwidth. (Courtesy of Beckman Instruments, Inc.)

a photocell, however, the receiver is a bolometer positioned inside a parabolic reflector which focuses the radiant energy on the measuring filament. A second filament compensates for ambient temperature and voltage changes across a Wheatstone bridge. The bridge output is transmitted to a null balance amplifier in the recorder. Smoke density is presented in percentage light transmission which may be correlated to Ringlemann numbers. Dust density is presented in grains per cubic foot. Calibration is made easy by the use of standard light absorption grids.

Colorimeter

Gauging color in the laboratory usually is accomplished by holding a container filled with liquid in front of a light and comparing the color with a group of color standards. The process analyzer uses the same method for gauging color and substitutes photocells for the human eye. In a previous paragraph on turbidity measurement, the analyzer in Figure 9.68 measured turbidity in the same manner and may be used as a colorimeter to measure darkness or absorbence due to color. When used as a colorimeter, the signal may be transmitted to a readout device, or, in the case of color alarms, the signal may be amplified to drive an alarm relay.

Occasionally, however, it is desirable to monitor color due to one specific component in the stream. The calibration curve in Figure 9.73 indicates that if the light wavelength is limited to a narrow band near 340 millimicrons, the instrument will be sensitive only to changes in chlorine content. This is because chlorine absorbs strongly at that wavelength. The analyzer in Figure 9.68 can be sensitized in this manner by the proper choice of light source, filters and photocell. Sensitivity of the unit may be increased or decreased within limits by varying the wavelength or sample cell length or both.

Viscosity

The viscosity of a fluid is its resistance to deformation under shear. Viscosity measurement is essential in many instances to control the flowability of fluids and has become more widely used to determine the molecular weight of fluids during the manufacturing process. Among other uses is the determination of solids concentration and/or size in slurry streams. The unit of absolute viscosity in the cgs system is the poise with the practical unit being the centipoise (cp). Kinematic viscosity is the ratio of the absolute viscosity to the density of the given fluid. Its unit in the cgs system is the stoke with the practical unit being the centistoke (cs).

Numerous types of viscosity analyzers are available to furnish both laboratory and process analysis. Some of these units are discussed in this section.

Ultrasonic

A relatively new method of viscosity measurement is the ultrasonic probe. Figure 9.74 shows the ultrasonically vibrated probe which, coupled with the electronic computer, measures the product of viscosity and density (cp x g/cc). Acting as a transducer, the probe shears the liquid under measurement and measures the damping effect of the liquid. The actual sensing element is a thin strip of metal. One end of the strip is a magnetostrictive alloy, and the other end, which is shown exposed to the process, is stainless steel. The two pieces are butt-welded and clamped close to the geometric center with the stainless section protruding through a flat circular diaphragm. The computer sends out a short pulse of current to a coil surrounding the magnetosensitive section inside the probe. The resulting magnetic field excites the blade and causes it to vibrate longitudinally.

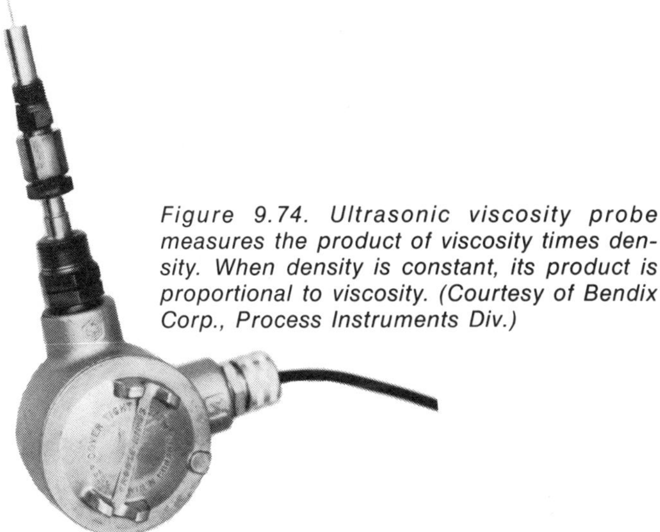

Figure 9.74. Ultrasonic viscosity probe measures the product of viscosity times density. When density is constant, its product is proportional to viscosity. (Courtesy of Bendix Corp., Process Instruments Div.)

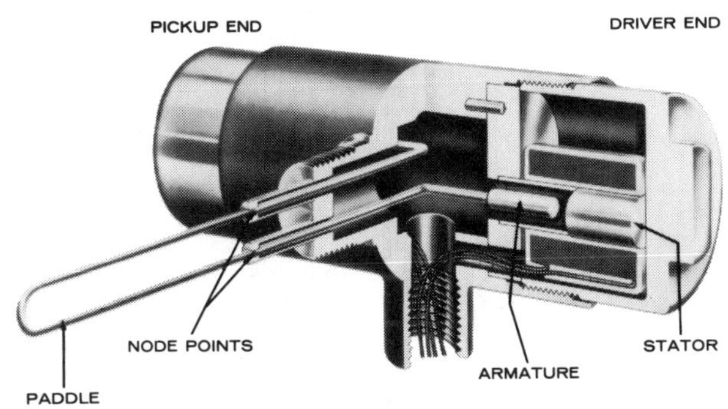

Figure 9.75. Vibrating reed viscometer. A drive coil sets up 120-hz. vibration to probe. Resistance to vibration is proportional to fluid viscosity. (Courtesy of Automation Products, Inc.)

The blade then vibrates at its own natural frequency with a maximum amplitude of motion at the tip. The computer translates the damping action of the viscous fluid on the probe blade into a useful electronic signal proportional to viscosity when the density remains constant. A thermocouple and automatic temperature compensator are required to correct the output signal to a value at a designated base temperature. The output of both the computer and the compensator are calibrated at two points on the curve.

Output of the unit is 0 to 100 millivolts DC and can be used for indicating or controlling viscosity of both Newtonian or non-Newtonian fluids. Available ranges vary from 0-5 to 0-50,000 cps/g/cc. Periodic calibration with standard Newtonian oils is recommended. The hermetically sealed probe will operate at temperatures to 650°F and pressures to 1,000 psig. Response time is less than 1 second, and accuracy is ±2% full scale. Temperature compensation is possible over a range of ±25°F from a reference point.

Vibrating Reed

The vibrating reed viscometer, which has become quite common, is used to measure viscosity continuously. The detector consists of a driver section, a vibrating probe and pickup circuitry (Figure 9.75). The drive coil uses 60 hz. power to set up 120 hz. vibration in the drive armature. The vibration is transmitted through a stationary (welded) mode point to the probe rod where the amplitude of probe vibration is proportional to fluid viscosity. The resistance to the shearing action caused by the probe vibration increases with an increase in viscosity and results in a decrease of vibrational amplitude. The resulting amplitude is transmitted through a second mode to the pickup assembly. A permanent magnet inside the pickup coil induces a 120 hz. signal, the magnitude of which is proportional to the viscosity. Output of the unit is converted to the desired millivolt signal as required. Viscosity ranges are available to 1×10^5 cp with accuracies of ±1%. Additional components are available for on-off or proportional-control.

Since viscosity varies with pressure and temperature, these must be controlled. Temperature compensation is available with the units when required. Product buildup on the probe may be a problem in some installations, and equipment vibration in the region of 120 hz. may also be a source of error.

Rotational

The rotational viscometer (Figure 9.76) measures the torque required to rotate an element in a liquid and relates this torque to viscosity. The oscillating system, consisting of a synchronous motor (14), potentiometer slide wire (11) and pointer (8), is suspended from a torsion wire (6) to eliminate friction in the bearings (12, 20) which vertically align the system. The synchronous motor (14) drives the measuring body (18), rotating in the liquid to be measured, through a magnetic coupling (16, 17). The viscous drag on the metering body causes a torque reaction over the oscillating system. This torque is offset by a calibrated measuring spring (10). The pointer (8) furnishes a visual reading, and the potentiometer (11) varies the circuit resistance proportionally to the angular change due to torque. A variable capacitor is preferred in low-viscosity ranges over a variable resistor. Signal output is thus converted to viscosity units. Viscosity ranges are available from 0-10 to 0-4,200,000 cp for both Newtonian and non-Newtonian fluids.

The detector head may be installed on open vessels, inline or in a bypass system. Temperature should be constant for consistent viscosity accuracy. At least one manufacturer offers temperature compensation while others control temperature by steam jacketing. Pressure ratings of some models are as high as 7,000 psig. The unit is adaptable to a wide variety of applications because of its nonclogging and easy to clean features.

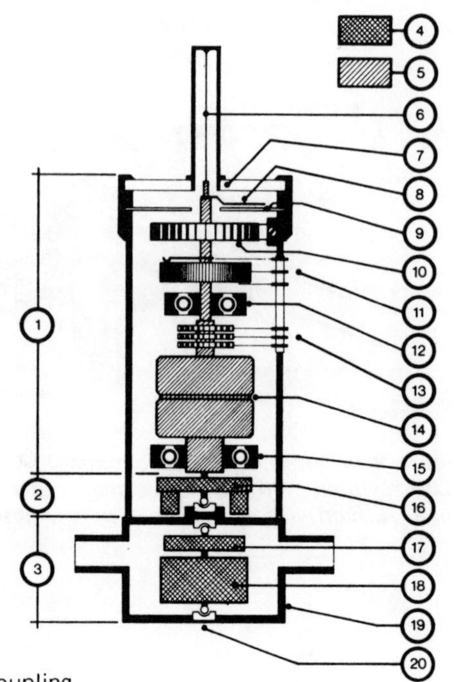

Description:

1 Viscometer
2 Magnetic coupling
3 Measuring system
4 Rotating parts
5 Oscillating parts (345°)
6 Torsion wire
7 Transparent window
8 Pointer
9 Scale (0 to 100 units)
10 Measuring spring (0 to 40 cm g)
11 Measuring potentiometer
12 Upper ball bearing
13 Spring power connections
14 Synchronous motor
15 Lower ball bearing
16 Magnet
17 Nickel rod (armature of coupling)
18 Measuring body (rotating in liquid)
19 Measuring cell
20 Lower body bearing (st. steel/sapphire)

Figure 9.76. Rotating viscometer. The torque required to rotate a body in the measured fluid is proportional to viscosity. (Courtesy Polyscience Corp.)

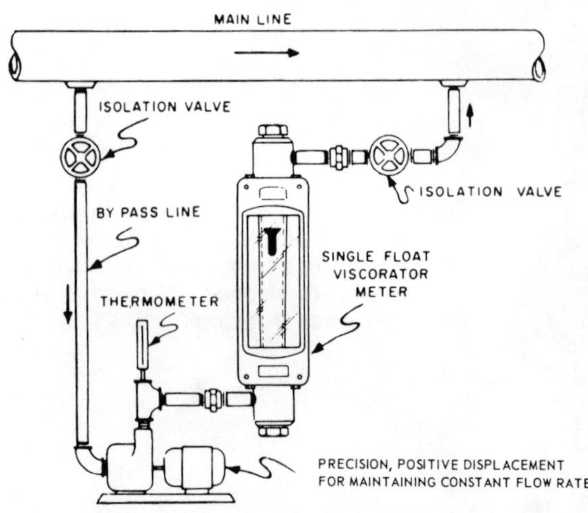

Figure 9.77. Single float viscometer used for Newtonian and non-Newtonian fluids. (Courtesy of Fischer and Porter Co.)

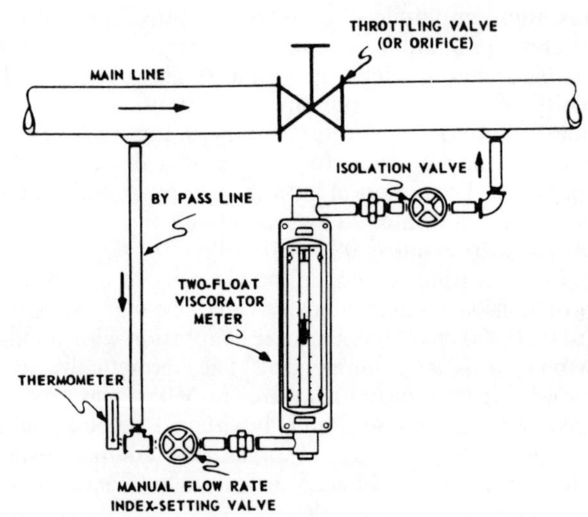

Figure 9.78. Two-float viscometer. A flow rate float is used in meter in place of constant flow rate equipment. Flow is manually set and readout is local only. (Courtesy of Fischer and Porter Co.)

Float Types

The float viscometer is similar in construction to the rotameter flowmeter. In the flowmeter, the forces acting on the float are flowrate, specific gravity of the float and liquid and viscosity of the metered fluid. The float is so designed, however, that the viscous drag area is relatively small and the float is viscosity insensitive. In the viscometer, the viscous drag area is increased, the flow is held constant, and the float responds to forces due to viscosity and density (kinematic viscosity).

The single float viscometer (Figure 9.77) is furnished with a constant flow pump (or other flow control) and provides a continuous direct reading. Recommended flow is 0.75 to 20 gpm. Temperature compensation is not provided. Recording, transmitting or controlling mechanisms may be attached to the meter. The glass tube viscometer is rated at 90 psig at 450°F while the steel tube model is rated at 650 psig and 450°F and may be purchased with auxiliary recorder for both viscosity and temperature recording. It can be used for non-Newtonian fluids with viscosities under 400 cp and can handle Newtonian fluids up to 10,000 cp. Accuracy is ±4% of scale and reproducibility is ±1%.

The two-float viscometer (Figure 9.78) is the economy model of the single float. By subtracting the constant flow equipment and adding a flow-rate float in the meter, one has

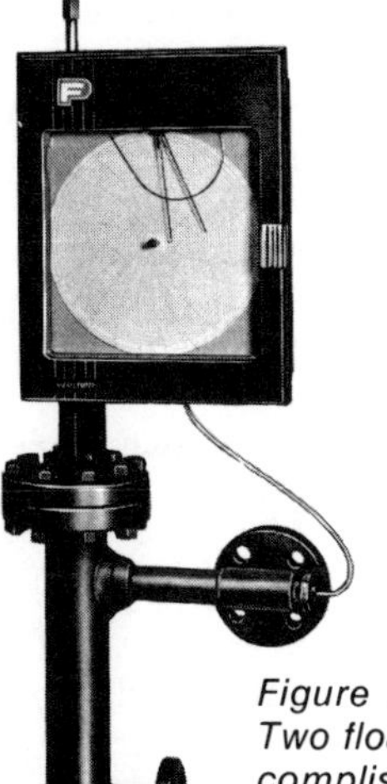

Figure 9.79. Concentric float viscosity meter. Two floats are used, one inside the other, to accomplish flow control and viscosity measurement. (Courtesy of Fischer and Porter Co.)

a local indicator only, whose flow must be manually set each time a reading is taken. Pressure rating is 300 psig at 450°F. This meter is suitable for Newtonian fluids with viscosities up to 250 cp and with accuracies from ±2 to ±4%, depending upon the range.

A concentric flow viscometer (Figure 9.79) uses two floats, one inside the other, to accomplish flow control and viscosity measurement. Shoulders on the outer (differential pressure) float and the inner wall of the steel tube act as a differential pressure regulator. Pressure drop across the meter is determined by the weight of the float. Positioned inside the differential pressure float is the viscosity float. With a constant differential across the meter, flow is constant and viscosity may be read, fulfilling the conditions of a single-float meter.

Advantages of the concentric unit are the elimination of a constant flow pump and increased permissive flow rate. Pressure drop across the valve should be approximately 4 psi, and flow rates from 2 to 30 gpm are acceptable. Magnetic coupled transmitting equipment is available and temperature compensation is possible, permitting operation at 50°F above a reference temperature. Pressure rating is 650 psig at 450°F. The concentric viscometer can measure viscosities in the range of 0.5 to 550 centistokes. Accuracy is from ±2 to ±4%, depending upon the range, and reproductibility is ±1%.

ANALYZER TERMINOLOGY

Absorb: to take in, soak up, as a sponge.

Adsorb: to hold by adhesion to the surface of a solid body.

Alpha ray: a ray consisting of the positively charged particles (H_2 atoms) emitted by a radioactive source.

Anode: The positive electrode of an electrolytic cell.

Attenuation pot: a variable resistor used for reducing (calibrating) an output signal.

Backflush: to flow in a reverse direction.

Bar graph: a recording of a peak signal, made with the chart drive off, resulting in a straight line across the chart, and followed by a movement of the chart after return of the pen to zero and before the next peak signal.

Beta ray: a ray consisting of the negatively charged particles (electrons) emitted by a radioactive source.

Bolometer element: a resistance element—a platinum strip—which changes resistance when exposed to radiant heat.

Capacitance detector: also called a microphone or pneumatic detector. A detector block is divided into two cells by a flexible metallic diaphragm. This diaphragm and a fixed-metal plate form a capacitor in an electrical circuit. Gas in the filled chamber absorbs infrared radiation, expands due to a rise in temperature and moves the diaphragm, thereby varying the circuit capacitance.

Carrier gas: an inert gas, most frequently helium, used to sweep the sample gas through the column and detector.

Catalytic combustion: the burning of a component in the presence of another, as in the case of oxygen and hydrogen.

Cathode: the negative electrode of an electrolytic cell.

Chopper: a beam interrupter, usually a rotating disc, with one or more sections capable of blanking the beam as the disc turns.

Chromatogram: the recording of component peaks as they are eluted from the chromatographic analyzer.

Chromatograph: an analyzer which physically separates and quantitatively identifies two or more components by distributing them in two phases, one phase being a stationary bed of adsorbent surface and the other phase a carrier gas which percolates through and along the stationary bed.

Coalescer: a unit for collecting liquid in a fluid stream to form droplets large enough to be removed.

Collimate: to make straight or parallel (as a beam of light or particles).

Column (chromatographic): a tube filled with finely ground material which may or may not be plated with a chemical, the purpose of which is to act as the stationary bed and medium for separating components in a sample.

Conductivity: the ability to conduct or transmit an electrical signal.

Curie: unit quanity of radioactivity equivalent to 3.7 x 10^{10} disintegrations per second.

Density: mass of a substance per unit volume.

Dessicant: a drying agent.

Dispersive spectrometry: infrared analysis by the use of prisms, mirrors and slits to scatter first the light from a source and then to eliminate stray and unwanted wavelengths, leaving for use only light rays of a specific wavelength.

Electrode: a conductor used to establish electrical contact with a nonmetallic part of a circuit.

Electrolyte: a nonmetallic electric conductor in which current is carried by the movement of ions.

Electromagnetic: of, relating to or produced by an electric current.

Elute: to extract from or to sweep away.

Fluidized: to suspend in a rapidly moving stream of gas for transportation.

Galvanic: dissimilar substances (as metals) capable of acting together as an electric source when brought in contact with an electrolyte.

Gamma ray: highly penetrating radiation from a radioactive source, similar to x-rays but of higher frequency.

Geiger tube: a gas-filled counting tube with a cylindrical cathode and axial wire electrode for detecting the presence of cosmic rays or radioactive substances by means of the ionizing particles that penetrate its envelope and set up momentary current pulsations in the gas.

Half-life: the time required for half of the atoms of a radioactive substance present to become disintegrated.

Hygrometer: an instrument to measure moisture content.

Hygroscopic: readily taking up and retaining moisture.

Infrared: thermal radiation of wavelengths longer than those of visible light.

Injection valve: valve, sometimes called a sample valve, used to inject automatically a predetermined volume of sample into the carrier stream.

Ionization chamber: a gas-filled metal tube with an axial rod or plate electrode. With 200 or 500 volts potential, X-rays entering eject electrons photoelectrically from the gas atoms: other X-rays are scattered and create "recoil" electrons which collide with other atoms, knocking out more electrons and creating a cloud of electrons and ions.

Liquid junction: a restricted flow of electrolyte from the reference electrode into the sample solution to achieve electrical continuity.

Magnetostrictive: the property of a ferromagnetic body to undergo a change of dimension with a change in its state of magnetization.

Memory unit: also called peak-picker; a capacitance electronic circuit which monitors a chromatographic peak output, storing a charge equal to the maximum amplitude of the peak signal and holding this charge for use by controllers, computers, etc., until another peak signal occurs.

Microphone detector: see *Capacitance detector.*

Microwave: a very short electromagnetic wave (between 1 and 100 cm long).

Newtonian fluids: liquids which, according to Newton's law of motion, when subjected to a shear stress, undergo deformation, whereby the ratio of flow (shear rate) to force (shear stress) is constant.

Oscillator: a device for producing alternating current; a radio-frequency or audio-frequency generator.

Paramagnetic: a substance subject to magnetization, but whose susceptability to magnetization varies little with magnetizing forces.

Peak-picker: see *Memory unit.*

pH: The negative logarithm of the effective hydrogen ion concentration or hydrogen ion activity in gram equivalents per liter; used in expressing both acidity and alkalinity on a scale whose values run from 0 to 14.

Photocell: an electronic tube which converts light energy to electrical energy.

Pneumatic detector: See *Capacitance detector.*

Polarographic: analysis based on current-voltage curves obtained during electrolysis of a solution.

Precut column: a section of the chromatographic column which may be switched out of the main carrier flow stream by the column switching valve. With the "unwanted" components in the rest of the column, the precut section is switched, and auxiliary carrier flow is used to flush the "unwanted" components out while the rest of the components are swept through the detector.

Prism: a transparent body bounded in part by two plane faces that are not parallel, used to deviate or disperse beams of light.

Resolution: the degree to which two adjacent peaks are separated or resolved.

Roentgen: the unit of X-radiation or gamma radiation that produces in one cubic centimeter of dry air under standard conditions ionization of either sign equal to one electrostatic unit of change.

Sample valve: See *Injection valve.*

Scintillation detector: a device consisting of a fluorescent material which absorbs incident X-rays and emits light pulses in the visible or ultraviolet region. These pulses are detected with a photomultiplier tube.

Selective detector: one which responds to only one component.

Sensitizing: designing a system so that it is preferentially and highly responsive to a particular condition, phenomena or objective.

Sparge column: one in which gas vapor is bubbled through a liquid bed.

Specific gravity: the ratio of the density of a substance to the density of another substance (as pure water or air) regarded as a standard.

Spectrophotometer: an instrument for measuring the relative intensities of light in different parts of the spectrum.

Spectrum: the record of plotting transmittance verses wavelength when a sample of a given element or compound is subjected to electromagnetic radiation.

Stationary bed: the solid support of a chromatographic column, normally an inert porous solid, whose function is to absorb the components of the sample as they are swept down the column by the carrier gas. The solid support may or may not be coated with a chemical. The inner wall of the column may be coated with a chemical and serve as the solid support.

Thermal conductivity: the ability of a substance (a gas) to conduct heat.

Thermistor: an electrical resistor made of a material whose resistance varies sharply in a known manner with the temperature.

Thermocompensator: a resistance thermometer element placed in the measured medium and connected in the electrical circuitry of a measuring unit in order to compensate for changes in temperature by a corresponding change in circuit resistance.

Thermopile: a number of thermocouple elements combined so as to multiply the effect of generated currents or for determining the intensities of incident radiation.

Toroid: doughnut shaped.

Transducer: a device that is actuated by power from one system and supplies power in any other form to a second system.

Transmittance: the fraction of radiant energy that, having entered a layer of absorbing matter, reaches its farther destination.

Ultrasonic: having a frequency above the human ear's audibility range of about 20,000 hz.

Viscosity: the property of resistance of a fluid to deformation under shear; the resistance to flow by release of counteracting forces.

10 Miscellaneous Instruments

Gareth Eugene Ivey, William G. Andrew

Miscellaneous groups of instruments that are not classified as a major categorical group are included in this chapter. An arbitrary classification was made primarily on the basis of the amount of space devoted to their discussion. The lack of coverage in no way minimizes their importance. Neither does it indicate that the quantity used is small compared to those in the major categorical groups. On the contrary, on some jobs there are more transducers used, for example, than any other type device.

This chapter does not attempt to cover all instrument items that are not listed in a major category. However, those discussed, together with the major groups, comprise well over 95% of the instrument items normally used in the processing industries under consideration in this book. Items covered in this section include annunciators, several types of transducers and converters, vibration switches, speed sensors, tachometers, torque devices, solenoid valves and program controllers.

Annunciators

Function

The annunciator is a central alarm system for announcing abnormal or changing conditions in plant operation. When operating parameters (pressure, level, flow, etc.) vary beyond prescribed limits, the annunciator receives an actuating signal and provides a visual and audible display of the alarm.

In recent years, the annunciator has become more important as an alarm or monitoring system. The complexity of modern process systems requires that the operator be alerted to any significant change in the process conditions. Increased reliability with the application of solid-state techniques in annunciator circuitry has also contributed to its use as an operating tool.

The design of different configurations of annunciator systems vary, depending on the manufacturer. The most widely used systems are the electromechanical relay and the electronic types. Annunciators are normally designed and built to allow future expansion or revision with a minimum of time and expense.

Terminology in common use for discussing and specifying annunciator systems and consistent with ISA-RP18.1 is given below.

Terminology

Abnormal operating conditions: the monitored variable is not within the specified operating limits. This condition is also known as off-normal.

Abnormal visual indication: the nameplate or bullseye that is lighted during the abnormal operating condition.

Acknowledgment: recognition of a change in operation.

Alert: the state of the system after a point has been actuated.

Alert-Lock: the state of a system, having memory and return alert, after a condition-sensing contact has returned to normal, but the reset has not been actuated.

Auxiliary switching: an accessory to the annunciator system which is capable of actuating an external device.

Backlighted nameplate: a translucent nameplate which is illuminated from behind which can be marked for identification.

Bullseye: a lens mounted in front of an illuminating source.

Clear: to allow a new sequential series to be monitored even if points in the original series are still in an abnormal operating condition.

Conditioning-sensing: a condition where an external contact or signal starts an annunciator alert sequence (may also be field-contacts).

Fail safe: a design in which an automatic indication is given when a system component fails.

Flasher: a subunit of the annunciator system used to cause the abnormal visual indicator to turn on and off during particular portions of the system sequence.

Hazardous locations: locations where hazardous fumes, dust or gases exist or are present in case of accidental leaking, which present fire, explosion or other dangers to either people or equipment.

Hermetically sealed: an enclosure, considered permanent, such that no gas or liquid may escape or enter.

High-low: a system which can indicate abnormally low, normal and abnormally high.

Intrinsic safety: safe operation obtained by a system design which ensures a low energy level at all times such that no device is capable of causing ignition of any mixture.

Low power drain: a system that uses minimum power under normal conditions.

Memory or lock-in: an electrical component to the annunciator system which holds the abnormal signal until acknowledgment, even if the condition-sensing contacts return to their normal state during the time interval involved.

Normal: an operating condition during which the process lies within specified limits.

Normally open annunciator operation: in the normal operation of the annunciator, the condition-sensing contacts are in the open position.

Normally closed annunciator operation: in the normal operation of the annunciator, the condition-sensing contacts are in the closed position.

Off-normal: see *Abnormal operating conditions.*

Plug-in card: an unenclosed electronic package designed for a particular function which can be mounted into a connector base.

Plug-in module: an electromechanical or electronic package designed for a particular function which can be mounted into a connector base.

Point: same as *Station.*

Pulse: electrical energy for a short period.

Reset: the process of returning the annunciator system to its normal state after acknowledgment of an abnormal condition.

Return alert: the annunciator sequence which causes the annunciator system to indicate that a change in condition has occurred after return of the conditioning contacts to normal. The operator is expected to acknowledge this condition by turning off the lighted nameplate.

Sequence: various states of operation of annunciator system (see Table 10.1 for sequence).

Station: the unit of the annunciator system necessary to cause system operation upon a change of the condition-sensing contact. An annunciator system may have active, spare or future stations.

Test: a check of operational readiness of the system, which can be manually simulated.

Sequences

Various stages of operation of the annunciator, or a description of the order of detailed events of the annunciator and associated components is called the sequence.

Several different sequences are available from most annunciator manufacturers. Table 10.1 shows a list of sequences commonly used. A typical basic sequence follows.

1. The conditioning-sensing contact goes off-normal.

   a. The horn sounds.
   b. The nameplate lamps light.

2. The operator responds by depressing the silence switch which silences the horn. The nameplate lamps remain on.
3. The conditioning-sensing contacts return to normal and the nameplate lamp goes out, and the point is ready for realarming.

Table 10.1 reveals that the many available sequences are divided into four classes. Class 1 contains the simple standards and operational sequences. Class 2 sequences include the alert step. Class 3 illustrates sequences with the high-low option, and Class 4 sequences are simultaneous multiple alert situations.

Table 10.2 shows some available sequences with visible and audible actions for each type. Among these are some first out sequences (5 and 6). First out sequences allow the first alarm point actuated by the condition-sensing contacts to be identified in order that the problem may be corrected. In many situations the first abnormal condition causes other monitored conditions to go off-normal. The first out sequence indicates the point normally by a flashing signal.

Sequences available from most manufacturers include:

1. Nonlatching, self-reset
2. Momentary, self-reset, latching before acknowledge, nonlatching after acknowledge
3. Momentary, manual reset, latching before and after acknowledge
4. Flashing, nonlatching, self-reset
5. Flashing, momentary, manual reset, latching before and after acknowledge
6. Flashing, momentary, self-reset, flashing and latching before acknowledge, steady on and nonlatching after acknowledge
7. Flashing first out, momentary, manual reset, latching before and after acknowledge
8. Flashing first out, momentary, manual reset
9. Flashing momentary first out, subsequent points nonlatching
10. Flashing return alert after first acknowledge

Return alert sequences provide additional information announcing that a previously abnormal condition has returned to normal. A variation of frequency or color may be used to identify this step. High and low sequences provide an indication that the abnormal point is either on the high or low side of the measuring point. Nameplates are properly

<table>
<tr><th colspan="12">Table 10.1 Table of Common Sequences</th></tr>
<tr><th>Sequence</th><th>Signal Device</th><th>Normal</th><th colspan="2">Alert</th><th colspan="2">Condition-Sensing Returns to Normal Before Acknowledge</th><th colspan="2">Acknowledge Reset</th><th>Condition-Sensing Returns to Normal</th><th>Return to Normal Reset</th><th>Remarks</th></tr>
<tr><td>ISA-1</td><td>Visual
Audible</td><td>Off
Off</td><td colspan="2">Flash
On</td><td colspan="2">Flash
On</td><td colspan="2">On
Off</td><td>Off
Off</td><td>—
—</td><td>Flasher memory</td></tr>
<tr><td>ISA-1A</td><td>Visual
Audible</td><td>Off
Off</td><td colspan="2">On
On</td><td colspan="2">On
On</td><td colspan="2">On
Off</td><td>Off
Off</td><td>—
—</td><td>Memory</td></tr>
<tr><td>ISA-1B</td><td>Visual
Audible</td><td>Off
Off</td><td colspan="2">Flash
On</td><td colspan="2">Off
Off</td><td colspan="2">On
Off</td><td>Off
Off</td><td>—
—</td><td>Flasher</td></tr>
<tr><td>ISA-1C</td><td>Visual
Audible</td><td>Off
Off</td><td colspan="2">On
On</td><td colspan="2">Off
Off</td><td colspan="2">On
Off</td><td>Off
Off</td><td>—
—</td><td></td></tr>
<tr><td>ISA-1D</td><td>Visual
Audible</td><td>Dim
Off</td><td colspan="2">Flash
On</td><td colspan="2">Flash
On</td><td colspan="2">On
Off</td><td>Dim
Off</td><td>—
—</td><td>Memory—flasher—continuous lamp test</td></tr>
<tr><td>ISA-1E</td><td>Visual 1
Visual 2
Audible</td><td>Dim
On
Off</td><td colspan="2">On
On
On</td><td colspan="2">Dim
On
Off</td><td colspan="2">On
Off
Off</td><td>Dim
On
Off</td><td>—
—
—</td><td>Continuous lamp test—normal indication</td></tr>
<tr><td>ISA-1F</td><td>Visual 1
Visual 2
Audible</td><td>Dim
Dim
Off</td><td colspan="2">On
On
On</td><td colspan="2">On
On
On</td><td colspan="2">Off
On
Off</td><td>Dim
Dim
Off</td><td>—
—
—</td><td>Memory—(alarm silence indication)—continuous lamp test</td></tr>
<tr><td>ISA-2</td><td>Visual 1
Visual 2
Audible</td><td>Off
Off
Off</td><td colspan="2">Flash
Off
On</td><td colspan="2">Flash
Off
On</td><td colspan="2">On
Off
Off</td><td>Off
Flash
On</td><td>Off
Off
Off</td><td>Memory—flasher—return alert (light distinction)</td></tr>
<tr><td>ISA-2A</td><td>Visual
Audible</td><td>Off
Off</td><td colspan="2">Flash
On</td><td colspan="2">Flash
On</td><td colspan="2">On
Off</td><td>Dim-Flash
On</td><td>Off
Off</td><td>Memory—flasher—return alert (light distinction)</td></tr>
<tr><td>ISA-2B</td><td>Visual
Audible 1
Audible 2</td><td>Off
Off
Off</td><td colspan="2">Flash
On
Off</td><td colspan="2">Flash
On
Off</td><td colspan="2">On
Off
Off</td><td>Flash
Off
On</td><td>Off
Off
Off</td><td>Memory—flasher—return alert (sound distinction)</td></tr>
<tr><td>ISA-2C</td><td>Visual
Audible</td><td>Off
Off</td><td colspan="2">Flash
On</td><td colspan="2">Flash
On</td><td colspan="2">On
Off</td><td>On
Off</td><td>Off
Off</td><td>Memory—flasher—return alert</td></tr>
<tr><td>ISA-2D</td><td>Visual
Audible</td><td>Off
Off</td><td colspan="2">On
On</td><td colspan="2">On
On</td><td colspan="2">On
Off</td><td>On
Off</td><td>Off
Off</td><td>Memory—return alert</td></tr>
<tr><td>ISA-3</td><td>Visual H
Visual N
Visual L
Audible</td><td>Off
On
Off
Off</td><td>Off
Off
Flash
On</td><td>Flash
Off
Off
On</td><td>Off
Off
Flash
On</td><td>Flash
Off
Off
On</td><td>Off
Off
On
Off</td><td>On
Off
Off
Off</td><td>Off
On
Off
Off</td><td>—
—
—
—</td><td>High-low—memory—flasher</td></tr>
<tr><td>ISA-3A</td><td>Visual H
Visual N
Visual L
Audible</td><td>Off
On
Off
Off</td><td>Off
Off
Flash
On</td><td>Flash
Off
Off
On</td><td colspan="2">Off
On
Off
Off</td><td>Off
Off
On
Off</td><td>On
Off
Off
Off</td><td>Off
On
Off
Off</td><td>—
—
—
—</td><td>High-low—flasher</td></tr>
<tr><td>ISA-3B</td><td>Visual H
Visual N
Visual L</td><td>Off
On
Off</td><td>Off
Off
Flash</td><td>Flash
Off
Off</td><td colspan="2">Off
On
Off</td><td>Off
Off
On</td><td>On
Off
Off</td><td>Off
On
Off</td><td>—
—
—</td><td>Same as ISA-3A, except no audible alarm</td></tr>
</table>

Source: Instrument Society of America

Table 10.1 continued

Table 10.1 continued

Sequence	Signal Device	Normal	Alert		Condition-Sensing Returns to Normal Before Acknowledge		Acknowledge Reset		Condition-Sensing Returns to Normal		Return to Normal Reset	Remarks
ISA-4	Visual 1	Off	Flash	Off	Flash	Off	On	Off	On	Off	Off	Sequence—flasher—memory (two-color indication). *If reset is operated, a new sequence can be started using visual 1 and 2 in the flashing state.
	Visual 2	Off	Off	Flash	Off	Flash	Off	On	Off	Off	Off	
	Audible	Off	On	On	On	On	Off	Off	Off	Off	Off	
ISA-4A	Visual	Off	Flash	On	Flash	Off	On	On	Off		—	Sequence—clear. *After acknowledgement the next point to alert will flash.
	Audible	Off	On	Off	On	Off	*Off	Off	Off		—	

displayed in order to indicate the abnormal point. A third indication may also be incorporated to show a normal condition.

The selection of the desired sequence depends on several factors. In many cases it is determined primarily by plant custom. If the design group or the operating group is accustomed to a particular sequence, little thought is usually given to changing the choice. Logic is displayed in that approach, because uniformity of operation is desirable. However, other sequences should be considered for each system to determine whether they can provide better information and operating assistance for the plant involved.

Types of Displays

Three basic types of annunciator displays exist:

1. Bullseye indicator
2. Lighted window
3. Graphic display

The simplest indication is the bullseye lamp (Figure 10.1). This simple light indicator is normally 1 inch in diameter and red, white or green. The bullseye is used where a limited number of indicators are required on a panel and identification presents no problem. Explosionproof annunciators often incorporate this type of indicator because of the packaging and space advantages it offers.

Lighted windows, the most widely used displays, are manufactured in various sizes and shapes. Figure 10.2 is a typical example, and Figure 10.3 shows a typical specification sheet for an annunciator of this type. Nameplates are available in several translucent colors; the most widely used color is translucent white with black letters.

Graphic annunciator panels (Figure 10.4) are conveniently adaptable to process changes and visually show the location of alarm points in the process. In the graphic annunciator shown, replaceable translucent drawing sheets are illuminated from behind by several lamps representing annunciator points. The drawing becomes the graphic panel. Color is incorporated in the symbols and graphic lines to code the graphic presentation. The graphic drawings may be as simple or detailed as required to show alarm locations or symbols. The drawings are easily revised when a process change is made. Sizes of the drawings vary depending on the manufacturer.

Types of Annunciators

Two basic types of annunciators are described below:

1. Electromechanical
2. Electronic

The electromechanical type, introduced long before solid-state electronic units were available, are still in wide use throughout industry.

Electromechanical Annunciator

The electromechanical annunciator uses modular fail-safe plug-ins, no-drain plug-ins, flasher plug-ins, auxiliary relays and many other items (see Figure 10.5).

Table 10.2. Sequences Including First Out Sequence of Annunciator Systems

SEQUENCE TYPES		CONDITIONS					
No.	**Description**	**Outputs**	**Normal**	**Alert**	**Acknowledge**	**Recover**	**Reset**
1	Non-flashing automatic reset	Visible Audible	Off Off	On On	On Off	Off Off	
2	Non-flashing manual reset	Visible Audible	Off Off	On On	On Off	On Off	Off Off
3	Flashing, automatic reset	Visible Audible	Off Off	Flash On	On Off	Off Off	
4	Flashing, manual reset	Visible Audible	Off Off	Flash On	On Off	On Off	Off Off
5	First-out flashing, automatic reset	1st Visible Oth. Visible Audible	Off Off Off	Flash On On	On On Off	Off Off Off	
6	First-out flashing, manual reset	1st Visible Oth. Visible Audible	Off Off Off	Flash On On	On On Off	On On Off	Off Off Off
7	Ring-back flashing, manual reset	Visible Audible	Off Off	Flash On	On Off	Slow Flash On	Off Off

Source: Deltron, Inc.

Figure 10.1. The bullseye indicator is simple, easy to install and used normally where only a limited quantity is desired. Lens size is 1-inch diameter with many colors available. A hole, 1 1/32-inch diameter is required for mounting the bullseye. A minimum spacing of 1 3/8 inches between adjacent bullseyes must be observed. Bullseye pilot lights are available for panel mounting. (Courtesy Pan alarm Division of the Riley Co.)

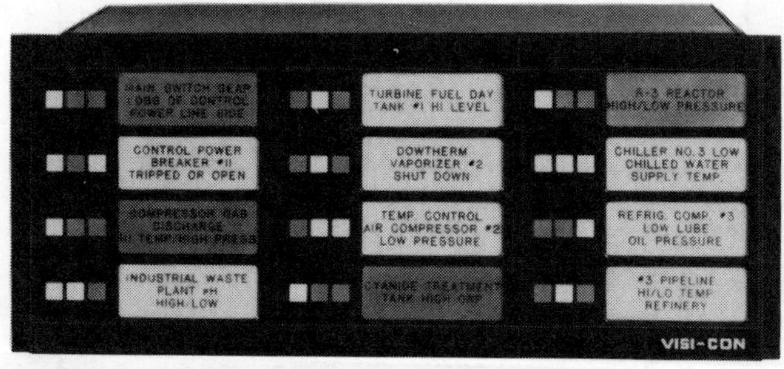

Figure 10.2. Typical 12-point lighted window annunciator cabinet. (Courtesy of VISI-CON Annunciator/Monitoring System)

Figure 10.6 shows a typical schematic for this type system. All relays are shown in the deenergized state. The system is designed for basic sequence plus flashing light. When an off-normal condition occurs, the horn blows and the proper alarm point produces a flashing light. Acknowledgment silences the horn and causes the light to burn steadily.

A typical process malfunction, causing the condition-sensing contacts to close, produces the following actions:

1. Relay A, Line 5 is deenergized.

   a. The horn and flasher are energized from Line 8 contact *A* through terminal *S*.
   b. The lamps flash through flasher contact *F*, B-N.O. and A-N.C.

2. The operator acknowledges the condition by actuating the silence button shown on Line 3.

   a. Relay *B* is deenergized.
   b. Relay *B* remains deenergized because B-N.O. Line 3 has opened.
   c. The lamps remain on through the circuit at Line 3 through the silence button, B-N.C. between Line 3 and Line 2, through A-N.C. Line 2, then to the lamps.
   d. The flasher has stopped and the horn silenced because the B-N.O. contact Line 9 has opened.

3. The condition-sensing contacts return to their corrected position. (N.O.).

	NO. ____ PROJECT ____ CLIENT ____	ANNUNCIATORS			
					SPEC. NO.
	LOCATION ____ CLIENT NO. ____	BY	DATE	ITEM	SHT. of

GENERAL

1. Tag No. ____ Location: ____
2. Cabinet Size: ____ Rows High By ____ Columns Wide ____
3. Mounting: Flush Panel ☐ Surface ☐
4. Cabinet Style: Plug-In Light Boxes ☐ Swing Door ☐ Remote Logic Cabinet ☐ Watertight Door ☐
5. Rating: General Purpose ☐ Weather proof ☐ Explosion proof ☐ Class ____ Group ____ Division ____
6. Power Supply: 117V 60Hz ☐ 125 Vdc ☐ 12 Vdc ☐ 24 Vdc ☐ Other: ____
   Location: Integral ☐ Remote ☐ ____

DISPLAY

7. Backlighted Nameplates: White Translucent ☐ Other: ____ Size: ____
8. Alarm Points Per Lightbox: One ☐ Two ☐ Three ☐ Four ☐
   Lamps Per Alarm: One ☐ Two ☐ Three ☐ Four ☐
9. Bullseye Type: Number of Lights: ____ Color: ____
10. Other Display: ____

LOGIC

11. Logic: Electro-Mechanical Relay ☐ Solid-State Electronic ☐ Mercury Bottle ☐ Fluidic ☐
12. In Display Cabinet ☐ Remote Cabinet ☐ Strip Chassis ☐
13. General Purpose ☐ Weather proof ☐ Explosion proof ☐ Class ____ Group ____ Division ____ Intrinsically Safe ☐
14. Field Contact Voltage: 117 Vac ☐ 12 Vdc ☐ 125 Vdc ☐ Other: ____
15. On Alarm, Actuating Contacts: Open ☐ Close ☐ Field Selectable ☐ Form ____

FEATURES

16. Required Features: Lock-In of Momentary Alarms ☐ Auxiliary Contacts ☐ Sequential Alarm Circuit ☐
17. Ring-Back Circuit: Via Alarm Audible Signal ☐ Via Other Audible Signal ☐
18. Fail-Safe Circuit to Signal Own Failure ☐ Operational Test ☐ Lamp Test ☐
19. Flasher: Remote ☐ In Cabinet ☐
20. Acknowledge Common ☐ Unit ☐ Light ☐ Audible ☐ PB Location in Cabinet ☐ Remote ☐ Others ☐
21. Reset Common ☐ Unit ☐ Light ☐ Audible ☐ PB Location in Cabinet ☐ Remote ☐ Others ☐

21A. Test Common ☐ Unit ☐ Light ☐ Audible ☐ PB Location in Cabinet ☐ Remote ☐ Others ☐

SEQUENCE

22.

STAGE	VISUAL SIGNAL	AUDIBLE SIGNAL
Normal		
Alert, Initial		
Alert, Subsequent		
Acknowledge, Int.		
Acknowledge, Subs.		
Return to Normal		
Reset		
Test		

ISA Sequence Number: ____ (See RP 18.1)

OPTIONS

23.	Horn: Model No.		30. Cabinet: Model No.	
24.	Bell: Model No.		31. Relay: Model No.	
25.	Dimmer:		32. Flasher: Model No.	
26.	Color Caps:		33. Pushbuttons: Model No.	
27.	Power Supply Model No.		34.	
28.			35. Engraving Sheet No.	
29.	Manufacturer:		36. Flow Sheet	

Notes:

△					
NO.	REVISIONS	BY	DATE	APPVD	DATE

Figure 10.3 Specification sheet for lighted window annunciator (© 1978 by the Instrument Society of America. Reprinted and modified with permission).

a. Relay *A* is energized.

b. Relay *B* is energized from Line 3, silence button, *B* coil and A-N.O.

c. The lamps go out since A-N.C. Line 2 opens.

Electronic Annunicator

Electronic solid-state annunciators are available from several manufacturers. They are very reliable and adaptable to almost any application in petrochemical plants. All solid-state circuitry makes these units compatible with modern fast-action process control equipment. Small modules make troubleshooting and replacement easy in such a system. These solid-state types are packaged in smaller units with less weight than the electromechanical type. Figure 10.7 is a schematic of a typical solid-state annunciator. Each circuit **provides for one alarm point. Capable of driving two 3-watt lamps and an alarm/flasher module, it receives its flash**

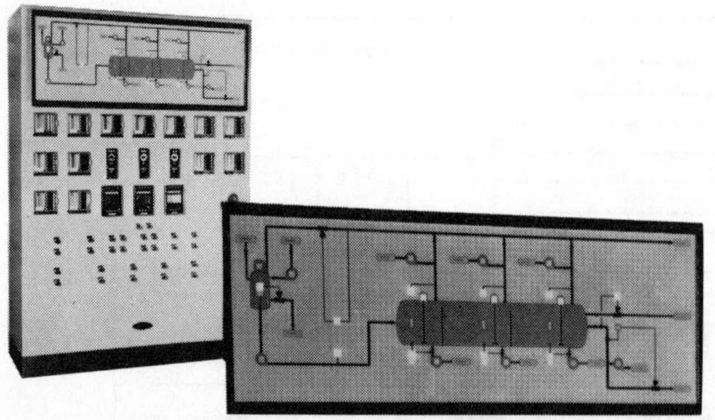

Figure 10.4. Graphic annunciators display schematic of main lines and equipment of the process with proper location of alarm devices. (Courtesy of Panalarm Division of the Riley Co.)

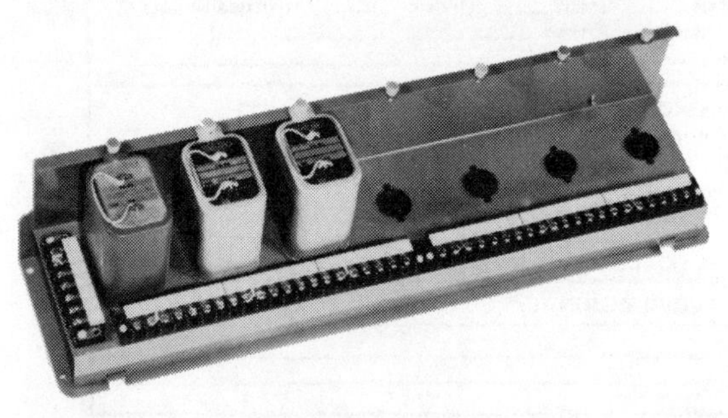

Figure 10.5. Plug-in relay cabinet for electromechanical annunciator which typifies modular concept. (Courtesy of Panalarm Division of the Riley Co.)

signal from the flasher bus supplied by the alarm/flasher module. The unit receives plus and minus 12 volts DC from the annunciator power supply, which also provides the proper voltage for the lamps and the audible alarm.

Intrinsically safe annunciator systems are designed to limit the amount of electrical energy that can pass through the annunciator and its wiring, normally or abnormally, and reach the field contacts. Intrinsically safe electric barriers limit power to terminals which have wires leading to equipment mounted in hazardous areas. Non-intrinsically safe wiring must be physically prevented from inadvertent shorts to safe wiring by means of mechanical barriers or terminations in a separate compartment, and by electrical circuit isolation means such as electromechanical relays. Field wiring must conform to intrinsically safe code requirements if system integrity is to be maintained. Horns, bells, relays and other related equipment can be used, provided their power is derived from an intrinsically safe power supply.

The function of the electronic solid-state annunciator is the same as the electromechanical type. The nature of solid-state equipment is such that it should be more maintenance-free. However, the electromechanical type is not a high maintenance item.

Alarm Printouts

Computer usage in the industry has given rise to the use of automatic printing of alarm points and conditions. Analog values, switch number, time and date are usually shown, often in red-colored print, as an intermediate entry to regularly logged data. In some instances, a separate automatic typewriter is used for alarm logging.

Data loggers—systems designed to read, log and alarm sensors for a designated machine or process unit—are provided with an adding-machine-type printer. Some data

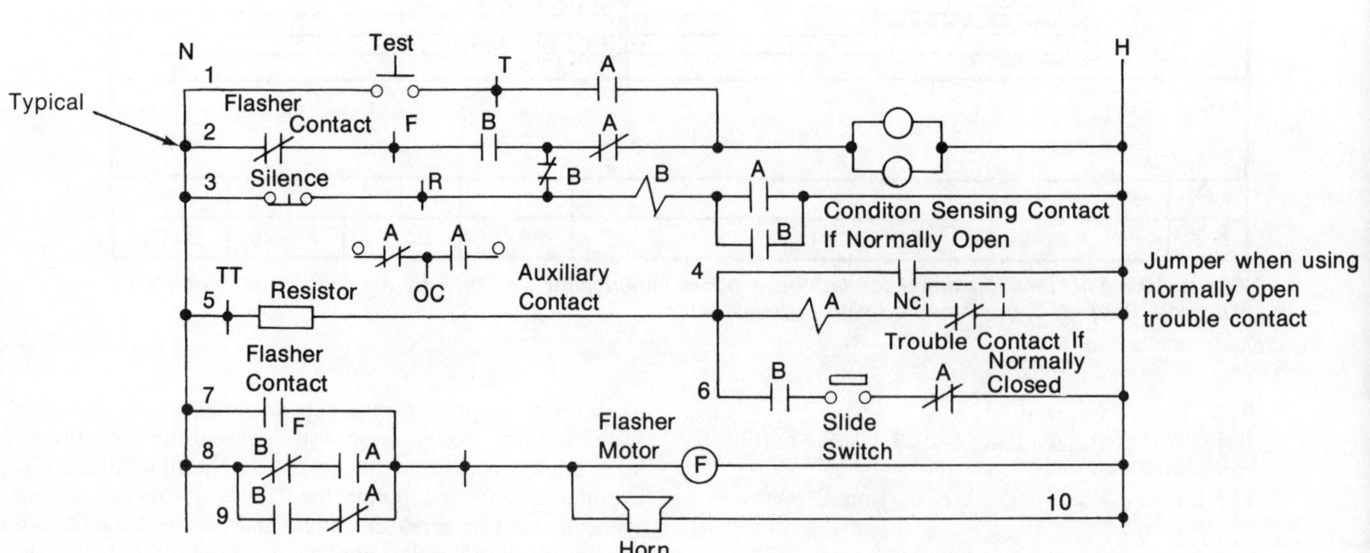

Figure 10.6. Electrical schematic for an electromechanical annunciator system.

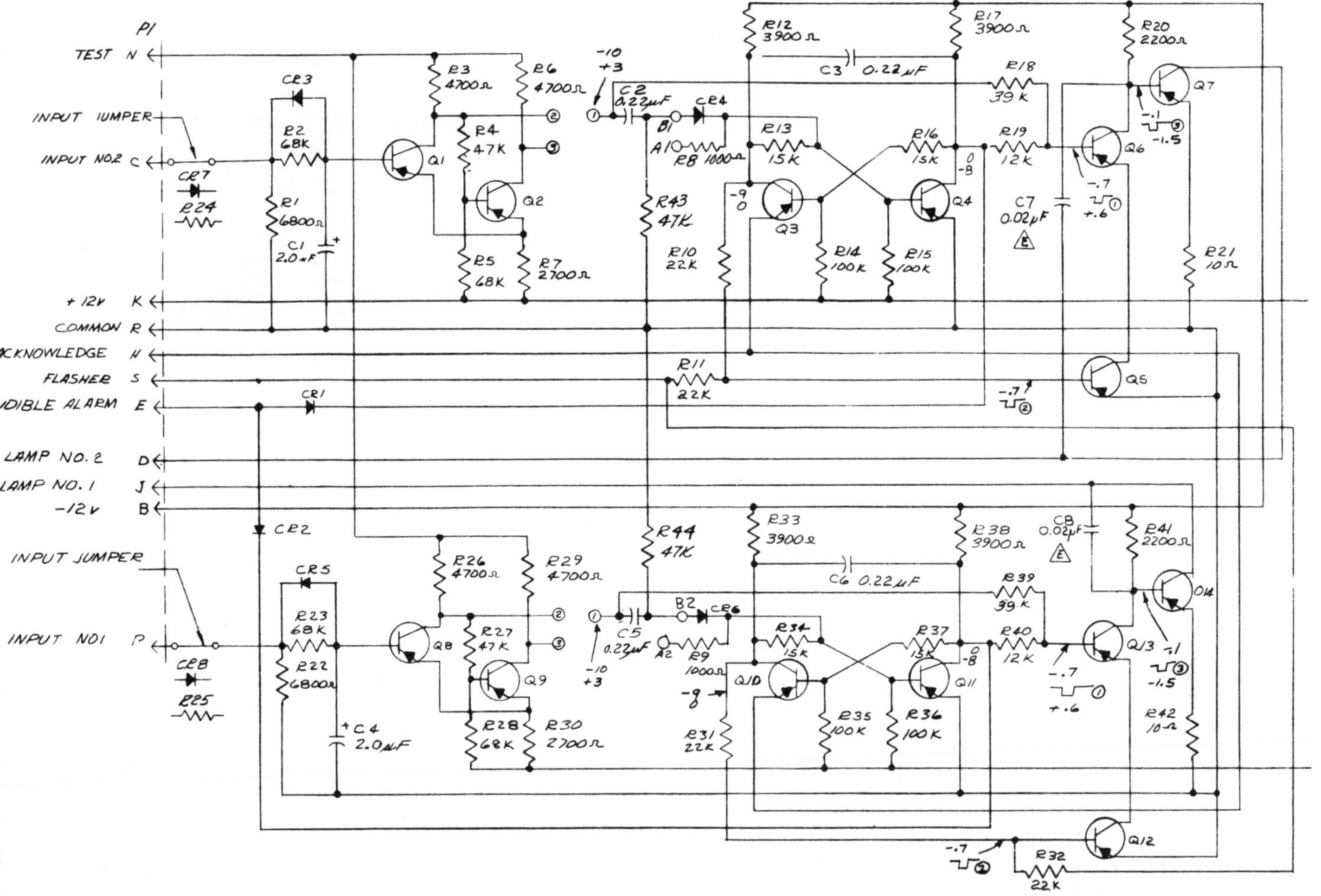

Figure 10.7. Electrical schematic diagram of electronic annunciator. (Courtesy of Rochester Instrument Systems)

loggers are microprocessor-based and are reprogrammable from an integral keyboard. See Chapter 14 for a detailed discussion of data acquisition.

Selection and Specification

Several items should be considered when writing specifications for an annunciator.

1. Ambient temperature range and voltage of lamps: damage may occur to visual indicators if the units require high voltage on a continuous basis. A normal range is less than 10 watts with a maximum of 12 watts per point. Window cabinet temperature of 120°F should be maximum. (Temperatures may vary slightly from this because of design and construction.)
2. Power requirements and line voltage availability: on AC power the units should operate on 60-Hz line voltage over a range of 90 to 135 volts. If DC power is used, requirements should be listed.
3. Contact current rating: auxiliary contacts for operation of external devices must be determined and specified.
4. Isolation: condition-sensing contact circuits are to be isolated from the AC supply voltage by a transformer. Condition-sensing contact leads may be shorted to ground or power without damaging the alarm unit.
5. Condition-sensing contacts: voltages of N.O. and N.C. contacts, operating mode and current rating of contacts should be specified.
6. Size of enclosures and number of points or stations: the number of points purchased should include spares for future use. The size is determined not only by the number of points but also by the configuration that fits the available space. (Note: It is usually cheaper in the long run to purchase a cabinet with all of the alarm modules or relays included, rather than have the expense of another purchasing activity and higher prices due to escalation.)

7. Area Electrical Classification: in addition to general purpose equipment, annunciators are available in enclosures rated NEMA 1, NEMA 4, NEMA 12. Explosionproof ratings include N.E.C. Class I, Group D, Division I and Division II; and Class II, Group E, F & G, Division I and Division II. Single-point units and up to 20-point systems are manufactured.
8. System purging requirements: sometimes required in order to meet an electrical area classification.
9. Mounting, dimensions and panel thickness: not often a problem.
10. Power interlock: seldom required.
11. Type of indication: bullseye, lighted window or graphic.
12. Parts list recommendation: for spare parts purchasing.
13. Wiring diagrams: drawings should be furnished that are adequate for troubleshooting and installation and that follow recognized standards. All components should be recognized, voltage measurements noted and oscilloscope wave forms illustrated for pertinent points in the circuit.
14. Intrinsic safety requirements: if desired, the annunciator should be compatible with other instrument equipment.
15. Sequence of operation: see paragraph on sequence.
16. Electrical requirements:
    a. Electrical components such as wire size or terminal strips within the annunciator should be clearly identified.
    b. Transient voltage surges should not affect annunciator components.
    c. The unit should be prewired so that only power and input wiring need to be connected for system operation.
    d. Spare point addition or removal should not influence or affect other points in any way.
    e. Types of terminal strips, wire size and color code should be specified.
    f. Minimum period of field contact actuation should be noted.

Other detailed information may be specified or provided to the manufacturer by the customer such as:

1. Letter size and engraving
2. Color of indicator lamps or bullseyes
3. Memory capacity
4. Switching ability from N.O. to N.C. operation
5. Options
    a. Lower power drain
    b. Self-monitoring
    c. Test
    d. Auxiliary switching

By considering this information, specifications can be adequately written for an annunciator.

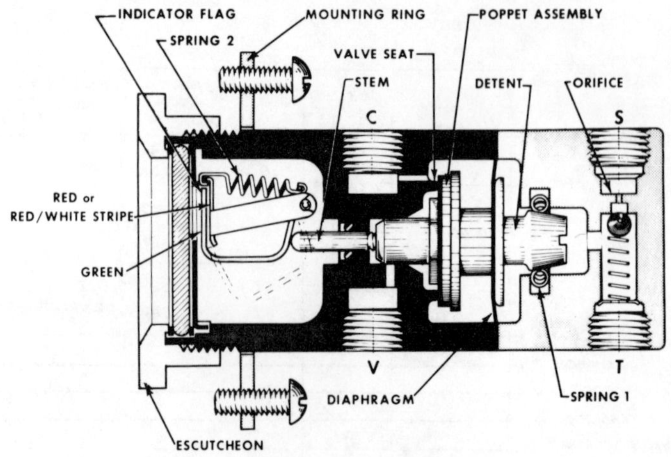

Figure 10.8. A pneumatic indicator-receiver dumps pneumatic pressure in a control line and indicates a "tripped" condition. (Courtesy of Robertshaw Controls Co.)

Pneumatic Shutdown Indicators

Many of the safety systems furnished with large engines and other rotating machines are pneumatically operated. A pneumatic source furnishes pressure through a small restricting orifice to a shutdown header. Each safety device connects to the header and dumps header pressure when actuating, causing a machine shutdown. Normally operating in the 5-25 psi range, indicators, relays, temperature- and pressure-sensors, and vibration sensors all snap-act to control, indicate, or shut-down. Indicators show a color change or change of words to show status change (Figure 10.8).

Enunciators

The enunciator alarm system is a verbal communication system that produces verbal messages to the operators. The enunciator alarm sequence is similar to the annunciator, except for the audible alarm. The sequence may be designed as follows:

1. The condition-sensing contacts open which causes a flashing light and a recorded message.
2. Acknowledgment stops the flashing light and repeats the message.
3. Correction of the problem causes the light to go out and resets the system.
4. The test push button gives a check on all parts of the audio system, except other taped messages.

A computer can be programmed to trigger the alarm messages. The alarm points are scanned for seeking out an alarm condition of the contacts. If a contact is in the alarm position, the logic circuit for this particular point actuates the sequence. Messages last about 10 seconds; therefore, if two alarms are actuated at the same time, one message must wait on the other. After a message is acknowledged, the second message will be played. The messages are recorded on magnetic tape cartridges which contain four parallel tracks with a message on each track.

Solid-state scanning logic checks each point at a nominal rate of 20 points per second. A number of different kinds of logic test functions are available with this system.

The cost of an enunciator system is over twice the cost of an annunciator system, generally because each system ends up as a custom designed system. Part of this cost is due to the communication network that serves the enunciator system. The communication system makes it possible to communicate with operators instantly.

Transducers and Converters

Transducers and converters include a variety of different equipment used in many different applications. They play an important role in modern instrumentation, linking electrical and pneumatic control systems together, changing signal levels in the application of computers or other electronic equipment to control problems and converting easily obtained sensing signals to other signal forms more easily transmitted or processed.

The terms *transducers* and *converters* are used somewhat interchangeably as they apply to process instruments. The device for converting electronic signals to pneumatic signals at control valves is almost always referred to as a transducer. Most other devices that convert the forms or levels of electronic signals to other electrical forms or levels or that switch from pneumatic to electronic signals or vice-versa are more likely to be referred to as converters.

One of the first converters to be used on a large-scale basis was the E/P (or MV/P) converter used to change a thermocouple voltage to a convenient pneumatic signal (Figure 10.9). It is more economical to use than the filled system temperature transmitter, especially when multiple measurements are required. Its use has declined, as electronic systems have become more popular, necessitating the use of its electronic equivalent, the E/I converter.

I/P transducers at control valves came into use when electronic controllers were introduced to the industrial market. Now they are an economic and operating necessity. Although electrically operated control valves have been developed, their use is minimal because of power deficiencies and lack of operating speed. Control valve transducers, therefore, are likely to be in use for a long time.

Many other devices could be classified by either term. Transmitters of every process variable measured convert from one energy form to another to accomplish the measurement and transmission. In processing signals for computer use, many signal conditioning devices might be classified by either term. The units to be discussed in this section, however, have a limited definition and are classified into five groups:

1. Current-to-pneumatic (I/P)
2. Pneumatic-to-current (P/I)
3. Voltage-to-current (E/I)
4. Voltage-to-pneumatic (E/P)
5. Current-to-current (I/I)

Current-to-Pneumatic Transducers

Inputs for current-to-pneumatic (I/P) transducers are normally 1-5, 4-20 or 10-50 milliamperes. Figure 10.10 shows a typical I/P unit. A high quality permanent magnet creates a field which passes through the steel body of the transducer and across a small air gap to the pole piece. A multiturn flexure-mounted coil is suspended in the gap. The input current flows through the coil creating an electromagnetic

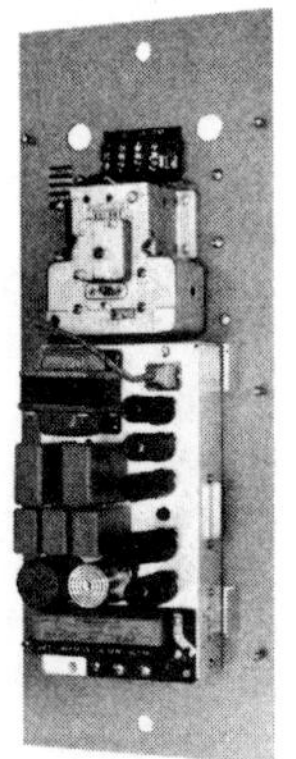

Figure 10.9. This E/P (or MV/P) converter is used to change a thermocouple voltage (or similar pneumatic output signal). Courtesy of Foxboro Co.

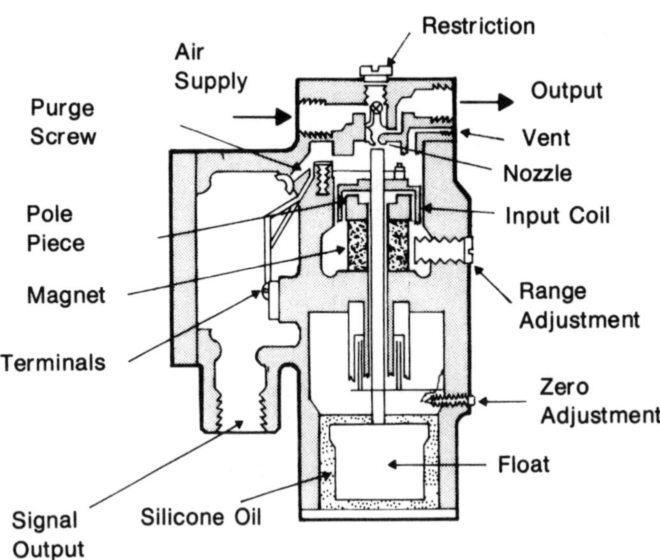

Figure 10.10. Electric to pneumatic transducer (I/P) converts low level DC currents to pneumatic signals. (Courtesy of Moore Products Co.)

force which tends to repel the coil and converts the current signal into a mechanical force.

A reaction nozzle is used to convert the mechanical force into a pneumatic output pressure. Air is supplied to a detection nozzle through a restriction, and the pressure in the nozzle varies according to the restrictive effect imposed by the nozzle seat. The reaction of the air jet impinging against the nozzle seat supplies the counterbalancing force to the coil motor. The nozzle back pressure is the transmitted output pressure which is the output of the transducer. This output is directly proportional to the input current.

The transducer is made insensitive to vibration by submerging the float in silicone oil. The coil is integrally mounted to the float. Sizing of the float is designed so that its buoyant force equals the weight of the assembly, leaving a zero net force.

The zero adjustment is accomplished by changing a leaf-spring force. Span is adjusted with a range adjusting screw which varies the gap between the screw and the magnet. This adjustment shunts some of the magnetic field away from the pole piece.

These shock-proof and vibration-proof transducers may be mounted in the upright position but not tilted more than 10% from the vertical centerline. The transducer may be mounted on or away from the actuating or control unit. I/P transducers can either signal a pneumatic positioner or can be used to direct load the control valve if a spring loaded actuator and volume booster are used.

Pneumatic-to-Current Converters

Pneumatic-to-electronic converters (P/I) are used when pneumatic signals must be converted to electronic signals (Figure 10.11). They find application when conversions to electronic controllers are made and existing pneumatic transmitters are used to save installation time and money. They are also used when pneumatic rather than electronic transmitters for a particular application are not available; when instrument air is not available at the controller location; when long transmission distances are involved (more than 500 feet); or when computer or telemetry inputs are required.

The P/I normally has an input of 3 to 15 psig pressure into a Ni-Span C pressure capsule and motion balance sensor. Accuracy of ±¼%, high response speed, small size, ability to operate in vibrational environments and a built-in power supply are desirable and obtainable characteristics of pneumatic-to-electronic converters.

This motion balance sensor is normally a linear variable differential transformer (LVDT) which converts the displacement of an input-pressure capsule into proportional electrical-output signals. All solid-state electrical systems are used by most manufacturers for P/I circuitry.

Figure 10.12 shows a simple schematic of the pneumatic-to-electronic converter, typical of the high quality devices made by several manufacturers. These units are not position sensitive and may be installed in any orientation. Pneumatic connections for converters are normally ⅛-inch npt female.

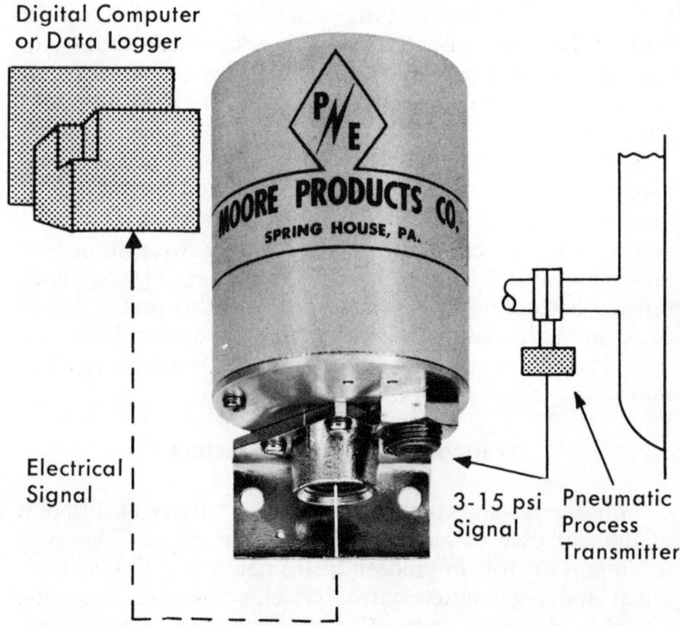

Figure 10.11. The pneumatic to current (P/I) converter uses a linear variable differential transformer (LVDT) to convert mechanical motion to an electrical signal. (Courtesy Moore Products Co.)

Voltage-to-Current Converters

Voltage-to-current (E/I) converters are widely used in the processing industries to convert voltage signals to current. Signal strength is usually in the millivolt range. Thermocouples are the most common millivolt generating sensing element. The millivolt signal must be converted to a current signal generally of 1-5, 4-20 or 10-50 milliamperes for inputs to controllers or other receivers. Analyzers, tachometers and other similar equipment may also require E/I converters for changing output signals into higher level transmission signals.

E/I temperature converters are designed to accept standard thermocouple inputs directly, converting them into high level DC current outputs. These units normally have built-in temperature compensation and thermocouple burn-out protection. High input impedance design of the unit permits the use of long thermocouple leads without error.

A millivolt-to-current converter amplifies the input signal by use of a solid-state chopper. The output of the chopper is fed to the input of an AC amplifier where the signal is amplified and fed to an AC to DC converter. The AC to DC converter output is amplified by a DC output amplifier. The standard outputs from these units are 1-5, 4-20 and 10-50 milliamperes DC. Figure 10.13 shows a typical E/1 converter with a voltage divider to reduce voltages when required.

Other voltage output devices that are included in the E/I converter group include strain gauges, various types of analyzers, speed devices, magnetic flow meters and many other voltage generating units.

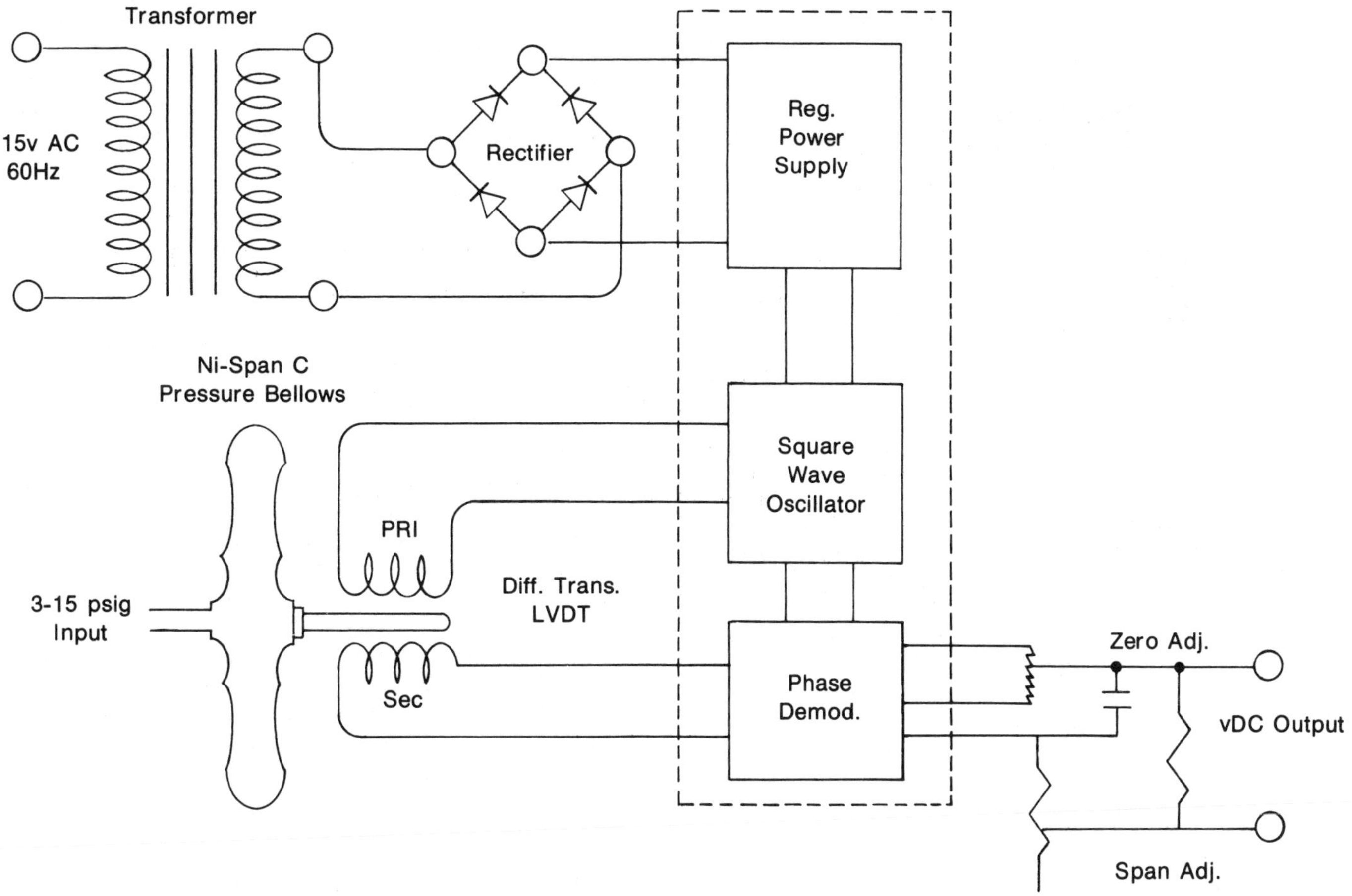

Transformer
115v AC
60Hz
Rectifier
Reg.
Power
Supply
Ni-Span C
Pressure Bellows
Square
Wave
Oscillator
PRI
3-15 psig
Input
Diff. Trans.
LVDT
Sec
Phase
Demod.
Zero Adj.
vDC Output
Span Adj.

Figure 10.12. Schematic diagram of the operation of a pneumatic to current (P/I) transducer.

Transformer
Voltage Input
+
+
+
DC Current Output
—

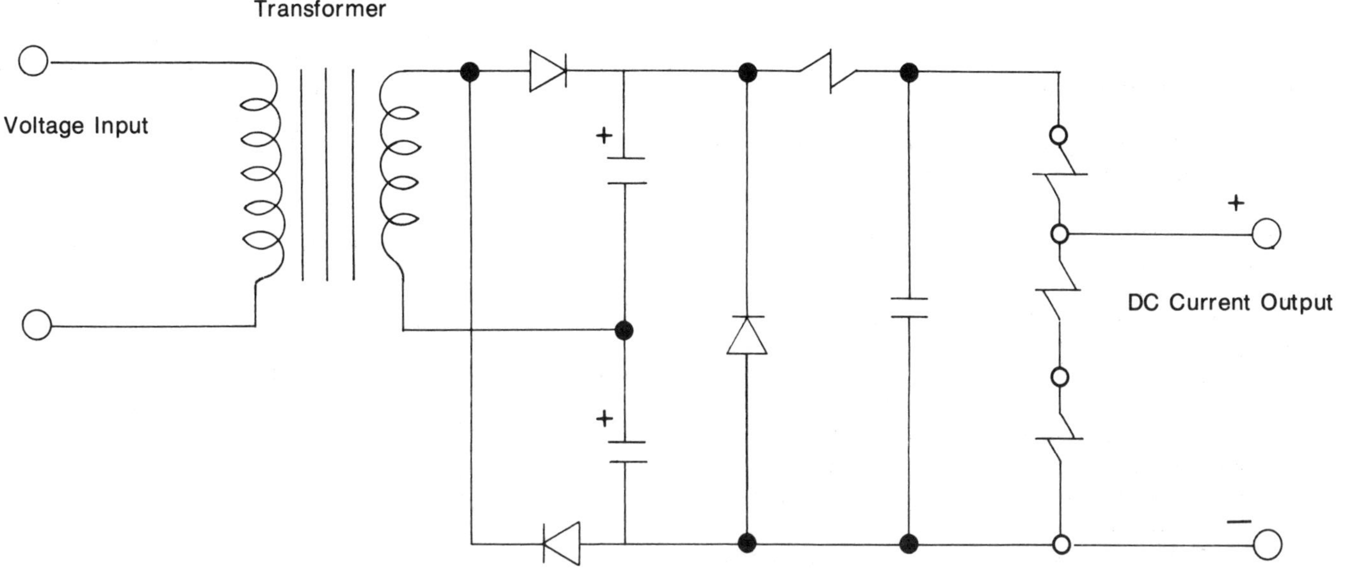

Figure 10.13. Simplified schematic diagram of a voltage to current (E/I) converter.

Figure 10.14. Voltage to pneumatic converters (E/P) are used for a wide variety of electrical measurements that are converted to air signals for pneumatic controls. (Courtesy of Foxboro Co.)

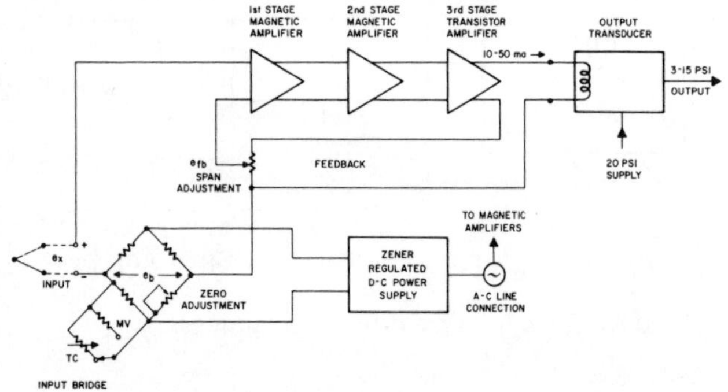

Figure 10.15. Electrical block diagram of a typical E/P converter. (Courtesy of Foxboro Co.)

Voltage-to-Pneumatic Converters

Voltage-to-pneumatic (E/P) converters are used in many control applications. Some of the more frequent uses are

1. Thermocouple outputs
2. Gas chromatograph outputs
3. Strain gauge outputs
4. Speed transmitter (tachometer) outputs
5. Electronic controller outputs
6. Other voltage generated outputs

It is obvious that the same measurements that are converted from voltage-to-current signals as described in the previous section can also be converted from voltage-to-pneumatic pressure signals.

A typical E/P device, shown in Figure 10-14, consists of a three-stage, solid-state amplifier and an output transducer (block diagram shown in Figure 10.15).

The input millivolt signal (e_x) is added algebraically to the output voltage of the input bridge (e_b) and the negative feedback signal (e_{fb}). The resultant signal is amplified to a 10 to 50 milliampere signal which drives the output transducer.

The input bridge performs two major functions:

1. Provides instrument zero elevation or suppression for measurement ranges not starting at zero millivolts
2. Provides reference junction temperature compensation for thermocouple temperature measurements

The feedback signal provides stable amplifier operation and an input impedance suitable for thermocouple and similar millivolt measurement signals. The amplifier output feeds the span adjusting potentiometer which is set to provide the proper amount of feedback for the millivolt range being measured. A change in the amount of feedback around an amplifier automatically changes the gain.

The amplifier consists of an input differential magnetic amplifier, a second-stage, self-saturating magnetic amplifier and a transistorized output stage. A closely regulated DC voltage source is provided by a Zener diode regulated power supply. This power supply provides DC excitation to the input bridge and bias voltages required by the amplifier. The output transducer is shown schematically in Figure 10.16.

The coils are mounted on an armature plate so that each is in a strong magnetic field of a permanent magnet. The direction of current through each coil determines whether that coil will be attracted to or repelled from the magnet. The permanent magnets are mounted with unlike poles upward and the coils connected so that the torques developed are additive. The polarity marks in dotted circles refer to those of the coils while the others denote the polarity of the permanent magnets. Arrows indicate the motion of the armature bar when a current passes through the coils.

Figure 10.17 is a more complete schematic diagram of the transducer, showing the coils, magnets, flexure (fulcrum), flapper, nozzle, zero-adjustment spring and feedback bellows. An increase in current in the coils causes the flapper to approach the nozzle. This results in a back

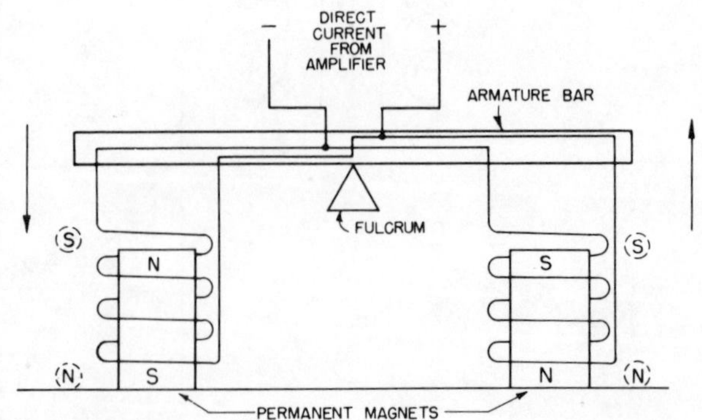

Figure 10.16. The output transducer for the E/P converter shown in Figure 10.14. (Courtesy of Foxboro Co.)

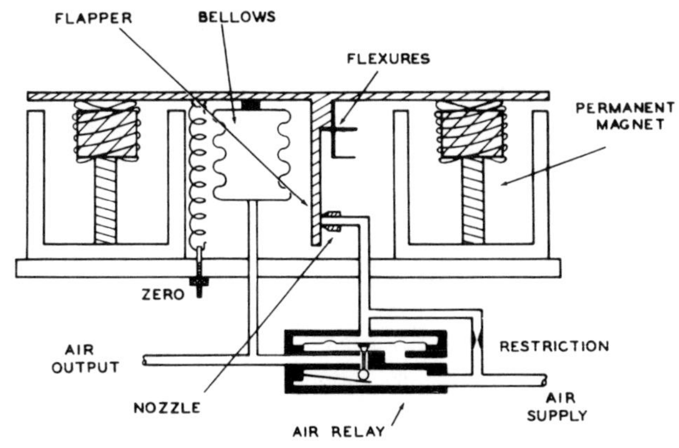

Figure 10.17. Schematic diagram of connections between the electrical and mechanical parts of the E/P converter. (Courtesy of Foxboro Co.)

pressure in the nozzle supply line which actuates the pneumatic relay. The relay action increases the output pressure enough so that the feedback bellows exerts a force on the armature bar to reposition the flapper relative to the nozzle.

Figure 10.18 shows a schematic diagram of a system with an E/P converter (Figure 10.19) used for low (or high) rpm trip on a turbine interlock system.

In case of undesirable fluctuation in turbine rpm, a low rpm trip system is designed to receive a millivolt signal from an electronic tachometer. Since the tachometer output is linear and proportional to the rpm, a chart is made of rpm versus millivolt output. (The pressure output is proportional to millivolt output and hence to rpm.) The pressure trip switch is adjusted to open at a value corresponding to the desired low rpm setting. As the pressure switch opens, the trip circuit is actuated and shuts down the turbine.

Current-to-Current Converters

The use of current-to-current (I/I) converters is not nearly as widespread as are the converters and transducers previously listed. They are used to change AC or DC signals, reducing or increasing signal level as desired. The most common applications are those converting 1 to 5 ma or 4 to 20 ma DC inputs to proportional 10 to 50 ma DC outputs. The I/I converter (Figure 10.20) may also be used simply as a current repeater which receives a 10 to 50 ma DC input signal and produces an electrically isolated signal of the same value. An inversely proportional 50 to 10 ma DC output is accomplished with the use of a reversing amplifier.

In addition to providing compatibility between systems of different signal levels, the I/I converters are also used occasionally when current inputs below or above the common signal levels are used for indication or readout.

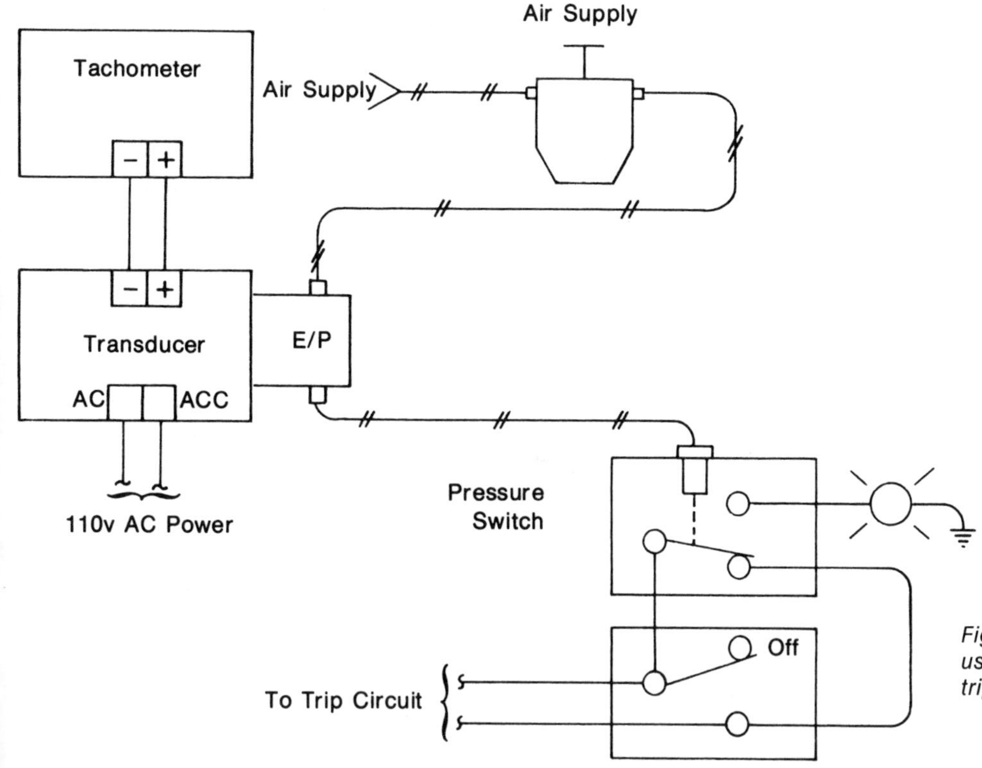

Figure 10.18. Loop diagram shows use of an E/P transducer as an rpm trip on a turbine interlock system.

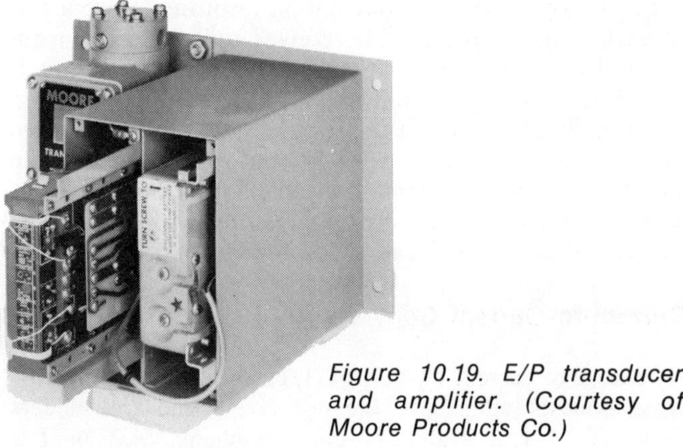

Figure 10.19. E/P transducer and amplifier. (Courtesy of Moore Products Co.)

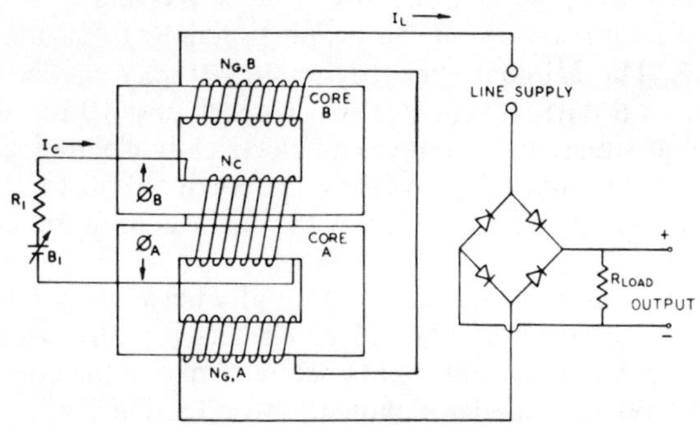

Figure 10.20. Schematic of a current to current converter used only as a repeater to isolate the signal. In the illustration, N_C = number of turns in control winding; N_G, A = number of turns in gate winding of Core A; N_G, B = number of turns in gate winding of Core B; I_C = control current; I_L = load current. (Courtesy of Honeywell)

Torque Devices

Torque is defined as the forces which act to change the linear motion or rotation of a body. The torque about a chosen axis is equal to the force times its moment arm,

$$T = LF$$

where F is the force
T is the torque
L, the moment arm, is the perpendicular distance from the axis of rotation to the line of action of the force.

Basically two types of torque measurement sensors are available: the in-line sensor used for rotating equipment and the stationary sensor used for nonrotating shafts. Figure 10.21 shows a round shaft with strain distribution and strain gauge bridge wiring. This configuration can be used on a stationary shaft with direct wiring or on a rotating shaft with slip rings or transformer transmission between the rotating and stationary sections of the system.

Rotating shaft torque can also be determined by measuring the angular displacement between two points of a shaft. Two gears with identical numbers of teeth can be mounted on the shaft to actuate two magnet pickups as shown in Figure 10.22. As the tooth of the gear passes the magnetic pickup sensors, alternating voltages are produced whose phase difference is proportional to the torque.

Program Controllers

Program controllers are used in batch type processes and in equipment automation schemes. They are used for control of chemical mixtures, for sequencing events in drying operations, for temperature control of reaction processes and for many other control schemes involving repeated operational sequences.

Digital programmers are designed for precise time schedule, simple setup, exact repeatability, positive process control and continuous visual display of switching functions. These units respond accurately to preselected program operation and repeat the operation on a scheduled basis.

Programmers may be self-contained and provide several output functions. Time ranges vary from a few seconds to hours on each step. The outputs are controlled through either pneumatic or electrical function switches. The most common function switches are SPDT rated at 10 amperes. The cycle function outputs can be set to operate or "not" operate on each step.

Programmers may be classified broadly as end point or time base controllers or a combination of the two. In many cases a time base is not adequate—a new step needs to begin only on the successful completion of a previous step. This would be called end point control. Many types of sensors or switches may signal the completion of one step, and only on their signal can the programmer advance to the following step. One advantage of end point control is that no time is lost before advancement to the next step.

The time base method of control is important where a definite time cycle is required. In many cases timers are used as sensors for end point programmers. All three types of programmers are discussed in the following paragraphs.

End Point Programmers

The drum programmer (Figure 10.23) is a pulse actuated, motor-driven, switching device which is preset for automatic sequence operation. The desired program is developed by inserting plugs into drum slots. A deliberate pulse signal, supplied by an external source such as a timer, relay, switch, thermostat or photocell, advances each step on the program pattern. Each successive step is interlocked with the end point of the preceeding step—the completion of one phase serves as the necessary trigger to start the next one.

The stepping drum programmer consists of a heavy gauge anodized aluminum drum, step driven by an electric motor and gear train through a Geneva mechanism, a series of roller-lever-actuated snap action switches, terminal strips for input and output connections, control circuitry for the

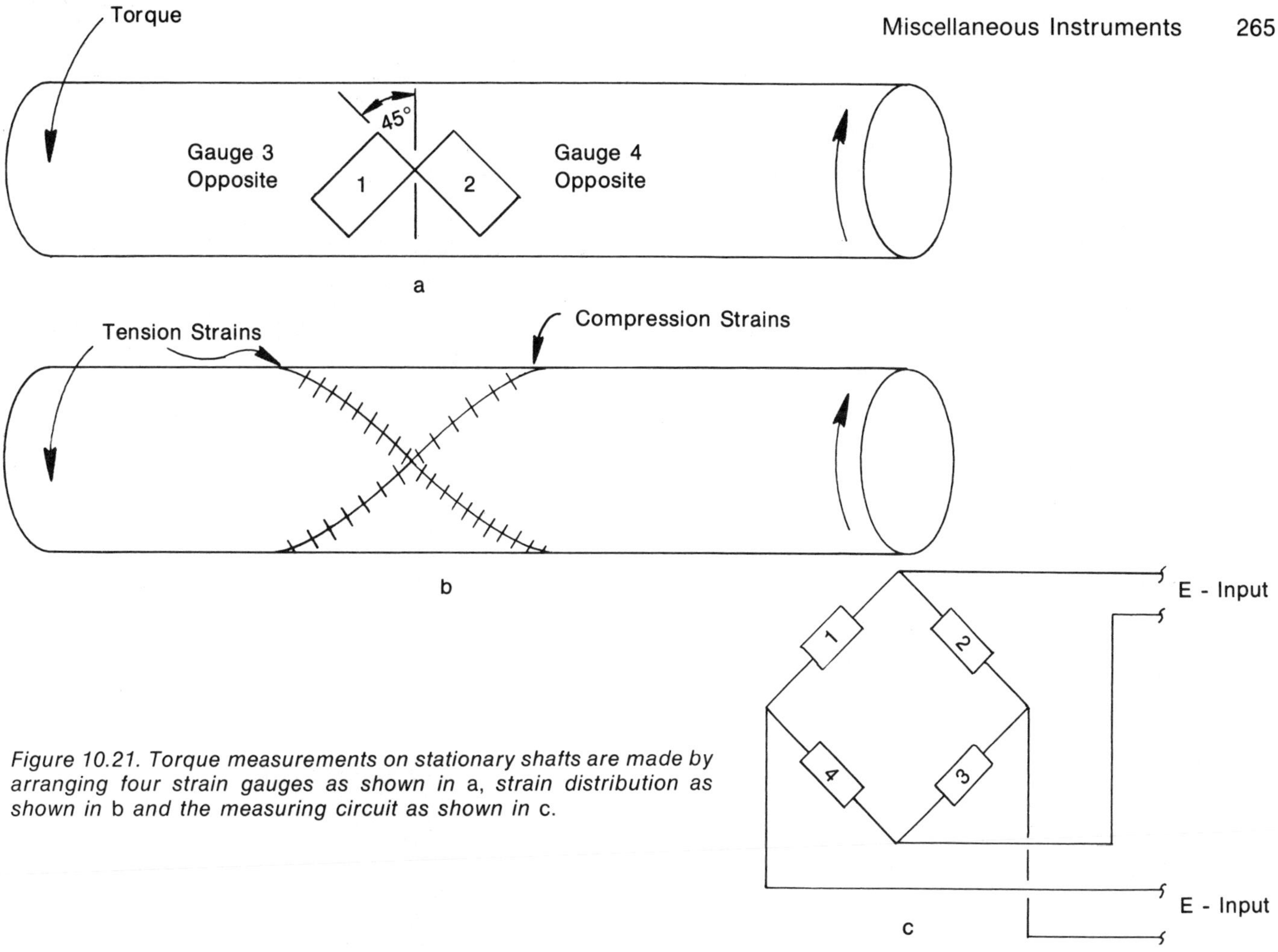

Figure 10.21. Torque measurements on stationary shafts are made by arranging four strain gauges as shown in a, *strain distribution as shown in* b *and the measuring circuit as shown in* c.

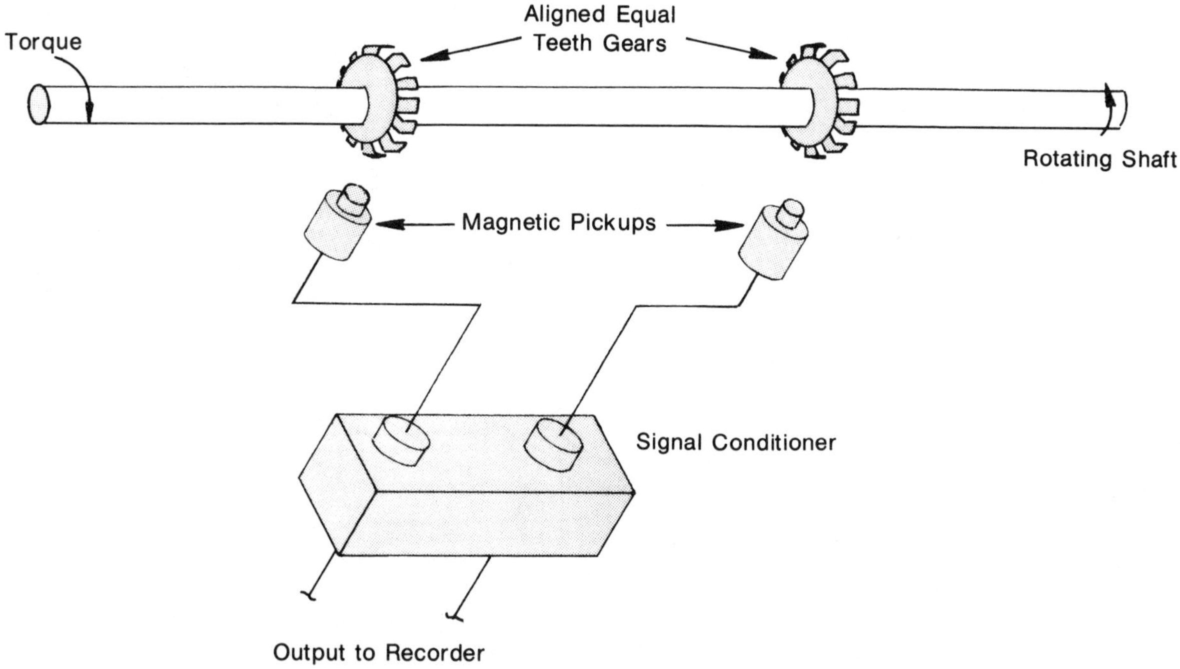

Figure 10.22. Rotating shaft torque is measured by the angular displacement of two magnetic pickup sensors whose AC voltages are out of phase in proportion to developed torque.

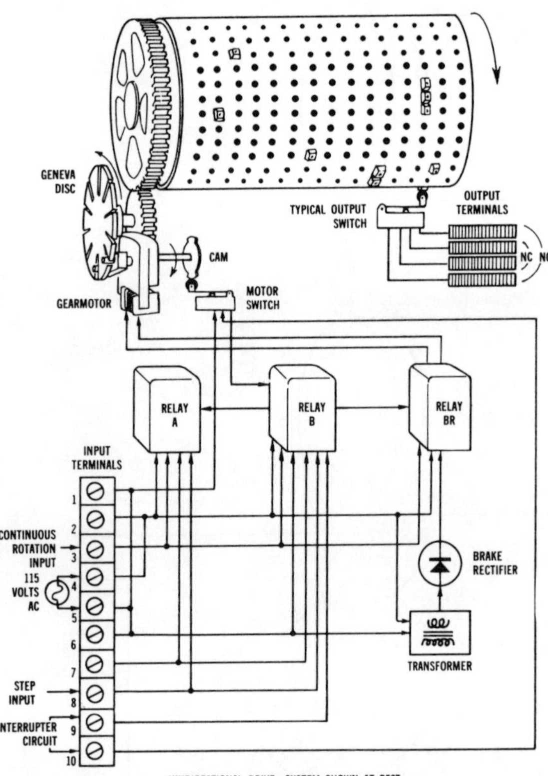

Figure 10.23. The stepping drum programmer is an example of the end point type, although timers may be used to advance a program. (Courtesy of The Tenor Co.)

Figure 10.24. Programs for future operation may be retained by changing drums. (Courtesy of The Tenor Co.)

Figure 10.25. The program is developed by cutting and punching the program master plastic sheet for the desired program. The program sheet is placed on the drum where it actuates the digital function switches or analog profile for operation of valves or other equipment. (Courtesy of Moore Products Co.)

drive motor and nylon plugs for inserting into the drum holes.

As the drum is stepped, the rows of holes pass over a line of switches, one for each column in the drum pattern. Insertion of a plug anywhere in the column actuates the switch as that point passes over the switch actuator. Programs can be retained for future operation by removing the entire drum (see Figure 10.24).

Figure 10.25 shows a programmer with a master chart which is punched for digital operation or pattern cut for analog operation. A batch process might be controlled by either method. Figure 10.26 shows a typical batch process, and Figure 10.27 shows a program chart for that operation. A description of the operation follows.

With the start of the program, initiated by a start button, limit switch 1 is energized to check on the position of the tank outlet valve. If the switch indicates that the valve is closed and the vat is therefore ready to accept materials, the programmer advances to Step 2.

Step 2 represents another check, a verification that the vat is actually empty. This is performed by energizing level control 1. If the level control indicates that the tank is indeed empty, the drum steps again.

Step 3 is a check to determine that there is sufficient material in the vessel containing product A to proceed with the program. This is accomplished by energizing level control 5. If the indication is correct, the drum steps again.

Up to this point, all steps have been manifestations of the end-point control concept around which the programmer is designed. Having ascertained that all is in readiness, the actual batching operation can begin. At *Step 4,* pump *A* and solenoid valve 1 are energized, pumping product *A* into the vessel. At the same time, the agitator is started; this unit will continue to run for the duration of the program. Finally, level control 2 is energized to indicate when the fill of product *A* has been completed. When the level control contacts close, the programmer advances to the next step.

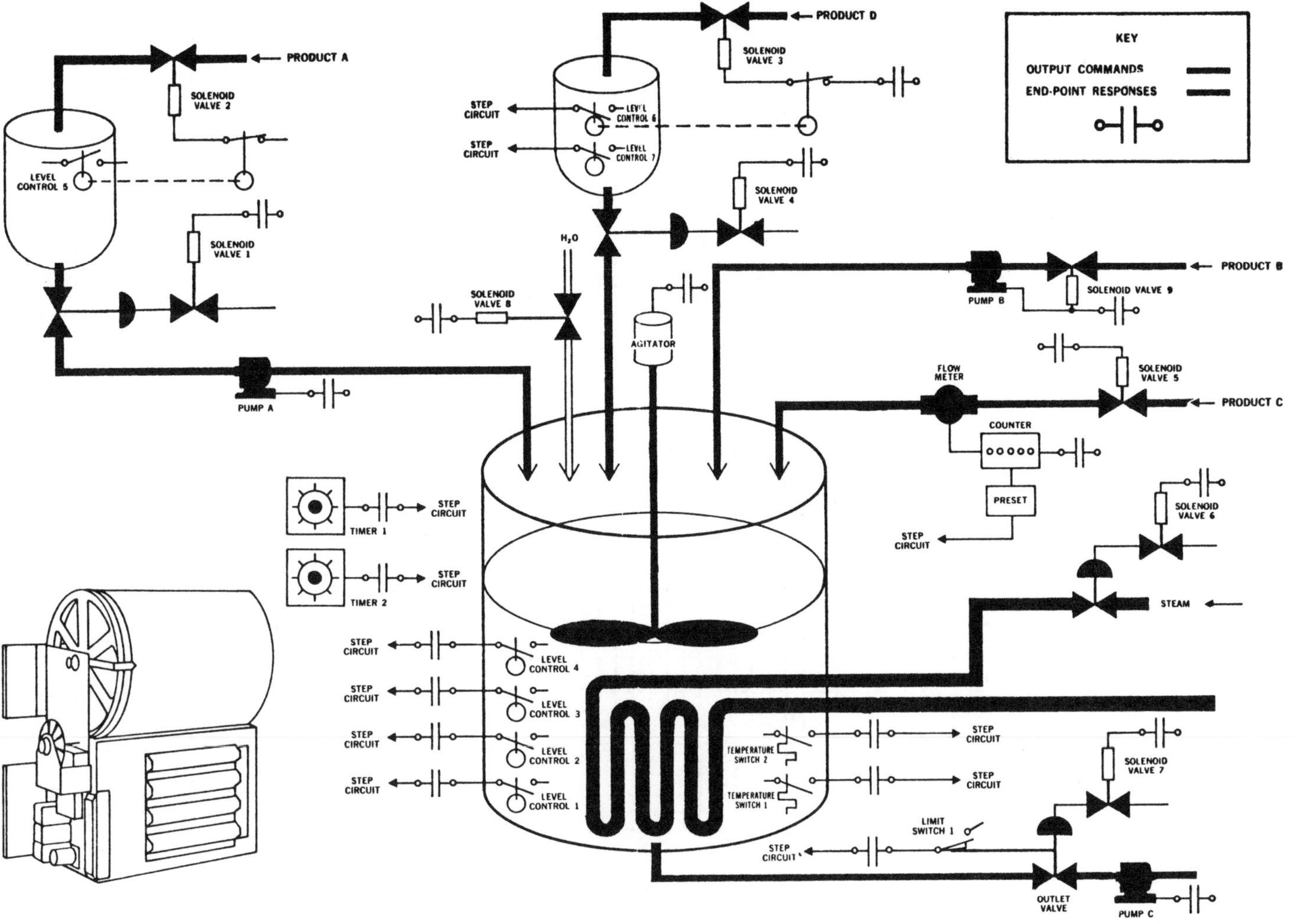

Figure 10.26. A typical programmed batch process controlled by a stepping drum programmer (see Figure 10.27 for program chart). (Courtesy of The Tenor Co.)

Step 5 provides for the addition of water to the vat and the refill of product *A* in its supply vessel. This is accomplished by opening solenoid valve 8 for water and solenoid valve 2 for product *A*. Solenoid valve 2 is programmed for the next four steps, but overfilling the product *A* vessel is prevented by an extra set of contacts on level control 5 which disables solenoid valve 2 when the correct level is reached. Solenoid valve 8 is closed when level control 3—also energized at Step 5—indicates that sufficient water has been added, and the programmer steps again.

Step 6 is a heating procedure in which solenoid valve 6 is operated, actuating a diaphragm valve which allows steam to enter the heating coil. Temperature switch 1 is also energized. The contents of the vat are heated until the contacts of temperature switch 1 close, at which time the programmer steps and the steam valve is closed.

Step 7 calls for the addition of product *B* to the vat. Accordingly, solenoid valve 9, pump *B* and level control 4 are energized, pumping product *B* into the vat until level control 4 contacts close, indicating that the step has been completed. The programmer steps, deenergizing the pump and associated solenoid valve and control.

Step 8 brings product *C* into the process. Solenoid valve 5 and a preset counter are energized. Product *C* travels through a flowmeter which delivers pulses to the counter. When the predetermined number of counts has been reached, the programmer is advanced, deenergizing solenoid valve 5.

Step 9 is a check to determine if there is sufficient material in the calibrated bin containing product *D*. If so, the programmer steps again.

Step 10 calls for the addition of product *D* to the vessel. This is accomplished by energizing solenoid valve 4 along with the low level control 7 in the bin. Solenoid valve 4 operates a slide gate valve and permits product *D* to enter the vessel. When the low level control indicates the bin is empty, the programmer steps again.

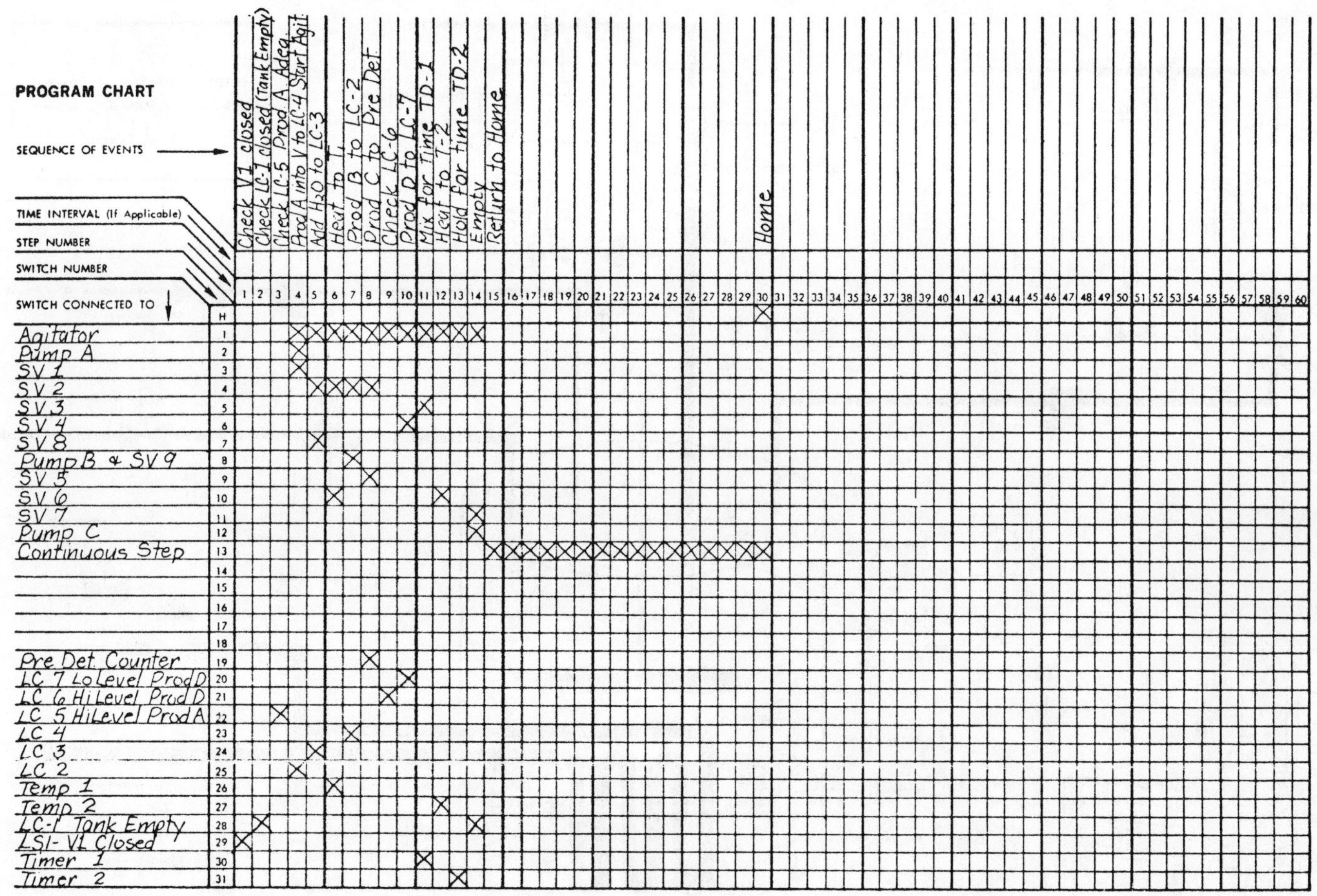

Figure 10.27. Program chart for a typical batch process (see Figure 10.26). (Courtesy of The Tenor Co.)

Step 11 is a timed process, calling for the agitation of all products in the vat for a time set on Timer 1. When Timer 1 times out, the programmer advances again. Solenoid valve 3 is energized to refill bin *D*.

Step 12 is another heating process, comparable to Step 6. Solenoid valve 6 again operates a diaphragm valve to allow steam to enter the heating coil. Temperature switch 2 is set to detect the temperature to which the contents are to be heated and causes the programmer to step when the temperature is reached.

Step 13 is another timer step prior to dumping the contents of the vessel. Timer 2 is energized, and the contents of the vat are agitated for the preset length of time. When Timer 2 times out, the programmer steps again.

Step 14 dumps the end-product contents. Solenoid valve 7 is energized, pump *C* is started, and level control 1 is energized to indicate when the vat is empty. When level control 1 contacts close, the programmer steps again.

Step 15 is the beginning of the homing procedure, in which the programmer steps continuously until it reaches the "home" position, where it remains until it receives its next start signal.

End-Point and Real-Time Programmers

A new design in program controllers is the combined end-point and real-time programmer (see Figure 10.28). This unit has been developed in a 7-day, modular, multiple load circuit unit that includes removable, plug-in, solid-state modules. Figure 10.29 shows a removable program module which can be changed in a few seconds. Alternate programs can be stored ready for instant use, or additional programs can be added without interruption of the existing program. Figure 10.30 shows a block diagram of this unit.

Time Program Controllers

A time program controller usually consists of a synchronous motor driving a camshaft which rotates continuously while the motor is energized. One or more adjustable cams may be run by the motor. The cam configuration determines the point of closing or opening a switch during each shaft revolution. Figure 10.31 shows a typical repeat cycle program timer. Cycle times may be changed by changing the motor or gear train.

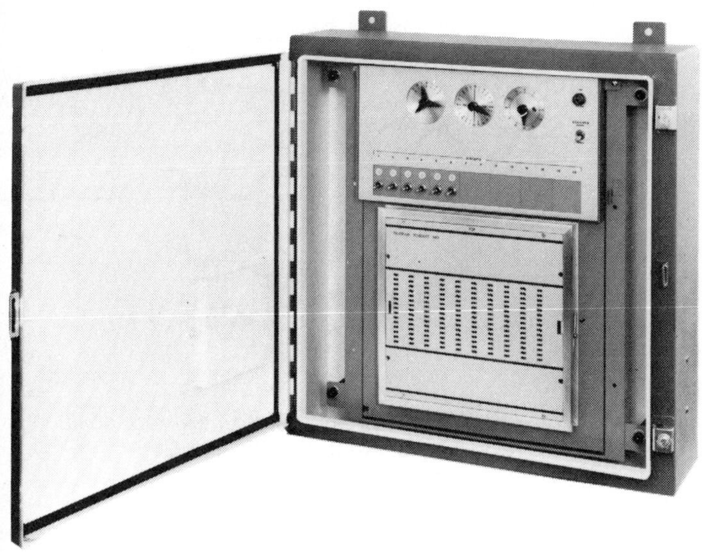

Figure 10.28. Typical combined end point and real time programmer. (Courtesy of The Edwards Co., Inc., a unit of General Signal Corp.)

Figure 10.29. Removable program module can be inserted or removed in a few seconds from the unit shown in Figure 10.28. (Courtesy of The Edwards Co., Inc., a unit of General Signal Corp.)

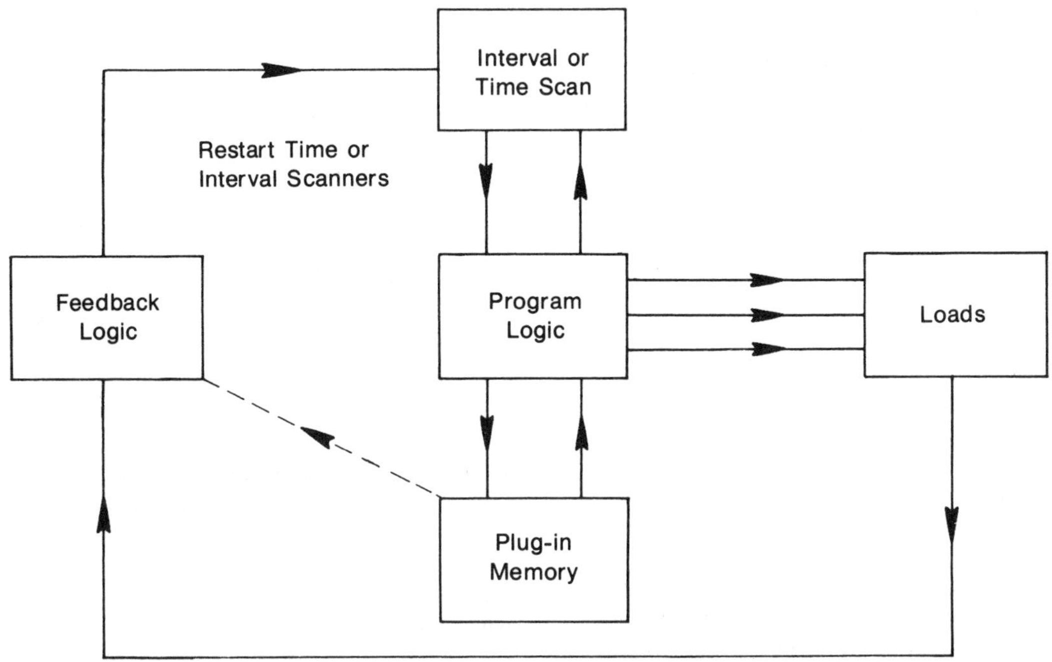

Figure 10.30. Block diagram of combined end point and real time programmer shown in Figure 10.27. (Courtesy of The Edwards Co., Inc., a unit of General Signal Corp.)

Figure 10.31. Repeat cycle program timer used to control a number of electrical circuits. The time duration of each function is determined by a "rise tab" or a "drop tab" which actuates a S.P.D.T. switch. (Courtesy of Eagle Signal Division of Gulf Western)

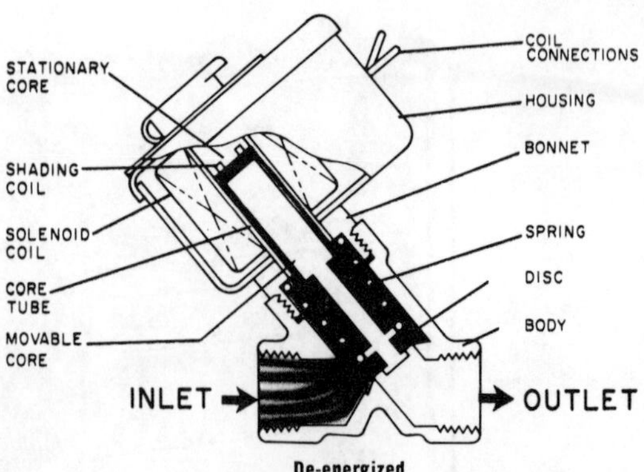

Figure 10.32. Solenoid valve in the deenergized position (power is off). (Courtesy of Automatic Switch Co.)

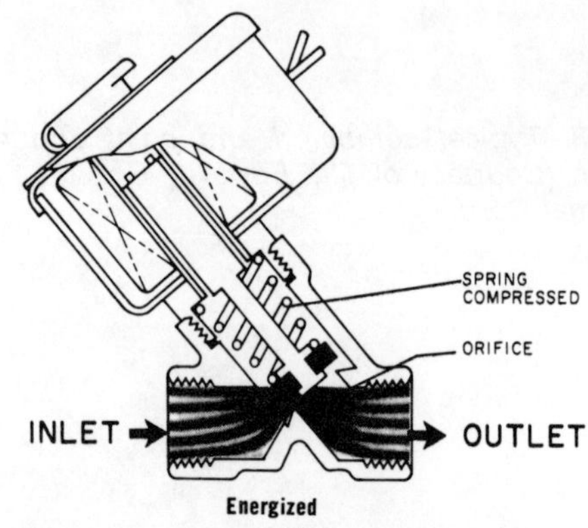

Figure 10.33. Solenoid valve in the energized position (power is on). (Courtesy of Automatic Switch Co.)

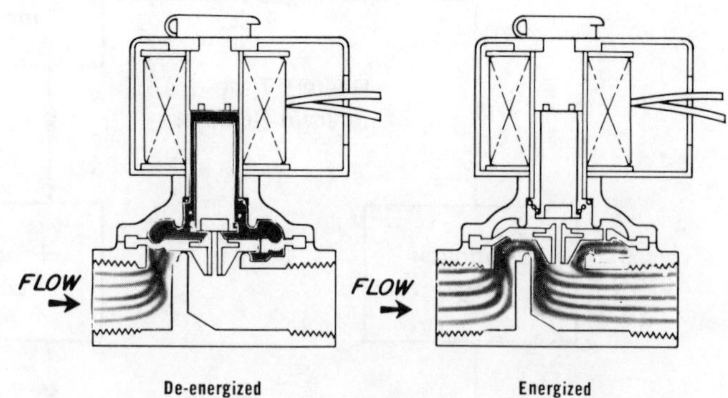

Figure 10.34. Two-way valve shown in both energized and deenergized positions. (Courtesy of Automatic Switch Co.)

A complete control panel system may be designed around a programmer. Such a system has many applications in today's modern plants. Regardless of the type programmer used, it can become a vital component in plant operation.

Solenoid Valves

A solenoid valve utilizes an electromagnetic coil to actuate an armature or valve stem in a magnetic field to control fluid flow. Solenoid valves are either fully opened or fully closed and are actuated by electric signals from remote locations.

When electrical power is supplied to the electromagnet, a magnetic field is built up in the valve coil which causes the solenoid plunger to be positioned in the coil. The plunger is connected to a valve disc which opens or closes the orifice dependent on the valve actuation—whether the valve is energized to open or energized to close (see Figure 10.32 and 10.33).

Types of Solenoid Valves

Four types of solenoid valves are discussed:

1. Two-way
2. Three-way
3. Four-way
4. Pilot-operated

Two-Way Valves

Two-way valves (Figure 10.34), normally used for shutoff purposes, have one inlet and one outlet connection. The two-way valves are designed and manufactured for either normally open or normally closed operation.

Three-Way Valves

Figure 10.35 shows a three-way solenoid valve that has three connections and two orifices. One orifice is closed when the other is open. These valves are used for operation of diaphragm valves, cylinder operators or other fluid loading applications requiring alternate loading (pressuring) and unloading (exhausting) of pressure chambers. Other applications include the selection or diversion of fluid streams as determined by valve actuation (see Figure 10.36).

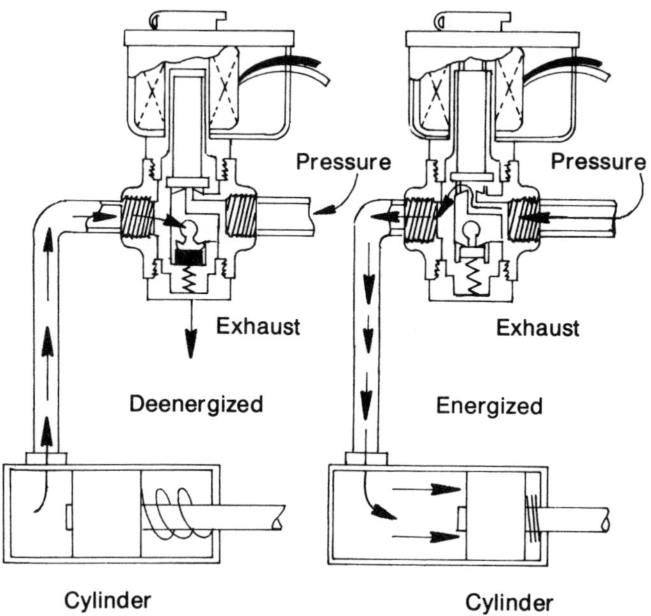

Figure 10.35. Three-way solenoid valve shown in energized and deenergized positions in operation of single acting spring return cylinder.

Four-Way Valves

Operation of double-acting cylinders may be accomplished with a four-way solenoid valve, as shown in Figure 10.37. These valves have four connections: one pressure, two cylinder and one exhaust. Energizing the solenoid causes one end of the cylinder piston to be pressurized while the opposite one is exhausted by opening to the atmosphere. When deenergized, the opposite actions occur to the cylinder piston. The four-way (and three-way) valve may be direct or indirect, and it may utilize single or dual solenoid construction.

Pilot-Operated Valves

Figure 10.38 shows a pilot-operated valve which incorporates a pilot and bleed orifice and utilizes the line pressure for operation. When electrical power is applied to the solenoid, the pilot orifice opens and releases pressure from the upper port of the valve piston or diaphragm to the outlet side of the valve. This action results in an unbalanced pressure which causes the line pressure to lift the piston off the main orifice, opening the valve. Deenergizing the solenoid closes the pilot orifice and applies full line pressure to the top of the piston through the bleed orifice, thereby providing a seating force for tight closure.

Single Solenoid Pilot-Operated Valve

The single solenoid pilot may be incorporated on two-way, three-way and four-way valves. In the deenergized position, the inlet is normally open to cylinder port *A*, and cylinder port *B* is open to exhaust (Figure 10.39). In the energized position, the inlet is open to cylinder port *B* and port *A* is open to exhaust. An unbalanced pressure causes the line pressure to lift the piston and open the valve. Single solenoids are used in applications where a loss of electrical power (thus reverting the solenoid to the deenergized position) does not impose a harmful effect on the process.

Double Solenoid Pilot-Operated Valves

The double solenoid pilot-operated valve is used to operate a double acting cylinder. The cylinder port on the same side as the solenoid being energized is open to the inlet port (Figure 10.40). The other is open to exhaust.

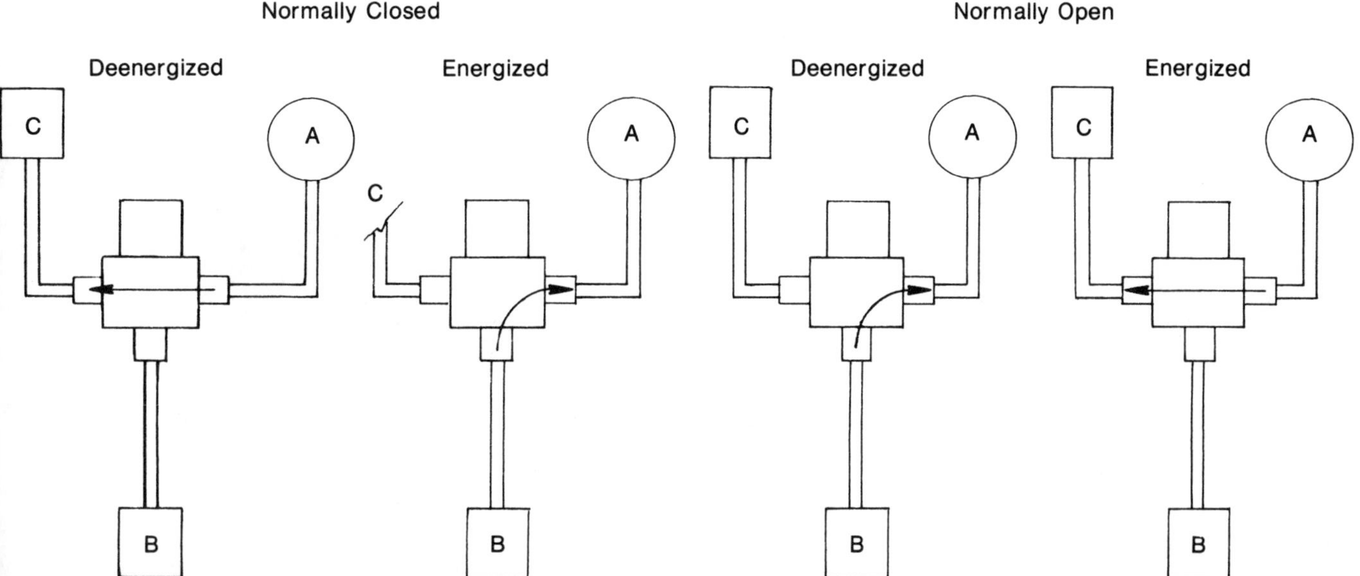

Figure 10.36. Fluid flow is controlled by three-way valves. At left an energized solenoid diverts flow from B to A whereas removal of power causes flow from A to C on normally closed valves. A normally open valve produces the reverse actions on coil energizing and deenergizing.

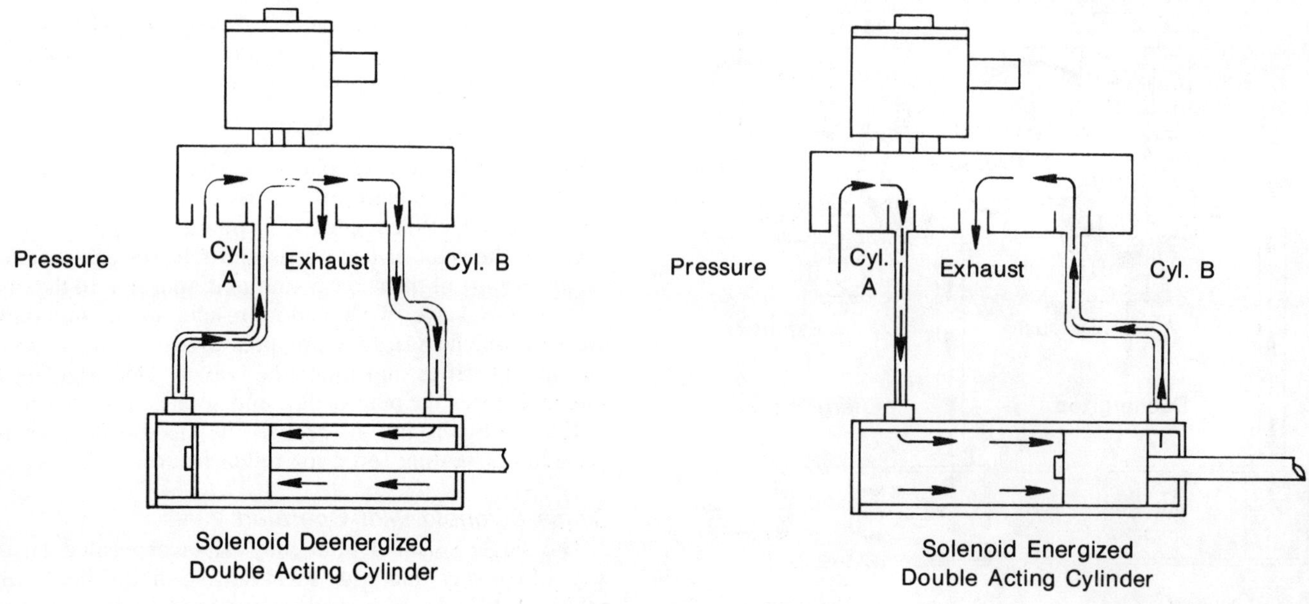

Figure 10.37. Four-way valves are used to operate double acting cylinders. At right when solenoid is energized, the left cylinder side is pressured from Port A while the right side is exhausted through Port B. At the left, the right cylinder side is pressured through Port B while the other side is exhausted through Port A.

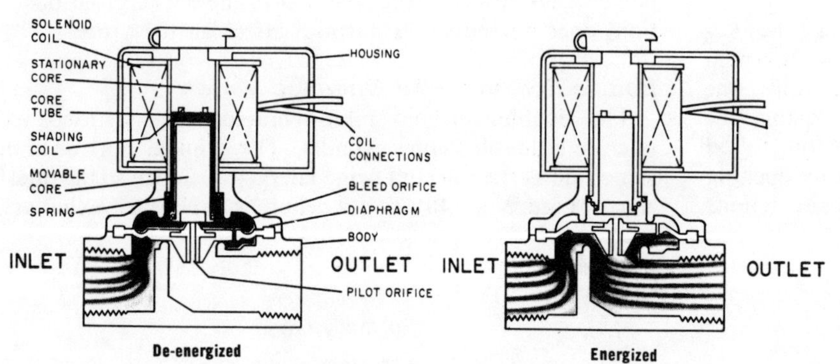

Figure 10.38. Pilot-operated solenoids utilize line pressure to furnish valve opening power by opening a pilot orifice when coil is energized. (Courtesy of Automatic Switch Co.)

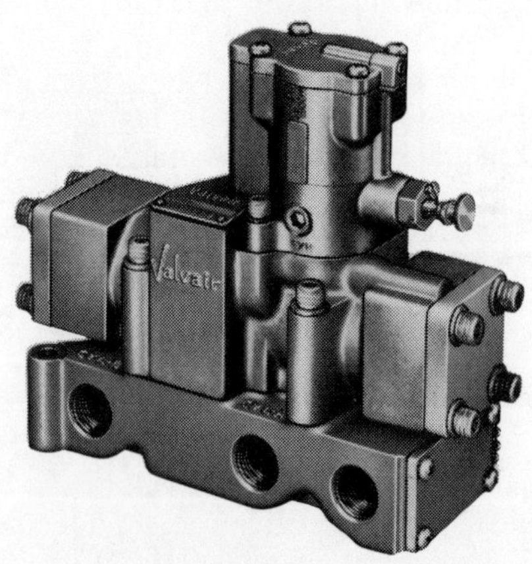

Figure 10.39. Pilot-operated valve using a single solenoid to operate valve. (Courtesy of Bellows-Valvair, Division of IBEC)

Figure 10.40. Double solenoids are used when reverse valve action on loss of electrical power is undesirable. (Courtesy of Bellows-Valvair, Division of IBEC)

Figure 10.41. General purpose solenoid valve with brass enclosure. (Courtesy of Automatic Switch Co.)

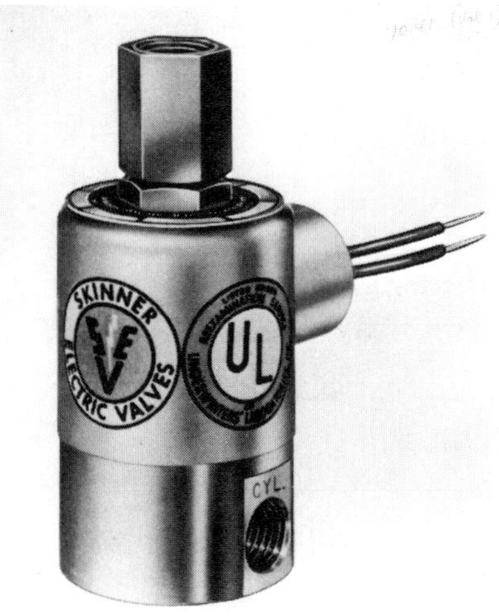

Figure 10.42. Explosionproof enclosure for stainless steel solenoid valve. (Courtesy of Skinner Electric Valve Division, Skinner Precision Industries, Inc.)

This unit is designed for momentary or sustained contact. One pilot must be deenergized before the other is energized. Double solenoids are used in applications where a loss of electrical power might have an undesirable effect by reversing the valve action. When using double solenoids, not only must an energized solenoid be deenergized, but the other solenoid must have power in order to reverse action.

Construction

Solenoid valve enclosures are designed for operation in general purpose or hazardous locations. A typical general purpose enclosure is made of pressed steel and is suitable for indoor locations where atmospheric conditions are normal; it protects the coil, but is not dust tight (Figure 10.41).

For hazardous locations enclosure materials are either die-cast aluminum or pressed steel; however, they may be fabricated from other materials (Figure 10.42).

The solenoid ports in contact with the line fluid are normally made of nonmagnetic stainless steel. Other materials are used if required for a particular service. The maximum operating pressure is largely dependent upon solenoid construction. For watertight and submersible applications, explosionproof enclosures are used and equipped with a gasket seal.

Gang-mounted solenoid valves are used to control multiple pneumatic cylinder operators. Common pressure and exhaust ports in the valve ends are butted together with intermediate gasketing to give high-density mounting with simplified piping and wiring. Normally-closed and normally-open operation can be reversed in some models by inverting the intermediate gaskets.

If flow-rates through 3-way and 4-way solenoids need to be throttled, such as for regulating the speed of pistons, integral adjustable flow control valves are available.

Valves rated for fuel gas service come in sizes ⅛ inch through 3 inches. Designs include those with manual reset feature and with proof-of-closure visual and electrical indication.

Some solenoid-style valves are available with pneumatic operators. Pressure ranges include 1.5-5 psi (fluidic-logic), 3-15 psi (instrument air), and 30-125 psi (pneumatic cylinder controls).

A wide range of fluidic-logic valves and control devices are now being marketed. Primarily being used in machine tool controls, this method of control has been slow to invade the petrochemical industries. But tried and proven filtering systems and careful system application have given fluidic-logic systems much better receptions lately.

Maximum Differential Pressure

The maximum differential pressure is the maximum differential pressure between the inlet and the outlet compartments at which the solenoid can safely operate. It may be much less than the maximum pressure rating of the valve body. The strength of the solenoid coil is a major factor in determining the maximum operating pressure differential for it must overcome the forces of the differential pressure across the valve ports.

Minimum Differential Pressure

The minimum differential pressure is the lowest operating pressure differential required for dependable operation. A two-way shutoff valve with floating diaphragm will tend to close if the valve is operated below that minimum differential pressure. To ensure complete transfer from one position to another for three- and four-way pilot valves, the minimum operating pressure must be maintained throughout the operating cycle. It is important to note that, on some models with no recommended minimum differential pressure requirement, the valves must nevertheless be piped such that certain ports *always* have pressures that are positive with respect to others.

Response Time

The time required to open the valve fully from a closed position is dependent upon the valve size, fluid temperature, pressure drop and electrical characteristics of the valve. It may vary from 3 to over 100 milliseconds, depending upon the above described conditions.

Table 10.3. Insulation Class and Temperature Ratings for Various Solenoid Power Output Groups

INSULATION CLASS AND MAXIMUM ALLOWABLE TOTAL TEMPERATURE①			
POWER OUTPUT GROUPS	A (105°C.)	F (155°C.)	H (180°C.)
Group I Basic Class A windings. (Class F & H for Higher Ambients or Fluid Temp.)③	Prefix — None Rise — 80°C. Excess margin for higher ambients or fluid temperatures — 105°C. — (80°C. + 25°C.) = 0°C.②	Prefix — FT Rise — 80°C. Excess margin for higher ambients or fluid temperatures — 155°C. — (80°C. + 25°C.) = 50°C.	Prefix — HT Rise — 80°C. Excess margin for higher ambients or fluid temperatures — 180°C. — (80°C. + 25°C.) = 75°C.
Group II (For Higher Valve Pressure Ratings & Higher Ambients or Fluid Temp.)③		Prefix — FB Rise — 105°C. Excess margin for higher ambients or fluid temperatures — 155°C. — (105°C. + 25°C.) = 25°C.	Prefix — HB Rise — 105°C. Excess margin for higher ambients or fluid temperatures — 180°C. — (105°C. + 25°C.) = 50°C.
Group III (For Higher Valve Pressure Ratings)		Prefix — FF Rise — 130°C. Excess margin for higher ambients or fluid temperatures — 155°C. — (130°C. + 25°C.) = 0°C.②	
Group IV (For Higher Valve Pressure Ratings)			Prefix — HP Rise — 155°C. Excess margin for higher ambients or fluid temperatures — 180°C. — (155°C. + 25°C.) = 0°C.②

Notes: ① As measured by the "Resistance Method."
② Equipment rated at an ambient temperature of 25°C. can be employed in areas where the ambient temperature reaches 40°C. **occasionally.**
③ Ambient temperatures are directly additive to coil rise — fluid temperatures are not.

Source: Automatic Switch Co.

Coils

The coils of solenoid valves normally are designed for continuous duty without overheating or failure. When they are not so designed, information on the valve should and normally does include statements to that effect. Coils are designed for various temperature parameters. High-temperature coils are constructed of fiberglass epoxy, silicone rubber and varnish and magnet wire rated at over 200°C. Coils are usually designed to operate at 15% under nominal voltage and 10% over nominal voltage. Power consumption and current ratings vary, depending upon the design and the manufacturer. Care must be taken to assure that coil inrush current, sometimes nearly eight times the value of holding current, does not exceed the contact rating of the actuating device.

Valve Sizing

Solenoid valve sizing is important in preventing unnecessary cost or substandard performance. The basic factors to be considered in sizing include the following:

1. Maximum and minimum flows to be controlled
2. Maximum and minimum pressure differentials
3. Specific gravity, temperature and viscosity of the flowing fluid

For calibration of the orifice size or c_v of solenoid valves, formulas and charts are furnished in manufacturers' catalogs.

Coils are constructed in accordance with UL, NEMA and other industry standards. Table 10.3 is a typical example of coil insulation systems and temperature limitations.

Miscellaneous Sensors

Vibration Monitors

Vibration is the periodic motion of a moving object sweeping back and forth repeatedly over the same path. It is a problem related to high-speed rotating and reciprocating equipment. In recent years there has been an increasing search for ways to reduce the cost of maintaining capital equipment. Vibration monitoring is one method used to detect equipment problems, allowing repairs to be made prior to total operating failure, reducing downtime and avoiding production losses. Systems have been designed and developed for measuring vibration of high-speed compressors, gas turbines, pumps, motors, generators and virtually every other type of equipment subject to vibratory shock damage. Vibration monitoring equipment (Figure 10.43) is normally designed to measure vibration in displacement; however, velocity and acceleration may also be taken into account in fixing normal operating and maximum displacement limits. Vibration measurement units are usually given in mils (inch X 10^{-3}) and represent peak-to-peak values. In the metric system amplitudes of vibration are stated in microns.

Vibration may be monitored on both attended and unattended equipment for determining the early stages of bearing or other failures (see Figure 10.44 and 10.45). One

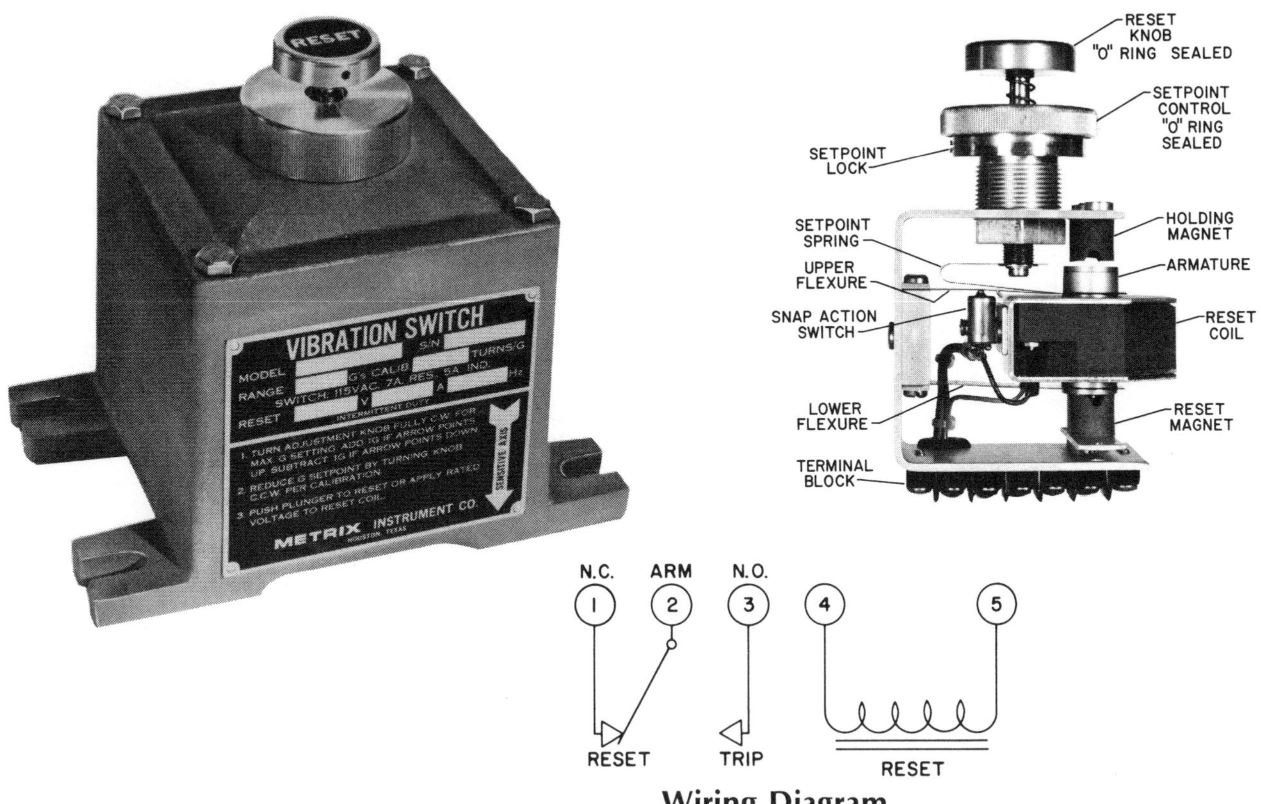

Figure 10.43. Adjustable inertia switch which provides vibration and shock protection for rotating or reciprocating equipment. (Courtesy of Metrix Instrument Co.)

Figure 10.44. Metrix model 007 vibration sentinel used as a vibration monitor and malfunction detector on engines, compressors, motors, pumps, generators and other equipment. (Courtesy of Metrix Instrument Co.)

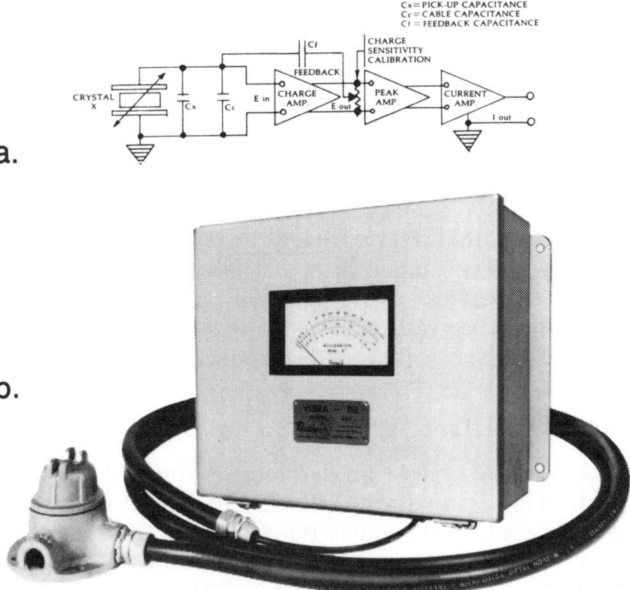

Figure 10.45. (a) Block diagram for Robertshaw model 367 vibra-tel vibration system; and (b) Robertshaw model 367 vibra-tel continuous vibration monitoring system. (Courtesy of Robertshaw Controls Co.)

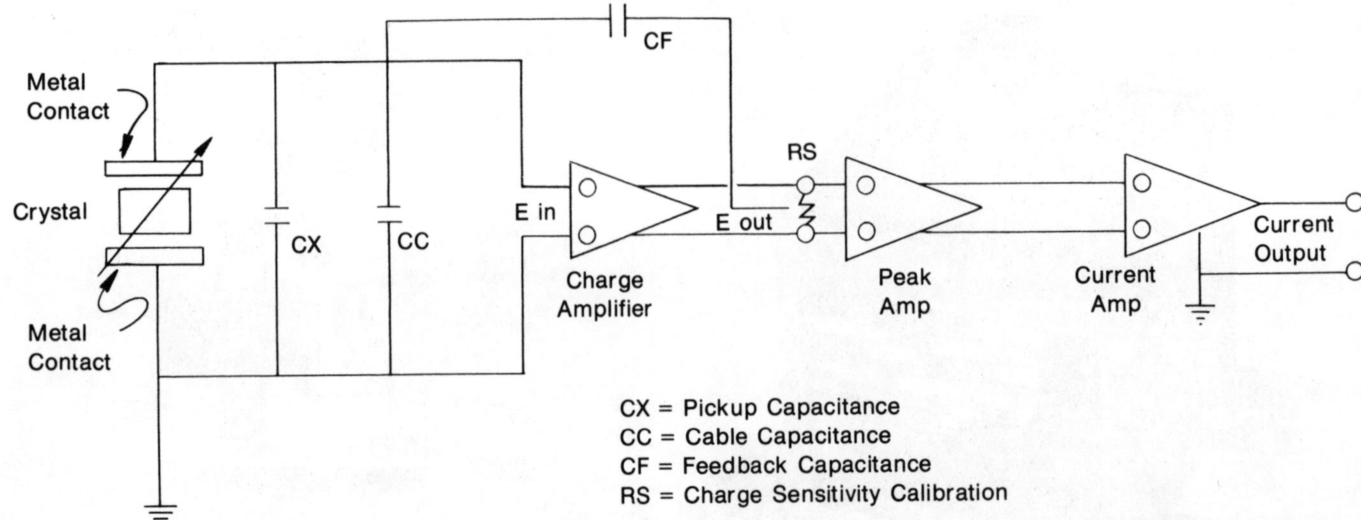

Figure 10.46. Schematic of a piezoelectric pickup and amplifier used for vibration measurement.

damaging component of vibration is acceleratory shock; therefore, acceleration sensitive vibration must be measured to determine accurately the destructive forces present. Acceleratory shock is determined by the frequency and amplitude of displacement and a constant, K, where K depends upon the units of amplitude and frequency. The formula for acceleratory shock is

$$G = AF^2K$$

where G is acceleration in G units
A is amplitude in centimeters or inches
F is the frequency in cycles per second or revolutions per minute.

If the amplitude is at peak value, measured in inches, and frequency is measured in cycles per second, then K is equal to 0.1.

An examination of the formula $G = AF^2K$ reveals that frequency has a greater effect on the destructive force than amplitude. Both, however, should be monitored. Although its effect is less, amplitude or displacement can be great enough to destroy a machine even at low frequency. Monitoring sensors, then, should have a velocity (or similar type) pickup that will sense displacement electronically.

Piezoelectric Type

Piezoelectric charge accelerometers are used on high-speed equipment to monitor vibration frequencies. They are manufactured from various types of crystals sandwiched between metal contacts in a capsule form (Figure 10.46). A voltage is produced by separation of charge on the opposite faces of the crystal when it is subjected to acceleration forces. A charge Q is developed when acceleration is applied to the mass, m, and crystal sandwich. The crystal is subjected to a varying force, F, which develops a charge, Q, proportional to the acceleration, (a), and the crystal constant, (k). The result is $Q = kF = kma$.

Piezoelectric charge pickups have some advantages over other sensors:

1. Self-generating output
2. Electrical output directly proportional to acceleration
3. Small size
4. Wide frequency response
5. System compatible with long transmission lengths
6. No phase shift in the normal operating range

Electromechanical Switches

The basic mechanism of this type switch consists of an armature suspended on a frictionless flexure pivot and held in position by a permanent magnet as shown in Figure 10.47. The armature pivots around the flexure and is forced in one direction by the adjusting spring and attracted

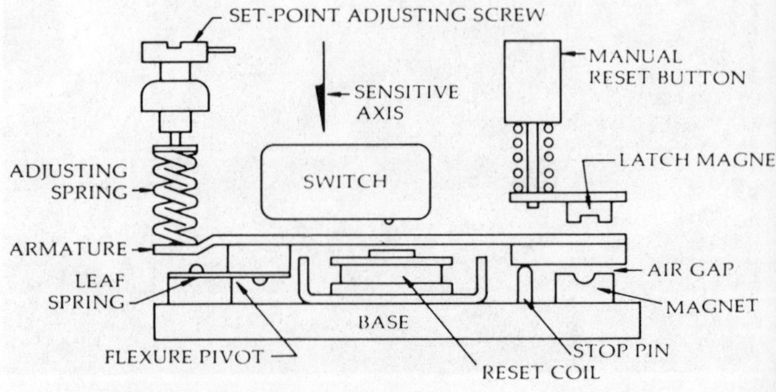

Figure 10.47. Electromechanical switch where inertial force and spring work against a permanent magnet; the variable force is the inertial force caused by vibration.

in the other direction by the magnet. When the entire assembly is subjected to vibrations perpendicular to the base, the product of peak acceleration and armature mass produces an inertial force aided by the spring and tries to pull the armature away from the magnet. When the peak acceleration exceeds the preset level, the armature leaves the stop pin and moves to the latching magnet. Movement of the armature actuates a switch that opens or closes an electrical circuit. After the switch has been tripped, it can be reset remotely by energizing the DC reset coil momentarily or by depressing the reset button mounted on the switch.

Strain Gauge Accelerometers

Strain gauge accelerometers, widely used as vibration monitors, measure a change in electrical resistance which is proportional to the force due to acceleration. This accelerometer incorporates a bending arm on which is mounted strain gauges that measure the strain on the arm as it bends. Strain is equal to the elongation of the arm divided by its length. The strain can be calculated by use of Young's modulus of elasticity:

$$E = \text{Stress}/\text{Strain}$$

From this expression, it can be seen that strain is directly proportional to stress. A curve representing this relationship is shown in Figure 10.48.

The resistance of a metallic conductor at a constant temperature varies directly with the length and inversely with the cross-sectional area. This resistance is expressed as a ratio called gauge factor. Symbolically, the gauge factor becomes

$$G = (\Delta R/R)/(\Delta L/L)$$

where G = gauge factor
- R represents the total gauge resistance in ohms
- L represents the total gauge length
- ΔL represents the total conductor length change, same units as L
- ΔR represents the total gauge resistance change, ohms

The principle is incorporated in the strain gauge which consists of a very fine wire loop or foil bonded to a carrier of paper, plastic or epoxy. The wire is about 0.001 inches in diameter and normally has a resistance of 120 or 350 ohms. A semiconductor may be used instead of wire or foil. The semiconductor strain gauge consists of semiconductors bonded to a mass. The deformation of the mass changes resistance when subjected to a strain. Because it has a much greater resistance change than the wire gauge, it provides a much greater millivolt output.

The resistance element is composed of four strain gauges which form a Wheatstone bridge. Two of the gauges are loaded in compression and two in tension. Figure 10.49 shows a typical bending sensor. The accelerometer measures movement in one plane only as indicated in Figure 10.50.

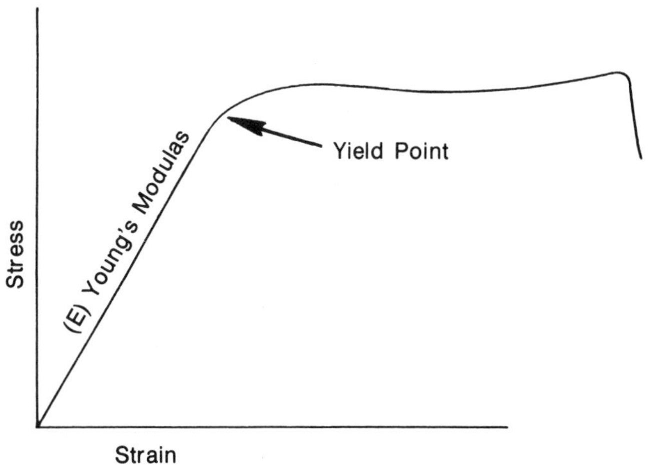

Figure 10.48. Young's modulas of elasticity curve showing the linear relationship of stress to strain until a material reaches its yield point.

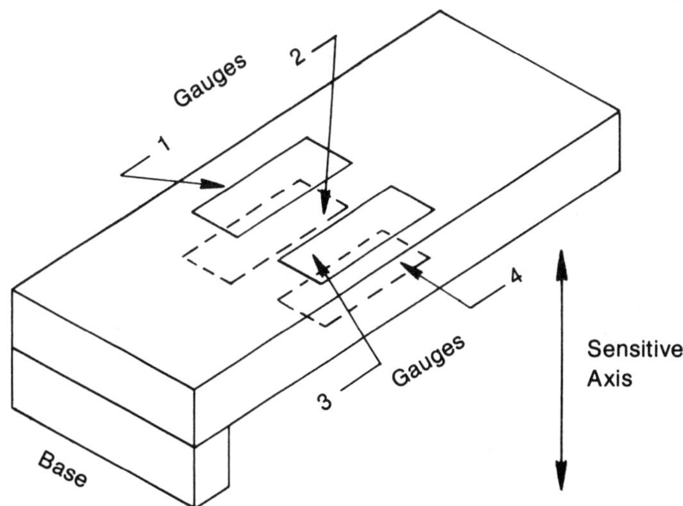

Figure 10.49. A bending full bridge accelerometer sensor is used as a vibration monitor.

Since one end of the cantilever is secured to the base of the accelerometer, the opposite end is left free to move when a force is exerted on it. The base of the accelerometer is mounted perpendicular to the sensitive axis as shown in Figure 10.50. As a force is exerted on the cantilevered mass, the resistance of the gauges changes giving an accurate proportional output.

The output of a semiconductor accelerometer may be calculated if the following information is known:

1. Gauge resistance
2. Full-scale resistance change or gauge factor
3. Excitation voltage

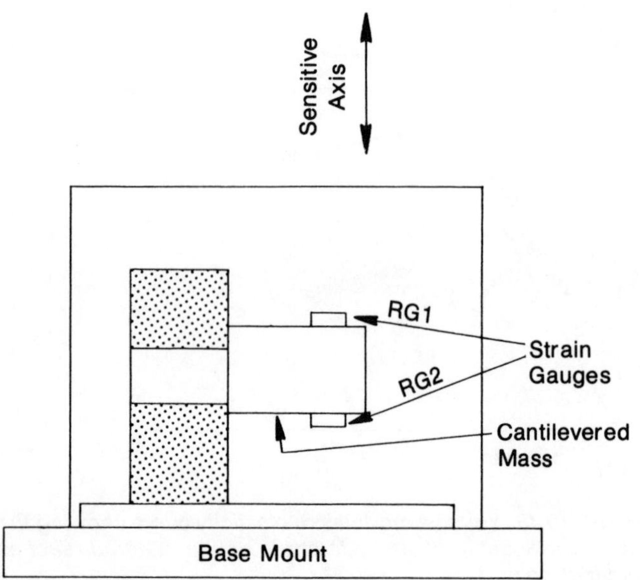

Figure 10.50. Bending accelerometer. Acceleration in the sensitive axis causes a resistance change in the strain gauges which are wired in the form of a Wheatstone bridge. The millivolt output is calibrated in G units (acceleration).

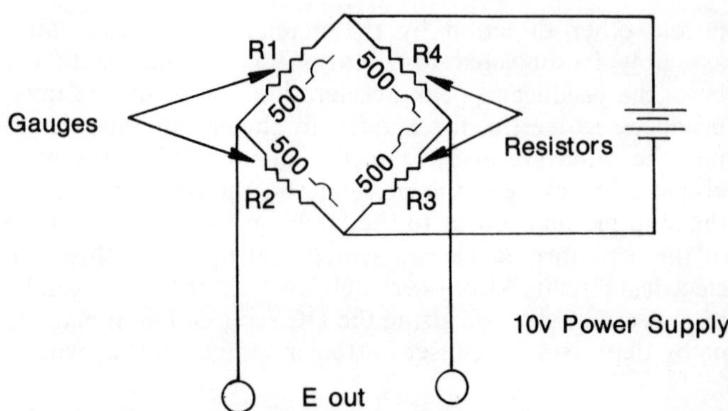

Figure 10.51. Wheatstone bridge arrangement of the semiconductor accelerometer shown in Figures 10.49 and 10.50.

Figure 10.51 shows $RG1$ and $RG2$ as 500-ohm semiconductors that form half the bridge with $R3$ and $R4$ (also 500-ohm resistors) used for bridge completion resistors. Full-scale deflection of $RG1$ and $RG2$ mounted on the beam may be determined, if ΔRG is known, by the following relationship.
[assume: $\Delta RG/RG = 0.1$, then $\Delta RG = 0.1\ (500) = 50$]

$$E_o = E \quad \frac{RG1 + RG2}{(RG1 + \Delta RG) + (RG2 - \Delta RG)} - \frac{R4}{R3 + R4}$$

$$E_o = 10 \quad \frac{500 + 50}{(500 + 50) + (500 - 50)} - \frac{500}{1000}$$

$$E_o = 10 \quad \frac{550}{1000} - \frac{500}{1000}$$

$E_o = 0.5$ volts or 500 millivolts

Semiconductor strain gauges may be used without temperature compensation; however, it is desirable if large temperature changes are expected. Semiconductors are regarded as the best in the state of the art for transducers sensed by strain gauge techniques.

Proximity Type

Monitoring and measurement of radial vibration and axial thrust positions on large machine trains are made successfully with proximity pickups. Shaft vibration is measured with respect to the bearing housing. Turbines and centrifugal compressor vibrations are measured with two proximity pickups located in the bearing or immediately adjacent to and on the same diameter as the bearing journal. One of the two probes should be mounted horizontally and the other vertically. Proximity probes are also used to measure wear on the thrust bearings and to monitor axial movement of shafts.

One proven type of proximity probe works on the eddy current principle. The motion measurement is derived from the distance between the gap and the shaft. Voltage outputs depend upon the gap size and are not affected by oil, steam or air in the gap. Although the eddy current probes are sensitive to gap distance, conductivity and permeability of the shaft, once the probe is calibrated with the machinery material, gap size is the only remaining sensitive factor since the conductivity and permeability factors are removed through calibration.

The vibration monitoring equipment for this type probe should measure and record average peak-to-peak amplitude, in mills of displacement. Indicating lights, alarms, shutdown and reset features are normally required on this equipment. The vibration monitor should have an automatic self-checking circuit and an external "OK" light or fault indicator; it should measure only one probe on each bearing. Figure 10.52 shows a block diagram of a proximity measuring probe and monitor.

Figure 10.53 shows a method of recording vibration points in order that a history file may be kept on each measurement point. A Polaroid camera is used to photograph the vibration at a specified rpm. This photograph is kept on file and used for reference and comparison of future photographs of the same point. If the file is properly kept and proper analysis is made, problem areas can be detected and costly repairs and shutdowns prevented.

Incipient Failure Detection

A new technology in measuring vibrations acoustically is being developed by the Boeing Company. Called *Incipient Failure Detection* (IFD), it is intended to detect antifriction bearing failures in rotating or reciprocating machinery. The normal method of failure detection used by plant personnel is simply to listen to the bearings' sound. Unfortunately, the unaided ear may not be able to detect failures sufficient-

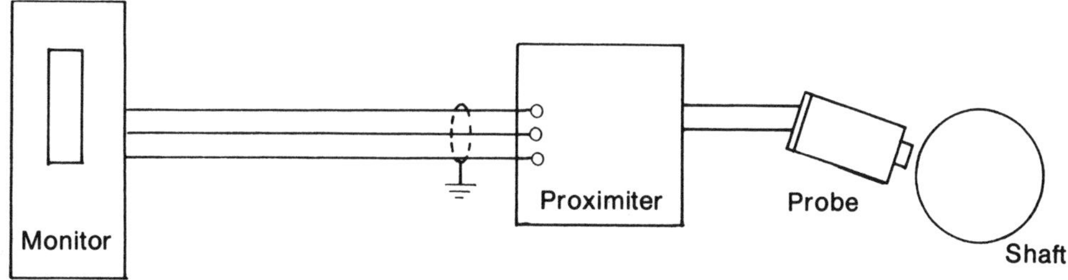

Figure 10.52. A typical hookup for a proximity switch used for vibration monitoring.

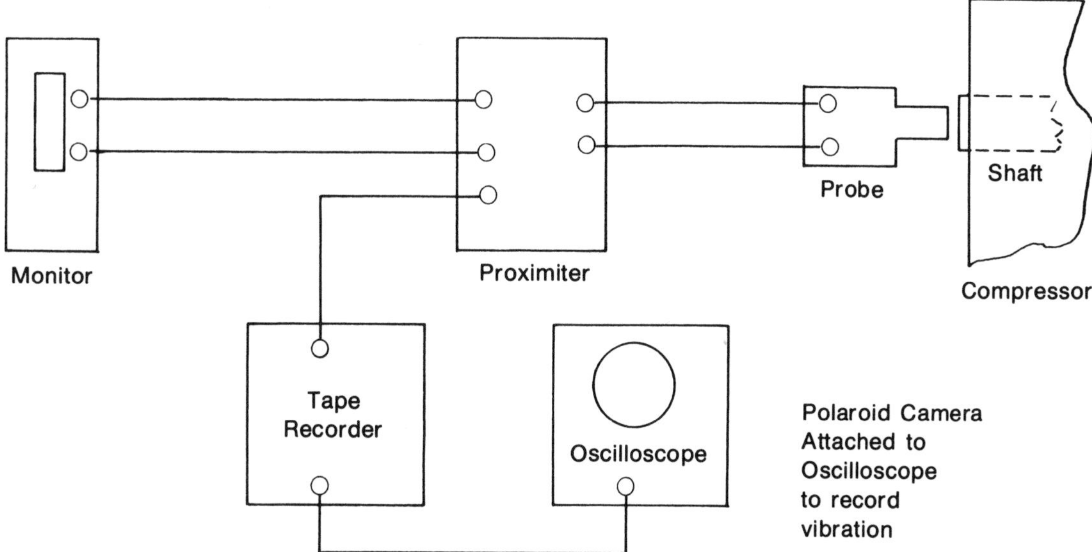

Figure 10.53. Permanent records of vibration at critical points can be the key to early detection of problem areas.

ly early to prevent untimely shutdowns and subsequent projection losses. Shutdowns may be avoided if spare machinery is available, but in many situations, spare equipment is not economically feasible. The need for early detection of incipient failures is readily apparent in order to schedule preventive maintenance functions.

The IFD system measures the acoustic emission bursts that are characteristic of defects or flaws in structural materials subjected to high repetitive loads. It uses acoustic transducers to convert acoustic signals to electrical energy. Because the fundamental frequencies generated by spalls (flaws) in antifriction bearings are in the same frequency range as other sources of industrial noise (rotating parts, friction sources, etc.), the IFD system measures some of the higher resonant frequencies of the failure source rather than base frequencies. The typical piezoelectric accelerometer or acoustic transducer normally used for this purpose is generally more sensitive in this high frequency region. Figure 10.54 shows a system used for detecting and counting acoustic emissions associated with the spall excited bursts. A spall in the bearing shock-excites the transducer at its resonant frequencies, resulting in an electrical signal similar to that shown in figure 10.55, which shows a single burst of energy. Figure 10.55 shows a single small spall. Several repetitive excitations are shown. The continuous background between the spall sinusoids result from continuous background noise due to friction.

The need for the high frequency detection technique is illustrated in Figure 10.56. Figure 10.56 is a spectrum, generated by using a low frequency measuring technique, of a bearing with a defect in its outer race. The data on the flaw was buried in the background noise. The spectrum for the high frequency shock excitation technique shown in Figure 10.56 revealed the flaw with a high signal-to-noise ratio. It was revealed at 735 Hz, the same value indicated by a mathematical calculation using an equation for determining spall frequencies.

The IFD technique allows detection of acoustic emissions at the earliest formation of the spalls, thus giving maximum time for maintenance functions without emergency shutdowns. Unfortunately, interference signals also exist at these high vibration frequency levels, just like they exist at

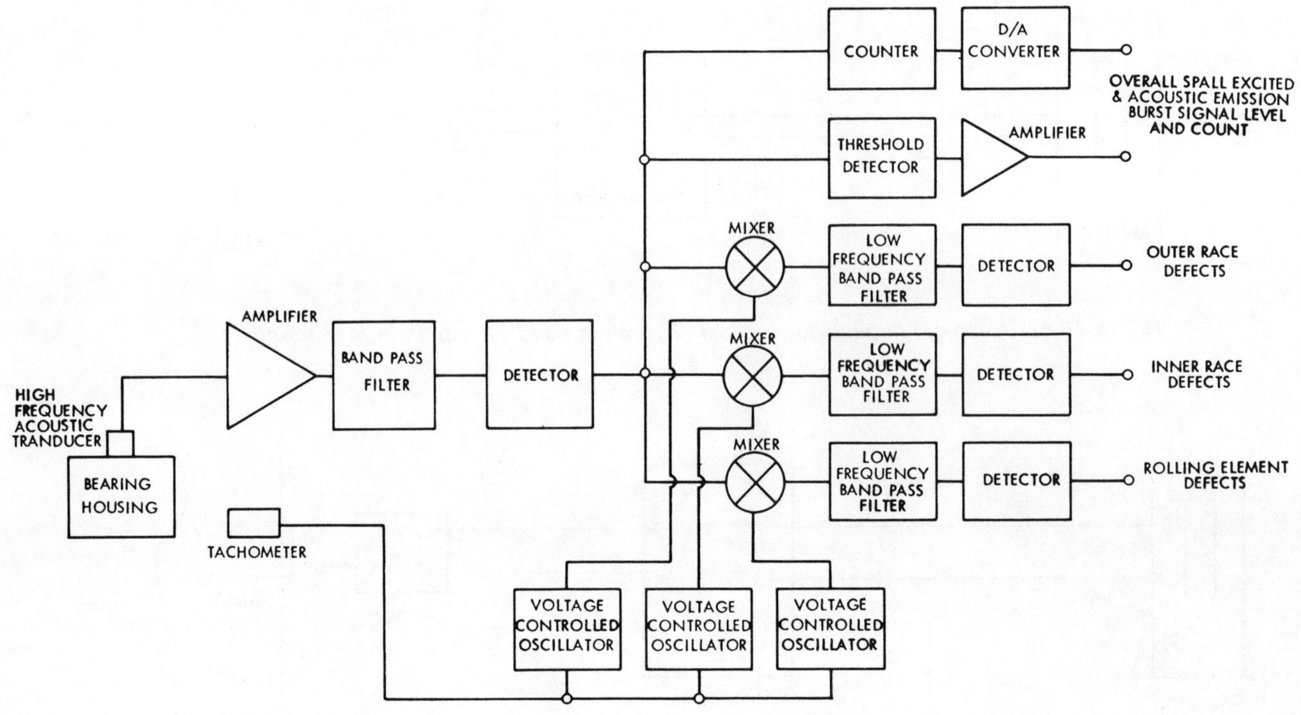

Figure 10.54. Incipient failure detection (IFD) system by the Boeing Company utilizes acoustic measurements to detect damage and potential damage to many types of equipment.

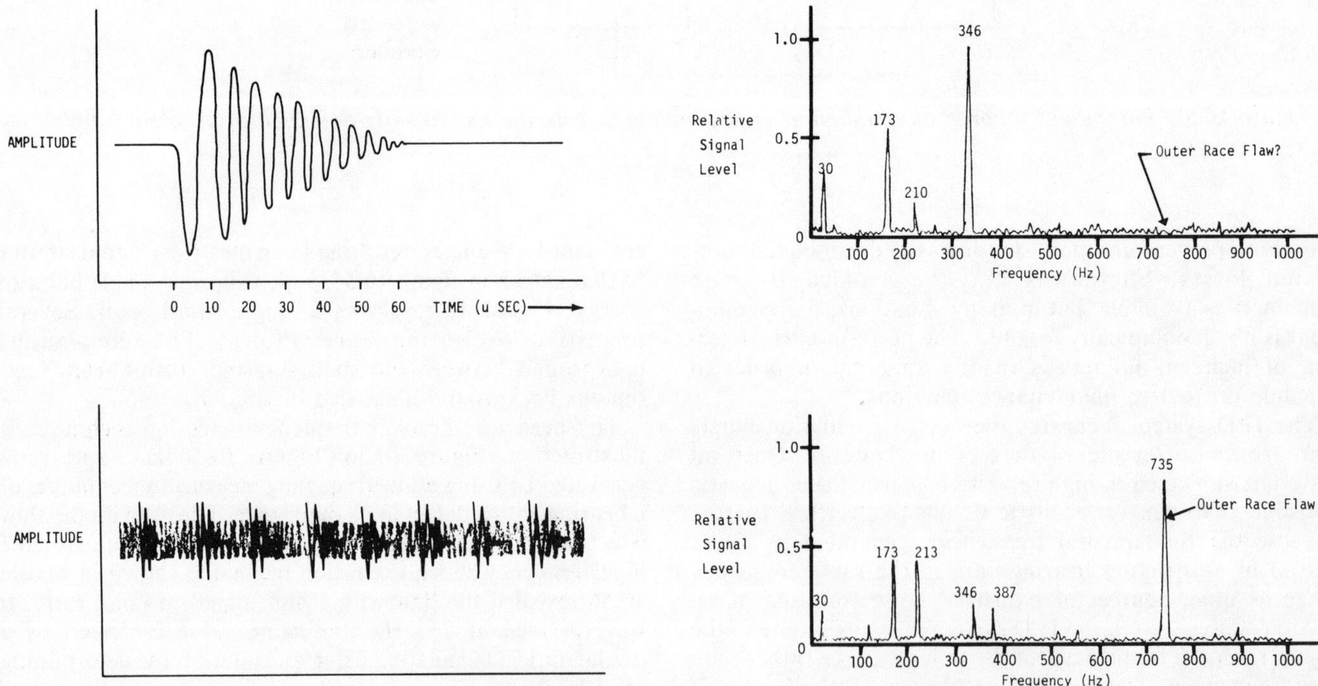

Figure 10.55. Acoustical and frequency measurement signatures that can be used in IFD systems. (Courtesy of the Boeing Co.)

Figure 10.56. High frequency flaws can be detected by using a high signal-to-noise ratio as indicated by the lower spectrum. (Courtesy of the Boeing Co.)

low vibration frequencies. However, these interference signals (at high frequencies) are greatly reduced in amplitude so that higher signal-to-noise ratios are realized. Among the extraneous signal sources are gears, other bearings, vibrating parts, fluid noise, loose solid particles in the machinery and mechanical devices such as motors, solenoids and relays.

Electrical interference is also a factor, coming from AM radio signals, power line transients and lightning transients. These can usually be made insignificant by use of preamplifiers or differential transformers.

The IFD system is relatively new. Some of the techniques employed will probably change as development progresses. The economics possible through improved operation of high and low speed heavy machinery, however, make the continued development of such techniques worthwhile.

Speed Sensors

The design of speed sensing devices covers several different principles of operation, including electromagnetic induction, voltage generation and magnetic sensing. The need for accurate speed indication has been brought about by the development and use in recent years of high-speed equipment required by modern processes. Accurate hand speed indication is not sufficient for industrial plant speed measuring problems. Following are some speed sensors now available.

Magnetic Pickup Sensors

The magnetic pickup sensor consists of a cylindrical permanent magnet behind a soft iron pole piece around which a coil has been wound (Figure 10.57). When a ferrous metal is placed in front of the pole piece, the magnetic flux is high and the flux drops off as the ferrous metal is passed. A voltage is generated which is proportional to the rate of change of flux in the pole piece and proportional to the speed with which a ferrous mass makes the flux build up or collapse.

The ferrous material is made up in a gear form with the desired number of teeth. The voltage wave form varies depending upon the teeth shape, thickness and spacing. The voltage amplitude is inversely proportional to the clearance between the pickup tip and the surface of the gear.

Magnetic pickups provide an accurate indication or record of equipment speed (rpm). The frequency of voltage buildups (and collapses) is linearly proportional to rpm. The pickup output frequency is equal to

$$\text{rpm} \times \text{No. of Gear Teeth} \; / \; 60$$

By using a 60-tooth gear, the frequency in hertz is equal to the rpm of the gear shaft. A gear of 120 teeth would make the frequency twice the rpm of the shaft.

The magnetic pickup may be operated under conditions where oil, water or noncorrosive liquids are present. Ambient temperature operation is normally from -75° to 200°F.

The magnetic pickup tachometer system may be used on any type vibrating, rotating or moving surface. A wheel spoke, screw head, rivet head and many other types of ferrous materials may be used instead of gear teeth. Figure 10.58 shows an application where a gear is mounted on a variable speed motor. This application may be used also for a closed loop control system. The electronic frequency-to-current converter receives an electrical input at a frequency directly proportional to the rpm of the motor. This signal is converted to a proportional output current, which is directly proportional to the input frequency, for operation of a rpm indicator or for a control application.

Figure 10.59 shows a magnetic pickup used as a speed switch for alarm or shutdown functions on rotating equipment. The pickup output is fed to an amplifier which amplifies the signal for operation of a relay. For use as a

Figure 10.57. Magnetic pickup sensor measures speed by generating a voltage whose frequency is proportional to rpm. (Courtesy of Airpax Electronics/Controls Div.)

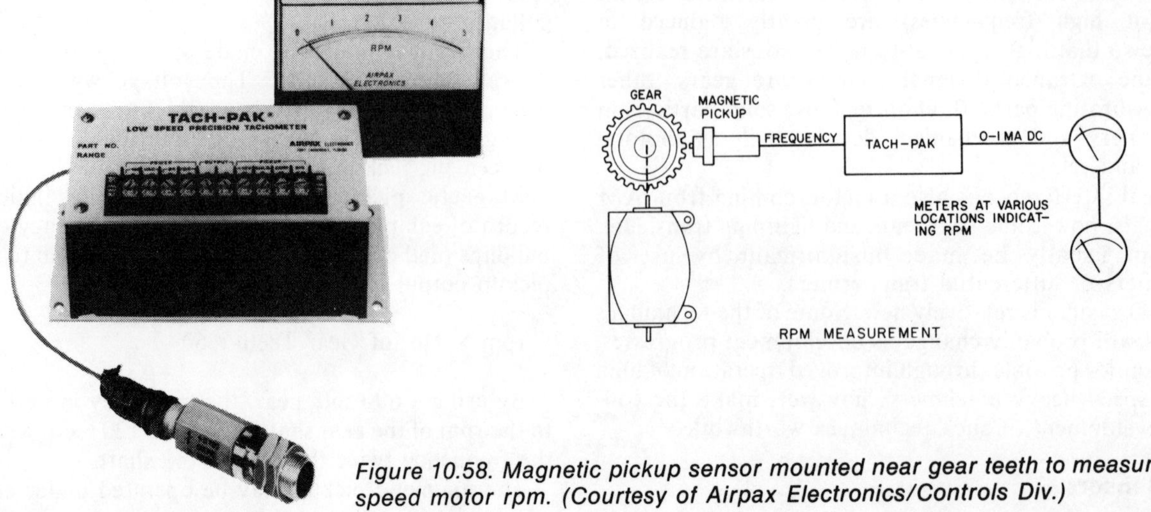

Figure 10.58. Magnetic pickup sensor mounted near gear teeth to measure variable speed motor rpm. (Courtesy of Airpax Electronics/Controls Div.)

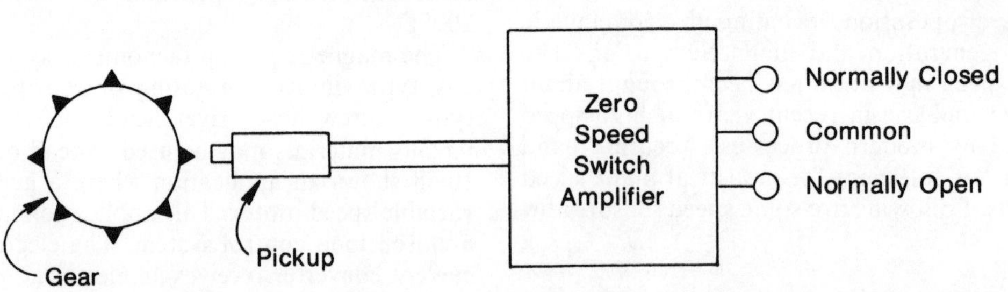

Zero Speed Switch Amplifier
Normally Closed
Common
Normally Open
Gear
Pickup

Figure 10.59. A magnetic pickup sensor used as a zero speed switch by adjusting amplifier to energize a relay at a specified low millivolt output.

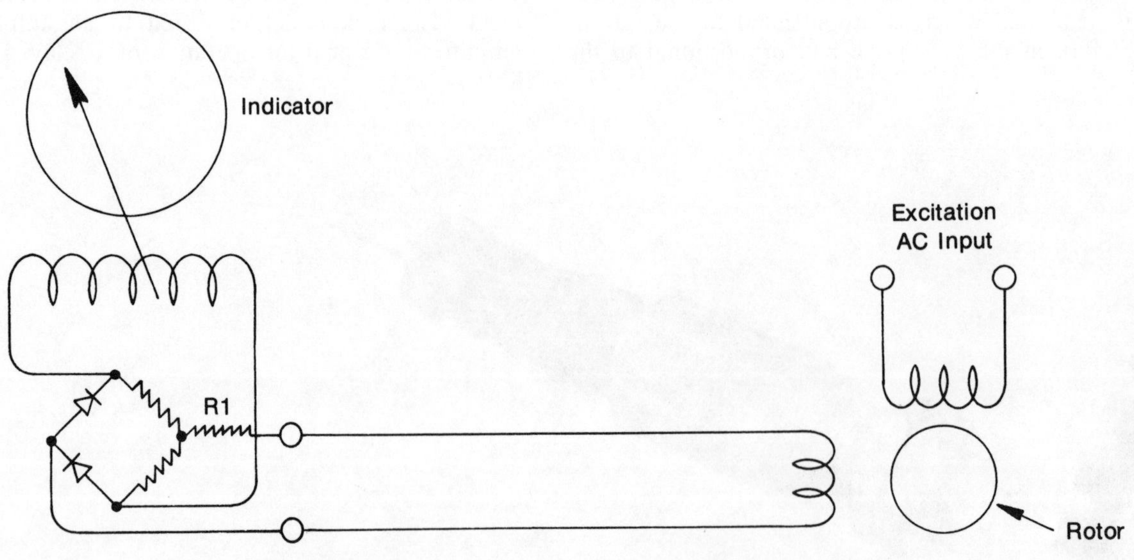

Figure 10.60. AC tachometers are used for measurement of rotor speed.

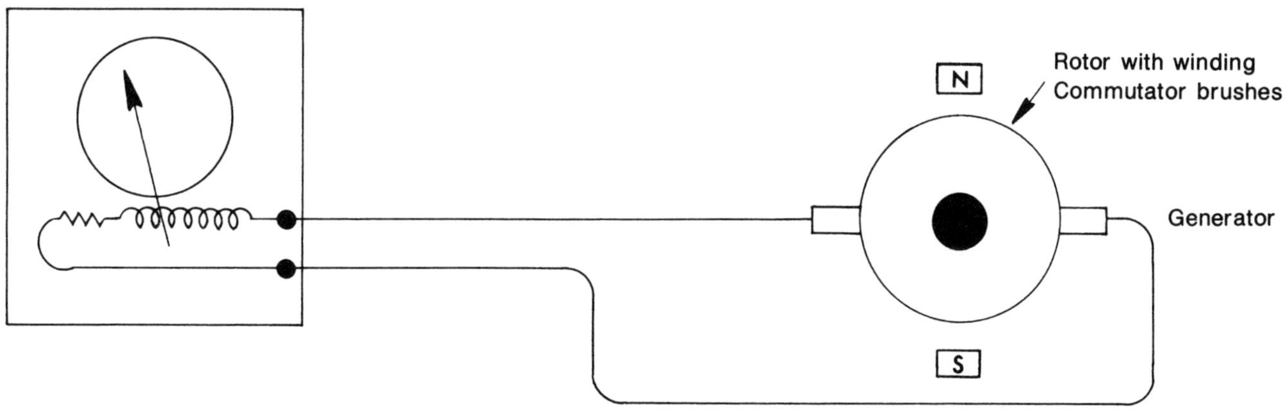

Figure 10.61. DC tachometer schematic: a permanent magnetic flux whose flux lines are cut by rotating shaft, generating an AC voltage which is converted to DC by commutator and brushes.

zero speed switch, the amplifier may be adjusted to energize the relay at a specified low millivolt output.

Magnetic pickup applications in the petrochemical industry have increased in the past few years. Since this pickup generates a signal whose frequency is proportional to shaft speed with excellent accuracy, it is a valuable tool for industrial applications.

Tachometer Generators

The tachometer generator is an electromechanical device that generates a voltage output proportional to shaft speed. This instrument may be designed for readout in revolutions per minute, feet per minute, yards per minute, miles per hour, frequency or any other unit that can be converted into a function of rotational motion.Types of speed tachometers include AC, DC and AC frequency response.

AC Tachometer. Components of the AC tachometer generator system consist of a primary winding, a secondary winding that is located physically 90° to the primary winding, a rotor and a high resistance indicator. Since the windings are 90° to each other, the output winding is basically zero when the primary winding is excited. When the rotor is turned, a voltage is induced in the secondary winding with a magnitude proportional to the rotor speed.

Figure 10.60 shows an AC tachometer system. A resistance, R_1, is added in the output line as a calibration resistor for the measuring circuit. For example, the generator maximum output may be from 1 to 50 volts which can be measured, in either case, with the readout meter by selecting the proper calibration resistor, R_1. Generator speed may be measured from 500 to 10,000 rpm.

DC Tachometer. The DC tachometer generator differs in design characteristics from the AC tachometer. The magnetic flux is originated by a permanent magnet, and the output winding is attached to the rotor. An increase in rotor speed causes a proportional increase in the output voltage. Figure 10.61 shows a diagram of a DC tachometer installation. DC tachometers are designed for operational speeds of 10 to 5,000 rpm.

The polarity of the generated voltage depends upon the direction of rotation. The voltage induced in the rotor winding by the magnetic field is sinusoidal; however, the commutator and brushes on the rotor convert the AC voltage to DC voltage.

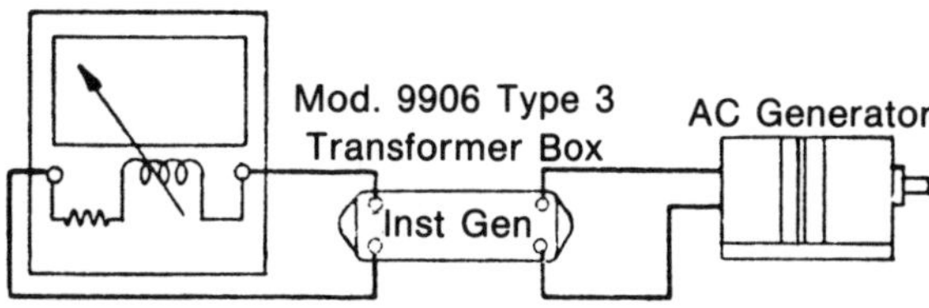

Figure 10.62. Hookup for AC frequency response tachometer used normally with a bearingless generator.

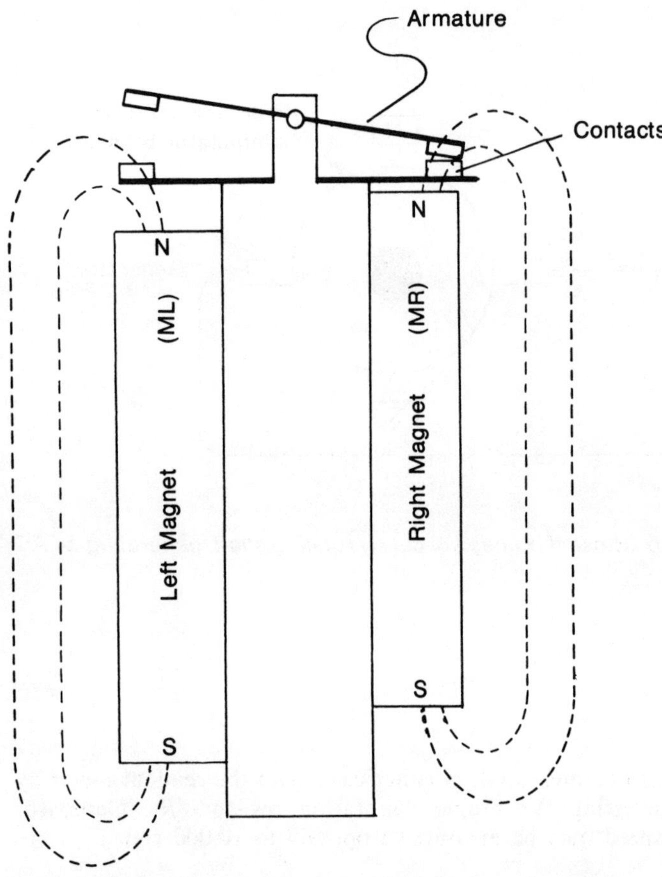

Figure 10.63. The proximity switch shows the closer location of right-hand magnet to armature holding switch closed until a ferrous metal approaching armature robs that side of magnetic flux, allowing the switch to open.

AC Frequency Response Tachometer. The AC frequency tachometer consists of a DC indicator, a frequency response network and an AC tachometer generator. This system is normally used with a bearingless tachometer generator. Figure 10.62 shows the circuit diagram for a saturable core transformer AC tachometer. This system operates in the speed range of 500 to 5,000 rpm with a conventional bearing generator or 500 to 100,000 rpm with a bearingless generator.

Proximity Sensors

Proximity sensors are used in vibration measurements and in speed and automatic sequence measurements and operations. These sensors are designed to detect ferrous material; therefore, they can operate on continuous or on-off functions.

Proximity switches are normally built for duty operation. Contacts are sealed in a hydrogen-filled atmosphere to reduce contamination of contacts. The armature and contact assembly is made for snap action of the contacts. Figure 10.63 shows the static operation of the switch. The magnet on the right-hand side is mounted closer to the armature to hold the right-hand contact normally closed. Since the right-hand magnet (*MR*) is mounted closer to the armature than the left-hand magnet (*ML*), it has more flux lines holding than the left-hand side. The flux lines will remain stronger on *MR* until a ferrous material is brought into the proximity of the armature which robs the flux from that side. As the *MR* flux is reduced to a value lower than the *ML* flux, the left contacts snap closed and, at the same time, open the right contacts.

Sensing distance for contact closure is a function of mass, area and position with respect to the sensing surface. This distance varies, depending on the individual switch design.

11 Control Valves

William G. Andrew

A control valve is the most commonly used final control element used to regulate material flow in a process. In a process loop it is the only *controlled* variable resistance in the system. Other resistances vary because of flow changes in the system or because of coating of pipe and equipment wall surfaces. These resistance variations are undesirable, and they must be compensated for by the control valve.

In most cases a control valve is expected to modulate continually in response to a control signal to keep a process variable steady. Because wide rangeability is inherent in control valve design, there is considerable flexibility in selection of its size. There are also several choices to be made relative to body design, flow characteristic, actuator type and trim design.

The following sections define valve terms in common use and describe valve capacity, rangeability, body designs, trim characteristics, mechanical features and valve accessories.

Valve Terminology

Balanced valve: a body design in which the same pressure acts on both sides of the valve plug, thus reducing the actuator force required to open the valve. This characteristic is inherent in the design of a double port valve but is also available in single port valves.

Bellows seal: a physical barrier or seal which uses a bellows for sealing against leakage around the valve plug stem.

Bonnet assembly: an assembly including the part through which a valve plug stem moves and a means for sealing against leakage along the stem. It usually provides a means for mounting the actuator.

Capacity: rate of flow through a valve under stated conditions.

Cavitate: the formation and subsequent collapse of voids or cavities in a valve resulting from increased fluid velocity through the restricted areas of the valve. It occurs in liquids when the valve operates near the vapor pressure of the liquid.

Characteristic: relation between flow through the valve and percent rated travel as the latter is varied from zero to 100%.

Corrosion: the reactions between materials of the valve and the fluids handled which cause valve deterioration.

C_v: flow coefficient, the accepted unit of measurement to define the capacity of a valve. It is defined as the number of gallons per minute of water at room temperature which will pass through a given flow restriction with a pressure drop of one psi.

Dead band: the amount the diaphragm pressure can be varied without initiating valve plug motion.

Diaphragm actuator: a fluid-pressure operated spring or fluid-pressure opposed diaphragm assembly for positioning the actuator stem in relation to the operating fluid pressure or pressures.

Equal percentage flow characteristic: an inherent flow characteristic which, for equal increments of rated travel, will ideally give equal percentage changes of the existing flow.

Erosion: A wearing action on valve trim and body, resulting from high velocity fluids and fluids containing solids particles. It is prevalent in steam service, in flashing service, in some chemical services and where high pressure drops occur.

Extension bonnet: a bonnet with an extension between the packing box assembly and bonnet flange.

Flow characteristic: see *Characteristic*.

Flow coefficient: see C_v

Galling: a high friction condition between the valve stem and stem guides, bushings or packing, resulting from a lack of lubricity between the moving parts.

Guide bushing: a bushing in a bonnet, bottom flange or body to align the movement of a valve plug with a seat ring.

High recovery characteristic: a term applied to the design characteristic of a valve which allows a high percentage ratio of downstream to upstream pressure.

Inherent flow characteristic: flow characteristic when constant pressure drop is maintained across valve.

Inherent rangeability: ratio of maximum to minimum flow within which the deviation from the specified inherent flow characteristic does not exceed some stated limits.

Installed flow characteristic: an inherent flow characteristic which can be represented ideally by a straight line on a rectangular plot of flow versus percent rated travel.

Leakage: quantity of fluid passing through an assembled valve when valve is in fully closed position under stated closure forces, with pressure differential and temperature as specified.

Linear flow characteristic: an inherent flow characteristic which can be represented ideally by a straight line on a rectangular plot of flow versus percent rated travel.

Low recovery characteristic: a term applied to the design characteristic of a valve which causes a low percentage ratio of downstream to upstream pressure.

Modulate: the function of a controller which causes a valve to respond to an infinite number of positions between the closed and full open positions.

Normally closed: applying to a normally closed control valve assembly, one which closes when the actuator pressure is reduced to atmospheric.

Normally open: applying to a normally open control valve assembly, one which opens when the actuator pressure is reduced to atmospheric.

ΔP: the pressure drop across a valve. The condition must be specified. For example: ΔP for sizing; ΔP at normal flow; ΔP at valve closure; etc.

Packing box assembly: the part of the bonnet assembly used to seal against leakage around the valve-plug stem, including various combinations of all or part of the following: packing, packing nut, packing follower, lantern ring, packing spring, packing flange, packing flange studs or bolts and packing flange nuts.

Plug: a moveable part which provides a variable restriction in a port.

Rangeability: the ratio of maximum to minimum usable sizing coefficient.

Rated C_v: the value of C_v at the rated full open position.

Rated travel: linear movement of the valve plug from the closed position to the rated full open position.

Seat: that portion of a seat ring or valve body which a valve plug contacts for closure.

Seat ring: a separate piece inserted in a valve body to form a valve body port.

Stem: a rod extending through the bonnet assembly to permit positioning the valve plug.

Trim: the parts (except the body) of a valve which come into contact with the flowing fluid.

Turndown: the ratio of maximum to minimum flow requirements.

Valve body: a housing for internal valve parts having inlet and outlet flow connections.

Valve characteristics: see *Characteristics*.

Valve plug: see *Plug*.

Valve plug guide: that portion of a valve plug which aligns its movement in either a seat ring, bonnet, bottom flange or any two of these.

Valve stem: see *Stem*.

Yoke: a structure by which the diaphragm case assembly is supported rigidly on the bonnet assembly.

Valve Capacity

Valve manufacturers have adopted a common basis for rating capacities of control valves. The yardstick accepted for this purpose is termed *flow coefficient* (C_v) and has been in use since the middle 1940s. The valve coefficient C_v is defined as

$$C_v = q/[(\Delta P/G)^{1/2}]$$

where q is the volumetric flow rate through the valve in gallons per minute, ΔP is the pressure drop across the valve in psi (including inlet and outlet losses) and G is the specific gravity of the flowing fluid. Stated another way, it is the number of gallons per minute of water at room temperature which will pass through a given flow restriction with a pressure drop of 1 psi. For example, a control valve which in the full open position passes 25 gpm of water with a 1 psi pressure drop has a maximum flow coefficient of 25. The flow is similarly determined at various increments of valve lift, and the C_v at each increment is obtained. A *plot* of these values reveals the "*characteristic curve*" of the valve. The curve is made by plotting the percent of maximum travel against the percent of maximum flow. Figure 11.1 reveals a typical flow curve for a percentage valve plug. The characteristic curve thus obtained reflects the control characteristic of that particular plug and determines whether it or another type plug is best suited for a particular application.

Valve Rangeability

Rangeability of a control valve may be defined as the ratio of maximum controllable flow to the minimum controllable flow. Stated another way, it is the ratio of maximum to minimum *usable* sizing coefficient. It is obtained by dividing the minimum usable sizing coefficient in percent into the maximum usuable sizing coefficient in percent. Figure 11.2 shows how rangeability is obtained. More properly, this term should be called *inherent* rangeability.

Rangeabilities of control valves vary, depending somewhat on the body style. Rangeabilities of globe valves vary generally from 30:1 to 50:1. Throttling ball valves may have rangeabilities in excess of 100:1; standard butterfly valves have rangeabilities between 10:1 and 20:1; and pinch

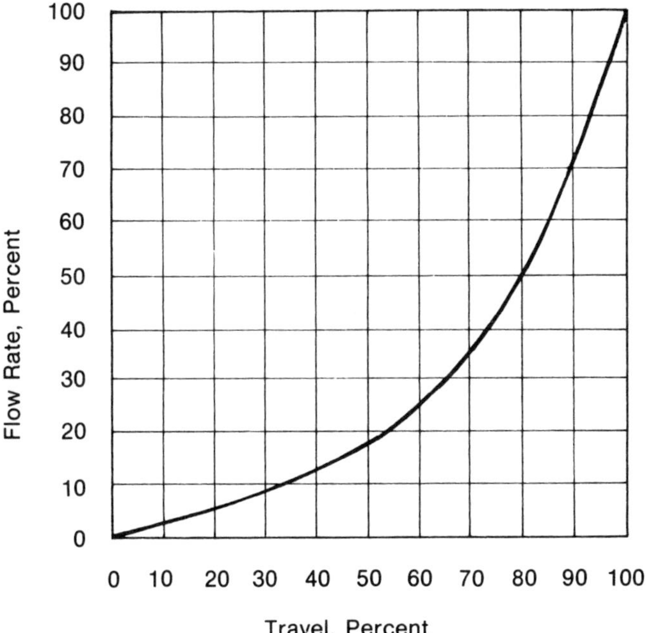

Figure 11.1. An equal perccentage characteristic curve. Such curves are determined by plotting valve opening versus flow rate.

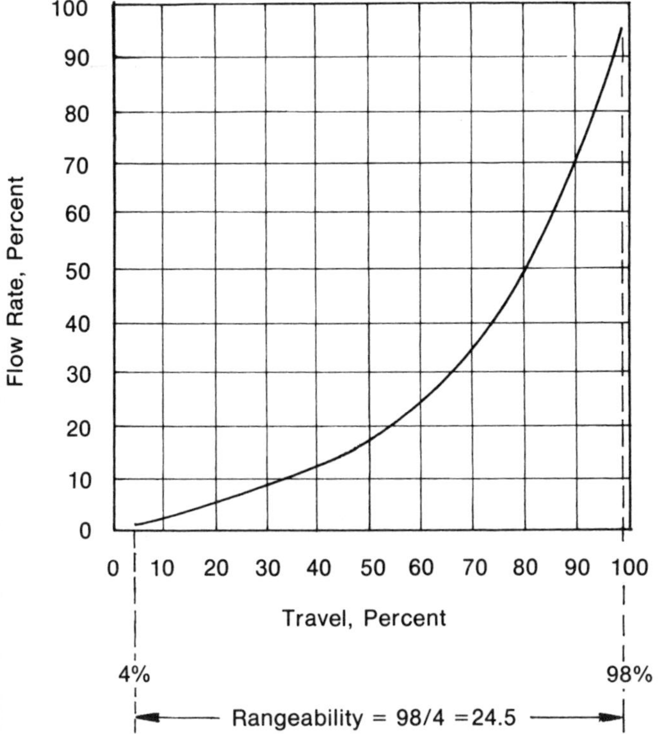

Figure 11.2. Rangeability determined for an equal percentage valve, using practical flow limits.

and diaphragm (Saunders-Type) valves have rangeabilities as low as 5:1.

In selecting a valve for a control application, the *installed rangeability* or *operating rangeability* is just as important as inherent rangeability. Operating rangeability may be defined as the relationship between rangeability and pressure drop. It can be expressed by

$$Ro = (q_1/q_2)\sqrt{(\Delta P_2/\Delta P_1)}$$

where q_1 is initial flow, q_2 is final flow, ΔP_1 is the initial pressure drop across the valve, and ΔP_2 is the final pressure drop across the valve.

For example, the flow requirement may decrease from 100 to 25% while the pressure drop increases from 16 to 100%. The required installed rangeability in this case would be

$$R = (100/25)\sqrt{(100/16)} = 10$$

The importance of looking closely at the loop requirements cannot be minimized.

Body Designs

No attempt will be made to show all body designs now in use. Those shown cover the majority of applications in the processing industries.

Globe Bodies

Globe valves are the most common type in use today and have been for many years. They may be divided into several categories, including single port, double port and three-way. Split body and angle valves are classified as special type globe valves.

Single Port

Single port valves are simple in construction; frequently used in sizes 2 inches and below; provide tight shutoff for metal-metal or composition seating; have wide rangeabilities; but may have high unbalanced forces on the plug requiring large actuators. They can be constructed to have the valve plug move into or out of the port with increasing actuator loading pressure. Figure 11.3 shows a typical design for a reversible plug, single port body, unbalanced design.

The need for low leakage, balanced valves has brought about some notable designs for balancing single seated valves. Figures 11.4 and 11.5 are typical examples of how balance has been achieved. It is accomplished by porting through the body or plug so that the same pressure (inlet or outlet) acts on both sides of the valve plug.

Double Port

Double port valves were developed to balance the forces normally acting on single port valves (Figure 11.6). Generally they have higher flow capacities and require smaller stem

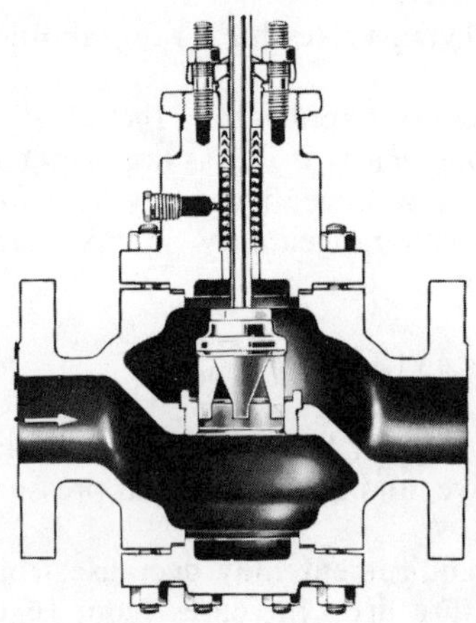

Figure 11.3. Single port globe body with reversible plug. To move in or out of seat with increasing signal pressure. (Courtesy of Fisher Controls Co.)

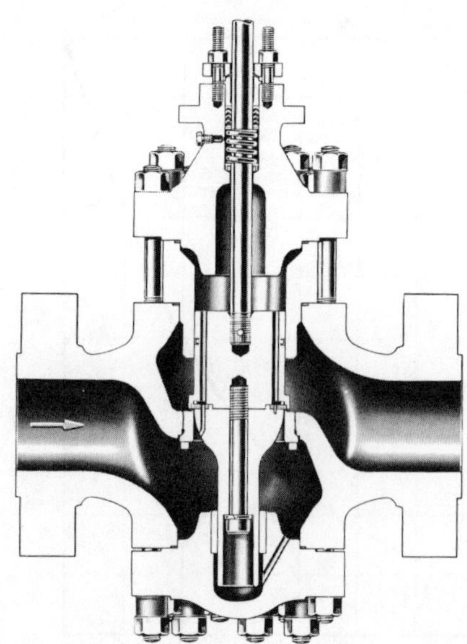

Figure 11.5. Balance is achieved on this single port valve by passageway from inlet to top of plug. (Courtesy of Fisher Controls Co.)

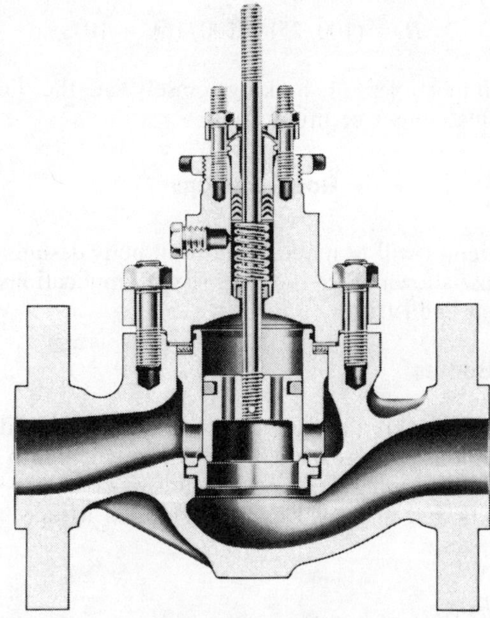

Figure 11.4. Single port cage balanced plug with outlet pressure acting on both sides of plug. (Courtesy of Fisher Controls Co.)

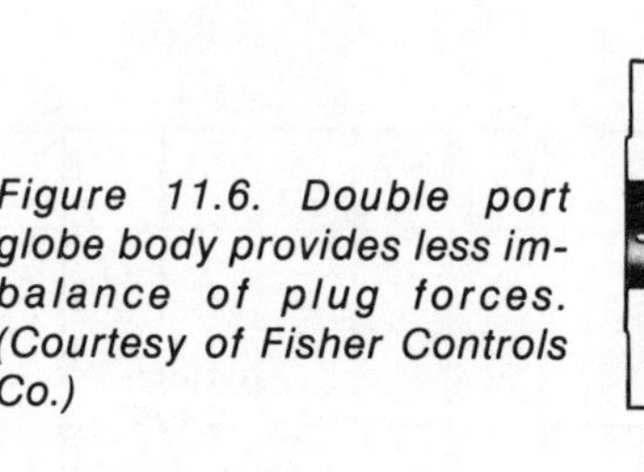

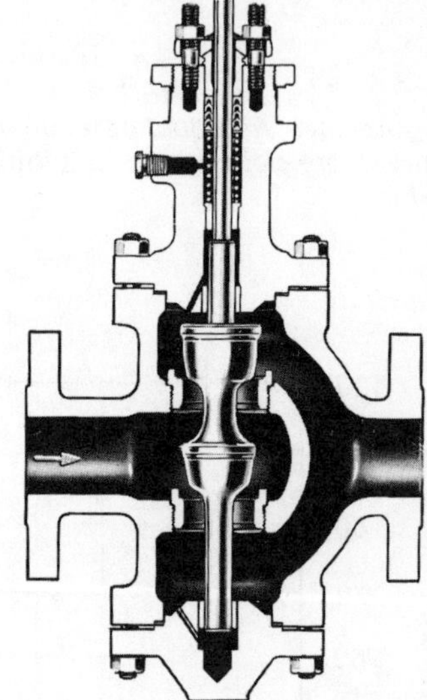

Figure 11.6. Double port globe body provides less imbalance of plug forces. (Courtesy of Fisher Controls Co.)

forces compared to the same size single port valve. They are frequently specified for sizes larger than 2 inches but should not be used when leakage is objectionable. Special construction may provide composition seating for tight shutoff when required. Reversible plug design is available to open or close the valve with increasing loading pressure.

Three-Way

Three-way valves are designed to blend (mix) or to divert (split) flowing streams. They can replace two straight-through valves in many applications. In blending service, there are two inlet ports and one outlet, whereas in diverting service there are one inlet and two outlet ports. Total flow is proportioned only, not controlled, in either service.

The three-way valve used for blending is a modified single port body (Figure 11.7). When used for diverting, it is a modified double port body (Figure 11.8). Most three-way valves have the characteristic of unbalanced forces on the valve plug and require large operators. They are usually installed with the flow tending to open the valve plug discs to prevent "slamming" of the valve plug.

Balancing a three-way valve is accomplished by using four seats as shown in Figure 11.9. In essence, it is done by using dual double seats in one body.

Split Body

A special type globe body, consisting of two body halves with a seat ring clamped between them, is called a split body valve (Figure 11.10). Its design is applicable only to single seated valves. Its construction minimizes erosion effects, allows parts to be replaced easily and is relatively inexpensive.

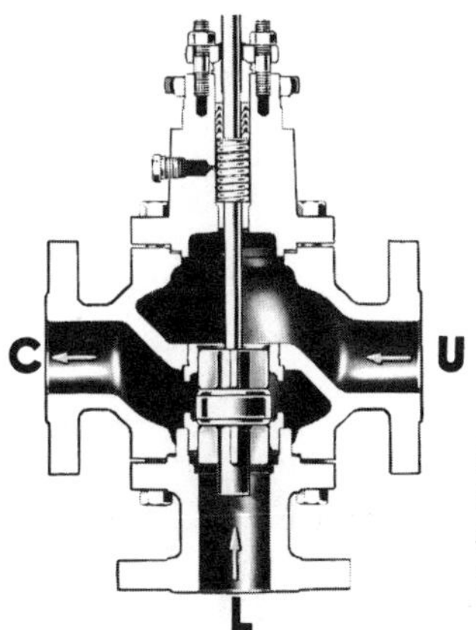

Figure 11.7. Three-way valve (mixing) utilizing modified single port body. (Courtesy of Fisher Controls Co.)

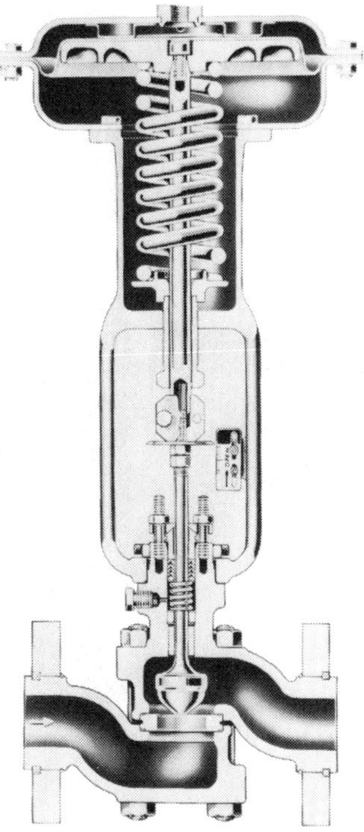

Figure 11.10. Split body valve with separable flanges allows easy disassembly and economical construction. (Courtesy of Fisher Controls Co.)

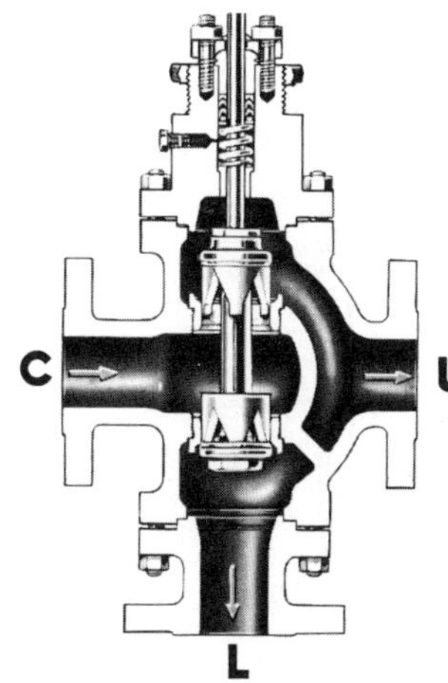

Figure 11.8. Three-way valve (diverting) utilizing modified double port body. (Courtesy of Fisher Controls Co.)

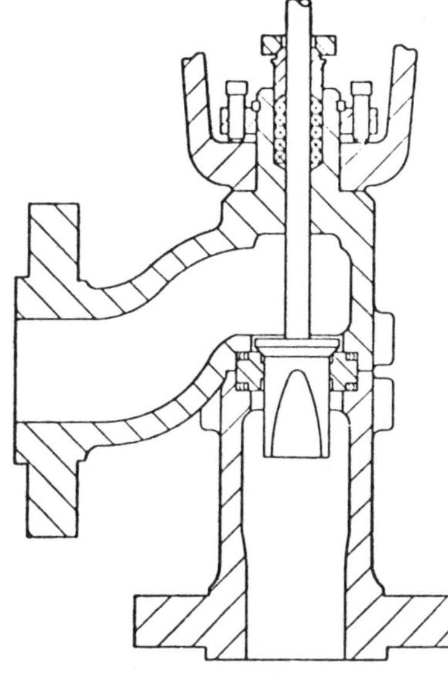

Figure 11.11. Angle valve with split body construction is easily removed and reinstalled. (Courtesy of ITT Hammel Dahl/Conoflow)

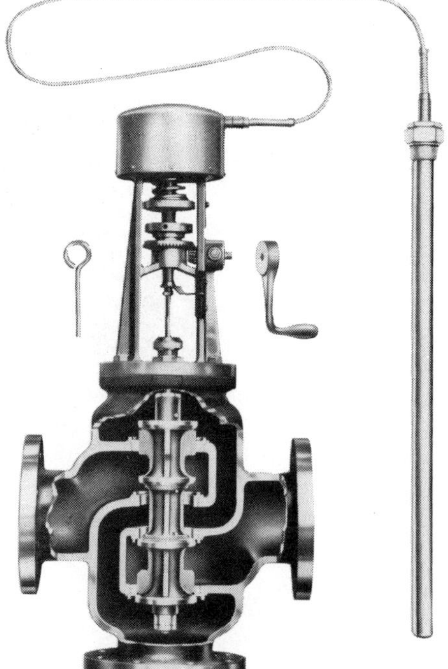

Figure 11.9. Three-way valve with balanced plug, using dual double seats. (Courtesy of Robertshaw Controls Co.)

Angle

Angle valves, nearly always single ported, are often used where space is at a premium (may eliminate a 90° ell in the piping). They are applicable to services requiring high pressure drops or where the effects of turbulence, cavitation or impingement present problems. Several styles of angle construction are available. All have good control characteristics, high rangeabilities and high pressure and temperature ratings. They are easily removed from the line and can handle sludges and erosive materials.

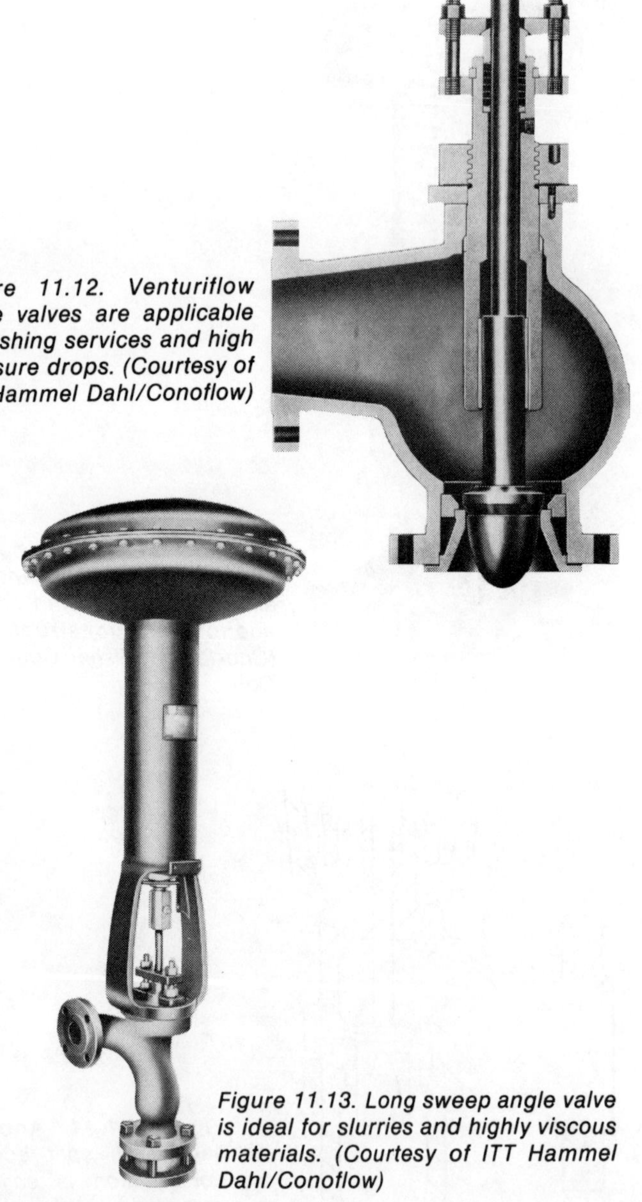

Figure 11.12. *Venturiflow angle valves are applicable to flashing services and high pressure drops. (Courtesy of ITT Hammel Dahl/Conoflow)*

Figure 11.13. Long sweep angle valve is ideal for slurries and highly viscous materials. (Courtesy of ITT Hammel Dahl/Conoflow)

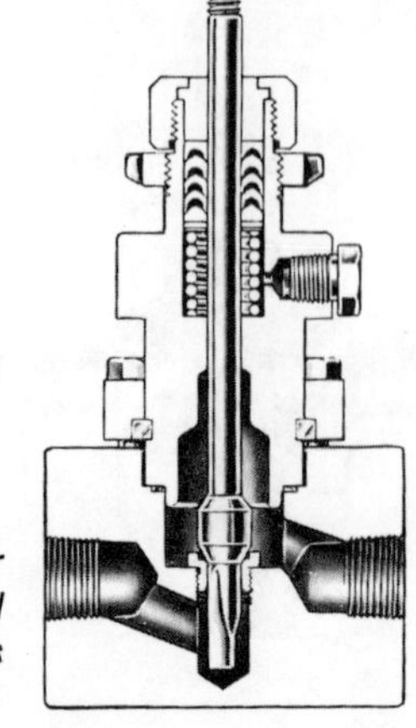

Figure 11.14. Needle valve of bar stock body construction for small flows. (Courtesy of Fisher Controls Co.)

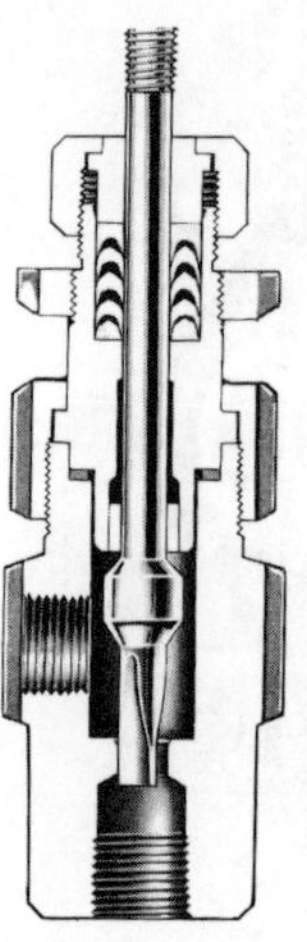

Figure 11.15. Needle valve, angle style barstock body. (Courtesy of Fisher Controls Co.)

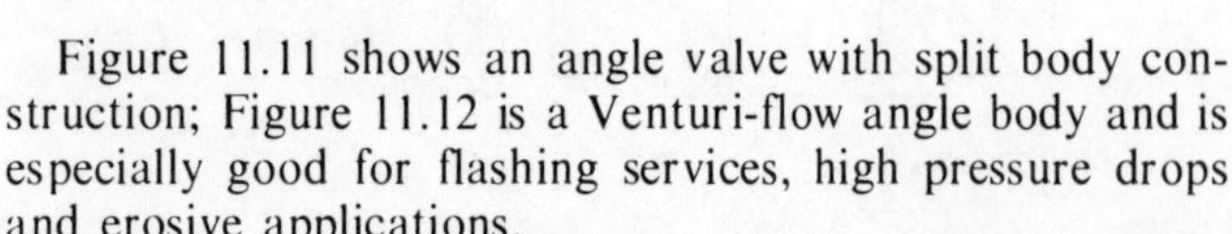

Figure 11.11 shows an angle valve with split body construction; Figure 11.12 is a Venturi-flow angle body and is especially good for flashing services, high pressure drops and erosive applications.

A long sweep angle design (Figure 11.13) has a slightly higher capacity than other angle patterns, uses a long-radius bend at the valve inlet while the outlet simulates a Venturi. Its construction is ideal for slurries and highly viscous fluids. However, this design is seldom used.

Needle

Designs that may be included in this category include the barstock body designs (Figures 11.14 and 11.15) and forged bodies for high pressure applications requiring small flows and high rangeabilities (Figure 11.16). The latter design is

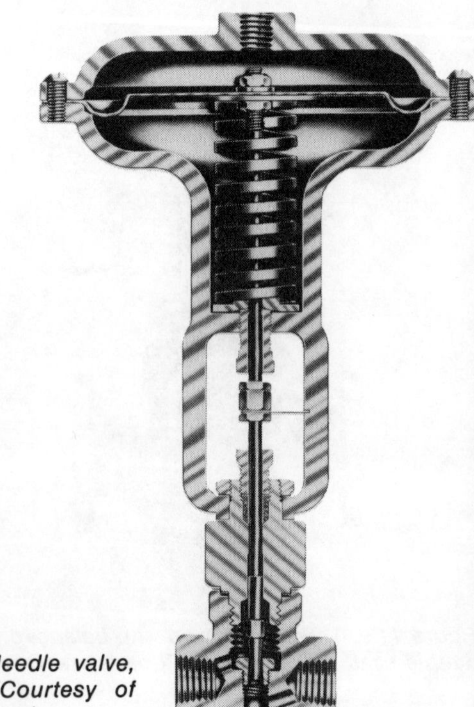

Figure 11.16. Needle valve, forged body. (Courtesy of Badger Meter, Inc.)

especially good for pilot-plant facilities, for control of liquid catalyst or additive flows to various processes and for pressure letdown services to analytical equipment.

Ball

One type of ball valve employs a cage to carry a solid ball into the mouth of the body opening (Figure 11.17). This valve is used in the pulp and paper industry. Its design ensures a self-cleaning action; it provides tight shutoff and has wide rangeability for accurate flow control.

Figures 11.18 and 11.19 show other variations of ball-type designs made for hard-to-handle fluids, such as paper stock, polymer slurries and other fluids with entrained solids. These high-recovery (low-pressure loss) valves have good control characteristics and high rangeabilities.

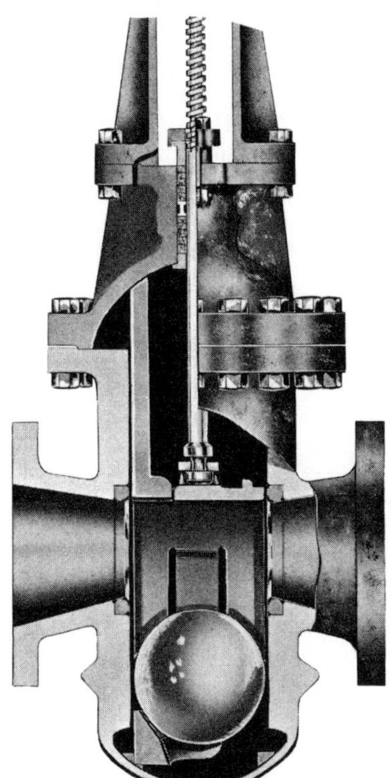

Figure 11.17. Ball valve with solid ball and cage. It provides tight shutoff and high rangeability. (Courtesy of Bell and Howell, Control Products Division)

Figure 11.18. Partial ball (Vee-ball) body design for hard to handle fluids as paper stock and polymer slurries. (Courtesy of Fisher Controls Co.)

Figure 11.19. Partial ball body design possessing good control characteristics. (Courtesy of Masoneilan)

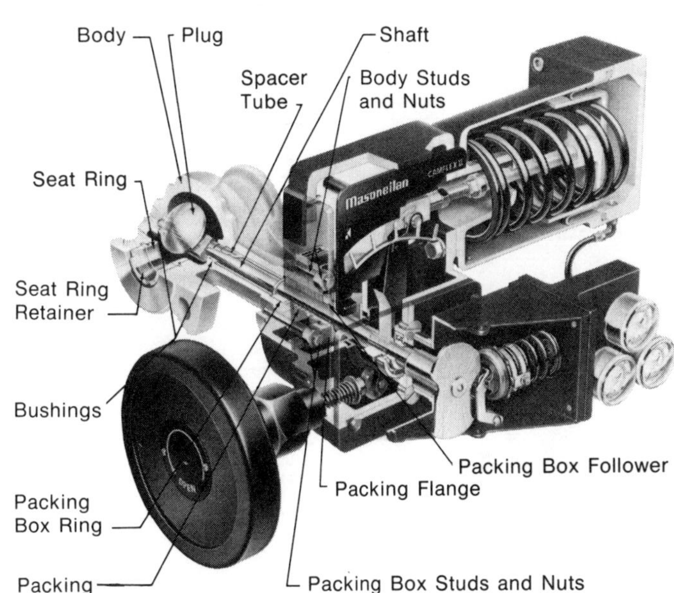

Figure 11.20. The Camflex valve has a center of rotation eccentric to the center line of the seat. (Courtesy of Masoneilan.)

Eccentric Rotating Plug

The "Camflex" valve by Masoneilan is a rotating plug valve that has a center of rotation eccentric to the centerline of the seat (Figure 11.20). When the plug rotates to close the valve port, the plug face moves into the seat with a cam-like motion. Design is such that little or no rubbing action occurs after contact is made between plug and seat, and the stem elastically deforms to give a tight shutoff. Some valve designs permit installation of reduced-trim seats without replacement of the plug. Valve flow characteristics are between equal percentage and linear, but are nearly linear.

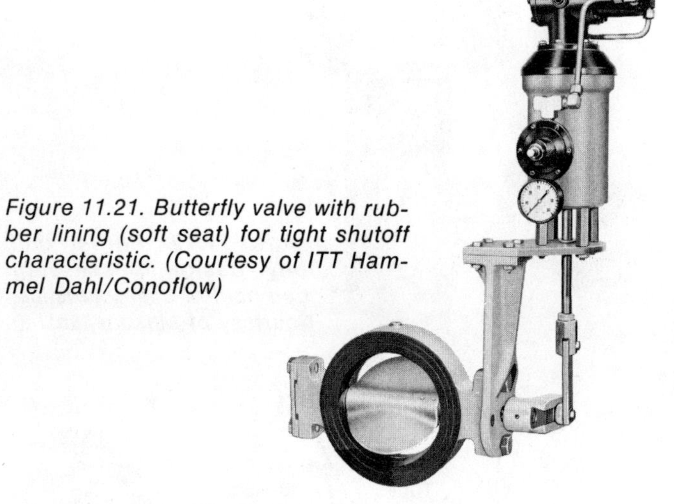

Figure 11.21. Butterfly valve with rubber lining (soft seat) for tight shutoff characteristic. (Courtesy of ITT Hammel Dahl/Conoflow)

Butterfly

A butterfly valve consists of a shaft supported vane or disc capable of rotating within a cylindrical body. In early industry use, butterfly valves were specified primarily for low pressure drop applications at low static pressures, where control was not critical and where high leakage rates could be tolerated. In the last few years, butterfly designs have been upgraded for high pressure drops, high static pressures and tight shutoff. Tight shutoff is accomplished through use of soft composition seats for seating the metal vanes (Figure 11.21).

Butterfly valves are economical, especially in larger sizes, because of their simple design and high capacity. They require a minimum space for installation and often reduce pumping costs because of their low pressure drop characteristic.

One of the disadvantages of the butterfly valve is the high operating torque requirement due to fluid flow through the valve (Figure 11.22). The Continental Division of Fisher Controls now markets a fishtail design that offers lower torque and improved flow characteristics (Figure 11.23). Butterfly valves commonly have been used for throttling control between 10° and 60° openings because torque con-

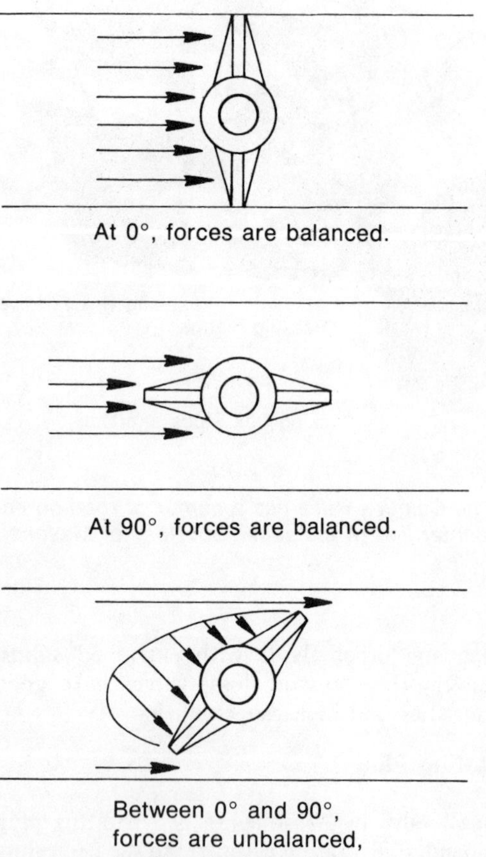

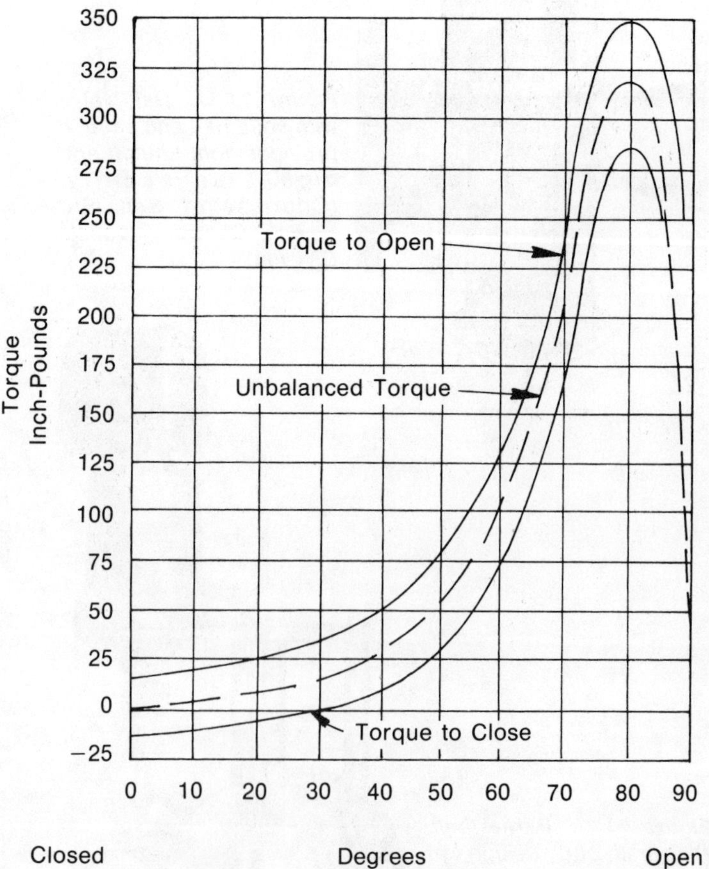

Figure 11.22. Force is plotted against valve opening to show the torque characteristics from closed to open to closed for butterfly valve.

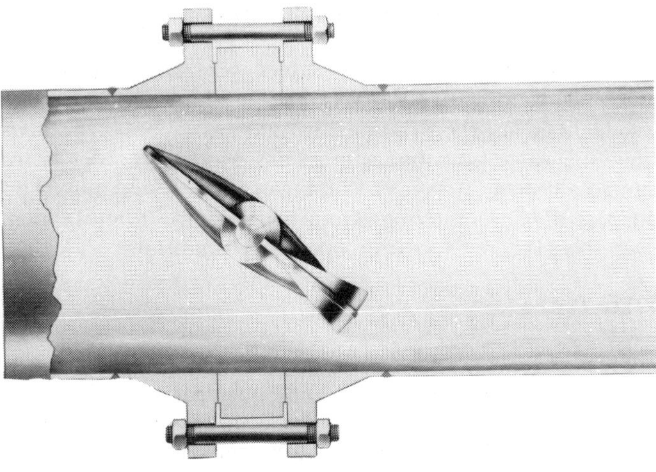

Figure 11.23. Butterfly valve with fishtail design disc which reduces operating torque and provides improved stability. (Courtesy of Fisher Controls Co.)

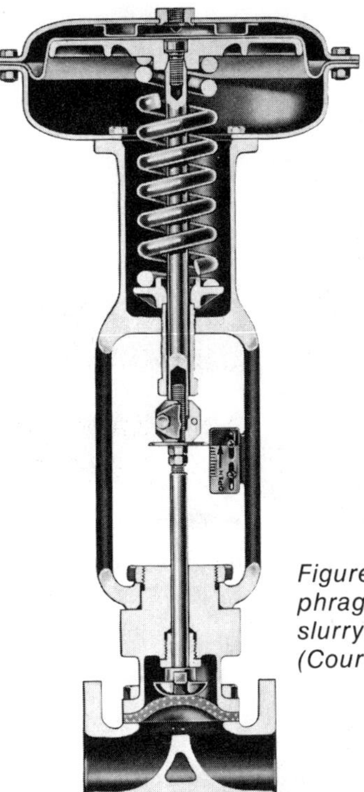

Figure 11.24. Saunders-type (diaphragm) valve bodies are used in slurry and highly viscous services. (Courtesy of Fisher Controls Co.)

ditions cause instability beyond this range. In its improved design, Fisher added a tail to the trailing edge of the disc, tending to retard the flow at low angles of opening and controlling the flow at high angles where the normal disc edge has been shadowed by the hub. As a result, the fishtail disc exhibits equal percentage characteristics from a low angle through a 90° opening. This design gives a turndown ratio as high as 100:1, allows high pressure drops with a reduction in tendency to cavitate and reduces the dynamic torque.

Several manufacturers sell butterfly valves which cam into place similar to the Camflex valve. Tight shutoff is available through some designs rated up to 720 psi, and temperature ratings are approximately −100°F to 450°F.

Diaphragm

The diaphragm valve consists basically of a body, bonnet, and flexible diaphragm (Figure 11.24). It is more often referred to as a Saunders-Type valve. Closure is made by forcing a flexible dome-like diaphragm against a weir.

Well-suited for slurries and viscous fluids, the diaphragm valve has high capacity; its cost is relatively low. The diaphragm seals the working parts of the valve from the process fluid and is the only wearing part of the valve. The Saunders-Type valve exhibits relatively poor control characteristics and has a low turndown ratio.

Pinch

Pinch valves are designed for slurries including metallic ores, fibers, sand, coal, sugar, pulp and paper stock and chemicals. They are made of a sleeve molded of rubber or synthetic material, with flanged or clamp ends for pipe connections and with a pinching mechanism for control (Figure 11.25). Control is accomplished by air or hydraulic pressure applied to the sleeve for closure.

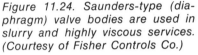

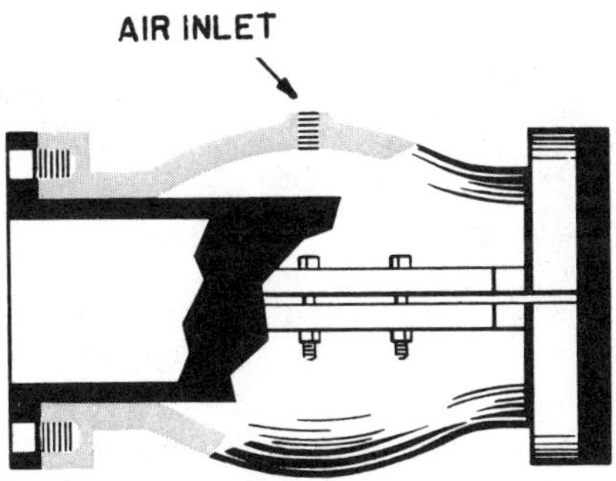

Figure 11.25. Pinch valves are used for heavy slurry services including metallic ores, coal and paper stock. (Courtesy of Red Valve Company, Inc.)

These inexpensive valves have high capacities and inherent self-cleaning action, but they have poor control characteristics and low rangeabilities.

Drag

The drag valve, a unique concept in control valve design by Control Components, Inc. (Figure 11.26), utilizes a patented multiple, disc technique (Figure 11.27) which divides the incoming flow into a series of smaller streams with tortuous flow paths. The paths are engineered to maintain the fluid velocity through the valve at or near line velocity. Control is obtained by positioning the plug (Figure 11.28) inside the stack of discs to change the flow area.

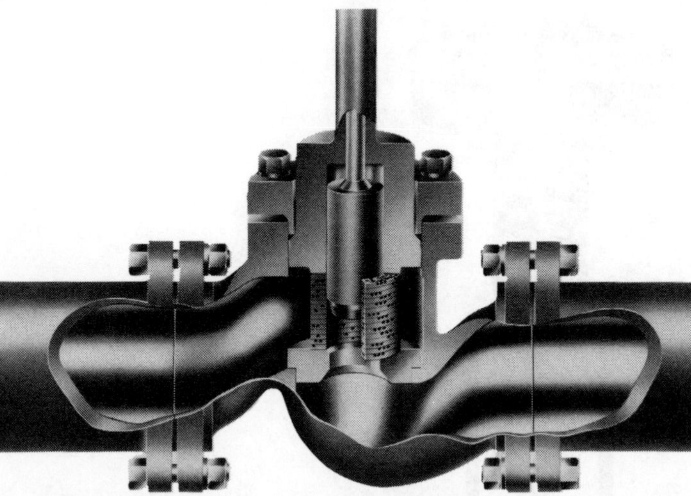

Figure 11.26. The drag valve uses a multiple disc cage trim that provides numerous fluid flow paths through the valve. (Courtesy of Control Components, Inc.)

Figure 11.27. Flow patterns and velocities are engineered by stacking grooved discs in selected arrangement. (Courtesy of Control Components, Inc.)

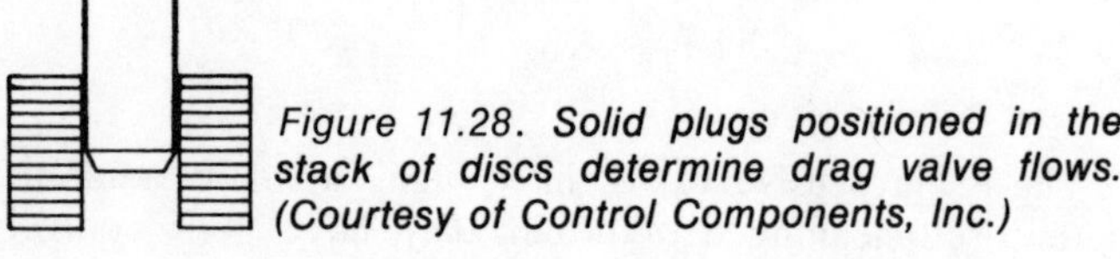

Figure 11.28. Solid plugs positioned in the stack of discs determine drag valve flows. (Courtesy of Control Components, Inc.)

The valve was developed for difficult control applications—high pressure drops, high temperatures and pressures, flashing services and erosive applications. Although it probably was not designed with this specifically in mind, a significant advantage it offers is the reduction of noise inherent in its design. It is more expensive than a standard valve.

Flow Characteristics

The flow characteristics of valves were discussed briefly under the heading of "Valve Capacity." Flow characteristic was defined as the relationship that exists between valve flow and valve position. Almost any kind of characteristic can be obtained by proper shaping of the seat and plug. The purpose of characterizing is to provide control loop stability over the expected range of operating conditions.

Trim Types

Flow characteristics fall into three major types (Figure 11.29): quick opening, linear and equal percentage.

Many variations of these types occur because of inherent valve design or because changes are engineered into the plug and seat design. The three major types are discussed below as well as some of their modifications.

Quick Opening

A quick opening characteristic provides for a maximum change in flow rate at low stem travel while maintaining a linear relationship through most of the stem travel. In Figure 11.29 about 90 percent of valve capacity is obtained at 30 percent valve opening, and a straight line relationship exists to that point.

Quick opening valve plugs are used primarily for on-off service or in self-actuated control valves or in regulators. They are also suitable for systems with constant pressure drops where linear characteristics are needed.

Linear

A valve with a linear flow characteristic produces flow directly proportional to the valve lift. Fifty percent of valve lift produces 50% of valve flow, etc. This proportional relationship produces a constant slope so that each incremental change in valve plug position produces a like incremental change in valve flow if the pressure drop is constant. Linear valve plugs are commonly specified for liquid level control and for control applications requiring constant gain.

Equal Percentage

An equal percentage flow characteristic is one in which equal increments of stem travel produce equal percentage changes in existing flow. For example, when the flow is small, the change in flow (for an incremental change) is small; when the flow is large, the change in flow (for an incremental change) is large. The change is always proportional to the quantity flowing before the change. Equal percentage valve plugs are used on pressure control applications where only a small percentage of the system drop is available for the control valve.

Modified Parabolic

The modified parabolic flow characteristic curve falls between the linear and equal percentage types (Figure 11.30). This type of characteristic is used on applications

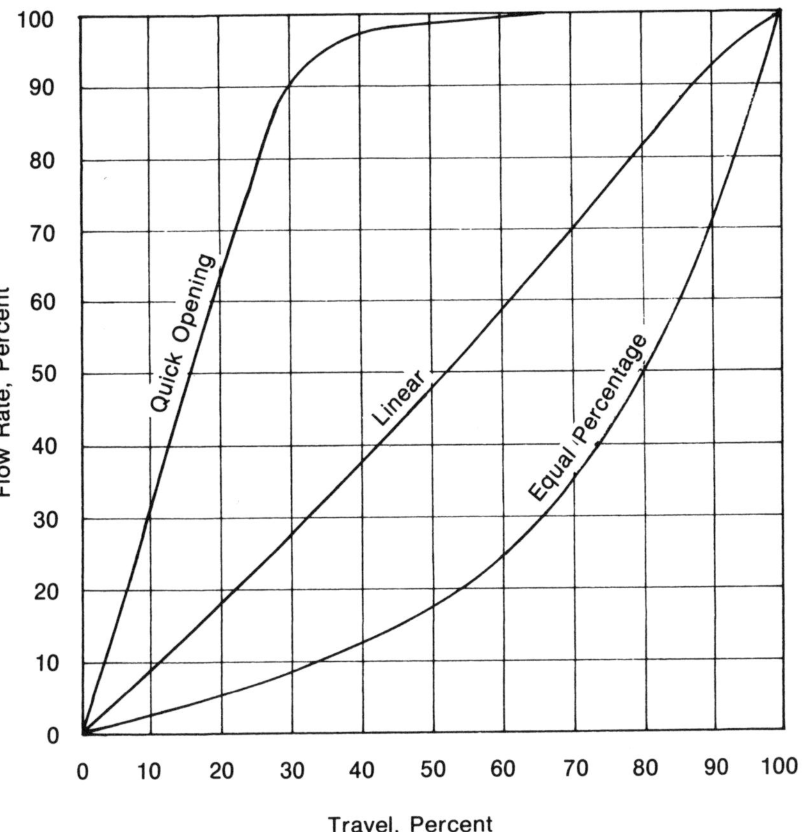

Figure 11.29. Three major types of flow characteristic curves.

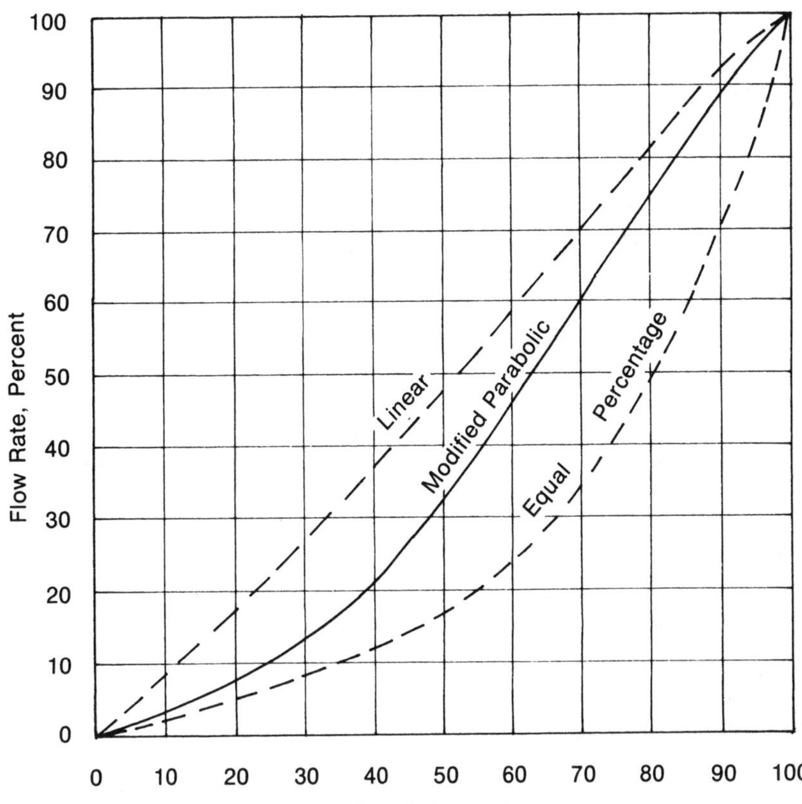

Figure 11.30. Modified parabolic flow characteristic curve falls between the linear and equal percentage types.

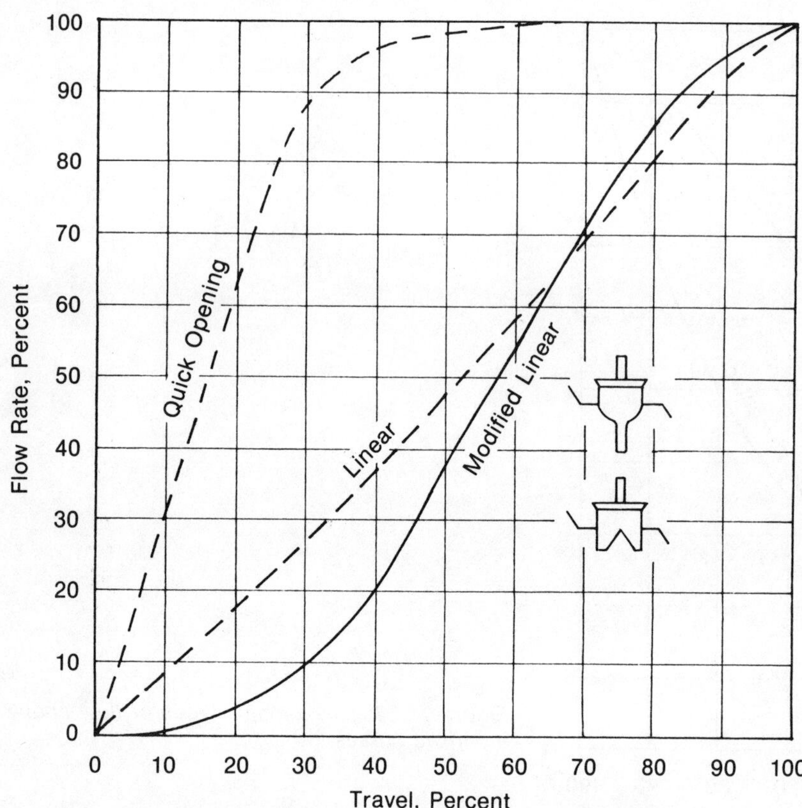

Figure 11.31. Modified linear characteristic curve falls between the linear and quick opening types.

where most of the system pressure drop is available at the control valve.

Modified Linear

The modified linear flow characteristic curve falls between the linear and the quick opening types (Figure 11.31). On the low and high flow rates, the valve sensitivity is low. Large plug motion changes produce small flow changes. This may work at a disadvantage on the high end where large motion changes also produce small flow changes.

Trim Design

The shaping of plugs and seats to obtain the desired flow characteristic would logically be a function of trim design. This section, however, covers other design concepts relating to valve trim that affect not only the characteristic curve but also how the valve responds to problems such as erosion, cavitation, vibration, high pressure drop, noise and other similar problems.

The term *trim* applies to the parts of a valve (except the body housing) that come into contact with the flowing fluid. Another term often used is *wetted parts*.

Material

Stainless steels 304 and 316 are widely used as seats, plugs, guides, bushings and other trim parts. Carbon steel is suitable also and is used in many applications. Bronze is used for water, air and steam services. Monel, Hastelloy and other alloys are used as required where corrosion, contamination or other considerations dictate.

Valve seats, plugs and guides often require harder materials because of erosive effects, flashing service, higher pressure drops or high temperatures. Stainless steels 17-4PH and 440C and K Monel are used frequently for these applications. Stellite No. 6, Colmonoy No. 6, chrome and tungsten carbides are used for small plugs, seats, inserts, guides and bushings or for facing larger areas in services that are both erosive and corrosive.

Plugs

Valve flow characteristics are determined primarily by valve plug shapes or patterns. Figure 11.32 shows some typical plugs for linear trim for both single port and double port valves. Figure 11.33 has typical shapes for equal percentage trim, and Figure 11.34 shows plugs for quick opening characteristic valves.

Seats

The seat or seat ring is that portion of the valve trim or body that the plug contacts for closure. The seat ring may be screwed or welded to the body (Figures 11.35 and 11.36).

Metal-to-metal contact between plugs and seats is standard practice. They can be machined accurately enough to

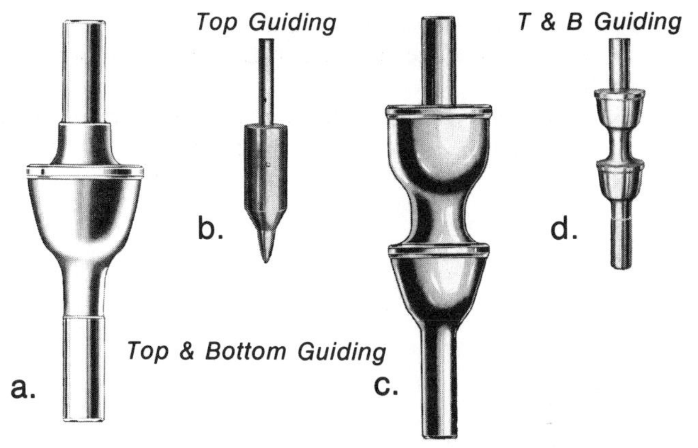

Figure 11.32. Typical examples of linear plugs, single and double port. The method of plug guiding is shown for each plug. (Courtesy of Fisher Controls Co. and Masoneilan)

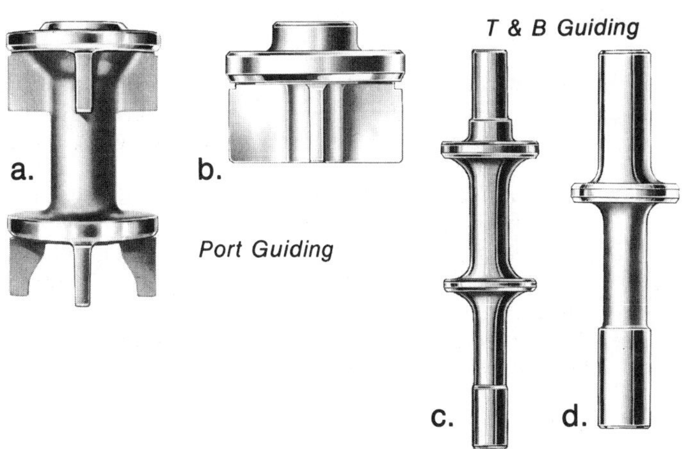

Figure 11.34. Typical quick opening plugs with type guiding shown. (Courtesy of Fisher Controls Co.)

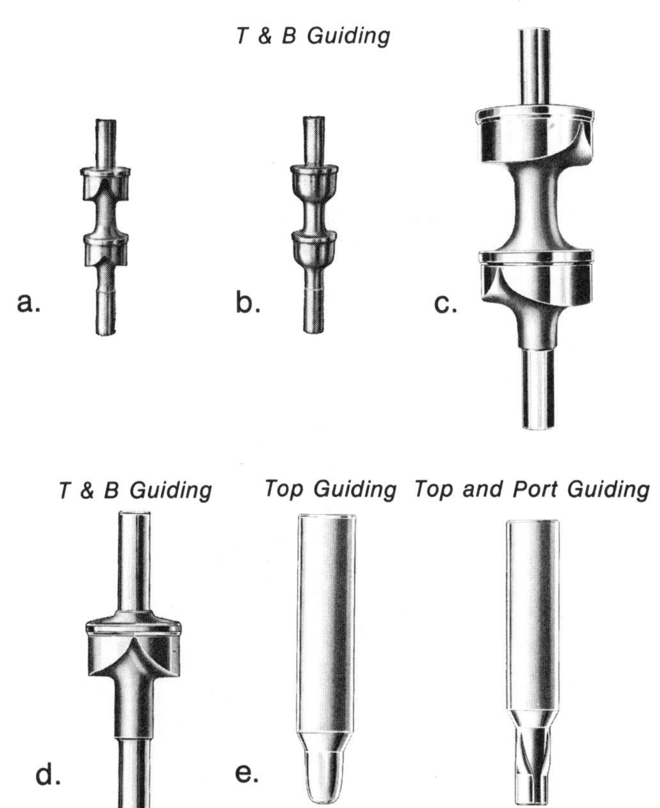

Figure 11.33. Typical examples of equal percentage plugs showing type of guiding. (Courtesy of Fisher Controls Co. and Masoneilan)

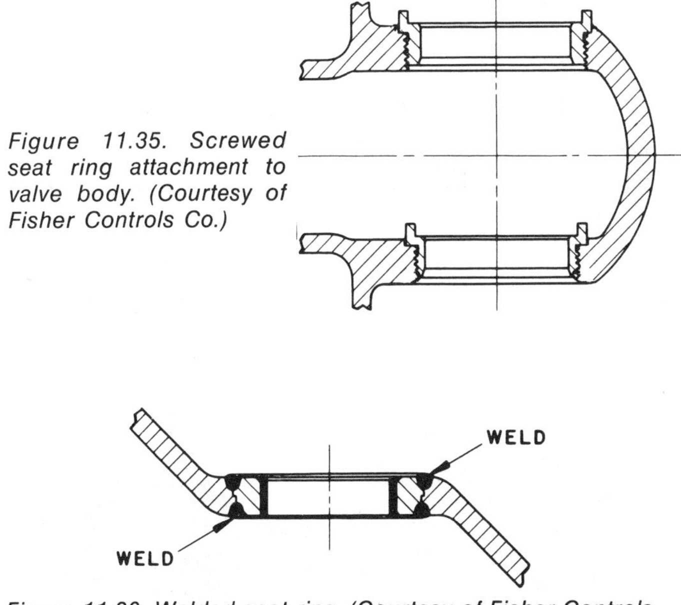

Figure 11.35. Screwed seat ring attachment to valve body. (Courtesy of Fisher Controls Co.)

Figure 11.36. Welded seat ring. (Courtesy of Fisher Controls Co.)

prevent high leakage rates. However, when tight shutoff is required, soft seats (Figure 11.37), made of Teflon, hard rubber or other resilient composition materials, are used to provide the necessary tight closure. The resilient part may be an insert in the seat.

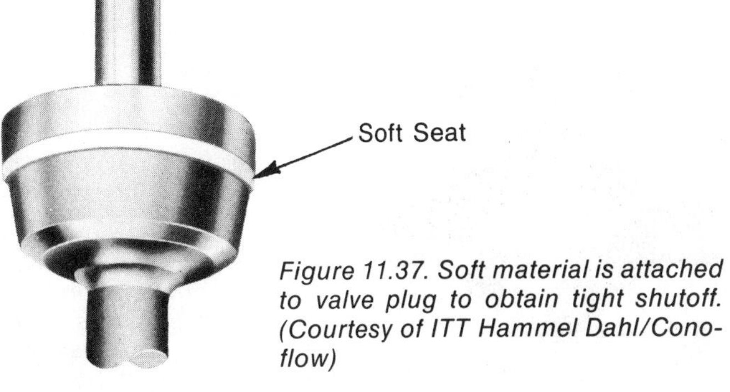

Figure 11.37. Soft material is attached to valve plug to obtain tight shutoff. (Courtesy of ITT Hammel Dahl/Conoflow)

Guides

Proper operation of control valves depends on a positive relation between the valve plug and seat. Accurate guiding of the valve plug assembly is accomplished in several ways.

1. *Top guiding* (Figure 11.38). The plug is aligned by a guide bushing in the bonnet or body.
2. *Top and bottom guiding* (Figure 11.39). The plug is aligned by guide bushings in the bonnet and bottom flange.
3. *Port guiding* (Figure 11.40). The plug is aligned in the body port or ports only.
4. *Top and port guiding* (Figure 11.41). The plug is aligned by a guide bushing in the bonnet and by the body port.
5. *Stem* (Figure 11.42). The plug is aligned by a guide bushing acting on the valve plug stem.

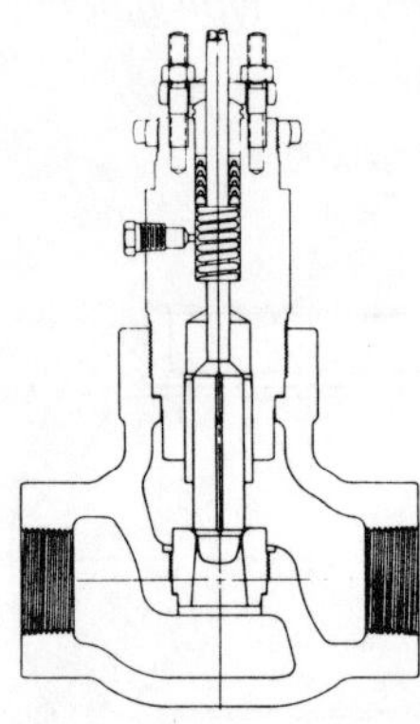

Figure 11.38. Top guided plug. Guiding of plug into valve seat is accomplished by a top guide only. (Courtesy of Fisher Controls Co.)

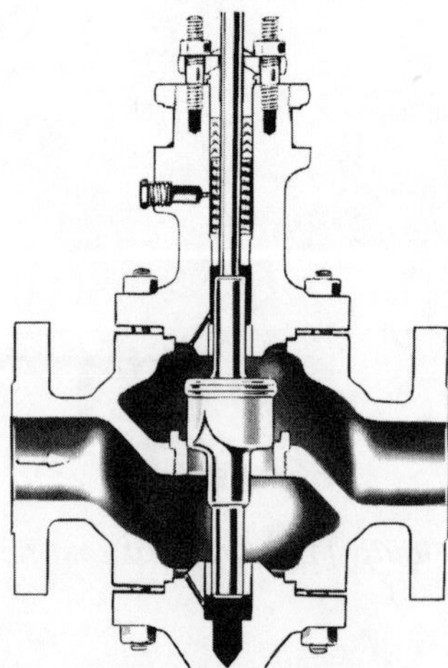

Figure 11.39. Solid sturdy support is available from guide bushings at both the top and bottom of the body in top and bottom guided plugs. (Courtesy of Fisher Controls Co.)

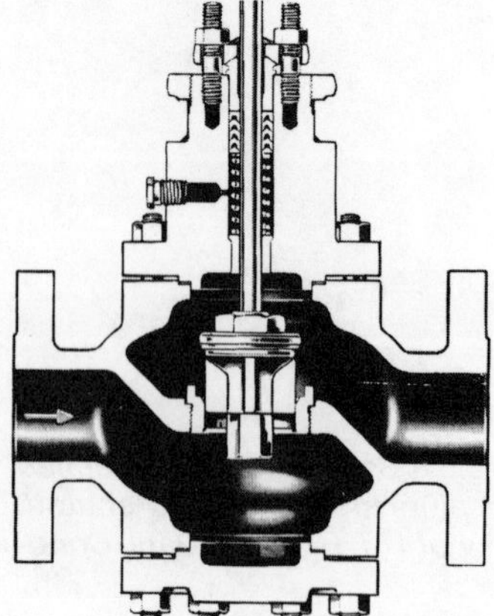

Figure 11.40. Port guided plugs rely primarily on the seat ring in the valve port for guiding. (Courtesy of Fisher Controls Co.)

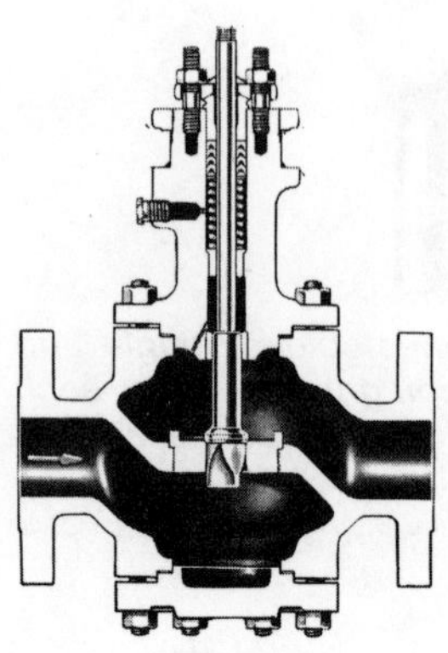

Figure 11.41. A bushing in the bonnet and the valve port are the supports in this top and port guided valve. (Courtesy of Fisher Controls Co.)

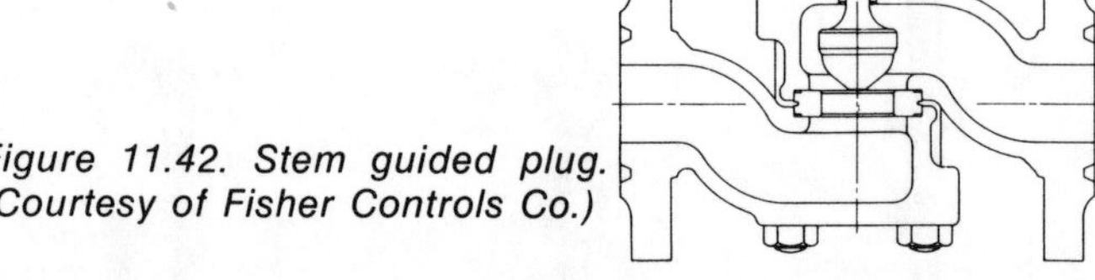

Figure 11.42. Stem guided plug. (Courtesy of Fisher Controls Co.)

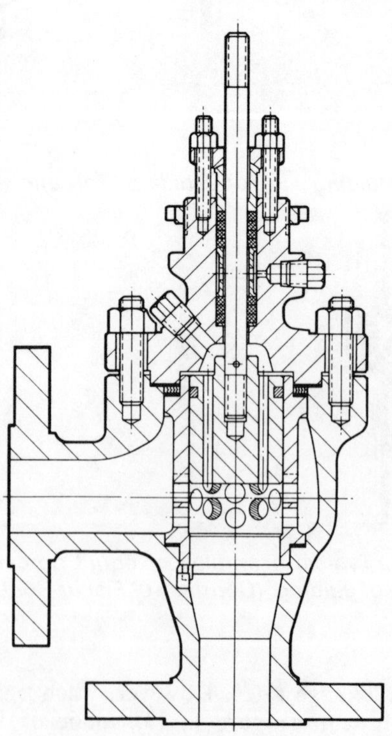

Figure 11.43. The cage in the valve body is ported to control flow as the plug position varies. It also guides the plug. (Courtesy of ITT Hammel Dahl/Conoflow)

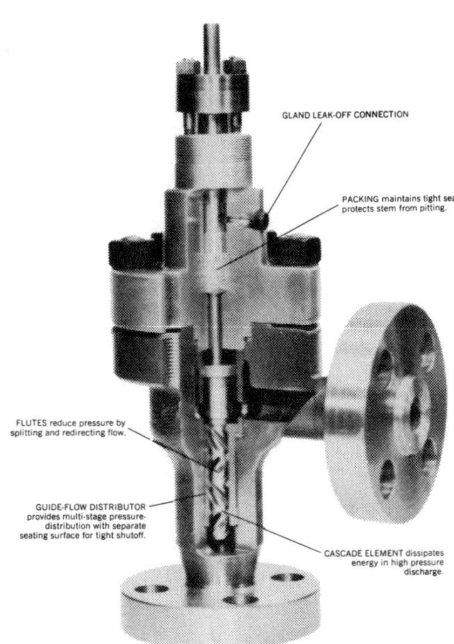
GLAND LEAK-OFF CONNECTION
PACKING maintains tight seal, protects stem from pitting.
FLUTES reduce pressure by splitting and redirecting flow.
GUIDE-FLOW DISTRIBUTOR provides multi-stage pressure-distribution with separate seating surface for tight shutoff.
CASCADE ELEMENT dissipates energy in high pressure discharge.

Figure 11.44. Flow energy is dissipated in the turbo-cascade trim design shown. (Courtesy of Yarway Corp.)

Cage

To help solve problems associated with high pressure drops and their attendant velocity and noise problems, various designs of cage trim have been developed. Cage trim valves usually provide higher capacity for the same size valve body than other globe valves, making them economically attractive.

Cage trim design employs a cage which fits into the valve body and serves as a guide for the plug. Excellent guiding is inherent in its design. The valve plug may be balanced or unbalanced; balanced design allows high pressure drops with low unbalanced forces. The design of cage porting results in longer trim life under high pressure drops than standard ports and trim.

Cage trims may be ported for almost any flow characteristic. Reduced capacities are obtained by the port configuration or a reduced opening in the end of the cage.

ITT Hammel-Dahl uses a multiple hole porting technique (Figure 11.43) in its "Flash-Flo" design approach to convert upstream potential energy to heat through friction within the throat of the seat ring. This absorption of energy from the flowing medium helps prevent cavitation and erosion.

Yarway's Turbo-Cascade valve (Figure 11.44) and Masoneilan's 10,000 Series labyrinth, Lo-db trim valves (Figure 11.45) accomplish about the same result through the dissipation of mass flow energy within the valve itself. Fisher Controls' E-Body Series has similar characteristics.

Control Components, Inc. has produced a unique design in cage valves with its "drag" valve concept (Figure 11.46). It utilizes a multiple disc technique to divide the incoming flow stream into a series of smaller streams (Figure 11.47). Each small flow stream is then directed through individual passages, making a series of right angle turns to control both pressure and velocity. A predetermined (engineered) grouping of discs (with flow passages etched) are stacked and positioned around a valve plug. Flow is controlled by plug position as its movement opens or closes flow passages along its travel.

Mechanical Features

Discussions to this point on control valves have centered around features affecting flow control of fluids through the

Figure 11.45. Labyrinth lo-db cage trim is another energy absorbing trim for high pressure drop applications. (Courtesy of Masoneilan)

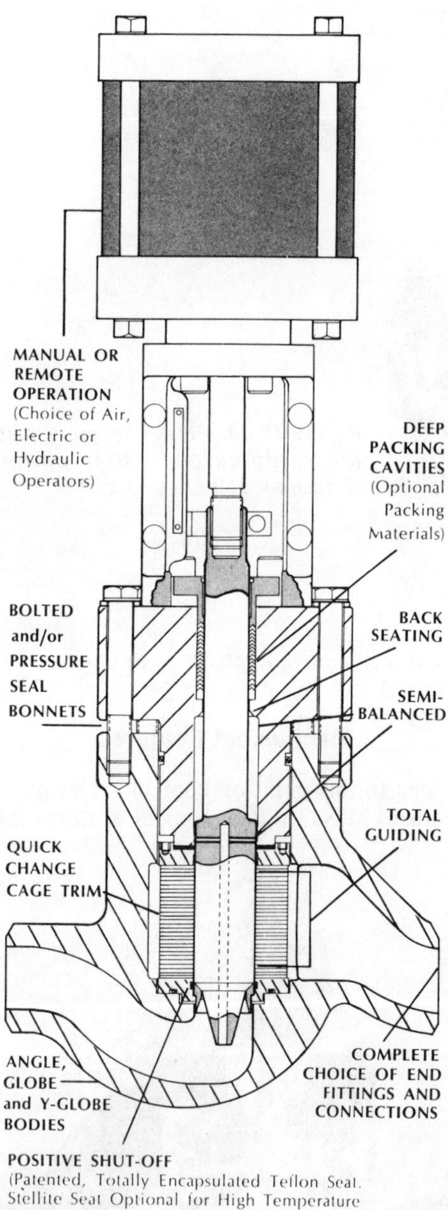

Figure 11.46. Drag valves with multiple plates to form trim use the cage trim and guide principle. (Courtesy of Control Components, Inc.)

valves. Mechanical aspects have been discussed only as they related to the control problem. Features are listed below which must be considered for proper mechanical functioning of the valve. Some of the features listed are not pertinent to all body designs discussed in the preceding part of the chapter.

End Connections

The most common methods of installing valve bodies in pipelines are screwed pipe threads, bolted flanges and weld ends (Figure 11.48). There are other methods among which are tubing fittings, Grayloc fittings and special high pressure types for ratings of 5,000 psi and above.

Tubing connections are sometimes used for sizes under 1 inch; screwed connections are most common from ½ through 1½ inches. Flanged and welded types are used extensively on 2-inch and larger lines. The piping specifications usually determine the type connection to be used.

Bonnet Assembly

A bonnet assembly is the part of the valve through which the valve plug stem moves and is the means for sealing against leakage along the stem. Figure 11.49 shows a typical bolted flange bonnet with a standard packing box. This type bonnet is good for temperatures from 0° to 450°F.

Extension Bonnet

In cryogenic or low temperature services, the packing box is kept at reasonable temperatures by extension bonnets (Figure 11.50). The bonnets may be cast or fabricated and are supplied in any required length. They may be insulated, if necessary.

Finned Bonnet

Many packing materials are limited to temperatures of 450°F or less. In services with higher temperature requirements, high temperature packings are used, or finned bonnets (Figure 11.51) are added to provide a large heat radiating area. This may increase the temperature range for standard packing by several hundred degrees.

Figure 11.47. Stacked discs of the drag valve removed from the valve body. (Courtesy of Control Components, Inc.)

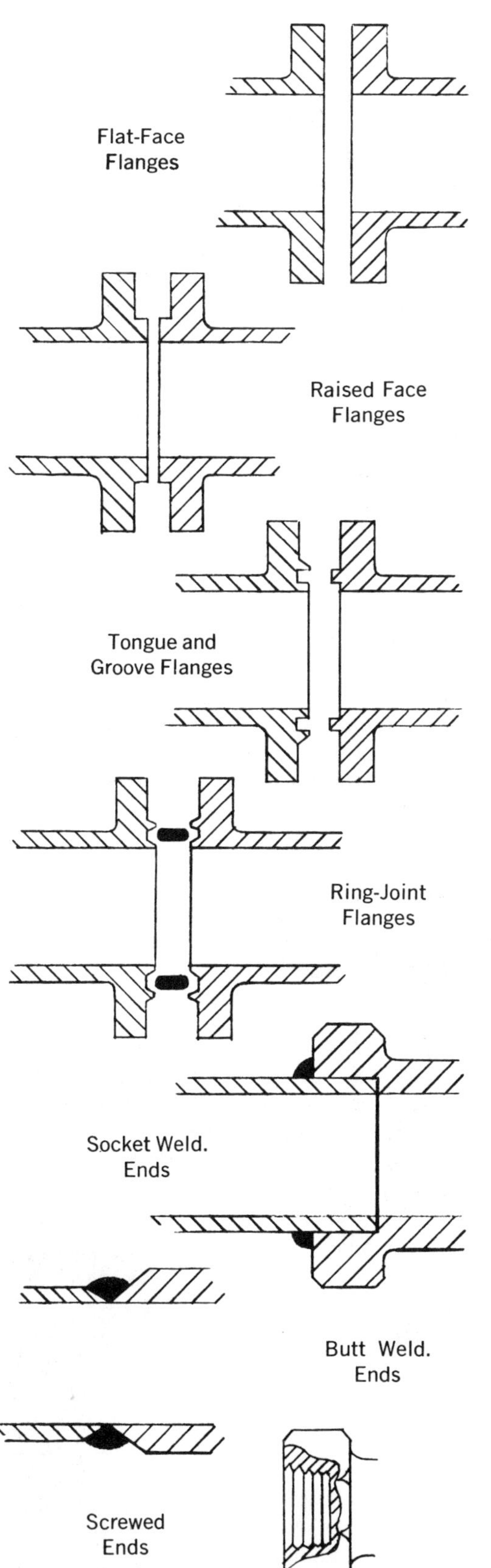

Figure 11.48. End connection methods used for installing valves in piping systems. (Courtesy of ITT Hammel Dahl/Conoflow)

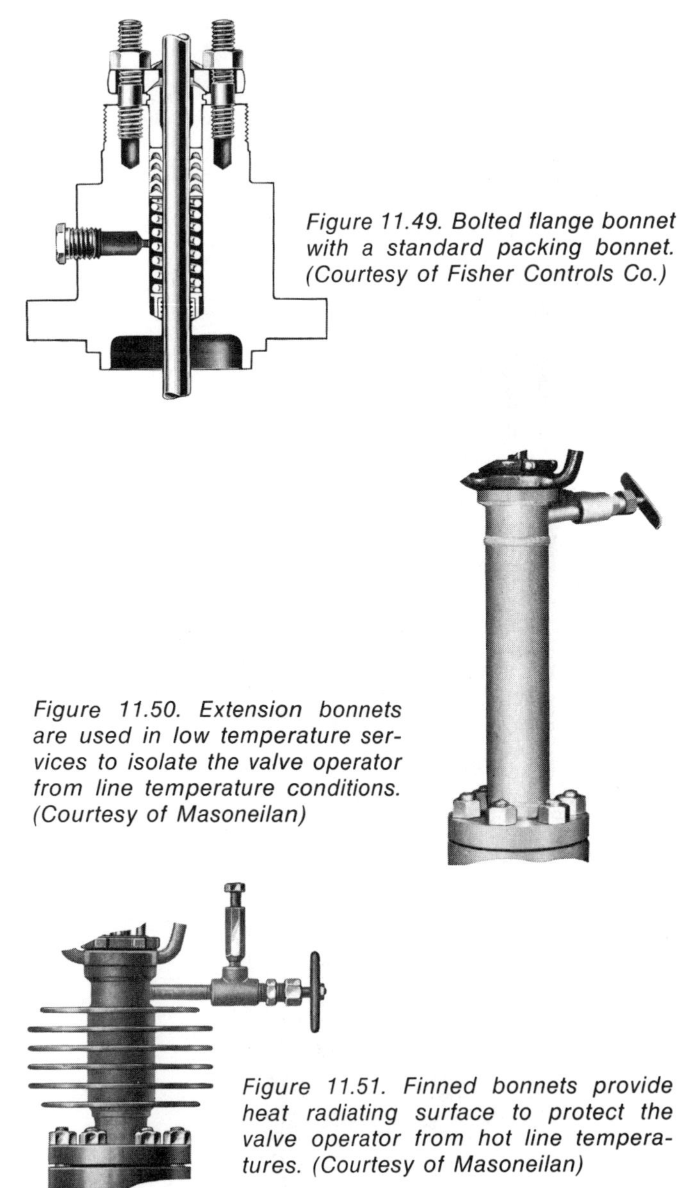

Figure 11.49. Bolted flange bonnet with a standard packing bonnet. (Courtesy of Fisher Controls Co.)

Figure 11.50. Extension bonnets are used in low temperature services to isolate the valve operator from line temperature conditions. (Courtesy of Masoneilan)

Figure 11.51. Finned bonnets provide heat radiating surface to protect the valve operator from hot line temperatures. (Courtesy of Masoneilan)

Bellows Seals

Figure 11.52 shows a bonnet assembly used on applications where no leakage along the stem can be tolerated. Bellows seals are used on installations where the process fluid is pyrophoric, toxic, explosive, highly expensive, or difficult to contain, such as hydrogen gas. Two types of packless sealing are made: one in which process fluid is sealed inside the bellows and one in which fluid is sealed outside the bellows.

Packing and Lubrication

A standard type of packing box, shown in Figure 11.53, includes a lubricator and isolating valve. Successful functioning of the control valve requires free movement of the

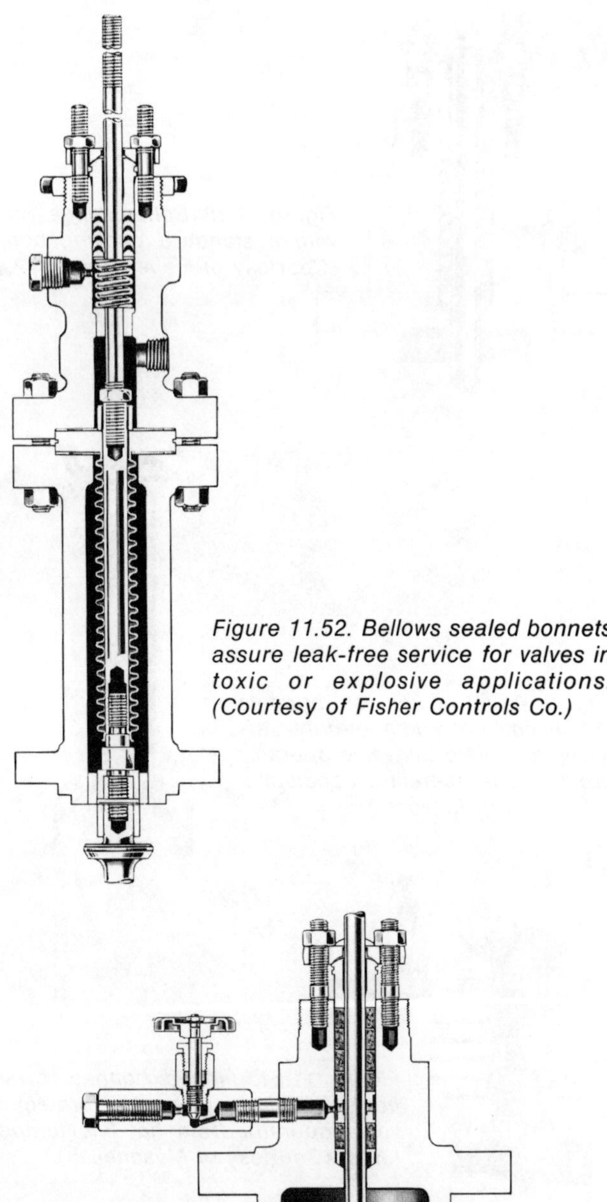

Figure 11.52. Bellows sealed bonnets assure leak-free service for valves in toxic or explosive applications. (Courtesy of Fisher Controls Co.)

Figure 11.53. Standard packing box with lubricator and isolating valve. (Courtesy of Fisher Controls Co.)

stem through the packing box with little friction; yet the packing must be tight enough to prevent objectionable leakage. Certain types of packing require lubrication. Valve lubricants are usually furnished as sticks which are placed in the lubricator and forced through the passage with the screw.

Standard packing materials include Teflon, Teflon-impregnated asbestos and graphited asbestos. For temperatures above 450°F, semimetallic packing is available. However, it causes considerable friction, and valve stems should be hardened or chrome plated to prevent excessive scoring and wear. Lubrication is required.

Actuators

Control valve actuators may be operated pneumatically, electrically, hydraulically, manually or by a combination of electrical, pneumatic and hydraulic forces. Pneumatic operation is the most widely used method. The forces actuators must overcome are the unbalanced force caused by the pressure drop across the valve, friction between and weight of moving parts and stem unbalance (insignificant except for high pressure drops). The discussion here is limited to actuators for modulating services and therefore excludes solenoid valves and other electrical and mechanical operators used primarily for on-off service.

Pneumatic Types

Pneumatic operators may be classified into two basic types—spring and diaphragm and springless (cylinder or piston). The spring and diaphragm is the most frequently used type.

Spring and Diaphragm

Spring and diaphragm operators may be *direct* acting or *reverse* acting. A direct acting operator (Figure 11.54) is designed so that air pressure (usually 3-15 psi) on the top of the diaphragm moves the stem downward, closing the valve. This action is termed *fail-open* (air-to-close). This force is opposed by compression of the spring, and loss of the operating medium (usually air) allows the compressed spring to open the valve.

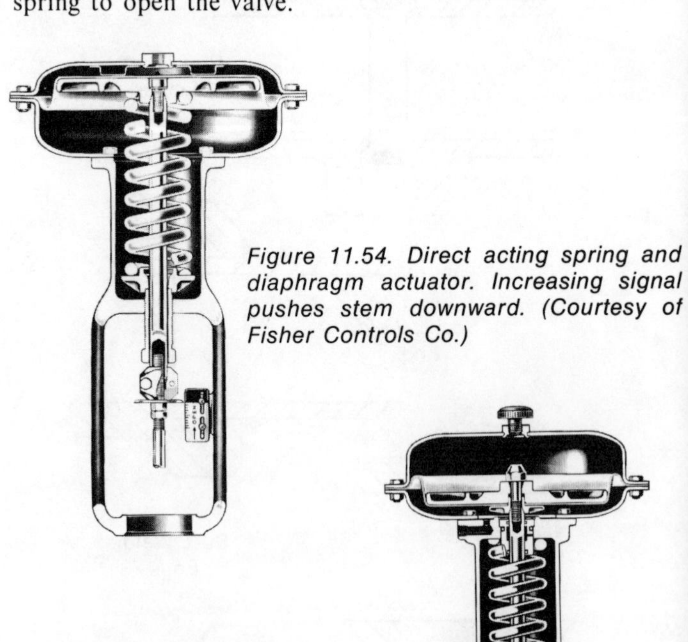

Figure 11.54. Direct acting spring and diaphragm actuator. Increasing signal pushes stem downward. (Courtesy of Fisher Controls Co.)

Figure 11.55. Reverse acting spring and diaphragm actuator. Increasing signal pulls stem upward. (Courtesy of Fisher Controls Co.)

On a reverse acting diaphragm actuator, air pressure below the diaphragm moves the stem up, opposing the spring action (Figure 11.55). This action is termed *fail-closed* (air-to-open). On air failure the spring closes the valve. Diaphragms shown in Figure 11.54 and 11.55 are not reversible. Some designs are made so that the diaphragm can be reversed to obtain the desired action (Figure 11.56).

Springless

Springless operators include Saunders-Type valves and pinch valves whose closures are obtained by compressing elastic diaphragms. However, this section is intended to cover pneumatic cylinder or piston operators. Cylinder or piston operators are increasing in usage because of the need for increased power and fast action. Increased power results from their ability to use higher pressure supply air. These operators sometimes include built-in valve positioners.

Figure 11.57 shows how the ITT Hammel Dahl/Conoflow Cylinder Conomotor operates. It derives its power from differential pressure across the piston. The piston is forced upward by a constant pressure from a reducing regulator, adjustable to suit the stem load. The chamber above the piston is dynamically loaded through a positioner which operates on the force-balance principle. In the top loading, direct action positioner form (illustrated), an increase in instrument air pressure increases the pressure in the chamber above the piston, moving it downward. This extends the range spring until the positioner forces are brought back into balance, at which point the positioner stabilizes the pressure on top of the piston to hold the new position. A decrease in instrument air pressure reverses the procedure. Higher supply pressures provide greater power and faster stroking speeds.

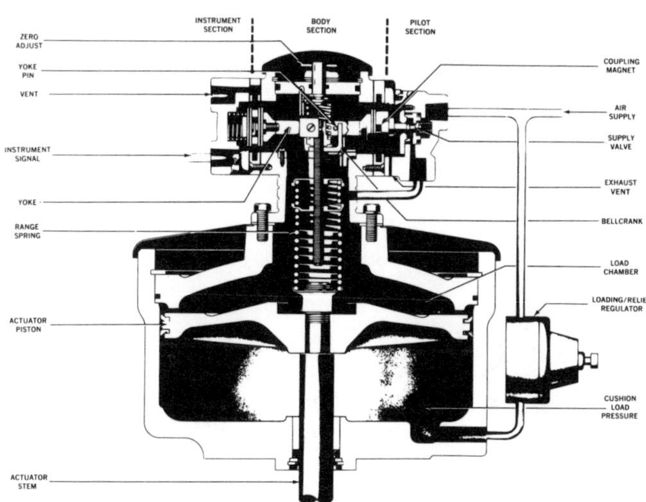

Figure 11.57. Pneumatic cylinder operator. The piston is loaded through the top mounted positioner which operates on the force balance principle. (Courtesy of ITT Hammel Dahl/Conoflow)

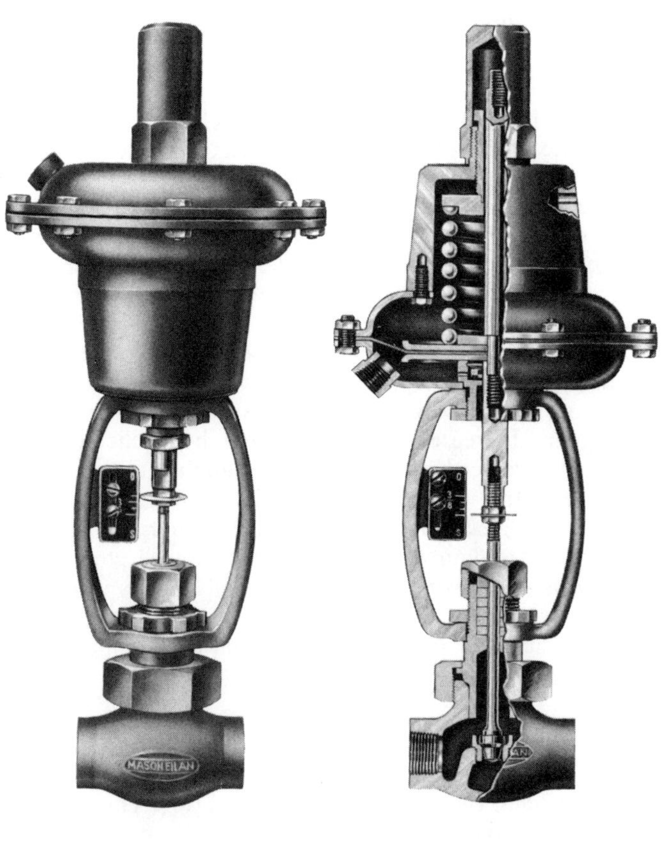

Figure 11.56. Reversible diaphragm actuator. At left, air to top of diaphragm opens valve. When the diaphragm is reversed, applied air opens the valve. (Courtesy of Masoneilan)

To provide fail-closed or fail-open modes, cylinder operators can be furnished with spring-return features. For fail-safe operation on electric power loss and for air supply loss, bottled gas with appropriate regulators and trip valves is sometimes employed. Valves are sometimes required to maintain the position they were in when supply pressure or signal is lost. Such a state is known as "fail-last position," and can be accomplished by trapping the last signal pressures within the cylinder or piston assembly.

Electric Types

Electric operators with proportional or infinite positioning control have limited use in the process industries. Their primary use has been in remote areas, such as tank farms and pipeline stations, where no convenient air supply is available. Slow operating speeds, maintenance problems in hazardous areas and economics have prevented wide acceptance for throttling applications. However, several companies have offered electrically powered units. Figure 11.58 shows an electrically operated butterfly valve which can be supplied with an automatic amplifier-relay control package for use with a remote command potentiometer. The remote potentiometer is part of a Wheatstone bridge arrangement with the feedback potentiometer in the actuator. Changes in the command potentiometer cause the actuator to reposition to a point where bridge balance is reestablished.

Figure 11.58. Butterfly valve with electric operator for remote infinite positioning control. (Courtesy of Hills-McCanna)

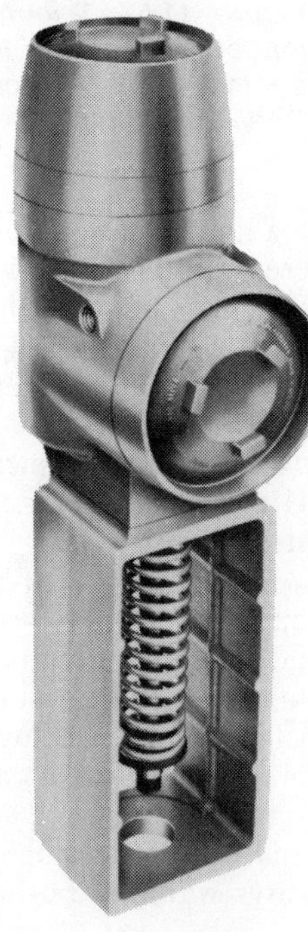

Figure 11.59. Electrohydraulic valve actuator with integral hydraulic unit self-contained. (Courtesy of ITT General Controls)

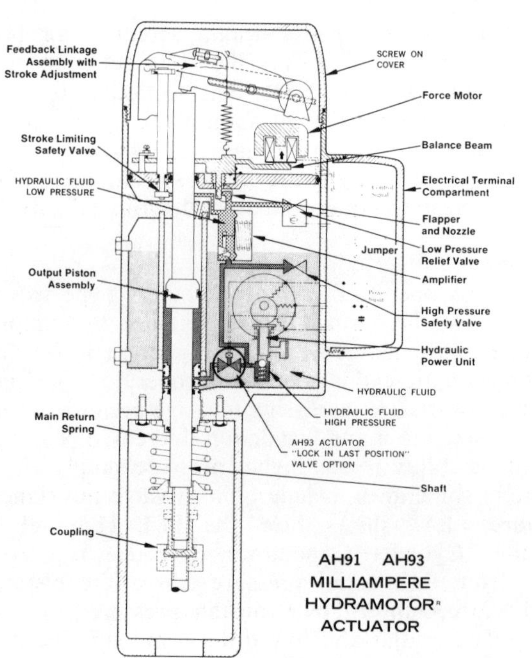

Figure 11.60. Schematic diagram of the electrohydraulic valve actuator unit shown in Figure 11.58. (Courtesy of ITT General Controls)

Electrohydraulic Types

Two approaches have been made in designing electrohydraulic actuators: the hydraulic unit as an integral self-contained system with the actuator and the hydraulic unit as a separate system, designed for one or perhaps several control valves.

Integral Unit

For several years valve manufacturers and users have recognized the need for a dependable control valve that would operate directly from the low power electronic signals of electronic controllers. ITT General Controls has introduced their Milliamp Hydramotor Actuator (Figures 11.59 and 11.60), which features a self-contained, sealed hydraulic power unit with stem thrusts to 3,000 pounds. It has a stroking speed as high as ⅜ inch per second.

The hydramotor unit is a force balance system composed of a balance beam, a flapper-nozzle, a force motor and a feedback loop (Figure 11.60). When the force motor coil is energized by a standard (4-20 or 10-50 ma DC) signal, an upward force is exerted on the force motor coil proportional to the input current. This force is transmitted to the balance beam, causing the flapper-nozzle to restrict oil flow through the nozzle and build up the pressure in the hydraulic amplifier. Increased pressure in the hydraulic amplifier results in greater hydraulic pressure supplied to the cylinder and the output piston assembly lifting the actuator.

The feedback linkage assembly, through its feedback spring, exerts an upward force on the balance beam, moving it against the pull of the force motor coil. The flapper-nozzle relationship adjusts to bring the system into equilibrium with the actuator position proportional to the input control signal.

Separate Hydraulic Unit

An electrohydraulic actuator offered by Masoneilan International, Inc., is shown in Figure 11.61. This unit consists of a power cylinder, filter, servovalve and sensing element. The servovalve and its manifold mount directly on the power cylinder. A full flow filter is installed just ahead of the servovalve. A valve stem position sensing element is also provided to give a feedback signal proportional to the plug position. This valve-mounted unit is used in conjunc-

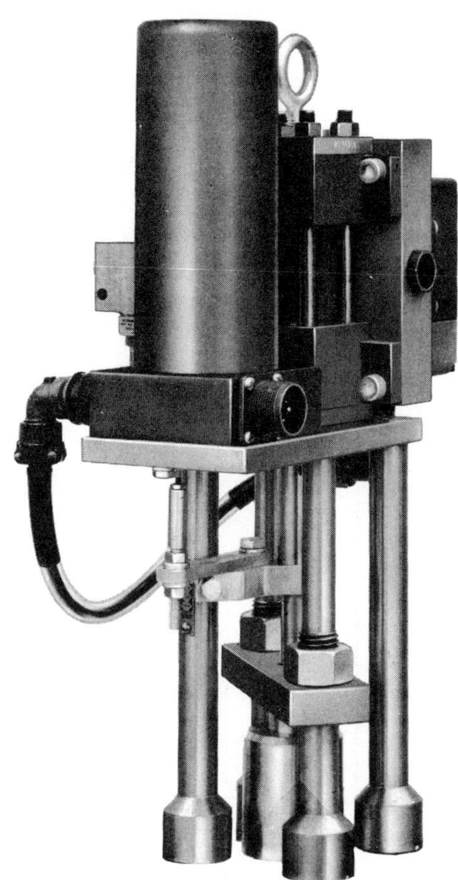

Figure 11.61. Electrohydraulic operator meets high thrust requirements and operates at frequencies as high as 20 Hertz. (Courtesy of Masoneilan International, Inc.)

tion with a standard remote controller and receives its power from an accessory hydraulic system which may or may not be serving other valve actuators.

This actuator system is capable of operating frequencies as high as 20 cycles per second, attaining maximum stem velocities up to 50 inches per second.

The hydraulic system used with the actuator is a packaged unit consisting of a reservoir, motor, pump and accumulator. The hydraulic piping system must be clean, not containing particles larger than 10 microns in size. It operates at pressures up to 3,000 psi, providing power and speed for exacting control requirements.

Positioners

A valve positioner is basically a relay that senses both an instrument signal and a valve stem position. The primary function is to ensure that the valve plug position is always directly proportional to its controller output pressure. For example, if the positioner receives a 35% signal, it will supply sufficient pressure to the actuator to cause it to stroke 35% of its travel. Effectively it may be described as a closed-loop controller that has the instrument signal as input, an output to the valve diaphragm and feedback from the valve stem.

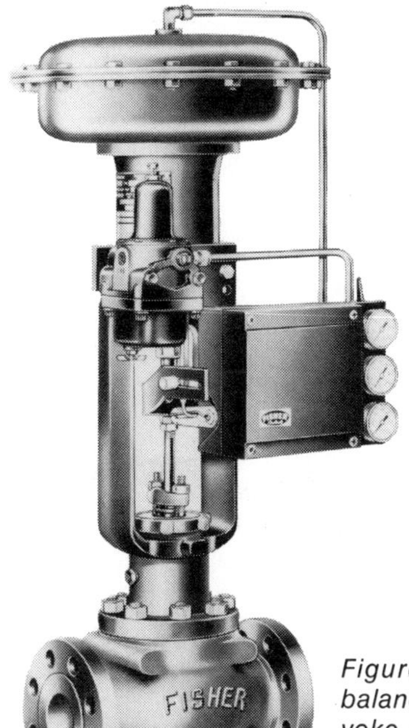

Figure 11.62. Pneumatic force balance positioner is mounted on yoke of valve topworks. (Courtesy of Fisher Controls Co.)

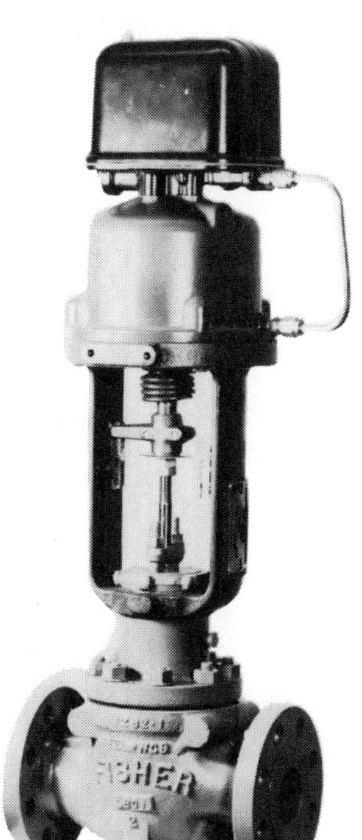

Figure 11.63. Pneumatic positioner mounted on top of piston actuator. (Courtesy of Fisher Controls Co.)

It is used to split range valves, to reverse the signal to a valve, to overcome friction forces within a valve on high pressure drop applications and on applications requiring fast, accurate control. It is normally mounted on the control valve.

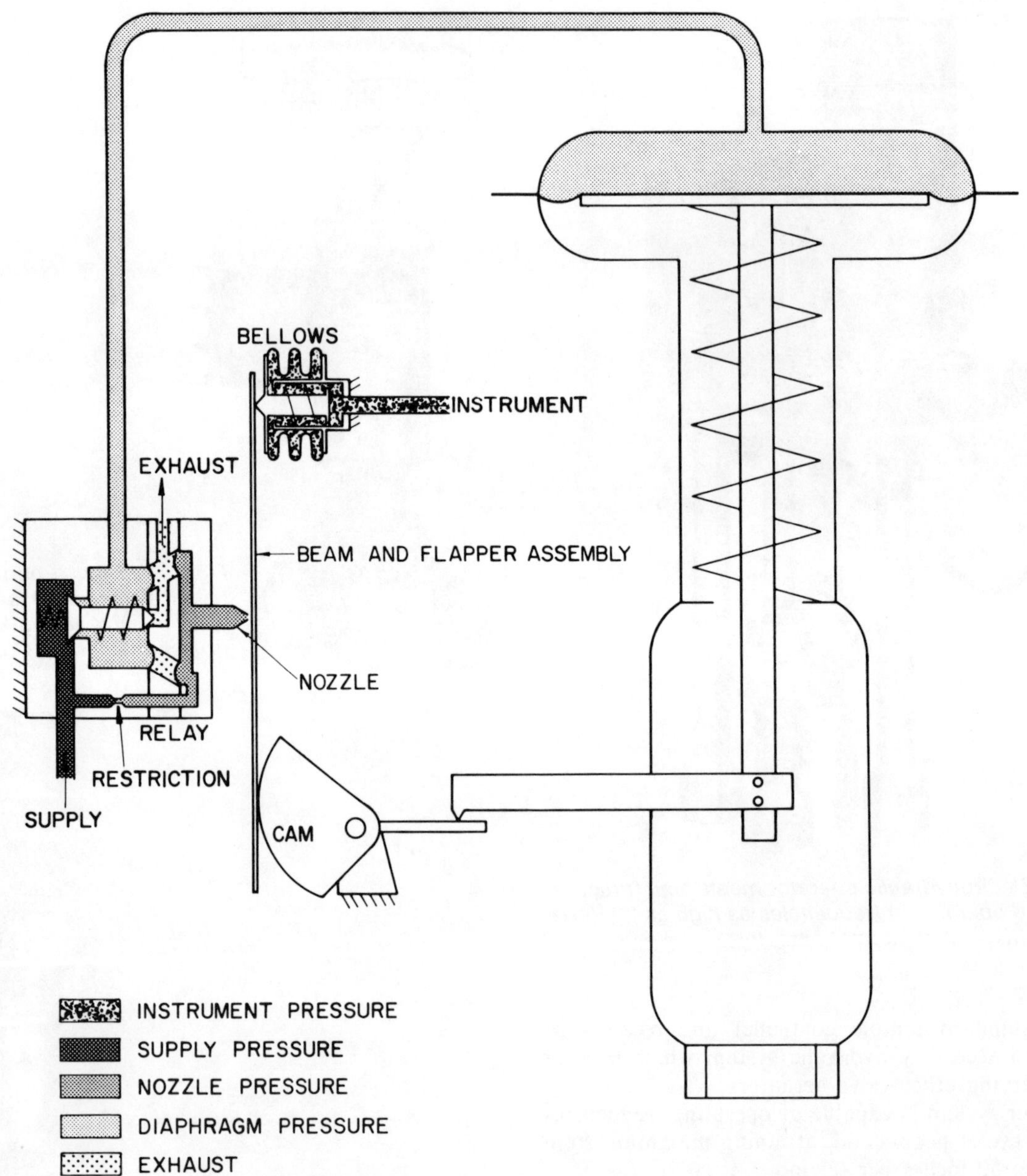

Figure 11.64. A schematic of the motion balance pneumatic positioner reveals how it operates. (Courtesy of Fisher Controls Co.)

Pneumatic

There are many valve positioners available, but two basic design approaches have been made: motion balance and force balance. Positioners usually are mounted on the side of diaphragm actuators (Figure 11.62) and on top of piston (Figure 11.63) and rotary actuators. They are mechanically connected to the valve stem or piston so that the stem piston can be compared with the piston dictated by the controller. Typical designs are shown and described.

Motion Balance.

Figure 11.64 shows a schematic of a motion balance positioner, and its physical appearance is seen in Figure 11.65. It consists basically of (a) a bellows to receive the instrument signal, (b) a beam fixed to the bellows at one end and, through linkage, to the valve stem at the other end, and (c) a relay whose nozzle forms a flapper-nozzle arrangement with the beam. As the bellows moves in response to a changed instrument signal, the flapper-nozzle arrangement moves, either admitting air to, or bleeding air from, the

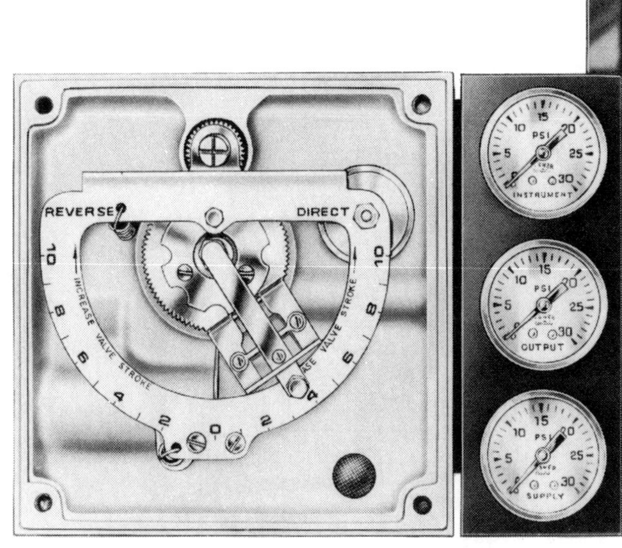

Figure 11.65. Motion balance pneumatic positioner with outside cover removed. (Courtesy of Fisher Controls Co.)

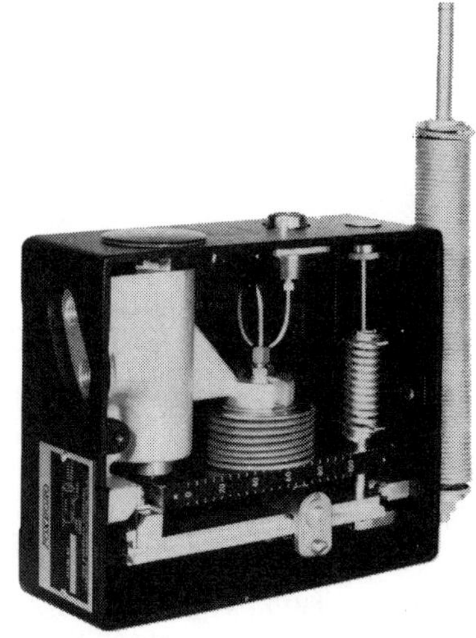

Figure 11.66. Force balance pneumatic positioner with cover removed. (Courtesy of Foxboro Co.)

diaphragm until the valve stem position corresponds to the instrument air signal. At this point the positioner is once again in equilibrium with the changed instrument signal.

Force Balance

A force balance positioner is shown in Figure 11.66. The output pressure from a controller is applied to the bellows (12) of the positioner. For any given controller output, there is a corresponding position of the controlled element (in this case, a cylinder). At the slightest change in signal pressure, full supply pressure is available to ensure immediate and accurate positioning of the cylinder.

To illustrate the positioner operation, when it is mounted for air-to-lower action (note piston in Figure 11.66) and is at a null position (mechanical and penumatic balance), an increase in signal pressure will expand the bellows (12), raise the bellows bar assembly (5) and allow the spring-loaded, pilot valve stem (8) to rise. High pressure air is passed through the pilot valve to the top chamber of the cylinder. Exhaust air from the lower cylinder chamber also passes through the pilot valve to the exhaust port. Because the piston rod is attached to the positioner through an actuating arm (30) and spring (18) arrangement, the rod's downward movement is transmitted to the lower bar assembly (6). The assembly pivots about two flexure strips (7) and moves in the same direction as the piston rod. The movement of the slide carriage assembly (4) opposes the movement of the bellows bar assembly (5) so that when the piston attains its proper position, the valve stem (8) is returned to the null position.

Built-in

Both positioners described in the preceding paragraphs are normally separately mounted on valve actuators. Several designs are available in which the positioner is an integral part of the actuator. Figure 11.67 shows a built-in force balance positioner for a cylinder actuator. It has a common air supply pressure piped to the top mounted loading regulator and positioner. The integral reducing-relief loading regulator provides a constant load on top of the piston. For most applications, the loading pressure is set at about one-half the supply pressure. The difference between positioner output pressure under the piston and the load above is the actuating force.

The actuator is supplied with a direct acting positioner (increasing signal retracts stem). The controller signal pressure acts within the positioner double diaphragm assembly. The area of the upper diaphragm is smaller than the area of the lower diaphragm. An increasing controller signal causes the diaphragm assembly to move down. This movement causes the positioner pilot valve to close the exhaust port and open the supply port, increasing pressure under the piston and retracting the piston rod. The piston moves until the range spring force is balanced by the force of the diaphragm assembly. Positioner output then stabilizes to position the piston rod in a linear relationship to the control signal.

Electropneumatic

Use of electronic control loops with air-operated valves has led to the development of electropneumatic positioners, a combination of an electronic-to-air transducer and a positioner (Figures 11.68 and 11.69). This is a force balance device and is supplied as direct or reverse acting. With direct action (increasing electrical input signal increases the air output pressure), an increase of the input signal causes the coil to produce a force on the beam, moving the flapper to cover the nozzle. The increase in nozzle back pressure causes the relay plug to close the exhaust in the diaphragm block and open the inlet, increasing positioner output pressure to the control valve actuator. The resultant valve stem motion extends the spring (through linkage) until the

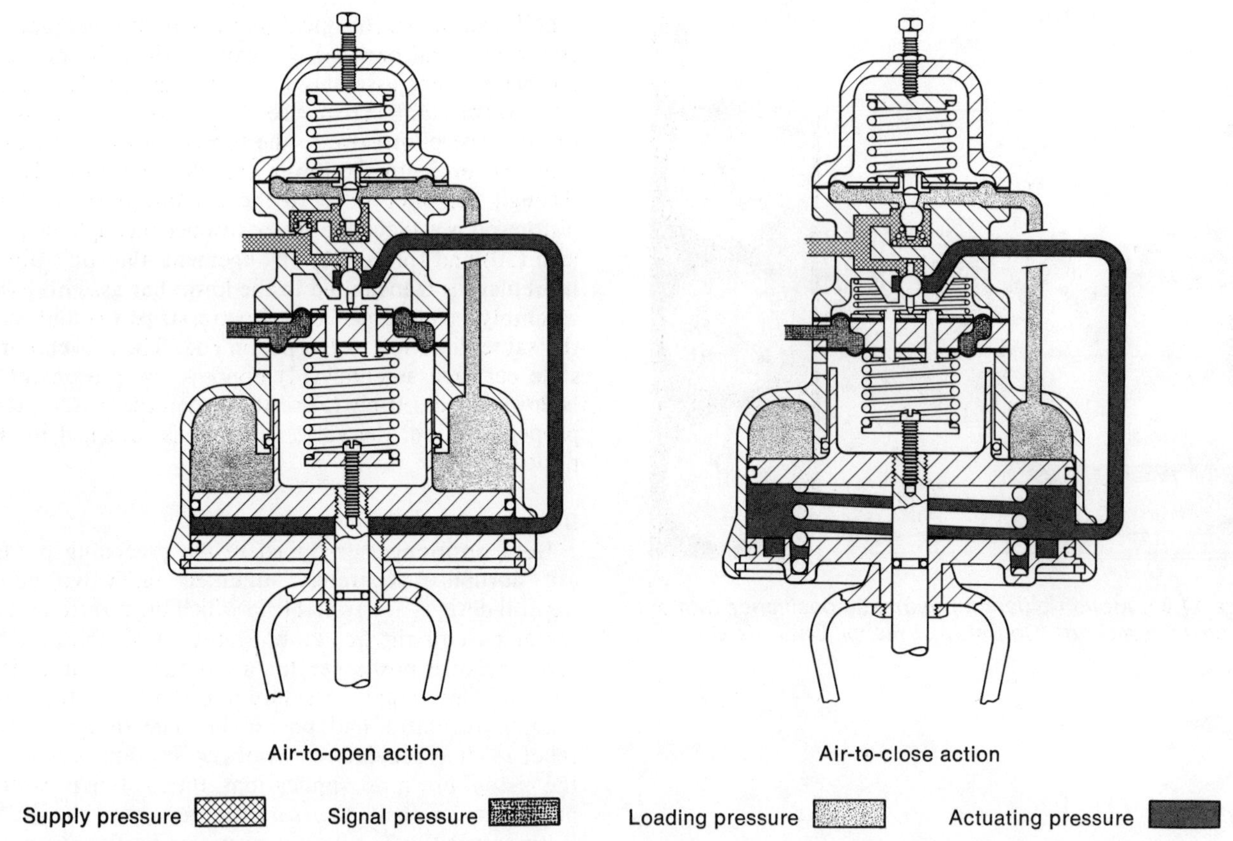

Figure 11.67. Built-in force balance positioner for a cylinder operator. (Courtesy of Masoneilan)

spring force is balanced by the coil force. As they equalize, the nozzle back pressure decreases, allowing the relay plug to close the inlet and open the exhaust. The system is in equilibrium, and the positioner output is stabilized at a value necessary to maintain the required plug position.

Positioner Features and Accessories

Stem travel (valve stroke) varies from a fraction of an inch to 3 inches and greater on control valves. Pneumatic positioners have mechanisms to adjust easily and quickly to the stroke length of the valve it serves (refer to Figure 11.65).

Range springs are changeable so that the positioner can be used for varying input ranges (3-15 psi or 6-30 psi). When split ranging is needed, smaller portions of the 12-psi range spread may be used for stroking the valve. For example, a controller operating two valves may close one as the signal changes from 3 to 8 psi and open the other from 9 to 15 psi. In this instance a dead band of 1 psi exists; both valves are closed from 8 to 9 psi.

Positioner actions are reversible to provide flexibility in choosing valve action; i.e., an increasing signal in controller output pressure may be reversed in the positioner to a decreasing signal output to the control valve.

Positioners may be supplied with guages and bypass cocks (Figure 11.70). The bypass cock in the "bypass" position sends the control signal directly to the valve; when the cock is turned to "positioner," the positioner is in service.

Three gauges may be provided: one to indicate the controller input signal, another to indicate the output signal to the valve and the third to indicate instrument air supply pressure.

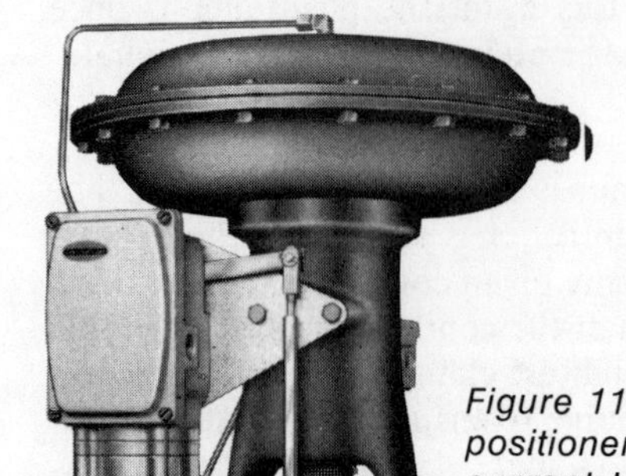

Figure 11.68. The electropneumatic positioner shown is a combination current-to-air transducer and a positioner. (Courtesy of Masoneilan)

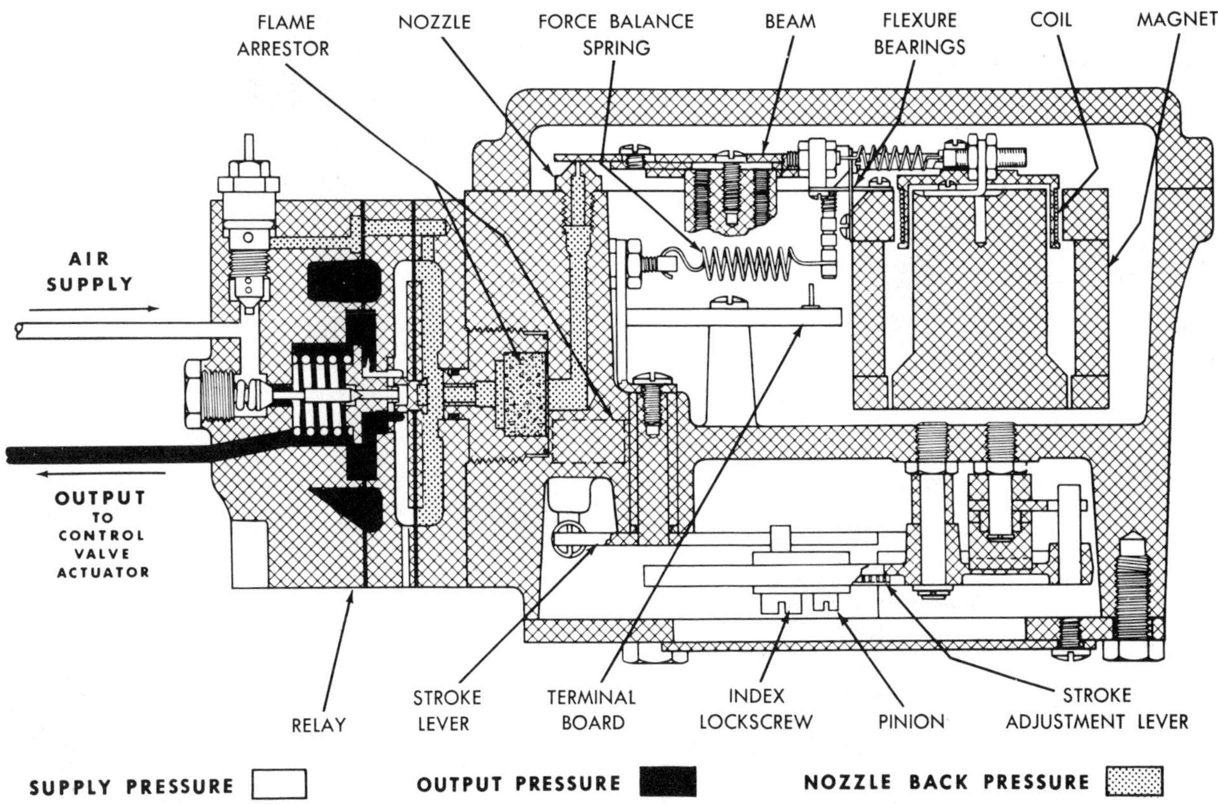

Figure 11.69. Electropneumatic positioner schematic diagram. (Courtesy of Masoneilan)

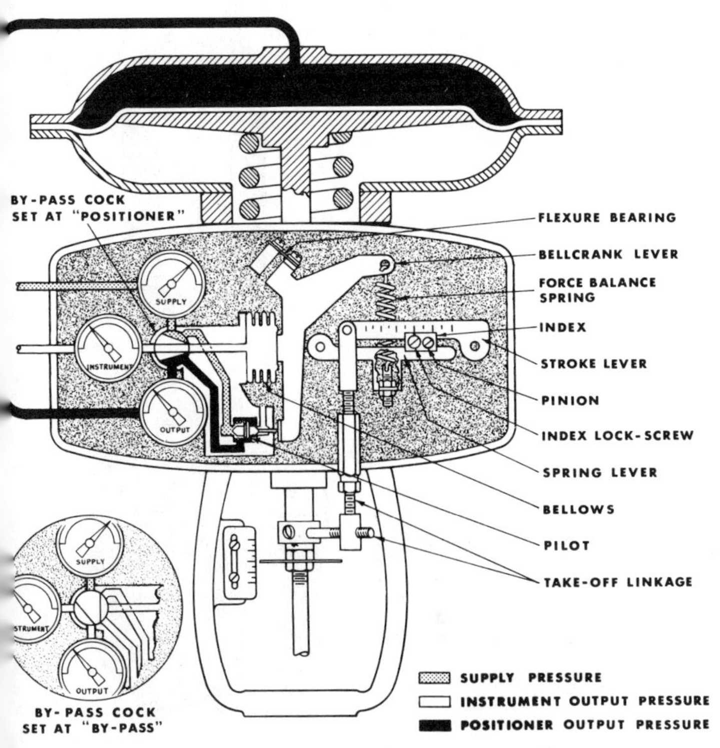

Figure 11.70. A schematic of the force balance valve positioner shows how the positioner operates; the inset shows the gauges and bypass. (Courtesy of Masoneilan)

Control Valve Accessories

Handwheels

Handwheels (Figures 11.71 and 11.72) may be supplied for manual operation of control valves for emergency use, during startup or in the event of air failure. They are used infrequently and primarily in critical services or when block and bypass valves are not provided.

Limit Switches and Solenoid Valves

Limit switches may be mounted on valves to operate signal lights for valve position indication, solenoid valve operation, alarms or relays (Figure 11.73). Similarly, solenoid valves may be provided for remote loading or unloading of valves (Figure 11.74).

Booster Relays

Booster relays are used to reduce lag time, resulting from long transmission lines, or when controller output capacity is insufficient for high demand devices, such as large diaphragm operators (Figure 11.75).

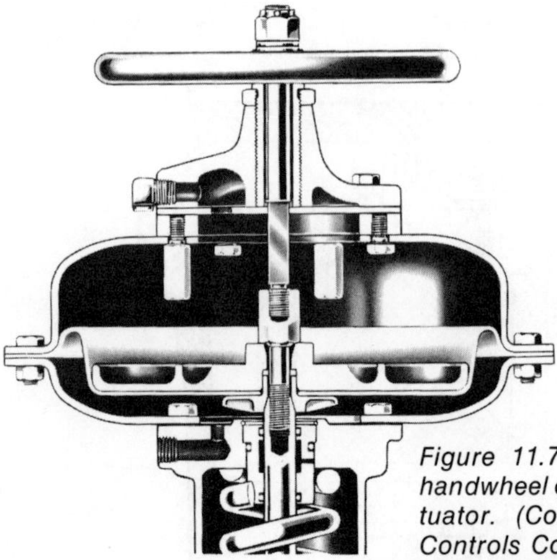

Figure 11.71. Top mounted handwheel on diaphragm actuator. (Courtesy of Fisher Controls Co.)

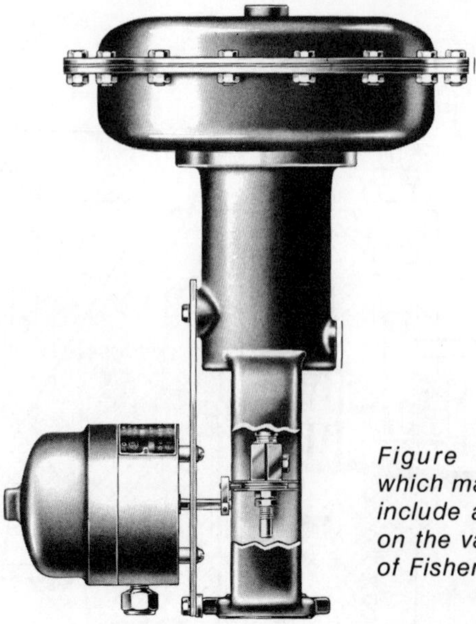

Figure 11.73. Auxilaries which may be valve mounted include a limit switch shown on the valve yoke. (Courtesy of Fisher Controls Co.)

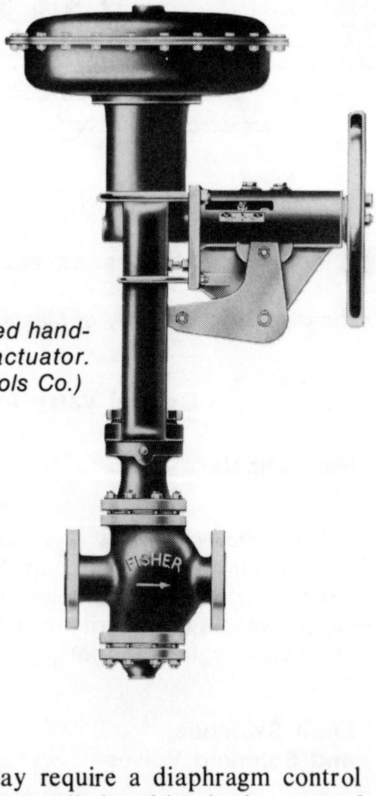

Figure 11.72. Side mounted handwheel on diaphragm actuator. (Courtesy of Fisher Controls Co.)

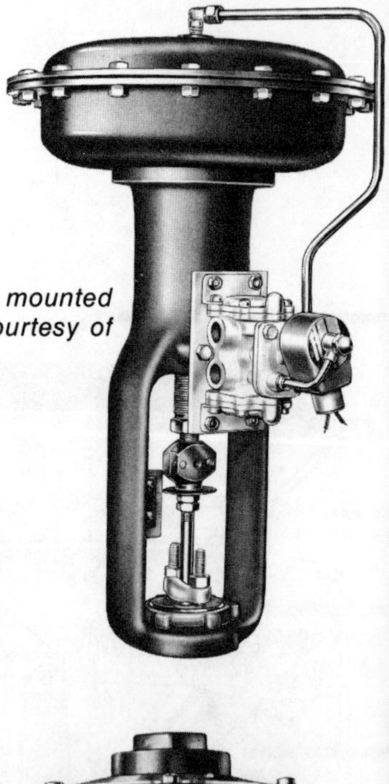

Figure 11.74. Solenoid valve mounted on diaphragm actuator. (Courtesy of Fisher Controls Co.)

Lockup Valve

Certain applications may require a diaphragm control valve to remain in its last controlled position in the event of air supply pressure failure. Figure 11.76 shows a valve used for the purpose, and Figure 11.77 shows a typical lockup installation.

Capacity Tank

Springless pneumatic actuators may stay in the last position on air failure but will more likely drift slowly to a closed or open position, depending on valve plug forces. When it is necessary to open or close a valve against line pressure, an airlock utilizing a capacity tank is used. Figure 11.78 shows a typical arrangement for this feature. This stored air pressure on the cushion-loading side of the piston provides

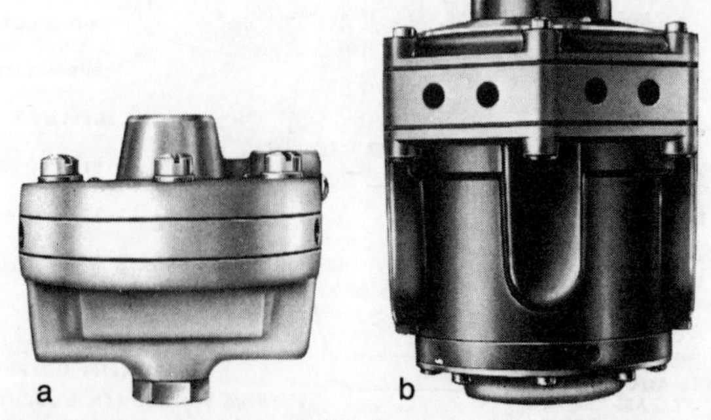

Figure 11.75. Pneumatic booster relays are used to reduce lag on long transmission lines or provide additional capacity for large operators. (Courtesy of Masoneilan)

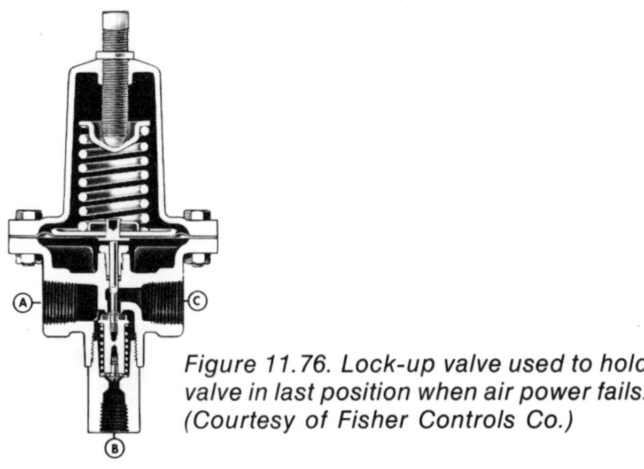

Figure 11.76. Lock-up valve used to hold valve in last position when air power fails. (Courtesy of Fisher Controls Co.)

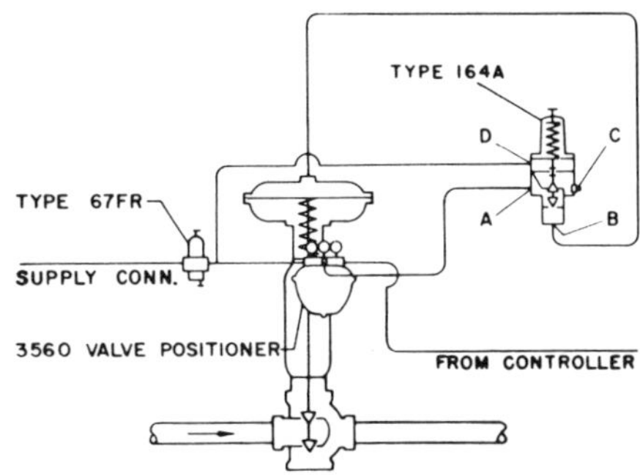

Figure 11.77. Schematic of a typical pneumatic lock-up installation. (Courtesy of Fisher Controls Co.)

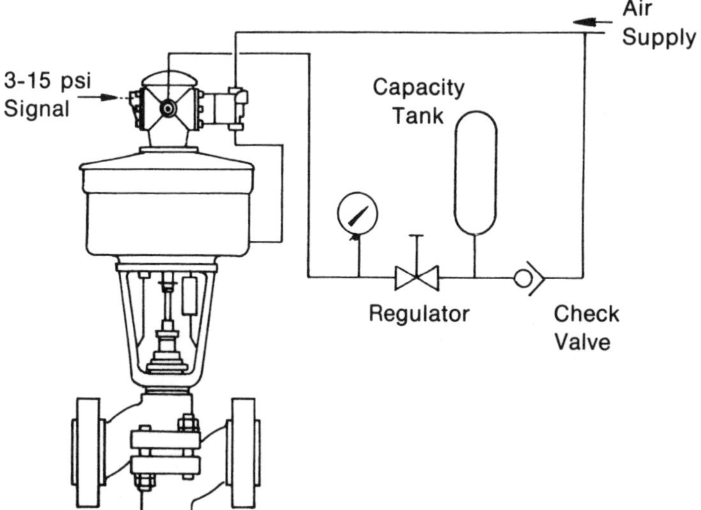

Figure 11.78. Schematic diagram of an airlock system with a capacity tank for valve operation following loss of air supply. (Courtesy ITT Hammel Dahl/Conoflow)

Figure 11.79. Electrical-to-pneumatic transducer mounted on pneumatic actuator. (Courtesy of Fisher Controls Co.)

positive valve opening or closing, regardless of the magnitude or direction of the forces involved when air supply failure occurs.

Transducers

Electronic control loops require an electrical-to-pneumatic transducer when the control valve is pneumatically operated. The transducer (Figure 11.79) may be mounted on the actuator. If the control valve is subject to vibration, however, it should be mounted on a nearby rigid support instead of the valve.

Jacketed Bodies

Jacketed valves are available for applications requiring heating of valve bodies. Jacketing provides more efficient and uniform heating than could be obtained by using copper tubing tracing. Steam, hot water or other heating media can be used where electric heating would not be desirable for safety reasons.

Airsets

The most common accessory is the air filter regulator. It is designed to furnish reduced air supply to the positioner at pressures ranging from 20 to 60 psi. The filter portion removes particulate matter large enough to clog instrument nozzles and orifices. The Fisher Controls filter in the 67FR removes particles of greater diameter than 0.0015 inch.

Ratio and Reversing Relays

A ratio relay is used to multiply or divide the pressure of an input signal. It might be used where split ranging of control valves occurs. For example, a 1:2 ratio relay could change a 3- to 9-psi control signal to a 3- to 15-psi signal.

Reversing relays are used when the action from a controller needs to be reversed. For example, split range valves may be operated from a controller where one valve is air-to-open while the other is air-to-close. A reversing relay can be used on one of the valves to achieve the desired action.

12 Pressure-Relieving Devices

William G. Andrew, Andrew Jackson Stockton

This chapter, intended to assist those engaged in specifying relief, safety and safety relief valves, provides pertinent information to make selection of the proper types easier and results in safer and more dependable installations. Fundamentals of valve construction, operation and purpose are given. Recommendations and specific selection guidelines are given in Volume 2, Chapter 6.

One of the prime responsibilities of the process plant designer is *safety*. An important aspect of this responsibility is the protection of equipment against failure from overpressure. Written specifications for the devices to provide this protection must be prepared. Equally important, provisions must be made for the safe disposal of the material released by the operation of the relieving devices.

Regulations or codes and recommended practices pertaining to pressure-relieving devices have been prepared by the American Society of Mechanical Engineers (ASME) and by the American Petroleum Institute (API). These codes and recommended practices have been adopted in some instances as mandatory by state and local regulatory agencies. They are certainly used as guidelines by these agencies and by operating companies in the process industries. Despite the codes and guidelines available, many judgments and decisions are entrusted solely to the designer. He must be equipped not only to solve an equation to determine a relieving quantity but also to analyze an operating system to determine the pressure effects that might develop from:

1. Chemical reactions
2. External fires
3. Explosions
4. Equipment failures
5. Operating errors
6. Thermal expansion
7. Control failures
8. Cooling failure
9. Reflux failure
10. Heat exchanger tube failure
11. Power failure
12. Misoperation
13. Combinations of the above

The following discussions should be helpful in arriving at the proper solutions to these problems.

Purpose

Pressure-relieving devices are installed for one or more of the following reasons:

1. To prevent the destruction of capital investment because of overpressure.
2. To provide safety for operating personnel.
3. To obtain favorable insurance rates on plant capital investment.
4. To conserve material loss that might occur with overpressure.
5. To prevent downtime that might result from overpressure.
6. To comply with applicable codes—local, state and national.

All these reasons could be capsuled into two primary factors: economy and safety. They may be expanded to include:

1. Avoidance of civil suits that might arise from the results of overpressure.

2. The prevention of pollution that may occur with overpressure and the subsequent release of vapors and/or liquids to the atmosphere.
3. Savings in equipment prices—without pressure-relieving devices, equipment would necessarily be designed to withstand higher pressures to which they might be subjected.

Definitions of Terms

To ensure a proper understanding of the various pressure-relieving devices and their associated terms, the following definitions are given. The first 18 definitions are based on API RP 520, Part 1, Paragraph 3.1.

Relief valve: an automatic pressure-relieving device actuated by the static pressure upstream of the valve; it opens in proportion to the increase in pressure over the opening pressure. It is used primarily for liquid service.

Safety valve: an automatic pressure-relieving device actuated by the static pressure upstream of the valve and characterized by rapid full opening or pop action. It is used for gas or vapor service. In the petroleum industry, it is used normally for steam or air.

Safety relief valve: an automatic pressure-relieving device suitable for use as either a safety or relief valve, depending on application. In the petroleum industry, it is normally used in gas and vapor service or for liquid.

Pressure relief valve: a generic term applying to relief valves, safety valves or safety relief valves.

Rupture disc: consists of a thin metal diaphragm held between flanges. Its purpose is to fail at a predetermined pressure, serving essentially the same purpose as a pressure relief valve.

Maximum allowable working pressure: as defined in the construction codes for unfired pressure vessels, the maximum allowable working pressure depends on the type of material, its thickness and the service conditions set as the basis for design. The vessel may not be operated above this pressure or its equivalent at any metal temperature. It is the highest pressure at which the primary pressure relief valve is set to open.

Operating pressure: the pressure, in pounds per square inch gauge, to which the vessel is usually subjected in service. A processing vessel is usually designed for a maximum allowable working pressure, in pounds per square inch gauge, which will provide a suitable margin above the operating pressure in order to prevent an undesirable operation of the relief device. It is suggested that this margin be as great as possible consistent with economical design of the vessel and other equipment, system operation and the performance characteristics of the pressure-relieving device.

Set pressure: the inlet pressure, in pounds per square inch gauge, at which the pressure relief valve is adjusted to open under service conditions. In a relief or safety relief valve on liquid service, the set pressure is to be considered the inlet pressure at which the valve starts to discharge under service conditions. In a safety or safety relief valve on gas or vapor service, the set pressure is to be considered the inlet pressure at which the valve pops under service conditions.

Cold differential test pressure: in pounds per square inch gauge, the pressure at which the valve is adjusted to open on the test stand. The cold differential test pressure includes the corrections for service conditions of back pressure and/or temperature.

Accumulation: pressure increase over the maximum allowable working pressure of the vessel during discharge through the relief valve, expressed as a percent of that pressure, or in pounds per square inch.

Overpressure: pressure increase over the set pressure of the primary relieving device. It is the same as accumulation when the relieving device is set at the maximum allowable working pressure of the vessel. Note: from this definition, it will be observed that when the set pressure of the first (primary) safety or relief valve to open is less than the maximum allowable working pressure of the vessel, the overpressure may be greater than 10% of the set pressure of the safety valve.

Blowdown: the difference between the set pressure and the reseating pressure of a relief valve, expressed as percent of the set pressure or in pounds per square inch.

Lift: the rise of the disc in a pressure relief valve.

Back pressure: pressure on the discharge side of safety relief valves.

Superimposed back pressure: the pressure in the discharge header before the safety relief valve opens.

Built-up back pressure: the pressure in the discharge header which develops as a result of flow after the safety relief valve opens.

Conventional safety relief valves: designed to have their bonnets closed (Figure 12.1). Their operating characteristics (set pressure, blowdown, capacity) are directly affected by the back pressure on the valve. Their performance,

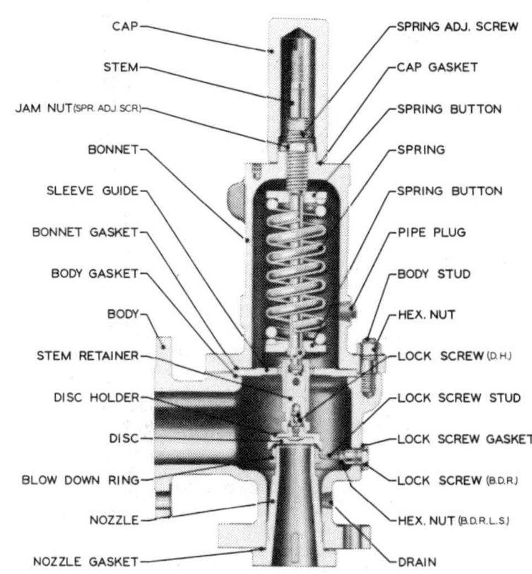

Figure 12.1. A conventional full nozzle safety relief valve. Note that the bonnet is plugged and is not open to the atmosphere. (Courtesy of Teledyne Farris Engineering)

therefore, is normally unsatisfactory under back pressure because of unbalanced forces acting on the disc and affecting the set pressure.

Balanced safety relief valves: designed to have their bonnets vented to the atmosphere or to a safe location. They are constructed so that back pressure has little or no effect on their operation at set pressure. There are two types: the piston (Figure 12.2) and the bellows (Figure 12.3). The piston type is designed so that the back pressure acts on both sides of the disc, cancelling out its effect. The bellows type is designed so that the back pressure is prevented from acting on top of the disc in the seat area while acting on both sides of the disc where the disc extends beyond the seat area.

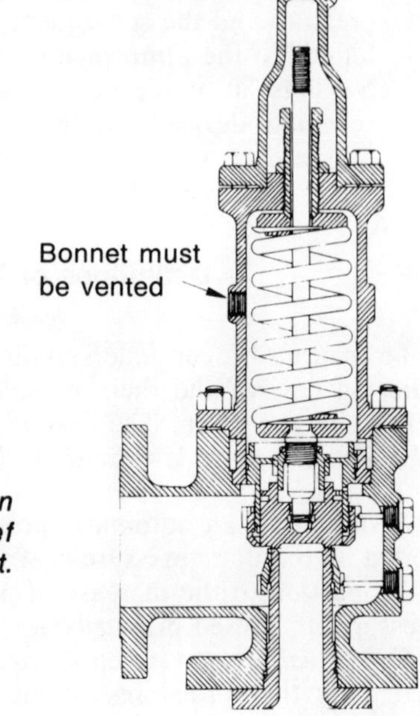

Figure 12.2. Piston balanced safety relief valve with vented bonnet. (Courtesy of API)

Types and Functions of Relieving Devices

Relief Valves

Relief valves (see Figure 12.4) have spring-loaded discs that close the inlet opening in the valve against the pressure source. The lift of the disc is in direct proportion to the over-

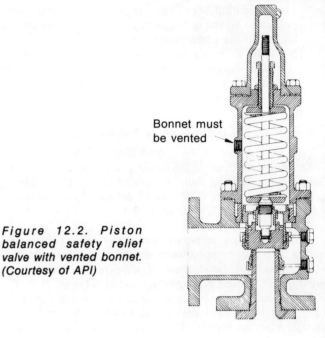

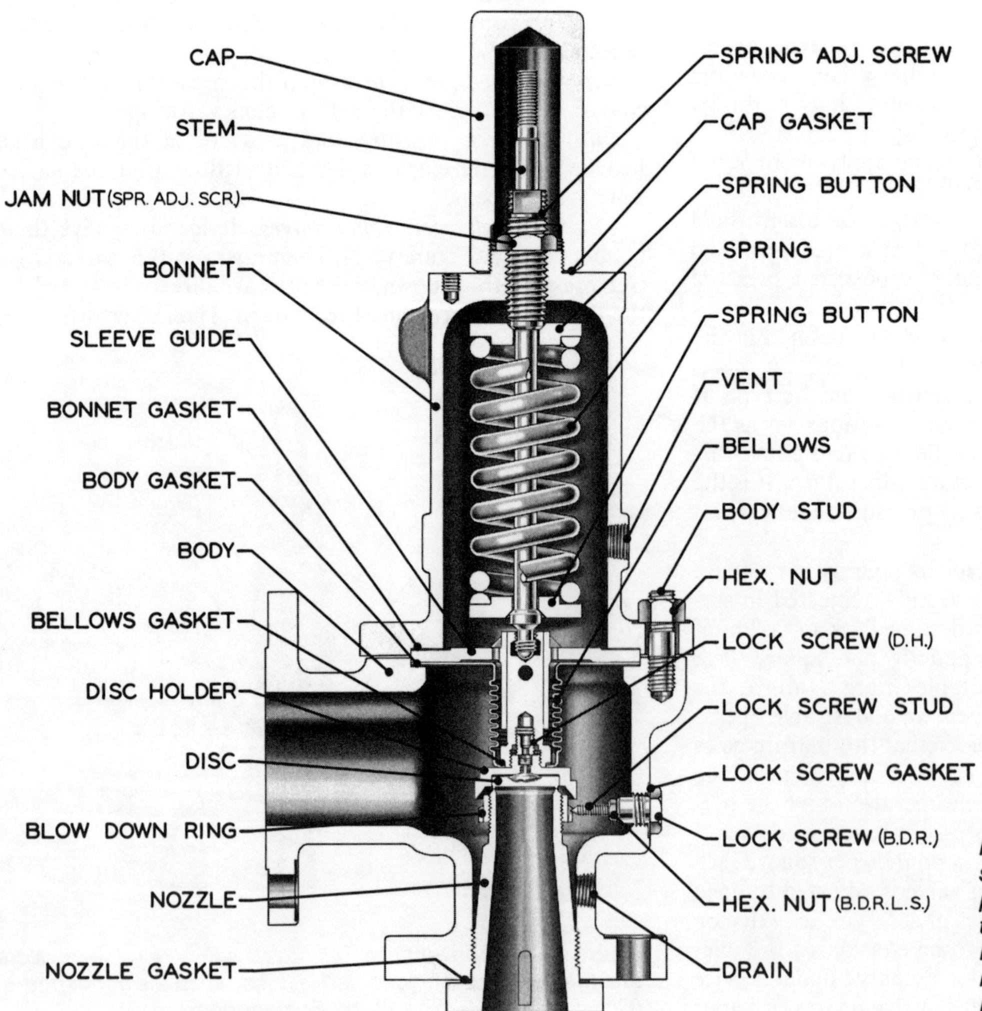

Figure 12.3. A bellows balanced safety relief valve prevents back pressure from acting on top of disc to affect opening at set pressure. Bonnet vent is open to atmosphere. (Courtesy of Teledyne Farris Engineering)

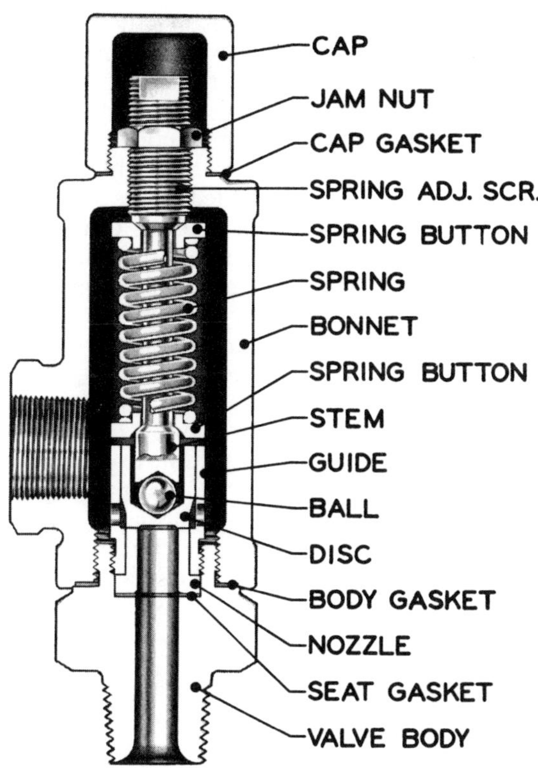

Figure 12.4. Typical spring-loaded relief valve whose lift (or opening) is directly proportional to the overpressure above set pressure. (Courtesy of Teledyne Farris Engineering)

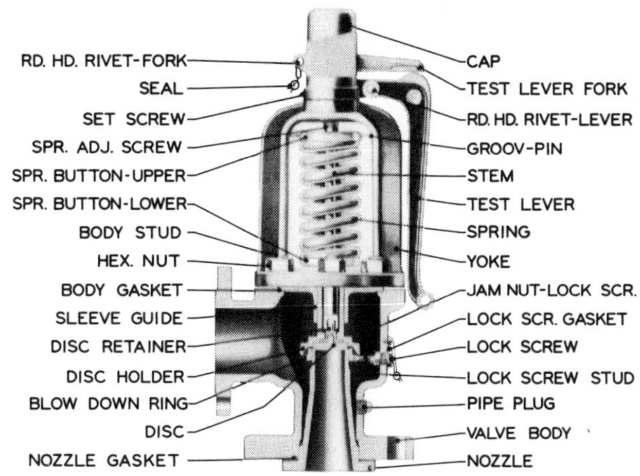

Figure 12.5. Safety valve—an increased disc and seat area exposed to inlet pressure produces pop action characteristic of safety valves. (Courtesy of Teledyne Farris Engineering)

pressure above the set pressure. When the inlet pressure equals the set pressure, the disc may rise slightly from the seat and permit a small amount of fluid to pass. As the higher pressure accumulates at the inlet, the spring is further compressed, permitting the disc to rise; thus, it provides additional area so as to allow an increasing flow of fluid.

The gradual rising of the disc with increasing upstream pressure throughout its useful range and the attainment of its full rated discharge capacity at 25% overpressure are characteristics of the relief valve. These features distinguish it from a safety valve whose disc attains its rated lift with low overpressure. It is further distinguished by not having a blowdown ring and a huddling chamber. It is used primarily for liquid service.

Safety Valves

Safety valves (and safety relief valves) are designed specifically to give full opening with little overpressure. They have spring-loaded discs that close the inlet opening in the valve against the upstream pressure and are characterized by rapid full opening, or pop action, produced by a huddling chamber which at set pressure, increases the disc and seat area exposed to inlet pressure to a point where the spring force can no longer overcome the inlet force (see Figure 12.5).

The flowing medium is directed to react against the disc and the spring force. This action utilizes kinetic energy (i.e., high velocity mass changing direction) to keep the valve in the open position. When the rate of flow is less than 25% of the valve capacity, the kinetic energy (mass) is not sufficient to keep the valve in its full open position, and the spring pushes the valve closed. Static pressure again pops the valve open, and again the low volume of flowing mass allows the valve to close. This repetition of opening and closing is characteristic of safety valves and is called *chattering*. It is an objectionable feature and occurs frequently when relief valves are oversized.

Characteristics such as chattering point to the need for accuracy in determining relief capacities. Knowledge of the relief devices is essential. For example, the discharge coefficient is not the same for valves of the various relief valve manufacturers—an assumed value might provide an erroneous size. The valve can be installed so that it limits the conditions on which sizing was based. Relief valves should be sized and installed so that they control (rather than being victims of) fluid release.

Properly sized, the relief valve continues to discharge until the inlet pressure recedes to 4 to 5% below the set point before reseating. The difference between set pressure and reseating pressure is known as the *blowdown*. Safety valves have an adjustable ring or rings for controlling blowdown.

Safety valves that meet the Power Boiler Code usually have body pressure limits equal to the inlet flange series, i.e., 300 psig for 300-psig inlet flanges, 600 psig for 600-psig flanges, etc.

Safety Relief Valves

The safety relief valve is a unique and extensively used relieving device in oil refineries and chemical process plants. It may be described as a safety valve with a closed bonnet

(Figure 12.1). Therefore, all the characteristics of the safety valve are inherent in the safety relief valve.

As the name implies, it may be used in either service—as a relief or safety valve. When used as a relief valve, the blowdown ring is backed off so that the huddling chamber effect is absent—the pop action is avoided—and the valve functions strictly as a relief valve. It may be used as a safety valve, except in situations where the temperature is high enough to alter the spring characteristics.

The versatility of the safety relief valve accounts for its extensive use, particularly in services where fluids must be contained or emission closely controlled.

Pilot-Operated Relief Valves

Pilot-operated relief valves have found increasing acceptance in relief service applications in the past few years. This trend is likely to continue and even to increase because of several significant features they possess:

1. Higher capacities for the same size valve body
2. Reduction of product losses
3. Reduction of downtime and maintenance
4. Capability of operating near the set relief pressure
5. Remote operating capability

The functions of pilot-operated safety valves are the same as for spring-loaded valves. They operate quite differently, however. The pilot-operated valve uses a floating piston as a main valve with process pressure on both sides. It is held closed by a larger area on top of the piston than on the bottom (Figure 12.6). When the set pressure is reached, a pilot valve relieves the pressure from the top of the piston, allowing quick action. Closure of the valve is accomplished by the pilot operator, which diverts pressure to the top of the piston whose greater area provides the necessary closing force at pressures close to the set pressure. Both the opening and closing of pilot-operated valves occur quickly and positively as opposed to direct spring-operated valves where process pressure first approaches the set pressure of the opposing spring force, then is equally balanced with that force and finally overcomes it and starts relieving. Firm closure of safety valves is accomplished by adjustment of blowdown rings which, although providing good closing forces, prevent closures close to the set pressure.

Since pilot-operated relief valves have more parts that are subject to failure (refer again to Figure 12.6 for a typical system), their acceptance has been slow. Proper evaluation of conditions where they may safely be applied, however, will lead to an increasing awareness of their economy and safety. They are acceptable for many clean services in gas transmission, at compressor stations and on unfired pressure vessels. A more detailed discussion follows later in this chapter.

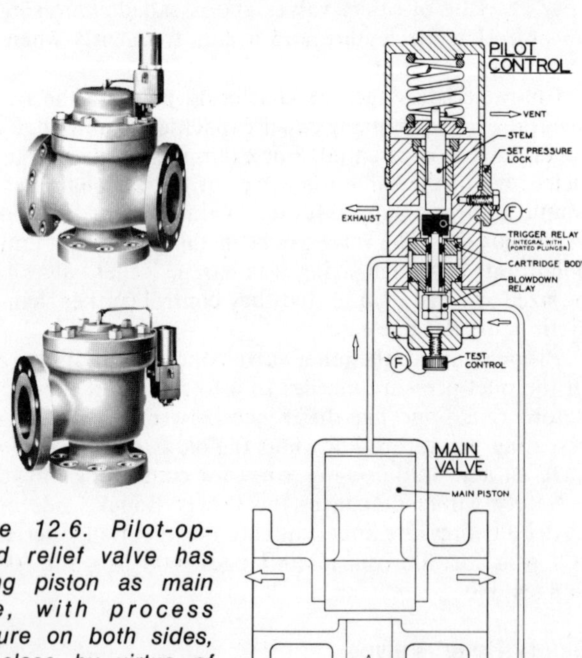

Figure 12.6. Pilot-operated relief valve has floating piston as main valve, with process pressure on both sides, held close by virtue of larger area on top of piston. (Courtesy of Teledyne Farris Engineering)

Rupture Discs

A rupture disc in an intentionally designed "weak spot" within a pressure system. Thus, it is expected to fail before other, more valuable equipment is damaged or destroyed. Requiring no moving parts, rupture discs are fabricated from carefully selected pieces of metal. They have defined limitations basic to their ultimate tensile or compressive strength and to creep, fatigue or corrosion resistance.

Metal rupture discs were introduced about 1930 and were used primarily in oil field service on oil and gas separators. Since this modest beginning, they have grown to usage on all types of chemical process vessels as well as many other functions such as quick opening valves.

Technical advancements in the last 10 years have greatly expanded the capabilities of the rupture disc for overpressure relief. A knowledge of the features and limitations of the various rupture disc designs are needed so that the correct one may be specified for a particular application. There are primarily five points to consider: type and thickness of metal, mechanical method of construction, operating margin, temperature and the types of loads the pressure system will impose on the disc during operation. These features will be discussed in greater detail later in the chapter.

Rupture discs may be grouped into four different design types: solid metal (Figure 12.7), composite (Figure 12.8), reverse-buckling (Figure 12.9) and shear type (Figure 12.10).

The most commonly used construction materials for rupture discs are aluminum, nickel, Monel, Inconel and austenitic stainless steel; however, discs are sometimes constructed from copper, silver, gold, platinum, tantalum and titanium. Metal foils, strips and sheets in soft, annealed condition are required. Today's wide variety of requirements

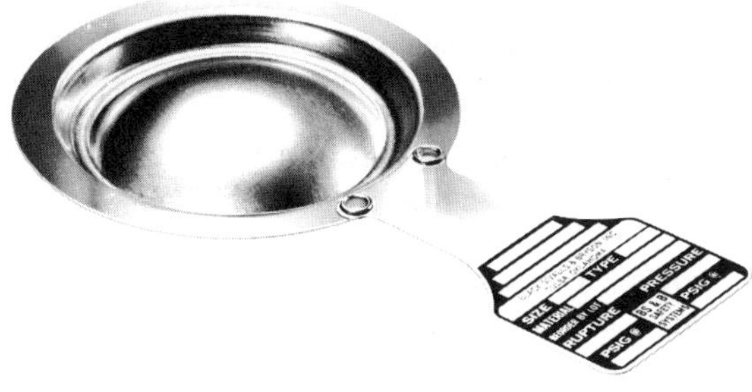

Figure 12.7. A solid metal rupture disc. (Courtesy of BS & B Safety Systems)

Figure 12.8. A composite rupture disc has a weakened metal disc backed by another metal or plastic disc that performs the sealing function. (Courtesy of BS & B Safety Systems)

necessitates the use of many increments of metal thicknesses ranging from approximately 0.002 to 0.060 inch. At least 70% of all metal rupture-disc applications involve the use of metal foil less than 0.010 inch thick.

Rupture discs are used for a variety of reasons, some of which are given below.

1. In polymer and slurry services (or other highly viscous materials) where material accumulation, plugging or freezing would make the use of safety valves impractical.
2. As secondary relief devices to release large volumes of fluids under fire or other emergency conditions after primary devices have been actuated.
3. As explosion protectors—they open a large area in a fraction of a second.
4. In series with relief valves to prevent leaking, to protect against corrosion or to avoid plugging or freezing of the relief valve.
5. Because they are more economical, especially in services requiring alloy materials.

ASME Code Requirements

The American Society of Mechanical Engineers set up a committee in 1911 to formulate standard rules for the con-

Figure 12.9. Reverse buckling disc before and after buckling. Knife action is used to shear disc when rupture is reached. (Courtesy of BS & B Safety Systems)

Figure 12.10. Shear rupture disc uses a hollow punch and a Bellville spring to assure an immediate rupture at desired burst pressure. (Courtesy of Ametek/Calmec)

struction of steam boilers and other pressure vessels. The Boiler and Pressure Vessel Committee, in the formulation of its rules and establishment of maximum design and operating pressures, considers materials, construction, method of fabrication, inspection and safety devices.

The National Board of Boiler and Pressure Vessel Inspectors is composed of chief inspectors of states and municipalities in the United States and provinces in the Dominion of Canada that have adopted the Boiler and Pressure Vessel Code. This Board, since its organization in 1919, has functioned to administer uniformity and enforce the rules of the Boiler and Pressure Vessel Code.

It should be pointed out that the state or municipality where the Boiler and Pressure Vessel Code has been made effective has definite jurisdiction over any particular installation.

Excerpts from ASME Unfired Pressure Vessel Code, Section VIII, 1968 Edition, are included in Volume 2, Chapter 6, so that the designer of pressure-relieving and venting systems will have at his disposal the code that authorities use in enforcing regulations concerning pressure vessels.

API Recommended Practices

In 1955, the American Petroleum Institute (API) first published a manual on *Recommended Practices for the Design and Installation of Pressure-Relieving Systems in Refineries,* issued as API RP 520. The published information was the result of several years' work by engineers in the petroleum industry and was intended to supplement the codes and regulations set forth in Section VIII of the ASME Boiler and Pressure Vessel Code.

A second and third edition of this manual have been issued, published in two parts: Part 1—*Design;* Part 2—*Installation.* The third edition was published in November, 1967.

Part 1 provides much useful information and discussion relative to pressure relief requirements, calculation methods, conditions requiring relief capacity, design of discharge systems, sizing procedures, etc. Included are illustrations, tables, charts and formulas to assist the designer in designing relief systems.

Part 2 covers requirements and suggestions pertinent to inlet and discharge piping, drain and bonnet or pilot vent piping, valve mounting positions and preinstallation handling and testing.

The increased complexity of modern processing units and the increased energy levels encountered in such units caused the petroleum industry to realize the need for additional guides for pressure relief and depressuring systems. As a result, in September, 1969, API RP 521 *Guide for Pressure Relief and Depressuring Systems* was issued to supplement information contained in API RP 520. This guide with its discussion of causes and prevention of overpressure, determination of relieving rates and the selection and design of disposal systems provides excellent guidelines for safety relief systems for modern plants.

API RP 520 and API RP 521 are essential tools for the designer of pressure-relieving systems along with Section VIII of the ASME Boiler and Pressure Vessel Code.

Safety and Relief Valve Design Features

To provide a better understanding of the function and operation of pressure relief valves, design and construction features are described and discussed. Illustrations are shown to emphasize construction details.

The basic purpose of a relief device is to relieve an overpressure condition of a system automatically and do it as economically and efficiently as practicable. The purpose also is to contain the pressure system as soon as the overpressure condition is relieved back to the normal condition. This is accomplished by a force-balance system acting on the closure of the relieving area (Figure 12.11). The orifice area of the pressure relief valve is selected to pass the required flow at specified conditions. This area is closed by a disc (Figure 12.12) until the set pressure is reached. The contained system pressure acts on one side of the disc and is opposed by a spring force on the opposite side. To complete this direct spring-loaded pressure-relieving valve, there must be a suitable body with inlet and outlet connections, a disc holder and other accessories to provide the performance characteristics desired.

Bodies

The valve body, usually an angle type (Figure 12.13), generally has the outlet sized larger than the inlet. This is necessary for expanding fluids such as gases, vapors, etc. A

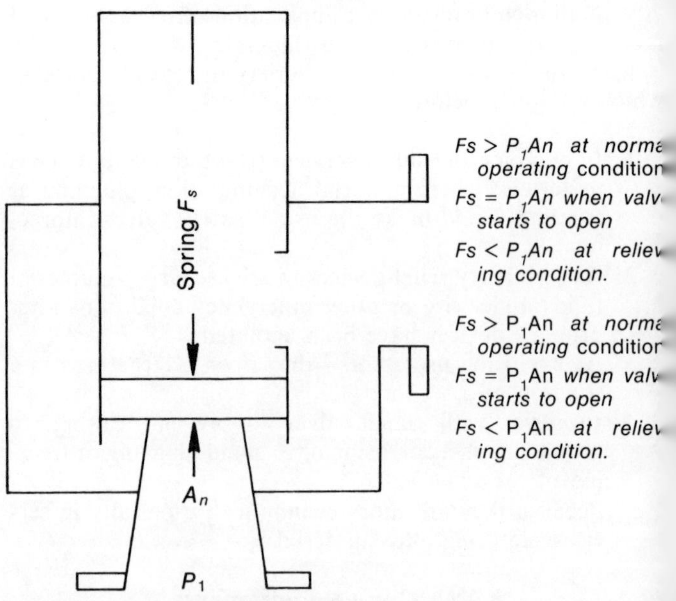

Figure 12.11. Pressure balance of a relief valve is determined by the spring (Fs), the inlet pressure (P_1) and the nozzle area (An).

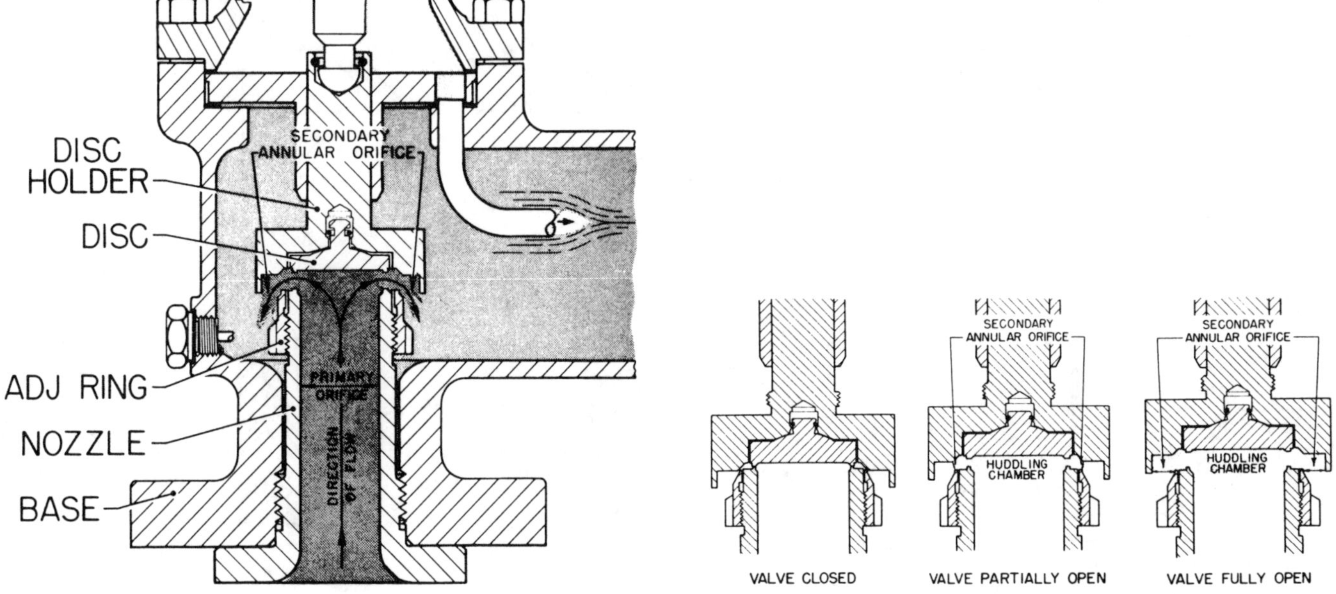

Figure 12.12. The valve disc seating against the nozzle closes the valve. (Courtesy of Dresser Industries, Inc.)

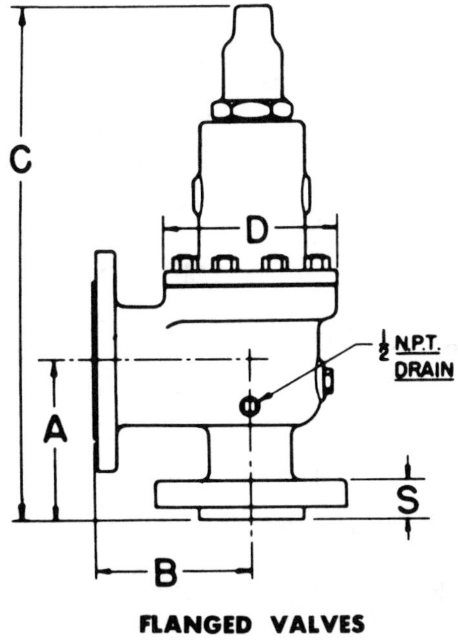

Figure 12.13. Relief valve body is normally an angle with larger outlet than inlet connection. (Courtesy of Dresser Industries, Inc.)

safety or relief valve must always relieve to a lower pressure (usually atmospheric). This lower outlet pressure requires an increase in pipe size sufficient to keep the back pressure under the allowable 10% of the inlet pressure which is necessary to ensure that the valve opens at its set pressure.

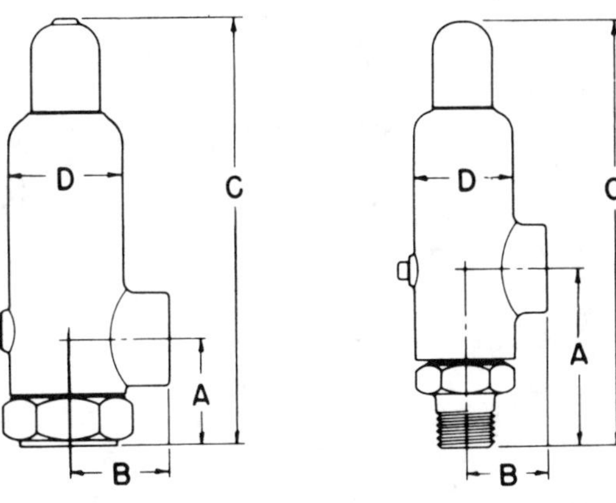

Figure 12.14. Inlet connections of screwed valves may be male or female; outlets are usually female. (Courtesy of Dresser Industries, Inc.)

The valve body must have an inlet connection capable of withstanding the pressure and temperature to which it will be subjected, not only under normal operating conditions but also at relieving conditions, which may be appreciably different. The body outlet connections and the bonnet assembly are likely to be designed for the lower pressures to which the valve will relieve.

Body connections may be screwed, flanged or welded. Screwed connections (Figure 12.14) are available in sizes up to 3 x 4 inches. Inlets may be male or female; outlet connections are usually female. Although the larger sizes (3 x 4) inches are available and used in some industries, many

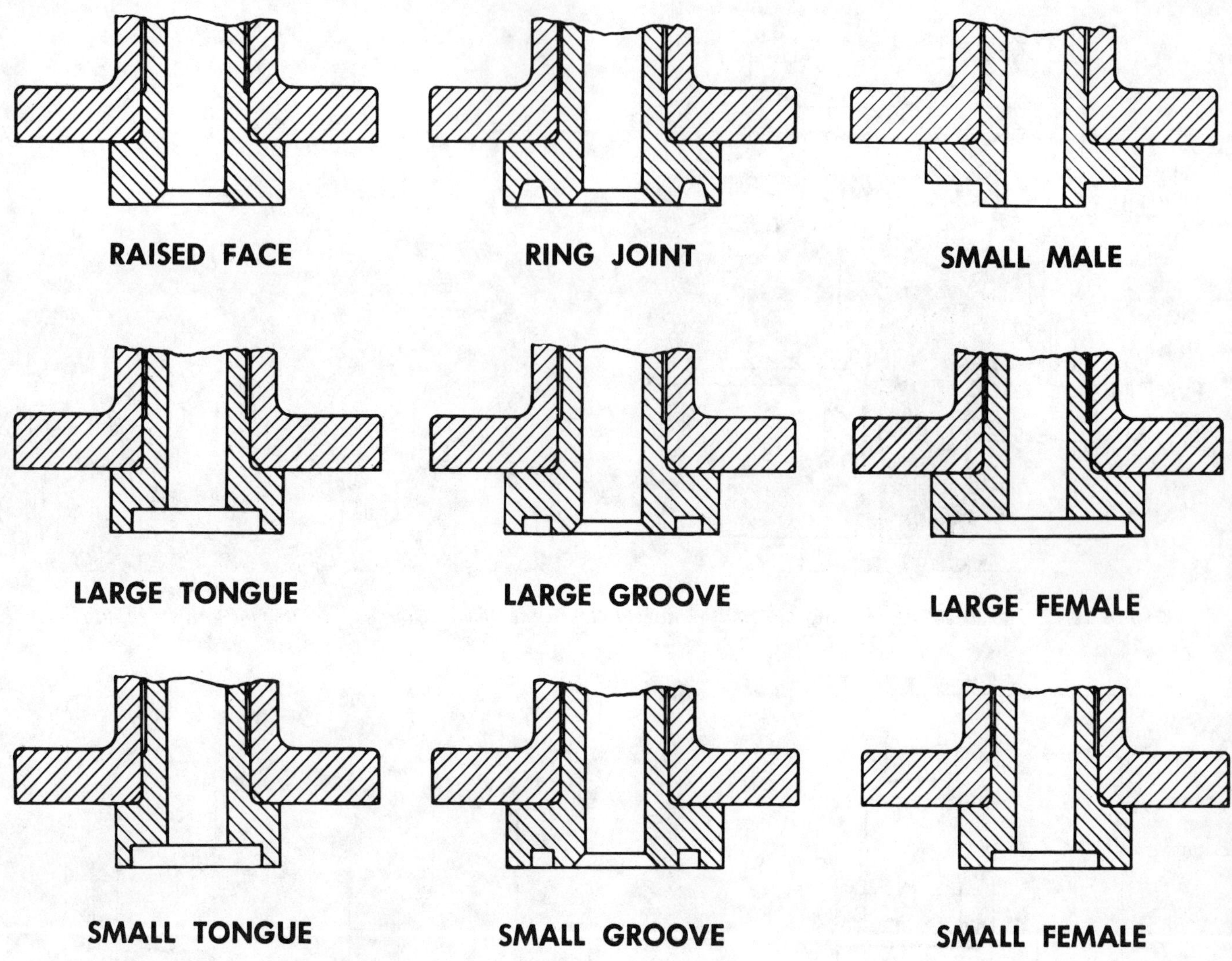

Figure 12.15. Typical flange connection methods for connecting safety relief valves. (Courtesy of J. E. Lonergan Co.)

plants in the chemical and petroleum processing field do not use screwed pipe larger than 2 inches. It is unusual to find sizes larger than 1½ x 2 inches in these industries.

Flanged connections are available from some manufacturers as small as ½ x ¾-inch size, but their use is not very common until larger sizes of 1 x 2 inches or 1½ x 2 inches are reached. Flange ratings are consistent with the pressure and temperature service conditions.

The inlet flange facings are machined on the flange of the nozzle of full nozzle steel valves. Raised face is the most common; however, other standard flange facings are also available (see Figure 12.15).

The outlet flange rating is compatible with the published maximum back pressure. Higher ratings are available on special applications. Ring joint outlets require total flange thickness heavier than raised face (standard) and are available when so specified (see Figure 12.16).

Welding nipple (Figure 12.17) and socket weld (Figure 12.18) connections are also available in smaller valve sizes but are not frequently used.

Body materials may be of cast iron, aluminum, bronze, steel, stainless steel, Hastelloy, Monel or other machinable metals. The great majority are iron, bronze and steel.

Pressure relief valve sizes are designated by a combination of their inlet and outlet sizes and a code symbol for their orifice size. Standard orifice sizes are given in the following section.

Nozzles

The arrangement of the nozzle in a pressure relief valve falls broadly into two types: full nozzle and seminozzle construction. Except when discharging, the only parts of a full nozzle valve that are wetted by the contained fluid are the

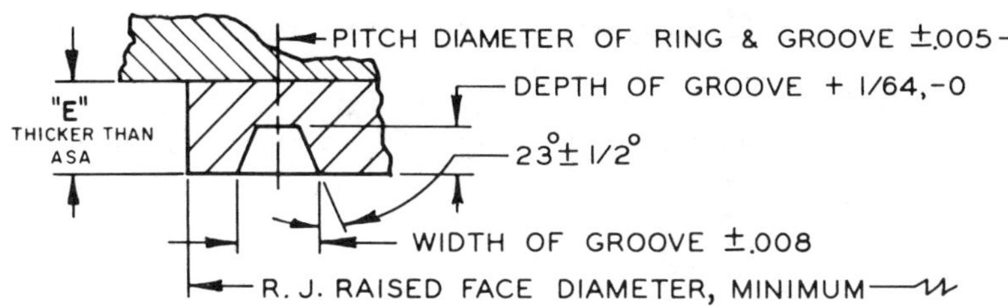

Figure 12.16. Ring joint outlets on relief valves can be furnished when specified. (Courtesy of Teledyne Farris Engineering)

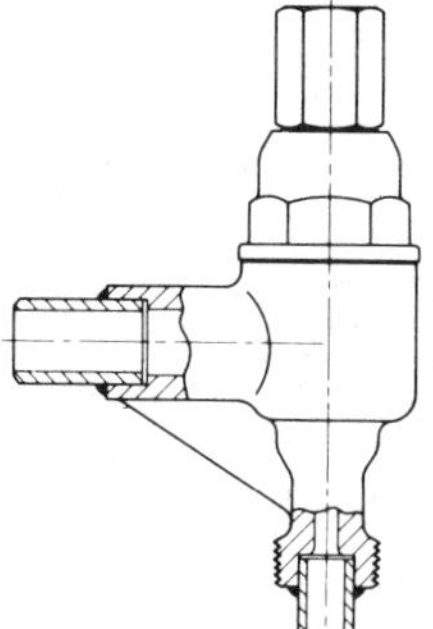

Figure 12.17. Relief valve using weld nipple connections. (Courtesy of Teledyne Farris Engineering)

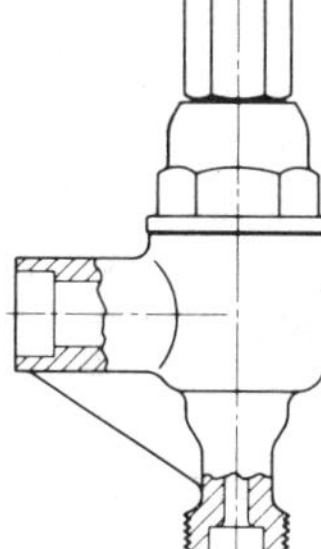

Figure 12.18. Socket weld connections for relief valve. (Courtesy of Teledyne Farris Engineering)

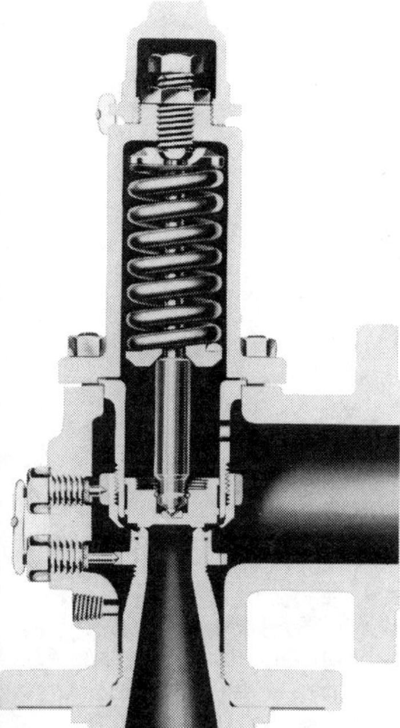

Figure 12.19. The nozzle and disc or disc insert are the only wetted parts of the valve until it relieves. (Courtesy of Crosby Valve and Gage Co.)

nozzle and disc or disc insert (Figure 12.19). In many applications that require special alloy materials, only the nozzle and disc need be made of the special alloy.

Seminozzle construction (Figure 12.20) is such that other parts of the valve are in contact with the line or vessel fluid. In most instances valves of standard material and construction are satisfactory.

A fact often overlooked in selecting relief valves is that they do operate infrequently. When this is assumed to be the case, considerable savings are achieved by the proper selection of materials for the nozzle and disc of full nozzle valves while using standard material for the body and bonnet.

Nozzle Orifice Sizes

Safety relief valve manufacturers have adopted standard orifice sizes to simplify relief valve selection and designation. Between 0.110 and 26.00 square inches are 15 different sizes designated by the letters *D* through *T* with the exception of *I*, *O* and *S*. Table 12.1 lists sizes and alphabetical codes.

Smaller orifice areas are available but are not designated by code; sizes are not standardized to the extent of those listed in Table 12.1. Larger sizes are also available but are

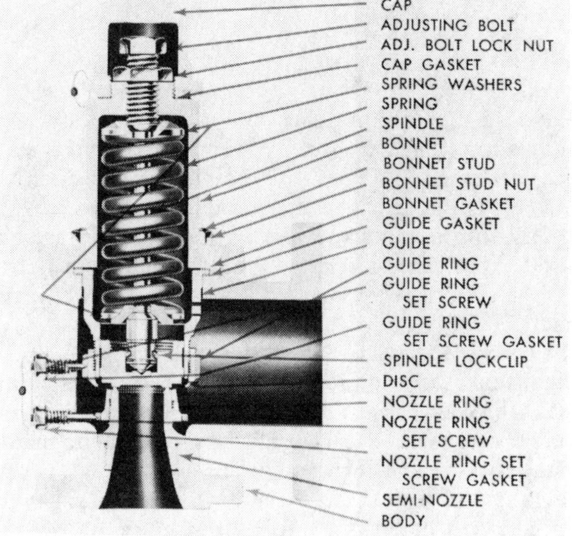

Figure 12.20. The inlet flange in the body of a seminozzle valve is wetted by the process fluid. (Courtesy of Crosby Valve and Gage Co.)

Table 12.1. Standard Relief Valve Orifice Sizes and Their Designation

EFFECTIVE ORIFICE AREA SQ. IN.	ORIFICE DESIGNATION
0.110	D
0.196	E
0.307	F
0.503	G
0.785	H
1.287	J
1.838	K
2.853	L
3.60	M
4.340	N
6.380	P
11.050	Q
16.0	R
26.0	T

not normally listed in the manufacturers' catalogs and bulletins. One manufacturer, for example, makes sizes up to 166 square inches which are housed in 20 x 24 inch bodies.

It may be pertinent to point out that there is an overlapping of inlet and outlet sizes associated with a particular orifice size. For example, a *J* orifice (1.287 square inch) may be installed in a 2 x 3 inch, a 2½ x 4 inch or a 3 x 4 inch body depending on inlet and outlet pressure conditions.

Springs

Standard springs for relief valves are normally carbon steel or alloy (tungsten) steel. Carbon steel springs are furnished for temperatures to 450°F and alloy (tungsten) steel springs for temperatures over 450° on closed bonnet valves. For steam (open spring) valves, the dividing point is 650°F.

For temperatures lower than −20°F, alloy steel springs (and bodies) are recommended to prevent breakage from impact.

For corrosive service, either alloy steel springs or special coating is advisable to prevent stress corrosion which might result in broken springs. Balanced bellows can be used to protect springs so that carbon steel springs and bonnets are satisfactory when bellows valves are used.

Valve springs are adjustable within a narrow range of settings. The adjustment may be 30% or more on low range springs and as little as 5% on higher spring ranges. This does not mean, however, that a valve setting may be changed that much on a particular valve for the original setting might have been at the very top or bottom of the spring range. It is likely that a new spring will be required when a valve set pressure is changed.

Spring washers (Figure 12.21) are fitted to a particular spring for proper alignment so that their change is also necessary when springs are changed.

The two important factors to remember in spring selections are temperature and corrosion.

Trim and Other Internal Parts

Some manufacturers consider that the term *trim* applies only to the nozzle and disc. Others include the disc holder in the term. Where full nozzle valves are not used, the base would be included in the term (Figure 12.22).

In the discussion on nozzles, the materials from which nozzles and discs are constructed were discussed. Standard materials for these and for other internal parts (disc holder, sleeve guide, stem retainer and blowdown ring) are alloy or stainless steels (Figure 12.23). During relieving conditions, these parts are subjected to the flowing medium, and corrosive applications may require Monel or Hastelloy.

Seating

As pressure relief valves approach their set pressure, there is increasing tendency for them to leak. This is inherent in their design as force balance mechanisms—the internal forces on the underside of the disc area approaches the opposing spring force on the top of the disc. The proneness to leak is overcome primarily by: 1. Precision machining of seating surfaces and aligning as accurately as possible. Surface finishes can be produced that deviate less than 5 microinches from a perfect finish. 2. The use of O-ring seat seals (Figure 12.24) which allows complete tightness at

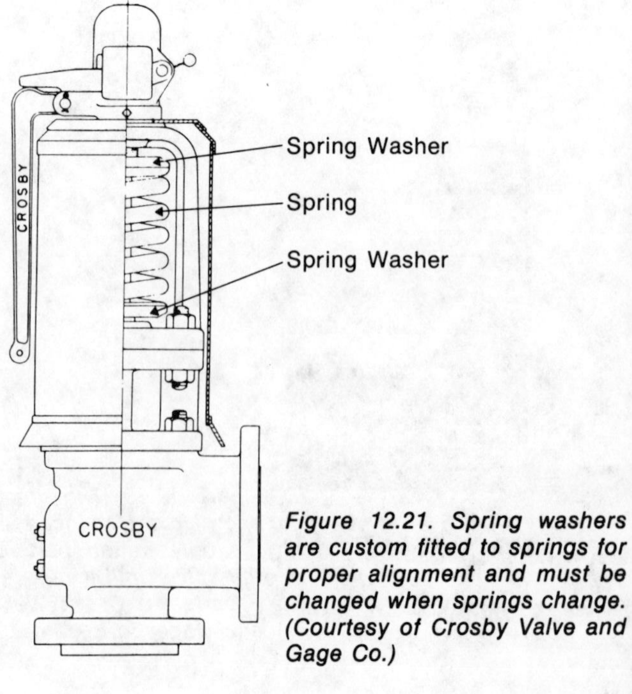

Figure 12.21. Spring washers are custom fitted to springs for proper alignment and must be changed when springs change. (Courtesy of Crosby Valve and Gage Co.)

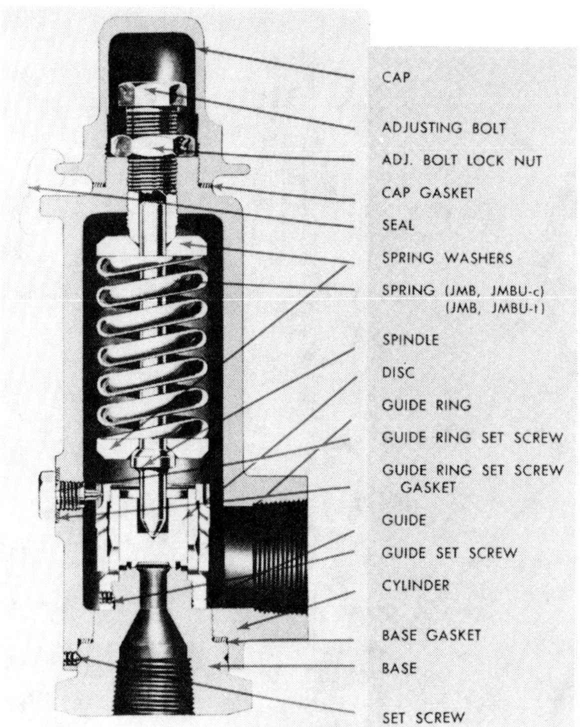

Figure 12.22. The base is considered as trim in the seminozzle or modified nozzle valves. (Courtesy of Crosby Valve and Gage Co.)

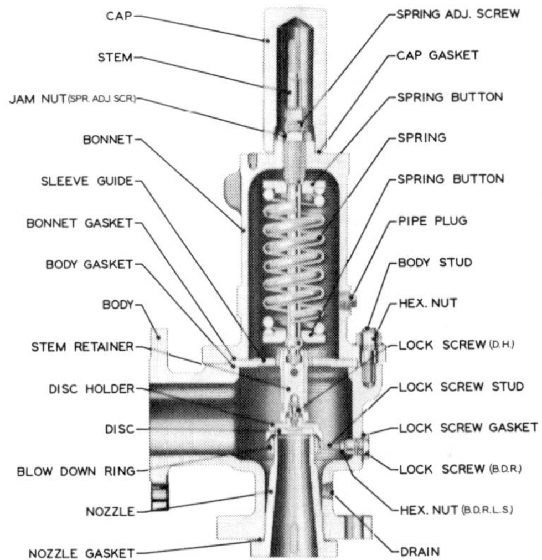

Figure 12.23. Internal parts such as disc holder, sleeve guide, stem retainer and blowdown ring are made of stainless steel or other alloy. (Courtesy of Teledyne Farris Engineering)

pressures much closer to the set pressure than is possible with standard metal-to-metal seats (see Figure 12.25).

Relief valve manufacturers have tried for many years to design a valve with resilient seats to overcome the leakage problem. The difficulty with the use of resilient seating material is its tendency to "cold flow" under the seat load imposed by the spring force. This gradually lowers the set pressure as the valve spring relaxes under these circumstances.

The use of an O-ring seat seal allows relief valves to maintain the set pressure and the desired tightness by using relatively rough metal-to-metal seating surfaces to carry the spring force while the O-ring is pressure loaded against a specially curved seating surface. With this construction, the resilient O-ring does not carry the seat load imposed by the spring force and is not subject to cold flow; yet it provides a high degree of tightness over an extended period of time.

Table 12.2 is a selection chart for O-ring materials for various services provided by Dresser Industries for their Consolidated Relief Valves.

Tightness

The terms used to define tightness are *commercial* and *bubble.* Bubble tightness means *no* leakage or bubbles of air at specified percentages of set pressure. Commercial tightness permits a stipulated maximum leakage at specified percentages of set pressure.

The test apparatus and procedures under which commercial tightness tests may be made are described in API RP 527, *Commercial Seat Tightness of Safety Relief Valves with Metal-to-Metal Seats,* first issued in September, 1964. Figure 12.26 shows the acceptable test arrangement.

Table 12.3 shows the maximum leakage rates permissible for conventional and balanced bellows valves for orifice size groupings shown. Tests are determined with the valve mounted vertically and the relief valve inlet held at 90% of set pressure immediately after the valve has popped. An exception to the above percentage is that for any valve set below 50 psig, the pressure shall be required to hold at 5 psig below the set pressure immediately after popping. In both cases cited above, the test pressure must be applied for a maximum of 1 minute for valve sizes through 2 inches; 2 minutes for 2½-, 3- and 4-inch sizes; and 5 minutes for 6 and 8-inch sizes. Air at approximately atmospheric temperature is used as the pressure medium.

API RP 527 applies only to relief valves with set pressures up to 1,000 psig. No leakage rates have been defined as permissible under this procedure for pressures above that. When there is a need for specifying tightness on valves of higher settings, complete specifications should be agreed upon between the customer and the manufacturer.

When better than *commercial* tightness is desired, this may also be specified by the customer. The same rate of leakage might be allowed at 95% of set pressure, for example, or a lesser leakage rate at the specified 80% of set pressure. Greater demands relative to tightness incur greater cost for the purchased item.

Common Causes of Leakage

Listed below are circumstances under which excessive valve leakage is likely to occur.

1. Set pressure too close to the normal operating pressure. As the differential between the operating and set pressure decreases, the force between the seating

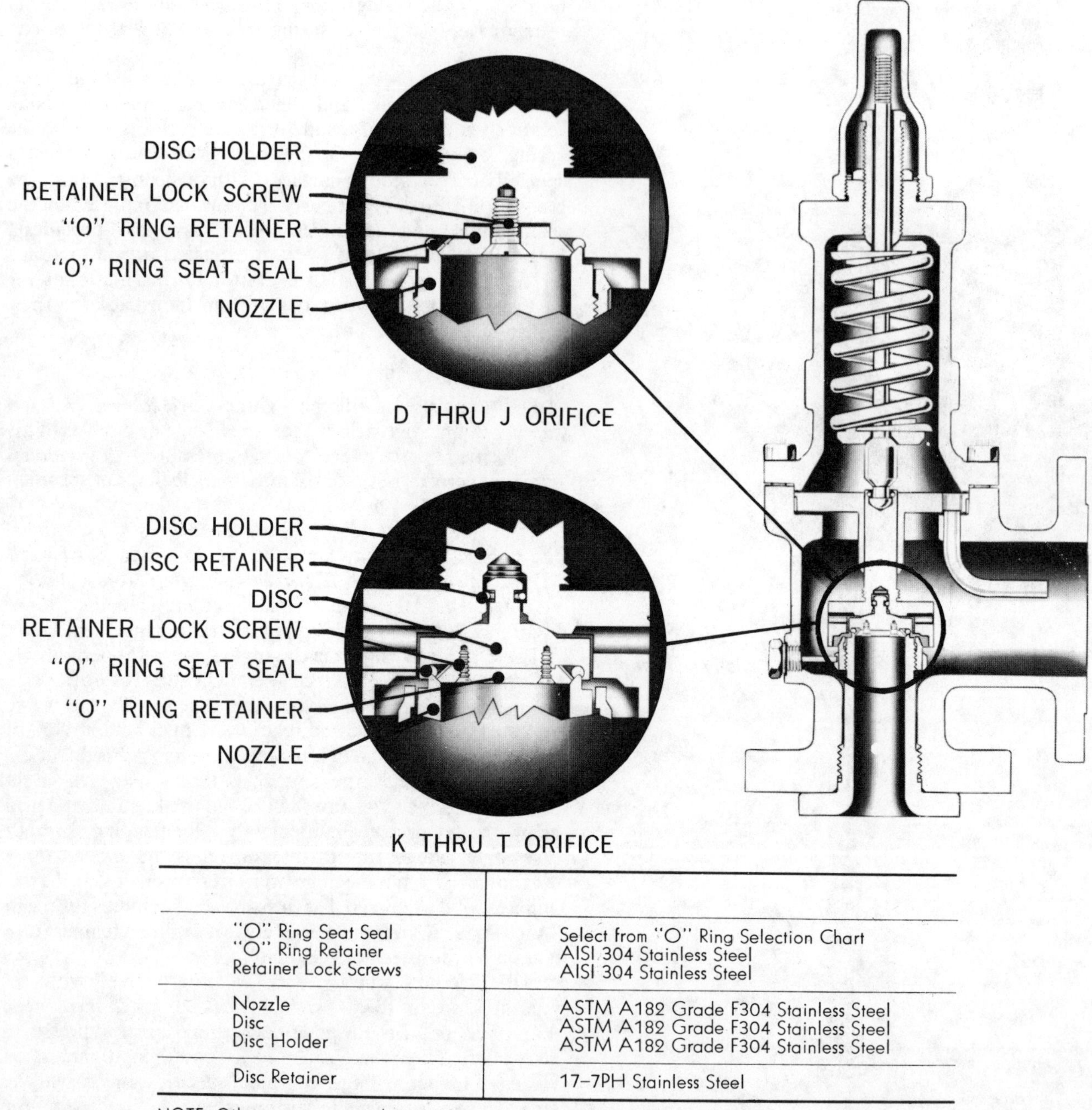

"O" Ring Seat Seal "O" Ring Retainer Retainer Lock Screws	Select from "O" Ring Selection Chart AISI 304 Stainless Steel AISI 304 Stainless Steel
Nozzle Disc Disc Holder	ASTM A182 Grade F304 Stainless Steel ASTM A182 Grade F304 Stainless Steel ASTM A182 Grade F304 Stainless Steel
Disc Retainer	17–7PH Stainless Steel

NOTE: Other part names and materials are as shown under conventional valve.

Figure 12.24. The use of O-ring seat seals reduces leakage problems considerably in comparison with metal-to-metal seats. (Courtesy of Dresser Industries, Inc.)

surfaces decreases also. Pressure peaks at these conditions produce leakage.

2. Failure of a valve to reseat properly after it has popped or after "simmering" occurs. A characteristic of a safety relief valve is that it must lift off the seat slightly and "simmer" before pressure is built up in the huddling chamber to cause the valve to pop. In many cases, the valve simmer takes care of the excessive pressure and the valve does not pop at all. At the simmer point, the seat load is zero between the seating surfaces, and the valve disc can float out of alignment. When the pressure goes down, the valve continues to leak until a high differential force causes it to seal more tightly.

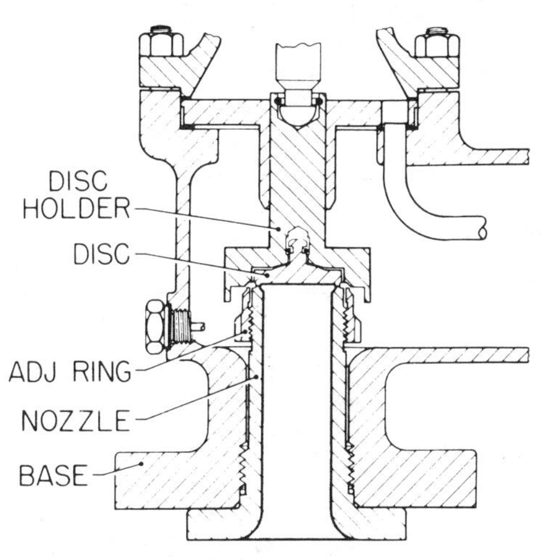

Figure 12.25. Metal-to-metal seats must be machined precisely and aligned accurately to seal properly.

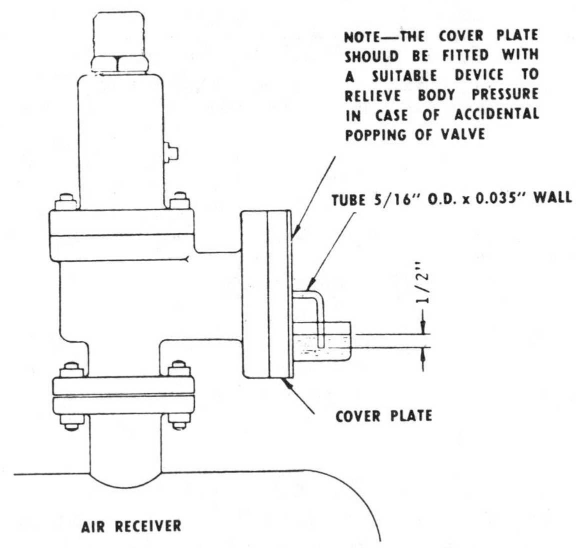

Figure 12.26. Test apparatus for seat tightness as specified in API RP 527. (Courtesy of Crosby Valve and Gage Co.)

Table 12.2. Service Recommendations for "O" Ring Materials

CHEMICAL SERVICE	VITON (−0 +400)	BUNA N (−0 +250)	BUTYL (−0 +250)	SILICONE (−100 +400)
Acetaldehyde			1	2
Acetate Solvents			1	2
Acetic Acid Vapors			1	2
Acetic Acid, 30%			1	2
Acetic Anhydride			2	1
Acetol			2	1
Acetone			1	2
Acetyl Chloride	2		1	
Acetylene	2		1	
Acids, Mild	1		2	
Acids, Concentrated	1		2	
Air	1		2	
Amines (Mixed)		2	1	
Ammonia, Aqueous		2	1	
Ammonia, Anhydrous (liquid)		2	1	
Ammonia Gas (Hot)			2	1
Aromatic Hydrocarbons	1	2		
ASTM #1 Oil	1	2		
ASTM #3 Oil	1	2		
Benzene (Benzol)	1	2		
Bleach Solutions	1		2	
Boric Acid	1		2	
Boron Fuels (HEF)	1	2		
Brine	1	2		
Butadiene	1	2		
Butane	1	2		
Bunker Oil	1	2		
Butyl Acetate			1	
Calcium Chloride	1		2	
Carbolic Acid (Phenol)	1		2	
Carbon Dioxide	1	2		
Carbon Monoxide (Hot)	1		2	
Carbon Tetrachloride	1			
Carbonic Acid	1		2	
Casing Head Gas	1	2		
Crude Oil	1	2		
Denatured Alcohol	1		2	
Detergent Solutions	1		2	
Dichlorobenzene	1			
Diesel Oil	1	2		
Diethanolamine			1	
Diethylamine			1	
Diethylene Glycol	1		2	
Dowtherm 'A'	1		2	
Dowtherm 'E'	1			
Ethanolamine		2	1	
Ethyl Acetate			1	
Ethyl Alcohol	1		2	
Ethyl Benzene	1	2		
Ethylene	1	2		
Ethylene Oxide	2		1	
Formaldehyde			1	
Formic Acid			1	
Freon 11 & 12	1	2		
Freon 22	2		1	
Fuel Oil	1	2		
Gasoline	1	2		
Glaubers Salt	1	2		
Glycerin	1	2		
Glycols	1		2	
Helium	1		2	
Heptane	1	2		
Hexane	1	2		
Hydrazine			1	
Hydrochloric Acid		Note A		
Hydrofluoric Acid	1		2	
Hydrogen Chloride Gas	1		2	
Hydrogen Gas	1		2	
Hydrogen Sulfide	1		2	
Isobutyl Alcohol	2		1	
Isopropyl Alcohol	2		1	
Kerosene	1	2		
Lacquer Solvents	1	2		
Lime	2		1	
L.P. Gas	1	2		
Lye Solutions	2		1	
Magnesium Chloride	1		2	
Methane	1	2		
Methanol (Wood Alcohol)		2	1	
Methyl Acetate			1	
Mineral Oil	1	2		
Naphtha	1	2		
Natural Gas	1	2		
Natural Gas (Sour)	1	2		
Nitric Acid		Note A		
Nitrogen	1		2	
Oxygen (Gas)	1		2	
Pentane	1	2		
Perchlorethyene	1			
Phenol	2		1	
Phosphoric Acid		Note A		
Propane	1	2		
Propyl Alcohol	2	1		
Propylene	1	2		
Propylene Glycol	2		1	
Propylene Oxide	2		1	
Silicon Oils and Greases	2		1	
Skydrol			1	2
Sodium Hydroxide			1	
Steam		Note B		
Sulfur Dioxide, Trioxide	1		2	
Sulfuric Acid		Note A		
Toluene	1	2		
Trisodium Phosphate	1		2	
Vegetable Oils	1		2	
Vinyl Chloride	1			
Water	1		2	
Zealites	1	2		

1 = 1st Choice
2 = 2nd Choice

NOTE A.—Use Parker Compound V-494-7 to 200°F.
NOTE B.—Use Teflon in Steam Service with Temperatures of 250°F. to 500°F.

Source: Dresser Industries, Inc.

Table 12.3. Maximum Permissible Leakage Rates for Relief Valves

Type of Valve	Manufacturer's Orifice Size	Max. Leakage Rate (bubbles/min.)	App. Leakage Rate (SCF/24 hr)
Conventional	F and smaller	40	0.60
	G and larger	20	0.30
Balanced bellows	F and smaller	50	0.75
	G and larger	30	0.45

3. The scratching of the metal-to-metal seating surfaces by foreign material such as pipe scale, welding beads, sand, dust, etc., when a valve is open and flowing.
4. Corrosion of the seating surfaces.
5. Piping strains caused by normal thermal expansion or by incorrect pipe installation.
6. Applications where heavy vibrations occur. Vibrations cause the spring force to increase and decrease alternately. On the decrease part of the cycle, the differential force between the contained medium and the spring may actually approach zero, resulting in leakage.
7. Nozzle icing conditions resulting from the refrigerant effect of the flowing fluid when a valve relieves. Ice sometimes actually forms on the seat. Reseating is improper. For valves subject to frosting when discharging, a knife seat is advisable. (Knife seats are satisfactory also for fluids containing fine solids in suspension.)

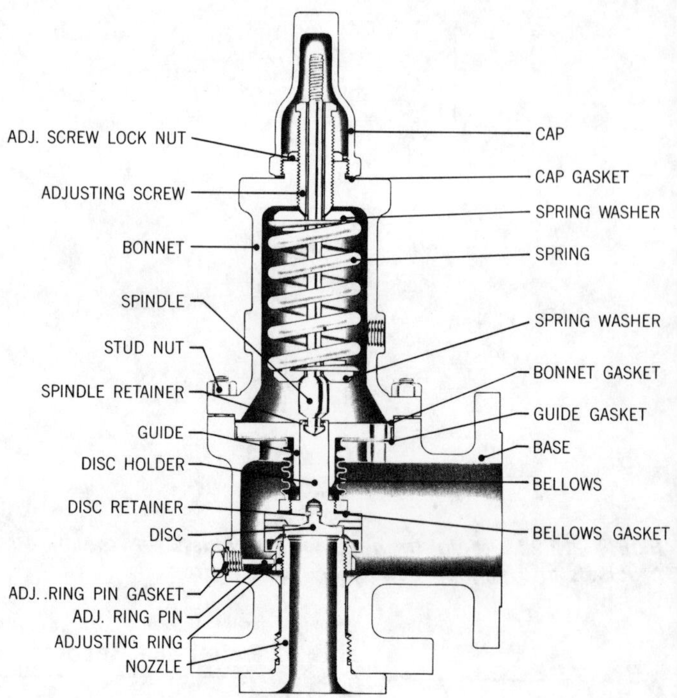

Figure 12.27. Bellows seals overcome the effect of back pressure on the set pressure of relief valves by covering the disc whose area the back pressure would affect. (Courtesy of Dresser Industries, Inc.)

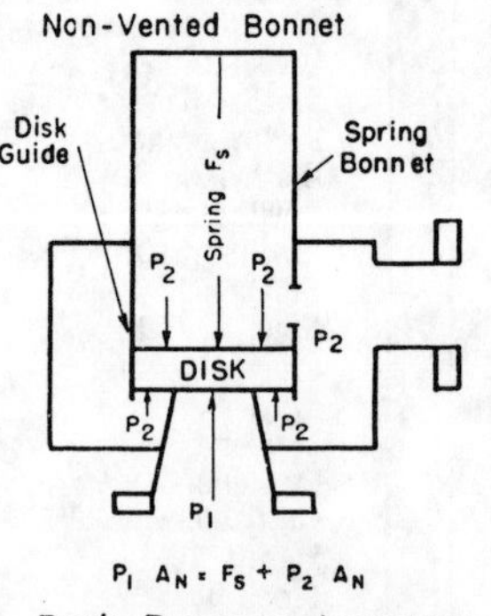

Figure 12.28. Effect of back pressure of conventional safety relief valves. (Courtesy of API)

Bellows Seals

Bellows seals (Figure 12.27) were developed to overcome the effect that back pressure exerts on the valve set pressure in pressure-relieving systems. A secondary advantage is that they also protect valve moving parts from corrosion.

When variable back pressure exists, it is not possible for a conventional valve to relieve consistently at any set pressure. The relieving pressure varies with the magnitude of the back pressure.

Figure 12.28 shows how back pressure (downstream pressure) acts on a disc to effectively increase the set pressure of the valve. It can be seen clearly that the pressure opposing the process pressure, P_1 is the spring force, F_s, plus P_2, the downstream (back) pressure. Conventional valves may be used when the back pressure is constant or when the variation in back pressure does not exceed 10% of set pressure. When back pressure is constant, it is simply subtracted from the set pressure to determine the net spring setting.

When back pressure is variable or when its value exceeds 10% of the setting, balancing bellows should be used. The inside of the bellows remains atmospheric at all times, keeping the back pressure from exerting its force on the unbalanced seat area. When balanced bellows are used, the bonnet should be vented to the atmosphere so that the bellows can breathe freely; otherwise (with plugged vent), pressure buildup in the bonnet and bellows restricts bellows movement.

The use of a balanced bellows does introduce another variable factor into valve sizing calculations. This bellows "flow factor" is labeled differently by the various manufacturers, but each one furnishes curves for its determination based on actual tests. The curves are plots of the factor (% of rated capacity) versus back pressure percentage. Factors are established for liquids and for vapors or gases. They vary a great deal among manufacturers. Figure 12.29 shows a flow factor correction curve for Farris valves at 10% over-

Variable or Constant Back Pressure Sizing Factor
K_V
10% Overpressure
BalanSeal Valves Only — Vapors and Gases

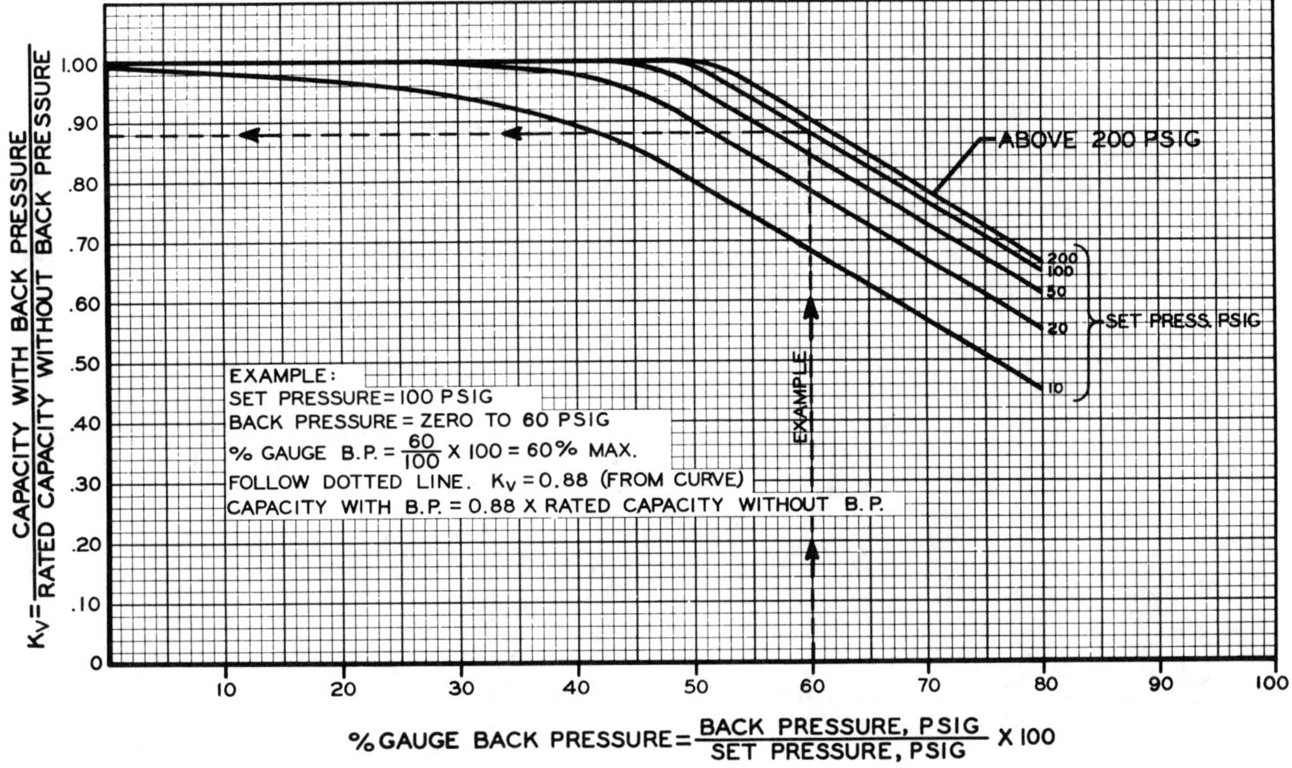

Figure 12.29. Flow factor correction curve used for Farris bellows valves at varying percentages of back pressure. (Courtesy of Teledyne Farris Engineering)

pressure. It may be noted that for set pressures 100 psig and higher, back pressures up to 50% of set pressure have no effect on capacity ratings of balanced bellows valves. At low set pressures, the effect becomes quite pronounced. For example, at 10 psig the factor is 0.8 at 50% back pressure.

Conventional top guided safety valves are readily adaptable to the addition of balanced bellows. The addition is sometimes desirable to prevent corrosion of the guiding surfaces that might result from the lading fluid contamination.

From a historical standpoint, it might be well to mention that an early effort to overcome back pressure effects was the use of balanced pistons (Figure 12.30). However, with the successful development of the more effective bellows seals, their use has been discontinued—at least by the major manufacturers.

Blowdown Rings

Blowdown rings accomplish two basic purposes:

1. Along with the disc holder (Figure 12.31), they form a "huddling" chamber that provides the "pop" action necessary for safety valves.
2. Adjustment of the blowdown ring(s) varies the difference between the set pressure and reseating pressure of the valve.

Pop action on safety valves is needed to prevent the condition known as chattering which occurs when the valve starts to open. Where a relief valve on liquid service opens slightly and the pressure drops sharply with the release of a relatively small volume of fluid, safety valves in vapor service exhibit different characteristics. There is a tendency for the valve to cycle, alternately opening and closing. This often causes damage to the valve seats resulting in leakage.

The solution was the design of the disc holder and blowdown rings to take advantage of the stream pressure and kinetic forces to obtain a greater lifting force as soon as the valve started opening. The resulting "pop" action is the distinguishing characteristic between relief and safety valves.

Blowdown is defined as the percentage difference between the pressure at which the valve starts to open (set pressure) and the pressure at which it reseats. Maximum blowdown occurs when the blowdown ring(s)—there may be one or two—are brought toward the disc (Figure 12.32). As the blowdown ring is adjusted away from the disc, blowdown decreases.

Normal blowdown on a valve is 5% of set pressure. It is normally set by bringing the ring all the way up to the disc, then backing off the number of turns suggested by the vendor. This recommendation comes from the vendor because there are few test facilities where it may be checked by customers. Blowdown per turn is usually based on tests using methane. Other fluids have different blowdown characteristics.

Bonnets and Yokes

Bonnets are housings around the springs of relief, safety and safety relief valves. The valve spring may or may not be in the relieving stream. Figure 12.33 shows the bonnet for a screwed safety valve where the spring is in the flowing stream. Figure 12.34 shows the bonnet enclosing a spring that is outside the flowing stream.

Bonnets are not necessary where there is no need to contain escaping fluids to prevent their release in a particular area. In the petroleum and related industries, it is usually necessary to contain the liquids and vapors that might escape and direct their flow to flare headers or specific areas of release.

The majority of valves used in the processing industry are provided with bonnets. When balanced bellows valves are used, no leakage to the bonnet occurs; the bonnet vent in this case is left open so the bellows may "breathe" properly. On conventional valves, bonnet vents are plugged to contain flowing fluids.

Safety valve applications that do not require fluid containment use an open spring arrangement (Figure 12.35) which uses a yoke mount for spring mounting. It is a simple, economical design that is satisfactory for clean, nontoxic, nonflammable services.

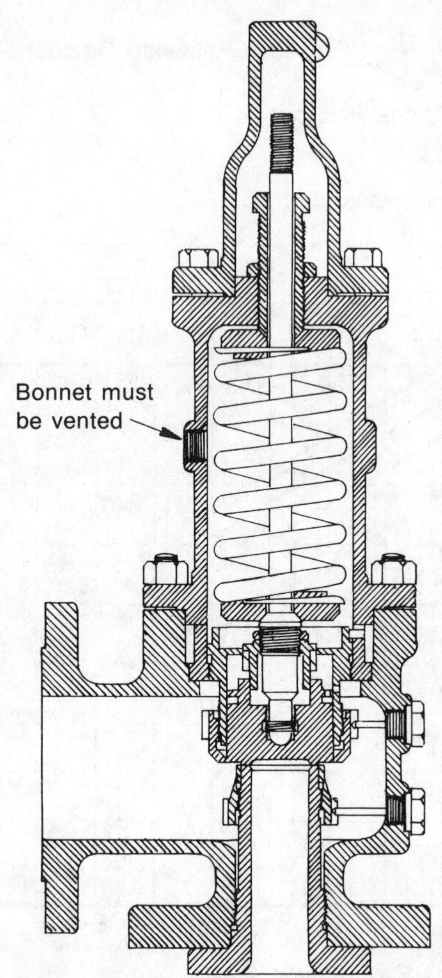

Figure 12.30. Safety valve with balanced disc and vented piston was an early effort to overcome the effects of back pressure. (Courtesy of API)

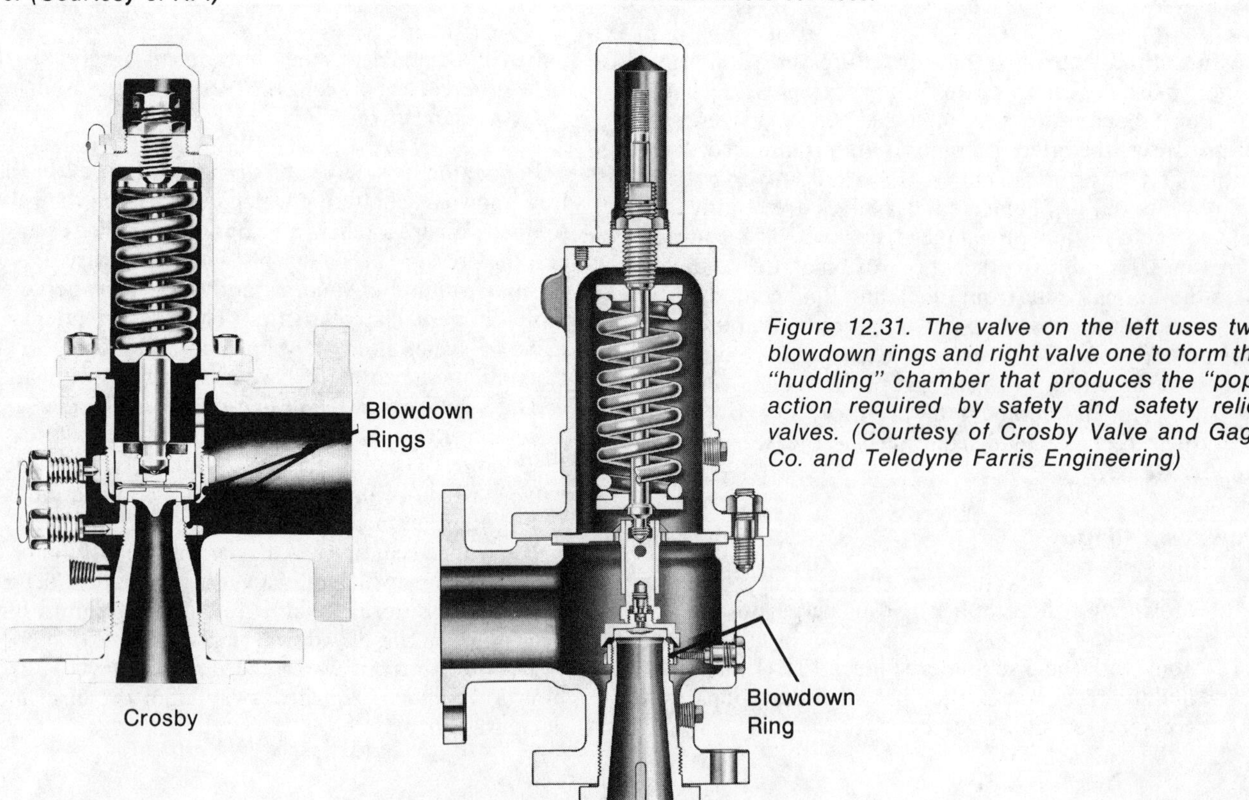

Figure 12.31. The valve on the left uses two blowdown rings and right valve one to form the "huddling" chamber that produces the "pop" action required by safety and safety relief valves. (Courtesy of Crosby Valve and Gage Co. and Teledyne Farris Engineering)

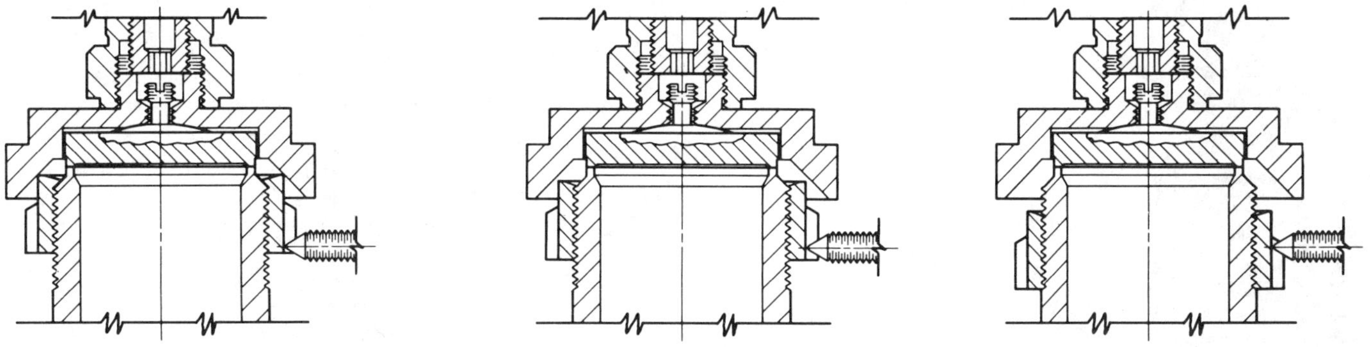

Figure 12.32. From left to right, the blowdown ring is brought completely up to the disc, backed away to provide "pop" action for vapor services and backed away completely for liquid relief service. (Courtesy of Teledyne Farris Engineering)

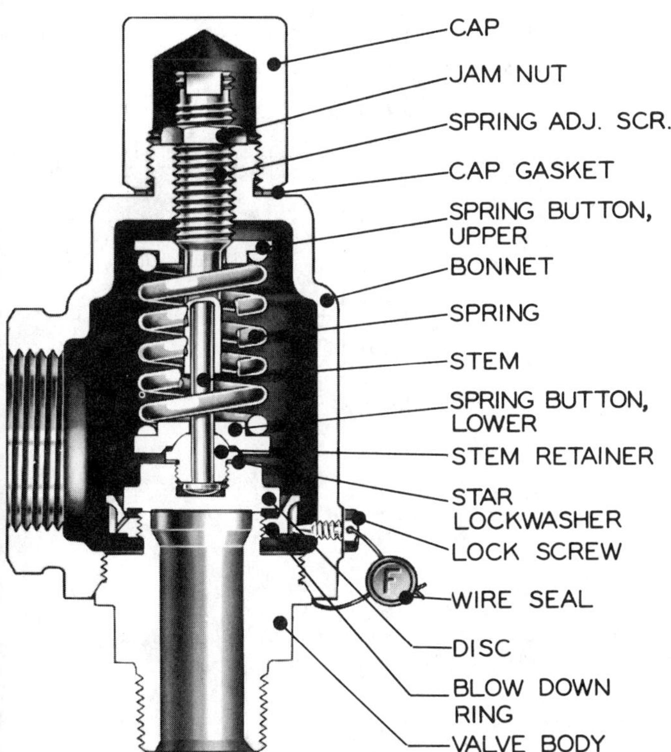

Figure 12.33. Relief valve bonnet encloses the spring inside the flowing relief stream. (Courtesy of Teledyne Farris Engineering)

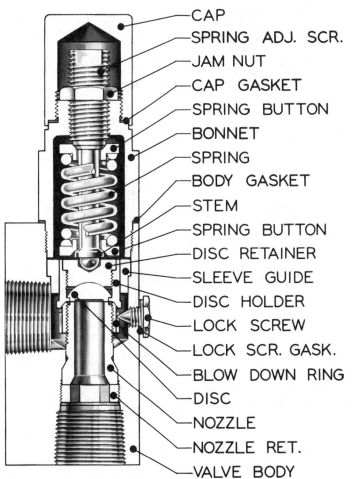

Figure 12.34. Relief valve bonnet encloses the bonnet outside the flowing relief stream. (Courtesy of Teledyne Farris Engineering)

Accessories

Several other items that are optional are discussed briefly under the heading of accessories.

Lifting Mechanisms

The purpose of a lifting mechanism is to open a valve when the pressure under the valve is lower than the set pressure. Lifting levers are supplied when periodic testing of the relief valve is desirable or mandatory. They are required on steam boilers and on air compressors and are quite common on other steam and air services. Their use on other process services varies with user practices.

Lifting mechanisms are furnished in three basic types: plain lever, packed lever and air-operated devices.

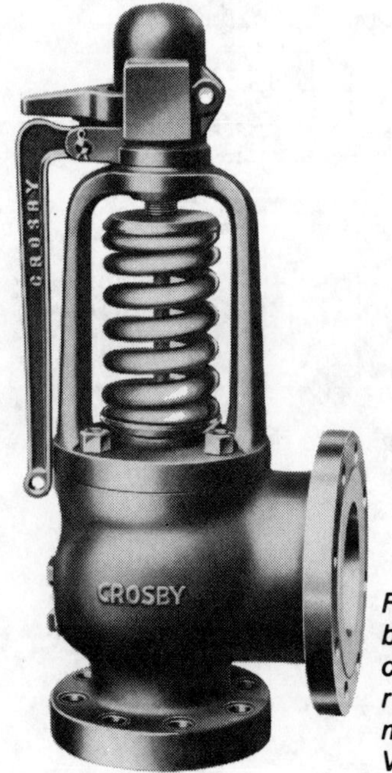

Figure 12.35. Enclosed valve bonnets are not necessary for clean services that may be released safely to the atmosphere. (Courtesy of Crosby Valve and Gage Co.)

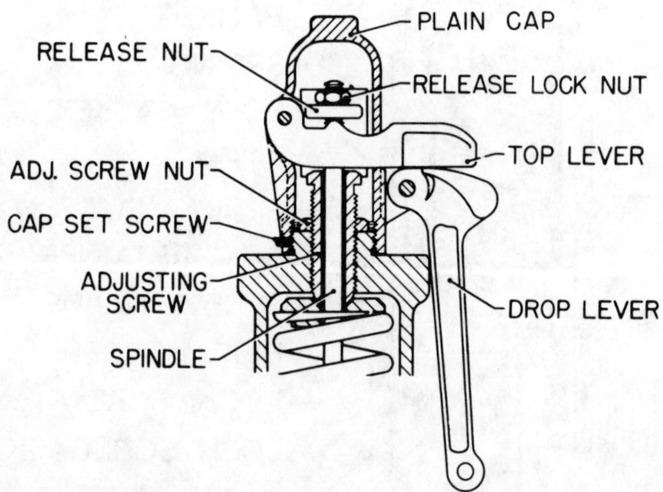

Figure 12.36. Plain lifting levers are used on services where escape of discharging gases to the atmosphere is not objectionable. (Courtesy of Dresser Industries, Inc.)

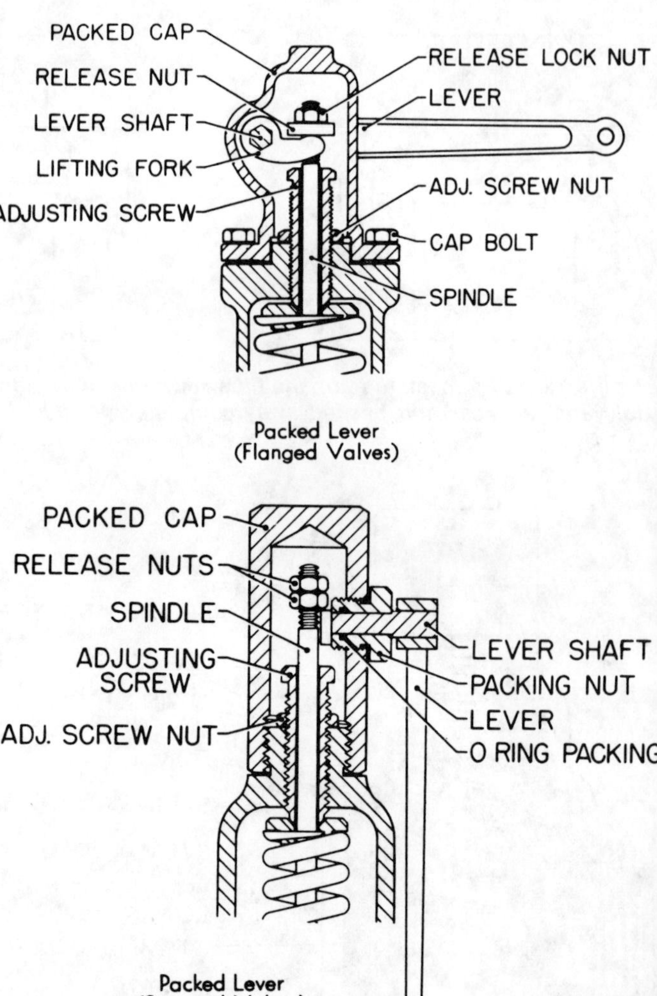

Figure 12.37. Packed lifting levers are used where back pressure is present and discharging vapor must be contained. (Courtesy of Dresser Industries, Inc.)

The plain lever (Figure 12.36) can be used where no back pressure is present and where the escape of discharging vapors is not objectionable.

Packed levers (Figure 12.37), as the name implies, are used on applications where leakage around the shaft cannot be permitted and where back pressure is present.

An air-operated lifting device (Figure 12.38) uses a diaphragm motor to obtain lifting power. Normal operation of the relief valve, however, is entirely independent of the lifting device. It is used when remote operation is desired.

In addition to its use for periodic testing, lifting mechanisms may also be used:

1. To lift the disc from the valve seat periodically to make sure the disc is not frozen as a result of corrosion, caking or other flowing media deposits.
2. To remove foreign particles which are sometimes trapped under the seat as the valve closes. This sometimes stops leaks by immediate cleansing of the flowing fluid, lowers maintenance cost and avoids shutdowns.
3. For venting purposes—piping and/or equipment.

Gags

The purpose of the gag (Figure 12.39) is to hold the safety relief valve closed while equipment is being subjected to a pressure greater than its set pressure. The valve may be left in place during a hydraulic test to avoid the cost of removal and reinstallation. The gag should be removed after using and never left on the valve.

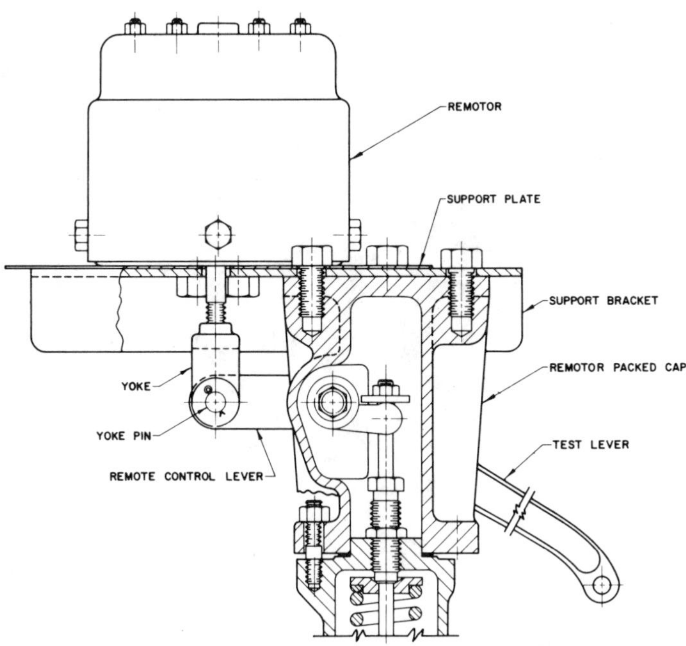

Figure 12.38. Air-operated lifting levers are used when remote operation of the lifting mechanism is desired. (Courtesy of Teledyne Farris Engineering)

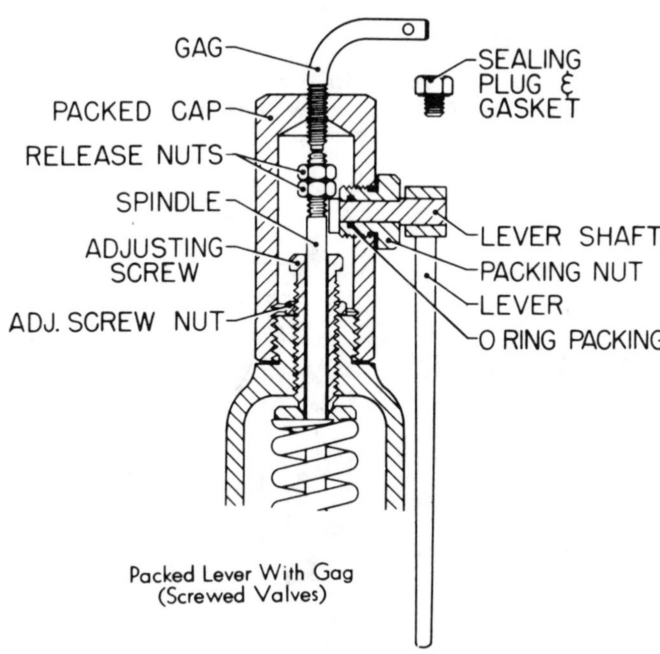

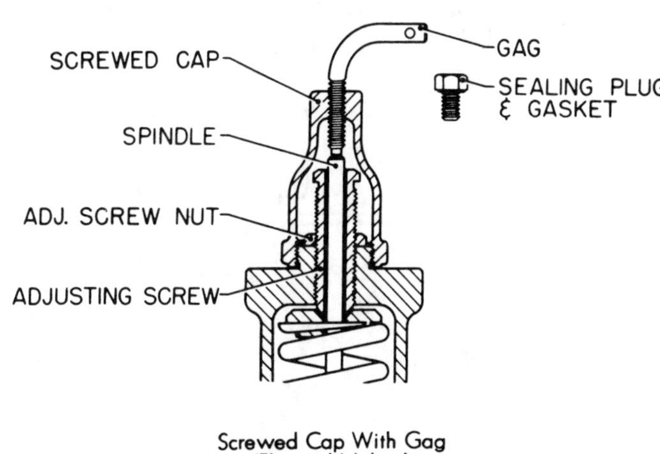

Figure 12.39. Gags are used when the need arises to hold the valve closed while operating or testing the system at pressures above the set pressure. (Courtesy of Dresser Industries, Inc.)

Caps

Relief valves are normally furnished with screwed caps (Figure 12.40) but are optionally furnished with bolted caps (Figure 12.41).

Jacketed Bodies

Where viscous or heavy residual fluids are used in process work, it is often necessary to apply heat to the piping to and from the safety relief valve as well as the valve itself in order to maintain flow and to avoid partial or total solidification of the flowing fluid. The jacketed safety relief valve (Figure 12.42) is suggested as a solution which keeps viscous fluids warm enough to flow, eliminates cumbersome steam tracing and avoids costly maintenance in removing or dismantling plugged valves. Previous to the development of this valve, steam tracing was the only method available to the user that would satisfy this requirement.

The steam jacketed valve not only is an improvement in heat transfer to maintain flow, but it also makes available a standard valve designed to accomplish this purpose.

Pilot-Operated Valves

The function of pilot-operated relief valves is the same as it is for direct-operated spring valves, except that the pilot-operated valves can provide a feature not inherent in spring loading—remote operation. There are several other features that distinguish them from spring-loaded valves, some of which are desirable while others are detrimental, at least for some services. Some of their salient features include:

1. Operability up to 98% of set pressure.
2. Positive opening and closing characteristics devoid of chatter.
3. Reduction of product losses, downtime and maintenance in many instances.
4. Higher capacities for the same size valve body.
5. Remote operating capability.

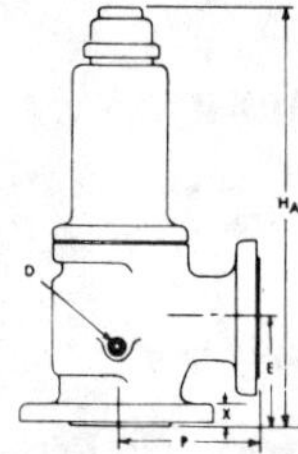

Figure 12.40. Screwed cap on bonnet protects against tampering with spring setting. (Courtesy of Crosby Valve and Gage Co.)

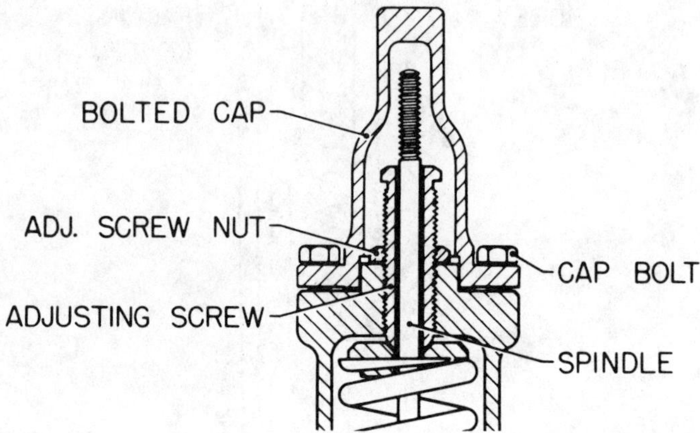

Figure 12.41. Flanged valves use bolted caps. (Courtesy of Dresser Industries, Inc.)

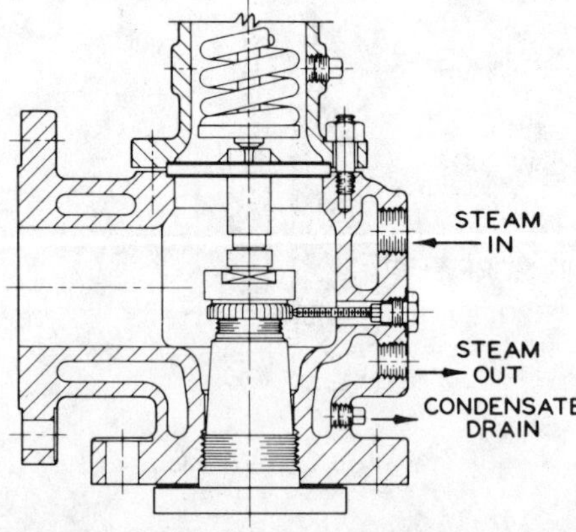

Figure 12.42. Relief valve bodies are steam-jacketed to prevent materials from solidifying or freezing or becoming too viscous. (Courtesy of Teledyne Farris Engineering)

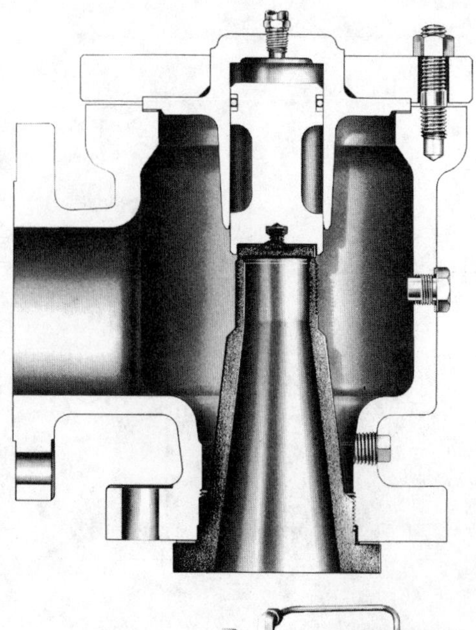

Figure 12.43. Pilot-operated relief valve uses piston main valve, with process pressure on both sides of piston. Larger area on top of piston keeps valve closed until top is vented by pilot. (Courtesy of Teledyne Farris Engineering)

Design Features

Pilot-operated valves use a floating piston as a main valve, with a larger area on top of the piston than on the bottom side at the nozzle bore (Figure 12.43). Since the static (process) pressure is the same on both ends of the piston, the force holding the piston closed increases as the process pressure increases. This feature prevents leakage inherent in spring-operated valves, especially as set pressure is approached by the process. At the set pressure, the pilot valve pops open and vents the top of the piston, allowing the process pressure on the bottom of the piston to force the valve open quickly. The valve operates in a full open position.

Closure of the pilot-operated relief valve is accomplished in different ways. One manufacturer uses a pilot valve (Figure 12.44) which utilizes a blowdown relay to open as the relieving system pressure is reduced past the set point. At the same time, a trigger relay (which opened at set pressure to exhaust the top of the piston) closes and diverts the system pressure again to the top of the piston, quickly closing the main valve smoothly and positively. The greater force due to the large top area keeps the valve closed tightly.

Another manufacturer uses a system in which a variable orifice is built into the pilot valve where the supply line (Item 10 in Figure 12.45) enters the pilot from the main valve. When the pilot opens, the flow through the supply line causes an immediate pressure drop across the orifice. The

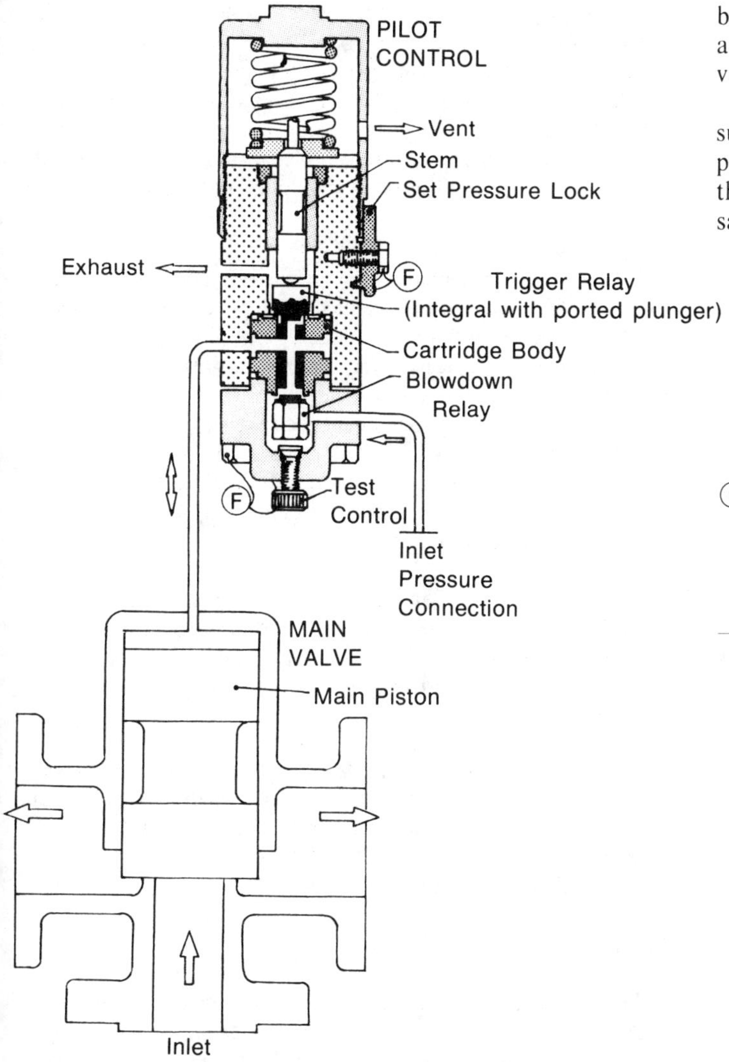

Figure 12.44. When the relieving system reduces below the set pressure, the blowdown relay opens and the trigger relay closes in the pilot valve, diverting system pressure to the top of the main valve piston to close valve. (Courtesy of Teledyne Farris Engineering)

amount of system blowdown is determined by adjusting the orifice size and thus the pressure drop across it. When the predetermined (orifice size) system blowdown pressure is reached, the pilot valve closes, and the full system pressure is immediately diverted to the dome. The piston is quickly moved down to close the main valve.

When pilot-operated relief valves discharge into a system where back pressure can build up to pressures higher than the set pressure, back flow may exist in the supply line, causing the valve to open. To prevent this possibility, check valves can be installed as shown in Figure 12.46. As the higher pressure is introduced into the dome through check valve *B*, check valve *A* prevents this back pressure from entering the pilot supply tube and main valve inlet. In normal operation, check valve *B* prevents dome pressure from bleeding into the discharge line. Back flow preventers such as these are optional features provided by all pilot-operated valve manufacturers.

The ability of the pilot-operated valves to operate within such narrow pressure bands is due to the snap action of the pilot relays in diverting process fluid pressure to and from the top side of the floating piston of the main valve. This same feature—the snap action—also provides the positive

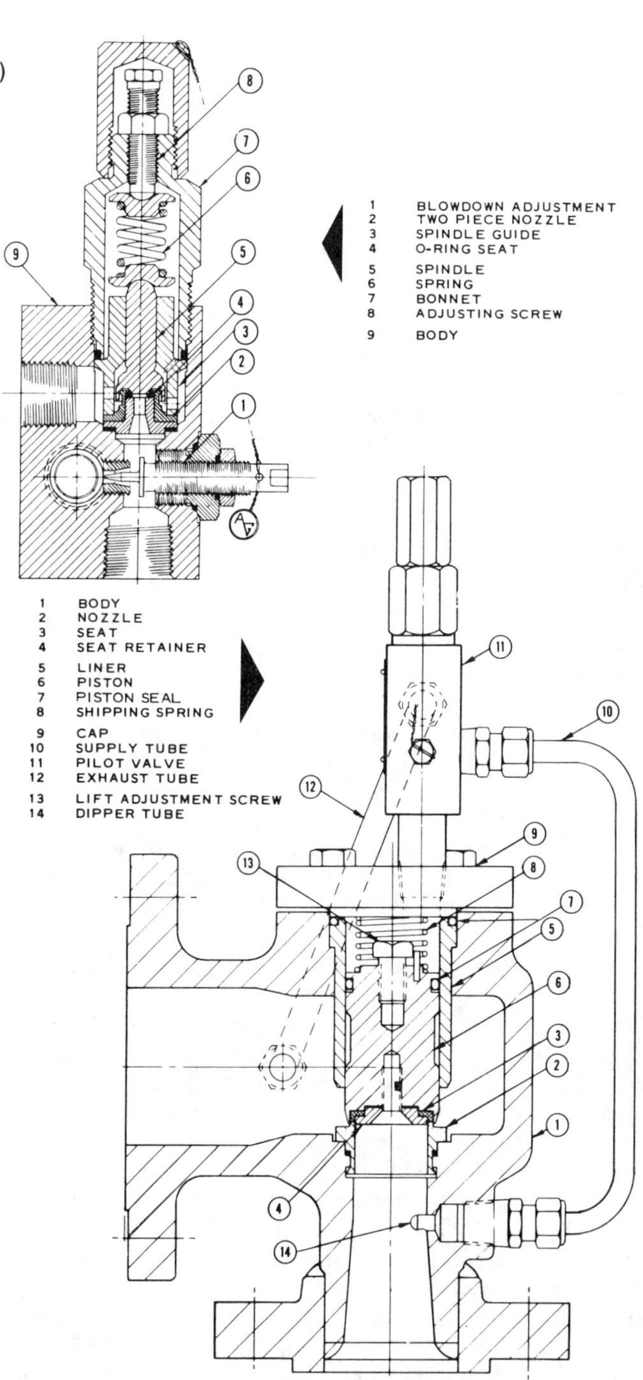

Figure 12.45. An adjustable orifice in the supply line to pilot valve determines the blowdown in this pilot-operated valve. (Courtesy of Anderson, Greenwood and Co.)

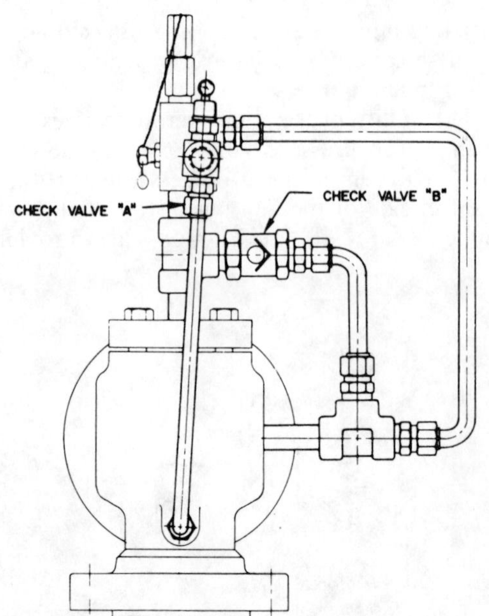

Figure 12.46. Back flow prevention by installing check valves A *and* B *in systems where back pressure may build up—for pilot-operated valves. (Courtesy of Anderson, Greenwood and Co.)*

seating force which eliminates chatter that is inherent in direct spring-operated relief valves.

Comparison with Direct Spring-Operated Valves

Several areas are discussed in comparing pilot-operated and spring-loaded valves, including leakage, capacity, maintenance, flexibility and safety.

Leakage problems are less severe with pilot-operated valves. The reasons have already been stated: the great force available for positive closure is available to maintain tightness until set pressure is reached and the pilot valve exhausts air pressure from the top of the main valve piston. Spring-operated valves, on the contrary, have decreasing differential pressure holding the valve closed as process pressure approaches the setting of the opposing spring.

Maintenance of direct spring-operated valves is likely to be higher, caused in part by the low differential pressures noted above which allow valve chatter near valve set pressures. The intermittent opening and closing of the valves (chatter or simmer) damage the seating surfaces of the disc and nozzle. To offset this problem partially, the blowdown rings of safety relief valves (spring-operated) are usually adjusted to a position where simmer and blowdown are both at acceptable values. The blowdown setting that reduces simmer to an acceptable value necessarily widens the pressure band between the opening and closing of the valve. Operating and set pressures may be closer together for pilot-operated valves—a good economical factor because it may allow smaller valves and lower pressure ratings on required equipment.

Capacities of pilot-operated relief valves are greater than conventional for the same size valve body as shown in Figure 12.47. This is true primarily because nozzles are not needed to generate the velocities necessary for pop action which is required for full opening spring-loaded valves in vapor service. Pilot-operated valves provide both high lift and full bore advantages to produce high capacity for a given inlet size.

Flexibility is provided with the pilot-operated valve in that it can be controlled remotely as well as used for standard relief service. Remotely mounted pneumatic devices or electrically operated solenoid valves may be used to trigger the pilot. Such manually operated systems are often used for emergency venting purposes.

Safety is a major consideration in every application of safety relief valves; in fact, it is the stated purpose for their use. It is in this area that pilot-operated valves have not met with the same acceptance as direct spring-loaded valves. The two significant factors are

1. More and smaller parts that may malfunction.
2. The small lead lines between the process and the pilot valve and between the pilot and main valves are subject to clogging of foreign material which may cause malfunctions.

These are valid objections, in some services at least. However, the objections must be weighed against other problems, such as leakage potential, waste (pollution) and economic factors which may offset such objections. Each situation must be evaluated separately.

Application

There are many applications in gas transmission services, at compressor stations and on unfired pressure vessels where pilot-operated valves may be used advantageously. At compressor stations along gas transmission routes, they provide emergency venting when equipment malfunctions or when hazardous conditions exist.

When operating pressures are changed in a process and equipment design limits are approached, the use of pilot-operated valves with their narrow pressure band are certainly an economic advantage. Figure 12.48 depicts quite well the comparative operating ranges of pilot- and spring-operated valves.

In processes containing toxic fluids where leakage presents extremely hazardous situations, pilot-operated valves should be considered. Their economy is unquestioned for many applications, and their safety matches the more widely used spring-operated valves in many services.

Rupture Discs

A rupture disc is a designed "weak spot" in a pressure system. In its simplest form, it is a thin metal or graphite membrane held between flanges and designed to burst at a predetermined pressure (Figure 12.49). Its purpose is basically the same as other pressure-relieving devices—to

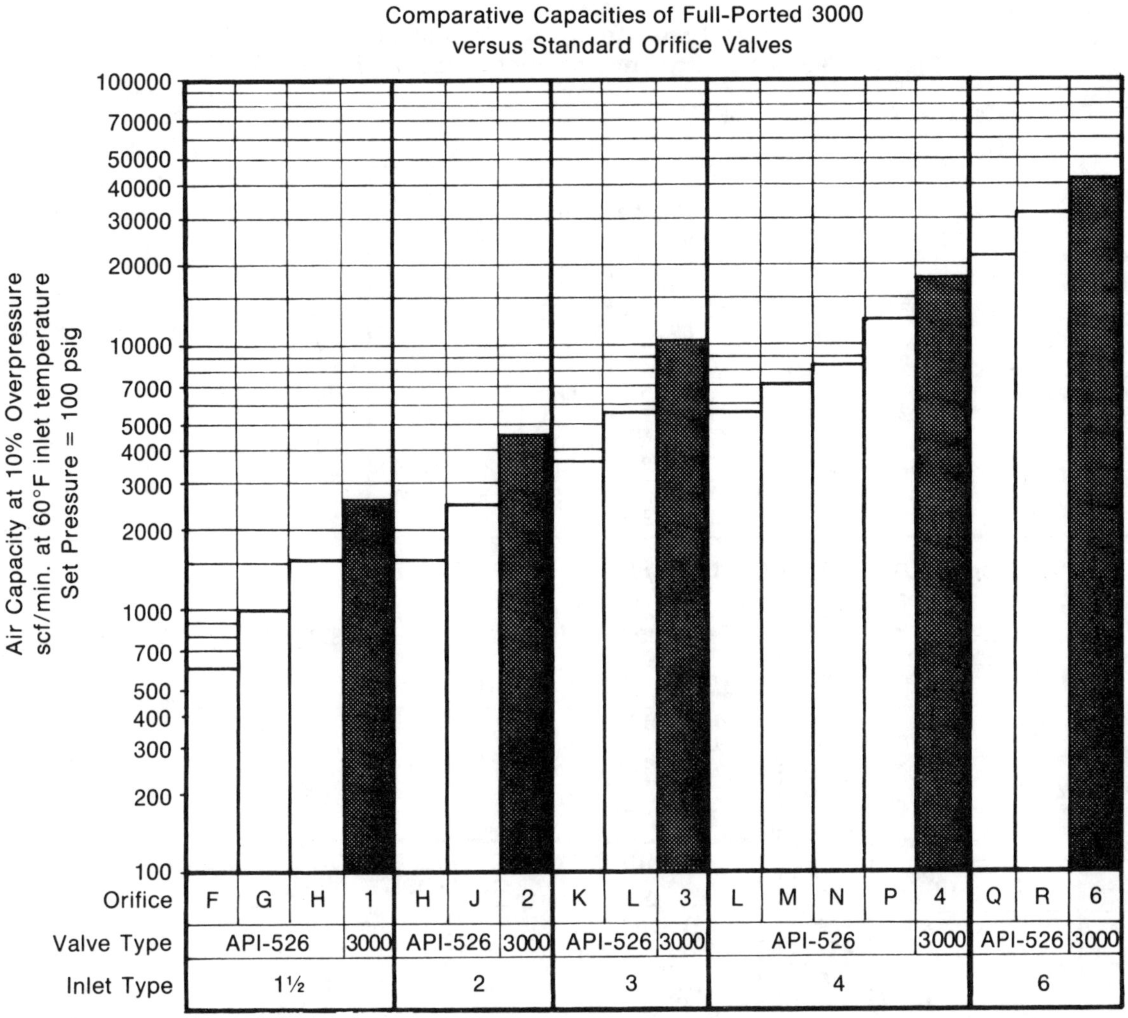

Comparative Capacities of Full-Ported 3000 versus Standard Orifice Valves

Air Capacity at 10% Overpressure scf/min. at 60°F inlet temperature Set Pressure = 100 psig

100000 70000 50000 40000 30000 20000 10000 7000 5000 4000 3000 2000 1000 700 500 400 300 200 100

Orifice	F	G	H	1	H	J	2	K	L	3	L	M	N	P	4	Q	R	6
Valve Type	API-526			3000	API-526		3000	API-526		3000	API-526				3000	API-526		3000
Inlet Type	1½				2			3			4					6		

Figure 12.47. Comparative capacities of pilot-operated valves with conventional relief valves. (Courtesy of Teledyne Farris Engineering)

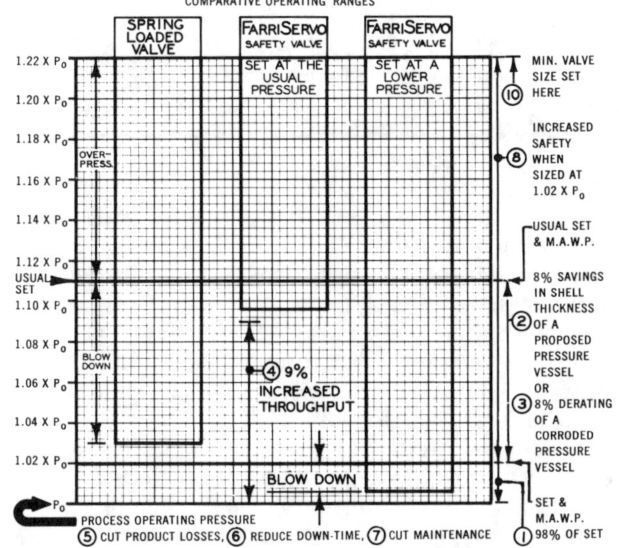

Figure 12.48. Comparative operating features of spring- and pilot-operated valves. (Courtesy of Teledyne Farris Engineering)

Figure 12.49. Rupture discs are thin metal or graphite membranes mounted between mating flanges and designed to fail at predetermined values. (Courtesy of BS & B Safety Systems)

prevent damage to equipment, to prevent unnecessary downtime and to provide safety. Rupture discs are used in services where pressure relief valves are impractical, as secondary relief devices for emergency operation, as explosion protectors and for economy.

Table 12.4 lists materials from which rupture discs are made. The most commonly used are aluminum, nickel, Monel, Inconel and austenitic stainless steel; however, other materials are used when needed. Metal foils, strips and

Table 12.4 Operating Limits of Metal and Composite Rupture Disc Material. Pressures Are Given in PSIG and Kg/Cm²; Temperatures in °F and °C.

SOLID METAL DISKS: Types B, BV, BR, BRV, BRSV

Disk Material	Unlined Disks			CTFE and FEP Lined Disks					Lead Lined Disks			
	Min. Pres.	Max. Pres.	Max. Temp.	Min. Pres. 1/side	Min. Pres. 2/sides	Max. Pres. 1/2 sides	Max. Temp. CTFE	Max. Temp. FEP	Min. Pres. 1/side	Min. Pres. 2/sides	Max. Pres. 1/2 sides	Max. Temp.
Aluminum	40 2.81	1000 70.3	250 121	90 6.33	140 9.9	1000 70.3	250 121	250 121	85 5.98	120 8.4	700 49.2	250 121
Silver	90 6.33	1500 105.5	250 121	140 9.9	190 13.4	1500 105.5	250 121	250 121	— —	— —	— —	— —
Nickel	145 10.2	8000 562.6	750 399	195 13.7	245 17.2	3000 211.0	250 121	400 204	210 14.8	245 17.2	1400 98.5	250 121
Monel	175 12.3	10000 703.2	800 427	225 15.8	275 19.3	3000 211.0	250 121	400 204	235 16.6	270 19.0	1400 98.5	250 121
Inconel	225 15.8	12000 843.9	900 482	275 19.3	325 22.9	5000 351.6	250 121	400 204	290 20.4	325 22.8	1400 98.5	250 121
316 S.S.	320 22.5	12000 843.9	900 482	370 26.0	420 29.5	5000 351.6	250 121	400 204	385 27.1	420 39.5	1400 98.5	250 121

COMPOSITE DISKS: Types D, DV, DR, DRSV, PLD, PLDV

Seal Material	D, DV, DR, DRSV Disks			PLD Disks			PLDV Disks				
	Top Section Material			Center Section Material			Center Section Material				
	Inconel, 316SS			316SS			316SS		Nickel		All Mtls.
	Min. Pres.	Max. Pres.	Max. Temp.	Min. Pres.	Max. Pres.	Max. Temp.	Min. Pres.	Max. Pres.	Min. Pres.	Max. Pres.	Max. Temp.
FEP (Teflon)	89 6.26	1000 70.3	400 204	130 9.1	1000 70.3	400 204	300 21.1	1000 70.3	230 16.2	600 42.2	400 204
TFE (Teflon/ Halon)	44 3.09	1000 70.3	500 260	87 6.12	1000 70.3	500 260	300 21.1	1000 70.3	230 16.2	600 42.2	500 260
Aluminum	50 3.52	1600 112.5	800 427								
Silver	110 7.7	2000 140.7	800 427								
Copper	170 12.0	2000 140.7	800 427								
Nickel	180 12.7	2000 140.7	1000 538								
Monel	220 15.5	2000 140.7	1000 538								
Inconel	285 20.0	2000 140.7	1000 538								
316 S.S.	400 28.1	2000 140.7	1000 538								
Hastelloy B or C	680 47.8	2000 140.7	1000 538								

MANUFACTURING RANGES FOR SOLID METAL OR COMPOSITE DISKS

Desired Pressure Rating	Test specimens must rupture within limits below of the pressure you specify		Desired Pressure Rating	Test specimens must rupture within limits below of the pressure you specify	
PSIG	Plus	Minus	Kg/CM²	Plus	Minus
2½— 3½	1 psig	1 psig	.176— .246	.070 Kg/CM²	.070 Kg/CM²
4 — 6	2 psig	1 psig	.281— .422	.141 Kg/CM²	.070 Kg/CM²
7 — 10	2½ psig	1½ psig	.49 — .70	.176 Kg/CM²	.105 Kg/CM²
11 — 16	3 psig	2 psig	.77 — 1.13	.211 Kg/CM²	.141 Kg/CM²
17 — 25	4 psig	2 psig	1.20 — 1.76	.281 Kg/CM²	.141 Kg/CM²
26 — 40	5 psig	3 psig	1.83 — 2.81	.352 Kg/CM²	.211 Kg/CM²
41 — 65	6 psig	4 psig	2.88 — 4.57	.422 Kg/CM²	.281 Kg/CM²
66 —100	9 psig	5 psig	4.64 — 7.03	.633 Kg/CM²	.352 Kg/CM²
101 —150	12 psig	6 psig	7.10 —10.55	.844 Kg/CM²	.422 Kg/CM²
151 —200	16 psig	9 psig	10.62 —14.06	1.125 Kg/CM²	.633 Kg/CM²
201 —350	23 psig	12 psig	14.14 —24.61	1.617 Kg/CM²	.844 Kg/CM²
351 —500	30 psig	15 psig	24.68 —35.16	2.11 Kg/CM²	1.055 Kg/CM²
501 & Up	6 pct.	3 pct.	35.23 & Up	6 pct.	3 pct.

Source: BS&B Safety Systems.

sheets in soft, annealed condition are required. At least 70% of all metal rupture disc applications involve the use of metal foil, less than 0.010 inch thick. As thickness requirements increase, the material is supplied in strip and sheet form. Figure 12.50 shows how disc material is stored. High cost precious metal and alloy materials come in much smaller sheets than are shown in the illustration, and storage schemes are different from those pictured.

Code Requirements

Rupture discs are accepted as primary relief devices under the ASME Unfired Pressure Vessel Code, Section VIII, Paragraphs UG-125 through UG-130; under Power Boiler Code, Section I, Part PVG, covering such equipment as Dowtherm boilers; and under Section III, Article 9, for nuclear vessels and equipment. A study of the paragraphs relating to rupture discs leaves the impression that they are accepted or tolerated; their use does not seem to be encouraged. API recommendations on relieving devices have very little to say about rupture discs.

Usage of rupture discs appears to be increasing; this increase has probably been brought about by increased production of plastics, other viscous materials, corrosive chemicals and high pressure processes. The use of relief valves on plastics or other viscous material streams is questionable because of plugging problems and flow characteristics of the process material. Highly corrosive

Figure 12.50. Storage facilities for rupture disc material. (Courtesy of BS & B Safety Systems)

Figure 12.51 Metal rupture disc—prebulged solid metal disc with angular seat, before and after rupture. (Courtesy of BS & B Safety Systems)

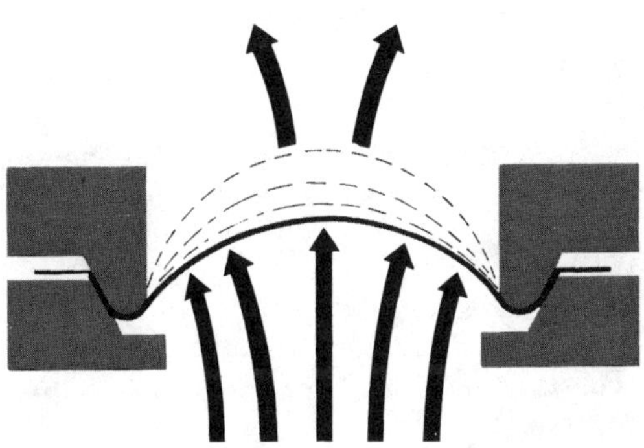

Figure 12.52. Conventional rupture disc with angular seating. Pressure loading is on concave side of disc, putting disc metal under tension. (Courtesy of BS & B Safety Systems)

chemical services that corrode operating trim cause valve operation to be questionable in some instances (expensive also). Some high pressure processes have relief requirements in ranges beyond the state-of-the-art of present relief valve development. Because of many questionable applications for relief valves, the rupture disc becomes an obvious solution for overpressure protection.

The codes make no mention of rupture discs as secondary relieving devices. This leaves the responsibility of specifying additional relieving devices to the discretion of the engineer who often uses a rupture disc set to rupture at the test pressure of a vessel. It can serve also as an emergency relief device where other relief equipment might possibly malfunction.

Types of Discs

Various characteristics of rupture discs need to be considered when selecting for different applications. They are classified into five basic types with variations.

Solid-Metal

Conventional solid-metal discs are shown in Figure 12.51, before and after rupture. The design utilizes a dome-shaped disc with angular seating arrangement. System pressure is applied to the concave side as shown in Figure 12.52. The prebulged shape is most common today as opposed to a flat disc design. Solid-metal discs may be lined with plastic material, such as Kel-F, Teflon or lead or coated with a corrosion retardant (Vinylite by BS&B) to meet corrosion requirements of some processes.

Another feature sometimes used on solid-metal discs is to score it mechanically with a cross or circular pattern to reduce the rupture pressure of a thicker piece of metal. Scored discs are more susceptible to fatigue failure from cycling pressures than conventional discs.

Composites

Because of some limitations of solid-metal discs, the composite disc design was developed. This involves the weakening of a solid-metal disc by a known amount with mechanical cutting and then creating a pressure seal by a separate plastic or metal membrane on the process (pressure) side. The composite design, compared to a solid-metal design, provides less margin between operating pressure and disc-burst rating and a lower bursting pressure for comparable metals of a given diameter. The composite

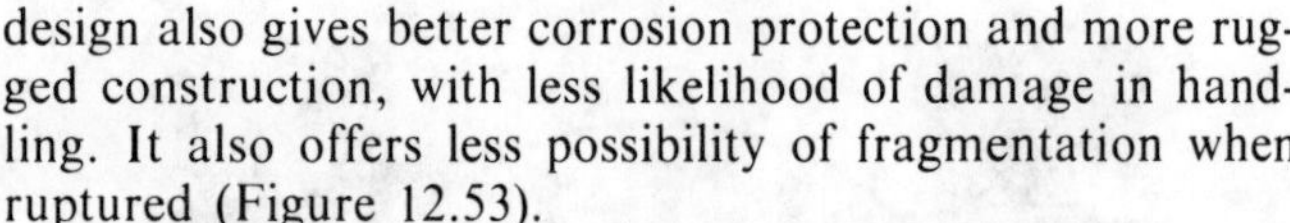

Figure 12.53. Composite rupture disc before and after rupture. (Courtesy of BS & B Safety Systems)

Figure 12.54. Composite disc, with vacuum support and plastic protective covers attached to both sides. (Courtesy of BS & B Safety Systems)

design also gives better corrosion protection and more rugged construction, with less likelihood of damage in handling. It also offers less possibility of fragmentation when ruptured (Figure 12.53).

The composite disc capability has recently been extended by the design of a disc with all of the composite benefits (similar construction) but with complete isolation of the metal by plastic. This design gives upstream and downstream corrosion protection and eliminates the need for a separate, exposed metal vacuum support on the process side. Figure 12.54 shows a composite disc of this design. The pie-shaped metal sections and plastic seals on both sides before and after rupture should be noted. The holder fittings for conventional and composite discs have angular seats. Figures 12.55 and 12.56 show types most often selected.

Reverse Buckling

Figure 12.57 shows a more recent design of rupture disc—a reverse buckling type. It provides greater accuracy and closer margin capability than the composite or the solid disc.

In the reverse buckling design, pressure is applied against the convex side of the disc, putting the metal in compression. As the compression load on the disc increases, a point is reached where the disc can no longer sustain the compression load. At this point, it buckles with an instantaneous, snap-acting movement.

The pressure at which the disc snaps to the reverse direction is determined by the inlet pressure-exposed diameter, type and thickness of disc metal and its radius of curvature. To take advantage of this predictable point of reverse buckling, a set of specially designed knife blades is permanently mounted in the downstream holding flange. The blades are made of hardened stainless steel and honed to a razor sharpness. They are heat-treated but left ductile enough to be resharpened.

The primary benefits from the reverse buckling design are

1. A reduction in the fatigue factor of the metal
2. A reduction in the creep characteristics of the metal
3. System operating pressures which can range up to 90% of set pressure over extended periods
4. The absence of fragmentation of disc particles when rupture and cutting open occur

Tests have shown that life expectancy can be 10 to 15 times that of older rupture disc designs. Tests also show that reverse buckling can be predicted within ±2% of nominal pressure rating.

Shear Discs

Figure 12.58 shows a shear rupture disc which uses a Bellville spring washer and a removable hollow punch. The Bellville spring washer possesses a unique design property that permits a decreasing force with an increasing stroke, allowing continuous operation of the disc up to 95% of burst pressure.

Impervious Graphite Discs

Figure 12.59 shows a graphite disc that is particularly useful in some highly corrosive services. This type utilizes a

(text continued on page 342)

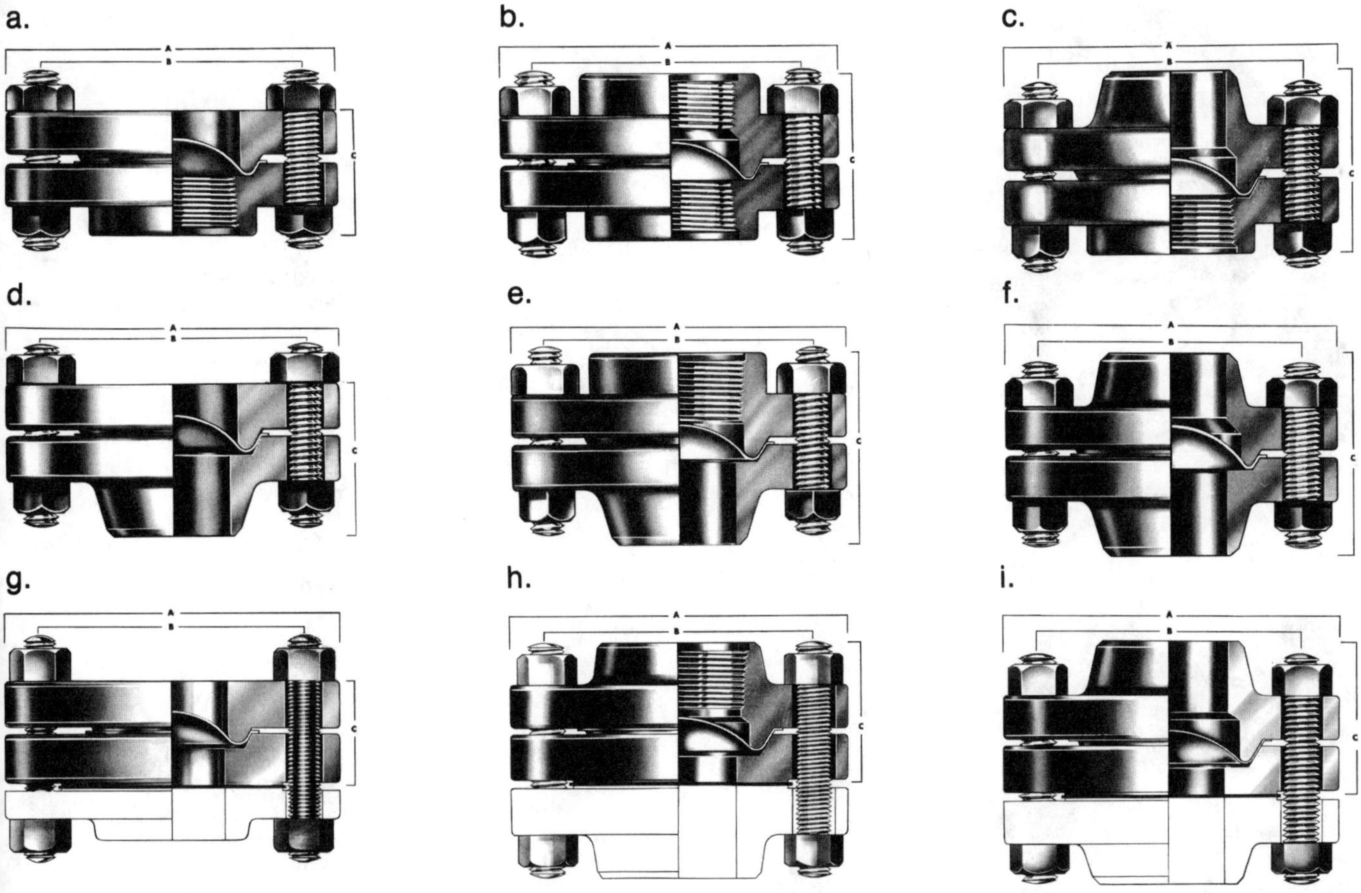

a. b. c. d. e. f. g. h. i.

Figure 12.55. Various holder assemblies for rupture discs. (Courtesy of BS & B Safety Systems)

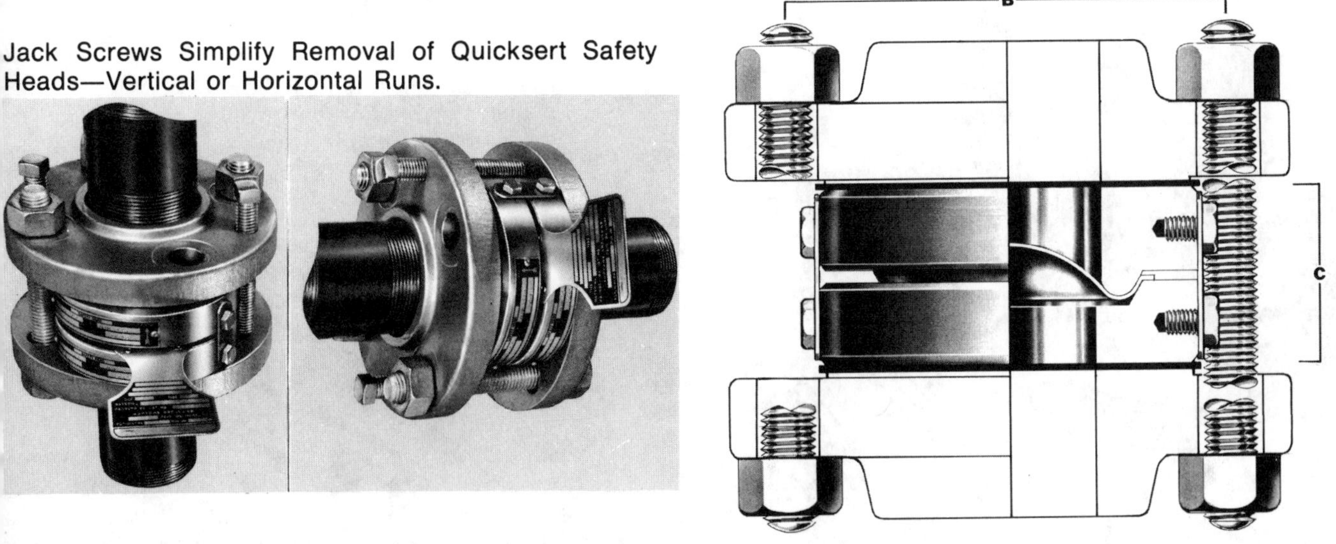

Jack Screws Simplify Removal of Quicksert Safety Heads—Vertical or Horizontal Runs.
A
B
C

Figure 12.56. An assembly used by BS & B for quick replacement of a rupture disc. (Courtesy of BS & B Safety Systems)

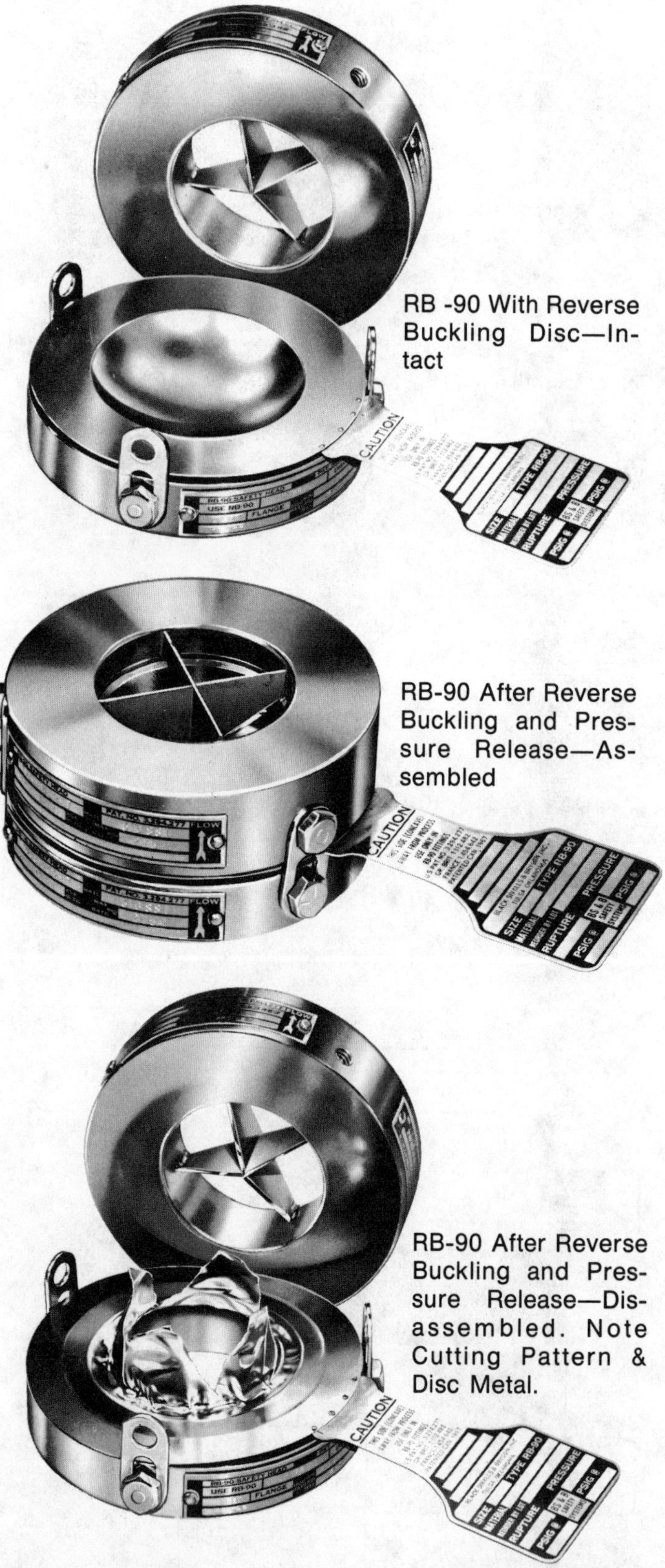

RB -90 With Reverse Buckling Disc—Intact

RB-90 After Reverse Buckling and Pressure Release—Assembled

RB-90 After Reverse Buckling and Pressure Release—Disassembled. Note Cutting Pattern & Disc Metal.

Figure 12.57. Reverse buckling design rupture disc allows close operating margin. (Courtesy of BS & B Safety Systems)

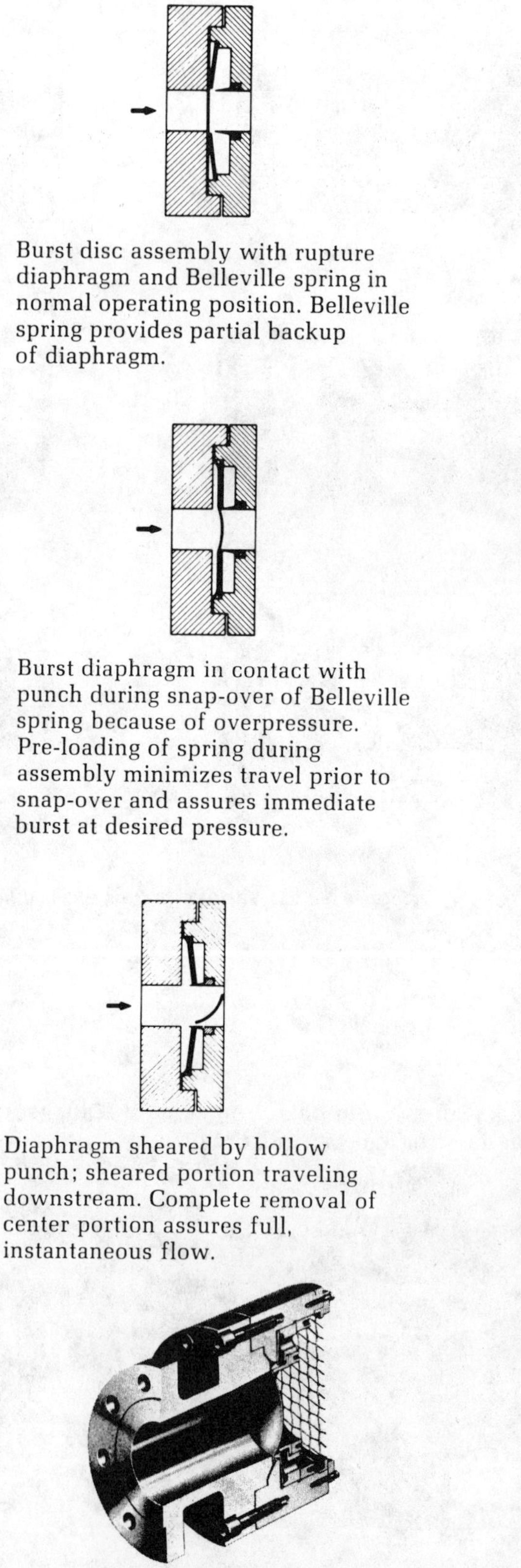

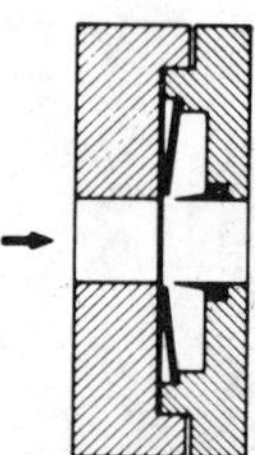

Burst disc assembly with rupture diaphragm and Belleville spring in normal operating position. Belleville spring provides partial backup of diaphragm.

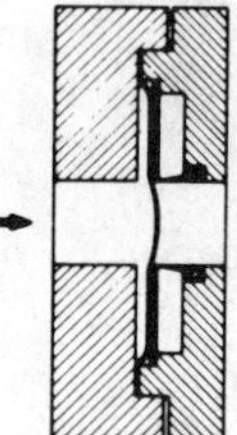

Burst diaphragm in contact with punch during snap-over of Belleville spring because of overpressure. Pre-loading of spring during assembly minimizes travel prior to snap-over and assures immediate burst at desired pressure.

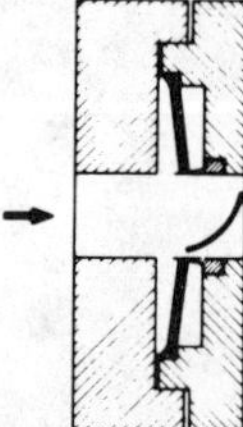

Diaphragm sheared by hollow punch; sheared portion traveling downstream. Complete removal of center portion assures full, instantaneous flow.

Figure 12.58. Operation of a shear disc manufactured by Ametek; it can operate to 95% of burst pressure. (Courtesy of Ametek/Calmec)

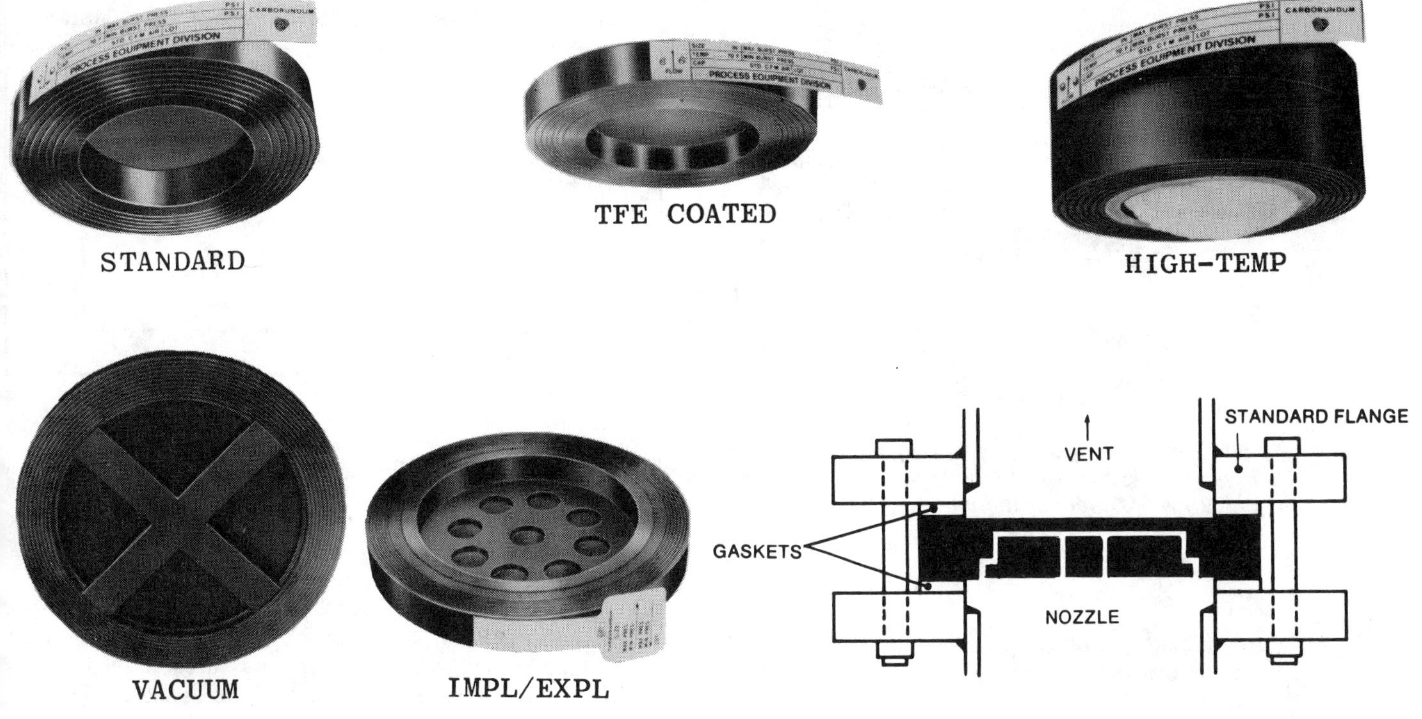

STANDARD DISKS

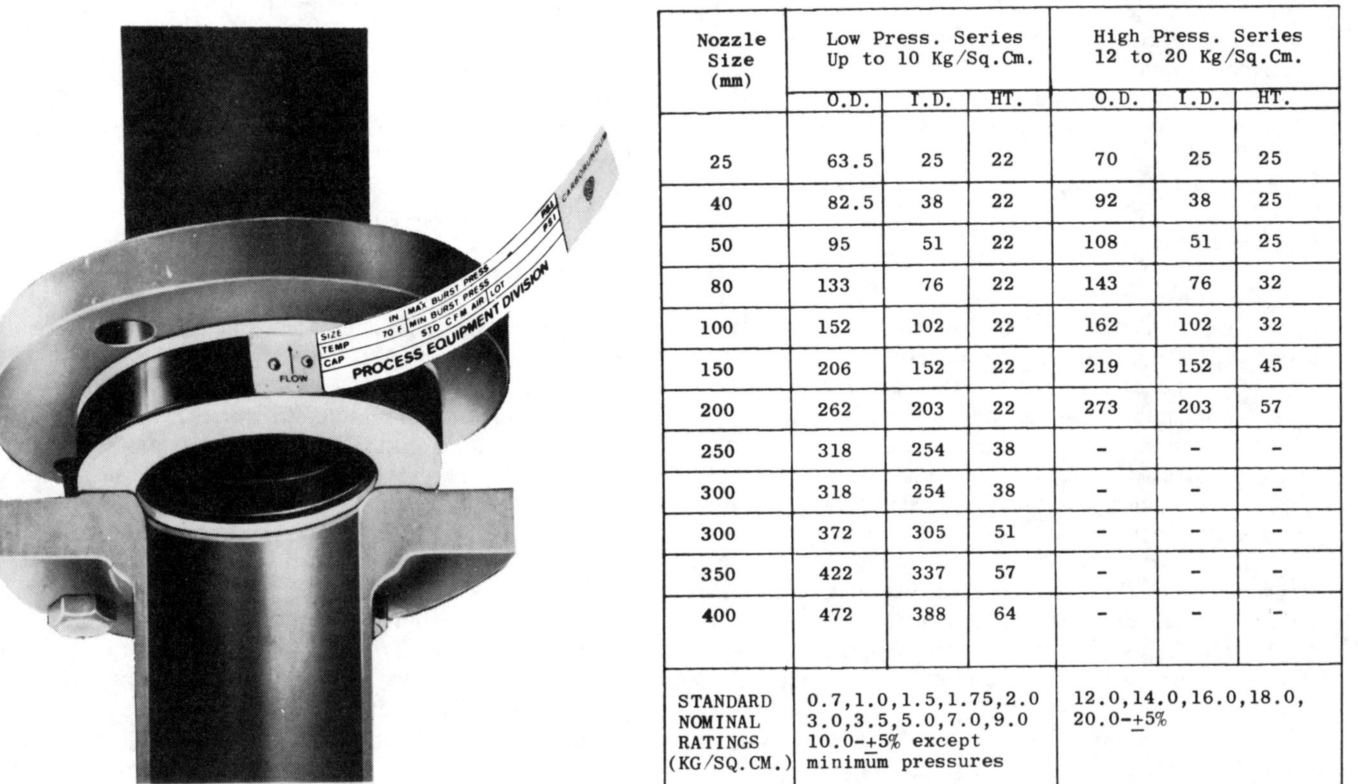

Nozzle Size (mm)	Low Press. Series Up to 10 Kg/Sq.Cm.			High Press. Series 12 to 20 Kg/Sq.Cm.		
	O.D.	I.D.	HT.	O.D.	I.D.	HT.
25	63.5	25	22	70	25	25
40	82.5	38	22	92	38	25
50	95	51	22	108	51	25
80	133	76	22	143	76	32
100	152	102	22	162	102	32
150	206	152	22	219	152	45
200	262	203	22	273	203	57
250	318	254	38	-	-	-
300	318	254	38	-	-	-
300	372	305	51	-	-	-
350	422	337	57	-	-	-
400	472	388	64	-	-	-
STANDARD NOMINAL RATINGS (KG/SQ.CM.)	0.7,1.0,1.5,1.75,2.0 3.0,3.5,5.0,7.0,9.0 10.0-+5% except minimum pressures			12.0,14.0,16.0,18.0, 20.0-+5%		

Figure 12.59. Impervious graphite discs are available for installation as shown between standard ASA flanges from 1- to 36-inch nominal pipe sizes. (Courtesy of The Carborundum Co.)

Figure 12.60. Vacuum supports are needed in services which cycle from vacuum to pressure conditions. (Courtesy of BS & B Safety Systems)

resin-impregnated graphite that makes it impervious to contained fluids. It resists the action of most acids, alkalis and organic solvents at temperatures up to 390°F.

Due to the nonfatiguing characteristic of graphite, no flexing failures occur, and burst ratings remain constant, regardless of temperature changes. Graphite discs can be used over an operating range of -400°F to almost 400°F.

When used in series with relief valves, extra care must be used in piping arrangements because of the fragmentation that occurs when they rupture. Graphite fragments caught in the valve may restrict fluid flow or prevent reclosing when the system pressure is reduced. Traps must be installed to catch all the fragments. With the use of vacuum supports, graphite discs are adaptable to vacuum service.

Vacuum Supports

A vacuum support is a specially constructed metal member mounted beneath a disc to prevent reversal during vacuum operation, but offering no resistance to adequate relief when the disc ruptures. There are many varieties; Figure 12.60 shows a typical low-pressure design. Vacuum supports are needed any time a process is likely to cycle from a positive to a vacuum condition.

Pressure cycling is extremely rough on solid-metal discs thin enough to bend easily. Great care is needed to support discs well enough to prevent reverse bending everywhere on the disc surfaces. Movement of the disc metal can create bends, ultimately causing fatigue cracks when cycling occurs. Disc life expectancy is increased proportionately to the degree of fit between the disc and its mating vacuum support.

Design criteria for vacuum supports include adequate opening with minimum pressure drop and minimum flow turbulance after the disc ruptures. Supports should retain all metal parts when a disc ruptures to prevent fragments from going downstream.

Table 12.5. Relative Performance of Disc Metals

Metal	Performance Value
Nickel	1,000
Thick aluminum (0.010 in. thick and up)	1,000
Inconel	700
Type 316 stainless steel	700
Monel	400
Type 347 stainless steel	400
Thin aluminum (0.005 in. thick and under)	7
Copper	2
Silver	2

Source: BS&B Safety Systems.

Pressure Pulsation and Fatigue

Pressure pulsations or pressure cycling always induces fatigue in thin metal discs. Disc life is thus shortened in pulsating service; the closer the operating pressure is to disc burst pressure, the shorter is the expected disc life. The most severe pulsating test on discs is cycling up to some given percentage of the disc's rated burst pressure and then releasing the pressure completely back to atmospheric. This induces a more rapid fatigue failure than does partial pressure unloading.

Extensive laboratory cycling pressure tests have been performed on solid-metal discs to provide a reliable guide for selection of the best disc materials for fatigue resistance. To compare performance, a value of 1,000 was assigned to the best performing metals, and related performance values were assigned to other metals. Table 12.5 shows the results of these tests and provides an excellent guide for material selection where severe cycling is encountered. One good solution to pressure cycling is the use of the reverse buckling, knife edge disc described in a previous section.

Temperature and Corrosion Effects

Other process characteristics which affect rupture disc selection are temperature and corrosion. Several allusions have already been made to the effects of corrosion. One should simply bear in mind that if corrosion occurs, very little would be required to change the burst pressure because rupture discs are so thin. With so many corrosion-resistant materials available, this does not pose a big problem.

Metals respond differently to changes in temperature (see Figure 12.61). Although the rupture pressure of stainless steel decreases rapidly with increase in temperature, an Inconel disc at the same rupture pressure at 72°F decreases far less rapidly. Table 12.6 shows how temperature affects disc burst ratings. The percentages given apply only to solid disc metals. Other disc types follow a similar pattern. This table illustrates the importance of having the correct temperature rating.

Estimating disc temperature is not always an easy task and can sometimes be misleading. If the material being processed and exposed to the disc is in the gaseous state and

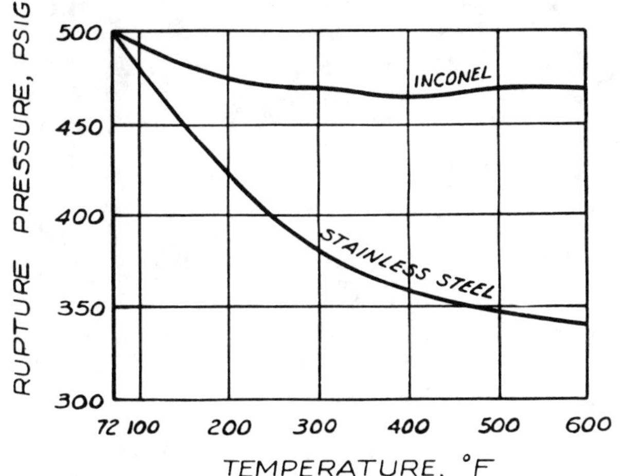

Figure 12.61. Rupture pressure variation with temperature. (Courtesy of Fike Metal Products Corp.)

Table 12.6. Effect of Temperature Change on Burst Rating of Solid-Metal Discs

Temp., °F.	Percent burst-pressure change from 72° F. Inconel	Monel	Nickel	316 Stainless steel	347 Stainless steel	Aluminum
0	+5	+9	+7	+16	+16	+14
32	+3	+5	+4	+10	+10	+10
72	0	0	0	0	0	0
150	−3	−7	−4	−9	−10	−6
250	−5	−11	−7	−14	−20	−19
350	−7	−15	−9	−18	−26	....
450	−7	−17	−11	−20	−30	....
550	−6	−18	−17	−21	−31	....

Source: BS&B Safety Systems.

if the material is normally a gas at atmospheric temperature, then the disc will be somewhere between these two temperatures.

When the process material is a condensing fluid, a temperature rating corresponding to its saturated temperature at the pressure rating of the disc should be used. This is assuming that the disc is exposed to the atmosphere on its downstream side.

Margin between Operating and Set Pressures

The term *margin* is defined as the percent difference between the rupture disc's rated bursting pressure and the operating pressure of the process during normal service. It might also be called a *safety factor.* Several years ago, a margin of 50% was common practice; that is, the operating pressure was supposed to be about 67% of the burst pressure setting. With the wider selection and use of materials and the many designs now available, it is no longer meaningful to use such a figure. Solid-metal discs, under some conditions, still need a 50% margin. Reverse buckling discs and shear discs, on the other hand, can operate with only a 10% (or less) margin.

Some disc metals that can be used with a closer margin than others are Types 316, 321 and 347 stainless steels, Monel, Inconel and nickel. These are better because they are basically tough, yet ductile and workable metals with predictable yield points. They can be operated successfully with no more than a 20% margin. Aluminum is good when limited to low temperatures and in the absence of corrosion. Copper and silver are relatively inferior because of low fatigue and creep resistance.

Thin metals are more subject to basic metallurgical defects and to handling and to installation damage than thicker ones. As a result, rupture discs ranging from 0.002 to about 0.008 inch thick should be operated with at least a 30% margin for a life expectancy of a year or more. When thicknesses exceed 0.008 inch, minor metallurgical defects and handling damage have neglible effects on performance, especially among the more ductile materials. They can, therefore, operate at lower margins.

The life expectancey of rupture discs follows an exponential curve. For example, a disc that would last for one month (rougher operating conditions than pure static pressure) at 10 to 15% margin of operation would easily last more than 12 months at 30% margin.

Another factor affecting the margin used on solid-metal discs is the set of pressure conditions imposed during normal service. If a rupture disc, made of one of the better metals, is subjected to cycling pressures from 0 to 70% of its burst rating and held static at that point, the disc could, in theory, hold for 10 years or more. Tests show that the same rupture disc, cycled from 0 to 85% of its burst rating, will fail after 10,000 to 12,000 cycles.

Because of the materials available and the varied conditions of operation, margins vary widely. Proper selection of rupture disc types and materials hinge on a good understanding of process conditions.

Flanges and Accessories

Flanges to grip the disc and incorporate it into a pressure system are made in a variety of sizes, materials, types and pressure ratings. Sizes vary from ½ to 44 inches with intermediate sizes corresponding to common pipe sizes. Materials are just as varied as for other pipe fittings. The most common is steel, but high priced alloy materials are also available. Following are some flange types and holders used.

1. *Bolted Type.* Figure 12.62 shows a standard bolted unit with screwed inlet and welded outlet. Connections can be threaded, welded or raised face as desired. Figure 12.63 shows a bolted flange with a BS&B Quiksert which allows a quick and easy changeout of rupture discs. The Quiksert shown alone in Figure 12.64 is a holder made especially for rupture discs and fits between standard flanges. Jack screws are available with flanges for ease of Quiksert removal.

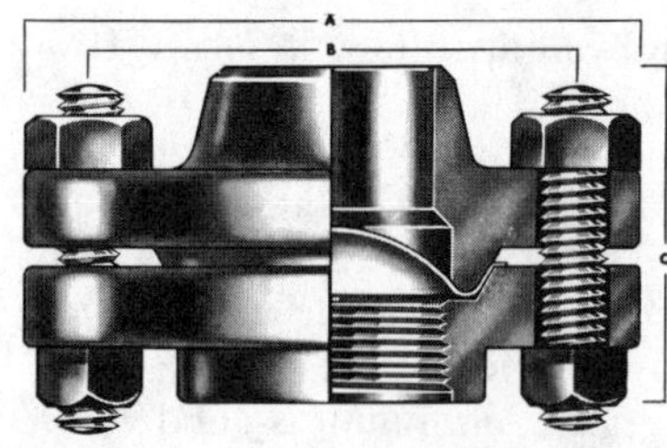

Figure 12.62. Flanged rupture disc holder with screwed inlet, welded outlet. (Courtesy of BS & B Safety Systems)

Figure 12.63. Quicksert preassembled safety head with jack screws inserted between ASA pipe flanges. (Courtesy of BS & B Safety Systems)

Figure 12.64. BS & B Quiksert rupture disc assembly that mounts between standard flanges. (Courtesy of BS & B Safety Systems)

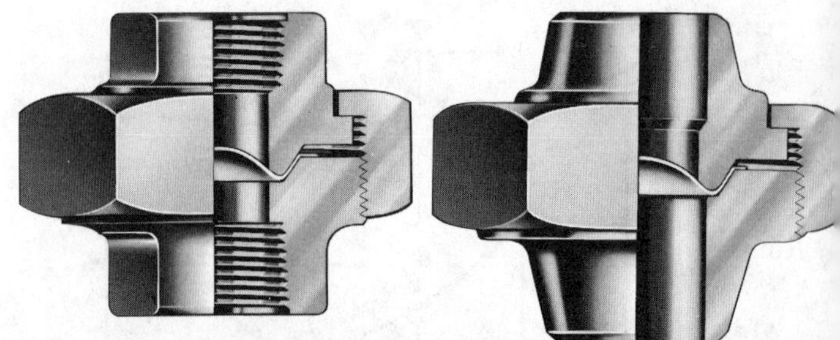

Figure 12.65. Typical union holders for rupture discs. (Courtesy of BS & B Safety Systems)

Figure 12.66. Screw rupture disc holders are used mostly for laboratory and aerospace equipment, compressed gas and cryogenic systems. (Courtesy of BS & B Safety Systems)

2. *Union Holders.* Figure 12.65 is a typical union holder that is used in sizes ½ to 2 inches (these sizes are standard). Connections may be screwed or welded. Reverse buckling discs are not used in union holders.
3. *Screw Type.* Screw safety heads are not widely used in the process industries (Figure 12.66). They do find application on high pressure laboratory equipment, aerospace equipment, bottled gases, cryogenic systems and compressed air or liquid systems. Figure 12.67 shows an assembly with a muffled outlet.

Sizes from 3/16 to over 1 inch are available in screw safety heads. They can be furnished in pressures up to 100,000 psi.

Accuracy

Rupture discs are quite accurate; most discs will rupture within ±5% of their stamped rating, and many will rupture within ±2% of their stamped rating, and that tolerance is often specified in their purchase.

When high accuracy is requested, more tests are made to determine that ruptures occur within the range or tolerance desired. Unlike material for most devices furnished, rupture discs are actually fabricated and tested to destruction on each order by the manufacturer. Actual tests determine the marked rating placed on the discs.

Figure 12.67. Muffled outlets are available for screw rupture disc holders. (Courtesy of BS & B Safety Systems)

Figure 12.68. The installation shown is typical of code-accepted use of rupture discs beneath relief valves in refineries and chemical plants. (Courtesy of BS&B Safety Systems)

High pressure discs are normally more accurate than low pressure ones. Discs made from the thinner materials are likely to have accuracies of ±5% whereas thicker materials may be ±3% or better.

Rupture Discs Versus Relief Valves

The question of whether to use a rupture disc or a relief valve is often raised. There is a great difference in economy with the initial costs greatly favoring a rupture disc. Since a relief valve is a reclosing device, overall economy usually favors a relief valve since shutdowns are usually avoided by their use.

One of the major code-accepted applications of rupture discs has been beneath relief valves. A typical refinery or chemical process plant application is shown in Figure 12.68. This type of application is expanding rapidly. The reverse buckling design with its 10% margin and accuracy of ±2% has accelerated their use. A major advantage of using a rupture disc beneath a relief valve is that it allows no leaks, and it isolates the valve from the process fluid, helping assure that the valve will work as intended when the disc ruptures.

By protecting the internal valve surfaces against corrosion, the rupture disc permits use of less expensive alloys in the valve trim. In many applications, a cheaper valve design may be used, since it is not normally exposed to the process media.

A rupture disc can be used on the outlet side of a relief valve. Whether to use a disc only or a disc and valve depends on the need for shutdown if an overpressure condition occurs. When used beneath a relief valve, it will open up on overpressure, relieve excess pressure and shut off. Operation can then continue until the next convenient shutdown for changeout of the disc.

The following criteria may be helpful in determining rupture disc use.

1. As a primary relief device on inexpensive and inert materials when shutdowns and material losses can be tolerated.
2. As a primary relief device for toxic or corrosive materials when leakage is intolerable and where they may be vented into a flare header system.
3. As a secondary relieving device where a relief valve is used as a primary device and where additional protection is needed for emergency purposes.
4. As explosion protectors; a large open area is made available in a very short period of time.
5. Upstream of a relief valve (previously discussed).
6. On polymer or other viscous services where relief valve operation would be questionable.
7. On high pressure applications beyond the range of relief valves.

Disc Sizing

Rupture disc sizing is simple. Most manufacturers use simplified formulas based on derivations from Crane Technical Paper No. 410. Crane has done extensive testing and developed a modified Darcy formula which takes into consideration correction factors to compensate for changes in fluid properties resulting from expansion. The Darcy formula is generally accepted for use in determining flow through piping and is derived by dimensional analysis, except for the friction factor which is empirically determined.

Sizing for Gases

BS&B uses the simplified flow formula taken from the Crane paper, simplified it further by assuming:

1. Rupture disc unit is taken as a short tube nozzle with a sharp edged entrance discharging to atmosphere.
2. Flow area is small compared to the tank or reservoir.
3. Critical flow conditions exist.

Starting with the accepted formula from the Crane paper.

$$q'm = 678\ Yd^2\ \sqrt{[\Delta P\ (P^1)/KT\ (Sg)]} \qquad (12.1)$$

and using a ratio of specific heats of 1.4 for gases, a limiting pressure drop of 0.66 P^1, and L/D of 75 pipe diameters for the safety head and a safety head size of approximately 2 inches, the above formula reduces to:

$$a = \{q'm\sqrt{[Sg\,(T)]}\}\,/(260P^1) \quad (12.2)$$

The required relief area, *a,* is taken as the smaller of the inlet or outlet pipe or safety head. Figure 12.69 is a nomograph for sizing gas flows using the above formula.

Sizing for Liquids

BS&B's equation used in developing the nomograph shown in Figure 12.70 for sizing rupture discs in liquid service is based on derivations from the same Crane paper. Starting with this accepted formula from Crane:

$$Q = 236\,d^2\,\sqrt{(\Delta\,P/K\rho)} \quad (12.3)$$

and substituting an L/D of 75 pipe diameters for the safety head and a safety head diameter of approximately 3 inches, it reduces to:

$$a = 0.0438\,Q\,\sqrt{(S/\Delta P)} \quad (12.4)$$

Because of assumptions, the above formula is limited to liquids having low viscosities (approaching that of water) and relatively high Reynolds numbers.

Sizing for Steam

The gas sizing formula can be used for steam, but there is much empirical data that gives good results for steam service. BS&B's equations used in developing the nomograph for sizing rupture discs in steam service (Figure 12.71), therefore, are simplified versions of various formulas and can be verified within approximately 10% by using Crane Technical Paper No. 410. For various conditions these are

Dry and Saturated:

$$a = W/30P^1 \quad (12.5)$$

Initially Superheated:

$$a = [W(1 + 0.00065\,\Delta t)]/30P^1 \quad (12.6)$$

Initially Wet:

$$a = [W(1 - 0.012y)]/30P^1 \quad (12.7)$$

Special Considerations

The above equations are not valid when extended lengths of upstream or downstream piping are connected to the rupture disc assembly or if flowing pressures are below that required for critical flow. In such instances, the pressure drop for the complete discharge piping must be calculated. Assume in such cases that the rupture disc assembly has a pressure drop equivalent to 75 pipe diameters (i.e.,- L/D = 75) and use Crane Technical Paper No. 410 as the basis for calculating the flow rate or size of this type system.

Nomenclature for Rupture Disc Sizing

a = required flow area in square inches
d = required diameter in inches
K = resistance coefficient
L = length in feet
P = rupture pressure or maximum allowable flowing pressure, psig (usually equals design pressure plus 10% of design pressure)
P^1 = pressure in pounds per square inch absolute
ΔP = differential pressure (usually equal to P)
Q = rate of flow in gallons per minute
$q'm$ = rate of flow in cubic feet per minute at standard conditions, 14.7 psia and 60°F
S = specific gravity of liquid relative to water, both at 60°F
Sg = specific gravity of gas relative to air, equals the ratio of the molecular weight of the gas to that of air
T = absolute temperature in degrees Rankine (460 + t)
t = temperature in degrees Farenheit
Δt = number of degrees superheat (Farenheit)
W = rate of flow in pounds per hour
y = percent moisture equals 100 minus steam quality
Y = net expansion factor for compressible flow through orifices, nozzles or pipe
ρ = density of fluid in pounds per cubic foot

Pressure-Vacuum Relief Valves

Storage tanks that operate at or near atmospheric pressure require relief protection on both the positive and negative (vacuum) sides of atmospheric pressure. The pressure changes that occur are small, usually in the order of inches of water column, but because they act on the large internal area of the tank, their effect can be great. Storage tank designs have been developed to eliminate or reduce venting requirements such as floating or lifting roof designs. The purpose of this section, however, is to discuss venting devices.

An open pipe is the simplest device that can be used for venting, but in many situations open vents allow too much product loss. The widely accepted method of pressure protection for storage vessels is the use of low pressure relief and vacuum breaker valves, or conservation vents, as they are often called. The use of the term *conservation vent* is obvious from its purpose and use in conserving product. Other aspects of the use of conservation vents are well worth noting, however. They prevent the release of toxic and/or corrosive vapors which are injurious to personnel and equipment. With the emphasis that is now placed on pollution control, the use of conservation vents is even more important.

(Text continued on page 350)

sizing for gas

This chart solves the following equation: $a = \frac{q'm\sqrt{Sg\,t}}{260\,P'}$

SAMPLE PROBLEM:

What size rupture disk will be required to relieve gas under the following conditions:

- Quantity: 4500 SCFM
- Specific gravity: 1.5
- Relieving temperature: 600° F.
- Vessel design pressure: 135 psig
- Requested burst pressure of rupture disk: 125-135 psig

q'm = 4500 SCFM
P' = 135 + 10% of 135 + 14.7 = 163.2 psia
Sg = 1.5
t = 600° F.

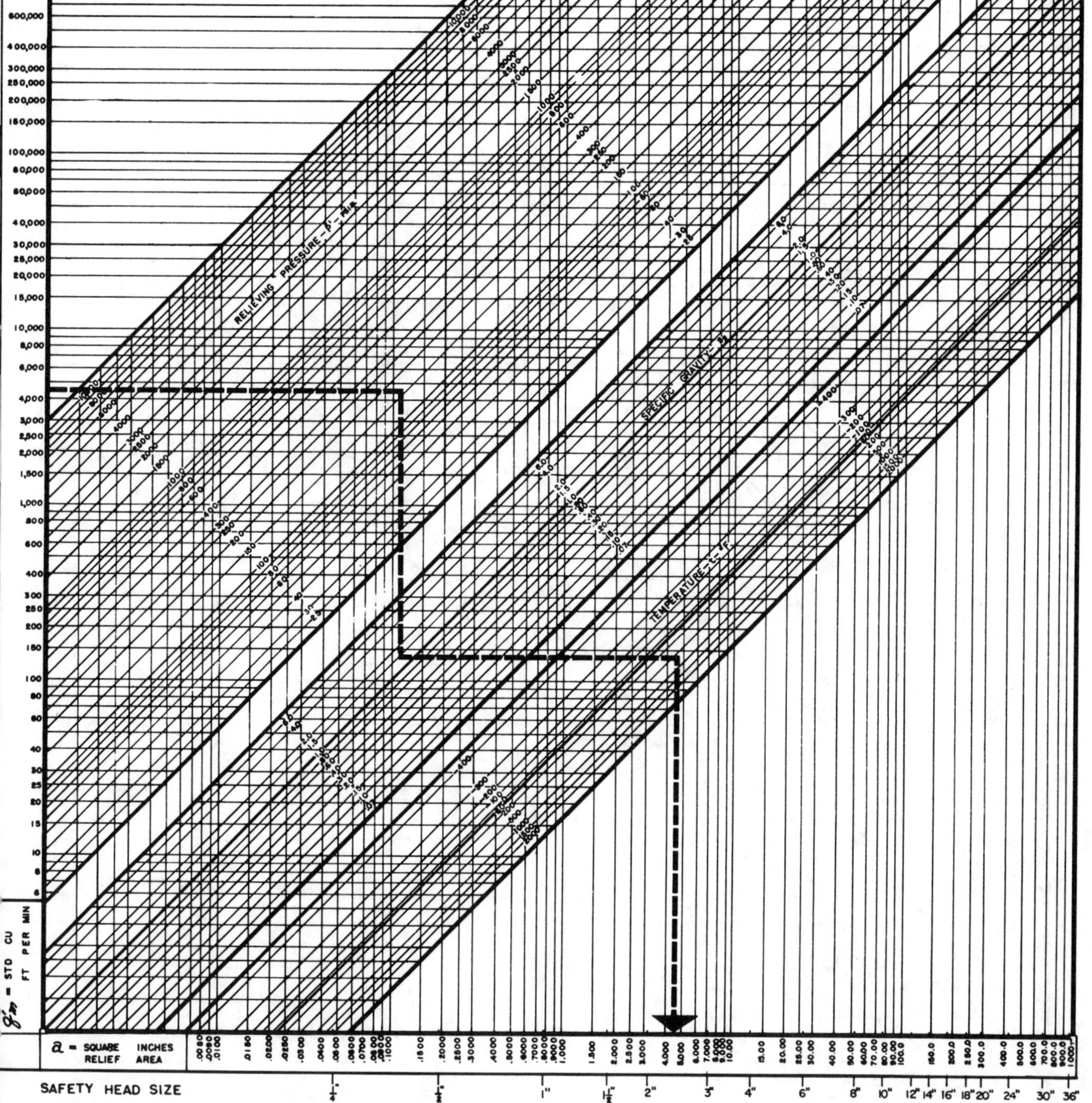

Figure 12.69. Nomagraph for sizing rupture discs for gas service. (Courtesy of BS & B Safety Systems)

sizing for liquid

This chart solves the following equation: $a = 0.0438 \frac{Q\sqrt{S}}{\sqrt{\Delta P}}$ or $Q\sqrt{\frac{S}{\Delta P}}$

SAMPLE PROBLEM:

What size rupture disk will be required to relieve liquid under the following conditions:

- Flow required: 6500 gpm
- Specific gravity: 1.5
- Vessel design pressure: 110 psig
- Requested burst pressure of rupture disk: 100-110 psig

Q = 6500 GPM
ΔP = 110 + 10% of 110 = 121 psi
S = 1.5

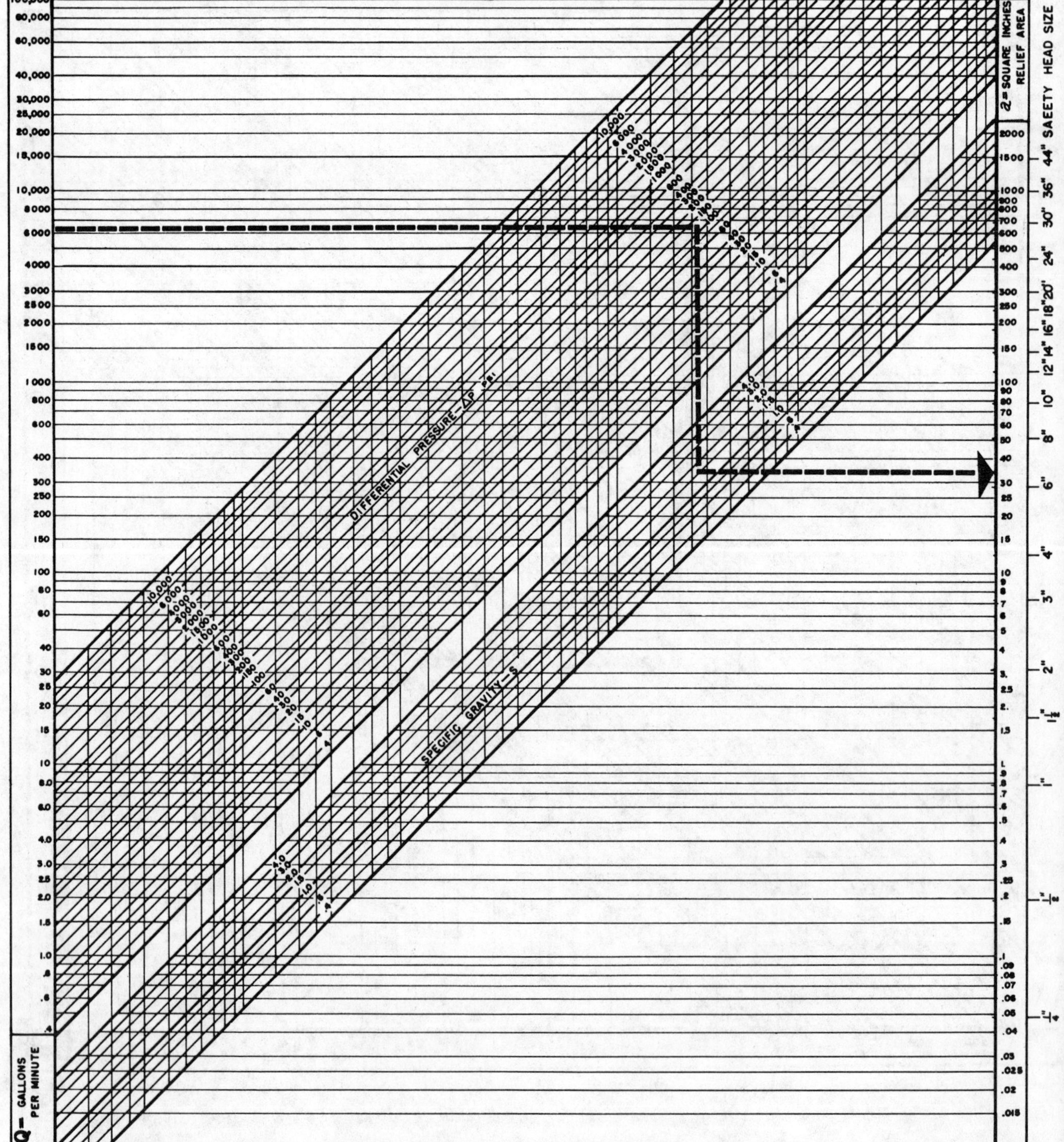

Figure 12.70. Nomagraph for sizing rupture discs for liquids. (Courtesy of BS & B Safety Systems)

sizing for steam

This chart solves the following equations:

$$a = \frac{W\,(1 + 0.00065\,\Delta t)}{30\,P'} \quad \text{and}$$

$$a = \frac{W\,(1 - 0.012\,Y)}{30\,P'}$$

SAMPLE PROBLEM:

What size rupture disk will be required to relieve steam under the following conditions:

Flow required: 4500 lbs/hr
Degrees superheat: 600° F.
Vessel design pressure: 100 psig
Requested burst pressure of rupture disk: 90-100 psig

W = 4500 lbs/hr
P' = 100 + 10% of 100 + 14.7 = 124.7 psia
Δt = 600° *F.*

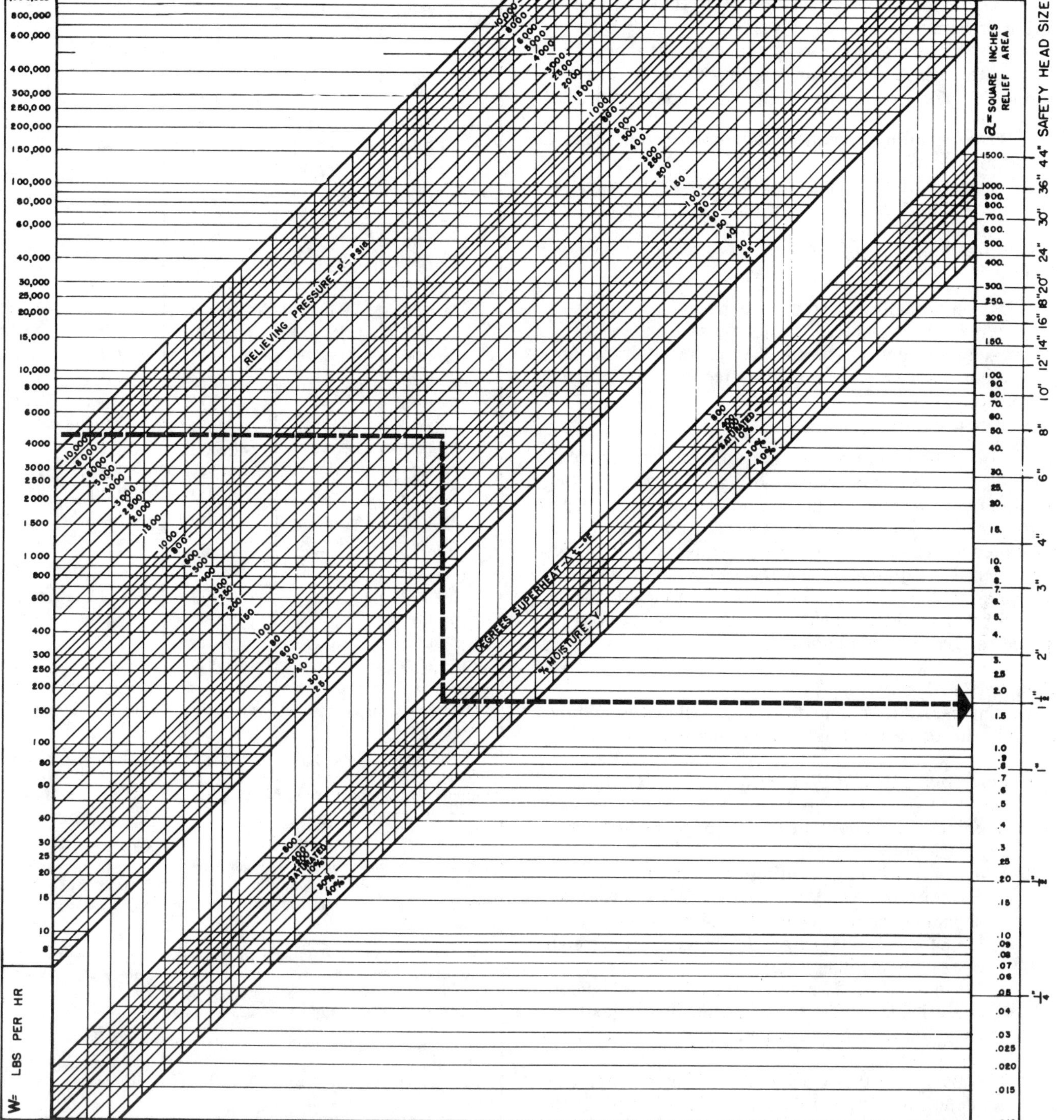

Figure 12.71. Nomagraph for sizing rupture discs for steam service. (Courtesy of BS & B Safety Systems)

API Recommendations

API RP 2000 is a guide for venting atmospheric and low pressure tanks. It covers the requirements for aboveground liquid storage tanks that operate from ½ ounce per square inch to 15 psig pressure. It recognizes five causes for determining venting requirements. Two of these causes result in "inbreathing" of the tank; three in "outbreathing." A lack of inbreathing capability produces a vacuum in the tank which could result in its collapse; a failure in outbreathing produces a positive pressure which could result in its rupturing or bursting.

For proper sizing of conservation vents, the inbreathing and outbreathing requirements must be determined separately. The inbreathing requirements considered are

1. Those resulting from maximum outflow from the tank (pump out)
2. Those resulting from contraction of vapors caused by maximum decrease in atmospheric temperature

Outbreathing venting requirements are those resulting from

1. Maximum inflow of product to the tank (pump in) and maximum evaporation caused by such inflow
2. Expansion and evaporation which result from maximum increase in atmospheric temperature (thermal outbreathing).
3. Fire exposure

The two inbreathing requirements and the first two outbreathing requirements are classified as "normal" venting requirements and are usually accommodated by a device that provides both pressure and vaccum relief. Requirements resulting from fire exposure fall under "emergency" venting requirements. Although the pressure-relieving capacity of the conservation vent is included in the overall relief needs, an additional device is sometimes required to meet fire conditions.

Venting Requirements

Venting capacities required for oil movement in and out of tanks are determined in the following manner.

1. *Pump out:* 560 cubic feet per hour of air for each 100 barrels (4,200 gallons) per hour of maximum emptying rate, including gravity flow to other tanks, for oils of any flash point.
2. *Pump in:* (including resulting evaporation)
   a. For oils with a flash point of 100°F or above, should be equivalent to 600 cubic feet of air per hour for each 100 barrels (4,200 gallons) per hour of maximum filling rate.
   b. For oils with a flash point below 100°F, should be equivalent to 1,200 cubic feet of air per hour for each 100 barrels (4,200 gallons) per hour of maximum filling rate.
3. Thermal inbreathing and outbreathing requirements are determined from Table 12.7, which is reproduced from API RP 2000.

Table 12.7. Thermal Venting Capacity Requirements

(Expressed in cubic feet per hour of air—14.7 psia at 60°*F*)

Tank Capacity* (Barrels) 1	Tank Capacity* (Gallons) 1	Inbreathing (Vacuum) All Stocks 2	Outbreathing (Pressure): Flash Point 100°F or Above 3	Outbreathing (Pressure): Flash Point Below 100°F 4
60	2,500	60	40	60
100	4,200	100	60	100
500	21,000	500	300	500
1,000	42,000	1,000	600	1,000
2,000	84,000	2,000	1,200	2,000
3,000	126,000	3,000	1,800	3,000
4,000	168,000	4,000	2,400	4,000
5,000	210,000	5,000	3,000	5,000
10,000	420,000	10,000	6,000	10,000
15,000	630,000	15,000	9,000	15,000
20,000	840,000	20,000	12,000	20,000
25,000	1,050,000	24,000	15,000	24,000
30,000	—	28,000	17,000	28,000
35,000	—	31,000	19,000	31,000
40,000	—	34,000	21,000	34,000
45,000	—	37,000	23,000	37,000
50,000	—	40,000	24,000	40,000
60,000	—	44,000	27,000	44,000
70,000	—	48,000	29,000	48,000
80,000	—	52,000	31,000	52,000
90,000	—	56,000	34,000	56,000
100,000	—	60,000	36,000	60,000
120,000	—	68,000	41,000	68,000
140,000	—	75,000	45,000	75,000
160,000	—	82,000	50,000	82,000
180,000	—	90,000	54,000	90,000

*Interpolate for intermediate sizes.

Notes:

1. For tanks with a capacity of more than 20,000 barrels (840,000 gallons), the requirements for the vacuum condition are very close to the theoretically computed value of 2 cubic feet of air per hour per square foot of total shell and roof area.

2. For tanks with a capacity of less than 20,000 barrels (840,000 gallons), the thermal inbreathing requirement for the vacuum condition has been based on 1 cubic foot of air per hour for each barrel of tank capacity. This is substantially equivalent to a mean rate of vapor space-temperature change of 100°F per hour.

3. For stocks with a flash point at 100°F or above, the outbreathing requirement has been assumed as 60% of the inbreathing capacity requirement. The tank roof and shell temperatures cannot rise as rapidly under any condition as they can drop, such as during a sudden cold rain.

4. For stocks with a flash point below 100°F, the thermal pressure-venting requirement has been assumed equal to the vacuum requirement in order to allow for vaporization at the liquid surface and for the higher specific gravity of the tank vapors.

Source: API

Table 12.8. Total Rate of Emergency Venting Required for Fire Exposure

(Wetted area versus cubic feet of free air per hour—14.7 psia at 60°F)

Wetted Area (sq ft)	Venting Requirement (cu ft/hr)	Wetted Area (sq ft)	Venting Requirement (cu ft/hr)
20	21,100	350	288,000
30	31,600	400	312,000
40	42,100	500	354,000
50	52,700	600	392,000
60	63,200	700	428,000
70	73,700	800	462,000
80	84,200	900	493,000
90	94,800	1,000	524,000
100	105,000	1,200	557,000
120	126,000	1,400	587,000
140	147,000	1,600	614,000
160	168,000	1,800	639,000
180	190,000	2,000	662,000
200	211,000	2,400	704,000
250	239,000	2,800	742,000
300	265,000	over 2,800*	

*For exposed wetted surfaces with more than 2,800 square feet, see Par. 4.21, 4.22, and 4.24.

Notes:

1. Interpolate for intermediate values.
2. The wetted area for the tank or storage vessel shall be calculated as follows: (a) *sphere and spheroid*—the total exposed surface up to the maximum horizontal diameter or to a height of 25 feet, whichever is greater; (b) *horizontal tank*—75% of the total exposed surface; (c) *vertical tank*—the total exposed area of the shell within a maximum height of 30 feet above grade.

Source: API

4. Emergency venting requirements for fire exposure are determined from Table 12.8, reproduced from API RP 2000. For tanks designed for pressures 1 psig and below, no additional venting is required even though the exposed wetted surfaces exceed the 2,800 square feet shown in the last row of the table. For tanks designed for pressures over 1 psig, the total rate of venting is determined from Table 12.8, except that when the exposed wetted area of the surface is greater than 2,800 square feet, the total rate of venting is calculated by the following formula:

$$CFH = 1{,}107 F A^{0.82} \tag{12.8}$$

where CFH are venting requirements, in cubic feet per hour of free air; A is the exposed wetted surface, in square feet, F is the environmental factor from Table 12.9.

The above formula is based on:

$$Q = 21{,}000 A^{0.82} \tag{12.9}$$

as given in API RP 520, *Design and Installation of Pressure-Relieving Systems in Refineries.*

The total venting requirements as determined from Table 12.8 above and Equation 12.8 are based on the assumption that the stored liquid will have the characteristics of hexane, since this will provide results which are within an acceptable degree of accuracy for almost all liquids encountered. If a greater degree of accuracy is desired or if liquids more volatile than hexane are stored, the total emergency venting requirement for any specific liquid may be determined by the following formula:

$$CFH \text{ (of free air)} = V\,[1{,}337/L\sqrt{(M)}] \tag{12.10}$$

where V is cubic feet of free air per hour from Table 12.8 or Equation 12.8
L is latent heat of vaporization of the specific liquid, in btu per pound
M is molecular weight of the specific liquid

Whether a pressure-vacuum relief is required, recommended or not needed depends on the applicable code or judgment. Standards within a company may vary, but

Table 12.9. Environment Factors to Be Used in Calculating Tank Venting Requirements for Fire Exposure

Type of Installation	Factor F*
1. Bare vessel	1.0
2. Insulated vessels† (these arbitrary insulation conductance values are shown as examples and are in British thermal units per hour per square foot per degree fahrenheit):	
a. 4.0	0.3
b. 2.0	0.15
c. 1.0	0.075
3. Water-application facilities on bare vessel ‡	1.0
4. Depressurizing and emptying facilities §	1.00
5. Underground storage	0.0
6. Earth-covered storage above grade	0.03

*These are suggested values for the conditions assumed in Par. 6.2. When these conditions do not exist, engineering judgement should be exercised either in selecting a higher factor or in providing means of protecting vessels from fire exposure as suggested in Par. 7.1.

† Insulation shall resist dislodgement by fire hose streams. For the examples a temperature difference of 1,600°F was used. In practice it is recommended that insulation be selected to provide a temperature difference of at least 1,000°F and that the thermal conductivity be based on a temperature that is at least the mean temperature.

‡ See Par. 7.3(c) for recommendations regarding water application.

§ Depressurizing will provide a lower factor if done promptly, but no credit is to be taken when safety valves are being sized for fire exposure.

Note: Paragraphs referred to are from API RP 520.

Source: API

API RP 2000 is likely to be used as a minimum requirement. It, too, leaves plenty of room for personal judgments, especially as to the method of venting. Several options can be exercised to obtain emergency venting, including fabrication methods for the tanks. The options available are well defined in API RP 2000.

From the information given, it is obvious that sizing of vents is based somewhat on empirical data. Capacity tables or curves are available on each type vent produced so the selection is a simple matter.

Available Designs

Pressure-vacuum valves can be classified into two basic types: pallet-loaded and pilot-operated. Pallet loading may be done with springs or with dead weights. Pallets are constructed differently and of different materials. Some pallets are made of metals, providing metal-to-metal seating; others use synthetic rubber or similar materials to provide tighter seals. Pilot operation on pilot-operated valves is obtained differently by various manufacturers. Following are descriptions and illustrations of these types.

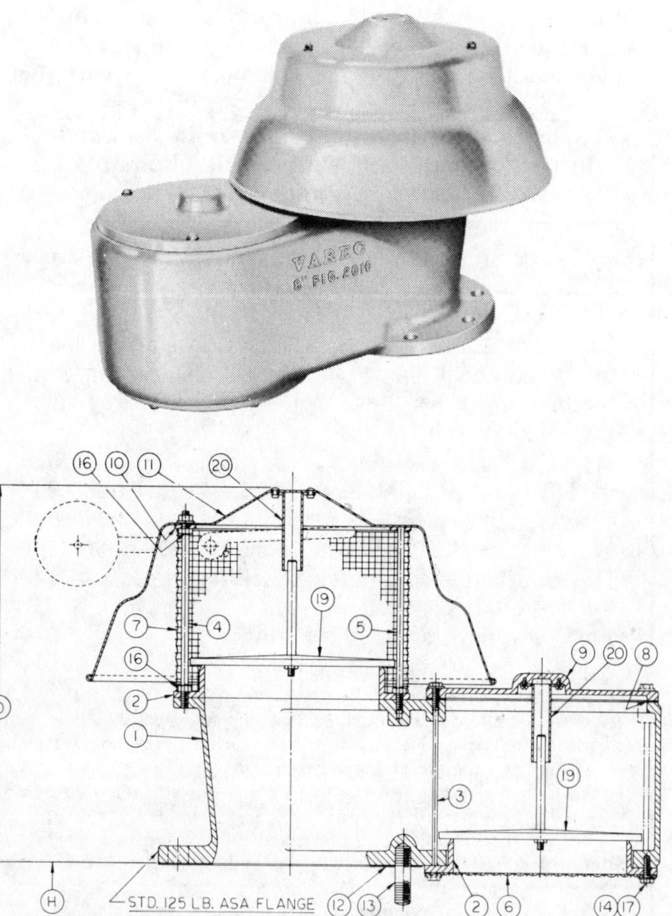

Figure 12.72. Pallet conservation vent used for both pressure and vacuum relief. (Courtesy of Varec, Inc.)

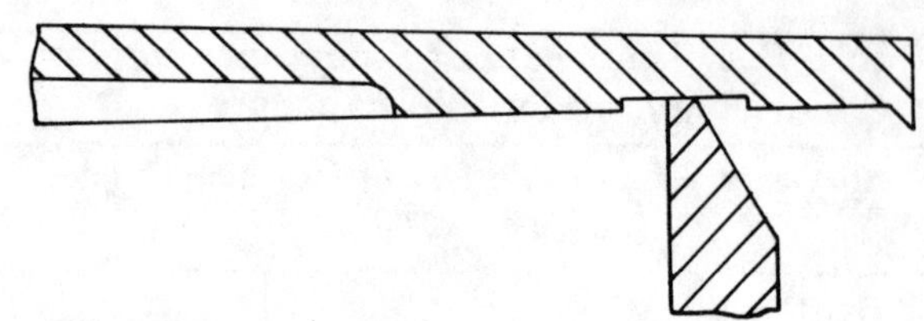

Figure 12.73. View of metal-to-metal seating on pallet conservation vent. (Courtesy of Varec, Inc.)

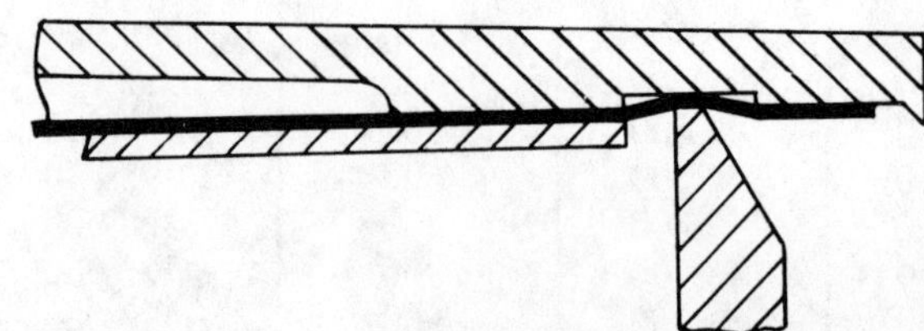

Figure 12.74. Better seals are obtained with "air-cushion" insert on pallet conservation vents. (Courtesy of Varec, Inc.)

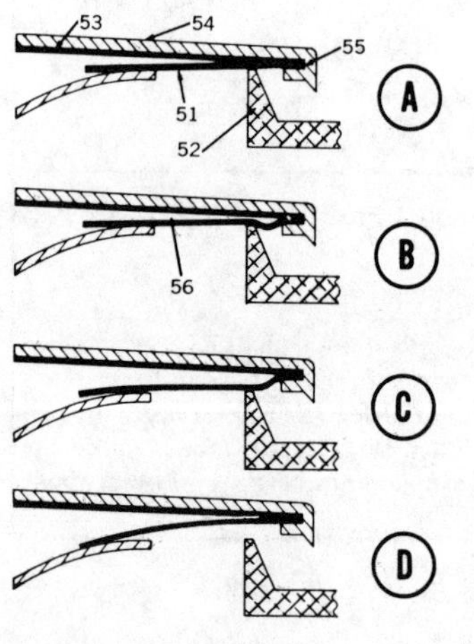

Figure 12.75. Diaphragm pallet uses double membrane diaphragm to achieve good sealing characteristics. (Courtesy of Varec, Inc.)

Pallet Type

Figure 12.72 shows a pallet conservation vent used for both pressure and vacuum relief. A section view of the valve shows the pressure pallet on the left and the vacuum pallet on the lower right section. The pallet shown is metal on a metal seat. Metal-to-metal seating (Figure 12.73) does not provide the seal tightness required of some applications. Better seals are obtained by an "air cushion" design (Figure 12.74) in which an "air cushion" insert fits into a groove near the perimeter of the pallet. The soft and flexible insert can be installed easily and quickly.

Even tighter seals can be obtained with a diaphragm pallet (Figure 12.75). The diaphragm is a double membrane type, and its sealing characteristics are illustrated by the cuts in Figure 12.75. The "balloon" diaphragm consists of members 51 and 53, one of which (53) is attached to the rigid pallet (54). The two pieces (51 and 53) are sealed

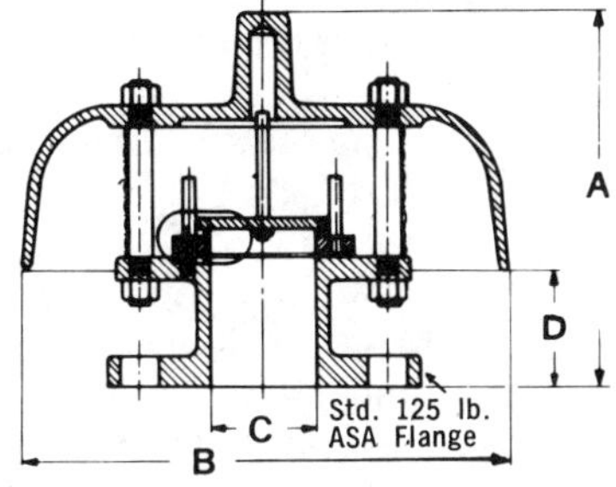

Figure 12.76. Picture and section view of a typical pressure relief only. (Courtesy of Varec, Inc.)

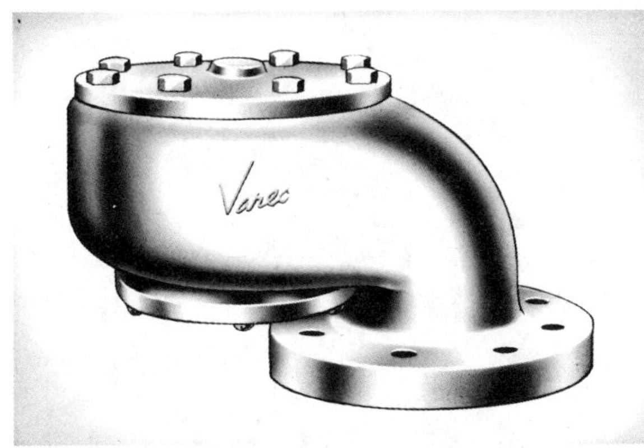

Figure 12.77. View of a vacuum only relief in a pallet design. (Courtesy of Varec, Inc.)

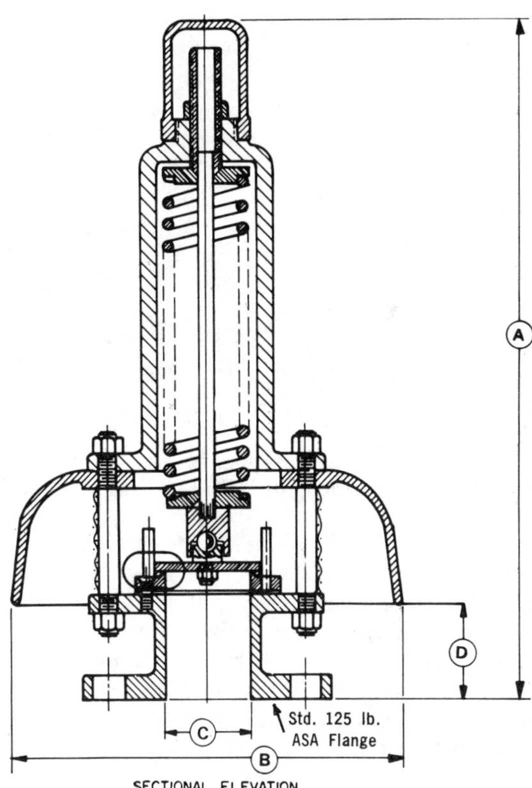

Figure 12.78. Spring-loaded pallets allow higher pressure relief values. (Courtesy of Varec, Inc.)

together at their outer edge but are free floating in groove 55.

At pressures below the valve setting, the weight of the pallet causes a tight seal on the valve seat (52) as shown in drawing *A*. As the tank pressure increases (drawing *B*), the effective weight of the pallet is reduced, but the tank pressure entering the space (56) causes the lower diaphragm member to be forced tightly against the seat maintaining a tight seal.

As the tank pressure increases further, the pallet lifts higher (drawing *C*) and the lower diaphragm member remains in contact with the seat until the tank pressure reaches the valve setting, when it lifts off the seat as shown in drawing *D*.

When pressure relief only is required, a relief valve may appear as shown in Figure 12.76. Similarly, a vacuum only relief is shown in Figure 12.77.

The pallets used on all the valves illustrated above are referred to as deadweight pallets—the only force holding the pallet closed is the weight of the pallet. Spring-loaded pallets can also be furnished (Figure 12.78), which permit higher pressure ranges than are obtained with deadweight pallets. Construction of the valve is the same.

One style of pallet type pressure-vacuum vent uses an "air cushion" type of Teflon expanding seal in conjunction with a magnetic latch. An Alnico permanent magnet holds the cover tightly sealed until set pressure is reached (Figure 12.79). When the force necessary to overcome the magnetic attraction is reached, the cover opens, venting instantaneously. The magnetic force, a function of the size and strength of the magnet, can be selected for pressure settings of ½ psi through 2½ psi in ½-psi increments.

API RP 2000 requires that at least one production model of every type and size of venting device be flow tested under specified conditions to determine valve capacities. This information must be presented in table or curve form. Flow capacity curves are usually drawn showing the required information. Figure 12.80 shows typical curves for pressure and vacuum relief.

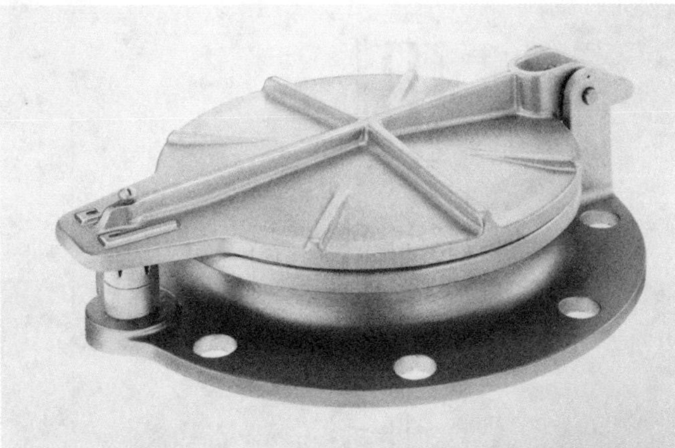

Figure 12.79. A permanent magnet latches the magnavent in its seal position. (Courtesy of GPE Controls, Shand & Jurs Div.)

Pilot-Operated Type

Pilot-operated pressure relief valves (Figure 12.81) are available in pressure ranges from 2½ psig to 10 psig. In Figure 12.81, under normal working conditions, an upward force is exerted on the main pallet (1) by the tank pressure acting over the area shown as A. The pressure in chamber 2, which is equalized with the tank pressure through hollow stem 7 and orifice 14, acts over an area V to exert a downward force on the pallet. Since area V is greater than area A, the pallet is held firmly in place against the seat ring (4). The vent is kept tightly closed at all pressures within the working range.

When the venting pressure is reached, the upward force of the tank pressure acting on diaphragm 5 in the control unit counterbalances the downward force of the calibrated spring (6) and balancing diaphragm (15), causing the vertical stem (7) to rise, opening the pilot valve (8) and venting chamber (2) to the atmosphere through the exhaust port (10). As chamber 2 is depressured, the main pallet is lifted by the pressure acting on its underside, and the excess vapors are vented outward through the weather hood.

When the tank pressure starts to drop below the set pressure, the valve (8) closes and chamber 2 is repressured through hollow stem 7 and orifice 14 causing the main pallet to close and remain closed until chamber 2 is again vented by action of the control unit.

A more recent type of pilot-operated valve, shown in Figure 12.82, uses a magnetically latched pilot to provide a snap action pilot opening, eliminating the flutter normally encountered in spring-operated valves and pilots. When the inlet pressure is lower than the setting, the magnetic poppet is held tightly against the pilot seat. Because the effective area inside the main valve bellows is greater than the seating

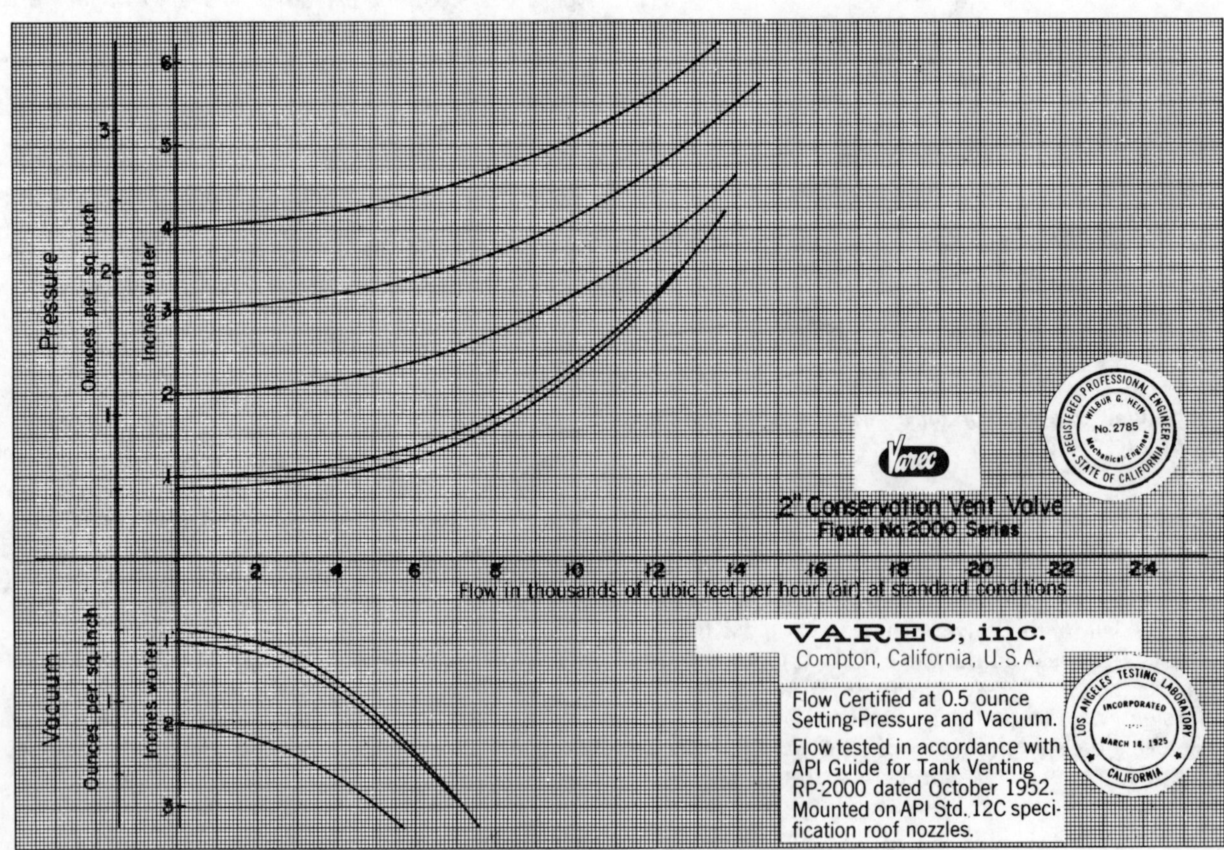

Figure 12.80. Two-inch flow capacity curves for combination pressure and vacuum relief valve. (Courtesy of Varec, Inc.)

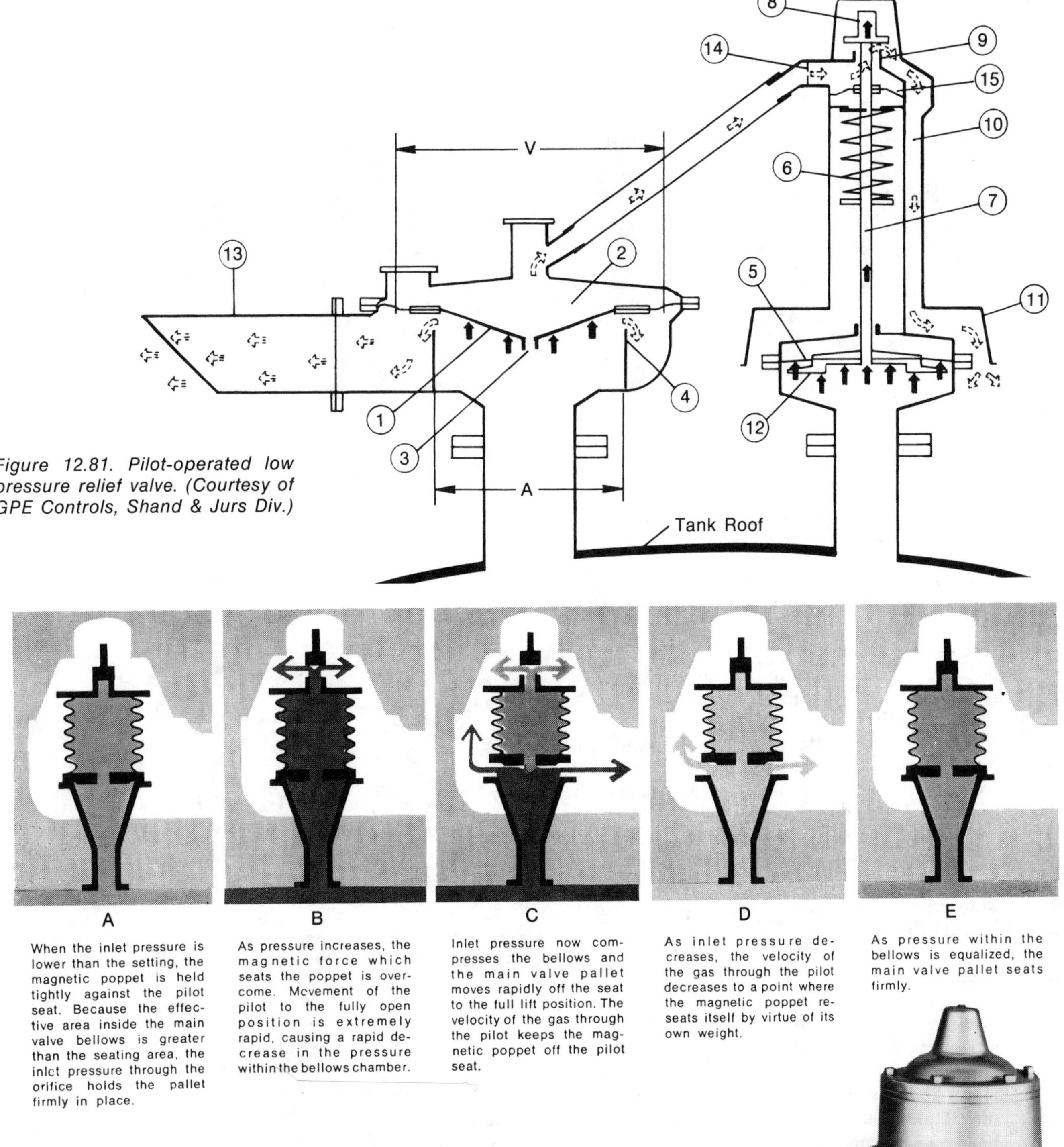

Figure 12.81. Pilot-operated low pressure relief valve. (Courtesy of GPE Controls, Shand & Jurs Div.)

A

When the inlet pressure is lower than the setting, the magnetic poppet is held tightly against the pilot seat. Because the effective area inside the main valve bellows is greater than the seating area, the inlet pressure through the orifice holds the pallet firmly in place.

B

As pressure increases, the magnetic force which seats the poppet is overcome. Movement of the pilot to the fully open position is extremely rapid, causing a rapid decrease in the pressure within the bellows chamber.

C

Inlet pressure now compresses the bellows and the main valve pallet moves rapidly off the seat to the full lift position. The velocity of the gas through the pilot keeps the magnetic poppet off the pilot seat.

D

As inlet pressure decreases, the velocity of the gas through the pilot decreases to a point where the magnetic poppet reseats itself by virtue of its own weight.

E

As pressure within the bellows is equalized, the main valve pallet seats firmly.

Figure 12.82. Low pressure pilot-operated relief valve uses magnetic poppet for pilot. (Courtesy of GPE Controls, Shand & Jurs Div.)

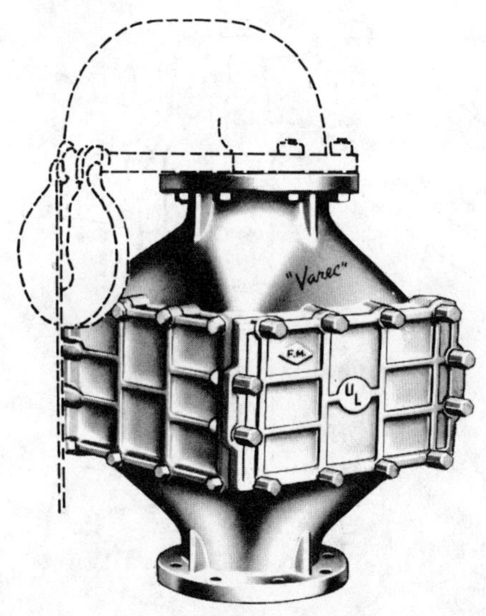

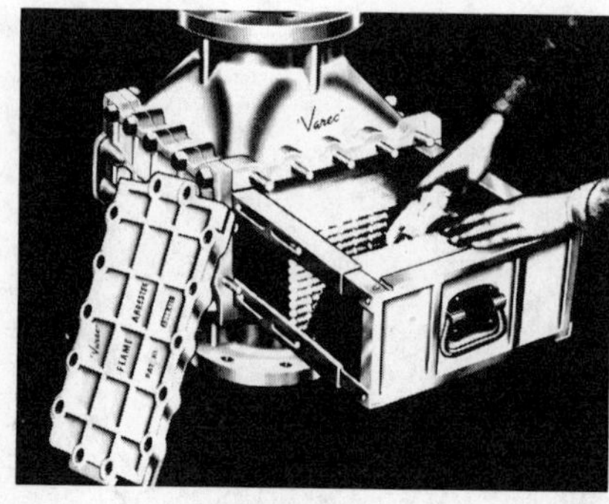

Figure 12.83. Typical flame arrestor for open vent tanks. (Courtesy of Varec, Inc.)

area, the inlet pressure through the orifice holds the pallet firmly in place.

As pressure increases, the magnetic force which seats the poppet is overcome. Movement of the pilot to the fully open position is extremely rapid, causing a rapid decrease in the pressure within the bellows chamber.

Inlet pressure now compresses the bellows, and the main valve pallet moves rapidly off the seat to the full open position. The velocity of the gas through the pilot keeps the magnetic poppet off the pilot seat.

As inlet pressure decreases, the velocity of the gas through the pilot decreases to a point where the magnetic poppet reseats itself by virture of its own weight.

As pressure within the bellows is equalized, the main valve pallet seats firmly.

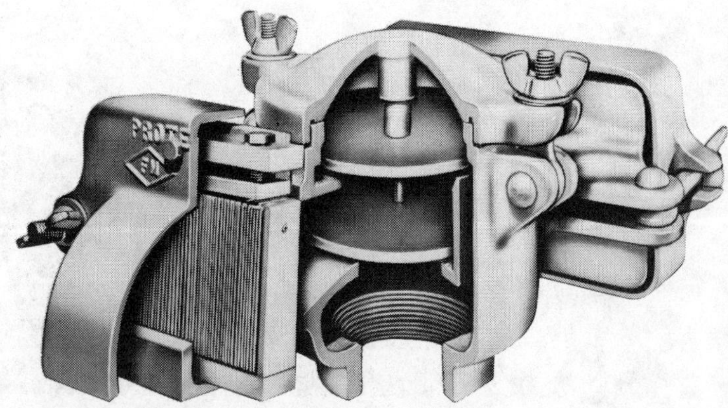

Figure 12.84. A combination pressure-vacuum valve with a flame arrestor. (Courtesy of The Protectoseal Co.)

Flame Arrestors

Flame arrestors are devices that are installed on free vents or conservation vents to prevent an external flame from entering a tank containing a flammable product. They consist of a bank of plates of sufficient size and spacing to cool and extinguish flames before they can pass through the unit. Figure 12.83 shows a flame arrestor that is used on open vents. Figure 12.84 pictures a pressure vacuum valve in combination with a flame arrestor. It should be noted that a flame arrestor of a given size has a lower capacity than the same size conservation vent. The flame arrestor, then, determines the line and vent nozzle size for a tank. Flame arrestors do not affect the venting requirements of a tank, but they do affect the capacity of vents.

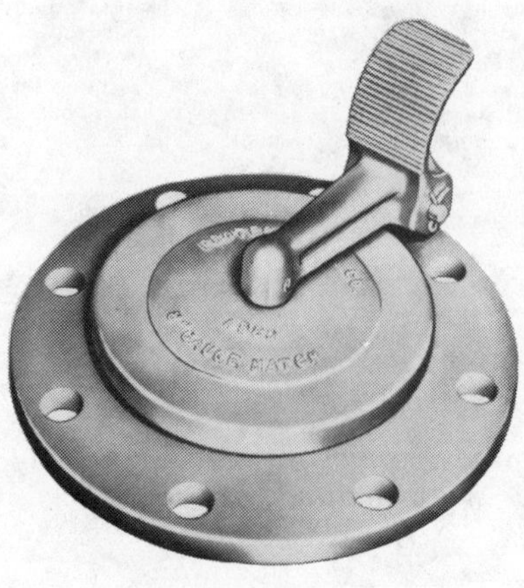

Figure 12.85. A gauge hatch which permits the cover to lift when abnormal internal pressure occurs. (Courtesy of The Protectoseal Co.)

Emergency Venting Methods

A previous discussion listed emergency venting requirements along with normal venting requirements. API RP 2000 lists methods by which emergency venting may be accomplished:

1. Larger or additional open vents
2. Larger or additional pressure vacuum valves or pressure relief valves
3. A gauge hatch (Figure 12.85) which permits the cover to lift under abnormal internal pressure
4. A manhole cover (Figure 12.86) which permits the cover to lift under abnormal internal pressure
5. Tank fabrication with weak roof-to-shell attachments
6. Other construction methods demonstrably comparable for pressure relief purpose

Materials of Construction

Conservation vents and flame arrestors are available in almost all materials that are compatible with tank construction and product storage requirements. Aluminum housings and parts are predominant because of weight and size when it is suitable for the product handled and for its environment. Stainless steels, cadmium plated steel, cast iron and corrosion resistant alloys are sometimes needed and used.

Figure 12.86. A typical manhole cover which lifts when abnormal internal pressures occur. (Courtesy of The Protectoseal Co.)

Nonmetallic materials are used for seating inserts, diaphragms, gaskets and seals.

13 Computer Theory and Mechanization

Richard H. St. John, Jr.

The electronic computer has the potential of becoming the most significant technological development in U.S. industry. The widespread application of digital computers for process control is the culmination of a 15-year struggle for acceptance, beginning in 1958 when the first industrial control computer was introduced. The role of the computer has progressed from the most elementary function of collecting and recording data, manually entered by the process operator to complete plant-wide automatic optimal control. This role is part of the evolutionary trend toward more continuous processing and less dependence on manual control.

Parametric information is a prerequisite for controlling a process. As processes become more sophisticated, the volume of information represents a significant time lag for operational analysis and decision making. The outstanding characteristics of a computer control system—the ability to acquire, assimilate, analyze and disseminate large amounts of information with great speed, accuracy and flexibility*—make the computer a logical tool for the automation of the process industries.

Status and Applications of Process Computers

Early process computers were largely experimental and very costly. Relatively few were installed prior to 1963. As technology progressed, reliability improved and the economics became more favorable, process computers experienced more widespread adoption. By mid-1968, over 3,000 digital process computer installations were reported throughout the world.† By mid-1971, this number had increased to over 25,000. There are four progressive trends in the functions of industrial process computers: data collection and reporting, on-line monitoring, open-loop advisory control and automatic closed-loop control.

Data collection and reporting involves the manual input to an off-line computer of process information for logging purposes and daily report generation.

On-line monitoring features the computer system directly connected to sensing devices which measure product qualities, raw material characteristics, temperatures, flows, pressures and other process conditions. In addition to logs and reports, the computer generates predictive trends and process alarm conditions.

Open-loop advisory control extends the on-line monitoring features to include computer generation of explicit operator instructions, such as process control set points, valve positions, motor selections, etc., but the operator performs the actual control functions.

Automatic closed-loop control allows the computer to adjust set points and make control decisions and selections automatically, thereby assuming many of the normal duties of the operator.

There are essentially two distinct categories of automatic closed-loop control: Supervisory Control and Direct Digital Control (DDC). When utilizing supervisory control of conventional process instruments, an on-line computer determines the value of the instrument set point, and an electronic or electromechanical interface mechanizes the set point change. The DDC approach can eliminate conven-

*Emanuel S. Savas, *Computer Control of Industrial Processes*, McGraw-Hill Book Company, New York (1965), p.4.

†U.S. Department of Labor, *Outlook for Computer Process Control*, Bulletin 1658 (1970), p.11.

tional instrumentation by utilizing the on-line computer to position the process control valves (or other final control elements) directly. Often, a Computer/Manual Station (C/M Station) or backup analog controller is provided to allow the operator to control the process in the event of computer failure.

Digital Computation

Evolution of Digital Computers

Electronic digital computers, barely a quarter century old, have reached their third generation in evolutionary growth. The first generation of electronic digital computers, the vacuum tube generation, began in 1946 with the ENIAC (Electronic Numerical Integrator and Computer). ENIAC was designed by J. Presper Eckert and John W. Mauchly and was built at the Moore School of Electrical Engineering, University of Pennsylvania. Eckert and Mauchly then formed a company which produced the first commercially available electronic computer, UNIVAC I (Universal Automatic Computer). This first generation of electronic digital computers is characterized by extremely large physical size, extensive power consumption, slow processing speed and frequent component failure.

The second generation, the transistor generation, began in 1956 with military application computers and in 1959 with commercial application computers. The transistorized computers are much smaller than their vacuum tube counterparts, consume only a small fraction of the vacuum tube power requirements, are very fast and have infrequent component failures.

The third generation of electronic digital computers are made up of monolithic integrated circuits. This type of computer construction began in 1965 and continues today. The IC computers generally require less power than those made of discrete components, are physically smaller and less prone to failure, mainly because of the greatly reduced component count. This generation of computers spawned the popular mini-computer.

The mini-computer is a miniature version of a full-scale digital computer. The mini can perform all of the basic operations of a digital computer. Speed and efficiency are usually sacrificed for size and economy.

Functional Parts

Modern electronic digital computers are composed of five basic units of functional parts: control unit, arithmetic unit, input, output and memory. The first two units are collectively known as the Central Processing Unit (CPU), and the last three are the basic peripherals.

The control unit is the nerve center of the computer. It sequences all events and directs the overall functioning of the other units of the computer by controlling the data flow between them. The control unit decodes program instructions and initiates the appropriate signals necessary to perform the specified function. Another important function assigned to the control unit is error detection for both hardware and program malfunctions. Often associated with the control unit is an operators control panel which contains all switches and status indicators necessary to operate the CPU.

The arithmetic unit performs numerical and logical calculations. This unit is composed of temporary storage registers used for operands and for the intermediate and final computational results of addition, subtraction, multiplication and division, as well as logical comparison. Information which is transferred from one section of the computer to another usually advances through the arithmetic unit.

The memory unit is a device which stores program instructions, intermediate results and data. Memory falls into two categories, internal and external. Internal memory is generally magnetic core, metal oxide semiconductor (MOS) or drum storage. External memory, often called auxiliary memory, is provided by magnetic tape, magnetic disc, magnetic card, magnetic drum or additional blocks of magnetic core. Memory can be accessed in two ways, random or sequential. The memory is called random-access if it takes just as long to obtain access to one address as it does to obtain access to another address. For sequential access memory, the computer must interrogate all locations between the current address and the one desired. An important distinction must be made between volatile and nonvolatile memory: information is lost by volatile memory when power is turned off but retained by nonvolatile memory.

The input section of the computer supervises the flow of information to the central processor by controlling input devices and scanning for interrupt signals. Prior to the execution of a program, instructions and data are loaded into computer memory via an input device such as a card reader, paper tape reader, magnetic tape, magnetic disc or magnetic drum. Many modern computers have the ability to perform input and output operations simultaneously with computer program processing.

The output section of the computer directs the transmission of information from computer storage to an output peripheral device. This process involves handling the output buffering operations as well as establishing data and control signal communications with the output device.

The following are the most common I/O peripherals.

Punched card equipment is a widely used method for both input and output of digital information. Standard cards contain 80 columns and 12 rows. Each column can be used to represent one character of data which takes the form of the presence or absence of punched holes in the 12 rows.

Punched tape is the least expensive (also one of the slowest) of the I/O techniques. Data is stored permanently in the form of punched holes. Generally there are eight rows of binary coded information plus a sprocket hole which serves either to advance the tape through the machine or to "clock" the word as it passes the read station. The eight rows run the length of the tape (which can be any length), and there are as many word column locations as there are sprocket holes.

Keyboard typer I/O equipment provides a convenient tool for operator conversation with the computer. The typer resembles an ordinary electric typewriter, except that the internal key signal codes are cabled to the computer.

Line printer equipment provides a "hardcopy," high-speed output. Modern line printers print a complete line of 80 to 160 characters in a fraction of a second and use either impact or nonimpact print mechanisms. Impact print mechanisms are generally one of five types: drum, chain, oscillating bar, type bar and type wheel. Nonimpact printers fall into three categories: electrostatic, ink jet and thermal matrix.

Magnetic tape equipment is the primary media in large-scale data processing facilities for reading data into the computer and recording results from the computer. The recording tape most commonly used has a plastic base coated with a magnetic oxide, ½ to 2 thousandths of an inch thick, from ⅛ to 3 inches wide and up to 3,600 feet long. Binary data is stored on the tape as spots of opposite magnetic polarity. Data is recorded in parallel rows (tracks), with a single character being made up of the information contained in one column—the width of the tape. The spacing between tracks is fixed by the recording head spacing, and the spacing between characters (packing density) is a function of the tape speed and the "record clock." One-half-inch wide tape, with either seven or nine tracks and 200 to 1,600 characters per inch, is the most commonly used type. Data is recorded sequentially and is erasable, but it is considered permanent in the sense that it is nonvolatile in the event of loss of power and can be retained indefinitely with proper reel handling and storage.

Magnetic disc storage involves the use of magnetic oxide coated discs which resemble phonograph records. Single or multiple discs are mounted on a horizontal or vertical shaft. Where multiple discs are used, they are separated from one another far enough to allow a read-write head to pass between them. Read-write heads may be fixed (one per track) or mounted on a movable comb-like access mechanism (one per surface). Data is stored in tracks circumferentially on the disc, generally on both sides. Disc diameters vary from 1 to 5 feet and there are commonly from 100 to 250 tracks. The speed of rotation is usually 1,800 to 2,400 revolutions per minute, and track access is random. Data may be erased and stored and is nonvolatile with proper handling.

Magnetic drum storage provides a relatively inexpensive method of storing large amounts of data. The drum is a precision metal cylinder with a coating of magnetic oxide on the outside surface. Data is recorded in circumferential tracks by fixed read-write heads floating above the surface of the drum on a thin layer of air. Drum speeds vary from 3,000 to 18,000 revolutions per minute. Drum diameters vary from 1 to 3 feet. Like other magnetic storage media, magnetic drum data is erasable and nonvolatile.

Magnetic card I/O equipment is a novel approach to providing on-line capabilities for rapid random access to an enormous amount of information. Magnetic cards are approximately 12 inches long, 3 inches wide and considerably thicker than magnetic tape. They are made of the same materials as magnetic tape. The cards are located in compartments similar to jukebox record slots. Information is recorded or retrieved from the card by extracting the card from its location, wrapping the card around a revolving drum and rotating the card under read-write heads.

Cathode ray tube display equipment provides an interactive system for the input/output of computer data. Information is displayed on the face of the tube by alphanumeric characters and/or special graphical symbols. The primary interactive devices associated with the CRT are alphanumeric keyboards and light pens, as well as special purpose switches.

Digital Computer Languages

Machine language is the basic repertoire of instruction and operation codes used by the computer. The computer operates on binary digits and responds to binary commands. For convenience to the programmer and to the operator, decimal or octal numbers are often used outside the computer, but these are converted to binary inside the machine. Machine instructions are made up of an operation code and an instruction address. The operation code defines the operation to be performed, and the address defines the location where the information can be found or stored.

Assembly language uses words called mnemonics to replace machine language numeric codes. The assembly program, or assembler, translates the assembly language program by creating the correct number coded in data format or instruction word format. The output of the assembler, known as the object program, consists of the assembled collection of data and instruction words.

Algorithmic language, commonly called procedural language, permits the programmer to use algebraic logic and English language procedure statements when preparing a computer program. Algorithms are the rules for solving mathematical problems. Procedure-oriented computer languages are designed for the efficient communication, between the computer and the programmer, of sequences of well-defined steps leading to problem solutions. These languages provide freedom of format and contain many equivalent machine language instructions for each procedural instruction. The task of translating the procedure language statements into machine language codes is accomplished with a special program called a compiler.

Data Acquisition

Digital data acquisition systems automatically monitor process instrumentation by recording and/or displaying the measured values. A data acquisition system consists of a scanner or multiplexer and an analog to digital converter.

The multiplexer samples all of the input signals at a predetermined time sequence. Data acquisition under computer control features programmable multiplexers with capabilities of priority interrupt or random scanning based on limit conditions and operator inputs. Accuracy, input sensitivity, signal conditioning and linearization and noise rejection are important specifications for multiplexers.

The analog-to-digital converter produces discrete representations of continuous inputs. The important characteristics of the A-D converter are defined in the following discussion.

Conversion time is the period required to generate a stable digital output after the command to digitize has been given.

Conversion error is the discrepancy between the digital output and the instantaneous value of the input.

Aperture defines the reference time which relates the output to the input. This generally equals the conversion time.

Quantum levels are the discrete step values that the output can assume. An n-bit A-D converter can assume 2^n unique voltage levels.

Linearity is a measure of the deviation from a straight line for a ratio of input to output over the entire input range.

Resolution is the ability to distinguish between adjacent input values.

Stability is a measure of maintaining fixed characteristics over a defined operating interval.

Hierarchical System Structure

Three distinct sizes of computer control systems are available to the process industry. Small mini-computer systems are available which perform dedicated control tasks on a limited number of loops. Medium-scale computer systems are available for process supervisory control and direct digital control functions, and their capabilities extend to limited unit optimization. Large, general purpose computers are available which are capable of handling optimization, management information and coordination for an entire company. Hierarchical systems interconnect several computers together for the purpose of information exchange. Figure 13.1 illustrates a hierarchical system interconnecting corporate offices with refineries, down to individual process areas. The system shown depicts dedicated mini-computers for each process area slaved to a plant control computer. The plant control computer is tied to the refinery general purpose computer, which is in turn linked to the corporate offices large-scale computer. The system culminates with a real time, hypothetical "profit meter."

Common Characteristics

The following characteristics are common to most of the digital computers available today.

1. All data are handled within the computer in binary form.
2. Machine operation is basically serial; one operation must terminate before another can begin.
3. Accuracy is dependent on the size of the computer and the numerical techniques of the programming, not on the quality of the computer components.
4. There is a natural ability to perform only a limited number of arithmetic operations, such as addition, subtraction, multiplication and division.
5. Logical operations and decisions are performed efficiently.
6. Problem scaling is eliminated due to the availability of floating point operations.
7. Programming techniques bear little direct relationship to the problem equation.
8. The system is capable of storing large amounts of data "on line" indefinitely.
9. Internal program control allows for automatically altering and controlling the topology of data flow.

Analog Computation

General Purpose Analog Computer

While digital computers operate on numbers directly with discrete step operation, analog computers operate on physical quantities which represent numbers. The manipulation of these quantities is continuous. There must exist on the analog computer a complete analogy between the physical quantities and events and the mathematical numbers and manipulations.

The introduction of the operational amplifier using DC voltages as the physical analog marked the beginning of the electronic analog computers. Lovell of Bell Telephone Laboratories is credited with the introduction of the operational amplifier during World War II. Early electronic analog computers used vacuum tube circuits. Like the digital computer, transistorized analog computers were developed in the early 1960s and are in widespread use today. Integrated circuit analog computers are available, but their popularity does not presently parallel that of the integrated circuit digital computer.

The analog computer has a multiplicity of uses in process work, beginning with the initial research and development and continuing with startup procedures and unit optimization. The analog computer is particularly advantageous in the early stages of process design because this work is intimately concerned with dynamic mechanisms. Process simulation on the analog computer is a useful engineering tool which results in a vast amount of parametric information. This information is economically obtained because simulated plant runs do not require the raw materials and operational manpower necessary for pilot or full-scale plant runs. Another area in which the use of the analog computer is particularly applicable is in the choice of process instrumentation and control. Various control schemes can be investigated easily, controller set points determined and instruments calibrated.

Modern analog computers are essentially a collection of electronic and electromechanical components which use DC voltages as variables in the solution of mathematical equations. Analog computers differ in two areas: (a) capacity, the number of computing elements, and (b) capability, the quality and versatility of computing elements. Small analog computers contain as few as 20 computing elements, whereas large general purpose analog computers contain several hundred computing elements. The quality of the analog computer is a measure of the signal-to-noise ratio of

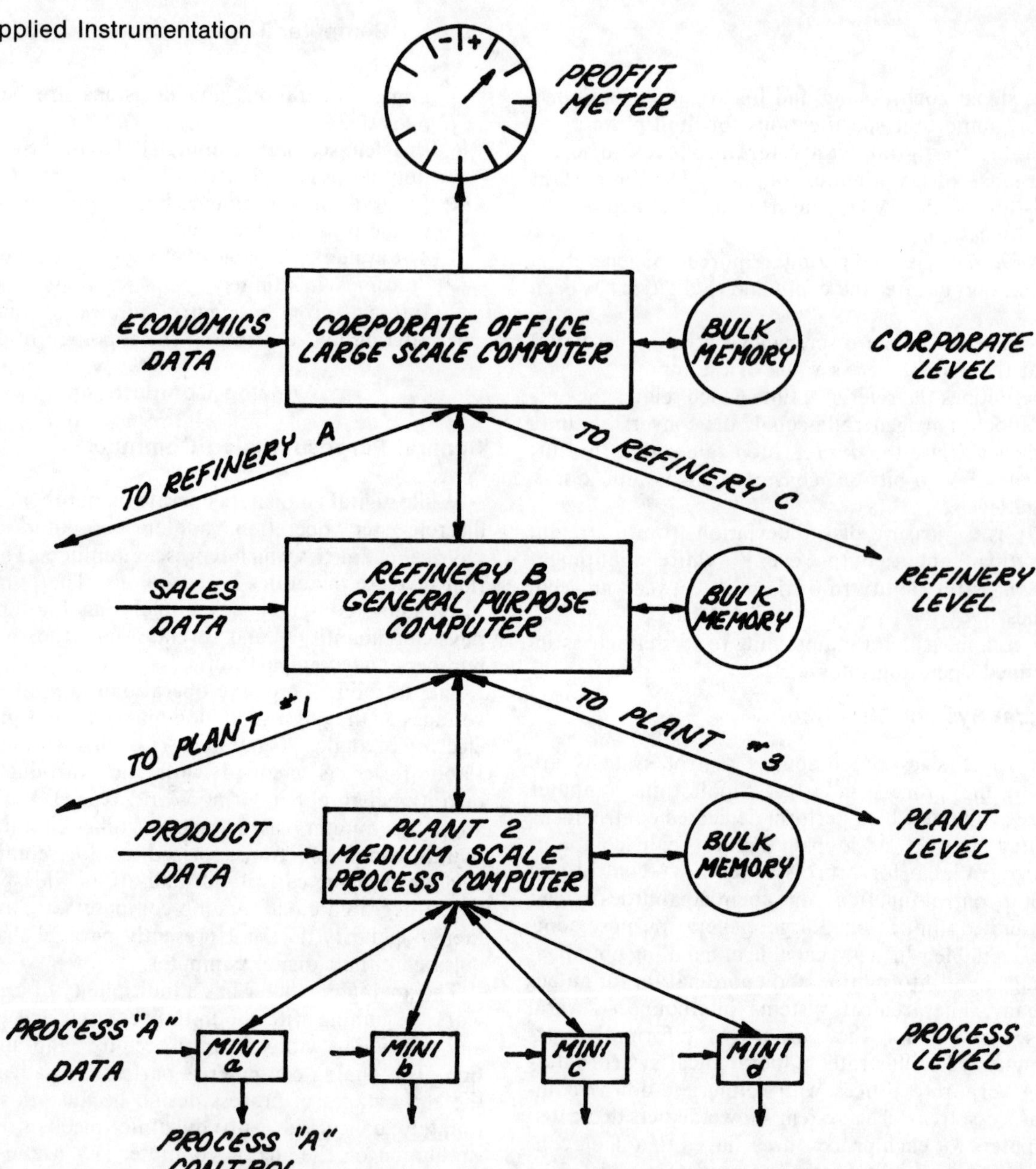

Figure 13.1. This hierarchical computer system illustrates the interconnection scheme of mini-computers at the process level, medium scale computers at the plant level, general purpose computers at the refinery level and large scale computers at the corporate level.

the computing elements; the higher the signal-to-noise ratio, the better the quality. The versatility of the computing elements describes the multipurpose operational features of the amplifiers.

There are two basic classifications of analog components, linear and nonlinear. The linear circuits are either attenuators or operational amplifiers. The nonlinear circuits are multipliers, resolvers and function generators. The mathematical operations performed by the linear circuits are multiplication by a constant, inversion, algebraic summation and continuous integration. The mathematical operations performed by nonlinear circuits are multiplication by a variable, generation of arbitrary functions and mechanization of logical constraints.

Linear Components

The basic circuit of the analog computer is the operational amplifier or op-amp. The linear modes of operation for the op-amp are inversion, multiplication by a constant, summation and continuous integration. For inversion and multiplication by a constant, the transfer characteristics of the op-amp can be closely approximated by:

$$e_o = (R_f/R_i)e_{in}$$

where e_o and e_{in} are the output and input voltages, and R_f and R_i are the feedback and input impedances (resistances). To perform the inversion operation, both resistors are made equal in magnitude so that the amplifier output voltages and input voltage differ only in polarity. If the resistors are not of equal magnitude, the result is multiplication of the input by a constant. To perform the algebraic summation of n variables, n input resistors must be connected to the summing junction of the op-amp. The generalized transfer equation for n variables is

$$e_o = -[(R_f/R_1)e_1 + (R_f/R_2)e_2 + \ldots + (R_f/R_n)e_n]$$

Integration is accomplished by replacing the feedback resistor with a capacitor. The transfer equation describing the output and input voltages is given by

$$e_o = -(1/RC)\int_o^t e_{in}dt$$

To obtain the integral sum with respect to time of n input voltages, n input resistors are used and the equation becomes

$$e_o = -\int_o^t [(e_1/R_1C) + \ldots + (e_n/R_nC)]\,dt$$

Attentuation and multiplication of a DC voltage by a positive constant less than unity is accomplished by using a potentiometer, or pot. This device is simply a fixed value resistor with a movable wiper arm. There are two types of pot circuits available for the analog computer, grounded and ungrounded. The grounded pot, a two-terminal device with one side of the pot permanently connected to ground, is generally used in two ways: obtaining a bias voltage which is less than the referenced voltage and multiplication of a problem variable by a constant less than unity. The transfer equation for the grounded pot is

$$e_o = K(e_{in}) \text{ where } O \leq K \leq 1.$$

For the ungrounded pot, all three terminal connections are available for programming. This permits a variety of useful operations to be performed, the most common of which is the attenuation of two variables. The transfer equation for the ungrounded pot is

$$e_o = e_l + K(e_2 - e_l) \qquad \text{where } o \leq K \leq 1.$$

In modern, large-scale analog computers, pots (K-values) can be set in two different ways—manually or by servomotors. Generally multiturn pots are used with a setting accuracy of better than ±.05%. Servomotors are an efficient way of quickly and accurately setting the pots.

Nonlinear Components

The basic nonlinear components for the analog computer are multipliers and function generators. Multipliers perform the mathematical operations of multiplication of two variables, division of two variables and taking the square root of a variable. Function generators generate the mathematical value, $f(X)$, where X is the independent variable or function generator input. The three basic types of function generators are fixed diode, variable diode and resolver. Common types of fixed diode function generators are X^2 and log X. Variable diode function generators are used to approximate any reasonably well-behaved, single-valued function by generating straight line segments to represent the curve $f(X)$. A resolver is composed of a sine and cosine generator and four multipliers. Connection options allow the resolver to perform (a) polar to rectangular coordinate transformation for up to two input radii at the same angle, (b) rectangular to polar conversion for one radius and angle and, (c) rotation of a set of rectangular coordinate axes.

Computation Accessories

Several additional circuits are generally available in the modern analog computer in addition to the linear and nonlinear components described above. These fall into two categories—passive elements and switching and limiting elements. The passive elements are typically resistors, capacitors and diodes. These elements are used to minimize the number of op-amps required for complicated transfer function simulation. The switching elements are generally function switches and comparators. Function switches are used to incorporate programming options efficiently. Comparators investigate the algebraic relationship between two inputs and determine the position of an output switch based on the following equations:

$$e_l + e_2 > 0, \text{ switch to position 1}$$
$$e_l + e_2 \leq 0, \text{ switch to position 2}$$

Limiters provide a setable output clamp or voltage limit for an input signal. The limiter output will track the input as long as the input is below (above) the upper (lower) set point. If the set limit is exceeded at the input, the output will automatically clamp at the limit voltage until the signal at the input falls below (above) the set point, at which time the output will again track the input.

Analog Computer Modes

The analog computer operator controls the mode of operation for the computer. The modes of operation fall into three categories: computational, check and special.

Computational Modes

There are four computational modes available for the analog computer: pot set, reset, operate and hold. The pot set mode is used to introduce the system parameters prior to problem solution. In the pot set mode, the reference voltage is removed from the amplifiers, and the potentiometers are connected to their respective loads and set. The reset mode, also known as the initial condition mode, connects the initial condition voltages to the output of the integrating op-amps. This charges the feedback capacitors to their initial values. In the operative mode, the initial condition is removed from the integrating op-amps, all inputs are connected and the time solution begins. The hold mode allows the computer to stop a problem solution at any time and begin again at the point where the solution was halted without reinitializing the system. This is accomplished by removing the inputs to the integrators without disturbing their outputs.

Check Modes

The most common check mode for the analog computer is the *static test.* In this mode of operation, a reference voltage is present at the output of all integrating amplifiers, regardless of their programmed initial conditions. This test is used to detect programming errors and component failures or malfunctions. The test procedure is divided into two parts, a program check and a circuit check. The program check determines whether the program actually represents the original equations. The circuit check determines whether the computer interconnections correspond to the program.

The *rate test* mode, available on the more sophisticated analog computers, is used to check the integration rate of all integrating op-amps. Computation accuracy of the integrators depends on a common rate of integration. To perform a rate test, the IC voltages are zero, and an input of all integrators is connected through a common pot to a common reference voltage. The integrator outputs are observed and the feedback capacitors adjusted if rate discrepancies exist.

Special Modes

The repetitive operation mode provides a way of presenting a problem solution repetively at a high speed. This enables trial-and-error parameter and initial condition adjustments with immediate feedback of results. High speed "Rep-Op" automatically cycles the computer between operate, hold, and reset modes. By monitoring the problem variables, the effects of parameter changes are easily and immediately observed.

The slave mode of operation is used when two or more computers are required for a problem solution. For this mode, one computer is designated "slave," the other "master," and their modes of operation are controlled by the master controls.

Analog Programming

The four basic steps in analog programming are

1. Mathematical modeling
2. Scaling
3. Equipment assignment
4. Documentation

A mathematical model is a collection of equations which defines and delimits a physical system. The model equations should be complete, with all approximations and assumptions stated. The equations should take the following forms:

1. First and second order ordinary differential equations solved for the highest derivative
2. Open-loop algebraic equations
3. Implicit equations
4. Closed-loop algebraic equations
5. Logical conditions

In addition to the equations, a listing of coefficients and initial conditions should be made. The analog representation of the mathematical equations forms the model of the program. It takes the form of a block or flow diagram of the problem, showing all of the major computing components and their interconnections.

Dynamic voltage range is one of the limitations of the analog computer. Since the coefficients of most models are not within this range, magnitude scaling is necessary. The steps in magnitude scaling follow.

1. Obtain the unscaled equations for each amplifier.
2. Substitute an equivalent scaled program variable for each problem variable. For example, if the problem contains a variable, X, whose maximum value can be as high as 500, this variable is replaced in the equation with the computer variable, $500\,(X/500)$, without unbalancing the equation.
3. Solve the equations for the scaled outputs in terms of the scaled inputs.
4. Adjust the amplifier gains so that all potentiometer settings are less than unity.

In addition to magnitude scaling, time scaling considerations are necessary. Time, the dependent variable, can be compressed or expanded, much like and accordian, by altering the time scale. This is done by increasing or decreasing the magnitude to the integrator feedback capacitors. The prime consideration for time scaling is that the dynamic response of the computer circuits should not be exceeded.

Modern analog computers are designed for operator convenience and ease of problem preparation. They are equipped with removable patch panels, allowing for remote problem preparation and storage. The holes in the patch panel are color coded and are annotated in accordance with their individual functions. Patch cords are available in assorted lengths and colors. Shorting plugs, or "bottle plugs," can be used to interconnect adjacent terminations, reducing patch cord clutter.

Analog Output Presentation

Generally, both an analog voltmeter and a digital voltmeter will be mounted on the operator's console. These displays are used to indicate the voltage level at the output of any circuit selected. Associated with these meters is an alphanumeric display indicating which circuit output appears on the voltmeter.

Oscilloscopes are often used for computer output display, especially during high speed, repetitive operation.

Strip chart recorders are employed when a record of the computer output is required. The recorders often contain multipen channels. The paper is drawn past the pens at a constant speed, and the pens deflect proportionally to their input voltages, resulting in a set of graphs of voltages as a function of time.

Another device used to record the computer outputs is the X-Y plotter. This device plots two variables simultaneously in the form of a rectangular coordinate graph. It differs from the strip chart recorder in that the two variables are plotted as a function of each other rather than as a function of time. Another difference is that the paper is held stationary and the pen moves.

Magnetic tape is also used to record the outputs of the analog computer. This tape is very similar to that used for digital recording, except that the information is recorded in analog form.

Common Characteristics

The following characteristics are common to most of the analog computers available today.

1. Variables are treated in a continuous form.
2. Accuracy is dependent on the quality of system components.
3. Computing operation is parallel for all system elements.
4. Selectable operating speeds are available for problem solution—from slow motion, to real time, to high speed, repetitive operation.
5. Efficient arithmetic operations, such as multiplication, addition, integration and nonlinear function generation, are obtainable.
6. Programming techniques involve the substitution of analog computing elements for corresponding physical system elements.
7. Simulation studies can include the use of actual system hardware and instrumentation.
8. Optimization studies are conducted efficiently as a result of the ease of parameter and coefficient changes.

Hybrid Computation

Hybrid Techniques

Hybrid computation includes all computing techniques which combine some of the features of digital computation with some of the features of analog computation. The term hybrid computer, however, is used primarily to describe computer systems involving the interface of a general purpose digital computer and an electronic analog computer. The first successful development of the hybrid computer occurred in the early 1960s, after both the transistorized analog computer and the transistorized digital computer were developed.

The spectrum of hybrid computing techniques can be divided into three parts: primary digital, balanced digital and analog and primary analog. The following is a descriptive survey of the more important characteristics of the hybrid spectrum.

Primary Digital

For the primary digital hybrid system, analog elements serve as peripheral equipment to the digital computer. These systems were developed to increase the speed of a pure digital system by performing some of the arithmetic calculations, such as multiplication and integration, using analog circuits.

Primary Analog

At the other end of the spectrum are the primary analog hybrid systems. These systems employ digital logic circuits to control pure analog solutions as well as using digital instruments, function generators and small, general purpose digital computers as peripheral equipment.

Balanced Digital and Analog

Modern hybrid computer systems are comprised of both general purpose digital computers and general purpose analog computers. Each component computer is capable of "stand alone" operation, but by interconnecting them, an even more powerful computer system is realized, the true hybrid computer system.

True Hybrid Systems

Bilateral Hybrid System

A bilateral hybrid computing system is one in which information can flow between the analog computer and the digital computer in a closed loop configuration. Figure 13.2 illustrates the general configuration of the bilateral hybrid system. In addition to the analog computer and the digital computer, the following major systems are identified with the bilateral hybrid system.

Multiplexer is a device for converting data from parallel form to serial form. The multiplexer accepts inputs from several analog elements, sequentially samples the output voltage of each element and produces over its single output line a sequence of pulses, each having an amplitude proportional to a corresponding input voltage. The output of the multiplexer serves as the input to the analog-to-digital converter.

Analog-to-digital converter is a device which converts the voltage pulses from the multiplexer into the binary-coded equivalents. The digital output of the A-D converter is a word which goes to the digital computer for processing.

Demultiplexer is a device which distributes sequentially generated digital data into an array of output registers, each of which controls a digital-to-analog converter.

Digital-to-analog converter is a device which converts data in binary form to an analog voltage proportional to the binary number.

Unilateral Hybrid Systems

The unilateral hybrid system functions in an open-loop form with information exchange either from the analog computer to the digital computer or from the digital computer to the analog computer. The direction of information flow depends on the configuration of the system. Figure 13.3 illustrates two basic configurations of the unilateral hybrid

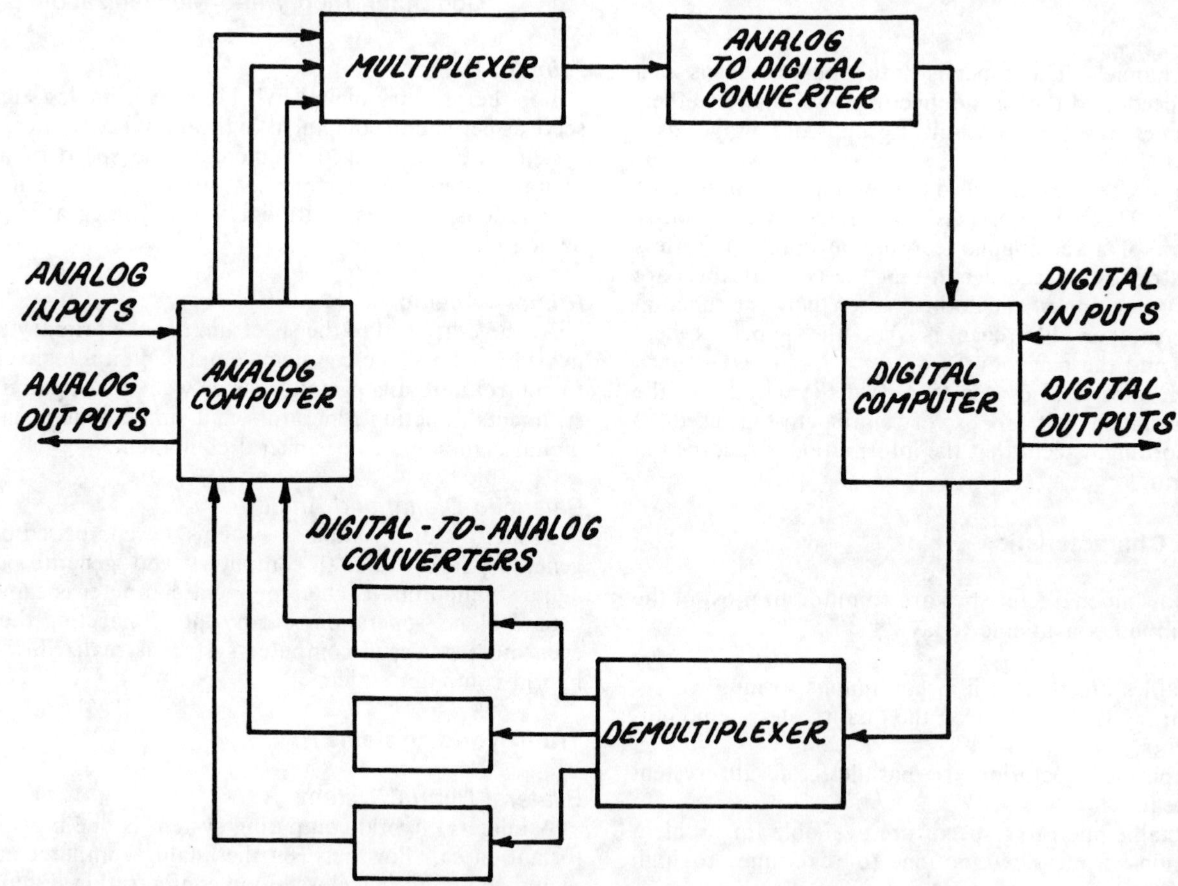

Figure 13.2. The major components of a bilateral hybrid computer system are interconnected with closed loop information paths between the digital computer section and the analog computer section.

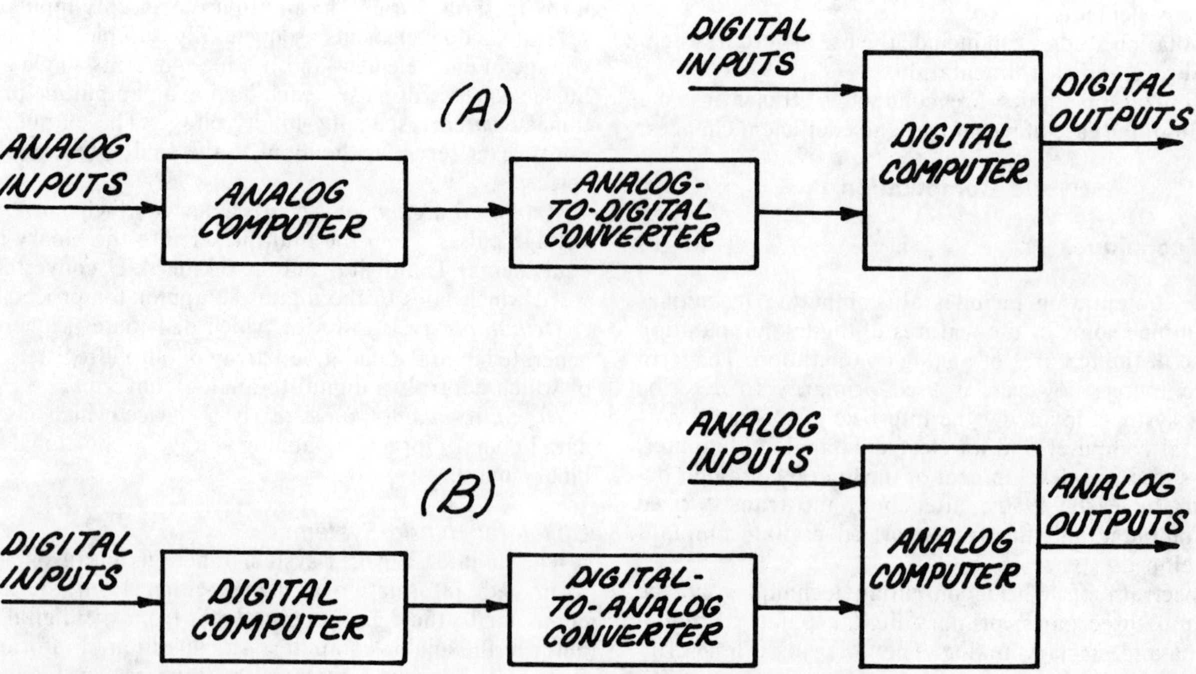

Figure 13.3. The unilateral hybrid computer system is an open loop system with information transfer: (a) from the analog computer to the digital computer or (b) from the digital computer to the analog computer.

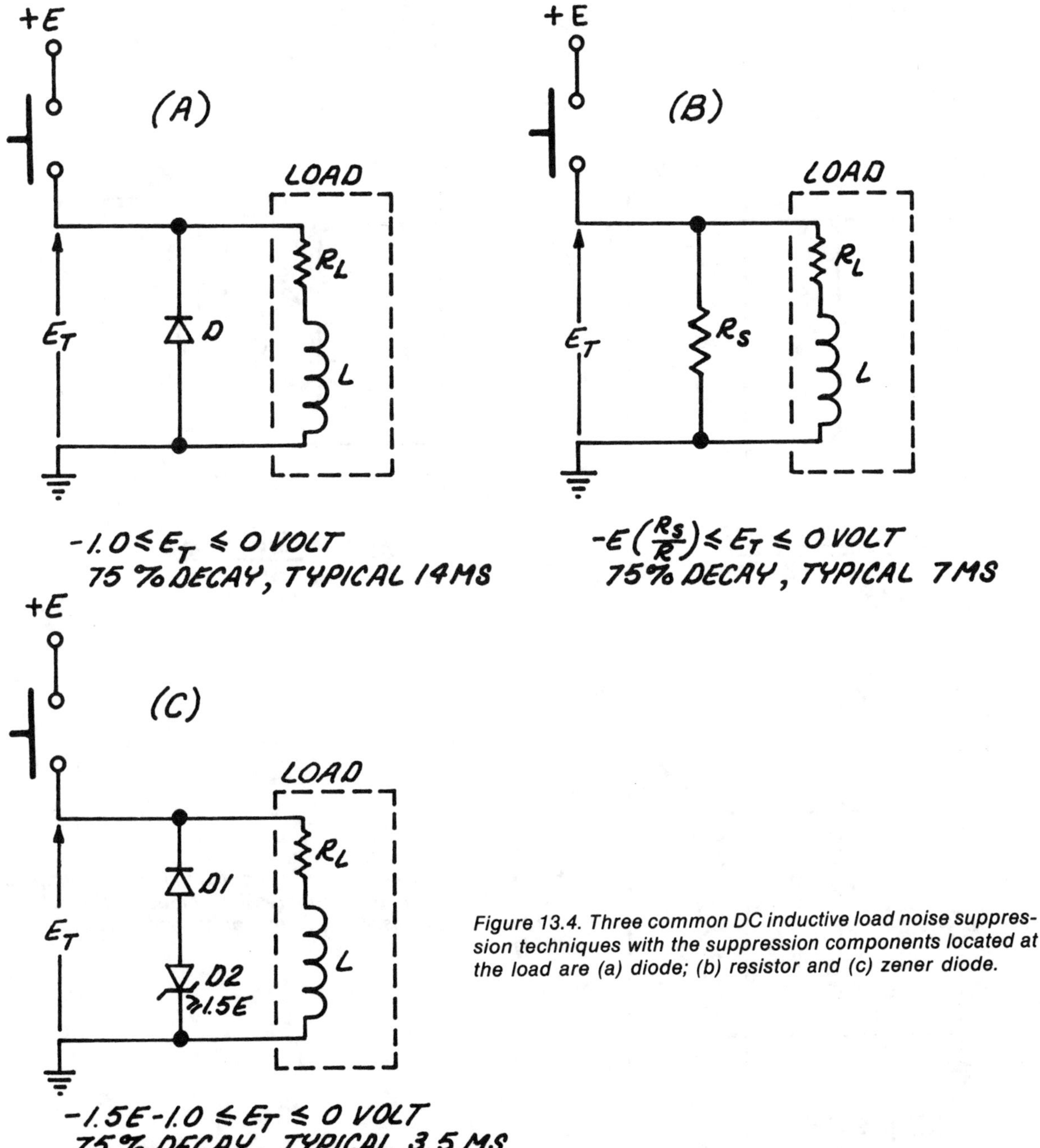

Figure 13.4. Three common DC inductive load noise suppression techniques with the suppression components located at the load are (a) diode; (b) resistor and (c) zener diode.

system. It should be noted that regardless of which configuration is used, only one type of converter, either analog-to-digital or digital-to-analog, is required at the interface between the two computers.

Areas of Application

Simulation

The hybrid computer permits the simulation of physical conditions for the efficient real-time analysis of physical systems. This simulation permits the study of the effects of parameter changes on the excitation-response characteristics of the systems. Hybrid models have been developed to investigate the response of process controls to random excitations, random parameter variations or random initial conditions. The process is simulated on the analog portion of the hybrid computer while the digital computer coveniently functions as the digital controller.

Another major area of hybrid simulation occurs in the simulation of a process start up and operator training. Because of the empirical nature of most sophisticated processes, optimum startup procedures are almost impossible to determine mathematically. Simulating the startup not only trains the operators on what to expect but also alerts

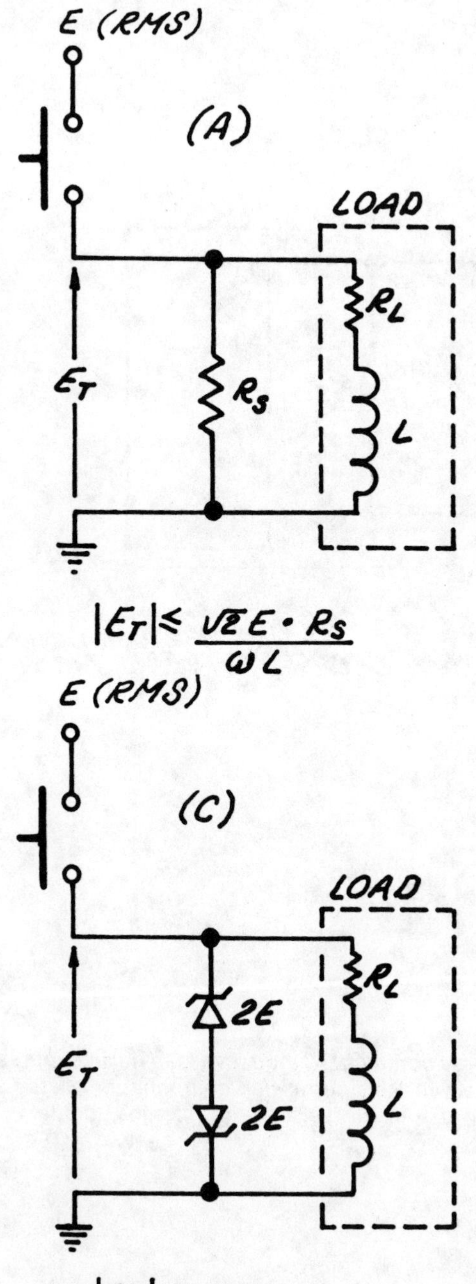

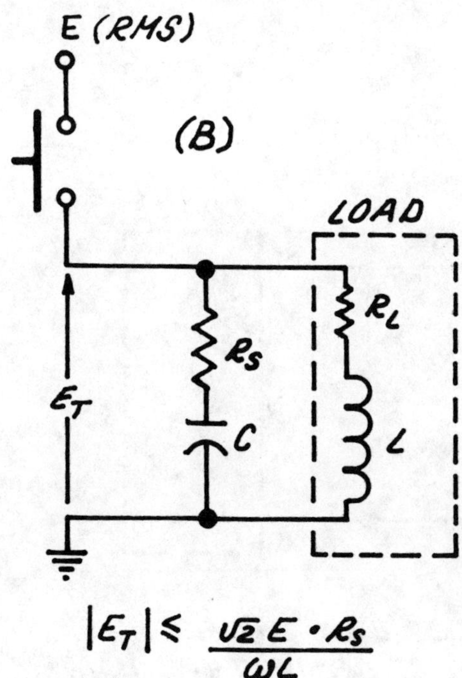

Figure 13.5. Three common AC inductive load noise suppression techniques with the suppression components located at the load are (a) resistor; (b) resistor-capacitor and (c) zener diode.

the startup team of potential process crises, enabling them to recognize and correct emergency conditions efficiently.

Optimization

System optimization is another major area for hybrid computing techniques. The criterion for optimization is that the values of a set of parameters or functions must be determined in such a way that a particular performance function, or cost function, is maximized or minimized. Hybrid computer optimization techniques are applicable to practically all areas of process control.

Motivation for Hybridization

Although there are many technical and economic reasons for using a hybrid computing system, the chief motivations for interconnecting digital and analog computers are:

1. To utilize the speed of the analog computer with the accuracy of the digital computer.
2. To permit the use of actual system components in a simulation study.

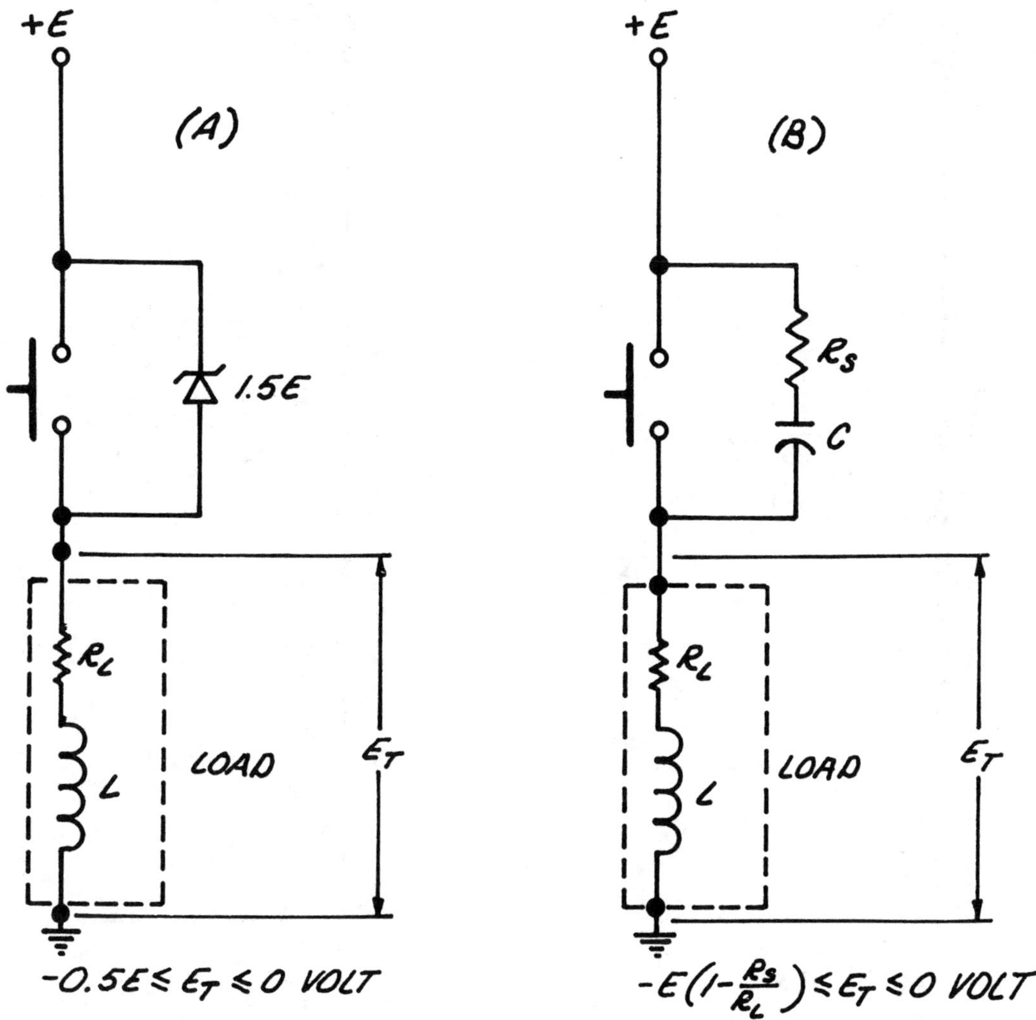

Figure 13.6. Two common DC inductive load noise suppression techniques with the suppression components located at the switch are (a) zener diode and (b) resistor-capacitor.

3. To increase the flexibility of an analog simulation by using digital memory and control.
4. To increase the speed of a digital computation by utilizing analog subroutines.
5. To permit the simultaneous processing of both discrete and continuous input data.

Data Transmission Interference

Causes of Electrical Interference

Electrical interference, commonly called noise, is any signal distortion which tends to mask the information carried by the signal. Although there are many terms used to describe noise, there are only four major ways electrical noise can enter a system:

1. Common impedance
2. Magnetic induction
3. Electrostatic coupling
4. Thermoelectric injection

Common impedance noise can occur whenever two circuits share conductors, impedance or power sources. Noise which exists in one circuit can be conducted to another circuit by a common neutral ground or power conductor. Another conduction mechanism is leakage path conduction, caused by moisture or inadequte insulation.

Magnetic induction or coupling is the transport mechanism where noise enters a circuit through the mutual inductance between conductors. The magnitude of the coupling depends upon the rate of change of noise source current. The most common cause of magnetic induction noise is capacitor switching transients, occurring when a switch is closed between a capacitive reactance and a low impedance. The impulse current which results is limited in amplitude only by the source impedance and the distributed resistance and inductance of the associated wiring.

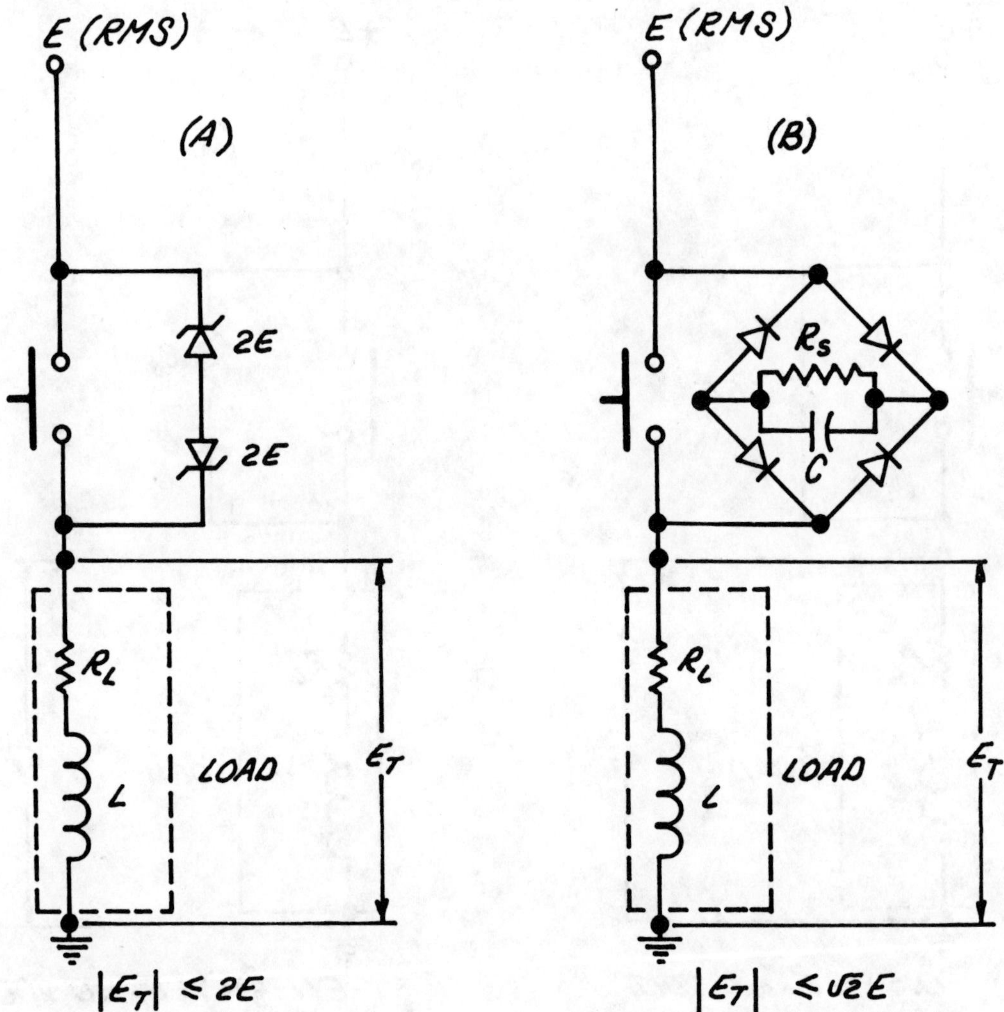

Figure 13.7. Two common AC inductive load noise suppression techniques with the suppression components located at the switch are (a) zener diode and (b) resistor-capacitor.

Electrostatic coupling, also called capacitive coupling is the transport mechanism where noise enters a circuit through the mutual capacity between circuits. The magnitude of the coupled noise depends upon the rate of change of voltage at the noise source. Electrostatic noise is most commonly caused by opening a switch, which attempts to interrupt the flow of current through an inductive reactance. Opening the switch causes a rapid voltage buildup across the inductance with a polarity that tends to prevent the interruption of current. The impulse voltage amplitude that results is limited only by the dielectric breakdown of either the switch air gap or the inductor insulation.

Thermoelectric injection noise enters a circuit as a result of the connection of dissimilar metal conductors. A low level noise voltage is inserted in the circuit if a temperature difference exists between the ends of the dissimilar conductors. Loose or corroded connections will also cause noise in a circuit due to thermal cycling and the resulting series noisy resistance.

Noise Reduction

The most successful approach to noise reduction is the suppression of the interference at its source. Often this is not the most economic or practical approach. Other reduction techniques are available. These include conductor and component shielding, twisting signal conductor pairs, signal line balancing, signal filtering and special grounding techniques.

The major sources of noise transients that can easily be suppressed are contactors, actuating solenoids, motors and high inrush-current loads. Inductive transients are the principal cause of system malfunction. Peak voltages exceeding 1 kilovolt are not unusual for unsuppressed switching transients. These transients are not a serious problem for conventional relay control systems because relays are insensitive to the high frequency noise they generate. Semiconductor logic, on the other hand, is quite sensitive to high frequency signals, even to the extent of permanent damage. The optimum place to suppress noise is at the source of the noise generation. Figure 13.4 illustrates several methods of noise suppression for DC inductive loads, giving typical ranges for transient voltages as well as a comparison of the decay switching time. Figure 13.5 illustrates three methods of noise suppression for AC inductive loads. When the suppression components cannot be placed near the source, they should be placed across the switch terminals. Figures 13.6 and 13.7 illustrate methods of noise suppression at the switch terminals.

Sometimes it is impractical, if not impossible, to suppress all of the noise sources in a process area. There are several good physical wiring practices that help to minimize the coupling of noise into control and monitor circuits. Twisting a two-wire signal transmission line every 2 inches can reduce magnetic induction by a ratio greater than 100:1 over parallel wires. Electrostatic coupled noise can be reduced by a ratio of over 6,000:1 when a twisted pair signal wire is shielded with an Aluminum-Mylar tape and grounded drain wire. The grounding of the shield drain wire and the signal common should be at the same point—and *only* at that point. Ground loops can result when a system is grounded at more than one point. The reason for this is that a potential difference often exists between ground points and this potential sets up current loops in the shields.

Common mode voltage, a voltage that is common to both lines of a signal pair, is a major cause of error for low level signal control. Line balancing (equalizing the IZ drop for each line of the signal pair) reduces the error induced by common mode voltage.

Filtering is another effective means of attenuating noise. Filters are designed to attenuate a certain range of frequencies and pass all others unchanged. Effective noise filtering requires careful attention to several difficult tasks. Among these are:

1. Defining the excitation—response characteristics of the filter
2. Specifying the permissible tolerance for signal deviation
3. Synthesizing a network of components which satisfy the excitation—response specifications
4. Realization of a physical network which closely approximates the network synthesis

The final selection of a noise filter must also be based on criteria involving cost, convenience, dependability and engineering judgement.

14 Data Acquisition Equipment, Microprocessors and Programmable Logic Controllers

Munawar Ahmad, Tom Lisch, Baxter Williams

Technology has recently taken quantum leaps forward. Microminiaturization of electronic circuits is just one example of such a leap. In 1971, microelectronic circuits were heading for the marketplace. Hand calculators with engineering functions and a few storage registers originally sold for $400; now they can be bought for less than $30. Programmable calculators, using programs stored on a magnetic strip smaller than a piece of chewing gum, solve complex equations inexpensively from an entry/readout device small enough to be carried in a shirt pocket. In the brief time since ENIAC (refer to Chapter 13), calculators have gone from room size to postage-stamp size.

Of course, the effect is even more dramatic when it is realized that true computer functions, such as storage and usage of massive amounts of data, control of pipeline pump and metering stations, and process control of battery limit units, are now available from a shoe-box full of postage-stamp-sized chips.

Data Acquisition Equipment

Collection of unit operating data is quite valuable to the safe and economic operation of the plant. A record of key temperatures in the unit is generally considered the minimum information required. Data is acquired and recorded in many different ways. Recording methods vary, ranging from readings taken from analog indicators and written by hand, to the logging of data collected and stored in a computer and printed by an automatic device. Several types of currently used systems are described.

Display-Only

One of the earliest and still most universally used data displays is the multipoint temperature indicator or recorder. Using toggle switches, rotary selectors, or push buttons to select the thermocouple to be monitored, the values are centrally displayed. The operator periodically triggers the switches and then writes each value on a report form. Indicator features now include digital readouts, electronic push-button type selectors, internal precision cold-junction compensation, self-contained calibration voltages, and adjustable alarm trips (Figure 14.1).

Analog Recording

In the case of multipoint recorders, logging is done automatically, but at a slightly higher initial cost. Most multipoint recorders print a process analog value as a dot with a corresponding point number printed beside it for identification (Figure 14.2). Options sometimes include several print colors to aid in distinguishing the adjacent points.

Some recorders are able to skip selected points without delaying the printing of succeeding points. Most are limited in the quantity of alarm trips that can be included, mainly

Figure 14.1. Indicator features now include digital readouts, electronic push-button type selectors, internal precision cold-junction compensation, self-contained calibration voltages, and adjustable alarm trips. (Courtesy of RdF Corporation.)

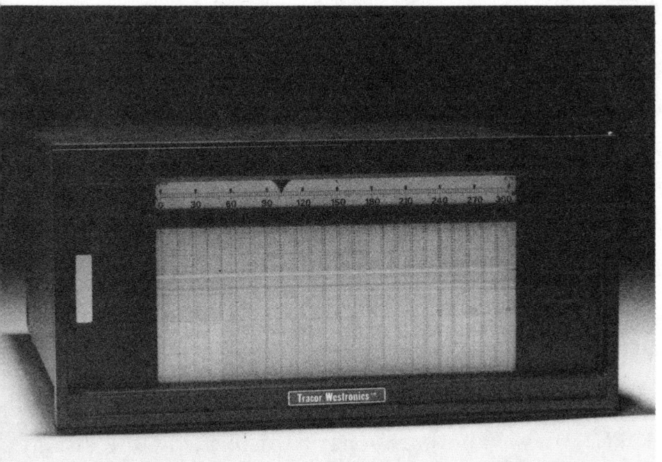

Figure 14.2. Most multipoint recorders print a process analog value as a dot with a corresponding point number printed beside it for identification. (Courtesy of Tracor Westronics.)

because of available room within the instrument case. Up-scale or down-scale thermocouple burnout printing can be specified.

Because the motions within a drive and print mechanism are complex, recorder maintenance must be considered. Early mechanisms were highly complicated, belts, gears, slide wires, and other moving parts being responsible for many ulcers. Recent designs have significantly reduced the number of parts, and have all but eliminated the recorder-related 2:00 a.m. call out.

Digital Recording

Digital temperature recorders are able to log temperatures at low cost (Figure 14.3). The first machines were not

Figure 14.3. Digital temperature recorders are able to log temperatures at low cost. (Courtesy of Kaye Instruments, Inc.)

much more than rebuilt adding machines. Many machines still incorporate narrow-tape printers because of their rugged, time-proven design. Other recorders use recent devices such as electrostatic or dot-matrix printers. Electrostatic types allow extremely fast print times. Dot-matrix print heads are very reliable, and are often seen in point-of-sale equipment in retail stores.

Many digital recorders are capable of printing out-of-limits data in color, usually red. Alarm conditions are thereby highlighted. Also, any point which enters an alarm condition is immediately printed. Data printed consist of date, time, and point number and its value.

Print intervals are selected by the operator, who can manually initiate a print of all points or a single point, or cause the scanning circuit to stay at a certain point.

Data Acquisition Systems

A data acquisition system is a stand-alone collection of equipment. The simplest such systems are data loggers.

Data Loggers

Only temperature measurements for data indication and recording have been mentioned so far. Any analog signal can be an input to a scanning digital recorder if proper signal conversions are made.

Recent equipment accepts a variety of electronic signals, including several types of thermocouples, resistance, 4-20 maDC, 0-1 vdc, frequency, and digital pulses (Figure 14.4). Circuitry converts each signal to a usable level, where linearization and scaling can be effected. Engineering units are also printed for each value.

Newer data loggers are capable of calculating values using input data before printing. Since microcomputer-

Figure 14.4. Recent data acquisition equipment accepts a variety of electronic signals. (Courtesy of Kaye Instruments, Inc.)

calculator circuitry is now available, sophisticated data loggers can calculate values such as mass flow and mass storage. Features of some equipment include a key switch to prevent inadvertant or unauthorized program modifications, internal battery backup, automatic restart after power interruption, full program listing ability, independently programmable high and low alarms for each input, accumulation (integration) of data for a given time period, channel skipping, expandable memory and input/output capacity, adjustable dwell time, limited programmable alphanumeric messages, and various outputs, such as 4-20 maDC, RS-232-C serial, BCD, and ASCII compatible types.

Some data logging equipment is designed for recording in unique circumstances. One manufacturer of water quality monitoring equipment provides data recording on standard Phillips cassettes using BCD format (Figure 14.5). The recording module can be mounted in a submersible housing for use at depths to 300 meters and operated from rechargeable batteries. It is used in unattended buoys, or aboard ships for profiling. The recorded cassettes are read by special equipment which can display, print on paper tape, and send data to an on-line computer by using an RS232C modem. A microprocessor-based logger can record averaged data, store and log peak values and their respective times, and output the data in serial format.

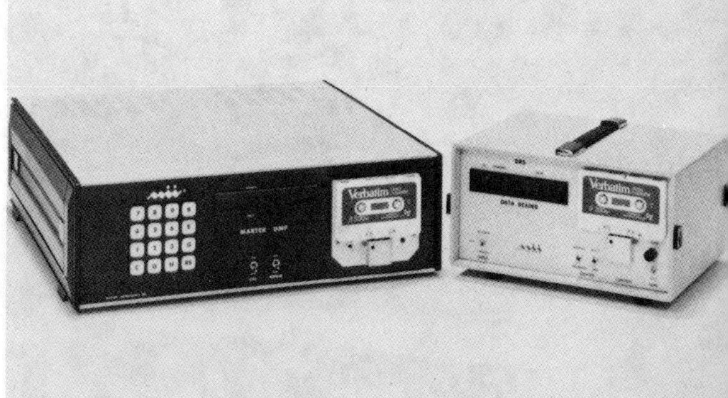

Figure 14.5. Data can be recorded on standard cassettes using BCD format. (Courtesy of Martek Instruments, Inc.)

Video-Based Display

The video-based display of data enables the control room operator to examine data of a large number of varying signals. Instead of examining a number of adjacent strip charts, historical values can be stored in an electronic memory and then displayed on a CRT device. One manufacturer displays as many as 32 current variables with 3 previous values for each, or up to 16 current variables with 8 previous values for each, all shown as horizontal bar graphs (Figure 14.6). Historical data is taken at preset time intervals, such as one minute intervals for temperature data and daily intervals for vibration data. Scales may also be displayed at the expense of reducing the number of displayed variables.

Variations of this display system include video-based annunciator systems with event recorders, and large multiplexed-input systems having dedicated CRT displays, data printers, page-selectable CRT monitors, programmable alarm relays, and host computer interfaces.

A good example of video-based display systems dedicated to specific services is tank gauging. In the tank farm, transmitters provide a signal developed from the take-up reel of a float-type level device. Resistance temperature detectors extend into the tanks to measure product temperature. These signals are fed into a local converter, which converts them to digital signals and sends them to a multiplexing device or to a scanning receiver. The receiver continually scans the field-mounted devices, stores data, and performs programmed calculations. A CRT terminal reads and displays tank data (Figure 14.7).

Additionally, most systems can alarm at operator-adjustable alarm values, compute gross tank volume from strapping tables stored in memory, compute net volume using temperature data and an API gravity value, indicate a faulty signal (parity check) or a malfunctioning transmitter, and display data in English or metric units. Output signals can feed to a large computer, to hard-copy printers, and to additional video terminals.

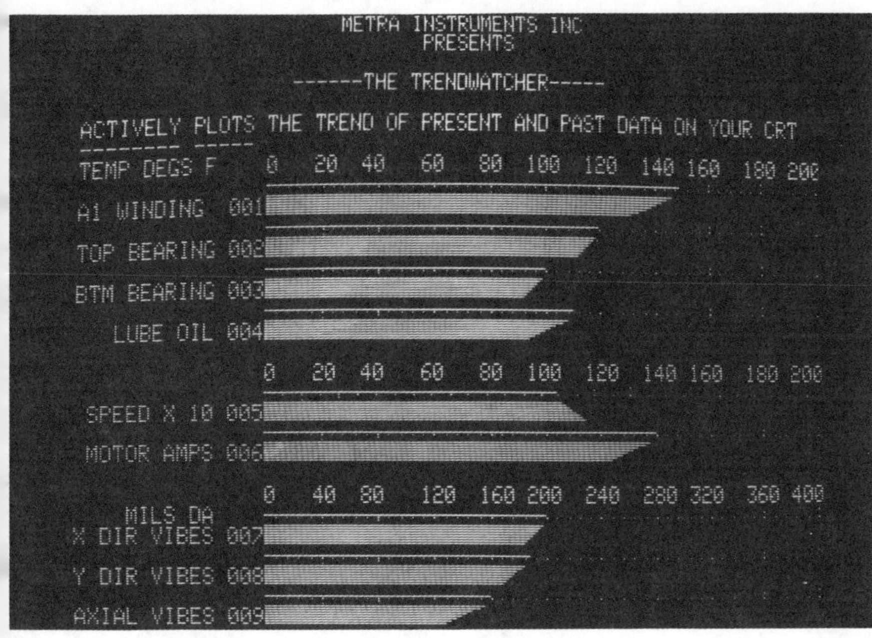

Figure 14.6. Historical data values can be displayed as horizontal bar graphs on a CRT. (Courtesy of Metra Instruments.)

Large Data Acquisition Systems

Data acquisition system (DAS) base hardware comes in a variety of sizes. Some manufacturers produce a "DAS-on-a-chip." Using microprocessor technology, the key elements—analog multiplexer, amplifier, analog-to-digital converter, voltage reference, and regulator—are combined on a single monolithic chip. But, while the heart of the system can be miniaturized, the pieces of hardware which complete the instrument must also be included.

Figure 14.7. The Telepulse 600 CRT terminal reads and displays tank data. (Courtesy of GPE Controls.)

Since the DAS chip is obviously intended for use with a microprocessor or microcomputer, memory, readout, and software are factors to be understood and considered when evaluating systems for application. A more detailed description of these devices is included in another section.

A computer-based DAS generally handles large quantities of inputs, usually thousands. Consisting of several cabinets full of terminations and input/output devices, and often accompanied by uninterruptable power supplies, data storage discs or tapes, CRT readouts, and other peripherals, systems of this size are imagined when a "data acquisition *system*" is mentioned (Figure 14.8). A computer-based DAS can successfully handle data for entire plant utility systems and for multi-train operating units. It would seem safe to say that the maximum size of such a system would be limited only by the imagination of the design engineer and the willingness of management to fund the project.

Because only a few more parts and pieces need to be added to a large DAS to provide a full-blown computer control package, most system designs incorporate features that allow upward compatibility. The forward-thinking designer assumes that a large plant will someday have advanced computer controls for optimization and energy savings, and designs the DAS accordingly. While more money can usually repair an awkwardly designed system, the more successful design is keyed to upward-compatibility of equipment, manufacturers who will obviously remain in business, and forward-looking software.

Upward-compatibility of equipment, or the equipment compatibility of a manufacturer's smaller designs with his larger, more advanced ones, can save many dollars in future expansion. Should succeeding models not be compatible, original hardware must sometimes be replaced in order to upgrade the system.

Software, or programming of the DAS, is a very important facet of any computer-based system. Most manufac-

Figure 14.8. A computer-based DAS generally handles large quantities of inputs. (Courtesy of S.I.P., Inc.)

turers have standard programs which are furnished with their packaged DAS. The salient features of this "canned" software are often what sells the system, although each is a custom system in that the data input sensor must be identified and programmed. Complete computer control of the process unit is effected when sufficient memory is programmed to send control signals to appropriate control elements, which, of course, changes the definition of the system to something other than a DAS.

An extremely important aspect of the programming of any computer-based system is the documentation of the programs. The permanent program, or "firmware," is usually stored in a nonvolatile section of memory. In microcomputer circuitry this is fixed in programmable read-only memory, or PROM, or in an erasable PROM, or EPROM. Some equipment, having no PROM-type programming, must store its base program on paper tape or magnetic memory, so that reprogramming can be quickly done should a loss of power wipe out the computer's intelligence. Most manufacturers provide the user with reprogramming backup by keeping the user's program on file. Program changes can be requested in detail over the telephone, modifications made, and a newly programmed PROM mailed for insertion by the user. With proper hardware, a modified temporary program can be stored in the DAS by a return phone call until the programmed PROM arrives. In any case, a listing (or printout) of the programming is always desirable for system analysis.

While an explanation of computer programming languages is beyond the scope of this text, it must be noted that there are several ways to "talk" to computers. The language that a specific piece of hardware is designed to understand is known as "machine language." There are as many machine languages as there are kinds of machines, and even more different kinds of modified languages. A good understanding of the machine's language will help an engineer realize the machine's full usage. Most DAS computers are programmed in machine language, with provisions being made through special keys or program modifications to allow the operator to add points and access data.

Fortunately, ways exist to converse with computers in languages that are similar to English. Known as high-level languages, such software used for programming automatically converts a keyboard input of P-R-I-N-T into machine language that causes an automatic typewriter to print data. One of the easiest languages for the layman to grasp is called BASIC. Developed at Dartmouth, BASIC is used in many consumer-oriented home computers.

Microprocessors

The microprocessor-based instrument or control system is a new concept to many operators or maintenance personnel. This chapter can help relieve the fear of the unknown that grips those who have not yet been given a conversational grasp of such a handy device. First, develop a simple idea analogous to the microprocessor.

A Simple Analogy

Picture a production superintendent sitting behind a desk. On one corner of the desk are baskets marked *in* and *out*. Just to the side is a wall full of pigeonholes. One of the superintendent's tasks is to continually check the in-box for intercompany memos and then process them. Finally, the mail clerk drops one in the basket. The superintendent looks at the memo, which says to place the data in the first few numbered pigeonholes, and then act on the basis of the instructions found in pigeonholes 46 through 100. The superintendent does this using only *one hand*. Finally, the resultant is placed in the out-box. The process is then repeated when a second memo is received.

The microprocessor is such a superintendent. It can execute instructions from the plant manager with lightning speed, assign duties to operators, check their progress, watch the entire unit, remember everything, pass on production reports at meetings, and request new data and tasks.

It does not tire of its duties, and will do them all at the same fixed, fast tempo. The more tasks it is given, though, the longer it takes to complete its duties. Depending on the machine's abilities, it might be capable of interrupting normal duties to handle an emergency, higher priority task, and then come back to the interrupted job.

The microprocessor's limitation usually is not enough pigeonholes (available memory spaces) to store information and instructions, but additional memory spaces can be purchased as add-on units.

Naturally, there are many different kinds of microprocessors. Some are faster than others; some have larger memories; and some use only simple four- or eight-letter words. (Watch out for the four-letter words!) Others can understand 16-letter words. Some can be given only a few instructions; others, many instructions. Some "trash-can" everything that is put into their memory spaces when an electrical power loss occurs; others do not. Some microprocessors have instructions permanently engraved in part of their memory; others have their "permanent" instructions written in chalk so that they can be erased and rewritten.

The analogy is obviously a simplification, but it is not far from the truth. The important thing to undertand is that microprocessors, programmable logic controllers, computers, calculators, and the like, are made for people to use. The intent of this entire chapter is to strip away the mystique of the bit, byte, and baud, to relieve fears, and to spur curiosity. An "A-Operator," regardless of age and educational background, can grasp the concepts involved. Any high school or college graduate can learn the principles to either sell such equipment or intelligently decide which equipment and features are best for a particular application. Once the concepts are realized, the engineer, manager, or superintendent begins to understand the enormous benefits and departmental effects. Budget requests, justifications, and educational programs can then be effected.

Excellent basic texts that provide detailed explanations of specific internal microprocessor circuitry are available. Electronics parts stores are excellent sources, as are technical bookstores. Treatment of the subject will be somewhat cursory, but will give the reader a conversational knowledge of basic hardware and systems.

Key Words in Microprocessor Terminology

A microprocessor is capable of doing several things in a prearranged order (preprogrammed). To understand what is happening from the operator's and the engineer's standpoint, it is first necessary to understand and use microprocessor terminology. Since microprocessors are new in the instrumentation field, it may be very difficult to understand these new gadgets that appear to be flooding the market and rapidly appearing in control rooms. Engineers are being bombarded by management, sales personnel, and clients who are all using these seemingly foreign terms.

Common Terms

ASCII: American Standard Code for Information Interchange. An eight-bit coding scheme for serial transmission of alphanumeric data. The first seven bits are representative of 128 standard ASCII characters. Bit 8 is a parity bit used for error checking.

Bit: a binary digit. A digital word representing a single character. It is the smallest unit of data in a digital microprocessor.

Byte: an eight-bit digital word, sometimes called an eight-bit byte.

Chip: a small piece of semiconductor material (silicon, germanium or other) containing one or more circuits. Often this term is used interchangeably with *integrated circuit* or *DIP.*

CMOS: Complementary Metal Oxide Semiconductor. An extremely low-power dissipation semiconductor device. The use of this type of circuitry allows smaller packaging of the microprocessor.

DIP: Dual In-line Package. The method of packing a chip in plastic or ceramic material with the input/output leads forming two lines.

LSI: Large Scale Integration. The accumulation of a large number of circuits on a single chip. In a microprocessor, there will be several hundred circuits on a chip.

Memory: the circuits in which the microprocessor data and/or instructions are stored. Common storage circuits are RAM, ROM, PROM, EPROM, magnetic tape, or disk. The size of the memory is defined by the number of digital words it will hold.

Microprocessor: a digital computer contained in one or more integrated circuits (usually an LSI CMOS chip).

Read: moving of data from one place to another.

Word: a set of bits (4, 8, 12, 16, 24, or 32) comprising the smallest addressable unit of information that can be manipulated at one time.

This list is not complete. It is only a few of the key words that sales personnel, management, and others interested in microprocessors are using.

Microprocessors and Related Components

A microprocessor is a digital computer contained in one or more integrated circuits. In fact, each chip is a complete system or subsystem in itself. This can be seen in the internal block diagram of a microprocessor (Figure 14.9).

Presently there are more than 24 manufacturers of microprocesssor units, all of which have several families (different microprocessors related to one particular series). Therefore, it is impossible to cover all units in this text. To review the different manufacturers and what they offer, see the *IC Update Master,* published annually by Cox Broadcasting Corporation. Manufacturers' literature is also helpful.

Much attention has been given to the microprocessor since its evolution was brought about by LSI circuitry. This miniaturization of circuitry enabled a complete microprocessor to fit into a single chip, making the microprocessor a very small and rugged package that can be placed almost

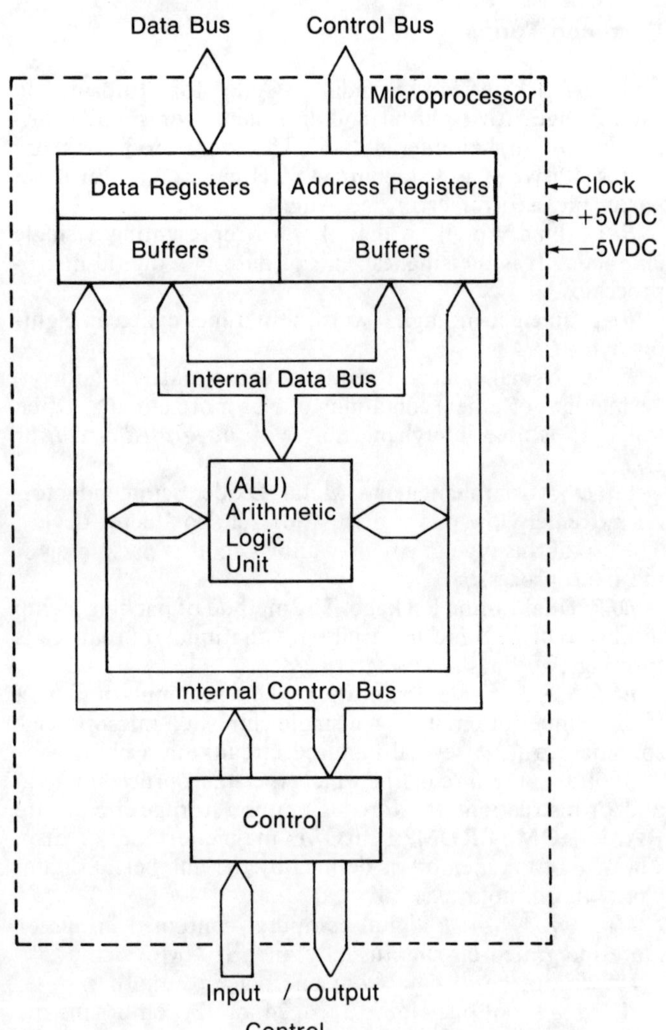

Figure 14.9. Block diagram of a single-chip microprocessor.

Figure 14.10. The Daniel flow calculator has the density table for a particular gas in memory. (Courtesy of Daniel Industries.)

anywhere. Cost was also minimized, allowing widespread application. A few examples are the Cadillac Seville automobile, hand-held calculators, microwave ovens, electronic watches, data acquisition systems, dedicated control instrumentation, and control sequencing equipment.

In the process control industry most manufacturers of instrumentation have used the microprocessor in one form or another. The Daniel flow calculator (Figure 14.10) has the entire density table for a particular gas in memory. The microprocessor, while reading flow rate, pressure, and temperature inputs, retrieves from memory the proper density correction for converting flow rate to mass flow. Because of the rugged characteristics of the chips, it is not necessary to put the Daniel unit in an air-conditioned, humidity-controlled room.

Single-Chip Microprocessors

The single-chip microprocessor (Figure 14.9), with program memory, data memory, and input/output (I/O) ports contained in one package, lends itself to dedicated control applications. These applications are limited by memory space (usually 1K), and the inability to diagnose the microprocessor's activities by not being able to look at the Data Bus in action. Actually, the present trend is toward using a single-chip microprocessor for solving small, well-defined process control problems.

Applications for microprocessors range from data processing centers to the dedicated control functions previously mentioned. In traversing this broad spectrum, it becomes apparent that two major categories of users exist. First are the consumers who are a high-volume market with items such as toys, calculators, and appliances. Second are the industrial users with many specialized applications of process control. This chapter is addressed to this second category of users.

Microprocessor Unit (MPU)

The MPU is sometimes referred to as the Central Processing Unit (CPU), and is the brains of the processing system. The MPU is supported by an external clock, RAM and ROM memory, an interconnecting bus system, and input/output (I/O) devices. The complexity of the system determines how many and what type of support devices are required.

The Motorola M6800 microprocessor is one of the chips widely used for process control applications. Figure 14.11a

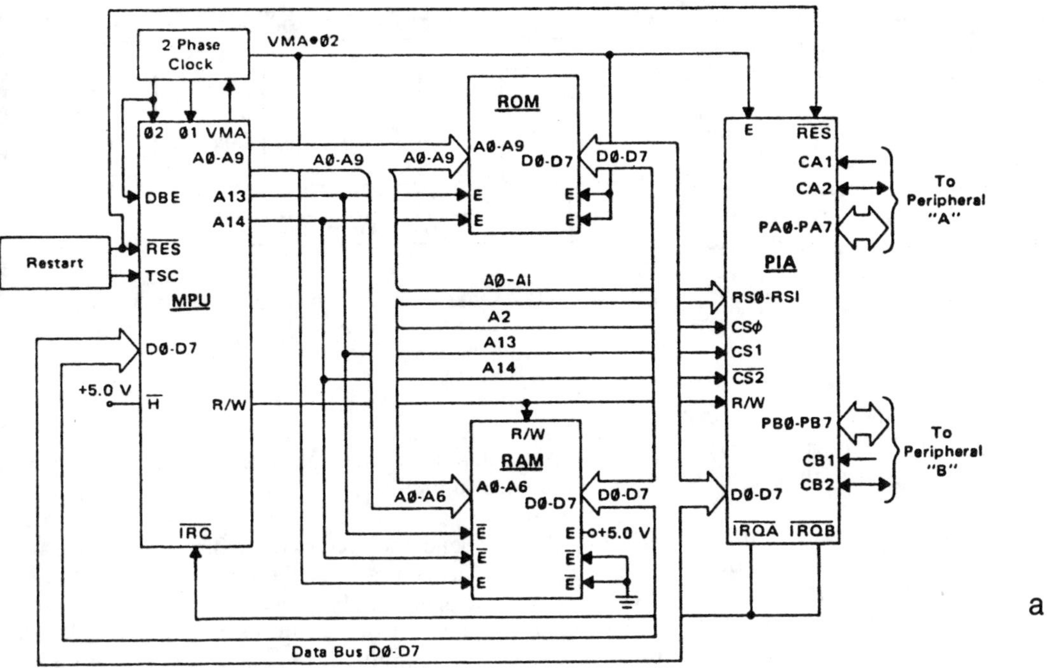

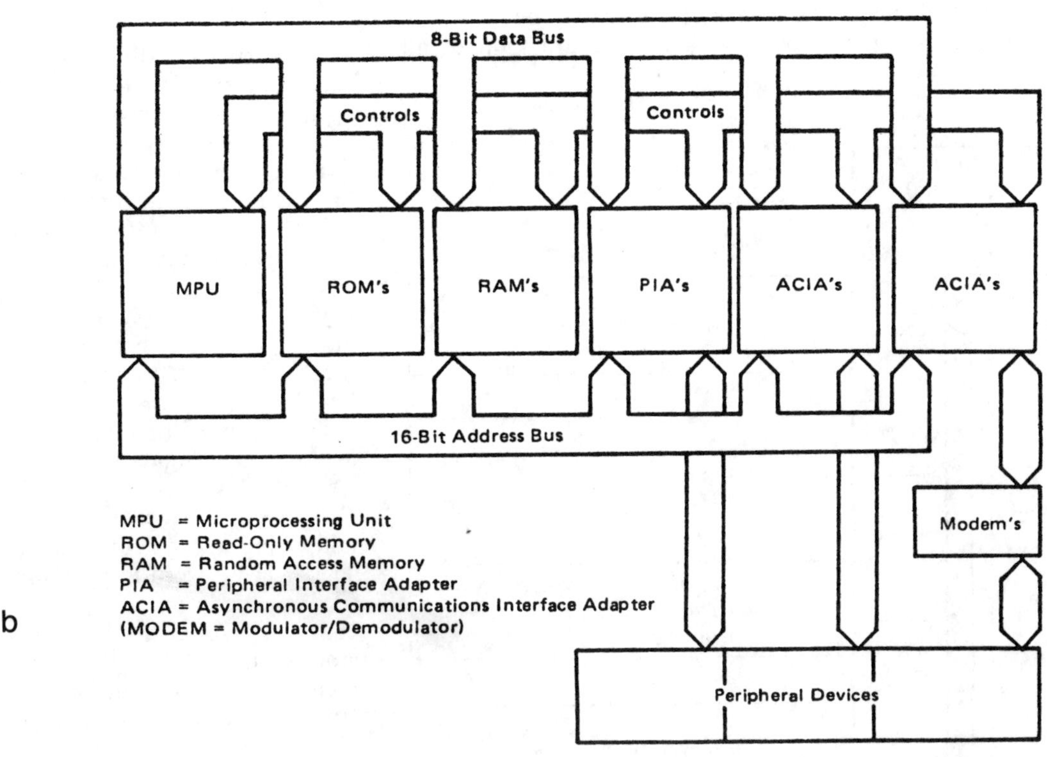

Figure 14.11. The basic architecture of a Motorola M6800 microprocessor is shown in (a), and an expanded version using the same MPU chip is shown in (b). (Courtesy of Motorola Semiconductor Products, Inc.)

shows the basic unit with the MPU, clock, memory and peripheral interface adapter, all of which are mounted on one printed circuit board. Figure 14.11b shows an expanded version using the same M6800 chip.

Microprocessors come in all shapes and sizes, and can be put into many different configurations. To clarify things further, the following sections examine a microprocessor unit in detail and look at the major devices used with a microprocessor. The components that make up an MPU system (Figure 14.12) are a clock, microprocessor, memory, peripheral devices, input/output devices, and the bus.

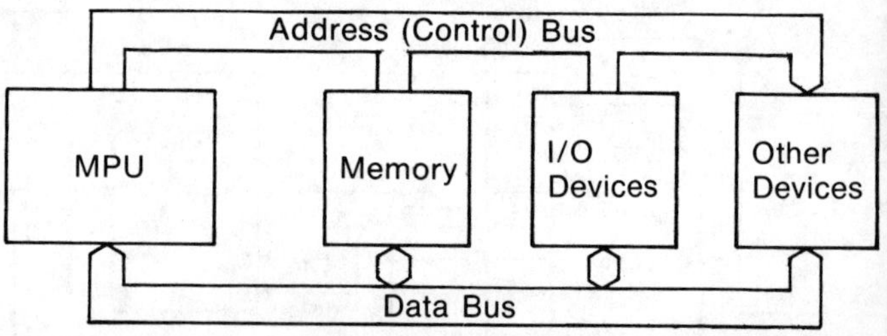

Figure 14.13. A basic microprocessor bus arrangement.

Clock

The clock's function is to generate all timing signal pulses for the microprocessor, memory, and the I/O devices. This pulse train is the common link that forces each system component to react at the same (or proper) time.

The timing function is important because it synchronizes the microprocessor to the memory. When asked, the memory puts data on the bus at the same time the microprocessor is ready to receive it. Refer to the analogy and watch the supervisor processing data. Things have been going fine all morning, and it is now time for lunch. While the supervisor is gone, an important document comes in that needs to be delivered to the boss immediately. But due to the poor timing of the document's arrival, the information is delivered to the boss late, which ultimately results in a plant shutdown.

The clock opens the data port by sending out a strobe signal. This signal alerts and holds the attention of all devices on the bus to the fact that data is being transferred from one device to another. Data is passed on the data bus when the control bus (Figure 14.13) addresses the devices, enabling them to send and receive data. The clock generates a timing pulse that opens a "port" (sometimes called a window) and holds it open just long enough for data to enter.

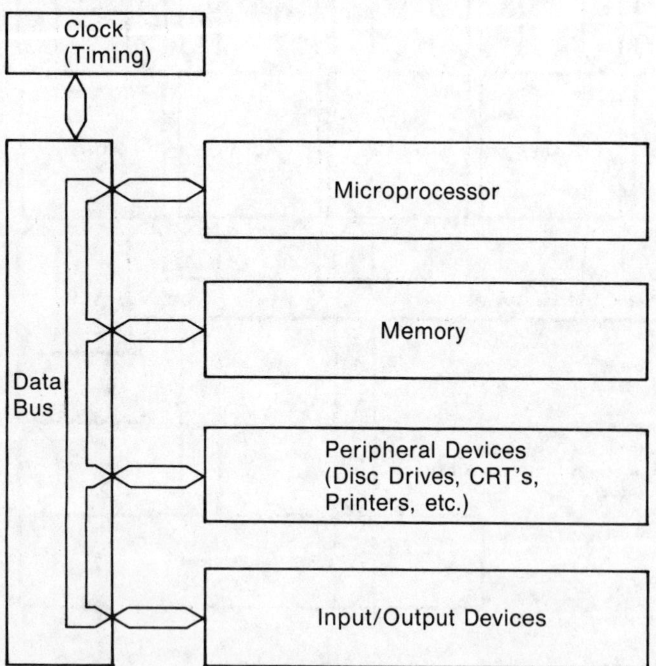

Figure 14.12. The components of an MPU system are a clock, microprocessor, memory, peripheral devices, input/output devices, and the bus.

The clock frequency also sets up the rate or speed of system operation. For example, if the clock rate is two microseconds, then it takes two microseconds to execute a command. Clock pulses also refresh the memory function by continually maintaining charges on dynamic memory while the power is on.

Memory

Memory is usually thought of as the ability to store and retain information for recall later. Here, memory is the *capacity* of a machine to store data and recall it.

Typical memory sizes are 1K, 4K, 8K, 16K, 32K, and 64K words of memory, with word lengths of either 4 bits, 8 bits, 12 bits, 16 bits, or 24 bits (Figure 14.14). Word length should not be confused with memory size.

Mathematically speaking, "kilo" means 1,000, but a kilo of memory equals 1,024 words of storage. To properly identify memory size, it is necessary to combine storage capacity and word length. For example, 4K by 16-bit memory indicates that it can store 4,096 words with a 16-bit word length.

Keep in mind that a 4-, 8-, 16-, or 24-bit word represents an American Standard Coding for Information Interchange (ASCII) character. It takes eight bits to represent an ASCII character. Larger words are used for gaining flexibility in addressing, speed of operation, and for scientific processing. The eight-bit word is generally used for instrument control (dedicated control) and for ASCII coding of alphanumeric characters. An eight-bit word is called a byte. For example, storing the word *CAT* in memory takes three bytes of memory, or three eight-bit words (C-A-T). Actually *CAT* is stored in machine language which is binary coded digits, or 1's and 0's (Figure 14.15).

All memories are classified as *volatile* or *nonvolatile*. A volatile memory is one that will not retain information if the power is removed. Anytime the power is interrupted, the contents of the memory are lost. The hand-held calculator has this class of memory. The only way to overcome this

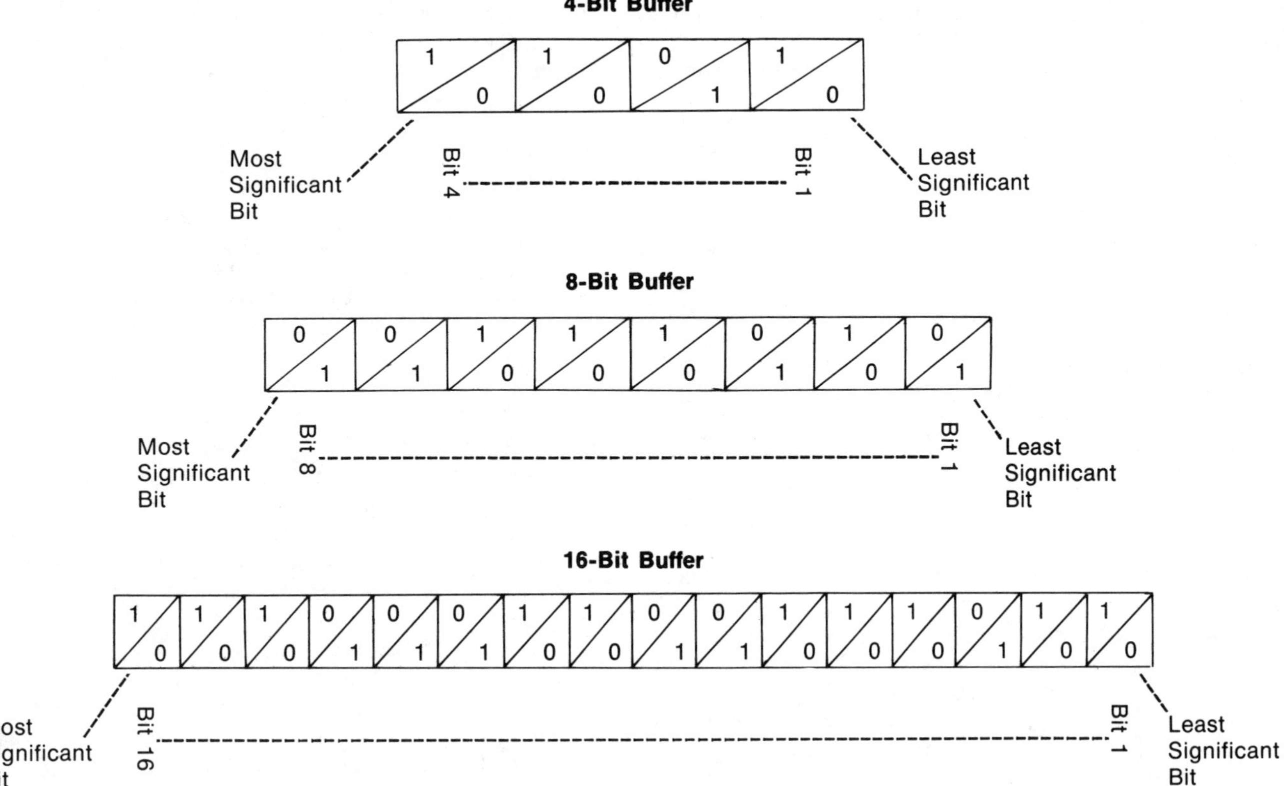

Figure 14.14. Typical output buffers containing a 4-bit, an 8-bit, and a 16-bit word.

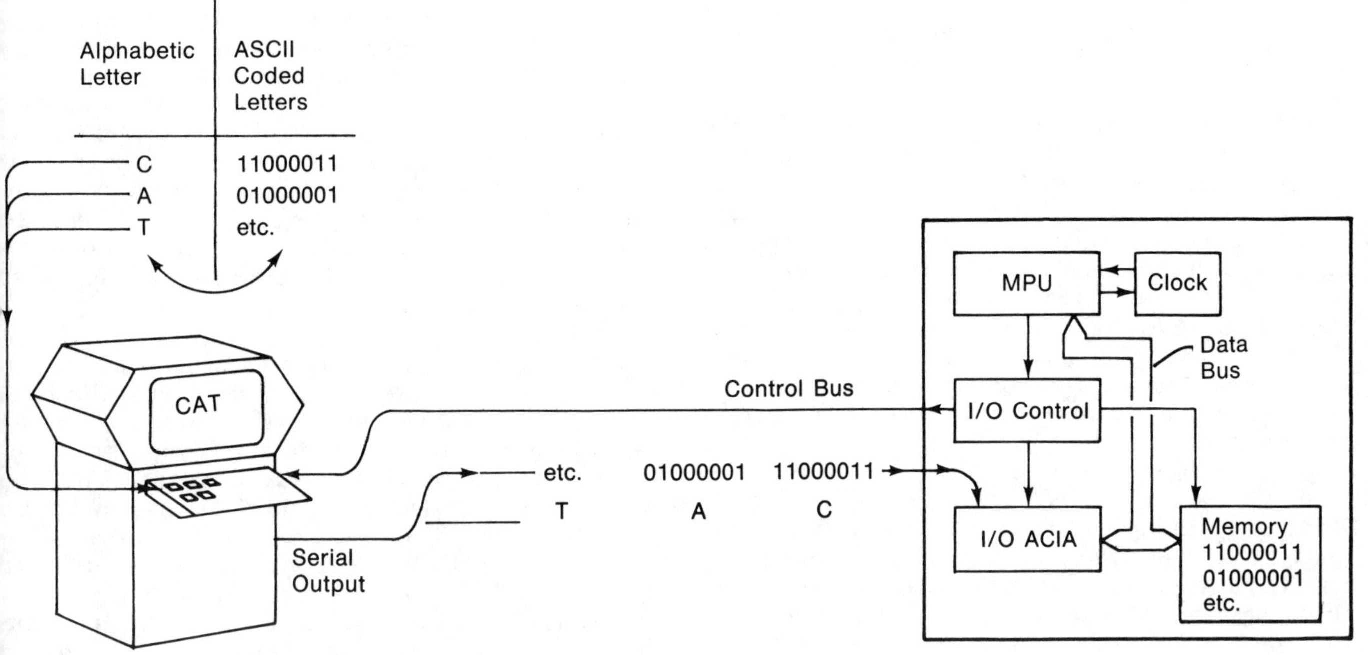

Figure 14.15. A graphic illustration showing the placing of the word "CAT" in memory.

disadvantage is to supply refresh clock pulses by means of a battery, thus making it nonvolatile.

Nonvolatile memory, once programmed, will retain its data after the power is removed. It is a type of semiconductor memory that, once programmed, must be replaced or erased and reprogrammed to modify. Magnetic ROM and PROM memory are examples of nonvolatile memory.

Each style of memory has a different access time. Access time is the time it takes the memory to deliver data once it is called for.

Types of Memory

There are several types of memory from which the engineer may choose. Smaller, less expensive, semiconductor memory is most frequently used with the microprocessor. Magnetic memory is also a commonly used method in which data is stored on a magnetic media such as tape, drum, disk, or diskette.

Semiconductor. Semiconductor memory is an inexpensive method of storing available data because of the technological advancements of solid-state circuitry. In one 24-pin, dual-in-line integrated circuit it is possible to store 8,192 bits of data. Some applications of semiconductor memory are data storage, code conversion, character generation, microprogramming, and program storage.

Metal Oxide Semiconductor (MOS) and Complementary Metal Oxide Semiconductor (CMOS) memories are widely accepted because of their low power consumption. Many of these chips can be assembled on a single PC board, forming a memory board. It is possible to purchase a memory board for any 4-bit, 8-bit, 12-bit, 16-bit, or 24-bit microprocessor that will store up to 16K words of memory.

Semiconductor memory uses a variety of solid-state latches, or flip flops, to retain information. It can be accessed for data in as few as 15 nanoseconds.

RAM (Random Access Memory). RAM is a read/write type of memory that can be randomly accessed—the user can go directly to any point in RAM without stepping through lower-numbered locations. RAM is similar to that used in calculators that have multistorage registers (memory) built into them. While at that point in memory, it is possible to read from or write into that location. For example, refer to the calculator with multistorage registers. To store a number in position three, and not in positions one or two, push "Store 3." The other registers remain empty, and three contains the data. The user has just randomly accessed memory, location three.

RAM may be either *static* or *dynamic*. In either case, it must have a backup battery to prevent loss of data because RAM is a volatile type of memory. Once again, byte size is selected when a microprocessor is purchased, so the user is forced to use words of that size. A typical memory access time ranges from 60 nanoseconds to 1 microsecond.

ROM (Read-Only Memory). Read-Only Memory chips are preprogrammed at the factory before they are put into the microprocessor. Semiconductor junctions within the chips are burned out with five times the normal voltage, permanently damaging the junctions in the chip, thereby becoming the data which is stored in that memory. The user cannot randomly access the ROM; that is, the microprocessor cannot store information in the ROM. All it can do is retrieve data from its memory.

This type of preprogramming is called *firmware*. The only way firmware can be changed in a ROM is by replacing chips with other preprogrammed chips.

ROM is a nonvolatile memory in which each word has a unique address. An exact address is needed to call up a specific word. Access times range from 45 nanoseconds to 2 microseconds.

PROM (Programmable Read-Only Memory). PROM is a later development of the ROM, and it is not preprogrammed at the factory. PROM can be programmed by the user who does not want to give the factory a proprietary software program, or who feels that money can be saved by personally developing the firmware rather than having the factory do it. However, there are many computer software companies that will work with a user to develop software and turn it into firmware for the PROM.

In using PROM, the key word to remember is "programmable." Programming is accomplished by inserting a special printed circuit card (available for most machines) into a socket on the microprocessor. A PROM chip is inserted into a plug-in socket provided on the card. Then, software programs are burned into the PROM chip permanently (Figure 14.16). This software program is now considered firmware, because it can no longer be changed.

EPROM (Erasable Programmable Read-Only Memory). The EPROM works in the same manner as the PROM, and is a later development of the ROM. *Erasable* is the key word to remember. In the event that a firmware program needs to be changed periodically, or new needs arise as processes change, an EPROM type of memory might be a good choice. It is erasable and reprogrammable. Erasing is done with ultraviolet light. A high-intensity, ultraviolet light is shined through the transparent window on the face of the chip, erasing it. As a matter of fact, if the chip was left lying in sunlight, it would eventually lose its firmware program. A special plug-in card is available to erase and reprogram the EPROM.

Magnetic Memory. Magnetic memory is an external memory. Its use requires an I/O (input/output) device, such as a peripheral interface adapter (PIA). Magnetic memory is used in addition to internal RAM or ROM to enhance the microprocessor's capabilities. Types of magnetic memories include cassette tapes, magnetic diskettes, and bubble memory. Magnetic memory is much slower (access time is in milliseconds) than solid-state memory, but is widely used because of the large volume of data it will hold. The cassette tape is suitable for small operations if time is not an important factor. The magnetic disk (floppy) is used if time is a critical factor. While the access time for a cassette tape is greater than 1 second, a floppy disk can be accessed in about 80 milliseconds.

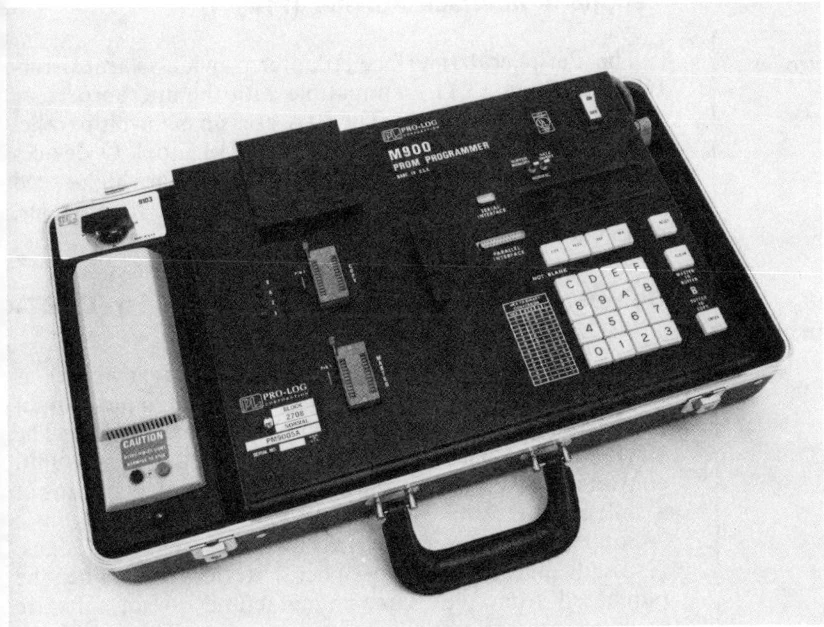

Figure 14.16. Software programs are burned into PROM chips using programmer devices such as this. (Courtesy of Pro-Log Corp.)

Bubble Memory. Today, as technological developments speed on, work on bubble memory is being done by Texas Instruments, Bell Telephone Laboratories, and IBM. It is a potentially low-cost, high-density memory. Actual chip size of TI's 92K-bit memory is 1.1 inches by 1 inch by .4 inch. Bubble memories have very large storage capacities, and are nonvolatile. The limiting factor is their speed of operation, which is very slow (approximately one to ten milliseconds access time).

Bubble memory is created in a substrate material when magnetic bubble domains are formed. By generating a magnetic field through using a pair of coils, bubbles representing bits are shifted around the chip in an endless loop (called the minor loop). When bubbles are shifted through another "major" loop, it acts as a unidirectional read/write register. This register transfers data into and out of memory by the presence of bubbles (binary 1's) or the absence of bubbles (binary 0's).

Summary of Memory Types

Two categories of memory used with microprocessors are solid-state semiconductor and magnetic. Semiconductor memories are either static (nonvolatile) or dynamic (volatile). Volatile memory requires refresh clock pulses to keep it functioning, or else it will lose its ability to store information. Nonvolatile memory will retain its information indefinitely. Common types of semiconductor memories are: ROM, RAM, PROM, and EPROM. Magnetic memory is the storage of data on magnetic media, such as floppy disks, magnetic tape cassettes, or the new bubble memory.

Bus

The bus plays a key role in microprocessors by being the freeway system which routes all traffic (information) to and from all devices. The bus is actually a number of parallel lines to which all devices are connected. Figure 14.17 shows a microprocessor using a single chip and the two-bus systems (Data Bus and Address Bus) which enable it to function as a mass flow computer/controller. The Data Bus moves data over parallel lines, one line for each bit of information. For example, if the microprocessor being used has an eight-bit word size, then it takes eight lines of bus to move the word. When data is placed on the bus, all devices connected to the bus see the data simultaneously. The Address Bus alerts a specific piece of hardware to act upon the data on the Data Bus, while all remaining units ignore the data.

When data is to leave the microprocessor proper, it is usually done through a PIA (Peripheral Interface Adapter), a UART (Universal Asynchronous Receiver Transmitter) sometimes called an ACIA (Asynchronous Communications Interface Adapter), and/or a modem (Modulator/Demodulator).

Input/Output

The rapid advancement of technology in the 1970s, and the arrival of the micros brought about improvements in peripheral devices. Operators' stations have gone from black-and-white CRT consoles to color CRTs, reverse video, and "smart" operators' consoles such as the Hewlett-Packard 2645A (Figure 14.18). The old-style teletype is now being replaced by a fancy printer that can print both frontwards and backwards. Technology has progressed from reel-to-reel tape units to cassette units and small, inexpensive floppy-disk memories.

Today's design engineer has many input/output devices from which to choose; and because of the high reliability of each one, decisions are hard to make. More important than choosing the device is being sure that the processor and the

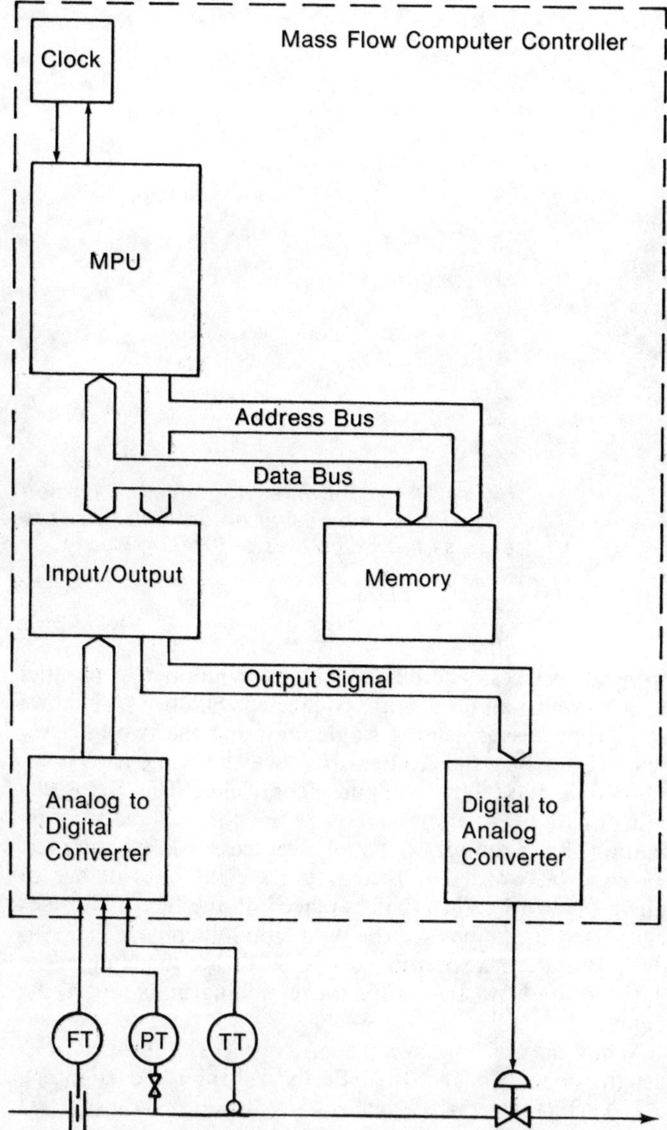

Figure 14.17. A microprocessor is diagrammed showing its use as a mass flow computer/controller.

peripheral devices can communicate with each other. This is accomplished by using the PIA, UART, and custom input/output devices.

The Electronics Industries Association (EIA), which represents most of the manufacturers, has developed a standard, EIA RS 232C (Figure 14.19), which specifies signals, signal levels, connectors and their wiring order. This makes it possible to specify a microprocessor and a peripheral device with an RS 232C connector, and to know that they will work as a unit.

With the use of a modem (Modulator/Demodulator), the microprocessor can communicate with other devices hundreds of miles away or only a few inches away, and do it over a single pair of wires.

Peripheral Interface Adapter (PIA)

The Peripheral Interface Adapter provides a transistor-transistor logic (TTL) compatible with the microprocessor and its peripheral device. The PIA uses an eight-bit parallel bus system to move data from the MPU to I/O devices (Figure 14.20). Some applications of the PIA are high-speed card reader, high-speed line printer, magnetic tape reader, and control of process instrumentation.

Universal Asynchronous Receiver Transmitter (UART)

The Universal Asynchronous Receiver Transmitter is used to adapt the microprocessor to serial input/output peripherals, such as an ASCII keyboard or a modem. The UART takes a parallel word from the microprocessor unit, converts it to serial bits, and sends it on its way by means of a shift register.

Some manufacturers use their own I/O interface designs. A disadvantage is that peripheral devices must then be purchased from that same manufacturer. Also, software packages are not usable in other equipment. This can greatly reduce the ability for future expansion.

The modem is used to modulate and demodulate digital data for transmission over a telephone line. It is now possible to have a microprocessor communicate, and even control, instrumentation at some remote location. It is also possible to have the factory troubleshoot and diagnose a microprocessor's malfunctions over the telephone. This is done by dialing the proper telephone number and allowing one computer to interrogate the other. It is even possible to reprogram a system remotely by using this hardware.

In any event, when purchasing a microprocessor investigate the types of input/output devices needed to ensure that the system components will function together.

Programming the Microprocessor

An engineer who is about to put in a system has three main alternatives to choose from: (1) buy a preprogrammed system; (2) purchase a system and hire an outside consultant (program analyst) to write the software program; or (3) do the programming him-/herself.

Several program packages are available as options for standard microprocessor systems such as the Motorola M6800 EXORciser (Figure 14.21). Some of these options are assemblers, compilers, linking loaders, FORTRAN, and BASIC. A complete list is available from each manufacturer. In fact, most manufacturers have a program library available to users for a nominal fee, or free library access for writing and submitting a new program to the library.

Even with many preprogrammed functions available in "canned" form, some custom programming is usually desirable. For example, if a unique control problem exists when using definable equations and control sequences, the microprocessor manufacturer may be asked to write this program. Programs developed in this manner are expensive, unless they are used in the consumer market. High-volume

Figure 14.18. Operator consoles can now use "smart" terminals such as the Hewlett-Packard model HP2645A. (Courtesy of Hewlett-Packard.)

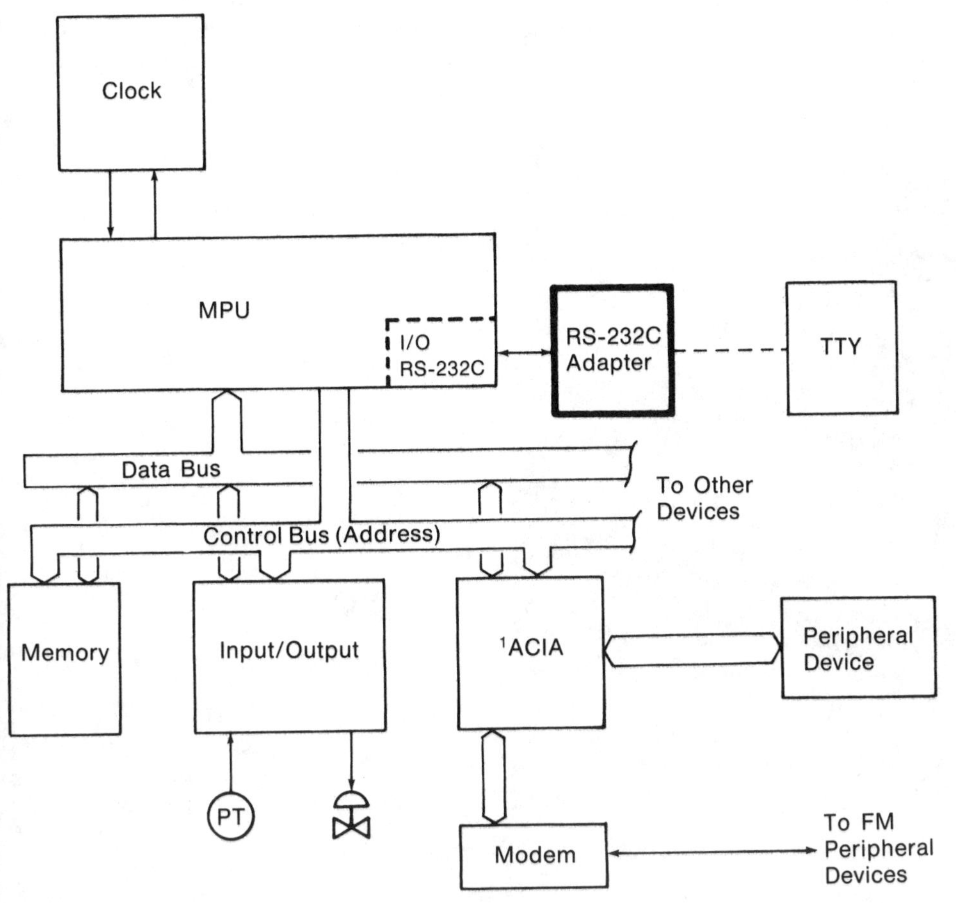

Figure 14.19. Using RS232C specification of connecting circuitry, peripheral devices are compatibly connected.

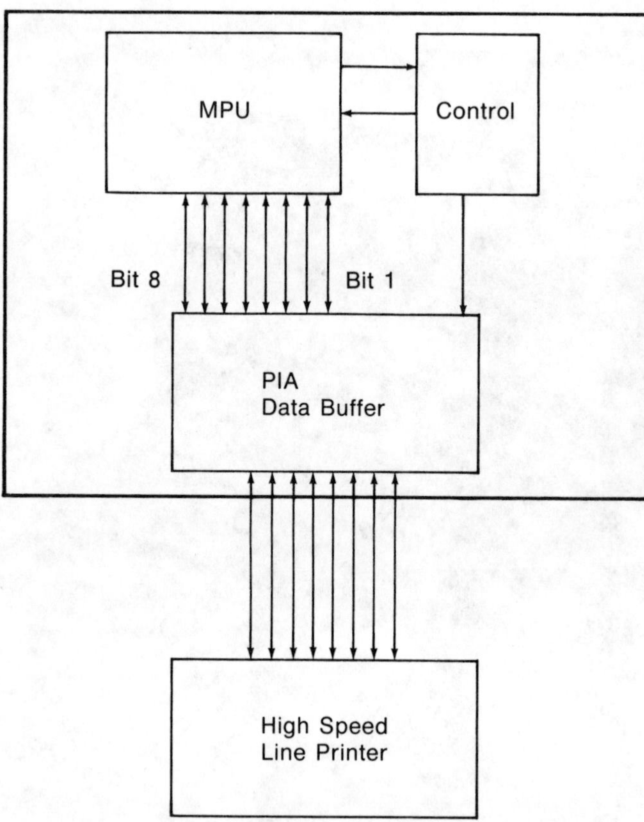

Figure 14.20. A Peripheral Interface Adapter (PIA) uses an 8-bit parallel bus system to move data from the MPU to I/O devices.

sales of preprogrammed games like chess and checkers, TV games, and calculators greatly reduce programming cost to the manufacturer. Remember, in order to get good results from the manufacturer, the program must be easily definable.

A more complex program can be handled by hiring a program analyst to write the program. In order for this approach to work effectively, the analyst and the engineer should work closely in developing a functioning program for the system. The engineer should be willing to share information freely with the analyst. Bear in mind that many analysts are proficient in machine language programming but have no knowledge of which factors are important in process control.

At times, the particular program needed will either contain classified information, such as company secrets, or will be too complex for outside help. Instrument manufacturers who have developed microprocessor-based equipment have found that, in order to achieve a working instrument, they also had to develop programs. The purchaser of a microprocessor unit may elect to use any of these program development methods or combine certain aspects of these methods.

Machine Language

All microprocessors, microcomputers, mini-computers, and full-blown computer systems use machine language. Machine language is the use of binary numbers (the 1's and 0's) to represent data and commands (Figure 14.14).

Early memories and computers used flip flops which had only two voltage levels, a high-level voltage that represented

Figure 14.21. The Motorola M6800 EXORciser accepts plug-in cards. (Courtesy of Motorola Semiconductor Products, Inc.)

a true state, or binary 1, and a low-level voltage that represented a false state, or binary 0. The same was true for the input switches and output switches—"on" was represented by binary 1, and "off" was represented by a binary 0. By locating the lights and switches in groups of fours, the operator could input and read the machine language with ease. As technology advanced, computers grew in size and sophistication, and input/output devices evolved to the extent that machine language is impractical for many microprocessor programming applications. Also, as larger words (8-bit, 12-bit, and 16-bit) became available, human error increased exponentially because of long strings of 1's and 0's.

Programming problems led to the use of octal and hexadecimal number systems and mnemonics. Octal and hexadecimal are numbered systems to base 8 and base 16, respectively. In either case, to use one of the program languages, a coder/decoder must be available in the MPU for converting data to machine language. Mnemonics (Figure 14.22) is the abbreviation of instructions used in microprocessor program language. Complete instructions are condensed into a three-letter code. This code is then translated by the MPU coder/decoder and compiler into machine language. In order to use mnemonics, instead of binary, octal, or hexadecimal programming, software and hardware must be added to the MPU to allow the translation to machine language.

Programming in machine language is useful and economical to the manufacturer who is manufacturing a dedicated control instrument that will not require additional programming. If operator reprogramming is or will be required, it is no longer economical to use machine language. The user should then choose either octal, hexadecimal, mnemonics or one of the higher-order program languages such as BASIC, FORTRAN, or COBOL.

In short, the purchaser of a microprocessor system should, prior to the actual purchase, consider and decide what approach to programming will be used. This requires an evaluation and a decision on the type of program language that is best.

Microprocessor Architecture

Architecture of the microprocessor refers to its internal construction, both physical and electronic. It is the microprocessor's format or arrangement of internal parts that interconnect each device. The architecture of single-chip microprocessors and microprocessors for control instrumentation will be discussed. Knowledge of the architecture of a microprocessor is a useful tool in understanding the internal logic of the unit.

Single-Chip Microprocessors

The single-chip architecture (Figure 14.23) of Intel's 8085A microprocessor has an eight-bit internal bidirectional data bus. This bus services all components through handling, manipulating, and generating data. This chip also contains an arithmetic logic unit (ALU) that adds, subtracts and performs Boolean algebra; a decoder/encoder for converting instructions from hexadecimal to machine language; a clock to handle timing and control functions; a serial I/O control for interfacing other devices; and the Interrupt Control, which "raises a flag" when a high priority task is to be performed.

Intel's 8085A also has a register array, which is a network of registers for doing "housekeeping." This array encompasses six registers which act as independent registers or as three pairs of registers. When operating as three pairs, one register can affect the other. The register array also includes Stack Pointers, a Program Counter, and Incrementer/Decrementer registers.

The Stack Pointer Register works on a last-in first-out basis. It is a 16-bit register, thereby able to address up to 64K of memory. The Stack Pointer keeps track of where the user is in memory. If a subroutine interrupts a program sequence, the Stack Pointer will return the user to that same spot, and the program can continue.

When running a program, the Program Counter advances through the program by automatically incrementing itself by one after each instruction is executed. The Program Counter Register always indicates the address of the next instruction that is to be fetched from memory.

The Incrementer/Decrementer Register is used in conjunction with the Program Counter Register. This register makes it easy to jump forwards or backwards by any number of program steps.

Another example of single-chip architecture is Motorola's M6800 (Figure 14.11). There are as many differences as similarities in this chip compared to the Intel 8085A. One outstanding difference is that the M6800 uses two accumulators instead of the regular array. Neither of the two microprocessors (Intel's 8085A and Motorola's M6800) are capable of stand-alone operation. Both need external memory and external I/O ports, and the M6800 must have an external two-phase clock for timing and control.

A microprocessor's mainframe architectural layout can be made very versatile. Figure 14.24 shows a schematic representation of a mainframe with an integral power supply. The design engineer can add additional boards or functions (Figure 14.25) by plugging them into a spare slot. Using the bus approach, wherein the mainframe contains lines of bus wired to each connector, the user need only confirm that (1) the accessories board meets the bus requirements of existing PC board signal lines, and (2) that logic levels are compatible with those already in the mainframe. Much more could be said about microprocessor mainframe architecture, but it is recommended that specific information be obtained from such manufacturers as Intel, Motorola, Texas Instruments, or others.

Microprocessors for Control Instrumentation

In the past, the number of full-blown computers in industry has gradually increased. At first it was thought that this was the way to control an entire plant. Some tried, and

ACCUMULATOR AND MEMORY INSTRUCTIONS

OPERATIONS	MNEMONIC	IMMED OP	IMMED ~	IMMED =	DIRECT OP	DIRECT ~	DIRECT =	INDEX OP	INDEX ~	INDEX =	EXTND OP	EXTND ~	EXTND =	IMPLIED OP	IMPLIED ~	IMPLIED =	BOOLEAN/ARITHMETIC OPERATION (All register labels refer to contents)	5 H	4 I	3 N	2 Z	1 V	0 C
Add	ADDA	8B	2	2	9B	3	2	AB	5	2	BB	4	3				A + M → A	↕	●	↕	↕	↕	↕
	ADDB	CB	2	2	DB	3	2	EB	5	2	FB	4	3				B + M → B	↕	●	↕	↕	↕	↕
Add Acmltrs	ABA													1B	2	1	A + B → A	↕	●	↕	↕	↕	↕
Add with Carry	ADCA	89	2	2	99	3	2	A9	5	2	B9	4	3				A + M + C → A	↕	●	↕	↕	↕	↕
	ADCB	C9	2	2	D9	3	2	E9	5	2	F9	4	3				B + M + C → B	↕	●	↕	↕	↕	↕
And	ANDA	84	2	2	94	3	2	A4	5	2	B4	4	3				A · M → A	●	●	↕	↕	R	●
	ANDB	C4	2	2	D4	3	2	E4	5	2	F4	4	3				B · M → B	●	●	↕	↕	R	●
Bit Test	BITA	85	2	2	95	3	2	A5	5	2	B5	4	3				A · M	●	●	↕	↕	R	●
	BITB	C5	2	2	D5	3	2	E5	5	2	F5	4	3				B · M	●	●	↕	↕	R	●
Clear	CLR							6F	7	2	7F	6	3				00 → M	●	●	R	S	R	R
	CLRA													4F	2	1	00 → A	●	●	R	S	R	R
	CLRB													5F	2	1	00 → B	●	●	R	S	R	R
Compare	CMPA	81	2	2	91	3	2	A1	5	2	B1	4	3				A − M	●	●	↕	↕	↕	↕
	CMPB	C1	2	2	D1	3	2	E1	5	2	F1	4	3				B − M	●	●	↕	↕	↕	↕
Compare Acmltrs	CBA													11	2	1	A − B	●	●	↕	↕	↕	↕
Complement, 1's	COM							63	7	2	73	6	3				$\overline{M}$ → M	●	●	↕	↕	R	S
	COMA													43	2	1	$\overline{A}$ → A	●	●	↕	↕	R	S
	COMB													53	2	1	$\overline{B}$ → B	●	●	↕	↕	R	S
Complement, 2's	NEG							60	7	2	70	6	3				00 − M → M	●	●	↕	↕	①	②
(Negate)	NEGA													40	2	1	00 − A → A	●	●	↕	↕	①	②
	NEGB													50	2	1	00 − B → B	●	●	↕	↕	①	②
Decimal Adjust, A	DAA													19	2	1	Converts Binary Add. of BCD Characters into BCD Format	●	●	↕	↕	↕	③
Decrement	DEC							6A	7	2	7A	6	3				M − 1 → M	●	●	↕	↕	4	●
	DECA													4A	2	1	A − 1 → A	●	●	↕	↕	4	●
	DECB													5A	2	1	B − 1 → B	●	●	↕	↕	4	●
Exclusive OR	EORA	88	2	2	98	3	2	A8	5	2	B8	4	3				A⊕M → A	●	●	↕	↕	R	●
	EORB	C8	2	2	D8	3	2	E8	5	2	F8	4	3				B⊕M → B	●	●	↕	↕	R	●
Increment	INC							6C	7	2	7C	6	3				M + 1 → M	●	●	↕	↕	⑤	●
	INCA													4C	2	1	A + 1 → A	●	●	↕	↕	⑤	●
	INCB													5C	2	1	B + 1 → B	●	●	↕	↕	⑤	●
Load Acmltr	LDAA	86	2	2	96	3	2	A6	5	2	B6	4	3				M → A	●	●	↕	↕	R	●
	LDAB	C6	2	2	D6	3	2	E6	5	2	F6	4	3				M → B	●	●	↕	↕	R	●
Or, Inclusive	ORAA	8A	2	2	9A	3	2	AA	5	2	BA	4	3				A + M → A	●	●	↕	↕	R	●
	ORAB	CA	2	2	DA	3	2	EA	5	2	FA	4	3				B + M → B	●	●	↕	↕	R	●
Push Data	PSHA													36	4	1	A → M_{SP}, SP − 1 → SP	●	●	●	●	●	●
	PSHB													37	4	1	B → M_{SP}, SP − 1 → SP	●	●	●	●	●	●
Pull Data	PULA													32	4	1	SP + 1 → SP, M_{SP} → A	●	●	●	●	●	●
	PULB													33	4	1	SP + 1 → SP, M_{SP} → B	●	●	●	●	●	●
Rotate Left	ROL							69	7	2	79	6	3				M	●	●	↕	↕	⑥	↕
	ROLA													49	2	1	A } □ ← ▭▭▭▭▭▭▭▭ ←	●	●	↕	↕	⑥	↕
	ROLB													59	2	1	B C b7 ← b0	●	●	↕	↕	⑥	↕
Rotate Right	ROR							66	7	2	76	6	3				M	●	●	↕	↕	⑥	↕
	RORA													46	2	1	A } → □ → ▭▭▭▭▭▭▭▭	●	●	↕	↕	⑥	↕
	RORB													56	2	1	B C b7 → b0	●	●	↕	↕	⑥	↕
Shift Left, Arithmetic	ASL							68	7	2	78	6	3				M	●	●	↕	↕	⑥	↕
	ASLA													48	2	1	A } □ ← ▭▭▭▭▭▭▭▭ ← 0	●	●	↕	↕	⑥	↕
	ASLB													58	2	1	B C b7 b0	●	●	↕	↕	⑥	↕
Shift Right, Arithmetic	ASR							67	7	2	77	6	3				M	●	●	↕	↕	⑥	↕
	ASRA													47	2	1	A } → ▭▭▭▭▭▭▭▭ → □	●	●	↕	↕	⑥	↕
	ASRB													57	2	1	B b7 b0 C	●	●	↕	↕	⑥	↕
Shift Right, Logic	LSR							64	7	2	74	6	3				M	●	●	R	↕	⑥	↕
	LSRA													44	2	1	A } 0 → ▭▭▭▭▭▭▭▭ → □	●	●	R	↕	⑥	↕
	LSRB													54	2	1	B b7 b0 C	●	●	R	↕	⑥	↕
Store Acmltr.	STAA				97	4	2	A7	6	2	B7	5	3				A → M	●	●	↕	↕	R	●
	STAB				D7	4	2	E7	6	2	F7	5	3				B → M	●	●	↕	↕	R	●
Subtract	SUBA	80	2	2	90	3	2	A0	5	2	B0	4	3				A − M → A	●	●	↕	↕	↕	↕
	SUBB	C0	2	2	D0	3	2	E0	5	2	F0	4	3				B − M → B	●	●	↕	↕	↕	↕
Subtract Acmltrs.	SBA													10	2	1	A − B → A	●	●	↕	↕	↕	↕
Subtr. with Carry	SBCA	82	2	2	92	3	2	A2	5	2	B2	4	3				A − M − C → A	●	●	↕	↕	↕	↕
	SBCB	C2	2	2	D2	3	2	E2	5	2	F2	4	3				B − M − C → B	●	●	↕	↕	↕	↕
Transfer Acmltrs	TAB													16	2	1	A → B	●	●	↕	↕	R	●
	TBA													17	2	1	B → A	●	●	↕	↕	R	●
Test, Zero or Minus	TST							6D	7	2	7D	6	3				M − 00	●	●	↕	↕	R	R
	TSTA													4D	2	1	A − 00	●	●	↕	↕	R	R
	TSTB													5D	2	1	B − 00	●	●	↕	↕	R	R

LEGEND:

- OP Operation Code (Hexadecimal);
- ~ Number of MPU Cycles;
- = Number of Program Bytes;
- + Arithmetic Plus;
- − Arithmetic Minus;
- · Boolean AND;
- M_{SP} Contents of memory location pointed to be Stack Pointer;
- + Boolean Inclusive OR;
- ⊕ Boolean Exclusive OR;
- $\overline{M}$ Complement of M;
- → Transfer Into;
- 0 Bit = Zero;
- 00 Byte = Zero;

Note – Accumulator addressing mode instructions are included in the column for IMPLIED addressing

CONDITION CODE SYMBOLS:

- H Half-carry from bit 3;
- I Interrupt mask
- N Negative (sign bit)
- Z Zero (byte)
- V Overflow, 2's complement
- C Carry from bit 7
- R Reset Always
- S Set Always
- ↕ Test and set if true, cleared otherwise
- ● Not Affected

Figure 14.22. Mnemonics is the abbreviation of instructions used in microprocessor program language. (Courtesy of Motorola Semiconductor Products, Inc.)

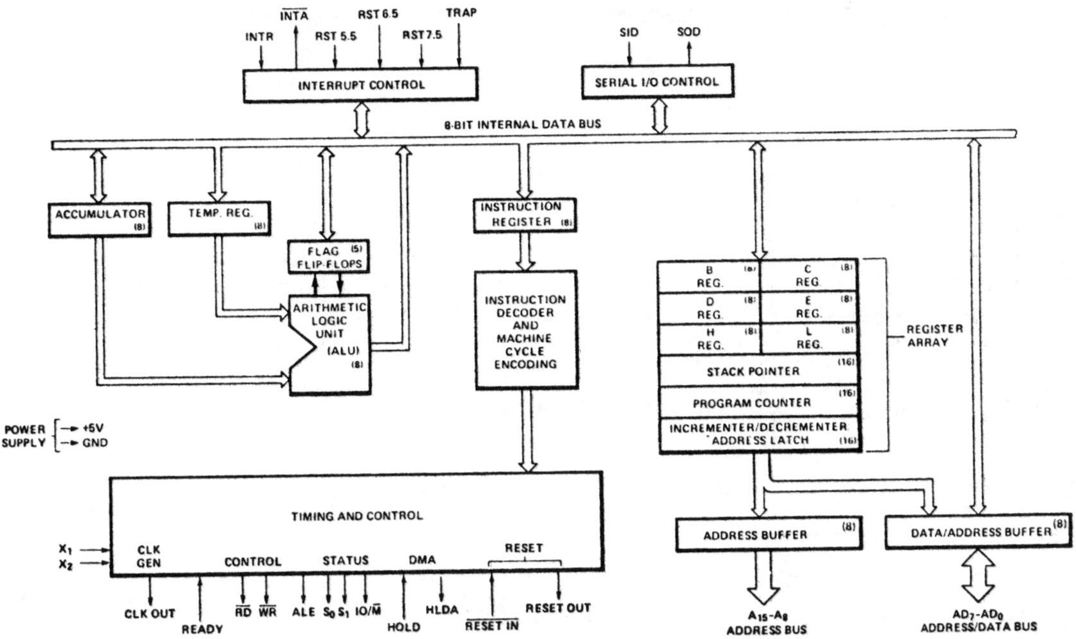

Figure 14.23. The single-chip architecture of Intel's 8085A Microprocessor has an 8-bit internal data bus. (Courtesy of Intel Corporation.)

had success until the computer failed. This caused mass confusion in the control room and the loss of many dollars. Many plants returned to dedicated hardware, putting only the most difficult control loops on the computer. The big computer became a data logger, collecting flow rates, pressures, and temperatures as data for making material balances. Some users tried implementing two computers, one as a backup for the first, but this proved quite costly.

Another variation was to use backup instruments with selectable computer/manual/auto modes. The operator, upon computer failure, was required to change the backup controller from computer to auto or manual mode. This control scheme worked effectively, but again was costly and required extra room for the computer.

The microprocessor is opening many previously closed doors because of its compact size, high speed of operation, low cost and high reliability. Eight control loops are now packaged into an enclosure (6 by 6 inches) that once housed a single-pen recorder. Figure 14.26 shows a block diagram of a microprocessor controller capable of handling 15 analog inputs, 8 of which may have 8 analog outputs for process control, 16 discrete inputs (contact closures), and 8 discrete outputs. Included in the same mainframe are the microprocessor's 8K of RAM memory with battery backup, display controls, 2 data links, and 24K of PROM memory. Typically, the 24K of PROM is preprogrammed by the manufacturer, while the 8K of RAM is configured by the user. This means that the field engineer or operator does not write the firmware program in PROM. The user is limited to using the controller program provided by the manufacturer, and to selecting the available configuration options needed to enable the controller to function for that particular control loop(s). Its data link is a dual bus output that is available for supervisory control or linked to a centralized operator's console.

A distributed control system (Figure 14.27) is another type of instrument architecture. The microprocessor, or Digital Control Element (DCE), is actually located in the field, at or near the final control element. Each DCE simultaneously functions as an independent controller. The controller provides a digital output on the data highway (bus) to the operator's station, where it can be addressed and reconfigured by the operator.

Honeywell's Total Distributed Control System TDC-2000 is available with a process computer, if need be. This system contains three levels of bi-directional buses: the I/O Bus which controls all incoming and outgoing signals to the microprocessor; the Microprocessor Bus, for busing signals within the microprocessor; and the Data Highway Bus, which is a dual (redundant) bus used to carry data to and from the operator station and the process computer.

In short, microprocessor architectures cover a very broad spectrum, from a single chip to very elaborate systems.

Microprocessor Applications in Instrumentation

Where are microprocessor's being used? How are they used? How are they applied? Are they difficult to program? These and other questions are best answered while examining different applications of microprocessors used in instrumentation—dedicated single-loop controllers; multi-

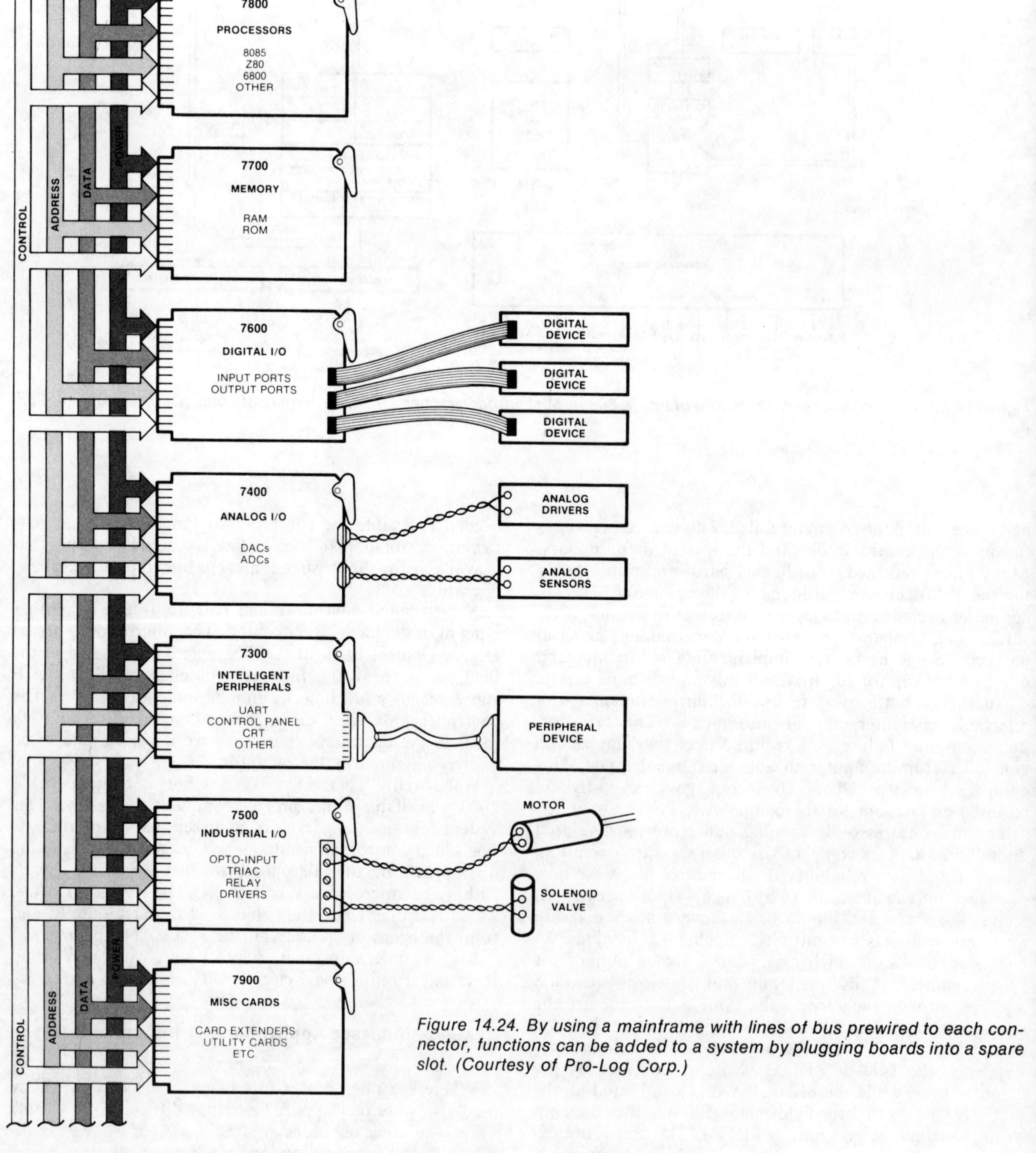

Figure 14.24. By using a mainframe with lines of bus prewired to each connector, functions can be added to a system by plugging boards into a spare slot. (Courtesy of Pro-Log Corp.)

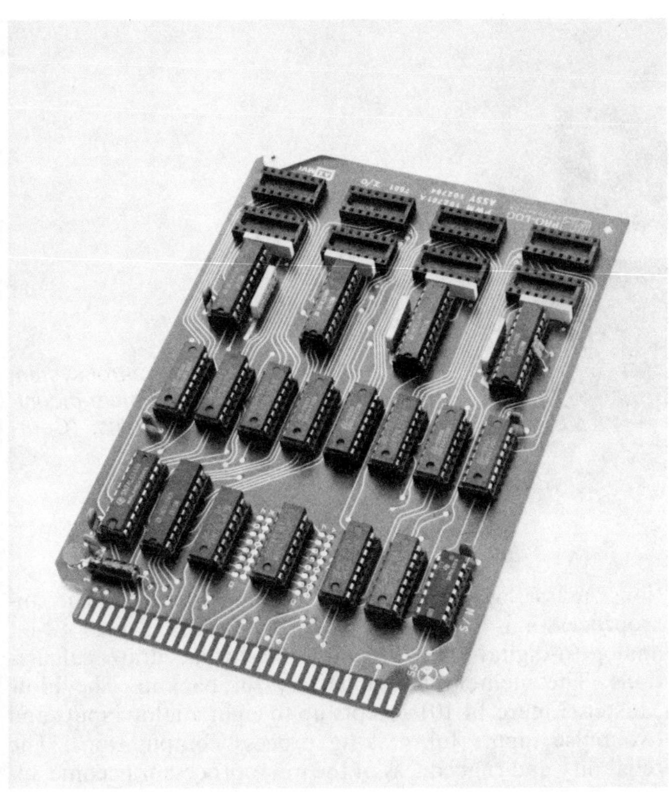

Figure 14.25. Plug-in boards such as this are used to expand a system easily. (Courtesy of Pro-Log Corp.)

loop controllers and data loggers; and, the use of modular components to build a console-operable system.

Dedicated Single-Loop Controllers

A dedicated single-loop controller using a microprocessor is illustrated in Figure 14.28. This controller contains a single microprocessor that is factory preprogrammed, except for the tuning constants, and that operates as any normal analog controller. It also contains RAM and PROM memories, process I/O modules, and a serial data transceiver for talking with a computer or data logger. Figure 14.29 is a duplicate of Figure 14.28, except for the addition of a backup microprocessor unit, which automatically takes action if the primary microprocessor unit fails.

Taylor Instrument Company offers the Micro-Scan 1300 Controller (Figure 14.30) which has many preprogrammed algorithms. Some of its standard features are remote analog adaptive gain, input for startup or sequential gain changes, digital 12-bit accuracy for control algorithms, and digital display of key loop parameters. The Micro-Scan 1300 Controller is packaged in a conventional case (3 by 6 inches).

The Daniel Flow Master was designed to solve mass flow, energy, and volume calculations using Intel's 8085A single-chip microprocessor. At the time of purchase, PROM's are selected for the measurement application. Its microprocessor includes a versatile executive program for control of the various subroutines available (i.e., math, energy calcula-

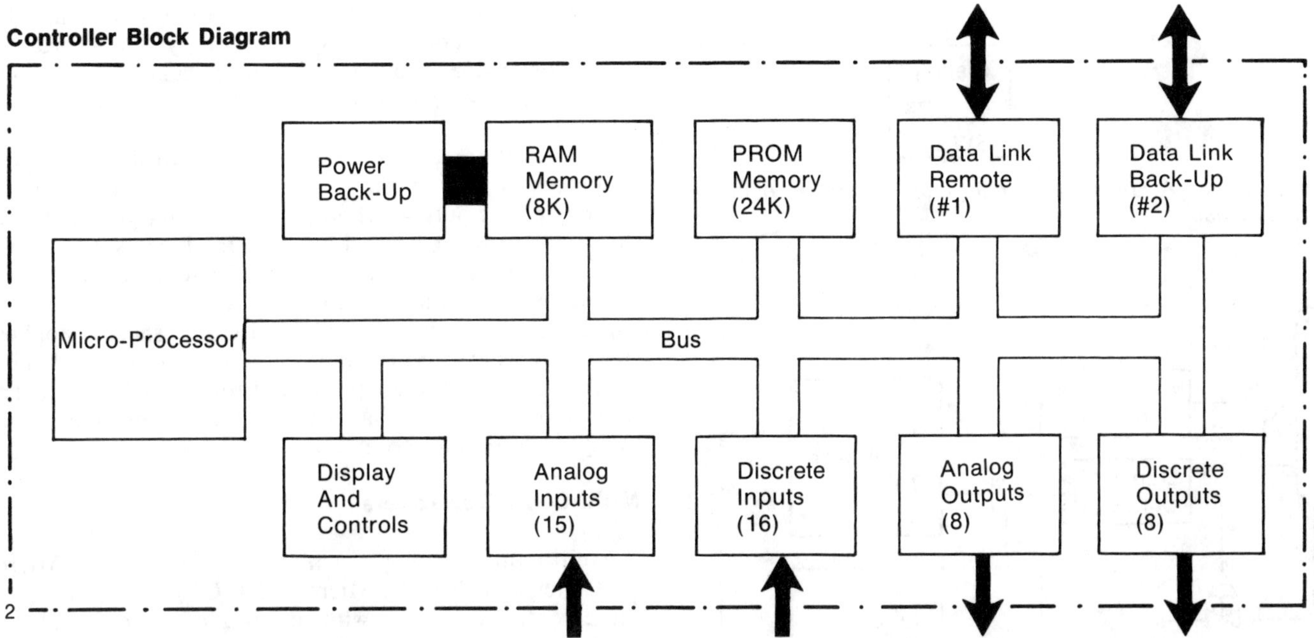

Figure 14.26. A block diagram of a Micon P-200 Controller shows its internal features. (Courtesy of Process Systems, Inc., Division of Powell Industries.)

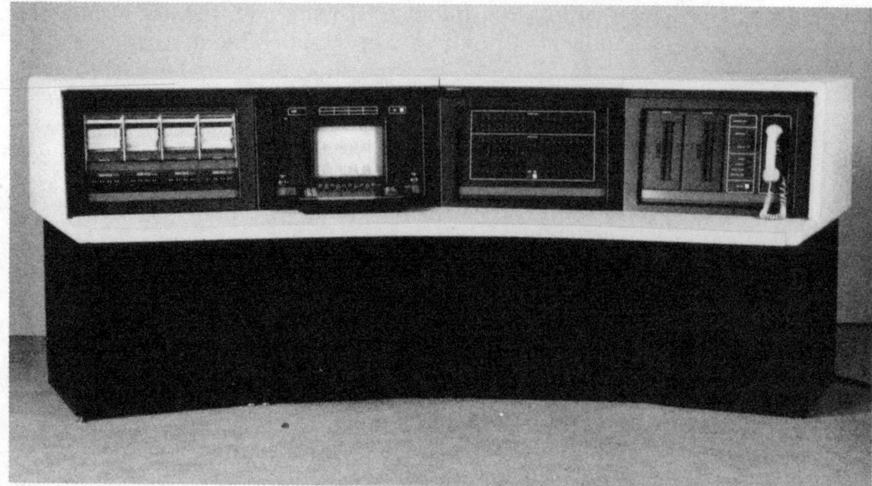

Figure 14.27. A distributed control system console communicates with remote-mounted microprocessor control units. (Courtesy of Foxboro Company.)

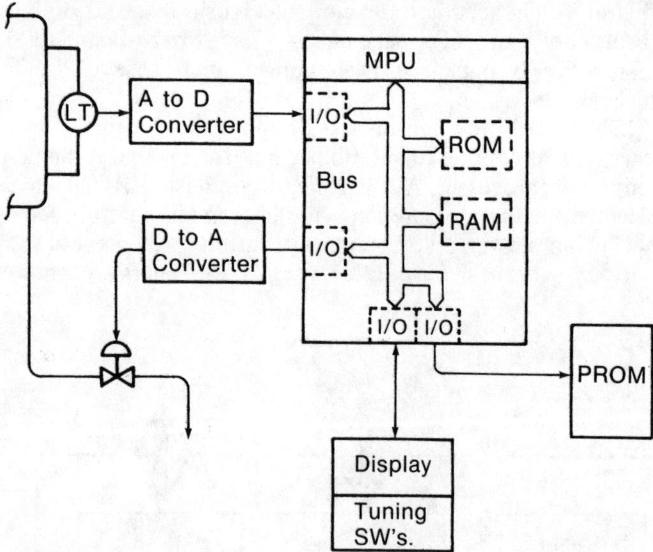

Figure 14.28. A dedicated single-loop controller using a microprocessor.

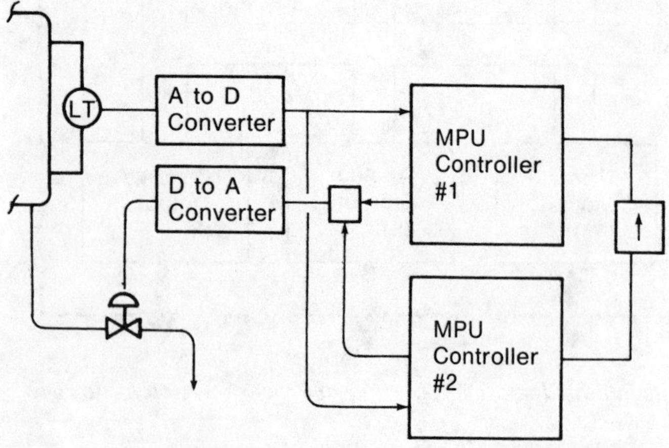

Figure 14.29. A back-up microprocessor unit automatically takes action if the primary unit fails.

tions, mass flow conversions, and input scaling). The microprocessor is capable of 80 instructions and uses 12-bit analog-to-digital (A/D) conversion for accurate calculations. The memory has a battery for backup. The Flow Master (Figure 14.10) accepts up to eight analog inputs and two pulse inputs for making process computations. The reliability and ruggedness of the microprocessor become apparent, since this unit is designed for placement in the field. No special heating or cooling equipment is required. It fits neatly into an enclosure rated for hazardous areas.

Solartron's Process Computer (Figure 14.31) is similar to the Flow Master. It also comes in an enclosure for field mounting. It is designed to make mass flow rate and totalized mass flow calculations, while correcting for temperature and density or specific gravity. It also provides pulse conversion and energy flow calculations. A comprehensive self-check is automatically performed at regular intervals.

So far, different forms of dedicated controllers using microprocessors have been discussed. Some controllers use a single input or output, while others use multiple inputs and outputs, and single control outputs. In all cases, these units are preprogrammed, making them limited to specific applications. This is not to say that they are deficient in any aspect, but they are intended for use with specific applications. Reprogramming and possible hardware changes would be required to adapt the unit for a new function. The multi-loop controller has similar limitations, but it has many more options from which to choose.

Multi-Loop Controllers

Multi-loop controllers, such as Doric's Control 80 Microcontroller and Powell's Micro P-200 Controller, offer versatile control equations with multi-function controls.

The Doric Microcontroller (Figure 14.32) is complete with up to four controller input/outputs. Its price starts at $1,000, as compared to the conventional single-loop analog controllers that sell for approximately $700. Controller modes, action, and tuning constants are program selectable from the front panel. This unit will accept thermocouples,

Figure 14.30. Taylor's Micro-Scan 1300 Controller has many preprogrammed algorithms. (Courtesy of Taylor Instrument Company.)

resistance temperature devices, analog inputs, and potentiometric inputs. It converts these internally (without an external signal transducer) to digital form. After converting the signals to digital, it performs one of the control functions selected by the user. There are also other options available such as alarms, input filtering, displays, relay outputs, anti-reset windup, and bumpless transfer. This unit is factory preprogrammed, and is limited to the function for which it was designed.

The Micon P-200 (Figure 14.33), developed by Powell Industries, is a dedicated multi-loop controller using microprocessor control techniques. This stand-alone unit is capable of performing a wide range of control functions—15 analog inputs, 8 analog outputs for control, 16 discrete inputs, and 8 discrete outputs. (Refer to Figure 14.26 for a block diagram of the unit.)

The P-200 has a preprogrammed PROM memory from which its many functions can be selected, such as link, control, math, limit, dynamic, and batch function. Each function has subgroups that are available for operating on any input. An individual loop may be acted upon by more than one group.

The Micon fits into a package (6 by 6 inches) and requires no programming by the operator, only loop configuration from its side-mounted keyboard.

The multi-loop controller promises sophisticated control, small size, moderate prices and high reliability. But what about the large control job? Can it control a plant? The Honeywell TDC-2000 (Figure 14.34) pioneered a change with its Total Distributed Control system by using the

Figure 14.31. Solartron's Process Computer is designed to calculate mass flow rate and totalized mass flow, while correcting for temperature, density, and specific gravity. (Courtesy of Solartron Division of Schlumberger.)

Figure 14.32. The Doric Microcontroller is complete with up to four controller input/outputs. (Courtesy of Doric Scientific Division of Emerson Electric.)

Figure 14.33. The Micon P-200 is capable of 15 analog imputs, 8 analog control outputs, 16 discrete inputs, and 8 discrete outputs. (Courtesy of Process Systems, Inc., Division of Powell Industries.)

Figure 14.34. Honeywell's TDC-2000 Total Distributed Control system uses a CRT and keyboard operator interface in the control room and remotely located controllers and process interface units. (Courtesy of Honeywell, Inc.)

architecture discussed previously. In pioneering the Total Distributed Control system, it has minimized the possibility of a complete system failure. This is accomplished through the use of a controller file (Figure 14.35) accepting up to 16 analog inputs and providing up to 8 analog control outputs. In addition to the inputs/outputs, the controller file contains a 24-hour historical memory card and data highway interface cards. Up to 64 controller files can be nested in terminal racks containing power supplies, field wiring terminations, battery backup, and auxiliary inputs. These controller files communicate with the operator console, continuously updating the operator's display, while each controller file is doing loop control independent of the operator's console and the operator.

This system is readily expandable due to its open-ended design approach. Additional operator consoles can be added, along with graphic display consoles. A large computer is also available.

The Micon P-200 is offered in an expandable operator console with remote nesting of controller assemblies. Similar systems, such as Fisher's ProVox, Foxboro's Videospec, and XOMOX's System 1100, are being sold. They also offer versatility of operational design.

When purchasing a large system, the user must prepare a program format. This includes configuring each microprocessor controller input, output, and the operator's console. Displays need to be identified, loops tagged and configured, and a tabulation of all words used, assembled, and put into

Figure 14.35. A TDC-2000 controller file accepts analog inputs, provides outputs, and connects to redundant data buses, thereby spreading system risk. (Courtesy of Honeywell, Inc.)

memory. If other options are used, they must also be programmed into the system.

As much programming as is practical will be done by the manufacturer, but that excludes the items just mentioned. They must be done by the design engineer or the operator, or be purchased as an option when the system is purchased.

Modular Components

The final form of application is the use of modular components to construct a system. Often, a microprocessor programmed to perform specific control functions is not available as an off-the-shelf manufactured item. The only alternatives are to (1) ignore the problem, (2) make do with what is available, or (3) create a system which will handle the problem. Whether a single-chip microprocessor, a mainframe microprocessor, or a microprocessor PC board is used, it is still necessary to consult individual manufacturers for these devices. A good way to start is to review chip manufacturing data (its architecture) and the pre-assembled PC board when available from suppliers. Often it is only a matter of purchasing pre-assembled boards, making the necessary interconnections, and the microprocessor is ready to be programmed.

Occasionally, a task may warrant a microprocessor system, complete with an operational keyboard and CRT, printer, floppy disks, and other peripheral devices. When choosing this path, it is wise to take advantage of the software packages provided with the system.

Programmable Logic Controllers

In the 1960s computers were considered by many as the ultimate way of increasing efficiency, reliability, productivity, and automation in industry because they presented the ability to acquire and analyze data at extremely high speeds, make decisions, and then disseminate relative information back to the control process. However, certain factors—high cost, complexity of programs, hesitancy on the part of industry to rely on a machine, lack of personnel trained in computer technology—greatly limited the application of the computer with respect to actual full-scale control functions in the process industry. Thus, computer applications through the 1960s were mostly in the area of data collection, on-line monitoring, and open-loop advisory control.

Traditionally, industrial processes and control systems used relays, timers, and counters—their control logic was depicted in the language of "ladder diagrams." Relay ladder diagrams are universally understood in the industrial world, whether in the process industry, manufacturing, assembly line, or electrical appliances and products. The need for applying the rapidly developing, solid-state technology to industrial and process control applications, and to be compatible with existing installations and control philosophies was realized. Because mini-computers require expensive software, specially trained personnel, and special peripheral equipment to interface with the industrial world, they were not readily accepted by industry. Thus, evolving from the concept of the traditional programmable controllers, such as the stepping drum type and the solid-state programmer with plug-in modules, the era of the Programmable Logic Controller (PLC) began in the mid 1960s. A new industry was born.

History of PLC's

Sweeping changes and rapid advances have taken place in U.S. industry in the last two decades. These have been characterized by new trends in labor and product markets. Production methods, processes, equipment, and machinery are continually being updated or replaced by more complex instrumentation and more sophisticated techniques. A special report, "Productivity, Electrotechnology to The Rescue," published in *IEEE Spectrum,* October, 1978, states:

> Today, with microprocessors and more sophisticated hierarchical programming concepts, it is theoretically possible, though not always cost-effective, to replace or parallel humans in many continuous-flow and batch-flow operations. For industries requiring a lot of assembly steps, like automobile making, however, even theoretical total automation seems a long way off.

Thus, it was originally in the automotive assembly line, faced with costly scrapping of controls due to changes during model changeovers, that the first PLCs were installed in 1969 as an electronic replacement of electro-mechanical relay controls. Here the PLC presented the best compromise of existing relay ladder schematic techniques and expanding solid-state technology. It eliminated the need of costly rewiring of relay controls, reduced downtime, increased flexibility, considerably reduced space requirements, and presented a more efficient system.

A Programmable Logic Controller, according to NEMA standards, is a digitally operating electronic apparatus that uses a programmable memory for the internal storage of instructions that implement specific functions such as logic, sequence, timing, counting, and arithmetic, to control machines and processes. The detailed description and operation of PLC's and their relation to standard relay control schemes will be explained in the next section.

Initially, due to development costs and relatively high costs of custom packaged chips and semiconductors, the installed cost of PLCs compared to equivalent relay systems was considerably higher. But, due to rapid developments in the semiconductor industry, development of solid-state memories, and LSI (Large Scale Integrated) chips that are now available at extremely low costs, PLC prices have shown a steady decline (See Figure 14.36). In 1978, the initial cost of a programmable controller installation became the same as the installed cost of an equivalent relay control system. As labor and maintenance costs continue to rise in industry, the cost of PLCs shows a downward trend through the 1980s. With the present trend, PLCs will continue to gain wider acceptance in industry—including the process industry.

Relay system costs are greatly affected initially by installation labor costs. Later, trouble-shooting and rewiring as required by changes in the control scheme affect the cost. Presently, wiring and installation varies from \$30 to \$50 per relay and can be higher. The cost of smaller PLCs can now be justified for systems involving as few as eight to ten relays. Solid-state PLCs will get less expensive as they get more compact and are widely accepted.

Basic Concepts of the PLC

A programmable controller, as previously defined, is essentially meant to replace relays, timers, and sequencers in traditional relay control systems, and is designed for installation and operation in industrial and process plants. There is a tendency to confuse PLCs with computers, minicomputers, and programmable process controllers that are used for numerical control and for position control. PLCs are used for sequence control. Unlike computer control, the PLC does not require very sophisticated programming, debugging, and maintenance techniques.

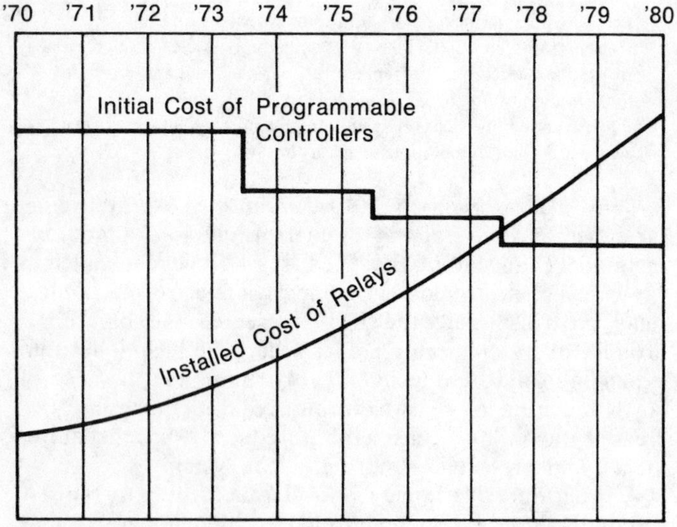

Figure 14.36. PLC prices have shown a steady decline, and now compare favorably with relay systems.

Typical PLC system architecture is shown in Figure 14.37. Models by Texas Instruments, Gould-Modicon, Allen-Bradley, Struthers-Dunn, Tenor Co., General Electric, Square D, Cutler-Hammer, and other manufacturers are available. The models have various levels of capabilities and complexities, and are continuously updated and improved to meet industry requirements. Figure 14.37 shows the CPU, memory, power supply, input/output section, and programming device. These main blocks and their functions, which are basically the same in all available PLCs, are explained in the following sections.

The Central Processing Unit (CPU)

The CPU is the heart of a PLC, computer, minicomputer, or microcomputer because it receives instructions from the memory and generates commands to the output modules. Input commands, device status, and instructions are converted to logic signals, "1" for input present, and "0" for no input signal in positive logic. These logic signals are then processed by the CPU. As in traditional ladder diagrams where the NO/NC contacts of the field devices activate relays and timers, PLCs process logic signals and activate output TRIACS that can be normally energized or de-energized. In a relay control system, NO/NC contacts from relays are available for use in the control scheme. Similarly in a PLC, internal and output coils have NO/NC contacts that can be used in the logic scheme. In contrast to hard-wired relay control systems, no wiring is needed for implementing the control logic in a PLC. All sequence con-

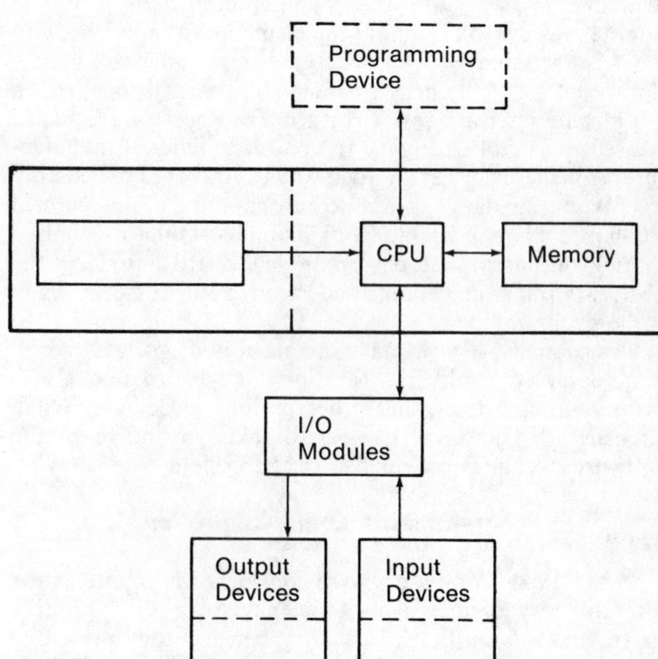

Figure 14.37. PLC systems have basic architectures that are usually similar.

trol logic is internal to the PLC, and is processed by the CPU as explained.

Available PLCs are based on various microprocessor chips which are preprogrammed with a main "executive program." The executive program enables the CPU to understand input command instructions and status signals, and provides logic processing capability. These capabilities (solving simple "and/or" logic, timing, sequencing, addition, subtraction, multiplication, division, and counting) vary from different models and manufacturers and often affect the hardware price. A careful study of the requirements of a control application should be made to decide the features currently needed and ones that may be needed in the future.

Memory

Memory in a PLC is where the central program is stored. The CPU utilizes program instructions stored in memory to tell itself to scan certain inputs and then to generate output commands. Memory capacities vary, and generally store 256, 512, 1024(1K), 2K or 4K words, depending on word size. Some models may have higher capacities. The memory size furnished in the PLC varies with the size of the control functions to be performed, and should be carefully selected only after evaluating present and future needs.

Various PLCs have different limitations on the number of horizontal and vertical contacts that can be programmed into each step. This affects net memory usage for a given ladder schematic (refer to the example under "Programming"). Also, the number of words of memory used per contact varies from model to model (Modicon uses two words per contact, whereas Square "D" uses one word per contact), and should be checked before selecting a particular memory size. Some PLCs use part of the advertised memory for the "executive program," which thereby reduces the available memory for the control program.

An important concept to understand in the operation of a PLC is "Memory Scan." A typical multi-node format network used by the Modicon Model 484 is shown in Figure 14.38. Some PLCs can be programmed only one horizontal rung at a time. Their maximum number of vertical elements is also limited. In PLCs such as Texas Instruments' Model 5 TI, there is no limitation on the number of series or parallel elements that can be used in a logic line or step.

As a typical example (Figure 14.38) the controller will solve each network of interconnected logic elements (NO and NC contacts, timers, counters, etc.) in their numerical sequence—the order in which they were programmed. The first network is scanned from the time that power is applied, first from top left to bottom left, and then continuing to the next vertical column to the right. Within a network, the logic elements are solved during the scan, then the coils are appropriately energized or de-energized to complete the scan. Since the scanning rate is very fast (4 milliseconds for 250 words, to 20 milliseconds for 4K memory for a Modicon 484), it appears that all logic is solved simultaneously. The result (change in coil state, numerical values, etc.) of each network scan is then available to all subsequent networks. Thus, all inputs and outputs are updated once per scan. The time from solving any individual network on one scan until that network is again solved on the next scan is defined as the *scan time* of the memory. It depends upon the complexity of the programmed logic and memory size. For memories with a longer scan time, a fast close/open input signal could possibly be missed by the PLC scan. In such a case, a push button would have to be held in longer. Though this seldom presents a problem in the average control systems, it should be considered.

Different forms of memory are available. The trend is towards using solid-state CMOS memories. Memory types include:

1. *Toroidal Core Read/Write Memory:* extremely flexible and easiest to reprogram, but is susceptible to voltage transients. In each scan, the program memory is read, stored in a register, and then rewritten in the original location.
2. *Stored-Field Read-Only Memory (ROM):* offers high noise immunity, and, for reprogramming, the original program can be erased by exposure to ultraviolet light. Portable erasing equipment is now available. These memories are also known as LEROM (Light Erasable ROM).
3. *Random Access Memories (RAM):* volatile, require battery backup in case of power loss. Generally, a nickel-cadmium cell(s) is provided on the memory board to ensure that the program is not lost. Some models offer a combination of RAM/PROM (Programmable ROM), in which the program is first developed on the RAM and then transferred to the PROM.
4. *Electrically Alterable ROM (EAROM):* nonvolatile, so battery backup is not required. Electrical alteration is possible via the "programmer," which can be either hand-held or a CRT type, where the program steps are displayed. PLCs with EAROM are more desirable because it is possible to add new program steps or revise the existing program in the processor while it is operating in the RUN mode. Gould, in their MODICON PLCs, and Square "D," in their SY/MAX-20 models, offer this feature.

Various models offer "scratch pad," or "trial" read/write memories in addition to the main memory. These enable the programmer to make changes, add to or delete from the program, debug the program, and then transfer it to the main memory.

An additional feature is "memory protect." A key interlock is provided to prevent unauthorized tampering with the stored program.

Power Supply

The power supply is an integral part of the PLC and is generally mounted in the mainframe enclosure. Line power at 120 volts, 60 Hz is converted to the appropriate DC voltages required by the solid-state circuitry and memory.

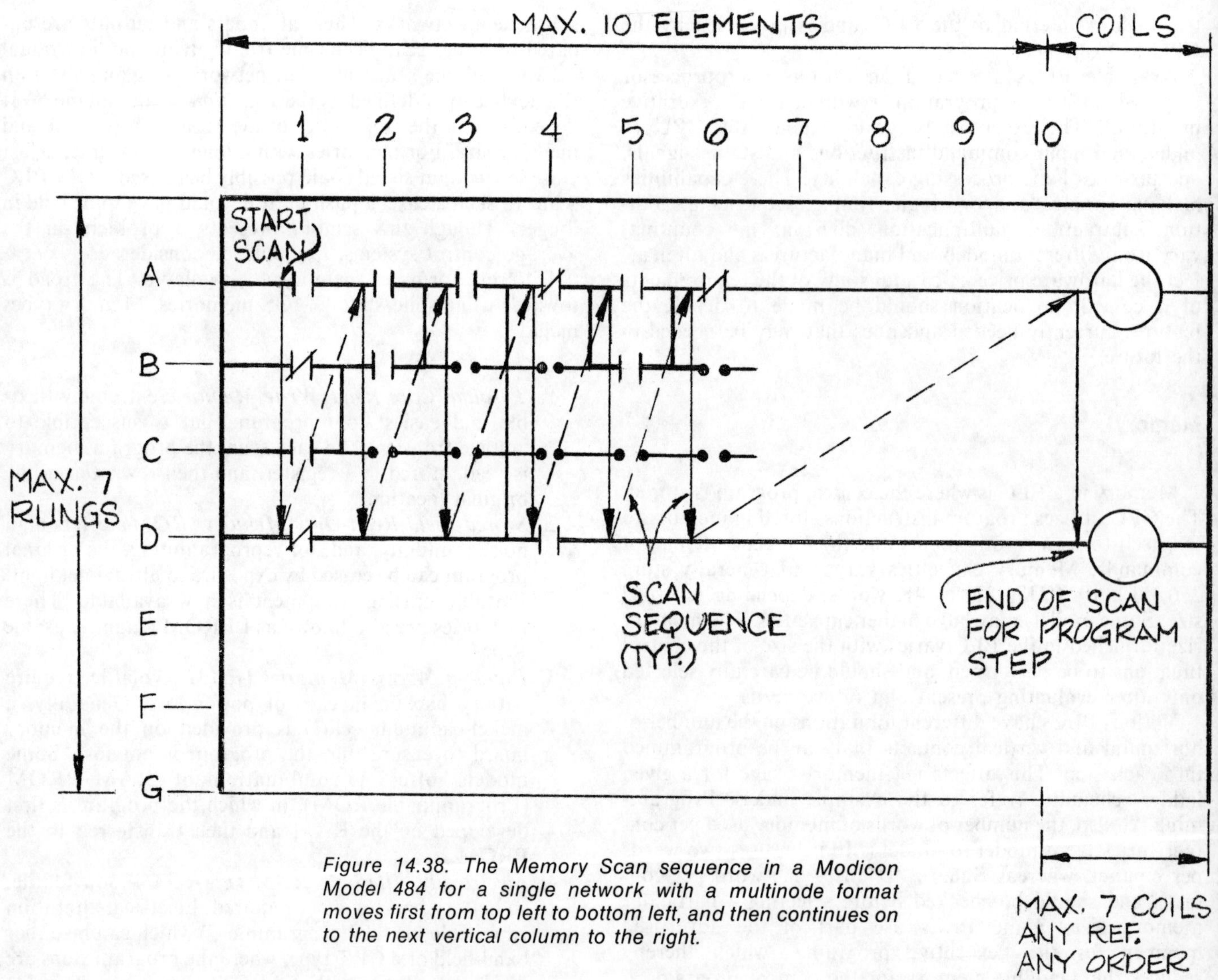

Figure 14.38. The Memory Scan sequence in a Modicon Model 484 for a single network with a multinode format moves first from top left to bottom left, and then continues on to the next vertical column to the right.

For volatile memories that require constant power to retain the stored program, DC cells are provided to ensure retention of the memory in case of main power failure. The power supply is designed to operate both the CPU and the basic number of inputs and outputs. For expanded input/outputs, an optional heavy-duty power supply usually has to be specified.

The power supply generally is designed for a controlled "power-down" sequence in case line power is lost. In this case, the CPU stops solving logic and retains the status of all coils, inputs, outputs, and registers. The outputs are all turned off. This eliminates the possibility of failure in an undetermined mode. In the "power-up" sequence, the stored inputs, outputs, status of coils and registers are checked, and then the memory-scan sequence is started.

Input/Output Section

One of the main characteristics that has made PLCs extremely attractive is that the input/output modules are designed to interface directly with industrial equipment. Input modules are generally available for interface with a wide variety of signal levels; for example, 120 VAC, 24 VDC, 48 VDC, 4-20 maDC, 5 VDC (TTL). Most manufacturers offer optically isolated inputs, which permit mixing of discreet and analog inputs, and prevent transients on the field wiring from affecting the internal logic. Input cards (modules) for each type of input signal are of plug-in construction and can usually be inserted or removed without a system shutdown. Most manufacturers now offer a status indicator light for individual inputs on the card.

Output modules are also available in the same wide variety of voltage ranges as are input modules. Each output is optically isolated and fused, and is available with output status indication. Field devices such as small motor contactors, valves, solenoids, and lights can be directly operated from the output modules. In some models the input/output section is directly connected to the mainframe, while in others it can be remotely located if the CPU is kept in a central location.

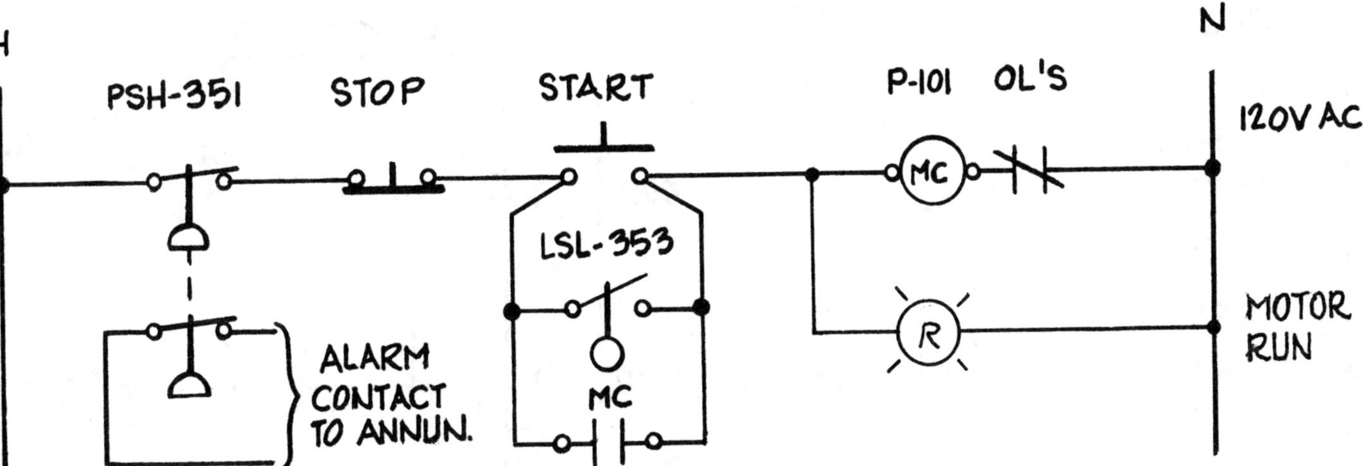

Figure 14.39. A typical relay system ladder diagram.

There is a basic difference between a PLC and a standard relay control system with regard to input/outputs. A section of a relay ladder diagram is shown in Figure 14.39. Its equivalent PLC input/output diagram is shown in Figure 14.40.

Note that all field switches are wired to input points identified by the PLC numbers 1001, 1002, etc. A closed-field contact, such as PSH-351, essentially energizes an internal PLC relay 1001, and all internal NO contacts referenced to 1001 in the logic will close (NC contacts of PLC relay 1001 will open). If PSH-351 opens, the PLC relay 1001 will de-energize and all references to NO contacts in the logic program will open (NC reference contacts will close). The logic operations during the scan are done on the internally programmed reference contacts shown in Figure 14.40b. If motor starting conditions are satisfied, output coil 0001 energizes and seals in. This causes the output TRIAC labeled 0001 to energize, which in turn energizes the motor contactor MC and starts pump P-101. The alarm output 0016 is wired to a 24 VDC output module as shown in Figure 14.40a.

Programming Devices

The programmer for a PLC is the device (usually, an external unit) that transforms the control scheme into useful PLC logic. The logic program is then stored in memory, where it is made available to the CPU for logic operations.

Various kinds of programming devices are available from PLC manufacturers. These range from a CRT Programming Panel, a hand-held calculator-like device, a thumbwheel input system, a cassette tape loader, or a hookup to a central computer or programmer through a telephone interface. For simplicity and compatibility with existing relay ladder schemes, most programming devices use either standard relay ladder symbols for NO/NC contacts, timers, counters, etc., or use Boolean terminology (AND, OR, NOT, etc.). Thus, there is no need to learn a sophisticated programming language or to redraw standard ladder diagrams in special format to program a PLC.

In addition to ease and flexibility of programming, most programming devices have a power flow light that enables the user to modify or troubleshoot the system. In CRT-based devices, visual display of the logic ensures the programmer of the accuracy of the punched-in program. Read/write or "scratch pad" memories are offered in some systems. These allow the programmer to add, delete, or modify the program before entering it in the main memory.

Some manufacturers offer, as a service to their customers, factory-written programs loaded into the PLC via a telephone interface. Programs can also be stored on cassette tapes and loaded into the PLC through a special interface unit.

Figures 14.41 and 14.42 show the various components of a typical PLC. Some different types of programming devices and other accessories are shown in Figure 14.43.

Programming

Programming a PLC can be simple and straightforward when approached in a systematic manner. As an example, convert a control scheme in relay ladder diagram form to a PLC ladder schematic for a Gould MODICON 484. A partial schematic for an oil heater burner ignition system is represented in Figure 14.44. The addresses of the input/output relays are assigned in a way similar to that shown in Figures 14.39 and 14.40. It is essentially a matter of converting the relay ladder control logic to PLC logic, using NO/NC contacts with the appropriate addresses to coincide with the input/output assignments. This is achieved by:

1. Draw a relay ladder schematic of the control scheme.
2. Assign an input address to each field input device (pressure switches, temperature switches, push-

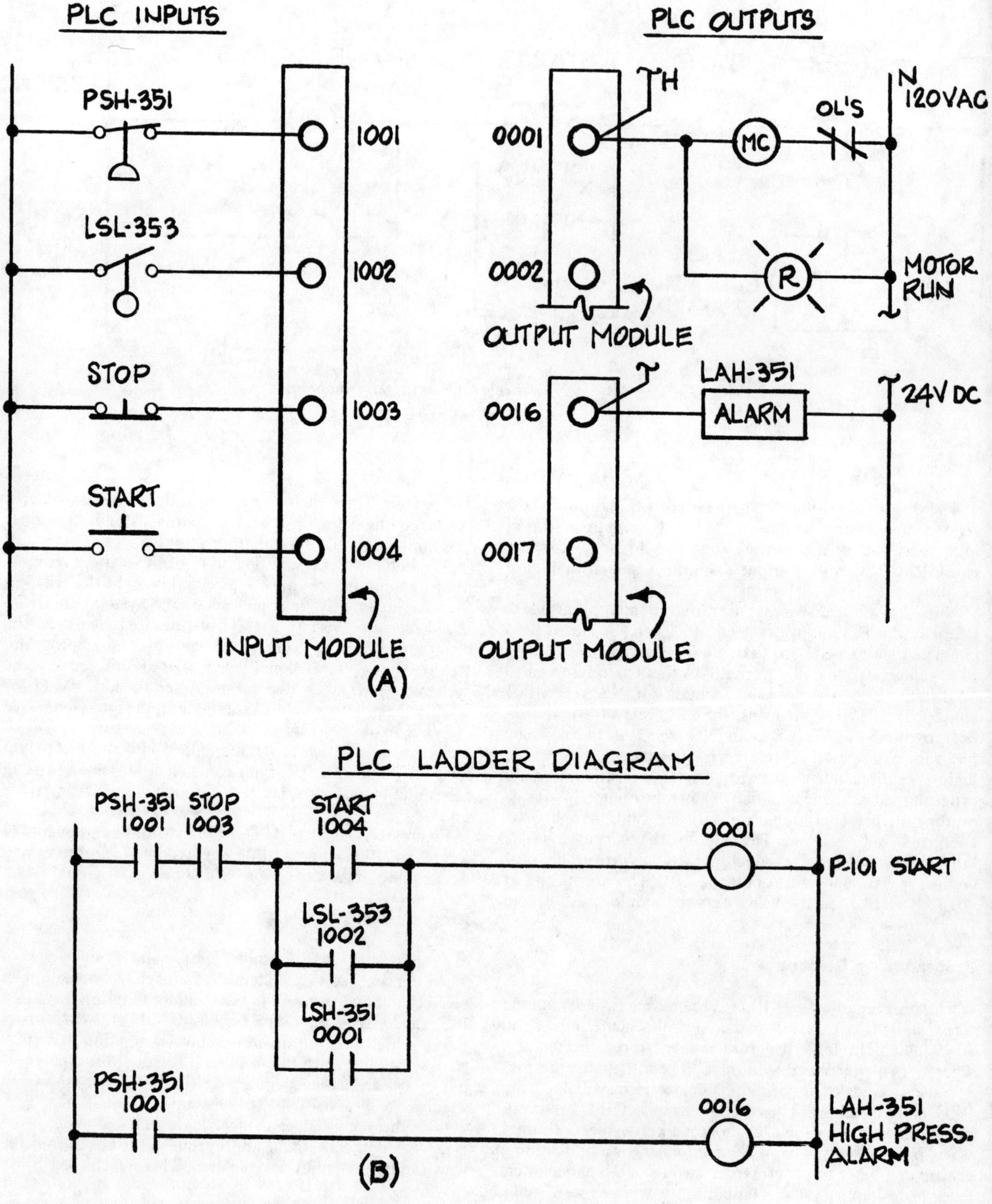

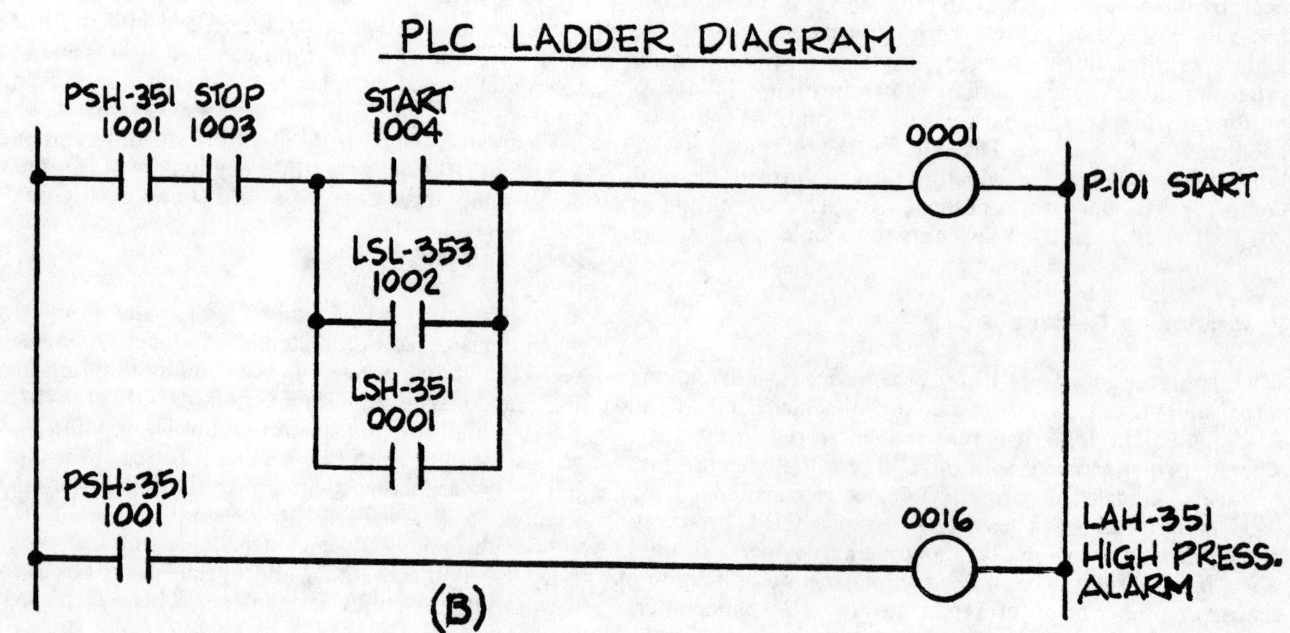

Figure 14.40. Diagrams showing (A) Typical PLC wiring to Input/Output modules and (B) a typical PLC equivalent ladder diagram for the system shown in Figure 14.39.

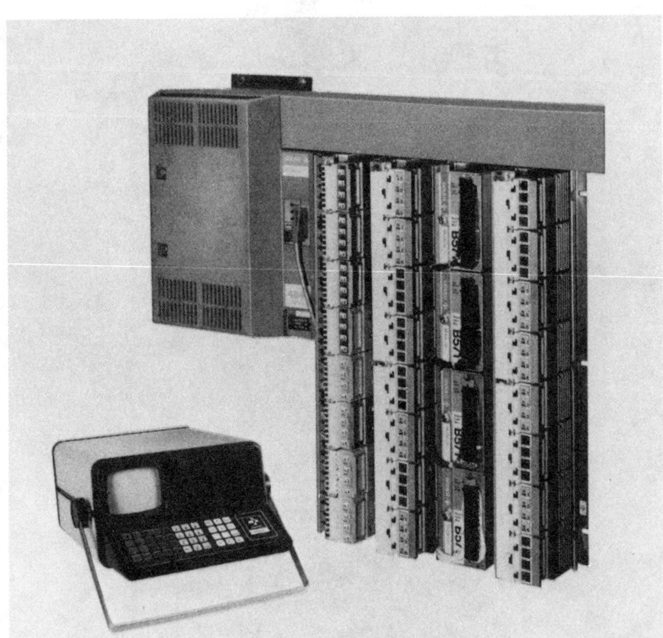

Figure 14.41. A Modicon Model 484 Programmable Controller system. (Courtesy of Gould, Inc.)

Figure 14. 42. A Texas Instruments Model 5TI Programmable Controller system (Courtesy of Texas Instruments, Inc.)

buttons, selector switches, etc.). These field contacts are wired to the Input Modules, per their respective address assignments. Generally, most PLCs allow any number of references (NO/NC contacts) to the input or output coils. Where an output is not required, internal "coils" are used. They are referenced to NO/NC contacts (e.g., coil No. 0258 in Figure 14.45).

3. Assign output relay coil numbers, where required, for solenoid valves, motors, indicator lights, alarms, etc. The output coil assignments and module types must be compatible with the system control voltage.
4. Then, enter the program into the processor using the CRT Programming Panel (Figure 14.41). Figure 14.45 shows how the screen display will look. It represents the PLC ladder schematic for Figure 14.44.

Once the input assignments have been made, the field wiring can be hooked up, and control logic changes carried out by addressing the particular program step. Some PLCs allow for on-line changes. Also, outputs can be locked into an energized/de-energized position, the program change made, and the output re-enabled.

Each type of programming device requires a different approach to actual program loading. A sample program done on a Texas Instruments 5 TI 2000 Series R/W programming device is shown in Figure 14.46. Programming can be done either directly from a ladder diagram or a Boolean logic diagram. The same program, done on a Modicon P180 CRT programming device, is shown in Figure 14.47. Even though the control functions are the same, a different amount of memory is used in each case.

Timers and counters can also be programmed easily. PLCs have a crystal-controlled clock signal that drives the timers. Maximum timing periods vary (999 seconds for a Modicon 484, and 54.6 minutes for a 5 TI). However, by cascading timers (counters), any desirable timing period (count) can be obtained. One form of a timer is shown in Figure 14.48.

Timer or counter functions are addressable from the programming panel. In Figure 14.48, 0030 is the timing period of 30 seconds. T1.0 denotes steps of 1 second each, and 4xxx gives the register address in the CPU. The timing function will start when a signal is given at T and when a logic "1" is present at the reset terminal. When the set period has elapsed, the Q output is energized. When the reset signal at terminal R turns to "0," the timer resets. When R = "1," the timer is enabled. $\overline{Q}$ signal is the inverse (opposite) of the Q signal. Counters operate in much the same way as timers except they increment the current count by one only when the energizing signal changes to "1."

In addition to the programming features, newer models of PLCs offer mathematical computing abilities (math packages) and enhanced instruction sets. These provide the ability to add, subtract, multiply, and divide, as well as other features. This enables the calculation of new values for process variables and set-points, comparison of field data with reference data stored in memory, and generation of statistical information for display or print out.

An extremely interesting approach to PLC programming is offered in the LDC-40 "line-o-logic" controller by

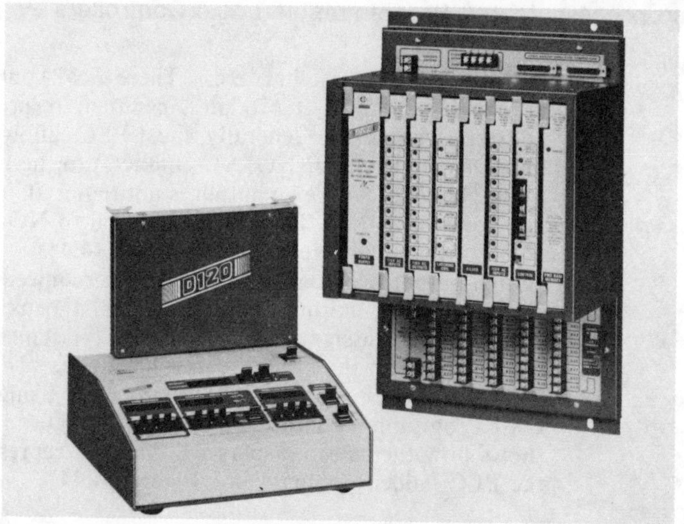

A. (Courtesy of Cutler-Hammer.)

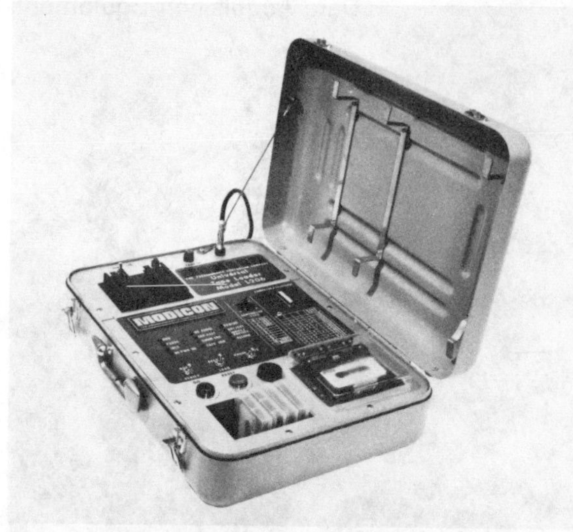

B. (Courtesy of Gould, Inc.)

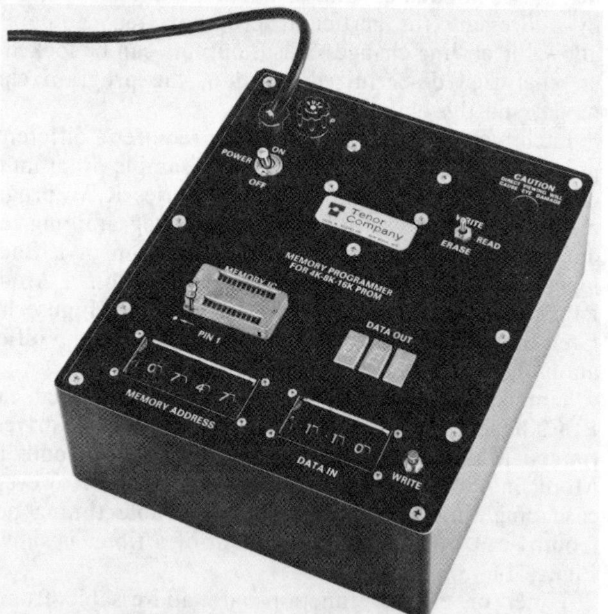

C. (Courtesy of Tenor Company.)

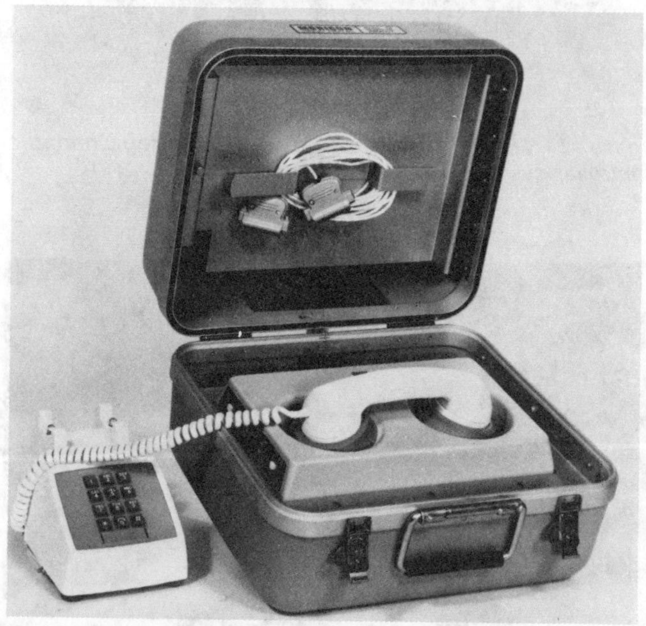

D. (Courtesy of Gould, Inc.)

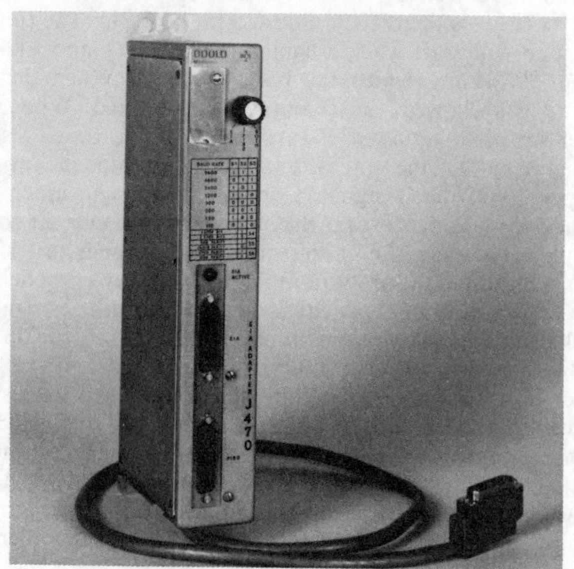

E. (Courtesy of Gould, Inc.)

Figure 14.43. Programming devices and accessories.

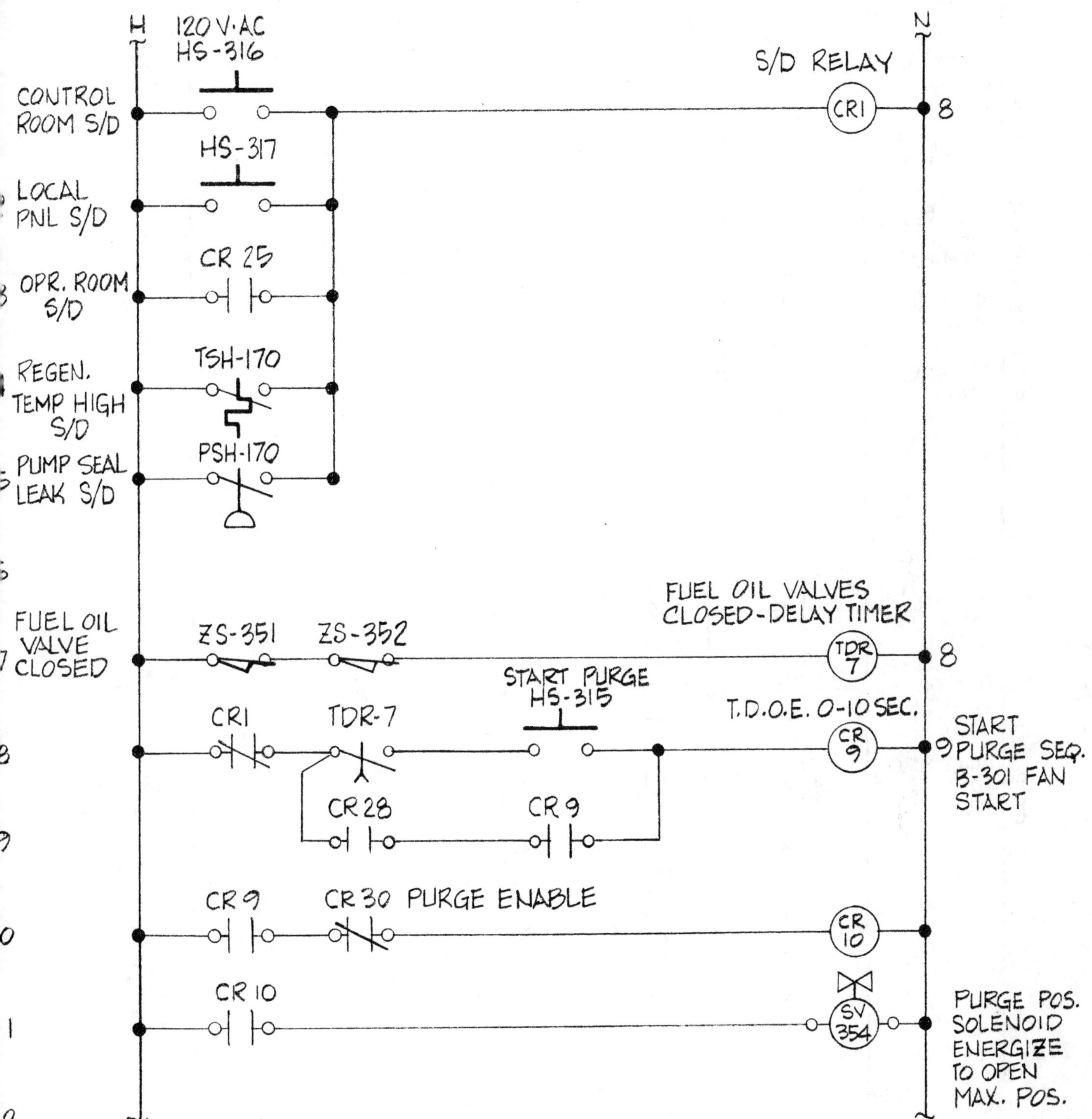

Figure 14.44. A partial schematic of an oil heater burner ignition system.

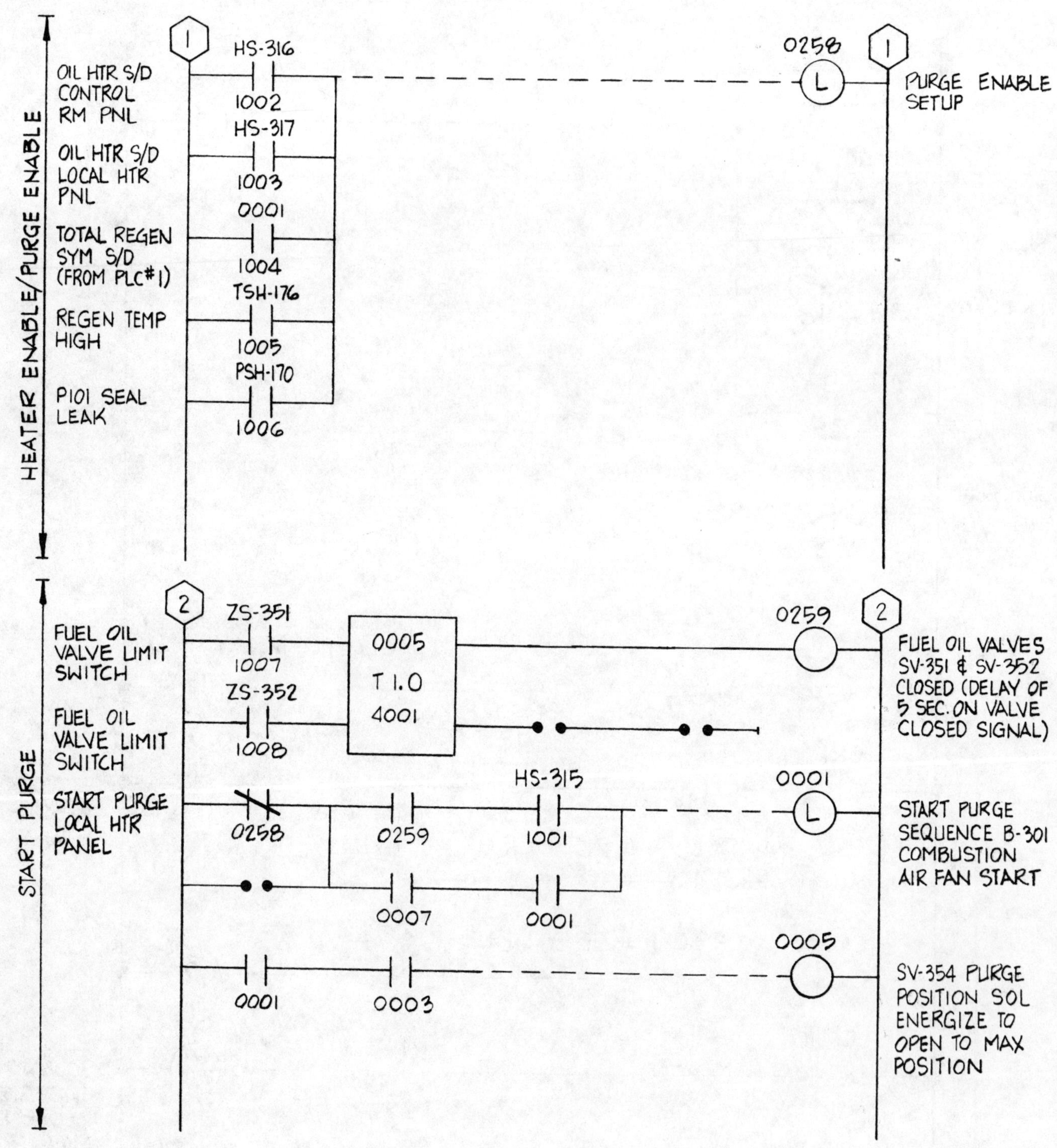

Figure 14.45. A PLC schematic of the oil heater schematic shown in Figure 14.44.

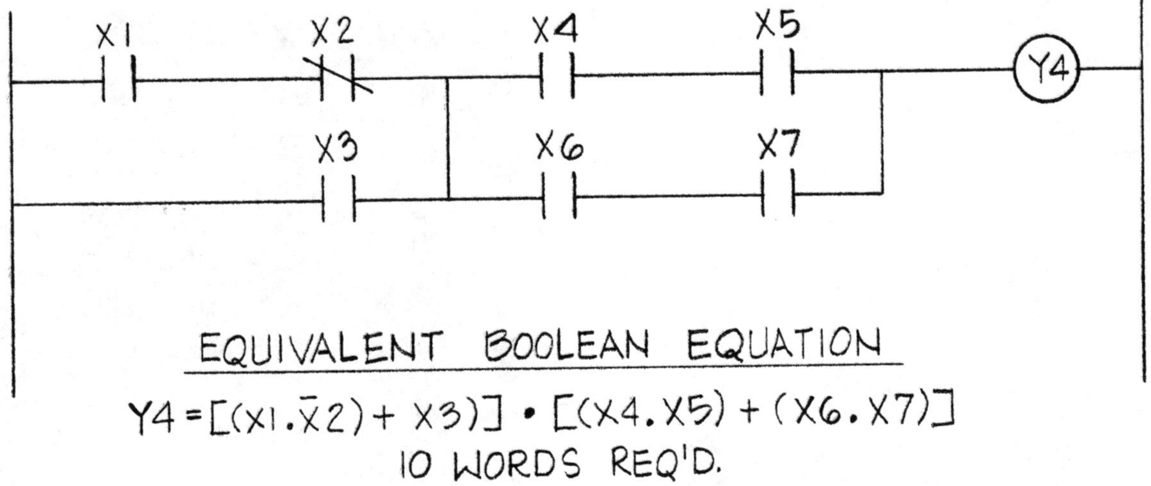

Figure 14.46. A program done on a Texas Instruments 5TI2000 Programmer.

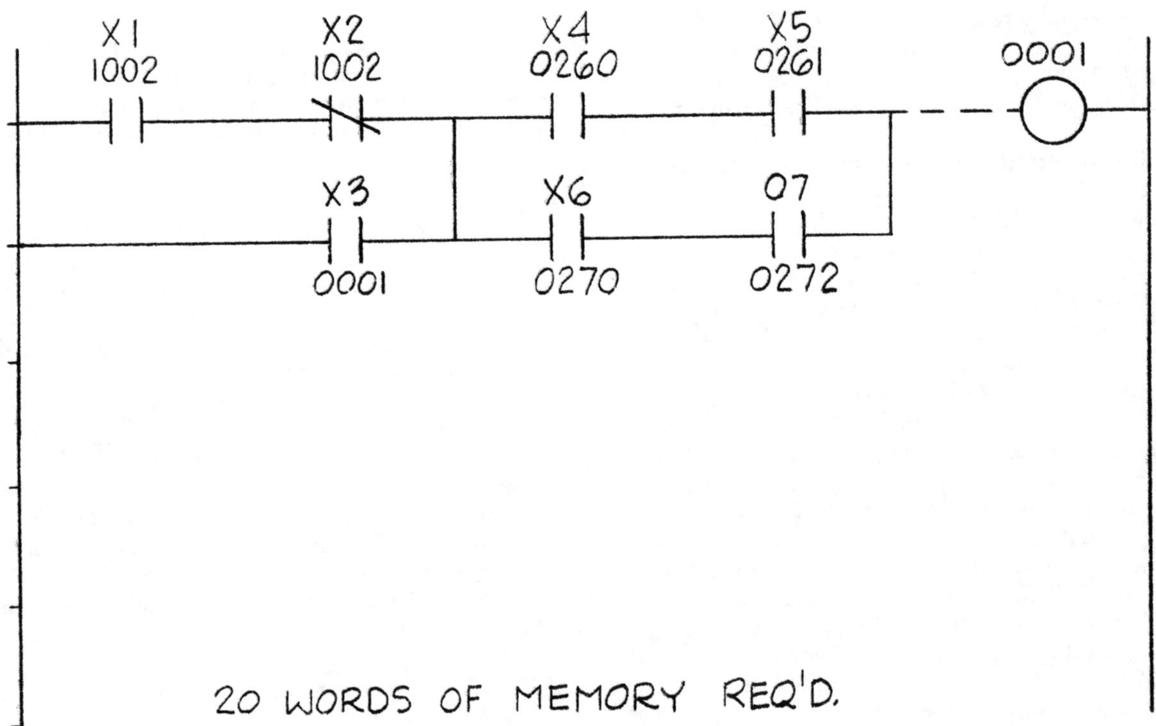

Figure 14.47. The logic of Figure 14.46 done on a Modicon P180 CRT Programmer.

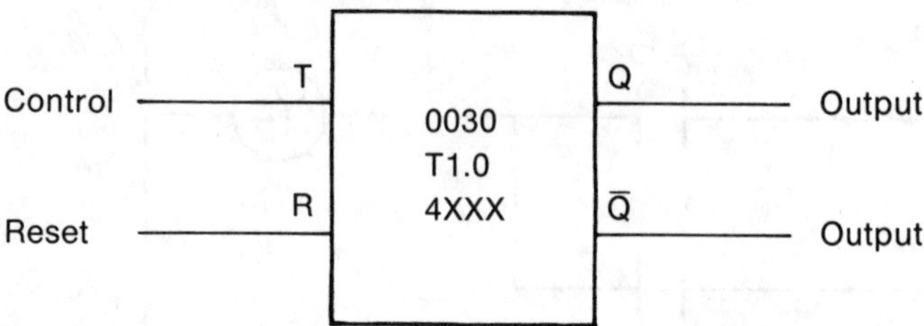

Figure 14.48. A schematic representation of a timer.

Automatic Timing and Controls Co. Although this controller is not a PLC as defined in this chapter, it is, according to the manufacturer, a "solid state device consisting of plug-in modules that are easily assembled and programmed to produce the appearance and function of standard relay ladder diagrams." An LDC-40 is shown in Figure 14.49. It is designed for 5 to 40 or more lines of ladder diagram logic. It is cost- and space-effective when compared to a PLC or conventional relay control systems. Solid-state modules are plugged into each line to coincide with the relay ladder diagram. Coil addressing is done by connecting cables on a patch panel behind the LDC-40. The input/outputs are connected directly to the terminals. Advantages of the LDC-40 are (1) that a programming device is not required, and (2) that the entire ladder circuit is displayed in the same form as the traditional relay ladder diagram.

Figure 14.49. When assembled, the LDC 40 front panel reproduces the original ladder diagram. Besides relay functions, the system provides timers and counters, either with or without a readout. (Courtesy of Timing and Controls Co.)

Application of Programmable Logic Controllers

Examples of actual PLC application in industry are listed, with a brief description of each control system.

Boiler control: A separate PLC is used for each of four boilers in a chemical plant to control the process of purging, pilot light-off, flame safety checks, all interlock and safety shut down checks, main burner light-off, temperature control, and valve switching for converting from natural gas to fuel oil. The PLC is programmed to be an energy management system for maximum efficiency and safety.

Compressor station control: A compressor station with multiple compressors is controlled by a PLC, which handles start-up and shutdown sequences and all safety interlocks.

Ethylene drying facility: In an ethylene drying facility, where moist gas is first removed from salt domes and then dried and pumped into the main pipeline, two PLCs are used for controlling the entire operation. One PLC is used to control heater combustion controls and shutdown system. The second PLC is used to control the drying and regeneration cycles in the dryer units. Both PLCs act as "slaves," and are tied into a "master" PLC located in a Central Operations Control Room. The "master" PLC has control over shutdown sequences, and also monitors some of the critical process variables.

Automatic welding: PLCs have been successfully applied to the control of automatic welding machines in the automotive industry. The use of aluminum in automobile bodies for weight reduction created a load distribution (AC power) problem because welding aluminum consumes more current per weld than welding steel. To eliminate this, controllers time-share automatic welding machines on a priority scheme that utilizes the data handling and arithmetic capability of the PLCs.

Coal fluidizing process: A PLC is installed on a fluidized bed to determine the amount of energy generated from a given amount of coal. A mixture of crushed coal and limestone is blown through jets over a heated bed. Burning rates and temperatures are monitored. The PLC controls the sequencing of the valves, and takes the place of a relay control system. The analog capabilities of the PLC enable jet valves to be controlled by the control system that is doing the sequencing. Control devices are also monitored on a CRT.

Material handling: In a storage/retrieval system controlled by a PLC, parts are loaded and carried through the system in totes (bins). The controller keeps track of the totes. An operator's console allows parts to be rapidly loaded or unloaded. A printer provides inventory printout, such as storage lane number, parts assigned to each lane, and the quantity of parts in a lane.

These examples show the variety of applications for PLCs in some phases of industry. Arithmetic and analog operations capabilities are being used in conjunction with normal sequence control, thus making PLCs attractive and cost-effective.

Future of PLCs

As the need for increased productivity, reliability, efficiency, and automation grows, industry is finding more ways to use sophisticated control systems. Though total automation and computer control are still in the future, the PLC offers a compromise between advanced control techniques and present-day technology. There is wider acceptance of PLCs as direct replacements of relay control systems. Also, a new breed of PLCs is gradually becoming available. This new breed, in addition to the sequencing, timing, and counting functions, has the capabilities of microprocessors (handling arithmetic functions, advanced process control, data acquisition, Binary/BCD conversion). Direct computer interfacing, and the hook-up of several PLCs for a distributed control system with a master control from another PLC or computer, is already a reality.

Though PLCs are not designed to (and will not) replace a computer, they are useful and cost-effective for small- to medium-sized control systems. With the capability of functioning in a hierarchical control system as a local controller, PLCs will retain their application in large, plant-wide control systems. Thus, in the next few years, PLCs should become as valuable to industry as hand-held calculators have become to education.

Index